Principles of RF and Microwave Design

For a complete listing of titles in the
Artech House Microwave Library,
turn to the back of this book.

Principles of RF and Microwave Design

Matthew A. Morgan

ARTECH
HOUSE

BOSTON | LONDON
artechhouse.com

Library of Congress Cataloging-in-Publication Data
A catalog record for this book is available from the U.S. Library of Congress.

British Library Cataloguing in Publication Data
A catalogue record for this book is available from the British Library.

Cover design by John Gomes

ISBN 13: 978-1-63081-649-0

685 Canton Street
Norwood, MA 02062

10 9 8 7 6 5 4 3 2 1

To my family, whom I treasure

Contents

Preface

I suppose the question I need to answer is, "Why another microwave engineering book?" There are already so many — Pozar, Rizzi, Collin; these were good enough for me, why not for everyone else?

The answer is threefold. First — and this is perhaps one of the best reasons for writing a book — I rather like the subject. I find that writing is one of the best ways to firm up one's own conceptual understanding of things. I recently discovered an interview of a fellow author, and a personal inspiration of mine, Dr. Stephen Maas, who admitted that he frequently goes back to his own books to find information. As he put it, there is really no substitute for a cleanly written reference that describes things in exactly the way the reader thinks about them. I could not agree more. My publications have become the go-to sources for me to refresh my memory of things that I once knew but have not used in a while. That is reason enough for me to write this book. I know not how many others may find it worth keeping on their shelves, but for myself, at least, it will always be close at hand.

The second reason for writing this book is that in the course of my career I have encountered a surprising number of young students and engineers — and even a few not so young — who have learned what steps are required to solve a particular microwave engineering problem, but lack the deeper insight to understand why. They know how to run a simulator and to optimize a given design, but then fail to notice the small, simple change in topology that could make the design work better. They can calculate the characteristic impedance of a transmission line, but then wonder how it can be real-valued without the line being lossy. They know how to draw a waveguide splitter, but fail to anticipate the resonance that appears due to a trapped higher-order mode. In short, they know what formulas to use, but not where those formulas came from, nor what implicit assumptions were made to derive them. To be clear, mathematics is an indispensable tool for us to shine a light on the truths hidden beneath the surface of physical laws, and it shall not be neglected in these pages, but neither shall we ignore the imperative to connect what the math teaches us with the clarity of an accurate mental picture. Math without intuition is clumsy. Intuition without math is impotent. We must have both in order to realize our full creative potential as engineers, and it is my hope to offer a way of looking at things that helps the next generation of electronics innovators do exactly that.

Finally, I have come to realize that most of the microwave engineers I have met fall into one of two groups: those who think primarily in terms of lumped elements, and those who think primarily in terms of transmission lines, as if one were a truer and more pure description of reality than the other. The former group is forever recasting problems — even those that pertain to waveguides and antennas — into their lumped-element equivalents. The latter group sees all as wave propagation, knowing in their hearts that all inductors are merely tightly coupled transmission lines in disguise, and believing (erroneously) that waveguides are transmission lines also, but perhaps less ideal in their characteristics than coax.

The truth, as I see it, is that all such models are merely useful idealizations, and that none has any greater claim to truthfulness or sophistication than another. Preferring one model over another is sort of like being right or left-handed — a simple preference that, in this case, may arise in part from our particular way of thinking about the world, or as a consequence of which teachers we had for which courses. As I have said in other writings, the most successful microwave engineer is the one that is adept at adjusting his or her viewpoint from one physical model to another as is best suited to a particular problem. It is for this reason that I reintroduce lumped-element concepts in Chapter 2 as a reduction of Maxwell's equations — even though most readers will surely have learned the former before mastering the latter. I then develop transmission lines, waveguides, and quasi-optical components in the same way, building up our repertoire of available circuit elements all on equal footing, each the master of their own particular corner of parameter space. Only then do I talk about engineered components such as couplers, baluns, filters, and active devices and describe their construction using all types of elements (lumped, distributed, and quasi-optical) side by side so that the reader may become accustomed to choosing the right approach for each individual application.

Acknowledgments

I am once again indebted to my friend and colleague, Dr. Shing Kuo Pan, for his meticulous review of the derivations found throughout this book. Thanks to him, a host of my embarrassing mistakes will never see the light of day!

Chapter 1

Fields and Waves

Although it is assumed that the readers of this book will already have a working familiarity with electromagnetic concepts, it is deemed useful to review those concepts here so that the derivations that follow in later chapters are well supported by sound, physical principles. It is further the intent of this chapter to acclimate the reader to my mathematical and conceptual style in a familiar setting, so that newer concepts introduced in later chapters are more easily grasped.

We therefore begin with an overview of the foundational laws of electromagnetics and the mathematical principles upon which all other technologies in this book will be built. Some notational conventions will be established, and the physical processes that lead to the macroscopic properties of the materials used to construct our circuits will be described.

1.1 MAXWELL'S EQUATIONS

The foundation of all electronic circuits — and arguably, by extension, of modern civilization[1] — is Maxwell's equations. Known by many equivalent formulations (integral, differential, time harmonic, relativistic), these classical equations govern the relationship and behavior of the electric and magnetic fields, which in turn describe the forces that charges, either static or in motion, exert upon one another.

1 The great physicist and teacher, Richard Feynman, famously stated, "From a long view of the history of mankind, seen from, say, ten thousand years from now, there can be little doubt that the most significant event of the 19th century will be judged as Maxwell's discovery of the laws of electrodynamics. The American Civil War will pale into provincial insignificance in comparison with this important scientific event of the same decade."

1.1.1 Large-Scale Form and Constitutive Relations

The integral (large-scale) form of Maxwell's equations is given below[2],

$$\oiint_S \mathbf{D} \cdot d\mathbf{S} = \iiint_V \rho dV \tag{1.1a}$$

$$\oiint_S \mathbf{B} \cdot d\mathbf{S} = 0 \tag{1.1b}$$

$$\oint_l \mathbf{E} \cdot d\mathbf{l} = -\frac{\partial}{\partial t} \iint_S \mathbf{B} \cdot d\mathbf{S} \tag{1.1c}$$

$$\oint_l \mathbf{H} \cdot d\mathbf{l} = \iint_S \mathbf{J} \cdot d\mathbf{S} + \frac{\partial}{\partial t} \iint_S \mathbf{D} \cdot d\mathbf{S} \tag{1.1d}$$

where $\mathbf{E}$ and $\mathbf{H}$ are the electric and magnetic field vectors, respectively. $\mathbf{D}$ is the electric flux density or *electric displacement*, which is related to the electric field via the dielectric permittivity, ε, of the constituent medium,

$$\mathbf{D} = \varepsilon \mathbf{E} \tag{1.2}$$

Similarly, $\mathbf{B}$ is the magnetic flux density, defined by the constitutive relation

$$\mathbf{B} = \mu \mathbf{H} \tag{1.3}$$

where μ is the magnetic permeability of the medium. These fields, in principle, are generated by charges and their motions, such that the charge density, ρ, and current density, $\mathbf{J}$, are referred to in these equations as *sources*. The integrals are taken on the surface S (with outward directed vector $\mathbf{S}$) enclosing an arbitrary volume V, or on the closed loop l (with directed arc length vector $\mathbf{l}$) enclosing an open surface S as selected to suit the boundary conditions of the problem under consideration. These subscripts may be omitted from the integrals in question when the region of integration is clear from context.

The electric and magnetic fields in turn exert a force, $\mathbf{F}$, on charged particles according to the *Lorentz law*,

$$\mathbf{F} = q\left(\mathbf{E} + \mathbf{v} \times \mathbf{B}\right) \tag{1.4}$$

2 For the purposes of this book, we will work almost exclusively with the SI unit system, which is most prevalent in engineering fields, as opposed to the Gaussian-CGS units, which in the past were popular in certain scientific disciplines such as theoretical physics and astronomy [1].

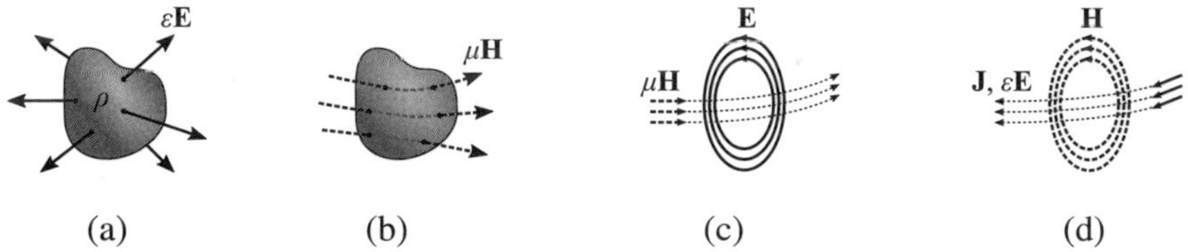

Figure 1.1 Visual representation of Maxwell's equations. (a) Gauss's law, eq (1.1a). (b) Nonexistence of magnetic charge, (1.1b). (c) Faraday's law, (1.1c). (d) Generalized Ampère's law, (1.1d).

where q is the electric charge and $\mathbf{v}$ is the velocity of motion of that particle, if any.

In general, the permittivity and permeability are anisotropic (direction-dependent) functions of space, time, and even field strength in nonlinear media; however, it is sufficient in most cases to consider them scalar constants on the temporal and spatial domain as well as the energy scale of a particular solution, as written above. Similarly, as with any classical field theory, it is now recognized that a correction (quantum electrodynamics) is needed to account for certain phenomena observed at the smallest scales, such as photon-photon interactions. However, for the purposes of all cases discussed in this book, the above formulation of Maxwell's equations, the constitutive relations, and the associated forces may be considered as exact and fundamentally true [2].

1.1.2 Physical Interpretation

A visual representation of Maxwell's equations is given in Figure 1.1. The first, (1.1a), illustrated in Figure 1.1(a), is known as Gauss's law and states that the total electric field strength emanating from a finite volume of space is equal to the net electric charge within that region. The second[3], (1.1b), illustrated in Figure 1.1(b), has a similar form, describing the total magnetic field strength emanating from a volume of space. As there is no magnetic charge, the magnetic field lines must be continuous throughout the region, and the number of lines entering it must be the same as the number leaving it.

The third, (1.1c), illustrated in Figure 1.1(c), is Faraday's law of induction, which states that the total electric field around any closed loop is associated with

3 There is no universally accepted name for the second of Maxwell's equations. Some call it "Gauss's law for magnetism," in recognition of its similarity to the original Gauss's law, others call it the "Absence/Nonexistence of Magnetic Poles/Charge" or the "Magnetic Nondivergence," and still others simply concede that it has no name.

an opposing increase of the magnetic field threading through that loop (opposing in this case because the time derivative of the magnetic field is negative when oriented according to the right-hand rule [3]).

Once again, there is a similar relation for magnetic field loops in (1.1d), illustrated in Figure 1.1(d), known as Ampère's law. In this case the center of the loop under consideration may be threaded by either a changing electric field, or an electric current, or both.

To summarize,

- The electric field emanates from electrical charge (1.1a) and circulates around changing magnetic fields (1.1c).
- The magnetic field, in turn, emanates from nowhere (1.1b) but forms closed loops threaded by electrical currents and changing electric fields (1.1d).

Note that in a *source-free region*, where $\rho = 0$ and $\mathbf{J} = 0$, the set of equations becomes symmetric in $\mathbf{E}$ and $\mathbf{H}$, in that one may be exchanged for the other without altering their relationships (with exception of a sign change in (1.1d) and some scaling factors associated with the impedance of the medium). This is a condition known as *duality* which plays an elegant role in the solution of many electromagnetic and circuit problems [4]. It shows that a field solution for given stimuli and boundary conditions applies equally well to the dual problem involving magnetic boundaries in place of electric boundaries, $\mathbf{E}$ in place of $\mathbf{H}$, and vice versa, with a reversal of polarity in the coupled fields resulting from the sign change. Duality will be discussed in the context of electrical networks in Section 5.1.7.

In the original form of Ampère's law, discovered experimentally, the second term involving the time derivative of the electric field was missed. This term (or its spatially localized form) is known as the *displacement current* and was introduced by James Clerk Maxwell to resolve certain inconsistencies in the remaining set of equations. One such inconsistency is the continuity of electric charge.

1.1.3 The Continuity Equation

Consider a surface, S, enclosing a finite volume of space with a small cutout bounded by a loop, l, as shown in Figure 1.2(a). We may apply (1.1d) to this geometry by integrating the magnetic field, $\mathbf{H}$, around the closed loop on the left side of the equation, and the physical current density, $\mathbf{J}$, plus the displacement current density, $d\mathbf{D}/dt$, through the surface on the right side of the equation. Let us further take the limit as the cutout defined by the loop, l, becomes vanishingly small. Since the magnetic field must be finite, the path integral on the left side

Figure 1.2 Derivation of the continuity equation for electric charge. (a) Evaluation of the path integral around a vanishingly small closed loop. (b) Summing the contributions from two coincident path integrals in opposite directions.

of (1.1d) approaches zero as the loop circumference diminishes. In this limit, the surface S also becomes closed, allowing us to substitute (1.1d) into (1.1a) such that

$$0 = \oiint \mathbf{J} \cdot d\mathbf{S} + \frac{\partial}{\partial t} \oiint \mathbf{D} \cdot d\mathbf{S} = \iint \mathbf{J} \cdot d\mathbf{S} + \frac{\partial}{\partial t} \iiint \rho dV \tag{1.5a}$$

$$\therefore \oiint \mathbf{J} \cdot d\mathbf{S} = -\frac{\partial}{\partial t} \iiint \rho dV \tag{1.5b}$$

This result, known as the *continuity equation*, makes explicit the conservation of electric charge. A net current entering a certain region of space must be associated with an accumulation of charge there, whereas a current exiting must lead to charge depletion.

If the reader is troubled by the concept of a path integral around a vanishingly small closed loop, an alternate derivation may proceed as follows. Instead of allowing the loop l to shrink in size, we evaluate the path integral in both directions, counter clockwise for surface S_1 and clockwise for surface S_2, as shown in Figure 1.2(b). By summing the contributions from both integrals, we obtain

$$\iint_{S_1} \mathbf{J} \cdot d\mathbf{S} + \iint_{S_2} \mathbf{J} \cdot d\mathbf{S} + \frac{\partial}{\partial t} \iint_{S_1} \mathbf{D} \cdot d\mathbf{S} + \frac{\partial}{\partial t} \iint_{S_2} \mathbf{D} \cdot d\mathbf{S}$$
$$= \oint_l \mathbf{H} \cdot d\mathbf{l} - \oint_l \mathbf{H} \cdot d\mathbf{l} = 0 \tag{1.6a}$$

$$\therefore \oiint_{S_1+S_2} \mathbf{J} \cdot d\mathbf{S} + \frac{\partial}{\partial t} \oiint_{S_1+S_2} \mathbf{D} \cdot d\mathbf{S} = 0 \tag{1.6b}$$

Figure 1.3 (a) Closed surface for determination of normal (perpendicular) fields at a conducting boundary. (b) Closed loop for determination of tangential fields at a conducting boundary.

$$\oiint_{S_1+S_2} \mathbf{J} \cdot d\mathbf{S} = -\frac{\partial}{\partial t} \iiint_V \rho dV \tag{1.6c}$$

To some, the truth of the continuity equation may seem rather obvious. Of significance, however, is that this result would not have been obtained without the displacement current term added by Maxwell, for in that case we would have found that the net current entering or leaving a finite region of space must always be zero. Put another way, if magnetic fields did not respond to changing electric fields as Maxwell suggested, it would be physically impossible for a non-neutral charge to accumulate anywhere in space (nor would it be possible to discharge any accumulated electrostatic potential leftover somehow by the initial conditions of the universe).

1.1.4 Boundary Conditions on a Perfect Conductor

To simplify the derivation of the behaviors of many kinds of circuit elements, it is useful to enumerate first the solutions of Maxwell's equations at the interface between conductive and nonconductive media (e.g., between metal and a dielectric insulator).

For many metals, it is a good approximation to consider their conductivity as infinite for the initial field solution and then account for their losses as a small perturbation. Such a medium will be a called a *perfect electric conductor* (PEC) or electric wall. It can be shown that both the electric and magnetic fields must vanish inside a perfect conductor, so we will be concerned primarily with the normal (perpendicular) and tangential components of the electromagnetic fields just outside the interface to the conductor.

Consider the closed surface drawn in Figure 1.3(a). Sometimes called a *pillbox*, we assume that it is vanishingly small in height and small enough (but finite) in diameter that the fields may be considered constant throughout. By applying

(1.1a), we find

$$\oiint \mathbf{D} \cdot d\mathbf{S} = (\mathbf{n} \cdot \mathbf{D})\, A = \rho_s A \tag{1.7a}$$

$$\mathbf{n} \cdot \mathbf{E} = \frac{\rho_s}{\varepsilon} \tag{1.7b}$$

where A is the area of the top of the pillbox, $\mathbf{n}$ is the unit surface normal, and ρ_s is the surface (as opposed to volumetric) charge density. Similarly, by applying (1.1b), we find for the magnetic field

$$\mathbf{n} \cdot \mathbf{B} = 0 \tag{1.8a}$$

$$\therefore \mathbf{n} \cdot \mathbf{H} = 0 \tag{1.8b}$$

To probe the tangential fields at the surface, we use a small closed loop that runs parallel to the interface lengthwise above and below the conducting boundary, as shown in Figure 1.3(b). In this case we assume that the length of loop parallel to the surface is small enough that the fields are constant along it, while the height of the loop is vanishingly small. Consider (1.1c) on this path, from which we can derive the tangential component of the electric field. The surface integral of $\mathbf{B}$ vanishes because the field is finite and the area enclosed by the loop is infinitesimal. The derivative of this surface integral also vanishes because we assume that the fields cannot change value infinitely fast. Therefore, we have

$$\oint \mathbf{E} \cdot d\mathbf{l} = (\mathbf{n} \times \mathbf{E})\, L = 0 \tag{1.9a}$$

$$\therefore \mathbf{n} \times \mathbf{E} = 0 \tag{1.9b}$$

where L is the small but finite length of the loop. Similarly, applying (1.1d) for the magnetic field,

$$\oint \mathbf{H} \cdot d\mathbf{l} = (\mathbf{n} \times \mathbf{H})\, L = \mathbf{J}_s L \tag{1.10a}$$

$$\therefore \mathbf{n} \times \mathbf{H} = \mathbf{J}_s \tag{1.10b}$$

where $\mathbf{J}_s$ is the current density at the surface, and we have used the fact that

$$\iint \mathbf{D} \cdot d\mathbf{S} = 0 \tag{1.11}$$

owing to the vanishingly small size of the area enclosed by the loop.

In conclusion, we find that the electric field must always be perpendicular to the surface of a perfect electric conductor with strength proportional to the charge concentrated at the surface, while the magnetic field must be entirely parallel to it, but perpendicular to the direction of current flow, with strength equal to the surface current density [5].

1.1.5 Boundary Conditions at a Dielectric Interface

Similar loop and pillbox test structures may be used to derive the boundary conditions from Maxwell's equations at an interface between two dielectric media instead of between a dielectric and a conductor. Let us consider the fields, $\mathbf{E}_1$ and $\mathbf{H}_1$, in the first dielectric medium having constituent parameters ε_1 and μ_1, and the fields in the second medium, $\mathbf{E}_2$ and $\mathbf{H}_2$, where the second medium has constituent parameters ε_2 and μ_2.

Since dielectric media are generally nonconducting, there is normally no current at the interface, nor any mechanism for a buildup of unbound electric charge. For this derivation, then, we assume that the interface is source free. The pillbox gives us the normal component of the electric field via (1.1a)

$$\oiint \mathbf{D} \cdot d\mathbf{S} = (\varepsilon_1 \mathbf{n} \cdot \mathbf{E}_1 - \varepsilon_2 \mathbf{n} \cdot \mathbf{E}_2)\, A = 0 \tag{1.12a}$$

$$\therefore \mathbf{n} \cdot (\varepsilon_1 \mathbf{E}_1 - \varepsilon_2 \mathbf{E}_2) = 0 \tag{1.12b}$$

Similarly, the normal component of the magnetic field is found by (1.1b)

$$\oiint \mathbf{B} \cdot d\mathbf{S} = (\mu_1 \mathbf{n} \cdot \mathbf{H}_1 - \mu_2 \mathbf{n} \cdot \mathbf{H}_2)\, A = 0 \tag{1.13a}$$

$$\therefore \mathbf{n} \cdot (\mu_1 \mathbf{H}_1 - \mu_2 \mathbf{H}_2) = 0 \tag{1.13b}$$

The tangential fields are found by evaluating (1.1c) and (1.1d) around a small loop at the surface

$$\oint \mathbf{E} \cdot d\mathbf{l} = (\mathbf{n} \times \mathbf{E}_1 - \mathbf{n} \times \mathbf{E}_2)\, L = 0 \tag{1.14a}$$

$$\therefore \mathbf{n} \times (\mathbf{E}_1 - \mathbf{E}_2) = 0 \tag{1.14b}$$

$$\oint \mathbf{H} \cdot d\mathbf{l} = (\mathbf{n} \times \mathbf{H}_1 - \mathbf{n} \times \mathbf{H}_2)\, L = 0 \tag{1.14c}$$

$$\therefore \mathbf{n} \times (\mathbf{H}_1 - \mathbf{H}_2) = 0 \tag{1.14d}$$

In conclusion, the tangential electric and magnetic field components are continuous across a dielectric boundary, while the normal field components change discontinuously at the boundary in proportion to the relevant constituent parameters (ε and μ) [5].

1.1.6 Differential Form

It is often convenient to work with Maxwell's equations in differential form instead of the integral form. The differential form of Gauss's law may be found by application of the divergence theorem (also called Gauss's theorem [6] or *Ostrogradsky's theorem*) to the surface integral of the electric flux density,

$$\oiint \mathbf{D} \cdot d\mathbf{S} = \iiint (\nabla \cdot \mathbf{D})\, dV = \iiint \rho dV \tag{1.15}$$

The symbol ∇, called the *del operator*, is given in Cartesian coordinates by

$$\nabla = \frac{\partial}{\partial x}\mathbf{x} + \frac{\partial}{\partial y}\mathbf{y} + \frac{\partial}{\partial z}\mathbf{z} \tag{1.16}$$

where $\mathbf{x}$, $\mathbf{y}$, and $\mathbf{z}$, are the unit vectors in the direction of the spatial coordinates x, y, and z, respectively. The del operator applied directly to a scalar function is referred to as the *gradient*, while the dot product of the del operator with a vector field is the *divergence* of that field, and the cross product is its *curl*.

Since (1.16) must hold for any arbitrary choice of the enclosing surface, S, the arguments inside the volume integrals must be equivalent at all points. We may therefore remove the integrals, and are left with

$$\nabla \cdot \mathbf{D} = \rho \tag{1.17}$$

Similarly, for the magnetic flux density, we have

$$\nabla \cdot \mathbf{B} = 0 \tag{1.18}$$

We may also apply Stokes' theorem [6] to the path integrals in (1.1c) and (1.1d), such that

$$\oint \mathbf{E} \cdot d\mathbf{l} = \iint (\nabla \times \mathbf{E}) \cdot d\mathbf{S} = -\frac{\partial}{\partial t} \iint \mathbf{B} \cdot d\mathbf{S} \tag{1.19a}$$

$$\therefore \nabla \times \mathbf{E} = -\frac{\partial \mathbf{B}}{\partial t} \tag{1.19b}$$

and

$$\oint \mathbf{H} \cdot d\mathbf{l} = \iint (\nabla \times \mathbf{H}) \cdot d\mathbf{S} = \iint \mathbf{J} \cdot d\mathbf{S} + \frac{\partial}{\partial t} \iint \mathbf{D} \cdot d\mathbf{s} \tag{1.20a}$$

$$\therefore \nabla \times \mathbf{H} = \mathbf{J} + \frac{\partial \mathbf{D}}{\partial t} \tag{1.20b}$$

The differential form of the continuity equation then follows directly by taking the divergence of (1.20b), noting that the divergence of the curl of any vector field must be zero [6], and then substituting from (1.17),

$$\nabla \cdot (\nabla \times \mathbf{H}) = \nabla \cdot \mathbf{J} + \frac{\partial}{\partial t} (\nabla \cdot \mathbf{D}) \tag{1.21a}$$

$$0 = \nabla \cdot \mathbf{J} + \frac{\partial \rho}{\partial t} \tag{1.21b}$$

$$\therefore \nabla \cdot \mathbf{J} = -\frac{\partial \rho}{\partial t} \tag{1.21c}$$

It is worth pointing out that Maxwell's equations are not mutually independent. Take, for example, the divergence of (1.19b),

$$\nabla \cdot (\nabla \times \mathbf{E}) = -\frac{\partial}{\partial t} (\nabla \cdot \mathbf{B}) = 0 \tag{1.22a}$$

$$\therefore \nabla \cdot \mathbf{B} = \rho_m \tag{1.22b}$$

where ρ_m is a placeholder for a fictitious magnetic charge, a dual counterpart to electric charge. Mathematically, this is consistent, but since no such charge is known to exist, we traditionally set this term equal to zero, and (1.22b) becomes equivalent to (1.18). It should be noted, however, that if such a charge is ever proven to exist, ρ_m may be reintroduced without any further disruption to the accuracy or self-consistency of Maxwell's equations.

1.1.7 Differential Boundary Conditions

In Section 1.1.4, boundary conditions were derived for fields adjacent to a PEC. A useful special case occurs when the PEC borders on a source-free region. In such a case, we may use the differential form of Maxwell's equations to derive constraints on the field derivatives as well.

We already know that the electric field must be perpendicular to the boundary (it has no tangential component), and the magnetic field must be parallel to the boundary (it has no surface normal component). We thus may write

$$\mathbf{E} = E\mathbf{n} \tag{1.23a}$$

$$\mathbf{H} = H\mathbf{u} \tag{1.23b}$$

where $\mathbf{u}$ is the tangential unit vector aligned with the magnetic field. Further, let $\mathbf{v} = \mathbf{u} \times \mathbf{n}$. According to Ampère's law[4], then,

$$\nabla \times \mathbf{H} = \frac{\partial \mathbf{D}}{\partial t} \tag{1.24a}$$

$$\frac{\partial H}{\partial v}\mathbf{n} - \frac{\partial H}{\partial n}\mathbf{v} = \varepsilon \frac{\partial E}{\partial t}\mathbf{n} \tag{1.24b}$$

$$\therefore \frac{\partial H}{\partial n} = 0 \tag{1.24c}$$

However, Faraday's law tells us

$$\nabla \times \mathbf{E} = -\frac{\partial \mathbf{B}}{\partial t} \tag{1.25a}$$

$$-\frac{\partial E}{\partial v}\mathbf{u} + \frac{\partial E}{\partial u}\mathbf{v} = -\mu \frac{\partial H}{\partial t}\mathbf{u} \tag{1.25b}$$

$$\therefore \frac{\partial E}{\partial u} = 0 \tag{1.25c}$$

It is important to realize that this last result does not imply the electric field has no derivative tangential to the surface, only that it has no derivative tangential to the surface except in the direction perpendicular to the magnetic field (that is, $\mathbf{E}$ varies in the $\mathbf{v}$ direction, not the $\mathbf{u}$ direction).

At a dielectric interface, neither normal nor tangential field components are excluded, but we may still derive relationships pertaining to the continuity of certain derivatives near its surface. If we apply Gauss's law separately in region 1 and region 2 proximate to the boundary, and assume that there is no charge present,

$$\nabla \cdot \mathbf{E}_1 = 0 \tag{1.26a}$$

$$\nabla \cdot \mathbf{E}_2 = 0 \tag{1.26b}$$

and then subtract the two, we have

$$\nabla \cdot (\mathbf{E}_1 - \mathbf{E}_2) = 0 \tag{1.27}$$

4 By considering the source-free region just outside the conductor, we may assume that $\mathbf{J} = 0$. Right at the boundary, however current flows as an infinitely thin sheet; thus, the volume density approaches ∞. This allows $\mathbf{H}$ to change discontinuously, from a finite value outside the conductor to zero inside.

Table 1.1

Boundary Conditions of Electromagnetic Fields

Interface	Condition Type	$\mathbf{E}$ Condition	$\mathbf{H}$ Condition
Dielectric	Normal	$\mathbf{n} \cdot (\varepsilon_1 \mathbf{E}_1 - \varepsilon_2 \mathbf{E}_2) = 0$	$\mathbf{n} \cdot (\mu_1 \mathbf{H}_1 - \mu_2 \mathbf{H}_2) = 0$
Dielectric	Tangential	$\mathbf{n} \times (\mathbf{E}_1 - \mathbf{E}_2) = 0$	$\mathbf{n} \times (\mathbf{H}_1 - \mathbf{H}_2) = 0$
Dielectric	Differential	$\mathbf{n} \cdot \left(\frac{\partial \mathbf{E}_1}{\partial n} - \frac{\partial \mathbf{E}_2}{\partial n} \right) = 0$	$\mathbf{n} \cdot \left(\frac{\partial \mathbf{H}_1}{\partial n} - \frac{\partial \mathbf{H}_2}{\partial n} \right) = 0$
PEC	Normal	$\mathbf{n} \cdot \mathbf{E} = \frac{\rho_s}{\varepsilon}$	$\mathbf{n} \cdot \mathbf{H} = 0$
PEC	Tangential	$\mathbf{n} \times \mathbf{E} = 0$	$\mathbf{n} \times \mathbf{H} = \mathbf{J}_s$
PEC	Differential	$\frac{\partial E_n}{\partial u} = 0$	$\frac{\partial H_u}{\partial n} = 0$

$\mathbf{n}$ is the unit vector normal to the surface (see Figure 1.3).
$\mathbf{u}$ is the unit vector tangential to the surface and aligned with $\mathbf{H}$.
Dielectric regions are presumed source-free.

Since the tangential components of $\mathbf{E}$ are continuous across the interface, the difference $\mathbf{E}_1 - \mathbf{E}_2$ above contains only a normal component. Thus,

$$\mathbf{n} \cdot \left(\frac{\partial \mathbf{E}_1}{\partial n} - \frac{\partial \mathbf{E}_2}{\partial n} \right) = 0 \tag{1.28}$$

Therefore, while the normal component of the electric field may change discontinuously across a dielectric boundary, its derivative in the direction normal to that boundary does not.

Similarly, for the magnetic field, we may apply the magnetic nondivergence law to obtain

$$\mathbf{n} \cdot \left(\frac{\partial \mathbf{H}_1}{\partial n} - \frac{\partial \mathbf{H}_2}{\partial n} \right) = 0 \tag{1.29}$$

A summary of the boundary conditions for electromagnetic fields appears in Table 1.1.

1.2 STATIC FIELDS

An important limiting case in electromagnetics is that which occurs when all boundary conditions have ceased changing, and all source charges and currents are stationary. We may thus rewrite Maxwell's equations (and the continuity equation) for this special case with the time derivatives replaced by zero,

$$\nabla \cdot \mathbf{D} = \rho \tag{1.30a}$$

$$\nabla \cdot \mathbf{B} = 0 \tag{1.30b}$$

$$\nabla \times \mathbf{E} = 0 \tag{1.30c}$$

$$\nabla \times \mathbf{H} = \mathbf{J} \tag{1.30d}$$

$$\nabla \cdot \mathbf{J} = 0 \tag{1.30e}$$

Note that, just as a stationary charge may exist in the absence of any current, a stationary or time-invariant current may exist in the absence of any isolated charge. This is true in the context of macroscopic electric and magnetic fields, even though we know in reality that an electrically neutral current is composed typically of negative charges (electrons) moving against an immobile background of neutralizing positive charge (the conductor lattice ions), or of equal parts positive and negative charges (electrons and holes) moving simultaneously in opposite directions (as in a semiconductor).

Therefore, we see that the static field assumption has effectively decoupled the electric and magnetic fields, with (1.30a) and (1.30c), along with the constitutive relations in Section 1.1.1, describing the electric field as emanating from isolated charge alone, and the magnetic field in (1.30b) and (1.30d) arising as a consequence solely of current. The two fields may therefore be calculated independently, the former as a problem in *electrostatics* and the latter as a problem in *magnetostatics*.

1.2.1 Electrostatics

Equations (1.30a) and (1.30c) reveal the electrostatic field to be curl free everywhere with divergence only where non-neutralized charge exists. It is therefore convenient to make use of a mathematical identity that states that the curl of a gradient of any scalar function in three dimensions is identically zero. That is,

$$\nabla \times (\nabla \varphi) = 0 \tag{1.31}$$

for any scalar function of three dimensions, φ (note that the truth of this statement is limited to three-dimensional quantities only because the curl, unlike the gradient itself, cannot be defined for any but three dimensions). We may then write the electrostatic field as the gradient of a scalar function,

$$\mathbf{E} = -\nabla\varphi \tag{1.32}$$

where φ is known as the *electrostatic potential.* The absence of curl in $\mathbf{E}$ is then guaranteed by (1.31), while substitution of (1.32) into (1.30a) yields

$$\nabla \cdot \mathbf{D} = \nabla \cdot (\varepsilon\mathbf{E}) = -\nabla \cdot (\varepsilon\nabla\varphi) = -\varepsilon\nabla^2\varphi = \rho \tag{1.33a}$$

$$\nabla^2\varphi = -\frac{\rho}{\varepsilon} \tag{1.33b}$$

The del-squared notation is shorthand for the divergence of the gradient and is known as the *Laplacian operator.* Although mathematically the sign of φ is arbitrary, we have used the negative sign in (1.32) to ensure that our electric field potential is highest (most positive) near positive charges and lowest (most negative) near negative charges. Equation (1.33b) is known as *Poisson's equation.* If the region is source-free, we have

$$\nabla^2\varphi = 0 \tag{1.34}$$

which is known as *Laplace's equation.*

The value of defining an electrostatic potential is that we may now solve for the electrostatic field by first solving for the potential using, say, Laplace's equation, which is a single differential equation in three dimensions, as opposed to (1.30a), which is a system of three differential equations in three dimensions.

Consider the electrostatic potential field created by a single point charge, Q, located at the origin. Laplace's equation therefore applies at all points except the origin, and in spherical coordinates has the form

$$\nabla^2\varphi = \frac{1}{r^2}\frac{\partial}{\partial r}\left(r^2\frac{\partial\varphi}{\partial r}\right) + \frac{1}{r^2\sin\theta}\frac{\partial}{\partial\theta}\left(\sin\theta\frac{\partial\varphi}{\partial\theta}\right) + \frac{1}{r^2\sin^2\theta}\frac{\partial^2\varphi}{\partial\phi^2} = 0 \tag{1.35}$$

This may appear quite intimidating at first, but the selection of spherical coordinates was not arbitrary. Due to the symmetry of our problem, we know that our potential field will depend only on the distance from the origin, r, and will not depend on θ or ϕ. The last two terms therefore drop out, and we are left with

$$\frac{1}{r^2}\frac{\partial}{\partial r}\left(r^2\frac{\partial\varphi}{\partial r}\right) = 0 \tag{1.36a}$$

$$\frac{\partial \varphi}{\partial r} = C_1 r^{-2} \tag{1.36b}$$

$$\varphi(r) = C_2 - C_1 r^{-1} \tag{1.36c}$$

To determine the constants C_1 and C_2, we must consider the boundary conditions. First, let us integrate the Poisson's equation over a spherical volume of radius r surrounding the origin,

$$\iiint \nabla^2 \varphi dV = -\iiint \frac{\rho}{\varepsilon} dV = -\frac{Q}{\varepsilon} \tag{1.37}$$

Since the Laplacian operator comprises a divergence, we may apply the divergence theorem to the integral on the left side,

$$\iiint \nabla^2 \varphi dV = \iiint \nabla \cdot \nabla \varphi dV = \oiint \nabla \varphi \cdot d\mathbf{S} = -\frac{Q}{\varepsilon} \tag{1.38a}$$

$$\oiint \frac{\partial \varphi}{\partial r} dS = -\frac{Q}{\varepsilon} \tag{1.38b}$$

$$\oiint C_1 r^{-2} dS = -\frac{Q}{\varepsilon} \tag{1.38c}$$

$$C_1 r^{-2} \cdot 4\pi r^2 = -\frac{Q}{\varepsilon} \tag{1.38d}$$

$$C_1 = -\frac{Q}{4\pi\varepsilon} \tag{1.38e}$$

$$\therefore \varphi(r) = C_2 + \frac{Q}{4\pi\varepsilon r} \tag{1.38f}$$

Note that the choice of C_2 is arbitrary. We have already satisfied all conditions required of the electric potential. It is customary to let the potential in a geometry such as this one fall to zero at a distance of infinity from the location of the charge. Therefore, we may simply let $C_2 = 0$. An illustration of this solution (and the electric field which is the negative gradient of this potential function) is shown in Figure 1.4.

Because Laplace's equation is a linear homogeneous differential equation, the principle of superposition holds. Importantly, this means that the solution of (1.38f) may be generalized to arbitrary charge distributions by integrating the individual

Figure 1.4 Illustration of electric field calculation using electrostatic potential around (a) a point charge, and (b) an arbitrary charge distribution. The closed loops represent equipotential contours, while the arrows represent electric field lines.

contributions of the infinitesimal charge elements. Thus,

$$\varphi = \frac{1}{4\pi\varepsilon} \iiint \frac{\rho}{r} dV \tag{1.39}$$

where r is the distance from any point within the volume V to the point at which the electric potential is to be calculated. A drawing of this calculated potential and electric field for an arbitrary volume of charge is given in Figure 1.4(b).

1.2.2 Magnetostatics

In contrast with the electrostatic fields, the magnetostatic field is found from (1.30b) and (1.30d) to be nondivergent, but having curl about its corresponding source quantity, the steady-state current. In this case, we make use of the fact that the divergence of a curl of any vector field is always zero, or

$$\nabla \cdot (\nabla \times \mathbf{A}) = 0 \tag{1.40}$$

for any three-dimensional vector field, **A**. It is customary to define the magnetic flux density as the curl of this *magnetic vector potential*,

$$\mathbf{B} = \nabla \times \mathbf{A} \tag{1.41}$$

The nondivergent property is thus guaranteed, and substitution into (1.30d) gives

$$\nabla \times \mathbf{H} = \nabla \times \left(\tfrac{1}{\mu}\mathbf{B}\right) = \nabla \times \left(\tfrac{1}{\mu}\nabla \times \mathbf{A}\right) = \tfrac{1}{\mu}\nabla \times \nabla \times \mathbf{A} = \mathbf{J} \tag{1.42a}$$

$$\therefore \nabla (\nabla \cdot \mathbf{A}) - \nabla^2 \mathbf{A} = \mu \mathbf{J} \tag{1.42b}$$

where the last step follows from the identity (C.12) in Appendix C.

Of course, this could hardly be described as a simplification of (1.30d), but we have some freedom in how to define the vector potential, since it is a quantity of our own specification, and multiple distinct vector fields may have the same curl. There are thus multiple ways to further constrain (1.42b), but for the purposes of maximum simplification in magnetostatics, we use the following,

$$\nabla \cdot \mathbf{A} = 0 \tag{1.43}$$

which is known as the *Coulomb gauge* for the magnetic vector potential [7]. The source equation for the magnetic vector potential then becomes

$$\nabla^2 \mathbf{A} = -\mu \mathbf{J} \tag{1.44}$$

which may be recognized again as Poisson's equation for each of the vector components of $\mathbf{A}$ in Cartesian coordinates.

We note the similarity between (1.44) and Poisson's equation for the electric field potential in (1.33b). In fact, the same final result from (1.39) may be imported here with a change of variables ($\varphi \rightarrow \mathbf{A}$, $\rho \rightarrow \mathbf{J}$, and $\varepsilon \rightarrow \frac{1}{\mu}$). That is,

$$\mathbf{A} = \frac{\mu}{4\pi} \iiint \frac{\mathbf{J}}{r} dV \tag{1.45}$$

For current, I, flowing through a thin wire, we may rewrite this as

$$\mathbf{A} = \frac{\mu I}{4\pi} \int \frac{d\mathbf{l}'}{r} \tag{1.46}$$

where $d\mathbf{l}'$ is the vector component of the wire geometry. In a true magnetostatic situation, the integral would have to be carried out over a closed path (i.e., in a loop of wire) to avoid having charge accumulate at one end or the other; however we allow for the nonclosed integral above for hypothetical situations such as an infinitely long wire carrying a DC current.

The magnetic flux density is then found as the curl of this vector potential,

$$\mathbf{B} = \nabla \times \mathbf{A} = \frac{\mu I}{4\pi} \int \nabla \times \left(\frac{d\mathbf{l}'}{r} \right) \tag{1.47a}$$

$$= \frac{\mu I}{4\pi} \int \left[\frac{1}{r} (\nabla \times d\mathbf{l}') + \nabla \left(\frac{1}{r} \right) \times d\mathbf{l}' \right] \tag{1.47b}$$

An important subtlety in dealing with this equation is that the curl operation above is carried out on the coordinates of the fields $\mathbf{A}$ and $\mathbf{B}$, whereas the path integral follows the primed coordinates of the wire, $d\mathbf{l}'$. These two coordinate systems vary independently, so the curl of $d\mathbf{l}'$ in the first term of the integral above is zero. The second term, the gradient of $1/r$, is given by

$$\nabla \left(\frac{1}{r} \right) = -\frac{\mathbf{r}}{r^2} \tag{1.48}$$

where $\mathbf{r}$ is the unit vector directed from the wire element (primed coordinates) inside the integral to the field point (unprimed coordinates). Substituting this back into (1.47b), we find

$$\mathbf{B} = \frac{\mu I}{4\pi} \int \frac{d\mathbf{l}' \times \mathbf{r}}{r^2} \tag{1.49}$$

This is known as the *Biot-Savart law* [7].

Since the magnetic potential, $\mathbf{A}$, is a vector, its mathematical utility is not as immediately obvious as the scalar electric potential, φ aside from the above analogy to known electrostatic solutions. We will make only sparing use of it in this book (mostly in the context of antennas in Chapter 9); however, it is interesting to note that some comparatively recent physical experiments have suggested the electric and magnetic field potentials to be in some ways more fundamental aspects of nature than the classic fields themselves, as it is possible for elementary particles under certain circumstances to respond to the presence of a potential, even when the field itself and the force it exerts are substantially absent [8, 9]. This viewpoint has not gone unchallenged, however [10].

1.2.3 Static Fields in a Coaxial Geometry

As a further example, let us consider the static fields inside a coaxial geometry, as shown in Figure 1.5. We assume two concentric, conducting cylinders, with radii a and b, separated by a dielectric medium. We assume a constant charge per unit length of $+Q$ on the inner conductor and a balancing charge per unit length of $-Q$ on the outer conductor. We further assume a constant current of $+I$ into the page on the inner conductor and a balancing current of $-I$ on the outer conductor. Note that the simultaneous presence of constant non-neutralized charges and currents on the same conductors leads to no contradictions in electrostatics and magnetostatics.

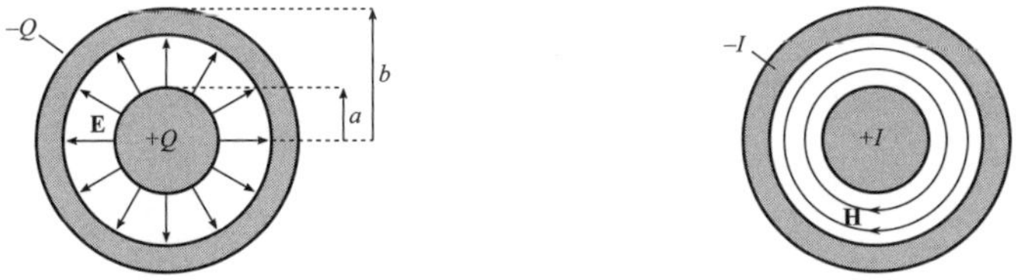

Figure 1.5 (a) Electrostatic and (b) magnetostatic fields in a coaxial geometry. Positive current flows into the page.

The physical situation is simply that of an endless stream of positive charge on the central conductor moving into the page, balanced by a continuous stream of negative charges on the outer conductor moving in the same direction. The two effects are simply uncoupled; the motion of charge, since it does not alter the charge distribution, has no impact on the electrostatic solution, while the existence of an unchanging charge distribution has no impact on the magnetostatic field.

To find the electrostatic field, we apply Laplace's equation in cylindrical coordinates to the region between the two conductors,

$$\nabla^2 \varphi = \frac{1}{r}\frac{\partial}{\partial r}\left(r\frac{\partial \varphi}{\partial r}\right) + \frac{1}{r^2}\frac{\partial^2 \varphi}{\partial \theta^2} + \frac{\partial^2 \varphi}{\partial z^2} = 0 \tag{1.50}$$

With this geometry, we may safely assume that there is no dependence on z or θ. Laplace's equation therefore simplifies,

$$\nabla^2 \varphi = \frac{1}{r}\frac{d}{dr}\left(r\frac{d\varphi}{dr}\right) = 0 \tag{1.51a}$$

$$r\frac{d\varphi}{dr} = C_1 \tag{1.51b}$$

$$\varphi(r) = C_1 \ln r + C_2 \tag{1.51c}$$

The electrostatic field is then given by (1.32),

$$\mathbf{E}(r) = -\nabla\varphi = -\mathbf{r}\frac{d\varphi}{dr} = -\mathbf{r}\frac{d}{dr}(C_1 \ln r + C_2) = -\frac{C_1}{r}\mathbf{r} \tag{1.52}$$

The constant of integration, C_1, may be determined by the electric boundary conditions on a perfect conductor, (1.7b),

$$\mathbf{n}\cdot\mathbf{E}(a) = -\frac{C_1}{a} = \frac{\rho_s}{\varepsilon} = \frac{Q}{2\pi a\varepsilon} \tag{1.53a}$$

$$C_1 = -\frac{Q}{2\pi\varepsilon} \tag{1.53b}$$

$$\therefore \mathbf{E} = \frac{Q}{2\pi\varepsilon r}\mathbf{r} \tag{1.53c}$$

For the magnetostatic fields, let us use (1.30d) in the space between the two conductors, noting that $\mathbf{H}$ may have only a $\boldsymbol{\theta}$-component, and again no dependence on z or θ (that is, $\mathbf{H} = H(r)\boldsymbol{\theta}$).

$$\nabla \times \mathbf{H} = \mathbf{z}\frac{1}{r}\frac{d}{dr}(rH) = 0 \tag{1.54a}$$

$$\mathbf{H} = \frac{C_3}{r}\boldsymbol{\theta} \tag{1.54b}$$

Using the tangential boundary condition for $\mathbf{H}$ from (1.10b),

$$\mathbf{n} \times \mathbf{H}(a) = \mathbf{J}_s \tag{1.55a}$$

$$\mathbf{r} \times \left(\boldsymbol{\theta}\frac{C_3}{a}\right) = \mathbf{z}\frac{C_3}{a} = \mathbf{z}\frac{I}{2\pi a} \tag{1.55b}$$

$$C_3 = \frac{I}{2\pi} \tag{1.55c}$$

$$\therefore \mathbf{H} = \frac{I}{2\pi r}\boldsymbol{\theta} \tag{1.55d}$$

1.3 WAVE SOLUTIONS

The most astounding outcome of Maxwell's work is the link between the speed of light and electromagnetic waves, and it was Maxwell who first posited that light itself was an electromagnetic wave. Consider the curl of (1.19b) in a source-free region,

$$\nabla \times \nabla \times \mathbf{E} = -\frac{\partial}{\partial t}(\nabla \times \mathbf{B}) = -\mu\frac{\partial}{\partial t}(\nabla \times \mathbf{H}) = -\mu\frac{\partial}{\partial t}\left(\varepsilon\frac{\partial \mathbf{E}}{\partial t}\right) \tag{1.56a}$$

$$\nabla(\nabla \cdot \mathbf{E}) - \nabla^2\mathbf{E} = -\mu\varepsilon\frac{\partial^2 \mathbf{E}}{\partial t^2} \tag{1.56b}$$

$$\nabla^2 \mathbf{E} = \mu\varepsilon \frac{\partial^2 \mathbf{E}}{\partial t^2} \tag{1.56c}$$

where (1.56c) follows since the divergence of $\mathbf{E}$ is zero in a source-free region. Similarly, by taking the curl of (1.20b) without sources, we have

$$\nabla \times \nabla \times \mathbf{H} = \frac{\partial}{\partial t}(\nabla \times \mathbf{D}) = \varepsilon \frac{\partial}{\partial t}(\nabla \times \mathbf{E}) = \varepsilon \frac{\partial}{\partial t}\left(-\mu \frac{\partial \mathbf{H}}{\partial t}\right) \tag{1.57a}$$

$$\nabla^2 \mathbf{H} = \mu\varepsilon \frac{\partial \mathbf{H}}{\partial t^2} \tag{1.57b}$$

Equations (1.56c) and (1.57b) are each forms of the wave equation in three dimensions. The geometry of the wave front in three dimensions can take on many forms, making it difficult to write down general solutions. It is therefore instructive to consider a fairly simple case, where the fields have no x or y dependence and are perpendicular to the z axis. We may then assume without loss of generality that

$$\mathbf{E} = E(z, t)\, \mathbf{x} \tag{1.58}$$

Substituting this into (1.56c), we have

$$\nabla^2 E(z, t)\, \mathbf{x} = \mu\varepsilon \frac{\partial^2}{\partial t^2} E(z, t)\, \mathbf{x} \tag{1.59a}$$

$$\frac{\partial^2}{\partial z^2} E(z, t)\, \mathbf{x} = \mu\varepsilon \frac{\partial^2}{\partial t^2} E(z, t)\, \mathbf{x} \tag{1.59b}$$

$$\frac{\partial^2 E}{\partial z^2} = \mu\varepsilon \frac{\partial^2 E}{\partial t^2} \tag{1.59c}$$

which has general solutions of the form

$$E(z, t) = E(z - v_p t) \tag{1.60}$$

where v_p is the *propagation velocity* or *phase velocity* and is given by

$$v_p = \pm \frac{1}{\sqrt{\mu\varepsilon}} \tag{1.61}$$

The reader may verify this is a solution by substituting (1.60) back into (1.59c).

In other words, under these conditions, the electric field may assume any profile whatsoever as a function of z, but that profile will evolve in time by a direct

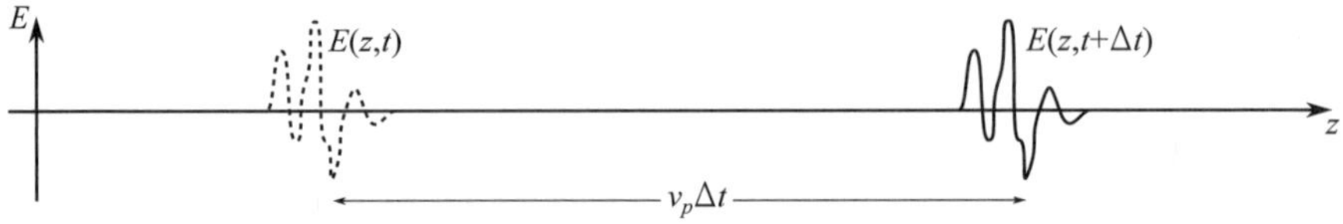

Figure 1.6 General solution for waves traveling in the z direction.

translation in the z direction at a constant rate of v_p, as shown in Figure 1.6. This is referred to as a linearly polarized *plane wave*, as the wavefront takes the form of an infinite plane (in this case, in the x-y plane) and the direction of the electric field vector is nonrotating. Plane waves are useful building blocks for complicated waveforms in many guided wave structures, as will be discussed in Chapter 7.

The magnetic field for this solution may be found by application of (1.20b),

$$\nabla \times \mathbf{H} = \frac{\partial \mathbf{D}}{\partial t} = \varepsilon \frac{\partial}{\partial t} E\left(z - v_p t\right) \mathbf{x} = -\varepsilon v_p E'\left(z - v_p t\right) \mathbf{x} \tag{1.62a}$$

$$= \mp \sqrt{\tfrac{\varepsilon}{\mu}} E'\left(z - v_p t\right) \mathbf{x} \tag{1.62b}$$

$$\therefore -\frac{\partial}{\partial z} H_y = \mp \sqrt{\tfrac{\varepsilon}{\mu}} E'\left(z - v_p t\right) \tag{1.62c}$$

$$\mathbf{H} = \pm \sqrt{\tfrac{\varepsilon}{\mu}} E\left(z - v_p t\right) \mathbf{y} \tag{1.62d}$$

Note that the positive sign in (1.62d) applies if v_p is positive, that is, if the wave is propagating in the $+z$ direction, and the minus sign applies if the wave is propagating in the $-z$ direction. The ratio of electric and magnetic field strength for any wave solution is known as the *wave impedance*. For this plane wave in unbounded space, that impedance is given by

$$\eta = \sqrt{\frac{\mu}{\varepsilon}} \tag{1.63}$$

In free space (vacuum), the permittivity and permeability have values given by $\varepsilon_0 \approx 8.854187817 \cdot 10^{-12}$ F/m and $\mu_0 = 4\pi \cdot 10^{-7}$ H/m, which are fundamental physical constants of nature. The wave impedance in that case is

$$\eta = \sqrt{\frac{\mu_0}{\varepsilon_0}} \approx 377\Omega \tag{1.64}$$

and the velocity of propagation is

$$|v_p| = \frac{1}{\sqrt{\mu_0 \varepsilon_0}} = c = 299,792,458 \text{ m/s} \tag{1.65}$$

which is exactly the speed of light in a vacuum.

The implications of this go far beyond just the realization that light is an electromagnetic wave, for there is nothing in these equations that suggests a medium through which the waves propagate. Physicists conducted numerous experiments in the nineteenth century trying to detect and characterize such a medium, which they called the *luminiferous aether*. Had the aether been detected, Maxwell's equations would have required modification to account for the velocity of this medium relative to some fixed coordinates or observational reference frame. Ultimately, all such experiments failed to show any evidence of the aether, and the concept was abandoned. Maxwell's original formulation stands, and the velocity at which electromagnetic waves propagate in free space is not dependent on the relative motion of the coordinate system. The speed of light through empty space simply *is*, for all observers. This is the basis of Einstein's Theory of Special Relativity.

1.4 SPECTRAL ANALYSIS

Although (1.60) shows that any arbitrary, translating waveform can be a valid solution to the wave equation (at least in one dimension), it is customary in modern circuit design to decompose a signal into spectral (frequency-domain) components, and to analyze the system behavior in terms of its response to those basis functions. The first step, spectral decomposition, is most commonly done using the Fourier transform [11],

$$A(\omega) = \mathfrak{F}\{a\}(\omega) = \int_{-\infty}^{\infty} a(t)\, e^{-j\omega t} dt \tag{1.66}$$

and its inverse,

$$a(t) = \mathfrak{F}^{-1}\{A\}(t) = \frac{1}{2\pi} \int_{-\infty}^{\infty} A(\omega)\, e^{j\omega t} d\omega \tag{1.67}$$

where $a(t)$ is any time-domain signal, such as a current, potential, or field quantity.

1.4.1 Maxwell's Equations in the Frequency Domain

Of particular importance when working in the frequency domain is that differential equations in time are transformed into algebraic equations in frequency, which are typically easier to solve. This is due to the property that differentiation in the time domain corresponds to a multiplication by the frequency variable in the frequency domain. Consider a function $b(t)$, which is defined as the derivative of another,

$$b(t) = \frac{d}{dt}a(t) \tag{1.68}$$

Using (1.67), we have

$$b(t) = \frac{d}{dt}\left(\frac{1}{2\pi}\int_{-\infty}^{\infty} A(\omega)\, e^{j\omega t} d\omega\right) = \frac{1}{2\pi}\int_{-\infty}^{\infty} j\omega A(\omega)\, e^{j\omega t} d\omega \tag{1.69a}$$

$$= \frac{1}{2\pi}\int_{-\infty}^{\infty} B(\omega)\, e^{j\omega t} d\omega \tag{1.69b}$$

$$\therefore B(\omega) = j\omega A(\omega) \tag{1.69c}$$

Applying this result to (1.1), we have for the large-scale form of Maxwell's equations in the frequency domain,

$$\oiint \mathbf{D} \cdot d\mathbf{S} = \iiint \rho dV \tag{1.70a}$$

$$\oiint \mathbf{B} \cdot d\mathbf{S} = 0 \tag{1.70b}$$

$$\oint \mathbf{E} \cdot d\mathbf{l} = -j\omega \iint \mathbf{B} \cdot d\mathbf{S} \tag{1.70c}$$

$$\oint \mathbf{H} \cdot d\mathbf{l} = \iint \mathbf{J} \cdot d\mathbf{S} + j\omega \iint \mathbf{D} \cdot d\mathbf{S} \tag{1.70d}$$

where $\mathbf{E}$, $\mathbf{H}$, $\mathbf{D}$, $\mathbf{B}$, ρ, and $\mathbf{J}$ are now all functions of frequency, ω. Similarly, by applying (1.69c) to (1.17) through (1.20b), we find the differential form of Maxwell's equations in the frequency domain,

$$\nabla \cdot \mathbf{D} = \rho \tag{1.71a}$$

Table 1.2
Maxwell's Equations in Various Forms

Name	Integral Form	Differential Form
	Time Domain	
Gauss's law	$\oiint \mathbf{D} \cdot d\mathbf{S} = \iiint \rho dV$	$\nabla \cdot \mathbf{D} = \rho$
Magnetic nondivergence	$\oiint \mathbf{B} \cdot d\mathbf{S} = 0$	$\nabla \cdot \mathbf{B} = 0$
Faraday's law	$\oint \mathbf{E} \cdot d\mathbf{l} = -\frac{\partial}{\partial t} \iint \mathbf{B} \cdot d\mathbf{S}$	$\nabla \times \mathbf{E} = -\frac{\partial \mathbf{B}}{\partial t}$
Ampère's law	$\oint \mathbf{H} \cdot d\mathbf{l} = \iint \mathbf{J} \cdot d\mathbf{S} + \frac{\partial}{\partial t} \iint \mathbf{D} \cdot d\mathbf{S}$	$\nabla \times \mathbf{H} = \mathbf{J} + \frac{\partial \mathbf{D}}{\partial t}$
The continuity equation	$\oiint \mathbf{J} \cdot d\mathbf{S} = -\frac{\partial}{\partial t} \iiint \rho dV$	$\nabla \cdot \mathbf{J} = -\frac{\partial \rho}{\partial t}$
	Frequency Domain	
Gauss's law	$\oiint \mathbf{D} \cdot d\mathbf{S} = \iiint \rho dV$	$\nabla \cdot \mathbf{D} = \rho$
Magnetic nondivergence	$\oiint \mathbf{B} \cdot d\mathbf{S} = 0$	$\nabla \cdot \mathbf{B} = 0$
Faraday's law	$\oint \mathbf{E} \cdot d\mathbf{l} = -j\omega \iint \mathbf{B} \cdot d\mathbf{S}$	$\nabla \times \mathbf{E} = -j\omega \mathbf{B}$
Ampère's law	$\oint \mathbf{H} \cdot d\mathbf{l} = \iint \mathbf{J} \cdot d\mathbf{S} + j\omega \iint \mathbf{D} \cdot d\mathbf{S}$	$\nabla \times \mathbf{H} = \mathbf{J} + j\omega \mathbf{D}$
The continuity equation	$\oiint \mathbf{J} \cdot d\mathbf{S} = -j\omega \iiint \rho dV$	$\nabla \cdot \mathbf{J} = -j\omega \rho$

$$\nabla \cdot \mathbf{B} = 0 \tag{1.71b}$$

$$\nabla \times \mathbf{E} = -j\omega \mathbf{B} \tag{1.71c}$$

$$\nabla \times \mathbf{H} = \mathbf{J} + j\omega \mathbf{D} \tag{1.71d}$$

Table 1.2 summarizes the various forms of Maxwell's equations articulated in this chapter.

By taking the curl of (1.71c) without sources, we may obtain the three-dimensional wave equation in the frequency domain,

$$\nabla \times \nabla \times \mathbf{E} = -j\omega \left(\nabla \times \mathbf{B}\right) = -j\omega\mu \left(\nabla \times \mathbf{H}\right) = \omega^2 \mu\varepsilon \mathbf{E} \tag{1.72a}$$

$$\nabla^2 \mathbf{E} + \omega^2 \mu\varepsilon \mathbf{E} = 0 \tag{1.72b}$$

or, as it sometimes written,

$$\left(\nabla^2 + k^2\right) \mathbf{E} = 0 \tag{1.73}$$

where

$$k = \omega\sqrt{\mu\varepsilon} \tag{1.74}$$

and k is known as the *wavenumber*.

This apparently time-independent form of the wave equation is often more generally called the *Helmholtz equation*, and is a widely studied differential equation in connection with a variety of physical phenomena. A similar expression for $\mathbf{H}$ is found by taking the curl of (1.71d) without sources,

$$\left(\nabla^2 + k^2\right)\mathbf{H} = 0 \tag{1.75}$$

1.4.2 Phasor Notation

Assuming that we restrict our analysis to real-valued quantities in time (which is an obvious requirement for physically realizable circuits), then it is possible to exploit the symmetry properties of the Fourier transform. Specifically, for any real-valued, time-domain signal, $a(t)$,

$$A(\omega) = \int_{-\infty}^{\infty} a(t)\, e^{-j\omega t} dt = \left(\int_{-\infty}^{\infty} a(t)\, e^{j\omega t} dt\right)^{*} = A^{*}(-\omega) \tag{1.76}$$

where the superscript asterisk denotes the complex conjugate. This property of the Fourier transform of real-valued signals is called *Hermitian symmetry*. It is therefore possible to reconstruct the entire Fourier spectrum, and consequently the time-domain waveform, of a real-valued signal from knowledge of just the positive half (i.e., for $\omega > 0$)

This reconstruction is especially simple for pure sinusoids of a known frequency, ω_0. Take a sinusoidal signal with amplitude a_0 and phase ϕ,

$$a(t) = a_0 \cos(\omega_0 t + \phi) \tag{1.77}$$

Its Fourier transform is

$$A(\omega) = \int_{-\infty}^{\infty} a_0 \cos(\omega_0 t + \phi)\, e^{-j\omega t} dt \tag{1.78a}$$

$$= \tfrac{1}{2} a_0 \int_{-\infty}^{\infty} \left(e^{j\phi} e^{-j(\omega-\omega_0)t} + e^{-j\phi} e^{j(\omega+\omega_0)t}\right) dt \tag{1.78b}$$

$$= \pi a_0 \left(e^{j\phi}\delta\left(\omega - \omega_0\right) + e^{-j\phi}\delta\left(\omega + \omega_0\right)\right) \tag{1.78c}$$

where δ is the *Dirac delta function*. Taking the coefficient of only the positive frequency component (and leaving off the factor of π for convenience), we have

$$a' = a_0 e^{j\phi} \tag{1.79}$$

This is known as the *phasor transform* of the sinusoidal signal in (1.77). It is a complex number with amplitude equal to the sinusoidal amplitude, and angle in the complex plane equal to the sinusoidal phase. Linear, frequency-domain calculations, such as the formulations of Maxwell's equations in (1.70) and (1.71), may be applied to phasor variables directly (using $\omega = \omega_0$) without loss of information or correctness. The time-domain waveform of the final result may then be recovered very simply with the relation

$$a(t) = \text{Re}\left\{a' e^{j\omega t}\right\} \tag{1.80}$$

This is obviously much simpler than going through Fourier transforms and is sufficient for the purposes of most passive electromagnetic circuit analyses. The phasor notation for electromagnetic quantities will be used extensively throughout this book, with one caveat: unless otherwise noted, many of the calculations and derivations will make use of *root-mean-square (rms)* phasor amplitudes to simplify expressions for energy and power. In those instances, the inverse-phasor transform would acquire an extra factor of $\sqrt{2}$,

$$a(t) = \sqrt{2} \cdot \text{Re}\left\{a' e^{j\omega t}\right\} \tag{1.81}$$

Let us now revisit the linearly polarized plane wave from Section 1.3. Expressed as a phasor, we have for the electric field

$$\mathbf{E}\left(z\right) = a_0\left(z\right) e^{j\phi(z)} \mathbf{x} \tag{1.82}$$

Note that by using the phasor notation, we have assumed a condition that the electric field at any point in space is sinusoidal in time. Since our more general solution in (1.60) comprised a linearly translating waveform, we know that our present solution must be sinusoidal in space as well. We may therefore drop the z dependence from a_0 and write ϕ as a linear function of z,

$$\mathbf{E}\left(z\right) = a_0 e^{j(\phi_0 - \beta z)} \mathbf{x} \tag{1.83}$$

Using (1.81) to recover the time dependence, we have

$$E(z,t) = \sqrt{2} \cdot \mathrm{Re}\left\{a_0 e^{j(\phi_0 - \beta z)} e^{j\omega t}\right\} = a_0\sqrt{2}\cos\left(\omega t - \beta z + \phi_0\right) \tag{1.84a}$$

$$= a_0\sqrt{2}\cos\left(-\beta\left(z - \tfrac{\omega}{\beta}t\right) + \phi_0\right) \tag{1.84b}$$

To satisfy (1.60) and (1.61) then, we must have

$$\beta = \frac{\omega}{v_p} = \omega\sqrt{\mu\varepsilon} = k \tag{1.85}$$

where we have assumed a wave traveling in the positive-z direction. β is known as the *phase constant* or *lossless propagation constant*. We may also at this point define a wavelength, corresponding to the spatial extent, λ, over which the complex angle of the phasor argument has increased by 2π radians, or

$$\beta\lambda = 2\pi \tag{1.86a}$$

$$\therefore \lambda = \frac{2\pi}{\beta} = \frac{2\pi v_p}{\omega} = \frac{v_p}{f} \tag{1.86b}$$

where f is the frequency under consideration.

1.5 WAVE-BOUNDARY INTERACTIONS

In Sections 1.1.4 and 1.1.5, we discussed the boundary conditions that apply to electromagnetic fields at the interfaces between media. Now that we have models for wave propagation within those media, it is useful to see how those waves interact with such boundaries.

1.5.1 Reflection from a Conducting Boundary

Consider a plane wave incident upon the boundary between free space and a perfect conductor. For simplicity, we set our coordinate system such that the boundary passes through the origin with the surface normal oriented in the $+\mathbf{z}$ direction, and the wave is incident at an oblique angle, θ_1 in the xz-plane. We articulate two canonical cases: one in which the electric field is polarized perpendicular to the plane of incidence, as in Figure 1.7(a), and another in which the electric field is

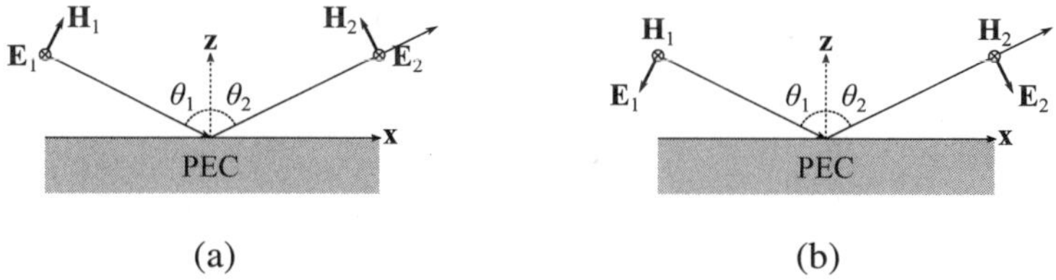

Figure 1.7 A plane wave incident upon a conducting boundary with (a) the electric field polarized perpendicular to the plane of incidence, and (b) the electric field polarized within the plane of incidence.

polarized within the plane of incidence, as in in Figure 1.7(b). The more general case of an arbitrarily polarized wave may be considered a superposition of these two.

In either case, since no field may exist inside the perfect conductor, we may write the total electric and magnetic fields as the sum of incident and reflected waves,

$$\mathbf{E}(\mathbf{p}) = \mathbf{E}_1 e^{-j\beta \mathbf{p}\cdot\mathbf{k}_1} + \mathbf{E}_2 e^{-j\beta \mathbf{p}\cdot\mathbf{k}_2} \tag{1.87a}$$

$$\mathbf{H}(\mathbf{p}) = \mathbf{H}_1 e^{-j\beta \mathbf{p}\cdot\mathbf{k}_1} + \mathbf{H}_2 e^{-j\beta \mathbf{p}\cdot\mathbf{k}_2} \tag{1.87b}$$

where $\mathbf{p}$ is any point in space (above the conductor) and $\mathbf{k}_1$ and $\mathbf{k}_2$ are unit vectors in the direction of propagation of the two plane waves. Since the phase constant, β is a property of the medium itself, it is the same for both waves. For points on the surface of the conductor within the plane of incidence, we may write $\mathbf{p} = x\mathbf{x}$; therefore,

$$\mathbf{E}(x) = \mathbf{E}_1 e^{-j\beta x \sin\theta_1} + \mathbf{E}_2 e^{-j\beta x \sin\theta_2} \tag{1.88a}$$

$$\mathbf{H}(x) = \mathbf{H}_1 e^{-j\beta x \sin\theta_1} + \mathbf{H}_2 e^{-j\beta x \sin\theta_2} \tag{1.88b}$$

In the first case, Figure 1.7(a), both $\mathbf{E}_1$ and $\mathbf{E}_2$ are tangential to the surface. The boundary conditions for a perfect electric conductor thus require that

$$\mathbf{E}_1 e^{-j\beta x \sin\theta_1} = -\mathbf{E}_2 e^{-j\beta x \sin\theta_2} \tag{1.89}$$

which can only be satisfied for all x if $\mathbf{E}_1 = -\mathbf{E}_2$ and $\theta_1 = \theta_2$. In the latter case, Figure 1.7(b), we have both tangential and normal components of the electric field. Isolating the tangential components, we have

$$\mathbf{x}\cdot\mathbf{E}(x) = \mathbf{x}\cdot\mathbf{E}_1 e^{-j\beta x \sin\theta_1} + \mathbf{x}\cdot\mathbf{E}_2 e^{-j\beta x \sin\theta_2} = 0 \tag{1.90a}$$

$$\therefore \mathbf{x}\cdot\mathbf{E}_1 e^{-j\beta x \sin\theta_1} = -\mathbf{x}\cdot\mathbf{E}_2 e^{-j\beta x \sin\theta_2} \tag{1.90b}$$

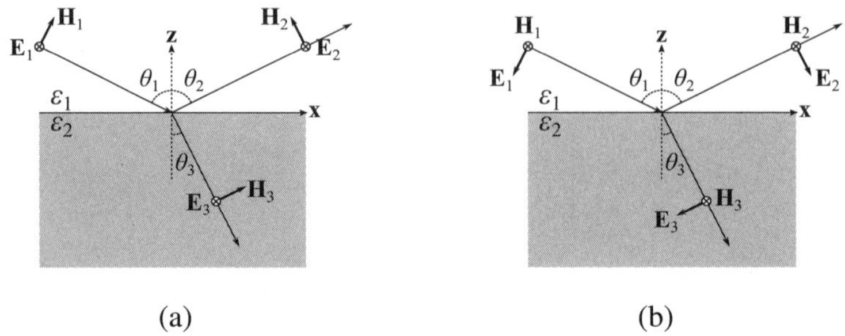

Figure 1.8 A plane wave incident upon a dielectric boundary with (a) the electric field polarized perpendicular to the plane of incidence and (b) the electric field polarized within the plane of incidence. We typically assume that the permeability, μ, is the same in both media.

and again we require that $\theta_1 = \theta_2$. This is the well-known law of *specular reflection* that was articulated by Hero of Alexandria in 10–70 A.D. [12], long before the basic laws of electromagnetics were understood.

Because the perfect conductor is lossless, simple conservation of energy requires that the amplitudes of the incident and reflected waves must be the same.

1.5.2 Reflection and Refraction from a Dielectric Boundary

If the boundary is not a perfect electric conductor, but rather the interface to another dielectric medium, as shown in Figure 1.8, then we must allow for the possibility of a third plane wave propagating in the new medium. Thus, points on the surface and in the plane of incidence may be written as

$$\mathbf{E}(x) = \mathbf{E}_1 e^{-j\beta_1 x \sin\theta_1} + \mathbf{E}_2 e^{-j\beta_1 x \sin\theta_2} \tag{1.91a}$$

$$\mathbf{H}(x) = \mathbf{H}_1 e^{-j\beta_1 x \sin\theta_1} + \mathbf{H}_2 e^{-j\beta_1 x \sin\theta_2} \tag{1.91b}$$

in the first medium, and

$$\mathbf{E}(x) = \mathbf{E}_3 e^{-j\beta_2 x \sin\theta_3} \tag{1.92a}$$

$$\mathbf{H}(x) = \mathbf{H}_3 e^{-j\beta_2 x \sin\theta_3} \tag{1.92b}$$

in the second medium. Note this time that we must use different phase constants, β_1 and β_2, in the two media. In the first case, Figure 1.8(a), the electric field is everywhere tangential to the surface, in the $\mathbf{y}$ direction ($\mathbf{E}_k = E_k\mathbf{y}$). These are conventionally called s-polarized waves.[5] The continuity of the tangential electric field across such a boundary gives

$$E_1 e^{-j\beta_1 x \sin\theta_1} + E_2 e^{-j\beta_1 x \sin\theta_2} = E_3 e^{-j\beta_2 x \sin\theta_3} \tag{1.93}$$

Since E_1, E_2, and E_3 are constants, this equation can only be satisfied for all x if the arguments of the exponentials are all the same; therefore,

$$\beta_1 \sin\theta_1 = \beta_1 \sin\theta_2 = \beta_2 \sin\theta_3 \tag{1.94}$$

From this, we once again find specular reflection in the first medium ($\theta_1 = \theta_2 = \theta_i$) and for the transmission to the second medium ($\theta_3 = \theta_t$) we have

$$\beta_1 \sin\theta_i = \beta_2 \sin\theta_t \tag{1.95a}$$

$$\omega\sqrt{\mu_1\varepsilon_1} \sin\theta_i = \omega\sqrt{\mu_2\varepsilon_2} \sin\theta_t \tag{1.95b}$$

$$\therefore \frac{\sin\theta_i}{\sin\theta_t} = \sqrt{\frac{\mu_2\varepsilon_2}{\mu_1\varepsilon_1}} = \frac{n_2}{n_1} \tag{1.95c}$$

where $n = c/v_p$ is known as the *refractive index* of the medium. This is *Snell's law* for refraction. Typically, for most dielectrics, we assume that $\mu_1 = \mu_2 = \mu_0$ and simply write $n = \sqrt{\varepsilon_r}$.

To determine the relative amplitude of the reflected and transmitted waves, we must also consider the continuity of the tangential component of the magnetic field, oriented in the $\mathbf{x}$ direction. Since $\mathbf{H}$ is not tangential to the surface, we must account for its projection,

$$H_1 \cos\theta_i e^{-j\beta_1 x \sin\theta_i} - H_2 \cos\theta_i e^{-j\beta_1 x \sin\theta_i} = H_3 \cos\theta_t e^{-j\beta_2 x \sin\theta_t} \tag{1.96}$$

Snell's law, more specifically (1.95a), allows us to to drop the exponentials, since all of their arguments are the same. Additionally, we may substitute E for H using the wave impedance, so that

$$\frac{E_1}{\eta_1} \cos\theta_i - \frac{E_2}{\eta_1} \cos\theta_i = \frac{E_3}{\eta_2} \cos\theta_t \tag{1.97}$$

5 The s stands for *senkrecht*, the German word for perpendicular, since the electric field is oriented perpendicular to the plane of incidence in this case [13].

Combining this with (1.93), we have

$$E_1 + E_2 = (E_1 - E_2)\,\frac{\eta_2 \cos\theta_i}{\eta_1 \cos\theta_t} \tag{1.98a}$$

$$r_s = \frac{E_2}{E_1} = \frac{\eta_2 \cos\theta_i - \eta_1 \cos\theta_t}{\eta_2 \cos\theta_i + \eta_1 \cos\theta_t} \tag{1.98b}$$

$$t_s = \frac{E_3}{E_1} = \frac{E_1 + E_2}{E_1} = 1 + r_s = \frac{2\eta_2 \cos\theta_i}{\eta_2 \cos\theta_i + \eta_1 \cos\theta_t} \tag{1.98c}$$

where r_s may be considered the reflection coefficient and t_s the transmission coefficient for s-polarized waves. Note that when $\mu_1 = \mu_2 = \mu_0$, as is often the case, the wave impedance and and index of refraction are related by

$$\frac{n_1}{n_2} = \frac{\eta_2}{\eta_1} \tag{1.99}$$

(a pernicious source of confusion, since the Roman letter n and the Greek letter η look so similar).

In the case of Figure 1.8(b), it is the magnetic field that is tangential to the surface, and the electric field is in the plane of incidence. These are known as p-polarized waves. The relevant equations are essentially the same, except that the roles of E and H are reversed, and consequently the wave impedances are inverted in the result, yielding the following expressions,

$$r_p = \frac{\eta_1 \cos\theta_i - \eta_2 \cos\theta_t}{\eta_1 \cos\theta_i + \eta_2 \cos\theta_t} \tag{1.100a}$$

$$t_p = \frac{\eta_2}{\eta_1}\,(1 + r_p) = \frac{2\eta_2 \cos\theta_i}{\eta_1 \cos\theta_i + \eta_2 \cos\theta_t} \tag{1.100b}$$

where r_p and t_p are the reflection and transmission coefficients for p-polarized waves, respectively, and Snell's law and the law of specular reflection remain the same.

These equations collectively are known as the *Fresnel equations* [14]. When expressed in terms of the indices of refraction, and substituting in Snell's law, we may derive a slightly simplified form

$$r_s = \frac{\eta_2 \cos\theta_i - \eta_1 \cos\theta_t}{\eta_2 \cos\theta_i + \eta_1 \cos\theta_t} = \frac{n_1 \cos\theta_i - n_2 \cos\theta_t}{n_1 \cos\theta_i + n_2 \cos\theta_t} \tag{1.101a}$$

Table 1.3
The Fresnel Equations and Snell's Law

Parameter	Expression
	s-polarized waves
Reflection coefficient	$r_s = \frac{\eta_2 \cos\theta_i - \eta_1 \cos\theta_t}{\eta_2 \cos\theta_i + \eta_1 \cos\theta_t} = \frac{n_1 \cos\theta_i - n_2 \cos\theta_t}{n_1 \cos\theta_i + n_2 \cos\theta_t} = \frac{\sin(\theta_t - \theta_i)}{\sin(\theta_t + \theta_i)}$
Transmission coefficient	$t_s = 1 + r_s = \frac{2\eta_2 \cos\theta_i}{\eta_2 \cos\theta_i + \eta_1 \cos\theta_t} = \frac{2n_1 \cos\theta_i}{n_1 \cos\theta_i + n_2 \cos\theta_t}$
	p-polarized waves
Reflection coefficient	$r_p = \frac{\eta_1 \cos\theta_i - \eta_2 \cos\theta_t}{\eta_1 \cos\theta_i + \eta_2 \cos\theta_t} = \frac{n_2 \cos\theta_i - n_1 \cos\theta_t}{n_2 \cos\theta_i + n_1 \cos\theta_t} = \frac{\tan(\theta_i - \theta_t)}{\tan(\theta_i + \theta_t)}$
Transmission coefficient	$t_p = \frac{\eta_2}{\eta_1}(1 + r_p) = \frac{2\eta_2 \cos\theta_i}{\eta_1 \cos\theta_i + \eta_2 \cos\theta_t} = \frac{2n_1 \cos\theta_i}{n_2 \cos\theta_i + n_1 \cos\theta_t}$
Snell's law	$\frac{\sin\theta_i}{\sin\theta_t} = \sqrt{\frac{\mu_2 \varepsilon_2}{\mu_1 \varepsilon_1}} = \frac{n_2}{n_1}$

θ_i is the angle of incidence and reflection. θ_t is the angle of transmission.
Where the expressions above are given in terms of the indices of refraction, it is assumed that $\mu_1 = \mu_2 = \mu_0$.

$$= \frac{\sin\theta_t \cos\theta_i - \sin\theta_i \cos\theta_t}{\sin\theta_t \cos\theta_i + \sin\theta_i \cos\theta_t} = \frac{\sin(\theta_t - \theta_i)}{\sin(\theta_t + \theta_i)} \tag{1.101b}$$

$$r_p = \frac{\eta_1 \cos\theta_i - \eta_2 \cos\theta_t}{\eta_1 \cos\theta_i + \eta_2 \cos\theta_t} = \frac{n_2 \cos\theta_i - n_1 \cos\theta_t}{n_2 \cos\theta_i + n_1 \cos\theta_t} \tag{1.101c}$$

$$= \frac{\sin\theta_i \cos\theta_i - \sin\theta_t \cos\theta_t}{\sin\theta_i \cos\theta_i + \sin\theta_t \cos\theta_t} = \frac{\tan(\theta_i - \theta_t)}{\tan(\theta_i + \theta_t)} \tag{1.101d}$$

where we have used a number of trigonometric identities which can be found in Section A.1. Equations (1.101b) and (1.101d) are known as *Fresnel's sine law* and *Fresnel's tangent law*, respectively. These results are summarized in Table 1.3.

One might expect that the square of the reflection and transmission coefficients represent relative power reflected and transmitted, respectively, and that conservation of energy would therefore require $|r|^2 + |t|^2 = 1$ for both polarizations. However, this is not the case. The square of the reflection coefficient does indeed give the fraction of reflected power, but the square of the transmission coefficient does not equal the fractional transmitted power, for two reasons. First, the transmission coefficient describes the ratio of the electric field amplitude, but since the

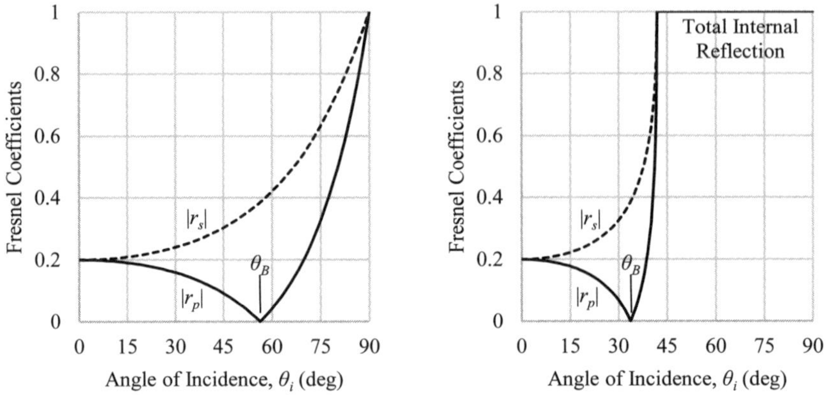

Figure 1.9 Fresnel reflection coefficients as a function incidence angle when (a) $n_1 = 1$ and $n_2 = 1.5$, which is typical for an air-to-glass interface, and (b) when $n_1 = 1.5$ and $n_2 = 1$, typical for a glass-to-air interface.

transmitted wave exists in a different medium than the incident wave, the ratio of electric to magnetic fields is different, and a correction proportional to the wave impedance must be applied. Additionally, the fact that a transmitted wave leaves the interface at a different angle than the incident wave means that the departing beam has a different cross-sectional area than the incident beam, due to angular projection (and coefficients on plane-wave solutions can only truly describe power densities rather than total power) [14]. The actual transmitted power ratio is therefore given by

$$T = \frac{\eta_1 \cos\theta_t}{\eta_2 \cos\theta_i} |t|^2 \tag{1.102}$$

The Fresnel reflection coefficients are plotted in Figure 1.9. Two special cases warrant some attention. For the p polarization, a specific critical angle is reached for which the reflection coefficient is exactly zero, and all the energy is transferred into the second medium (and, consequently, any reflection at this angle, whatever the polarization mix of the incident beam, is perfectly polarized). This occurs when

$$r_p = \frac{\tan(\theta_i - \theta_t)}{\tan(\theta_i + \theta_t)} = 0 \tag{1.103a}$$

$$\therefore \tan(\theta_i + \theta_t) = \infty \tag{1.103b}$$

$$\theta_i + \theta_t = \tfrac{\pi}{2} \tag{1.103c}$$

$$\cos\theta_i = \cos\left(\tfrac{\pi}{2} - \theta_t\right) = \sin\theta_t = \frac{n_1}{n_2}\sin\theta_i \tag{1.103d}$$

$$\tan\theta_i = \frac{n_2}{n_1} \tag{1.103e}$$

$$\theta_B = \theta_i = \tan^{-1}\left(\frac{n_2}{n_1}\right) \tag{1.103f}$$

and is known as *Brewster's angle.*

The second special case occurs when $n_1 > n_2$ and θ_i is sufficiently large, that is, the incident wave grazes the surface. We may thus find ourselves in a situation where

$$\sin\theta_t = \frac{n_1}{n_2}\sin\theta_i > 1 \tag{1.104}$$

This equation cannot be satisfied for any real angle θ_t. Clearly, our model of three propagating plane waves is insufficient to account for this scenario. In fact, what we need is an *evanescent wave* in the second medium to satisfy the boundary conditions. Such waves represent fields which decay exponentially from the surface and carry no power with them. Thus, all of the incident power is reflected. This phenomenon is known as *Total internal reflection (TIR)* and it occurs for incidence angles greater than a critical value,

$$\theta_i > \theta_c = \sin^{-1}\left(\frac{n_2}{n_1}\right) \tag{1.105}$$

1.6 LOSSY MATERIALS

So far, we have assumed that the dielectric permittivity, ε, and magnetic permeability, μ, are real valued and that all conductive media have infinite conductivity. This effectively makes these materials lossless; all of the energy stored in temporary electric and magnetic fields is eventually returned when the fields dissipate or propagates away as an electromagnetic wave without dampening (note, for example, that the amplitude of the plane wave solution in (1.60) does not diminish with time or distance propagated). In this section, we will examine the physical mechanisms of loss in realistic media and how these manifest themselves mathematically.

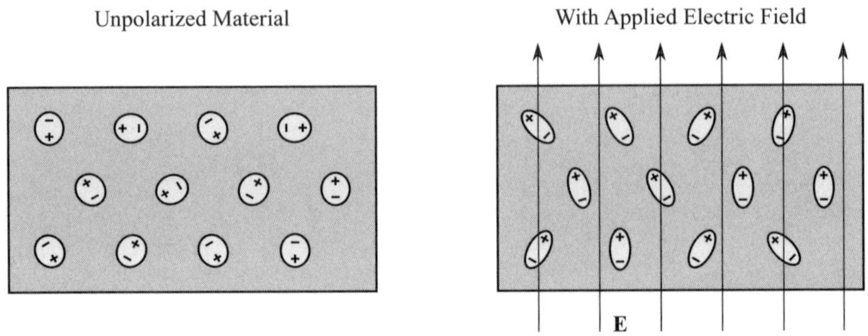

Figure 1.10 Illustration of dielectric polarizability in a material. (a) Molecular dipoles randomly oriented without an electric field. (b) Coorientation of dipole moments with an applied electric field.

1.6.1 Dielectric Polarization

It is tempting to think of the vacuum permittivity and permeability as simply characteristics of the medium of free space in the same way that the permittivity and permeability of other materials characterize those media, whereas the underlying physical processes that cause them are completely different. Space is not a medium, per se (recall that the luminiferous aether does not exist); rather, space is defined as the lack of any significant medium. In fact, there is a point of view that Maxwell's equations having ε_0 and μ_0 in the constitutive relations are the only true form, and that material permittivity and permeability are simply mathematical shortcuts for describing the behavior of macroscopic phenomena without having to explicitly account for the dipole moments and magnetic polarizability of individual molecules.

The vacuum permittivity is simply a natural constant that describes the potential energy density stored in a bare electric field. When matter is present, the ambient electric field causes the individual molecules within the material to become electrically polarized (if they are not already) and align them all in the same general direction. See Figure 1.10. This additional electric displacement energetically resists the creation of the electric field and thereby increases the apparent energy density stored within it. Specifically,

$$\mathbf{D} = \varepsilon_0 \left(1 + \chi_e\right) \mathbf{E} \tag{1.106}$$

where χ_e is called the *electric susceptibility* of the material. Since this molecular polarization is phenomenologically indistinguishable from ordinary vacuum permittivity on a macroscopic scale, it is customary to subsume both effects into a single parameter,

$$\varepsilon = \varepsilon_0 (1 + \chi_e) = \varepsilon_0 \varepsilon_r = \varepsilon' - j\varepsilon'' \quad (1.107)$$

where ε_r is known as the *dielectric constant* or *relative permittivity* of the material. In this case, we are using the phasor notation for a frequency-dependent effect, and the permittivity has both real and imaginary parts. The latter accounts for energy lost to heat as a consequence of dampened vibrations of the molecules, and must always be negative.

1.6.2 Magnetic Polarization

In a similar manner, an applied magnetic field may tend to align magnetic dipole moments of the molecules within a material. This net magnetic polarization, or *magnetization*, within the material increases the apparent energy stored in the magnetic field. Therefore, we may write

$$\mathbf{B} = \mu_0 (1 + \chi_m) \mathbf{H} \quad (1.108)$$

where χ_m is the *magnetic susceptibility*, and the effective material permeability becomes

$$\mu = \mu_0 (1 + \chi_m) = \mu_0 \mu_r = \mu' - j\mu'' \quad (1.109)$$

where μ_r is the *relative permeability*. Again, the imaginary part of μ accounts for losses due to damping, and is always less than zero.

1.6.3 Conduction

Long before Maxwell's equations were written down and the underlying physical principles of electromagnetics were understood, scientists had determined experimentally that the electric current through most conductive materials was proportional to the applied electric field. That is,

$$\mathbf{J} = \sigma \mathbf{E} \quad (1.110)$$

where σ is called the *conductivity*. This equation, called *Ohm's law* in honor of the German physicist, George Ohm, is essentially empirical rather than fundamental, but nonetheless explains the macroscopic behavior of a wide range of materials and

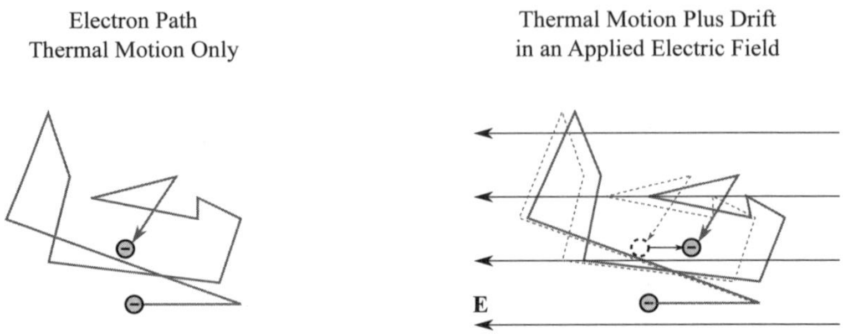

Figure 1.11 Illustration of a typical electron's path through matter in the Drude model. (a) Random scattering due to thermal excitation. (b) A drift component added to the electron's path from an applied electric field.

over a wide range of field conditions; in point of fact, electrical conductivity is one of the most widely varying of all measurable material parameters. Even ordinary materials like Teflon ($\sigma < 10^{-23}$ S/m) and silver ($\sigma \approx 6.3 \times 10^7$ S/m) span more than 30 orders of magnitude. Put another way, Teflon is to silver in terms of conductivity what the near-vacuum of interstellar space is to a black hole in terms of density.

The underlying physical cause of the simple yet pervasive direct proportionality between electric field strength and current was first satisfactorily explained by Paul Drude in 1900. The Drude model envisions free electrons set in motion by thermal kinetic energy within a conductor, scattering off of the fixed ions in the lattice of the embedding material at random intervals and in random directions, as shown for a typical electron in Figure 1.11(a). Even without an applied electric field, the net effect of a large number of electrons like this one undergoing random motions creates small (but measurable) fluctuations in current, called *Johnson-Nyquist noise*, and partially accounts for the static heard on radios when the station is out of range, or the snow visible on analog television sets (assuming the young readers of today remember such things) when no channel is present.

In Figure 1.11(b), due to an applied electric field, a small component of drift is added to the electron's trajectory during each interval between scattering events that is in the direction opposite to the field (because the electron has negative charge). The random kinetic velocity of the electrons due to thermal excitation is much larger than the drift imparted to them by acceleration in the electric field (by about 10 orders of magnitude [15]). As a result, the mean free time between collisions is decisively dominated by temperature and not by the field strength. The average drift velocity for a large ensemble of electrons is simply the acceleration due to Lorentz forces (which, according to (1.4), is proportional to field strength) times the mean time between collisions (which has effectively no dependence on field strength). Thus, the total current is directly proportional to field strength, leading to Ohm's law.

Later improvements and extensions based on quantum mechanics helped to explain other properties of materials related to their mobile electron population, such as thermal conductivity and heat capacity, but the simple Drude model for charge transport is sufficient for the purposes of circuit theory as described in this book.

Substituting (1.110) into (1.71d), we find

$$\nabla \times \mathbf{H} = \mathbf{J} + j\omega\mathbf{D} = \sigma\mathbf{E} + j\omega\varepsilon\mathbf{E} = \sigma\mathbf{E} + j\omega\left(\varepsilon' - j\varepsilon''\right)\mathbf{E} \tag{1.111a}$$

$$= \left(\sigma + \omega\varepsilon''\right)\mathbf{E} + j\omega\varepsilon'\mathbf{E} \tag{1.111b}$$

Thus, mathematically, conductivity appears in Maxwell's equations in the same way as dielectric losses. Material property datasheets often give only the real part of the dielectric constant and a *loss tangent* defined by the ratio of the real and imaginary parts of the total electric displacement (the final expression of (1.111b)), or

$$\tan\delta = \frac{\sigma + \omega\varepsilon''}{\omega\varepsilon'} \tag{1.112}$$

which embodies both the dielectric and conductive losses in one parameter.

1.6.4 The Skin Effect

In highly conductive media, the conductivity itself is usually a far greater source of loss than the complex permittivity or permeability. Let us consider a plane wave inside a good conductor, where the conductivity, σ, is the only source of loss taken into account. According to (1.110), there is a significant current density present, so we cannot use the source-free form of Maxwell's equations. We may still assume there are no isolated non-neutral charges present ($\rho = 0$), but we must allow for the presence of current density, $\mathbf{J}$. The wave equation therefore becomes

$$\nabla \times \nabla \times \mathbf{E} = -j\omega\left(\nabla \times \mathbf{B}\right) = -j\omega\mu\left(\nabla \times \mathbf{H}\right) = -j\omega\mu\left(\mathbf{J} + j\omega\mathbf{D}\right) \tag{1.113a}$$

$$= -j\omega\mu\left(\sigma + j\omega\varepsilon\right)\mathbf{E} \tag{1.113b}$$

$$\therefore \nabla^2 \mathbf{E} = j\omega\mu\left(\sigma + j\omega\varepsilon\right)\mathbf{E} \tag{1.113c}$$

The dielectric constants of good conductors such as metals are rarely found in physical tables, and for good reason: the conduction current greatly exceeds the displacement current (i.e., $\sigma \gg \omega\varepsilon$) within such materials at frequencies from DC to more than a terahertz, making the dielectric constant itself both inconsequential and very difficult to measure [2]. We may therefore safely neglect the displacement term above and write

$$\nabla^2 \mathbf{E} \approx j\omega\mu\sigma\mathbf{E} \tag{1.114}$$

A plane wave traveling in the z-direction and polarized in the x-direction would therefore have solutions of the form

$$E_x\left(z\right) = E_0 e^{\pm\gamma z} \tag{1.115a}$$

$$\gamma^2 = j\omega\mu\sigma \tag{1.115b}$$

Note that

$$\sqrt{j} = \pm\frac{1+j}{\sqrt{2}} \tag{1.116a}$$

$$\therefore \gamma = \frac{1+j}{\delta} \tag{1.116b}$$

where

$$\delta = \sqrt{\frac{2}{\omega\mu\sigma}} \tag{1.117}$$

The constant δ is known as the *skin depth*, and is the depth of penetration into the conductive material at which the amplitude of the wave has decayed by a factor e^{-1}.

It is also, per (1.116), the distance over which the wave phase angle changes by 1 radian (since the real and imaginary parts of γ are the same), making the wavelength inside the conductor equal to 2π skin depths. The skin depth in common metals varies from a few millimeters to less than a micron in the radio frequency (RF), microwave, and millimeter-wave frequency range, and is typically thousands to millions of times smaller than a wavelength in free space for the same frequency. The refractive index is therefore very large by a similar factor, and essentially guarantees that an external wave incident at even grazing angles will refract (via Snell's law) into the conductor at an angle virtually perpendicular to the surface. This ensures that the skin depth formula above applies to virtually any geometry, and is not limited to cases of perpendicular incidence.

An electromagnetic field at the surface of a conductor, then, will induce currents in the conductive medium that decay exponentially away from the surface.

The fields cannot sense the bulk material deep within the conductor, instead responding only to the first few skin depths below the surface. This leads to a net effective surface resistance which is simply given by one over the product of the bulk conductivity and the skin depth,[6]

$$R_s = \frac{1}{\sigma\delta} = \sqrt{\frac{\omega\mu}{2\sigma}} \tag{1.118}$$

The surface resistance, or *sheet resistance*, R_s, is usually given in units of $\Omega/\square$, pronounced "ohms per square."

Because of the skin effect, it is usually sufficient in the construction of practical circuit elements to deposit a very thin layer of a good conductor like copper, silver, or gold on top of the bulk geometry created with a cheaper, stronger, or lighter-weight material, which need not even be electrically conducting.

1.7 LORENTZ RECIPROCITY

Let us now apply what we have learned to prove a rather useful theorem in electromagnetic circuits. Consider a source current distribution, $\mathbf{J}_1$, creating in the surrounding medium the field quantities $\mathbf{E}_1$ and $\mathbf{H}_1$, and another current distribution, $\mathbf{J}_2$, creating fields $\mathbf{E}_2$ and $\mathbf{H}_2$. If we take the dot product of $\mathbf{H}_2$ with the curl of $\mathbf{E}_1$, and subtract from it the dot product of $\mathbf{E}_1$ with the curl of $\mathbf{H}_2$, we have

$$\mathbf{H}_2 \cdot (\nabla \times \mathbf{E}_1) - \mathbf{E}_1 \cdot (\nabla \times \mathbf{H}_2) = \mathbf{H}_2 \cdot (-j\omega\mathbf{B}_1) - \mathbf{E}_1 \cdot (\mathbf{J}_2 + j\omega\mathbf{D}_2) \tag{1.119a}$$

$$\therefore \mathbf{H}_2 \cdot (\nabla \times \mathbf{E}_1) - \mathbf{E}_1 \cdot (\nabla \times \mathbf{H}_2) = -j\omega\mathbf{H}_2 \cdot \mathbf{B}_1 - \mathbf{E}_1 \cdot \mathbf{J}_2 - j\omega\mathbf{E}_1 \cdot \mathbf{D}_2 \tag{1.119b}$$

We then apply to the left side of (1.119b) the following vector identity,

$$\nabla \cdot (\mathbf{A} \times \mathbf{B}) = \mathbf{B} \cdot (\nabla \times \mathbf{A}) - \mathbf{A} \cdot (\nabla \times \mathbf{B}) \tag{1.120}$$

which yields

$$\nabla \cdot (\mathbf{E}_1 \times \mathbf{H}_2) = -j\omega\mathbf{H}_2 \cdot \mathbf{B}_1 - \mathbf{E}_1 \cdot \mathbf{J}_2 - j\omega\mathbf{E}_1 \cdot \mathbf{D}_2 \tag{1.121}$$

6 An exception occurs in some cases at extreme cryogenic temperatures, in which the mean free path of the electrons may actually exceed the skin depth, leading to what is sometimes called the *anomalous skin effect* [16].

Further, since the choice of subscripts is arbitrary, we may also write

$$\nabla \cdot (\mathbf{E}_2 \times \mathbf{H}_1) = -j\omega \mathbf{H}_1 \cdot \mathbf{B}_2 - \mathbf{E}_2 \cdot \mathbf{J}_1 - j\omega \mathbf{E}_2 \cdot \mathbf{D}_1 \tag{1.122}$$

Subtracting (1.121) from (1.122) yields

$$\nabla (\mathbf{E}_2 \times \mathbf{H}_1 - \mathbf{E}_1 \times \mathbf{H}_2) = \mathbf{E}_1 \cdot \mathbf{J}_2 - \mathbf{E}_2 \cdot \mathbf{J}_1 \tag{1.123}$$

where we have used the fact that $\mathbf{H}_1 \cdot \mathbf{B}_2 = \mathbf{H}_2 \cdot \mathbf{B}_1$ and $\mathbf{E}_1 \cdot \mathbf{D}_2 = \mathbf{E}_2 \cdot \mathbf{D}_1$. This is true for linear, isotropic, time-invariant media, wherein the constituent parameters ε and μ are simple scalars. In some more exotic media, such as ferrites, these equalities may be violated (see Section 10.8).

Integrating (1.123) over volume and applying the divergence theorem, we have

$$\oiint (\mathbf{E}_2 \times \mathbf{H}_1 - \mathbf{E}_1 \times \mathbf{H}_2) \cdot d\mathbf{S} = \iiint (\mathbf{E}_1 \cdot \mathbf{J}_2 - \mathbf{E}_2 \cdot \mathbf{J}_1)\, dV \tag{1.124}$$

This is the *Lorentz reciprocity theorem* in integral form. If the source currents are taken to be localized (that is, they do not extend to infinity) and the volume of integration is taken over all space, then the surface integral on the left side vanishes, and we have

$$\iiint \mathbf{E}_1 \cdot \mathbf{J}_2 dV = \iiint \mathbf{E}_2 \cdot \mathbf{J}_1 dV \tag{1.125}$$

which is sometimes called the *Rayleigh-Carson reciprocity theorem.*

The implication of these results is that the relationship between a source current distribution and the field it creates is unchanged if the positions of the original source and the resulting field are swapped. This finds numerous applications in electromagnetic analysis. For example, it guarantees that the radiation pattern of antenna when used to transmit a signal is exactly equivalent to its reception pattern when used to detect that signal. This is exploited extensively in antenna measurements since one is often very much easier to measure in the lab than the other [17].

1.8 FROM FIELDS TO CIRCUITS

In order to synthesize electromagnetic circuits having desired characteristics, we must first derive a repertoire of standard circuit elements exhibiting relatively simple mathematical behaviors to use as building blocks. Otherwise, every design problem becomes a laborious exercise in vector calculus: the brute force and blind optimization of boundary conditions under Maxwell's equations to achieve a prescribed result. The development of even a basic component, such as a power splitter or filter, would be an impossibly open-ended problem. Progress would be prohibitively slow.

In Chapters 2–9, we develop this portfolio of standard elements in groups according to their physical size relative to the wavelengths at which they apply. We start with lumped elements, having physical dimensions very small compared to a wavelength, and yielding circuit behaviors that are the ratios of polynomials in frequency. Next we discuss transmission lines, which are small in cross-section, but potentially many wavelengths long, yielding circuit functions with a trigonometric rather than polynomial basis. Next, we discuss waveguides, having cross-sections comparable in size to a wavelength, and for which the interaction of various electromagnetic modes (some propagating, others evanescent) contributes to the overall frequency response. Finally, we introduce the reader to quasi-optical components (such as lenses and shaped mirrors), which are typically very large compared to a wavelength.

Each of these element classes will be recognized as a simplification of Maxwell's equations, based on the large or small dimension approximation. The effective interconnection and transmission of electromagnetic energy between elements of the same or different classes will be explored, as will their application to the design of many kinds of specific components and systems in Chapters 10–13.

Problems

1-1 How fast would a charged particle have to be traveling in a vacuum through perpendicular electric and magnetic fields of 1 V/m and 1 A/m, respectively, such that it experiences no acceleration?

1-2 A single electron (charge $q = 1.602176634 \times 10^{-19}$ C, and mass $m = 9.109 \times 10^{-31}$ kg) traces out a circular orbit 1m in diameter in free space under the influence of a perpendicular 1 A/m magnetic field. How fast is it orbiting?

1-3 A current density of 1 A/m^2 flows radially outward from a spherical boundary 1m in diameter. What is the rate of charge depletion from the region inside the sphere?

1-4 What is the strength of the electric field emanating from a flat plane carrying a surface charge density of 7 nC/m^2?

1-5 A 1A current flows through a metal tube 1 cm in diameter. What is the magnetic field at the surface of the tube?

1-6 An electric field penetrates the surface of a Teflon body ($\varepsilon_r = 2.1$) which is surrounded by vacuum. Assume the field, directed inward and perpendicular to the interface, is 1 V/m inside the Teflon, and 2 V/m outside. Calculate the charge density which must be present at the surface.

1-7 What is the form of the static magnetic field surrounding two parallel, infinitely thin wires with spacing s and equal currents, I, running in the same direction? With equal currents running in opposite directions?

1-8 Calculate the rate of accumulation/depletion of electric charge throughout a region of space where there is a radially divergent current given by $\mathbf{J} = J_0\mathbf{r}$.

1-9 Assume that the magnetic charge, ρ_m, in (1.22b) is real, and that it obeys a magnetic form of the continuity equation, $\nabla \cdot \mathbf{J}_m = -\partial\rho_m/\partial t$, where $\mathbf{J}_m$ is the magnetic current. Write down a modified form of the Faraday's law which accounts for this new magnetic current, but reduces to the present form of Faraday's law when $\mathbf{J}_m = 0$.

1-10 Prove, in Cartesian coordinates, the vector identity (C.12).

1-11 What is the magnetic vector potential in the center of a circular loop of wire 10 cm in diameter carrying 3 mA of current?

1-12 What is the wave impedance inside a dielectric medium (assume $\mu_r = 1$) for which the phase velocity is $v_p = 0.8c$?

1-13 What is the wavenumber for 100-GHz waves in free space? In glass ($\varepsilon_r = 4.7$)?

1-14 A spectrum of plane waves incident upon a pool of water ($n = 1.333$) at a 45° angle from the surface normal is unpolarized (it has equal amplitudes in the s and p polarizations). What is the polarization power ratio of the waves transmitted into the water? Of the reflected waves?

1-15 What is the relative power transmitted from air into Mylar ($\varepsilon_r = 3.1$) in the p-polarization at an angle $\theta_i = 30°$?

1-16 What is Brewster's angle for glass ($\varepsilon_r = 4.7$) under water ($\varepsilon_r = 80.2$)?

1-17 If one is lying in the bottom of a 6-foot-deep pool of water ($n = 1.333$ in visible light) and looking up, what is the apparent diameter of the hole in the surface of the water through which one is able to see the outside world? How wide is the field of view within which nearby objects above the surface can be seen?

1-18 Express the plane wave phase velocity in a medium in terms of the speed of light in a vacuum, c, and its index of refraction.

1-19 What is the electric susceptibility of polyimide ($\varepsilon_r = 3.4$)?

1-20 What is the skin depth of copper ($\sigma = 5.96 \times 10^7$ S/m, $\mu = \mu_0$) at 100 GHz? What is its surface resistance, neglecting roughness?

References

[1] Wikipedia. (2019, February) Gaussian units. https://en.wikipedia.org/wiki/Gaussian_units.

[2] S. Ramo, J. Whinnery, and T. Van Duzer, *Fields and Waves in Communication Electronics*. New York: Wiley, 1984.

[3] Wikipedia. (2017) Right-hand rule. https://en.wikipedia.org/wiki/Right-hand_rule.

[4] M. A. Morgan, *Reflectionless Filters*. Norwood, MA: Artech House, 2017.

[5] D. M. Pozar, *Microwave Engineering*, 4th ed. New York: Wiley, 2011.

[6] G. Arfken and H. Weber, *Mathematical Methods for Physicists*, 4th ed. San Diego, CA: Academic Press, 1995.

[7] D. Cheng, *Field and Wave Electromagnetics*, 2nd ed. Reading, MA: Addison-Wesley, 1996.

[8] Wikipedia. (2017) Aharonov-Bohm effect. https://en.wikipedia.org/wiki/Aharonov-Bohm_effect.

[9] Y. Aharonov and D. Bohm, "Significance of electromagnetic potentials in quantum theory," *Physical Review*, vol. 115, no. 3, pp. 485–491, August 1959.

[10] L. Vaidman, "Role of potentials in the Aharonov-Bohm effect," *Physical Review A.*, vol. 86, no. 4, October 2012.

[11] B. Lathi, *Linear Systems and Signals*. Carmichael, CA: Berkeley-Cambridge Press, 1992.

[12] Wikipedia. (2018) Specular reflection. https://en.wikipedia.org/wiki/Specular_reflection.

[13] Wikipedia. (2018) Plane of incidence. https://en.wikipedia.org/wiki/Plane_of_incidence.

[14] Wikipedia. (2018) Fresnel equations. https://en.wikipedia.org/wiki/Fresnel_equations.

[15] M. Schwartz, *Principles of Electrodynamics*. New York: Dover, 1987.

[16] T. van Duzer and C. Turner, *Principles of Superconducting Devices and Circuits*, 2nd ed. Upper Saddle River, NJ: Prentice Hall, 1999.

[17] C. Balanis, *Advanced Engineering Electromagnetics*. New York: Wiley, 1989.

Chapter 2

Lumped Elements

The traditional way to begin simplification of Maxwell's equations is to first consider the behavior of electrical (and magnetic) systems in their steady state condition, when a sufficient amount of time has passed since the boundary conditions and external stimuli have stopped changing. That is not to say that the circuits we thus derive cannot work predictably with changing inputs, only that the inputs must change slowly enough, or that the surrounding electromagnetic fields settle quickly enough, that the essential behavior of the circuit as a whole may be well approximated by an electrostatic (or magnetostatic) field solution.

In the time-harmonic case, this is equivalent to saying that the frequency of operation is very low, or that the geometric features that dominate the behavior of the circuit are small compared to a wavelength. This latter point, the compactness of critical features, is what earns the elements derived from such an assumption their common description as *lumped*.

It is very likely that any student reading this book will already be well familiar with lumped elements, as they are very often taught first in electrical engineering curricula. However, I believe that it is valuable at this stage to relearn some of the basic principles from a viewpoint of Maxwell's equations, in order to better appreciate their place alongside transmission lines and waveguides in the larger world of microwave and millimeter-wave engineering. Otherwise, students too often fall into the trap of believing that lumped elements are somehow more natural than other kinds of elements and that transmission lines should therefore attempt to approximate lumped elements in some way. In fact, both classes of elements are equally fundamental, and the most successful microwave engineer is the one that is adept at readjusting his or her viewpoint from one system to another as befits the situation.

Figure 2.1 (a) Voltage defined as the electric potential between two nodes in an electrically small circuit. (b) Voltages between nodes of a lumped-element network.

2.1 VOLTAGE, CURRENT, AND KIRCHHOFF'S LAWS

In Chapter 1, the direct sources of electric and magnetic fields were described as field quantities themselves, namely the charge density and current density. If we are to talk about these objects in the context of electrically small circuit elements, then it is useful to drop the distributed nature of this description and refer to scalar quantities instead, namely *voltage*, related to the electrostatic potential defined in Section 1.2.1, and (total) *current*, respectively.

2.1.1 Kirchhoff's Voltage Law

One assumption that we will make at this time is that our electronic system as a whole is designed to be electrically neutral. Were this not the case, then our circuit would tend to find a way to discharge itself to the outside world, either slowly via leakage in the surrounding medium, or too rapidly for our materials to withstand as an electrostatic discharge (the bane of electronics manufacturers everywhere). In effect, the surrounding environment then becomes an unavoidable part of our circuit, usually a part that is beyond our control.

To maintain neutrality then, it makes little sense to talk about isolated charge as a source of an electric field, but rather charge separation: two independent lumps of equal and opposite charge held apart at two points, or *nodes*, in our circuit. To the external viewer, this is measured as *voltage*, or the electrostatic potential spanning the separation between these two charges.

Consider the two nodes, a and b, connected to an arbitrary, but electrically small circuit, shown in Figure 2.1(a). To calculate the voltage between nodes, we need only to subtract the electrostatic potential, φ, at one node from that of the other. Effectively, this amounts to integrating the gradient of φ in (1.32) along some arbitrary path, l_k, leading from node a to node b,[1]

$$v_k = \varphi_a - \varphi_b = \int_a^b \mathbf{E} \cdot d\mathbf{l}_k \tag{2.1}$$

In general, we cannot assume that the voltage defined above will be unique, rather it will depend on the path of integration chosen. That is, the electric field potential is not strictly conservative. In the lumped circuit case, however, where the distance between nodes is very short, the electric potential is very nearly conservative, and may be considered so for practical purposes.

Consider, for example, the difference between voltages v_1 and v_2, taken as the integrals through two different paths in the space between the nodes,

$$v_1 - v_2 = \int \mathbf{E} \cdot d\mathbf{l}_1 - \int \mathbf{E} \cdot d\mathbf{l}_2 = \oint \mathbf{E} \cdot d\mathbf{l}_{1\&2} \tag{2.2}$$

Since the final integral above is taken around a closed loop, we may apply Faraday's law as follows

$$v_1 - v_2 = \oint \mathbf{E} \cdot d\mathbf{l}_{1\&2} = -\frac{\partial}{\partial t} \iint \mathbf{B} \cdot d\mathbf{S} \tag{2.3}$$

where the magnetic flux density on the right side is that which is threading the loop enclosed by the two paths, l_1 and l_2. In the electrostatic or low-frequency limit, the time derivative on the right vanishes (we assume in the spirit of the lumped-element paradigm that the path chosen does not stray too far from the rest of the circuit, lest the area subtended by the closed loop, and consequently the integral of $\mathbf{B}$, becomes arbitrarily large). We are then left with $v_1 = v_2$; the voltage is conservative.

When applied to a network of lumped elements, Figure 2.1(b), the same argument for the conservativity of electrostatic potential guarantees that the sum of the voltages around any mesh (closed loop of elements) is zero. In the context of Figure 2.1(b),

$$v_1 + v_2 + v_3 + v_4 = 0 \tag{2.4}$$

or, for that matter,

$$v_1 + v_2 + v_5 = 0 \tag{2.5}$$

This is known as *Kirchhoff's voltage law* [1].

---

1 In many books, the integral shown here is negated and with the limits reversed, but the end result is the same.

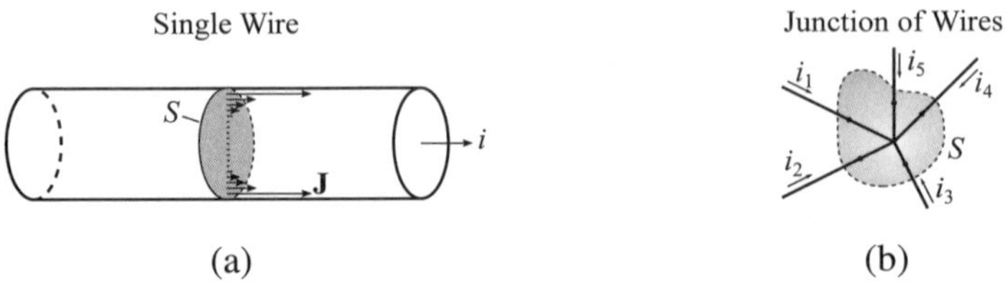

Figure 2.2 (a) Current density through the cross-section of a wire. (b) A junction of several wires.

2.1.2 Kirchhoff's Current Law

In almost all cases, the medium surrounding our circuit elements (e.g., air or vacuum) will be a very good insulator. As such, we can safely assume that currents can only flow inside our circuit elements or in the wires connecting them. Consider the rather common case of a circularly symmetric wire of metal, as shown in Figure 2.2(a). We know from Section 1.6.4 that the current will be flowing primarily at the surface, with exponentially less current toward the center of the wire (we have assumed that the wire is small compared to a wavelength in the surrounding space, but not necessarily small compared to a skin depth inside the conductor). The total current, i, is obviously just given by the integral of the current density throughout the cross-section of the wire,

$$i = \iint \mathbf{J} \cdot d\mathbf{S} \tag{2.6}$$

Consider now a junction of wires, as shown in Figure 2.2(b). The continuity equation (1.5b) gives us

$$\oiint \mathbf{J} \cdot d\mathbf{S} = -\sum_k i_k = -\frac{\partial}{\partial t} \iiint \rho dV \tag{2.7}$$

In keeping with the lumped-element approximation, we can assume that the time derivative on the right is zero. Therefore, the sum of all currents entering (or leaving) a junction is zero, or

$$\sum_k i_k = 0 \tag{2.8}$$

This is *Kirchhoff's current law* [1].

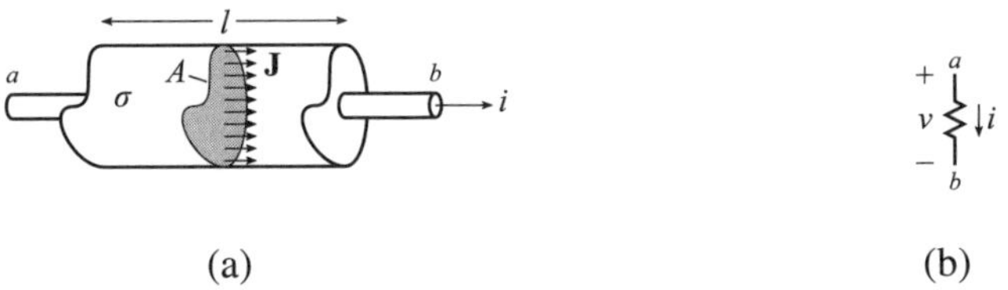

Figure 2.3 (a) Physical model of a resistor, and (b) schematic symbol.

2.2 LUMPED-ELEMENT DEVICES

Kirchhoff's circuit laws are the linchpins of lumped-element circuit theory; however, they are not enough. It would be fairly limiting if all we could do was impose voltages across isolated conductors, or shuttle currents around a mesh of empty wires. To make electronic circuits of any real value, we must construct elements that couple these two quantities to one another.

2.2.1 Resistors

The easiest way to do that is to pass our current through a piece of material for which the conductivity is not infinite (as we have implicitly assumed thus far for the wires in our circuit). See the geometry in Figure 2.3(a), comprising an extrusion of finite-conductivity material attached to wire leads at each end. In contrast with the high-conductivity wire, we may assume that the skin depth, given by (1.117), in this case is much larger than the diameter of the element. As such, the current distribution inside is approximately uniform, rather than concentrated at the periphery. The total current is found using (2.6) and then related to the electric field using Ohm's law (1.110),

$$i = \iint \mathbf{J} \cdot d\mathbf{S} = JA = \sigma EA \tag{2.9}$$

where A is the area of the cross-section. Likewise, we obtain the voltage between the two leads from (2.1)

$$v = \int_a^b \mathbf{E} \cdot d\mathbf{l} = El \tag{2.10}$$

where l is the total length of the conductive element. Combining these two results gives

$$v = EL = i\frac{l}{\sigma A} = iR \tag{2.11}$$

This is the more familiar form of Ohm's law expressed in the scalar quantities of this section instead of the field quantities in Chapter 1. Note that the total resistance is given by

$$R = \frac{l}{\sigma A} = \rho\frac{l}{A} \tag{2.12}$$

where $\rho = 1/\sigma$ is known as the *resistivity* of the material (not to be confused with the charge density, ρ, from Chapter 1). Equation (2.12) is known as *Pouillet's law*.

Not surprisingly, what we have just described is a resistor, shown by the usual schematic symbol in Figure 2.3(b).

2.2.2 Reactive Elements

Nothing we have done so far would allow our circuits to have any form of frequency dependence. This is because our basic assumption of low-frequency operation has zeroed out all time derivatives in Maxwell's equations. If we ever wish to make circuits that have rich frequency characteristics (say, a filter), we will need to find a way of reintroducing the time variability[2] from Maxwell's equations back into our networks.

Recall that another way of describing the lumped-element approximation is that the geometric features are very small compared to a wavelength, ideally, infinitesimally small. If we are indeed making the assumption that our circuits are infinitesimally small, then we must also be prepared to accept that some non-vanishing quantities must (notionally, at least) have infinite density. Any nonzero current, for example, passing through an infinitesimally narrow diameter wire must have infinite current density inside, while the total current itself is finite. The same can be said of the electric and magnetic flux densities that appear in the time derivatives of Maxwell's equations, assuming the boundary conditions are right. That will be the key to deriving the behaviors of reactive (time-dependent) components in the lumped-element regime.

2 *Time variability* and *time dependence* in this chapter refer simply to the fact that some of our circuit and field quantities may depend on derivatives with respect to time. It is not to be confused with the formal systems concept of *time invariance*, which is the property a system may have wherein its outputs depend only on the present value of its inputs and their derivatives, not on their past values.

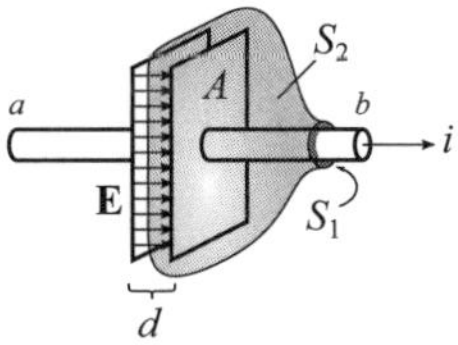

(a)

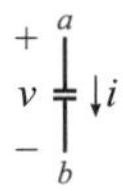

(b)

Figure 2.4 (a) Physical model of a capacitor, and (b) schematic symbol.

2.2.3 Capacitors

In reality, no field quantity can truly be infinite, but electrically small elements do require very concentrated fields in order to exhibit time-dependent effects. The electric field inside a tiny capacitor on a microchip, for example, can be hundreds of times stronger than the fields surrounding a high-voltage, overhead power line. (Capacitors were once called condensers for this very reason.)

Consider the two closely spaced plates connected to lead wires in Figure 2.4(a). The close spacing and small size of the plates are sufficient to guarantee that the electric field in the gap is essentially uniform. The voltage across the gap may be calculated again by (2.1),

$$v = \int_a^b \mathbf{E} \cdot d\mathbf{l}_k = Ed \tag{2.13}$$

where d is the distance across the gap. Now consider the magnetic field encircling the lead on the right, and apply Ampère's law,

$$\oint_l \mathbf{H} \cdot d\mathbf{l} = \iint_{S_1} \mathbf{J} \cdot d\mathbf{S} + \frac{\partial}{\partial t} \iint_{S_1} \mathbf{D} \cdot d\mathbf{S} \tag{2.14}$$

where S_1 is the small, disk-shaped surface that cuts through the cross-section of the wire (shaded gray in Figure 2.4(a)), and where the line integral on the left side is evaluated on its periphery, l. Note that since we assume wire leads have infinite conductivity (σ), but the total current (i) is finite, we must have

$$\iint_{S_1} \mathbf{D} \cdot d\mathbf{S} = \varepsilon \iint_{S_1} \mathbf{E} \cdot d\mathbf{S} = \frac{\varepsilon}{\sigma} \iint_{S_1} \mathbf{J} \cdot d\mathbf{S} = \frac{\varepsilon}{\sigma} i = 0 \tag{2.15}$$

on this surface. Therefore,

$$\oint_l \mathbf{H} \cdot d\mathbf{l} = \iint_{S_1} \mathbf{J} \cdot d\mathbf{S} = i \tag{2.16}$$

However, Ampère's law is required to hold for any surface bounded by the loop, l, including S_2, a balloon-shaped surface that avoids cutting through any conductor but passes through the gap between the plates (shaded lightly in Figure 2.4(a)). In this case, there is no current density, $\mathbf{J}$, to contribute to the integral, and we have

$$\oint_l \mathbf{H} \cdot d\mathbf{l} = \iint_{S_2} \mathbf{J} \cdot d\mathbf{S} + \frac{\partial}{\partial t} \iint_{S_2} \mathbf{D} \cdot d\mathbf{S} = \frac{\partial}{\partial t} \iint_{S_2} \mathbf{D} \cdot d\mathbf{S} \tag{2.17a}$$

$$\therefore i = \frac{\partial}{\partial t} \iint_{S_2} \mathbf{D} \cdot d\mathbf{S} \tag{2.17b}$$

Normally, we would assume that, in the lumped-element approximation, the time derivative on the right approaches zero. However, we have arranged the boundary conditions here to concentrate and intensify the field in a very small region penetrating the surface, S_2. Considering that the field in this region is nearly uniform, and zero outside of that region, we may simplify (2.17b),

$$i = \frac{\partial}{\partial t} \iint_{S_2} \mathbf{D} \cdot d\mathbf{S} = \frac{d}{dt} (\varepsilon E A) \tag{2.18}$$

where A is the area of the plates. Putting this together with (2.13),

$$i = \frac{d}{dt} (\varepsilon E A) = \left(\frac{\varepsilon A}{d} \right) \frac{dv}{dt} = C \frac{dv}{dt} \tag{2.19}$$

where C is called the *capacitance* [2].

Dimensional analysis reveals that the capacitance above has units of coulombs per volt, or of charge per electrostatic field potential. To make this more rigorous, let us consider the charge, Q, on the capacitor plate connected to terminal a. From the boundary conditions for a perfect conductor, (1.7b), we have,

$$Q = \rho_s A = \varepsilon E A \tag{2.20}$$

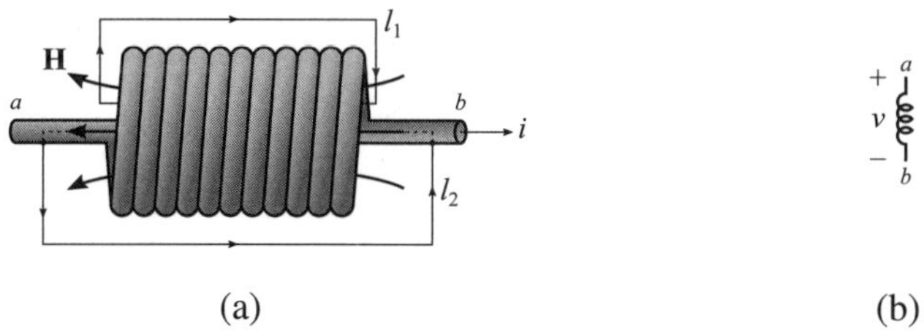

Figure 2.5 (a) Physical model of an inductor, and (b) schematic symbol.

Similarly, the charge on the opposite plate, with the field entering the surface instead of leaving it, is Q. Dividing the magnitude of this charge by the voltage in (2.13), we find

$$\frac{Q}{v} = \frac{\varepsilon EA}{Ed} = \frac{\varepsilon A}{d} = C \tag{2.21}$$

So the capacitance is also given by the amount of separated charge ($\pm Q$), divided by the electrostatic field potential that separates it. While the formula for capacitance in (2.19) is specific to the geometry in this example, the latter definition is true for all capacitors.

2.2.4 Inductors

Whereas the capacitor creates a time-dependent effect by creation of an intense, localized electric field, an inductor achieves a similar result by creating an intense, localized magnetic field. Since magnetic fields are created by currents, the logical way to do this is to wrap a wire around a tight region of space as a coil, or solenoid, as shown in Figure 2.5(a). It is advantageous to wind the wire as tightly as possible, so typically the wire will need to be insulated to avoid conductive contact between adjacent turns. This geometry ensures that a strong and largely uniform magnetic field will exist in the space inside the coil, while quickly becoming weak and divergent outside the coil.

The strength of the magnetic field may be calculated by applying Ampère's law to the closed path l_1, which passes through the center of the coil and returns outside of it,

$$\oint_{l_1} \mathbf{H} \cdot d\mathbf{l} = Hl = \iint \mathbf{J} \cdot d\mathbf{S} + \frac{\partial}{\partial t} \iint \mathbf{D} \cdot d\mathbf{S} = iN \tag{2.22a}$$

$$\therefore H = \frac{N}{l} i \tag{2.22b}$$

where N is the number of turns in the coil and l is the total length, as shown. Note that the electric displacement has been neglected in this region.

The voltage may be calculated by applying Faraday's law to another closed path, l_2, which follows the entire coil up the center of the wire, then returns back from lead a to lead b outside the coil.

$$\oint_{l_2} \mathbf{E} \cdot d\mathbf{l} = -\frac{\partial}{\partial t} \iint \mathbf{B} \cdot d\mathbf{S} = \mu N A \frac{dH}{dt} \tag{2.23}$$

where A is the cross-sectional area of the coil. Note that the magnetic field has penetrated the screw-shaped surface enclosed by l_2 a total of N times. As before, since the wire is assumed to have nearly infinite conductivity, we may assume that the electric field inside the conductor is effectively zero. Thus, the only part of l_2 that contributes to the line integral above is the part outside the coil between the two leads, which we recognize as simply the voltage across the part. Therefore, we have

$$v = \mu N A \frac{dH}{dt} \tag{2.24}$$

Combining this with (2.22b), we obtain

$$v = \mu N A \frac{dH}{dt} = \left(\frac{\mu N^2 A}{l} \right) \frac{di}{dt} = L \frac{di}{dt} \tag{2.25}$$

and L is called the *inductance* [2].

As with the derivation of capacitance, a number of simplifying assumptions were made above about the geometry and the stray fields in the space surrounding the part. These details have a minor impact on the actual value of the inductance or capacitance realized in practice, but the voltage-current relationship as a first-order derivative is quite accurate at low frequencies, or if the part is very small compared to a wavelength.

2.2.5 Transformers

Thus far, the lumped elements we have discussed have all been two-terminal devices. In this section, we discuss a four-terminal device known as a *transformer*. This device operates on the principle that the magnetic flux from one coil can be

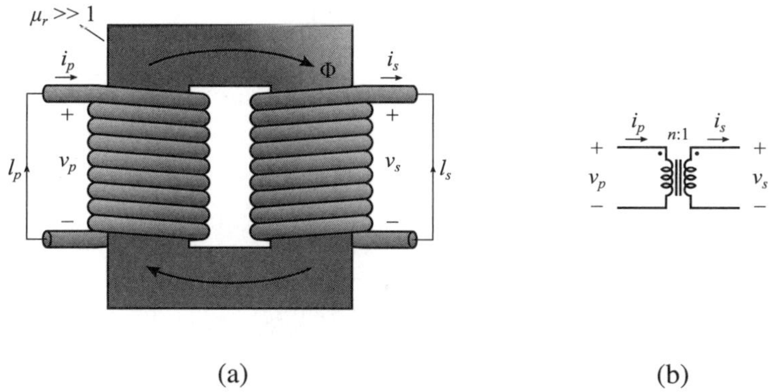

Figure 2.6 (a) Physical model of a transformer, and (b) schematic symbol.

linked to the magnetic flux of another. Typically, the two coils are wound around a common core with high permeability, usually a ferrite material, as shown in Figure 2.6(a), This helps to maximize the linkage of magnetic flux between the two coils by substantially confining the field inside the ferrite material. Although a square toroidal core is shown, in practice the core may take on many geometric forms.

The total magnetic flux in the core is simply the integral of the magnetic flux density over the cross-section,

$$\Phi = \iint \mathbf{B} \cdot d\mathbf{S} \tag{2.26}$$

We introduce this new quantity here because it is the one quantity that is constant throughout the full length of the core (not **B** or **H**, which get pinched, or concentrated, in areas where the core may narrow in cross-section, and may vary near corners or other discontinuities in the geometry).

The terminal voltages of both coils may be found much as we did for individual inductors, by applying Faraday's law to a closed path which follows each solenoid up from one terminal to the other and then returns outside of the transformer,

$$\oint_{l_p} \mathbf{E} \cdot d\mathbf{l} = v_p = -\frac{\partial}{\partial t} \iint_{l_p} \mathbf{B} \cdot d\mathbf{S} = -N_p \frac{d\Phi}{dt} \tag{2.27a}$$

$$\oint_{l_s} \mathbf{E} \cdot d\mathbf{l} = v_s = -\frac{\partial}{\partial t} \iint_{l_s} \mathbf{B} \cdot d\mathbf{S} = -N_s \frac{d\Phi}{dt} \tag{2.27b}$$

where N_p and N_s are the number of turns in the primary and secondary coils, respectively (we do not assume they are the same). Putting these together, we have

$$\frac{v_p}{v_s} = \frac{N_p}{N_s} = n \tag{2.28}$$

and n is known as the *turns ratio* of the transformer. The current relationship may be found by applying Ampère's law to a closed loop following the magnetic flux field lines around the core,

$$\oint_{\text{core}} \mathbf{H} \cdot d\mathbf{l} = \Phi R_\Phi = \iint \mathbf{J} \cdot d\mathbf{S} = N_p i_p - N_s i_s \tag{2.29}$$

where R_Φ is known as the *reluctance* of the core. The quantity evaluated above is known as the *magnetomotive force*, or *mmf*, and shall be denoted F_Φ. This completes a kind of analogy between ordinary circuits and magnetic circuits where the mmf takes on the role of voltage, magnetic flux takes the place of current, and the two are related through a counterpart to Ohm's law in which reluctance replaces the resistance,

$$F_\Phi = \Phi R_\Phi \tag{2.30}$$

If the core were uniform in cross-section, the reluctance would have value

$$R_\Phi = \frac{l}{\mu A} \tag{2.31}$$

where A is the cross-sectional area and l is the total length around the loop. Whether uniform or not, in an ideal transformer, the relative permeability, μ_r, is very large (often 10^3 to 10^5), such that the reluctance approaches zero.[3]

That being the case, we have

$$N_p i_p - N_s i_s = 0 \tag{2.32a}$$

$$\frac{i_p}{i_s} = \frac{N_s}{N_p} = \frac{1}{n} \tag{2.32b}$$

3 Effectively, this means that the $\mathbf{H}$ field inside the core is near zero, while the $\mathbf{B}$ field is finite. This is similar to an argument we made previously about perfect conductors, wherein the $\mathbf{E}$ field is near zero while the current density, $\mathbf{J}$, is finite.

Importantly, since the coupling of voltage from the primary to secondary coil relies on the magnetic flux changing (note the time derivatives in (2.27)), the above equations do not hold when the input signal is constant — the ratios above would be 0/0, or indeterminate. No real transformer can operate at DC [3].

2.3 ENERGY AND POWER

Before moving on to networks of elements, it is useful to talk about how energy is stored and dissipated in the devices that we have discussed so far. We shall find that the energy and power characteristics are completely determined by the behavior of the scalar voltage and current quantities measured at their terminals; the details of their internal structure, including bulk size, are unimportant.[4]

2.3.1 Power Dissipation in Resistors

In rigorous physical terms, power is defined as the rate of work done, where work is defined as a force acting over a distance. How can these concepts apply to an electronic system? We tend to think of most electronics as having no moving parts, so the concept of a force acting over a distance may at first seem not applicable. In all electronic systems, however, there are objects which are moving, namely the electric charges, and the force driving them is the Lorentz law, given in (1.4).

Consider an electric field, $\mathbf{E}$, moving a quantity of charge, q, against some external influence (in this case, the collisions of the charge carriers with the lattice). The work, U, done in moving that charge from point a to point b is given by

$$U = \int_a^b \mathbf{F} \cdot d\mathbf{l} = q \int_a^b \mathbf{E} \cdot d\mathbf{l} \tag{2.33}$$

In a resistor, this is a continuous process, with more charge passing through the resistor every moment. The quantity, q, may thus be considered a cumulative quantity (the total amount of charge that has passed through the resistor over a given period of time). To find the power, the rate at which the work is being done,

4 That is not to say that a physically larger resistor, for example, cannot dissipate more power without damage than a smaller resistor of the same value, but that given the same terminal inputs, the power delivered to them, however briefly, is the same.

we need only to take the derivative of U,

$$P = \frac{dU}{dt} = \left(\int_a^b \mathbf{E} \cdot d\mathbf{l} \right) \left(\frac{dq}{dt} \right) \tag{2.34}$$

Note that we have implicitly assumed the electric field itself is unchanging in the time that it takes an element of charge to traverse the distance from one end of the resistor to the other (an assumption that is compatible with the lumped-element approximations in this chapter). The two expressions in parentheses above should be recognized almost immediately as the terminal voltage and current, respectively. Therefore,

$$P = vi \tag{2.35}$$

and applying this to (2.11) we may further write

$$P = i^2 R = \frac{v^2}{R} \tag{2.36}$$

2.3.2 Energy Storage in a Capacitor

In principle, we should be able to derive the energy and power characteristics of a capacitor in much the same way, by tracking the movement of charge into the plates against the concentrated electric field which is accumulating there. Since we already know how a resistor dissipates energy, there is an easier way.

Consider the very simple circuit shown in Figure 2.7(a). We have a capacitor with capacitance C connected initially to a voltage source through a mechanical switch. It is not important at this time how the voltage source works, only that it imposes a fixed voltage, V_0, upon the terminals connected to it. Let us assume that at time $t = 0$, the switch is moved from one position to another, disconnecting it from the voltage source and connecting it to a resistor, R. The charge that was initially stored in the capacitor to create the electric field potential, V_0, is therefore discharged via current flowing out of the capacitor and into the resistor, until the charge on the capacitor is fully neutralized (see the plot in Figure 2.7(b)). We may calculate the amount of energy that was stored initially in the capacitor by integrating, over time, the power dissipated in the resistor.

Since the terminals of the capacitor are connected to those of the resistor, their voltages are the same,

$$v_C = v_R = v(t) \tag{2.37}$$

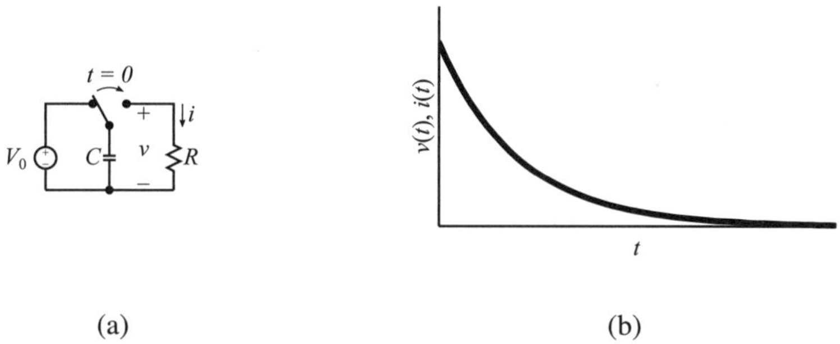

Figure 2.7 (a) A test circuit for determination of the energy stored in a capacitor. (b) Evolution of the terminal voltage/current as a function of time.

as required by Kirchhoff's voltage law. However, according to Kirchhoff's current law, the current flowing out of the capacitor is equal to the current flowing into the resistor, so

$$-i_C = i_R = i(t) \tag{2.38}$$

These two quantities must therefore evolve in time in such a way that satisfies both (2.11) and (2.19), the terminal characteristics of the two components. That is,

$$v = i_R R = -i_C R = -RC\frac{dv}{dt} \tag{2.39}$$

The solution to this relatively simple differential equation is

$$v(t) = V_0 e^{-\frac{t}{\tau}} \tag{2.40a}$$

$$\tau = RC \tag{2.40b}$$

which is what is plotted in Figure 2.7(b). The parameter τ is known as the *time constant* and is the time required for the voltage to settle to within $1/e$ of its final value in circuits like this one. Further, since the resistor ensures that v and i are always proportional to one another, we may write

$$i(t) = \frac{v(t)}{R} = \frac{V_0}{R}e^{-\frac{t}{\tau}} = I_0 e^{-\frac{t}{\tau}} \tag{2.41}$$

where I_0 is of course the initial current flowing through the resistor at the moment the switch is thrown. Finally, the energy delivered by the capacitor into the resistor is found by integrating (2.35) over time,

$$U = \int_0^\infty P dt = \int_0^\infty v i dt = \frac{V_0^2}{R}\int_0^\infty e^{-\frac{2t}{\tau}} dt = -\frac{V_0^2}{2R}\tau e^{-\frac{2t}{\tau}}\bigg|_0^\infty = \tfrac{1}{2}CV_0^2 \qquad (2.42)$$

The above is true for any capacitor of value C, regardless of its internal construction. However, using our simple, parallel-plate model for a capacitor in Figure 2.4(a), we may also use it to determine a more general expression for the energy stored in an electric field. According to (2.19) we have $C = \varepsilon A/d$, and by (2.13) we have $V_0 = Ed$; therefore,

$$U = \tfrac{1}{2}CV_0^2 = \tfrac{1}{2}\left(\frac{\varepsilon A}{d}\right)(Ed)^2 = \tfrac{1}{2}\varepsilon E^2 (Ad) \qquad (2.43)$$

where the final term in parentheses is recognized simply as the volume occupied by the electric field. The energy density is thus given by

$$u_E = \frac{U}{Ad} = \tfrac{1}{2}\varepsilon E^2 = \tfrac{1}{2}\mathbf{E}\cdot\mathbf{D} \qquad (2.44)$$

The final form involving the dot product of vector quantities is justified in our specific case since the electric field, **E**, and the electric flux density, **D**, are exactly parallel to one another (that is, we have assumed ε is a simple scalar). However, this need not always be the case if the medium is not isotropic. A more detailed derivation (which is beyond the scope of this book) reveals that the dot-product form shown above is true for the general case.

The total energy stored in an electric field of arbitrary configuration is given by

$$U_E = \tfrac{1}{2}\iiint \mathbf{E}\cdot\mathbf{D}dV \qquad (2.45)$$

Note that this section also provides an additional, very general definition for capacitance in terms of the total energy stored in an electrostatic field generated by a voltage,

$$C = \frac{2U_E}{V_0^2} = \frac{\iiint \mathbf{E}\cdot\mathbf{D}dV}{V_0^2} \qquad (2.46)$$

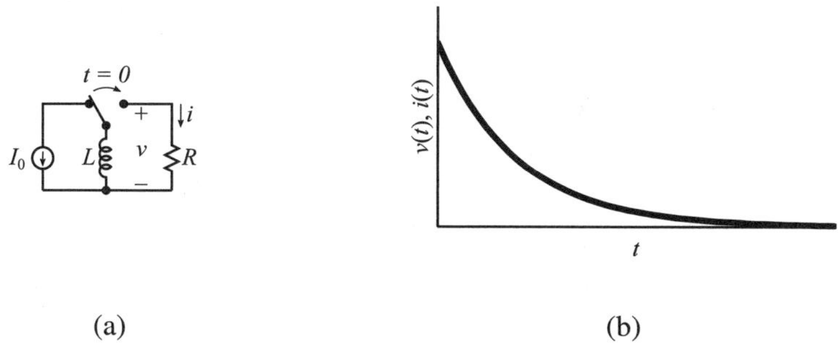

Figure 2.8 (a) A test circuit for determination of the energy stored in an inductor. (b) Evolution of the terminal voltage/current as a function of time.

2.3.3 Energy Storage in an Inductor

A similar derivation suffices to calculate the energy stored in the magnetic field inside an inductor. In this case, Figure 2.8(a), the inductor is first primed with a current source of value I_0. When the switch is thrown, the current flowing through the inductor is redirected through a load resistor. The voltage across its terminals must therefore satisfy both Ohm's law for the resistor, and (2.25) for the inductor,

$$v = iR = L\frac{di_L}{dt} = -L\frac{di}{dt} \tag{2.47}$$

The resistor current, i, has the solution

$$i = I_0 e^{-\frac{t}{\tau}} \tag{2.48a}$$

$$\tau = \frac{L}{R} \tag{2.48b}$$

Once again, the voltage and current both decay exponentially, only now with a time constant given by (2.48b), as shown in Figure 2.8(b). The total energy delivered by the inductor to the resistor is

$$U = \int_0^\infty P dt = \int_0^\infty vidt = I_0^2 R \int_0^\infty e^{-\frac{2t}{\tau}} dt = -\tfrac{1}{2} L I_0^2 e^{-\frac{2t}{\tau}}\Big|_0^\infty = \tfrac{1}{2} L I_0^2 \tag{2.49}$$

To relate this to the energy density of a general magnetic field, we may once again refer to our prototype element models. From (2.22b), we had at time $t = 0$,

$$I_0 = \frac{Hl}{N} \tag{2.50}$$

and from (2.25) we have

$$L = \frac{\mu N^2 A}{l} \tag{2.51}$$

Therefore, the total stored energy is

$$U = \tfrac{1}{2} L I_0^2 = \tfrac{1}{2}\left(\frac{\mu N^2 A}{l}\right)\left(\frac{Hl}{N}\right)^2 = \tfrac{1}{2}\mu H^2 \left(Al\right) \tag{2.52}$$

where again the last term in parentheses is the volume of the space occupied by the field. The energy density is then given by

$$u_H = \frac{U}{Al} = \tfrac{1}{2}\mu H^2 = \tfrac{1}{2}\mathbf{B}\cdot\mathbf{H} \tag{2.53}$$

and the total energy stored in an arbitrary magnetic field is

$$U_H = \tfrac{1}{2}\iiint \mathbf{B}\cdot\mathbf{H}dV \tag{2.54}$$

In this case, we have a very general energetic definition of inductance in terms of the magnetic field set up by a static current,

$$L = \frac{2U_H}{I_0^2} = \frac{\iiint \mathbf{B}\cdot\mathbf{H}dV}{I_0^2} \tag{2.55}$$

2.3.4 Energy Transfer in a Transformer

For a transformer, we may examine how the energy and power delivered to it flows by connecting a load resistor to the secondary terminals, as in Figure 2.9(a). The terminal voltages and currents are related as follows,

$$v\left(t\right) = v_p = nv_s = ni_sR \tag{2.56a}$$

(a) (b)

Figure 2.9 (a) A resistor connected to an arbitrary (AC) voltage source through an ideal transformer. (b) Equivalent circuit as seen from the source.

$$i_p = \frac{i_s}{n} \tag{2.56b}$$

Note that the primary voltage and current are proportional, just as the secondary voltage and current are, according to Ohm's law. The voltage source sees an equivalent resistance connected to it, R', where

$$R' = \frac{v_p}{i_p} = \frac{ni_s R}{i_s/n} = n^2 R \tag{2.57}$$

Thus, we may draw an equivalent circuit from the viewpoint of the voltage source which is shown in Figure 2.9(b).

The power delivered to the input terminals of the transformer may also be calculated as

$$P_p = v_p i_p = (ni_s R)\left(\frac{i_s}{n}\right) = i_s^2 R = P_s \tag{2.58}$$

Since the instantaneous power delivered to the load is equal to that supplied to the transformer, we may conclude that the (ideal) transformer neither dissipates nor stores energy; it merely transfers it, losslessly, from one pair of terminals to the next.

2.4 TERMINAL PARAMETERS IN THE FREQUENCY DOMAIN

Although the analyses of Section 2.2 are sufficient to fully describe the reactive elements in terms of their terminal characteristics, it would be most tedious if the analysis of every circuit we put together required solving a system of differential equations. We prefer to analyze our circuits in the frequency domain where, according to Section 1.4, derivatives in time are replaced with multiplicative factors in frequency.

Figure 2.10 (a) A series network of elements. (b) A parallel network of elements.

2.4.1 Impedance and Admittance

Converting our voltages and currents to their phasor form, we have for resistors,

$$v = iR \tag{2.59}$$

for capacitors,

$$i = j\omega Cv \tag{2.60}$$

and for inductors,

$$v = j\omega Li \tag{2.61}$$

Since the formulae for all two-terminal devices are now of the same basic form, we can generalize the concept of resistance into the frequency domain. That is, we define the impedance, Z, of a two-terminal element (or a two-terminal network of elements) as the ratio of voltage across its terminals to the current through it. For the basic two-terminal elements described thus far, we have

$$Z = \frac{v}{i} = \begin{cases} R \\ (j\omega C)^{-1} \\ j\omega L \end{cases} \tag{2.62}$$

When two or more elements are combined in series, as in Figure 2.10(a), their individual voltages add, while the current through each is identical. Therefore, the combined impedance of the group is simply the sum of the individual impedances.

The admittance, Y, will be defined as the inverse of this,

$$Y = \frac{i}{v} = \begin{cases} G \\ j\omega C \\ (j\omega L)^{-1} \end{cases} \tag{2.63}$$

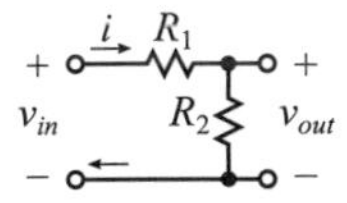

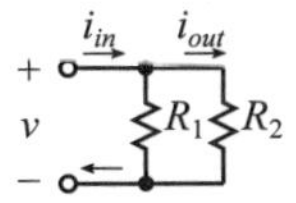

Figure 2.11 (a) Voltage divider. (b) Current divider.

where the conductance, $G = R^{-1}$. When two or more elements are combined in parallel, as in Figure 2.10(b), their individual currents add, while the voltage across each is identical. Therefore, the combined admittance of the group is simply the sum of the individual admittances. In contexts where either impedance or admittance quantities apply, the general term *immittance* is sometimes used instead.

One of the simplest and most immediate applications of series elements is the voltage divider, shown in Figure 2.11(a), a simply way to convert one voltage into a smaller one. Since the resistors are in series, the same current flows through both of them, where

$$i = \frac{v_{in}}{R_1 + R_2} \tag{2.64}$$

The voltage measured at the output is only that which develops across the second resistor, R_2. Therefore,

$$v_{out} = iR_2 = \frac{v_{in}}{R_1 + R_2} R_2 \tag{2.65a}$$

$$\therefore \frac{v_{out}}{v_{in}} = \frac{R_2}{R_1 + R_2} \tag{2.65b}$$

Similarly, the current divider in Figure 2.11(b) splits the current supplied by the external circuit across two resistors in parallel. The relevant relations in this case are

$$v = \frac{i_{in}}{G_1 + G_2} = \frac{i_{in}}{R_1^{-1} + R_2^{-1}} \tag{2.66a}$$

$$i_{out} = vG_2 = \frac{i_{in} R_2^{-1}}{R_1^{-1} + R_2^{-1}} \tag{2.66b}$$

$$\therefore \frac{i_{out}}{i_{in}} = \frac{R_2^{-1}}{R_1^{-1} + R_2^{-1}} = \frac{G_2}{G_1 + G_2} \tag{2.66c}$$

Figure 2.12 (a) A voltage source delivering power to a resistor. (b) A voltage source delivering the same instantaneous power to a device of arbitrary impedance.

2.4.2 Complex Power

The examples in Section 2.3 suggest that the instantaneous power entering or leaving an arbitrary, two-terminal network or device is simply the product of its real voltage and current, as given by (2.35). This can be proven in the following way.

For power leaving the device, one may simply connect a test resistor across its terminals such that the voltage and current, at least for that moment, remain unchanged. (For complex impedances, the required value of that resistor will change from one moment to the next, but that is not important to the argument being made here.) The power entering the resistor in that instant must be the same as the power leaving the device under test, so (2.35) describes that power. Note that the reference direction for current is always from the positive voltage terminal to the negative terminal. In the case just described, the current for the device under test will be flowing against this reference direction, making the computed power negative. This indicates that the device is delivering power instead of dissipating it.

For power entering the device, we may consider the two circuits shown in Figure 2.12. In Figure 2.12(a), a voltage source delivers a certain amount of power to a resistor. The resistor value is chosen such that the same current would be achieved if the voltage source was connected instead to our device under test, as in Figure 2.12(b), at a particular instant in time. Since the power being delivered by the voltage source must be the same in both cases, the power being dissipated in the device at that instant is the same as was dissipated in the resistor. Again, (2.35) describes that power.

Let us see how this applies to power computed in the frequency domain. Assume that we have a network or device driven with a sinusoidal voltage having peak phasor amplitude v and a current with peak phasor amplitude i. The time-domain voltage and current are then found using (1.80),

$$v(t) = \text{Re}\left\{ve^{j\omega t}\right\} = \text{Re}\left\{|v|\, e^{j(\omega t + \phi_v)}\right\} = |v| \cos(\omega t + \phi_v) \tag{2.67a}$$

$$i(t) = \text{Re}\left\{ie^{j\omega t}\right\} = \text{Re}\left\{|i|\, e^{j(\omega t + \phi_i)}\right\} = |i| \cos(\omega t + \phi_i) \tag{2.67b}$$

The instantaneous power flowing into the device is then

$$P'(t) = v(t)\, i(t) = |v|\, |i| \cos(\omega t + \phi_v) \cos(\omega t + \phi_i) \tag{2.68a}$$

$$= \tfrac{1}{2} |v|\, |i| \cos(2\omega t + \phi_v + \phi_i) + \tfrac{1}{2} |v|\, |i| \cos(\phi_v - \phi_i) \tag{2.68b}$$

Note that the instantaneous power dissipation comprises two terms. The first is an oscillatory term with frequency 2ω. It indicates that energy is alternately stored in the device and then released back to the external circuit, twice for every cycle of the input signal. Energy storage takes place inside the capacitors and inductors as intense, localized electric and magnetic fields. The total energy stored as a function of time is the integral of this term,

$$U = \int \tfrac{1}{2} |v|\, |i| \cos(2\omega t + \phi_v + \phi_i)\, dt \tag{2.69a}$$

$$= \tfrac{1}{4\omega} |v|\, |i| \sin(2\omega t + \phi_v + \phi_i) + U_0 \tag{2.69b}$$

The constant of integration, U_0, can be determined by recognizing that the stored energy can never be negative, nor can there be any permanent storage of energy in the final steady state. (Any long-term stored energy would eventually dissipate in the lossy elements, or else leak out to the external circuit. Even lossless networks cannot have zero external coupling, as that is required for energy to have been injected into the network in the first place.) Therefore, we may define U_0 so that the energy over time reaches exactly zero at its minimum points,

$$U_0 = \tfrac{1}{4\omega} |v|\, |i| \tag{2.70a}$$

$$\therefore U = \tfrac{1}{4\omega} |v|\, |i| \left[1 + \sin(2\omega t + \phi_v + \phi_i)\right] \tag{2.70b}$$

Note that U_0 may also be considered the average energy stored over time.

Returning to (2.68b) the second, time-independent term is seen to be the average power dissipated over time,

$$P_0 = \tfrac{1}{2} |v|\, |i| \cos(\phi_v - \phi_i) \tag{2.71}$$

One may observe immediately from (2.59)–(2.61) that the voltage and current through a resistor are in-phase ($\phi_v = \phi_i$), leading to finite power dissipation, while the voltage and current through a capacitor or inductor are in quadrature ($\phi_v - \phi_i = \pm\pi/2$), ensuring that the average power dissipation over time is zero for these elements.

To relate these concepts back to the phasor voltage and current, let us now define the *complex power* as

$$P = \tfrac{1}{2}vi^* = v_{rms}i_{rms}^* \tag{2.72}$$

where v_{rms} and i_{rms} are the root-mean-square phasor amplitudes, given for sinusoidal signals by taking the peak amplitude over $\sqrt{2}$. One may clearly see that the time average stored energy and dissipated power are given by

$$U_0 = \frac{|P|}{2\omega} \tag{2.73a}$$

$$P_0 = \text{Re}\left\{P\right\} \tag{2.73b}$$

It is important to not fall into the habit of thinking that the complex power, P, is the phasor amplitude of the time varying instantaneous power, $P'(t)$. The latter does not oscillate at the same frequency as the voltage and current, nor does it have a purely sinusoidal waveform, thus it cannot be represented as a phasor. Operations on phasor quantities yield phasor results only if the operation is linear (such as the sums and derivatives in Maxwell's equations). Note that impedance and admittance, though they operate on phasor quantities, are not themselves phasors, for the same reason.

2.5 RESONATORS

We may now talk about a particularly common grouping of elements, known as a *resonator*. Both of the networks shown previously in Figure 2.10 may be considered resonators, the first a *series RLC* type, and the second is a *parallel RLC* type.

2.5.1 Series RLC Resonator

Consider the case where the elements are in series. We know from Section 2.4.1 that the total impedance, Z_s, of the series combination is given by the sum of the

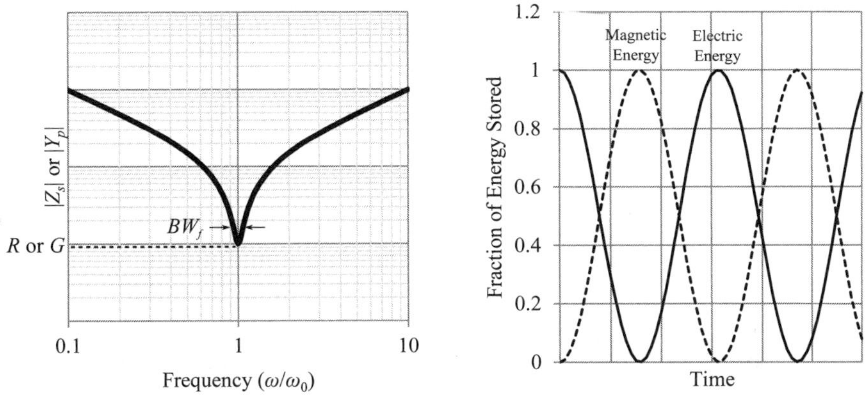

(a) (b)

Figure 2.13 (a) Magnitude of series impedance or parallel admittance versus frequency for series and parallel RLC resonators, respectively. BW_f is the half-power fractional bandwidth. (b) Fraction of energy stored in electric and magnetic fields versus time.

impedances of the individual elements,

$$Z_s = R + j\omega L + (j\omega C)^{-1} = R + \frac{1 - \omega^2 LC}{j\omega C} \tag{2.74}$$

The magnitude of this impedance function is plotted versus frequency in Figure 2.13(a). The resistor is usually assumed to be very small, and is often not included in the circuit intentionally, but is rather a parasitic element due to nonidealities in the reactive components. (Parasitic elements will be discussed in more detail in Section 2.8.) Note that there is a particular frequency for which the inductor and capacitor effectively cancel each other,

$$\omega_0 = \frac{1}{\sqrt{LC}} \tag{2.75}$$

At that particular frequency, the inductor and capacitor are said to be *in resonance*. The curve reaches its minimum point with a value of R (approaching a short circuit as $R \to 0$). At resonance, a constant amount of stored energy is passed back and forth each cycle between the capacitor and the inductor (between the electric field and the magnetic fields, as in Figure 2.13(b)), while the resistor dissipates a small amount. This is apparent since the average stored energy in the inductor,

$$U_L = \frac{|P_L|}{2\omega_0} = \frac{\left|\frac{1}{2}v_L i\right|}{2\omega_0} = \frac{\left|i^2 Z_L\right|}{4\omega_0} = \frac{\left|i^2 j\omega_0 L\right|}{4\omega_0} = \tfrac{1}{4}\left|i\right|^2 L \tag{2.76}$$

and in the capacitor

$$U_C = \frac{|P_C|}{2\omega_0} = \frac{\left|\frac{1}{2}v_C i\right|}{2\omega_0} = \frac{\left|i^2\right|}{4\omega_0 \left|Y_C\right|} = \frac{\left|i\right|^2}{4\omega_0^2 C} = \frac{\left|i\right|^2 LC}{4C} = \tfrac{1}{4}\left|i\right|^2 L \tag{2.77}$$

are the same. Furthermore, the average power dissipation in the resistor over time is given by

$$P_R = \text{Re}\left\{\tfrac{1}{2}v_R i^*\right\} = \tfrac{1}{2}\left|i\right|^2 R \tag{2.78}$$

The ratio of total energy stored to energy lost per unit time, multiplied by ω_0, is known as the *quality factor*, or Q, of the resonator. For the series resonator, this is

$$Q = \omega_0 \frac{U_L + U_C}{P_R} = \omega_0 \frac{2 \cdot \frac{1}{4}\left|i\right|^2 L}{\frac{1}{2}\left|i\right|^2 R} = \frac{\omega_0 L}{R} = \frac{1}{\omega_0 RC} \tag{2.79}$$

Thus, for the series resonator,

$$Q = \frac{X}{R} \tag{2.80}$$

where X is the reactance that is cancelled at resonance. Note that the Q is also an indicator of the half-power fractional bandwidth, BW_f, of the resonance. Consider the frequency, $\omega_0 + \Delta\omega$, at which the power dissipated in the resonator, with a constant voltage input, is half the power dissipated at resonance,

$$\frac{P_R\left(\omega_0 + \Delta\omega\right)}{P_R\left(\omega\right)} = \frac{\left|i\left(\omega_0 + \Delta\omega\right)\right|^2}{\left|i\left(\omega_0\right)\right|^2} = \frac{\left|Z\left(\omega_0\right)\right|^2}{\left|Z\left(\omega_0 + \Delta\omega\right)\right|^2} \tag{2.81a}$$

$$= \frac{R^2}{\left|Z\left(\omega_0 + \Delta\omega\right)\right|^2} = \frac{1}{2} \tag{2.81b}$$

$$\therefore 2R^2 = \left|R + \frac{1 - \left(\omega_0 + \Delta\omega\right)^2 LC}{j\left(\omega_0 + \Delta\omega\right) C}\right|^2 \tag{2.81c}$$

$$2R^2 = \left| R + \frac{1 - \omega_0^2 \left(1 + \frac{\Delta\omega}{\omega_0}\right)^2 LC}{j\left(\omega_0 + \Delta\omega\right) C} \right|^2 = \left| R + \frac{1 - \left(1 + \frac{\Delta\omega}{\omega_0}\right)^2}{j\left(\omega_0 + \Delta\omega\right) C} \right|^2 \tag{2.81d}$$

If we assume that the bandwidth is small ($\Delta\omega \ll \omega_0$), this may be simplified,

$$2R^2 \approx \left| R + \frac{1 - \left(1 + 2\frac{\Delta\omega}{\omega_0}\right)}{j\omega_0 C} \right|^2 = \left| R + j2\Delta\omega L \right|^2 = R^2 + 4\left(\Delta\omega\right)^2 L^2 \tag{2.82a}$$

$$= R^2 + \left(\tfrac{2\Delta\omega}{\omega_0}\right)^2 R^2 Q^2 \tag{2.82b}$$

$$\therefore BW_f = \frac{2\Delta\omega}{\omega_0} = \frac{1}{Q} \tag{2.82c}$$

2.5.2 Parallel RLC Resonator

Next, let us examine the parallel RLC resonator, shown originally in Figure 2.10(b). The admittance for this network has the same general form as the series RLC impedance, plotted in Figure 2.13(a), and is given by

$$Y_R = R^{-1} + j\omega C + \left(j\omega L\right)^{-1} = G + \frac{1 - \omega^2 LC}{j\omega L} \tag{2.83}$$

where $G = R^{-1}$. Noting the similarity with (2.74), all the results of the previous section may be imported here with a change of variables ($R \to G$, $v \leftrightarrow i$, $L \leftrightarrow C$). That is,

$$U_L = U_C = \tfrac{1}{4} \left|v\right|^2 C \tag{2.84a}$$

$$P_R = \tfrac{1}{2} \left|v\right|^2 G = \frac{\left|v\right|^2}{2R} \tag{2.84b}$$

$$Q = \omega_0 \frac{U_L + U_C}{P_R} = \omega_0 RC = \frac{R}{\omega_0 L} \tag{2.84c}$$

$$BW_f = \frac{2\Delta\omega}{\omega_0} = \frac{1}{Q} \tag{2.84d}$$

and, for the parallel resonator,

$$Q = \frac{B}{G} \tag{2.85}$$

where B is the susceptance that is cancelled at resonance.

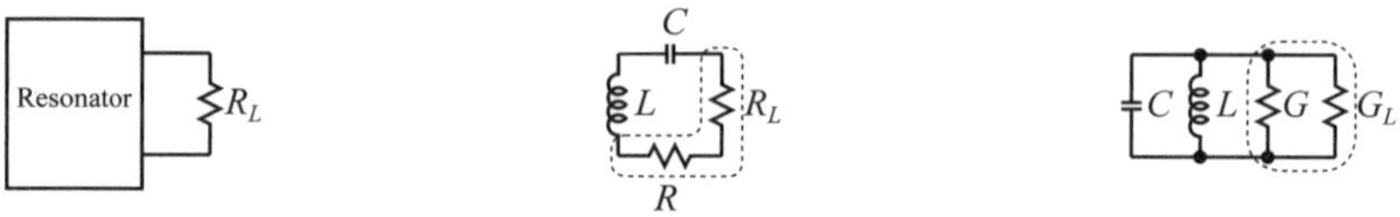

(a) (b) (c)

Figure 2.14 (a) A resonator loaded with an external resistance. (b) Series case. (c) Parallel case.

2.5.3 Loaded Q

Inevitably, a resonator will be coupled to an external circuit having its own termination impedance, as in Figure 2.14(a). This additional resistive load has the effect of lowering the Q of the resonator such that the half-power bandwidth is broadened, sometimes significantly.

In the case of a series resonator, Figure 2.14(b), the load resistor, R_L, adds in series with R in the resonator. According to (2.79) then,

$$Q' = \frac{\omega_0 L}{R + R_L} = \left(\frac{R}{\omega_0 L} + \frac{R_L}{\omega_0 L}\right)^{-1} = \left(Q^{-1} + Q_e^{-1}\right)^{-1} \tag{2.86}$$

where Q_e, called the *external Q*, is what the quality factor of the resonator would be if the resonator's resistance, R, were replaced by the external resistance, R_L.

For a parallel resonator, Figure 2.14(c), the load conductance, G_L, adds in parallel with the resonator conductance, G, such that

$$Q' = \frac{1}{\omega L\,(G + G_L)} = (\omega_0 LG + \omega_0 LG_L)^{-1} = \left(Q^{-1} + Q_e^{-1}\right)^{-1} \tag{2.87}$$

where, again, the external quality factor is that which would be obtained if the resonator's conductance (or resistance) were replaced by the external termination.

In both cases, we have found that

$$\frac{1}{Q'} = \frac{1}{Q} + \frac{1}{Q_e} \tag{2.88}$$

This is called the *loaded Q* of the resonator, and the above formula will apply generally to all types of resonators, not just those that are built upon lumped elements. The quality factors of Sections 2.5.1 and 2.5.2 are sometimes called *unloaded Qs* for clarity [4].

Table 2.1
Lumped-Element Resonators

Quantity	Symbol	Expression
Resonant Frequency	ω_0	$= \frac{1}{\sqrt{LC}}$
Total Stored Energy (at resonance)	$U = U_L + U_C$	$= \begin{cases} \frac{1}{2}\|i\|^2 L & \text{series} \\ \frac{1}{2}\|v\|^2 C & \text{parallel} \end{cases}$
Power Dissipation (at resonance)	P_R	$= \begin{cases} \frac{1}{2}\|i\|^2 R & \text{series} \\ \frac{1}{2}\|v\|^2 G & \text{parallel} \end{cases}$
Reactance Cancelled at Resonance (series)	X	$= \omega_0 L = \frac{1}{\omega_0 C}$
Susceptance Cancelled at Resonance (parallel)	B	$= \omega_0 C = \frac{1}{\omega_0 L}$
Unloaded Q	Q_0	$= \begin{cases} X/R & \text{series} \\ B/G & \text{parallel} \end{cases}$
External Q	Q_e	$= \begin{cases} X/R_L & \text{series} \\ B/G_L & \text{parallel} \end{cases}$
Loaded Q	Q'	$= \left(Q_0^{-1} + Q_e^{-1}\right)^{-1}$

A summary of the results for lumped-element resonators is given in Table 2.1.

2.6 SOURCES

Up to this point, we have used ideal voltage and current sources as a convenience for proving some mathematical truths. Real signal generators are not capable of imposing exact voltages and currents without constraint. If they were, simply connecting two voltage sources in parallel or two current sources in series would lead to a paradox. We never asked, for example, what would happen to the current source in Figure 2.8(a) after the switch disconnected it, where would its current go?

Figure 2.15 (a) Thévenin equivalent voltage source. (b) Norton equivalent current source.

In reality, all signal sources have some internal impedance. That impedance may appear in series with an otherwise ideal voltage source, as Figure 2.15(a), or in parallel with an ideal current source, as in Figure 2.15(b). The first is called the *Thévenin equivalent source*, and the latter is called the *Norton equivalent source*. What is perhaps surprising is that these two representations are exactly equivalent to one another, so long as $V_0 = I_0 R_s$. There is no way to tell one from the other simply by its terminal characteristics.[5] Consider the voltage and current that is developed across the terminals in Thévenin's case, for a load resistance, R_L,

$$i = \frac{V_0}{R_s + R_L} \tag{2.89a}$$

$$v = iR_L = \frac{V_0 R_L}{R_s + R_L} \tag{2.89b}$$

and in Norton's case,

$$v = I_0 \left(R_s^{-1} + R_L^{-1}\right)^{-1} = \frac{I_0 R_s R_L}{R_s + R_L} = \frac{V_0 R_L}{R_s + R_L} \tag{2.90a}$$

$$i = \frac{v}{R_L} = \frac{V_0}{R_s + R_L} \tag{2.90b}$$

which is the same.

What matters is the relative value of the internal source resistance compared to the range of expected loads. Let us imagine, for example, that we have an application where our source will be connected at times to loads that vary from 5Ω to 500Ω. If our source impedance is only 1Ω, then the source delivers nearly constant voltage to its terminals, while the current and power delivered vary significantly, as shown in Figure 2.16(a). This is true regardless of whether it is a Thévenin- or

5 However, one can see that the Norton equivalent circuit would dissipate power internally if its terminals were left open, while the Thévenin equivalent circuit would not, and vice versa if the terminals were short-circuited.

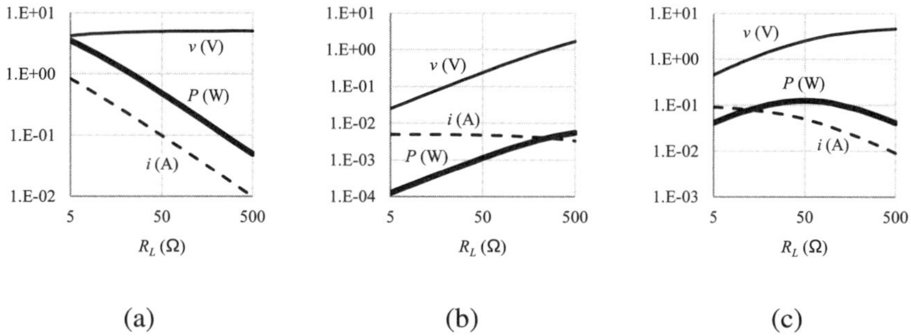

Figure 2.16 Terminal characteristics of a source having (a) 1Ω internal resistance, (b) 1 kΩ internal resistance, and (c) 50Ω internal resistance. In all cases, $V_0 = 5$V.

Norton-style circuit; the plot is the same. It is conventional to draw a Thévenin equivalent circuit in this case, however, as the terminal voltage that the generator approximates is more explicit.

However, if the source impedance is 1 kΩ, then the generator does a better job of maintaining nearly constant current over the given load range, as in Figure 2.16(b). In this case, the Norton equivalent is preferred, though either works.

Finally, it is interesting to note that a midrange source resistance does the best job of equalizing the power delivered to the load. See Figure 2.16(c), where the source resistance is 50Ω. In fact, when the source and load resistance are the same, the power delivered is maximized [1]. In this example, where $V_0 = 5$V, the power delivered is 125 mW.

One must be careful to remember that these are equivalent circuits only, describing the terminal behavior of a typical power generator, such as a battery, oscillator, or transmitter. The individual circuit elements within them do not necessarily correspond one-to-one to the physical processes that may be at work inside such a device. Consider, for example, that while the Thévenin and Norton equivalent sources determine the load impedance that maximizes delivered power, they do not always represent the true efficiency of that transference. This is made obvious by the fact that the two equivalent circuits do not achieve the same efficiency under most load conditions. Figure 2.17(a) shows the power delivered to the load (solid line) and the power dissipated in the source (dashed line) for a Thévenin equivalent generator with internal impedance of 50Ω. Figure 2.17(b) shows the same curves for a Norton equivalent generator. While the power delivered to the load in both cases

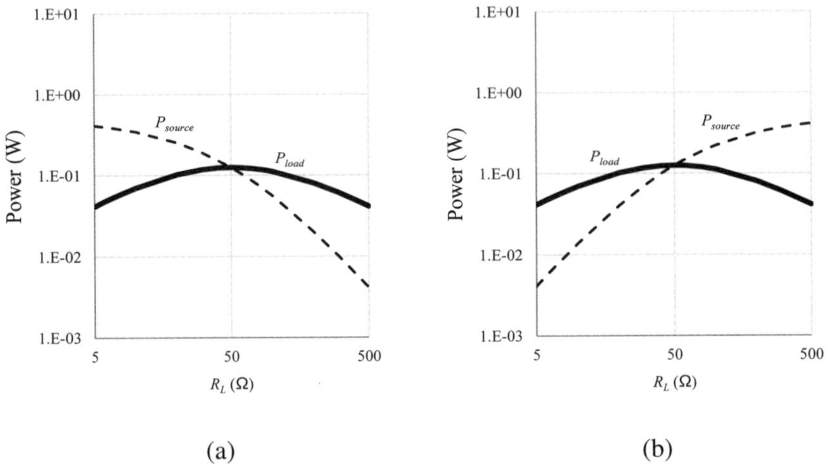

Figure 2.17 Power delivered to a load (solid line) and dissipated in the source (dashed line) for (a) a Thévenin equivalent generator, and (b) a Norton equivalent generator, where the source impedance in both cases is 50Ω.

is the same, the power dissipated in the source, and consequently the efficiency of the power transfer, is quite different.

2.7 TELLEGEN'S THEOREM AND ITS IMPLICATIONS

To illustrate the power of the lumped-element model, let us examine a rather remarkable theorem, which has found broad application not only in electronics, but by analogy to chemical, biological, and industrial processes as well.

Let us assume we have an arbitrary network of lumped elements and sources, as shown in Figure 2.18(a). The network has n nodes and b branches connecting them (note that more than one branch may connect the same two nodes if elements are in parallel). Define an incidence matrix, $\mathbf{A}$, size $n \times b$, such that the members of the matrix, a_{ij} are given by

$$a_{ij} = \begin{cases} 1 & \text{if current } j \text{ leaves node } i \\ -1 & \text{if current } j \text{ enters node } i \\ 0 & \text{otherwise} \end{cases} \tag{2.91}$$

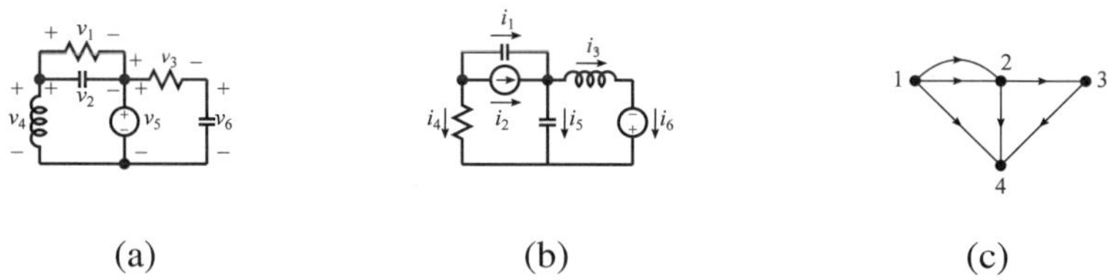

Figure 2.18 Illustration of Tellegen's theorem. (a) An arbitrary network from which voltages are taken. (b) A different, arbitrary network from which currents are taken. (c) Topology defined by the incidence matrix, with reference directions and node numbering shown.

In other words, the matrix $\mathbf{A}$ describes the linkage, or *topology*, of the network, but does not describe the network elements themselves. Let $\mathbf{K}$ be a column vector of the currents in all the branches of the network,

$$\mathbf{K} = \begin{pmatrix} i_1 \\ i_2 \\ \vdots \\ i_b \end{pmatrix} \tag{2.92}$$

Kirchhoff's current law for every node may then be written quite simply,

$$\mathbf{AK} = 0 \tag{2.93}$$

Further, let Φ be a column vector containing the potential of every node relative to some arbitrary reference,

$$\Phi = \begin{pmatrix} \varphi_1 \\ \varphi_2 \\ \vdots \\ \varphi_n \end{pmatrix} \tag{2.94}$$

The potential differences (or *voltage drops*) though each branch of the network may then be written,

$$\mathbf{V} = \begin{pmatrix} v_1 \\ v_2 \\ \vdots \\ v_b \end{pmatrix} = \mathbf{A}^T \mathbf{\Phi} \tag{2.95}$$

Since the voltage drops through each branch are derived by differencing absolute node potentials, Kirchhoff's voltage law is automatically satisfied by this definition.

Now consider the product, $\mathbf{V}^T\mathbf{K}$,

$$\mathbf{V}^T\mathbf{K} = \left(\mathbf{A}^T\mathbf{\Phi}\right)^T\mathbf{K} = \left(\mathbf{\Phi}^T\mathbf{A}\right)\mathbf{K} = \mathbf{\Phi}^T\left(\mathbf{A}\mathbf{K}\right) = \mathbf{\Phi}^T 0 = 0 \tag{2.96}$$

Therefore,

$$\sum_{k=1}^{b} v_k i_k = \mathbf{V}^T\mathbf{K} = 0 \tag{2.97}$$

This is *Tellegen's theorem*. On the surface, it may look like a simple expression of energy conservation, since the product vi at any branch we already know is the power dissipated by an element if positive, or delivered by a source if negative. So Tellegen's theorem, in the context presented here, simply tells us that the power delivered by all the sources is equal to the power absorbed by the remaining elements.

The true power of Tellegen's theorem comes from the realization of just how little we needed to assume about the list of voltages and currents in order to prove it. We said that the currents obey Kirchhoff's current law. We said that the voltages obey Kirchhoff's voltage law. We said nothing at all, however, about the elements. They need not be linear. They need not be passive. They need not be time-invariant. They need not be memory-less or reciprocal. They need not even be causal.

The list of voltages and currents need not have anything to do with each other at all; they may be taken from the circuit at totally different times, or even from totally different circuits, as shown in Figures 2.18(a, b). Nor are they required to be true voltages and currents, but could be any quantities that meet Kirchhoff's criteria (linear operators applied to voltages and currents, for example). Tellegen's theorem states that any two lists of numbers, whatsoever, that satisfy Kirchhoff's laws on a common directed graph (e.g., Figure 2.18(c)), regardless of their context or meaning, must satisfy (2.97) [5, 6].

2.7.1 Network Reciprocity

One of the more important results that can be proven using Tellegen's theorem is network reciprocity. Reciprocity in this context states that if a voltage source, v, in one branch of a network produces a current, i, in a different branch, then moving the voltage source to the second branch yields the same current in the first. Stated another way, the input and output locations of a reciprocal network are interchangeable.

Reciprocity in the more general context of Maxwell's equations was proven earlier in Section 1.7, subject to certain material property constraints. In principle,

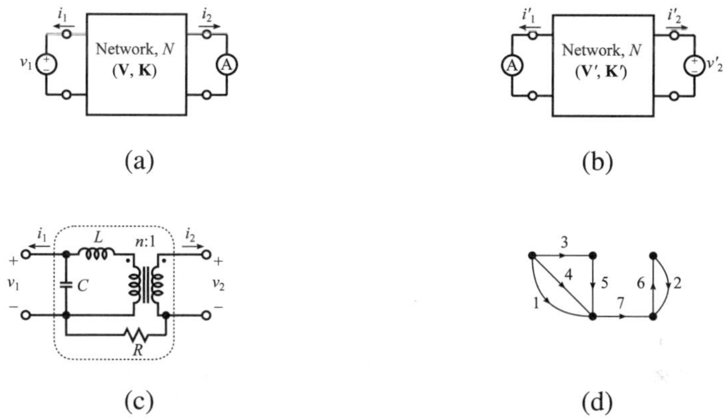

Figure 2.19 Illustration for the proof of network reciprocity. (a) A network with a voltage source at port 1, and output current measured at port 2. (b) The same network with a voltage source at port 2, and output current measured at port 1. (c) Example network with two port branches and five internal branches, and (d) its topological representation with the branches numbered.

then, we could apply that result here without further proof. However, since the lumped-element approximations have yielded for us a network theory that rests upon a far simpler set of rules (Kirchhoff's laws) than those that govern general electromagnetics (Maxwell's equations), it should be possible, and is instructive, to reconstruct reciprocity on that simpler basis alone.

Consider a network, N, with two branches set aside as input and output *ports*. For this proof, we shall connect a voltage source to port 1 (i.e., branch 1), and an *ammeter*, or current measurement device, to port 2, as indicated in Figure 2.19(a). The list of voltages and currents for the remaining internal branches shall be contained in the column vectors $\mathbf{V}$ and $\mathbf{K}$, respectively.

Consider now the same network, N, but with the positions of the voltage source and ammeter reversed, as in Figure 2.19(b). In general, the voltages and currents in each branch of the network will change, and the new values will be labelled with the primed parameters, $\mathbf{V}'$ and $\mathbf{K}'$. Since Tellegen's theorem applies to any two lists of voltages and currents on this network topology, even if not directly related, we may multiply the primed current vector with the unprimed voltage vector, and vice versa

$$v_1 i_1' + v_2 i_2' + \mathbf{V}^T \mathbf{K}' = 0 \tag{2.98a}$$

$$v_1' i_1 + v_2' i_2 + \mathbf{V}'^T \mathbf{K} = 0 \tag{2.98b}$$

At this time, we note that an ammeter, in order to measure current without otherwise altering the network, must have zero internal impedance. It therefore admits no voltage across it and we can say that $v_2 = v_1' = 0$. Substituting this into (2.98b) and subtracting the two expressions gives the following,

$$v_1 i_1' - v_2' i_2 = \mathbf{V}'^T\mathbf{K} - \mathbf{V}^T\mathbf{K}' \tag{2.99}$$

Let us now introduce the impedance operator, $\mathbf{Z}$, such that $\mathbf{V} = \mathbf{ZK}$. Were the network to contain only two-terminal lumped elements, like inductors, capacitors, and resistors, then $\mathbf{Z}$ would be a diagonal matrix containing the elemental impedances, since the voltage in each branch depends directly and only upon the current through the same branch. If the network contains four-terminal devices, like transformers, for example, we must be slightly more careful. The primary and secondary coils of the transformer may be considered as separate branches in our network, but coupled according to the terminal characteristics in (2.28) and (2.32). Putting these together, we have

$$\frac{v_p}{i_s} = \frac{v_s}{i_p} = z_x \tag{2.100}$$

These two ratios may be recognized as equivalent off-diagonal elements in the matrix $\mathbf{Z}$ (the actual value of these ratios, z_x, shall depend on the number of turns in the transformer and the embedding network; however, this is unimportant at this stage). The transformer is thus said to be bilateral, and (2.100) is essentially a statement about the reciprocal behavior of a transformer in isolation. The same can be said of individual two-terminal elements like resistors, capacitors, and inductors. What we are attempting to prove at present is that any compound network containing such elements is also reciprocal with respect to any two arbitrarily chosen ports.

As an example, consider the network shown in Figure 2.19(c), having two port branches and five internal branches populated with lumped-elements and a transformer. The impedance operator for the internal branches, numbered 3–7 in Figure 2.19(d), may be written as follows,

$$\mathbf{Z} = \begin{pmatrix} j\omega L & 0 & 0 & 0 & 0 \\ 0 & (j\omega C)^{-1} & 0 & 0 & 0 \\ 0 & 0 & 0 & z_x & 0 \\ 0 & 0 & z_x & 0 & 0 \\ 0 & 0 & 0 & 0 & R \end{pmatrix} \tag{2.101}$$

In any case, the arguments above guarantee that the matrix $\mathbf{Z}$ is symmetric, or $\mathbf{Z}^T = \mathbf{Z}$, with its diagonal elements given by two-terminal devices, and the off-diagonal elements made nonzero by transformers, if any. Applying this result to (2.99), we have

$$v_1 i_1' - v_2' i_2 = \mathbf{V}'^T \mathbf{K} - \mathbf{V}^T \mathbf{K}' = \left(\mathbf{Z}\mathbf{K}'\right)^T \mathbf{K} - \left(\mathbf{Z}\mathbf{K}\right)^T \mathbf{K}' \tag{2.102a}$$

$$= \mathbf{K}'^T \mathbf{Z}\mathbf{K} - \mathbf{K}^T \mathbf{Z}\mathbf{K}' = \mathbf{K}'^T \mathbf{Z}\mathbf{K} - \left(\mathbf{K}'^T \mathbf{Z}\mathbf{K}\right)^T = p - p^T = 0 \tag{2.102b}$$

where the last step follows because the product $p = \mathbf{K}'^T \mathbf{Z}\mathbf{K}$ is a scalar. Therefore, we have

$$v_1 i_1' = v_2' i_2 \tag{2.103}$$

and it follows that if $v_1 = v_2'$, then $i_1' = i_2$.

Similarly, if the ammeter was replaced by a current source, and the voltage source was replaced by a *voltmeter*, we could have derived the dual statement that $i_1 = i_2'$ implies $v_1' = v_2$.

The reader should note the similarity between (2.103) and the Rayleigh-Carson reciprocity condition for field quantities found earlier in (1.125).

It is worth pointing out here that there are some nonbilateral devices that disrupt reciprocity in the networks that use them. Generally, these will be nonlinear solid-state elements, such as diodes and transistors,[6] or nonreciprocal multiport devices such as ferrite circulators and gyrators. Some of these elements will be discussed in more detail in later chapters.

2.8 PARASITICS

Most of the lumped elements that we have described in this chapter so far represent the mathematical ideal of infinitely compact structures made with perfect materials: perfectly lossless and linear reactive elements (inductors and capacitors) and transformers having complete linkage of magnetic flux between the coils. Nothing in reality is so perfect. These imperfections can be modeled, however, using additional discrete parasitic elements based on the same mathematical foundation.

---

6 Diodes violate reciprocity for DC circuits, since their nonlinearity invalidates the use of the linear impedance operator, $\mathbf{Z}$. They may be *linearized* for small AC signals, however, in which case reciprocity holds. Transistors, as three-terminal devices, do not usually exhibit reciprocal behavior, even when linearized for small signals.

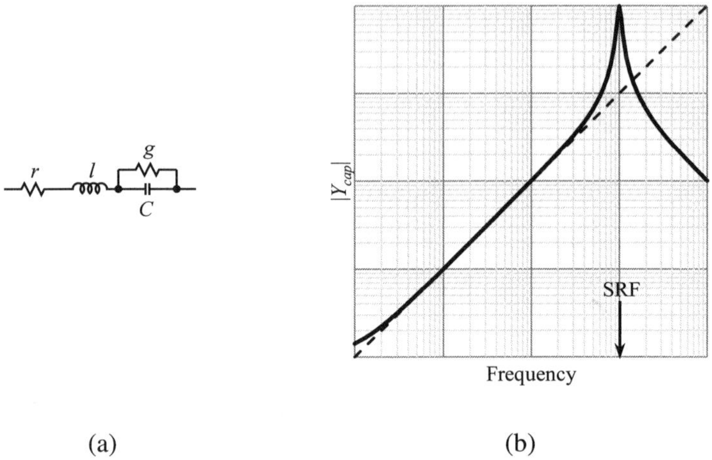

Figure 2.20 (a) Realistic model of a capacitor. (b) Magnitude of the capacitor's admittance versus frequency. The dashed line is for an ideal capacitor.

2.8.1 Realistic Capacitors

The idealized capacitor that was illustrated in Figure 2.4 assumed that there was no leakage between the plates. In reality, no insulator is totally insulating; there will always be some leakage. Any current that leaks through the medium between the plates essentially bypasses the capacitive action of the device. This can be modeled as a large-valued resistor (or small conductor) in parallel with the capacitor.

Furthermore, the metal used to make the plates and the lead wires was assumed to have infinite conductivity. In reality, no conductor is ever perfect (not even superconductors, which do have a small amount of resistance above absolute zero frequency). The finite conductivity of real-world materials means that any capacitor will inevitably have some series resistance associated with it. Finally, the flow of current through the leads and spreading out at the plates will involve a small amount of inductance which also appears in series with the main capacitor.

A more complete model of a real capacitor is thus shown in Figure 2.20(a). Parasitic elements, shown with lowercase labels, are assumed to be very small. The magnitude of the admittance of such a model is shown in Figure 2.20(b). It very closely approximates that of an ideal capacitor (dashed line) over most of the frequency range, but deviates at both the lowest and highest frequencies. In

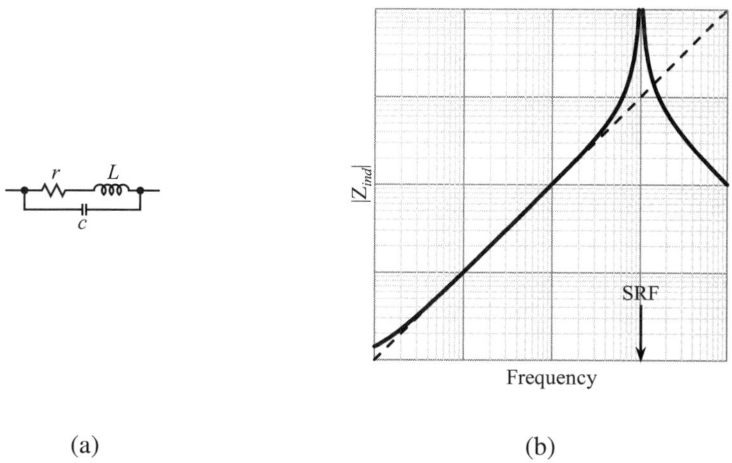

Figure 2.21 (a) Realistic model of an inductor. (b) Magnitude of the inductor's impedance versus frequency. The dashed line is for an ideal inductor.

particular, the leakage of the insulator prevents the capacitor from approaching the ideal open-circuit at DC, while the *equivalent series resistance* (ESR) and *equivalent series inductance* (ESL) prevent it from approaching the ideal short-circuit at high frequency.

In fact, the interaction of the primary capacitor with the parasitic inductor creates a *self-resonant frequency* (SRF) at which the capacitor's admittance becomes very large. While this can wreck havoc with carefully tuned devices such as filters, it is actually an acceptable condition when the capacitor is used for passing or shorting AC signals (as in a coupling or bypass capacitor).

2.8.2 Realistic Inductors

Similarly, the ideal inductor shown in Figure 2.5 made use of perfect electrical conductors, neglecting any series resistance in the coil. There is also a very small capacitance between the individual turns of the coil, which can be modeled as a parasitic capacitor in parallel with the main inductance.

Thus, a more accurate model for a realistic inductor is shown in Figure 2.21(a), where again the parasitic elements are labeled in lowercase. The magnitude of the impedance of this model is shown in Figure 2.21(b). Very similar to the

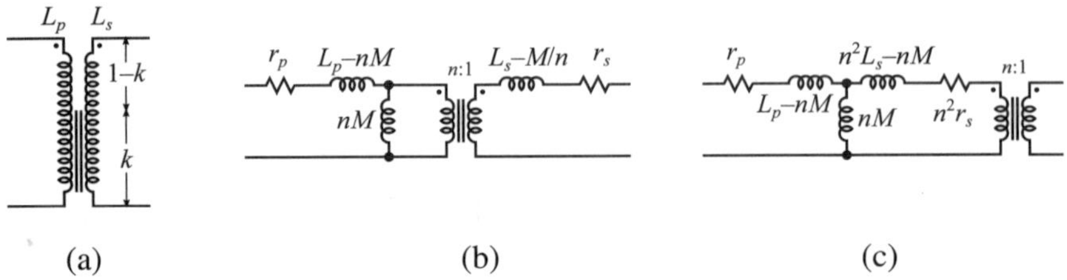

Figure 2.22 (a) Transformer model showing the coupling factor and the relationship between leakage and magnetizing inductance. (b) Circuit model with parasitic elements. (c) Parasitic elements moved to the side of the primary winding.

admittance of the capacitor, the parasitic elements cause the actual impedance curve to deviate from the intended ideal at very low and very high frequencies. Especially notable is the self-resonant frequency where the primary inductance interacts with the parasitic interwinding capacitance to create a pole. As was the case with the capacitor, this resonance may be disastrous for tuned circuits, but is acceptable when the inductor is used to block high-frequency signals (called a *choke*).

Even these models are imperfect, and commercial models may include additional parasitic elements for increased accuracy and frequency range, but the elements shown here are sufficient to capture the most significant and frequently cited features of true reactive elements.

2.8.3 Realistic Transformers

Transformers likewise are subject the series resistance and interwinding capacitance that degrades inductor performance, but in this case there is another effect that is even more significant: the imperfect linkage of magnetic flux between the two coils. Recall that in the derivation of Section 2.2.5, the permeability of the core around which the coils were wound was assumed to be very large, effectively infinite. This ensured that all magnetic flux threading through one coil also threaded through the other.

In reality, this coupling of the two coils' magnetic fields can never be perfect. Magnetic flux which threads only one of the two coils is known as *leakage flux*. In effect, this incomplete linkage causes some fraction of each coil to have isolated inductance that does not contribute to the transformer coupling. Each coil thus appears to have a parasitic inductor in series with it, as indicated in Figure 2.22(a). Let L_p be the total, independent inductance of the primary coil, and L_s be the

Table 2.2
Elements of a Realistic Transformer Model

Quantity	Symbol	Expression(s)
Turns ratio	n	$= \frac{N_p}{N_s} = \sqrt{\frac{L_p}{L_s}}$
Coupling factor	k	$= \frac{M}{\sqrt{L_p L_s}}$
Primary inductance	L_p	$= L_p^{mag} + L_p^{leak}$
Secondary inductance	L_s	$= L_s^{mag} + L_s^{leak}$
Mutual inductance	M	$= k\sqrt{L_p L_s}$
Magnetizing inductance:		
referred to primary	L_p^{mag}	$= kL_p = nM$
referred to secondary	L_s^{mag}	$= kL_s = \frac{M}{n}$
Leakage inductance, primary	L_p^{leak}	$= (1-k)L_p = L_p - nM$
Leakage inductance, secondary	L_s^{leak}	$= (1-k)L_s = L_s - \frac{M}{n}$

total, independent inductance of the secondary coil. Since these inductances are proportional to the square of the number of turns in each coil (see Section 2.2.4), the turns ratio may be written in terms of these quantities,

$$n = \frac{N_p}{N_s} = \sqrt{\frac{L_p}{L_s}} \tag{2.104}$$

The *coupling factor*, k, represents the fraction of each coil that contributes to the transformer coupling. Therefore, the leakage inductors have values $(1-k)\,L_p$ and $(1-k)\,L_s$. The inductances which do contribute to transformer coupling, kL_p and kL_s, are termed the *magnetizing inductance*, referred to the primary and secondary coils, respectively.

Recall also that real transformers cannot operate at DC. A changing magnetic flux is required, as a consequence of the time derivatives in Maxwell's equations. This is represented in a physical transformer model by a *mutual inductance*, which depends on the individual coil inductances and the coupling factor,

$$M = k\sqrt{L_p L_s} \tag{2.105}$$

In practice, the mutual inductance may appear in the circuit model on either side of the transformer. If on the primary side, the parasitic element will have value nM; if on the secondary side, it will have value M/n. The leakage inductances may then also be written

$$L_p^{\text{leak}} = (1 - k)\, L_p = L_p - nM \tag{2.106a}$$

$$L_s^{\text{leak}} = (1 - k)\, L_s = L_s - \frac{M}{n} \tag{2.106b}$$

A more complete model of a realistic transformer is thus shown in Figure 2.22(b). In some cases, it is customary to move all the parasitic elements to one side, as in Figure 2.22(c). The quantities associated with this model and their relationships are summarized in Table 2.2.

Problems

2-1 Nichrome has a resistivity of 1.1×10^{-6} Ω·m. What is the resistance of a thin wire of Nichrome that is 10 cm long and 100 μm in diameter?

2-2 How thin would a tantalum nitride (TaN) film having bulk resistivity of 200 $\mu\Omega$·cm have to be to exhibit a sheet resistance of 50 Ω/□?

2-3 Capacitors on an integrated circuit are made using a metal-insulator-metal (MIM) stack with the insulator formed by a thin film of silicon nitride (having dielectric constant $\varepsilon_r = 7.5$). If the film is 0.2 μm thick, what is the capacitance between two square plates that are 100 μm on a side?

2-4 What is the capacitance between two concentric spheres of radii a and b? Of a single sphere of radius a (that is, let $b \to \infty$)?

2-5 Imagine that a parallel-plate capacitor with an air gap between the plates holds a fixed charge, $\pm Q$. How is the energy stored by the capacitor changed if a slab of Teflon ($\varepsilon_r = 2.1$) having the same thickness as the gap is inserted between the plates? Does the capacitor attract or repel the slab?

2-6 Calculate the electric field strength between the plates of a microelectronic capacitor with plate spacing of 200 nm and holding a voltage of 5V, and the magnetic field inside the coil of a 10-turn microelectronic inductor 1 mm long and carrying 1A of current. Compare these to the electric and magnetic fields at the mid-point between the wires of a 300-kV power transmission line which are 1 meter apart and carrying a current of 1 kA each.

2-7 A thin wire that is 10 cm long is wound into an air-filled solenoid of 100 turns that is 1 cm in length. What is its inductance?

2-8 How much energy is stored in a 1F capacitor charged to 5V?

2-9 What is the energy density stored in the electric field between the two plates of a capacitor which are separated by 1 μm of silicon dioxide ($\varepsilon = 3.9$) if the capacitor is charged to 10V?

2-10 An inductor of unknown value is primed with a current of 1 mA, then switched to a 50Ω resistor. The current has dropped by 50% after 1 ms. What is the value of the inductor?

2-11 Resistors of 1Ω, 2Ω, and 3Ω are connected in series, in that order. What is the voltage across the 2Ω resistor if a total voltage of 5V is applied to the set?

2-12 Derive an expression for the complex power supplied by an AC current source to a parallel RLC circuit. For a given R and L at a given frequency, ω, what value of C maximizes the power delivered? What is that power, and what is the average stored energy in that case?

2-13 A parallel RC load, comprising a 1-pF capacitor and a 50Ω resistor, is driven by a Thévenin equivalent voltage source, with 1-V peak amplitude and internal resistance of 1Ω, through a series 1-nH inductor. What is the amplitude of the voltage swing across the RC load at 5 GHz?

2-14 What is the half-power bandwidth in absolute frequency of a 1-GHz resonator with $Q = 1,000$?

2-15 A series RLC resonator is found to have a 3-dB bandwidth of 2 MHz at a resonance of 1 GHz when loaded with a 1Ω resistor, and 3-MHz bandwidth at the same resonant frequency when loaded with a 2Ω resistor. What are the values of R, L, and C for this resonator?

2-16 A resonator with an unloaded $Q = 1,000$ is built into a circuit having a loaded $Q' = 100$. What is external Q_e?

2-17 A particular source when modeled as a Norton equivalent has an internal parallel resistance of 500Ω. What is the series resistance when modeled as a Thévenin equivalent source?

2-18 A Thévenin equivalent voltage source of 1V having internal resistance 1Ω is connected in parallel with a Norton equivalent current source of 1A having an internal resistance of 1 kΩ. What is the Thévenin equivalent source voltage and resistance of the combination?

2-19 Validate the reciprocity of a resistive tee-network comprising 1Ω and 2Ω in the series branches and 3Ω in the shunt branch by calculating the short-circuit current delivered at one terminal with an ideal voltage source applied to the other and vice versa.

2-20 What is the equivalent series inductance of a ceramic 1-pF capacitor having a self-resonant frequency of 8 GHz?

2-21 A 1-nH inductor is found to have a self-resonant frequency of 5 GHz and an unloaded Q of 1,000 when combined with an ideal capacitor in a 1-GHz series resonator. Derive a parasitic circuit model for this inductor.

2-22 What is the coupling factor of a pair of transformer coils for which the primary inductance is $L_p = 1$ nH, the secondary inductance is $L_s = 4$ nH, and the mutual inductance is $M = 1.85$ nH?

2-23 A 1:1 transformer has primary coil inductance of 1 μH and coupling factor $k = 0.99$. Calculate the frequency band over which the power delivered by a 50Ω source on the primary coil to a 50Ω load on the secondary coil is at least half what it would be without the transformer.

References

[1] J. Nilsson and S. Riedel, *Electric Circuits*, 5th ed. Reading, MA: Addison-Wesley, 1996.

[2] S. Ramo, J. Whinnery, and T. Van Duzer, *Fields and Waves in Communication Electronics*. New York: Wiley, 1984.

[3] R. Erickson and D. Maksimovic, *Fundamentals of Power Electronics*, 2nd ed. New York: Springer, 2001.

[4] D. M. Pozar, *Microwave Engineering*, 4th ed. New York: Wiley, 2011.

[5] P. Penfield, R. Spence, and S. Duinker, "A generalized form of Tellegen's theorem," *IEEE Transactions on Circuit Theory*, vol. 17, no. 3, pp. 302–305, August 1970.

[6] Wikipedia. (2017) Tellegen's theorem. https://en.wikipedia.org/wiki/Tellegen' s_theorem.

Chapter 3

Transmission Lines

Having previously described the behavior of electromagnetic fields at opposite ends of the size-frequency scale, first, as propagating waves in infinite free space, and then as confined and highly concentrated fields in infinitesimally small lumped elements, we now turn our attention to another network paradigm involving aspects of both. The structures that we describe in this chapter are assumed to be infinitesimally small (as lumped elements) in two spatial dimensions, while being very large in the third spatial dimension (the domain of wave propagation). "Long and skinny" is the defining characteristic of transmission lines, a powerful paradigm for electrical network design that literally extends the reach of lumped-element circuits to much greater distances and/or frequency regimes.

As both the large-scale and small-scale solutions of Maxwell's equations presented previously are equally valid, it is not surprising that one may construct the transmission-line solutions from either framework, with identical results. Both methods will be articulated here so that the reader may become accustomed to understanding the behavior of electromagnetic circuits from multiple viewpoints.

3.1 CONSTRUCTION FROM LUMPED ELEMENTS

To extend the infinitely compact domain of lumped elements into the long-and-skinny domain of transmission lines, we must first recognize that even distributed structures can be broken down into an effective ensemble of lumped-element parts. These parts may not closely resemble the idealized devices that we built in Chapter 2, but the description is still valid so long as the discretization is sufficiently fine to apply the electrostatic solutions.

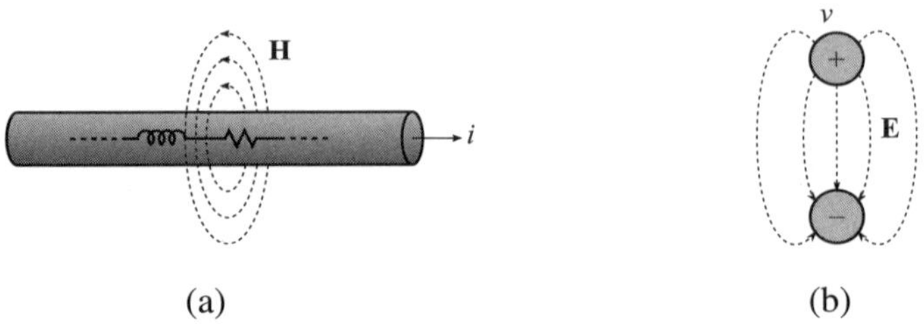

Figure 3.1 (a) Illustration of the magnetic field around a current-carrying wire, giving rise to inductance per unit length. Resistivity in the wire may also contribute some series resistance per unit length. (b) Illustration of the electric field between a pair of wires supporting a voltage, giving rise to capacitance per unit length. Conductivity in the embedding medium may also lead to some shunt conductance per unit length.

3.1.1 Distributed Inductance and Resistance

Consider, for example, that while a tight coil of wire was used in Section 2.2.4 to make an inductor, maximizing the effect of the magnetic field so that it was dominant over all other effects (e.g., the resistance of the wire or the capacitance between adjacent turns), even a straight wire has some inductance. This is because, according to Ampère's law, the passage of current through the wire creates a magnetic field encircling it, decaying radially as indicated in Figure 3.1(a). Faraday's law then states that this magnetic field induces an electromotive force, or voltage drop, across the length of the wire proportional to the time derivative of that current.

Put another way, it takes energy to create a magnetic field in the space around the wire, and that energy is released again when the field collapses. The reactive energy of creation and dissipation of magnetic fields is associated with inductance.

The wire may also have finite conductivity, in which case there is additionally a resistive component to the voltage drop along the line, according to Ohm's law. Both the inductance and resistance will depend upon the length of the wire (or segment of wire) under consideration. We thus say that a wire — or any long-and-skinny current carrying conductor, whatever its cross-sectional geometry — has a certain *inductance per unit length*, L', and *resistance per unit length*, R'.

3.1.2 Distributed Capacitance and Conductance

To complete a circuit, we cannot have a single wire extending over a long distance in isolation. There must be a return path for the current to flow back to the original

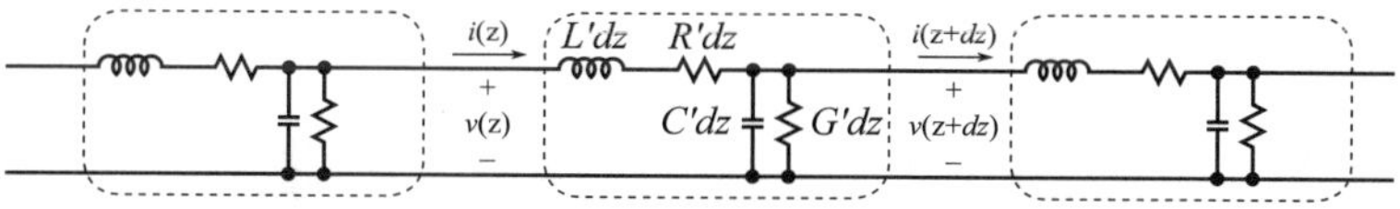

Figure 3.2 Model of a transmission line as a cascade of lumped-element cells.

source (e.g., a voltage or current generator). We thus envision a pair of wires running close by one another in parallel, carrying equal currents in opposite directions.

The electric potential between the two wires creates an electric field in the space between them, as shown in Figure 3.1(b), just as the plates of an ideal lumped capacitor. The creation of this electric field requires energy, which is released again when the field dissipates. The two wires are thus said to have a certain *capacitance per unit length*, C', between them.

Also, if the insulating medium between the wires has any leakage (i.e., the resistivity is not infinite), then there is a short-cut path for current to flow from one conductor to the other rather than down the full length of the line. It is convenient to use units of admittance in this case, so we call this a *conductance per unit length*, G', between the two wires.

3.1.3 The Telegrapher's Equations

We may therefore model a length of two parallel wires (or parallel conductors of any shape) as a cascade of an infinite number of tiny lumped-element cells, having inductance, resistance, capacitance, and conductance per unit length given by L', R', C', and G', respectively, as illustrated in Figure 3.2. Note that although the notional geometry of the transmission line comprises two wires, the inductance and resistance per unit length of both conductors have been combined in series in the top wire for convenience.

We assume that a current, $i\left(z\right)$, is flowing in the top wire at position z along the transmission line, and that a voltage, $v\left(z\right)$, is held between the two wires at the same position. The change in this voltage from one unit cell to the next is that which is dropped by the current flowing through the inductor and capacitor,

$$v\left(z\right) - v\left(z + dz\right) = i\left(z\right)\left(R'dz + j\omega L'dz\right) \tag{3.1}$$

and the current flowing in the top wire is reduced by an amount that is shunted through the capacitor and conductor,

$$i(z) - i(z + dz) = v(z + dz)(G'dz + j\omega C'dz) \tag{3.2}$$

Dividing these expressions by dz and taking the limit as $dz \to 0$, we have

$$-\frac{dv}{dz} = iZ' \tag{3.3a}$$

$$-\frac{di}{dz} = vY' \tag{3.3b}$$

where

$$Z' = R' + j\omega L' \tag{3.4a}$$

$$Y' = G' + j\omega C' \tag{3.4b}$$

These are known as the *Telegrapher's equations* in the time-harmonic form [1, 2]. By differentiating (3.3) with respect to z, we obtain

$$\frac{d^2v}{dz^2} = \gamma^2 v \tag{3.5a}$$

$$\frac{d^2i}{dz^2} = \gamma^2 i \tag{3.5b}$$

where

$$\gamma = \sqrt{Z'Y'} = \sqrt{(R' + j\omega L')(G' + j\omega C')} \tag{3.6}$$

This is a form of the Helmholtz equation in one dimension. Its solutions in the time-harmonic case are given by exponentials representing forward-traveling and backward-traveling waves,

$$v(z) = v^+ e^{-\gamma z} + v^- e^{\gamma z} \tag{3.7a}$$

$$i(z) = i^+ e^{-\gamma z} - i^- e^{\gamma z} \tag{3.7b}$$

The constants v^+ and i^+ are the phasor amplitudes of the forward-traveling voltage and current waves, respectively, while v^- and i^- represent the backward-traveling voltage and current waves. The reader may substitute (3.7) into (3.5) to verify that these are indeed solutions of the Helmholtz equation.

The choice of a negative sign for the i^- term is, at this point, arbitrary, but it will be a convenient one as it ensures that the reference direction for current is always in the direction of propagation for any given wave. Other conventions are valid, so long as they are applied consistently.

If the conductors are perfectly conducting ($R' = 0$) and the insulating medium between the conductors is perfectly insulating ($G' = 0$), then the line is said to be lossless. In that event, the parameter γ, known as the *propagation constant*, becomes purely imaginary,

$$\gamma|_{R'=G'=0} = \sqrt{(j\omega L')(j\omega C')} = j\omega\sqrt{L'C'} = j\beta \tag{3.8}$$

The imaginary part, β, is entirely analogous to the phase constant or lossless propagation constant for waves in free space derived in Section 1.4.2.

Substituting this back in (3.7),

$$v(z) = v^+ e^{-j\beta z} + v^- e^{j\beta z} \tag{3.9a}$$

$$i(z) = i^+ e^{-j\beta z} - i^- e^{j\beta z} \tag{3.9b}$$

Let us consider an example where v^+ and v^- are both real numbers. We may recover the time-domain voltage by application of the inverse phasor transform (1.80),

$$v(z,t) = \text{Re}\left\{\left(v^+ e^{-j\beta z} + v^- e^{j\beta z}\right) e^{j\omega t}\right\} \tag{3.10a}$$

$$= v^+ \cos(\omega t - \beta z) + v^- \cos(\omega t + \beta z) \tag{3.10b}$$

Imagine following the peaks of the sinusoidal waveforms as they evolve in time. This amounts to tracking a sequence of z-coordinates that maintain constant arguments inside the cosine functions as time increases. Since the terms inside the argument of the first cosine in (3.10b) for z and t have opposite signs, this means that z and t must increase simultaneously in order for the argument to remain constant. Hence, this is, as predicted, a forward-traveling wave. The second cosine in (3.10b) has the same signs for z and t. Hence, z must decrease while t increases in order to keep the argument constant; this is a backward-traveling wave. Note that we assumed v^+ and v^- were both real only for convenience. Had they been complex, the waveforms would have included a constant phase offset, but the direction of travel would have been unchanged.

This simple thought experiment also shows us how to calculate the velocity of propagation, v_p. By setting the argument of the forward-traveling wave to a constant, we have

$$\omega t - \beta z = C_1 \tag{3.11a}$$

$$\therefore z = \frac{\omega t - C_1}{\beta} \tag{3.11b}$$

$$v_p = \frac{dz}{dt} = \frac{\omega}{\beta} \tag{3.11c}$$

This is known as the *phase velocity*, since it follows the phase fronts of a sinusoidal wave. We will find later that more complex waveforms may have a number of different velocities associated with different aspects of the signal. Note the consistency between the findings of (3.11c) and those of (1.61), (1.85), and (1.86).

A parallel derivation finds the forward-traveling and backward-traveling current waves propagating in synchronization with the voltage waves. Furthermore, if the line were not totally lossless, then the propagation constant, γ, would have included a real term, indicating an exponential decay of the voltage and current amplitudes as a function of both space and time.

We have not yet determined how the voltage and current are related. This relationship is easily found by substituting (3.7) into (3.3a).

$$-\frac{dv}{dz} = iZ' \tag{3.12a}$$

$$-\frac{d}{dz}\left(v^+e^{-\gamma z} + v^-e^{\gamma z}\right) = \left(i^+e^{-\gamma z} - i^-e^{\gamma z}\right)Z' \tag{3.12b}$$

$$\gamma v^+e^{-\gamma z} - \gamma v^-e^{\gamma z} = i^+Z'e^{-\gamma z} - i^-Z'e^{\gamma z} \tag{3.12c}$$

As the line is uniform in the longitudinal direction, the above must hold for all z; therefore,

$$\frac{v^+}{i^+} = \frac{v^-}{i^-} = \frac{Z'}{\gamma} = \frac{Z'}{\sqrt{Z'Y'}} = \sqrt{\frac{Z'}{Y'}} = \sqrt{\frac{R' + j\omega L'}{G' + j\omega C'}} = Z_0 \tag{3.13}$$

This is known as the *characteristic impedance*, a property of the transmission line itself that applies to all waves traveling upon it, forward or backward. Note that had we chosen to use a positive sign before i^- in (3.9b), we would have been forced at this point to define a characteristic impedance that has the opposite sign in one direction compared to the other. While this definition could work in theory, it would be a difficult convention of which to keep track, since transmission lines need not always be so conveniently oriented along a Cartesian axis.

Typically, the line is assumed to be lossless ($R' = G' = 0$), or very nearly so, in which case the characteristic impedance from (3.13) is real-valued. A common choice is $Z_0 = 50\Omega$, which will be discussed in more detail in Section 3.3.6.

Some useful approximate expressions apply when the line is only slightly lossy, that is, when $R' \ll \omega L'$ and $G' \ll \omega C'$,

$$Z_0 = \sqrt{\frac{R' + j\omega L'}{G' + j\omega C'}} = \sqrt{\frac{L'}{C'}}\sqrt{\frac{1 + \frac{R'}{j\omega L'}}{1 + \frac{G'}{j\omega C'}}} \tag{3.14a}$$

$$\approx \sqrt{\frac{L'}{C'}}\left(1 + \frac{R'}{j2\omega L'} - \frac{G'}{j2\omega C'}\right) \tag{3.14b}$$

Therefore, we see that the characteristic impedance in the slightly lossy case acquires a small imaginary part, for which the conductor and dielectric contributions tend to compensate one another (that is, they have opposite sign). For the nearly lossless propagation constant, we have

$$\gamma = \alpha + j\beta = \sqrt{(R' + j\omega L')\,(G' + j\omega C')} \tag{3.15a}$$

$$= j\omega\sqrt{L'C'}\sqrt{\left(1 + \frac{R'}{j\omega L'}\right)\left(1 + \frac{G'}{j\omega C'}\right)} \tag{3.15b}$$

$$\approx j\omega\sqrt{L'C'}\left(1 + \frac{R'}{j2\omega L'} + \frac{G'}{j2\omega C'}\right) \tag{3.15c}$$

$$\therefore \alpha \approx \sqrt{L'C'}\left(\frac{R'}{2L'} + \frac{G'}{2C'}\right) = \frac{R'}{2Z_0} + \frac{G'}{2Y_0} \tag{3.15d}$$

where $Y_0 = Z_0^{-1}$. The *attenuation constant*, α, therefore comprises two summed terms, one associated with conductor resistance, and the other associated with dielectric loss.

A summary of the key results for transmission lines constructed from lumped-element concepts is given in Table 3.1.

3.2 CONSTRUCTION FROM FIELD EQUATIONS

The previous section derived transmission-line behavior as an extension of lumped elements, in essence, building structures of electrically significant size from our infinitesimal components by postulating an infinite number of them in one dimension. Now we take the reverse approach, deriving the same essential behavior as a reduction of the large-scale Maxwell's equations in three dimensions, confining them to an infinitesimally small space in two dimensions.

Table 3.1
Transmission Lines in Terms of Lumped-Element Parameters

Parameter	Symbol	Expression
Characteristic impedance	Z_0	$\sqrt{\frac{R'+j\omega L'}{G'+j\omega C'}}$
Propagation constant	γ	$\sqrt{(R'+j\omega L')(G'+j\omega C')}$
Attenuation constant (conductive)	α_c	$\frac{R'}{2Z_0}$
Attenuation constant (dielectric)	α_d	$\frac{G'}{2Y_0}$

3.2.1 Separation of Variables

Let us start with the three-dimensional wave solution derived for arbitrary source-free regions in Section 1.3,

$$\nabla^2 \mathbf{E} = \mu\varepsilon \frac{\partial^2 \mathbf{E}}{\partial t^2} \tag{3.16}$$

or in the frequency domain,

$$\nabla^2 \mathbf{E} + \omega^2 \mu\varepsilon \mathbf{E} = 0 \tag{3.17}$$

In order to constrain these equations to a long and skinny domain, it is useful to work with a specific geometry. For this example, let us use a coaxial cross-section [3], shown in Figure 3.3(a). We assume that the radii of the inner and outer conductors, a and b, are so small that they may be considered essentially infinitesimal, while the longitudinal axis, Figure 3.3(b), may be very long. For this problem, it will be convenient to work in cylindrical coordinates (r, θ, and z) where z is oriented with the longitudinal axis of the coaxial line. For now, we will also assume that the conductors are perfect ($\sigma = \infty$).

The boundary conditions for perfect conductors articulated in Section 1.1.4 state that the electric field must be perpendicular to the conducting walls, while the magnetic field must always be parallel to it. While it should be possible to construct a valid electric field solution with finite longitudinal (z) or azimuthal (θ) components that vanish on the appropriate boundaries, this would in general require the cross-section to be large, or the frequency to be very high such that the wavelength is small compared to the conductor diameters. We have assumed that

Figure 3.3 Geometry of a coaxial transmission line. (a) Cross-section. (b) Profile.

not to be the case here, so we can safely say that the electric field has radial (r) components only.

While the field is thus *oriented* in the radial direction, however, it may still have both radial and longitudinal *dependence*. The circular symmetry along with infinitesimally small diameter does, however, rule out any dependence on θ. Finally, the dependence on time is assumed to be sinusoidal, and will be hidden as we are going to work with phasor quantities. We may therefore write

$$\mathbf{E} = E(r, z)\,\mathbf{r} \tag{3.18}$$

Substituting this in (3.17) and applying the Laplacian operator in cylindrical coordinates, we have

$$\nabla^2 (E(r, z)\,\mathbf{r}) + \omega^2 \mu\varepsilon E(r, z)\,\mathbf{r} = 0 \tag{3.19a}$$

$$\left\{ \frac{1}{r}\frac{\partial}{\partial r}\left(r\frac{\partial}{\partial r} \right) - \frac{1}{r^2} + \frac{\partial^2}{\partial z^2} + \omega^2 \mu\varepsilon \right\} E(r, z) = 0 \tag{3.19b}$$

We thus have a second-order partial differential equation for E in the two variables, r and z. Note that we have constrained the boundary conditions along these two coordinate axes very differently: infinitesimally small in one dimension, while potentially very long in the other. We may thus expect the solution in r to be quasistatic, while the solution in z may exhibit wave behavior. It is logical, then, to attempt a solution to this partial differential equation using the *separation of variables* technique [4]. This means that we assume $E(r, z)$ may be written as the product of two independent functions, one depending only on r, and the other only on z, and see if we can find expressions for those functions which jointly satisfy (3.19b). Thus, let

$$E(r, z) = M(r)\,N(z) \tag{3.20}$$

and plug this into (3.19b),

$$\left\{\frac{1}{r}\frac{\partial}{\partial r}\left(r\frac{\partial}{\partial r}\right)-\frac{1}{r^2}+\frac{\partial^2}{\partial z^2}+\omega^2\mu\varepsilon\right\}M\left(r\right)N\left(z\right)=0 \tag{3.21a}$$

$$\therefore\frac{1}{rM}\frac{d}{dr}\left(r\frac{dM}{dr}\right)-\frac{1}{r^2}=-\left(\frac{1}{N}\frac{d^2N}{dz^2}+\omega^2\mu\varepsilon\right) \tag{3.21b}$$

Note that the left side is a function of r only and the right side is a function of z only. The quantity that both of these expressions represent, then, must not be a function of either, and may be considered a constant, C_0. Therefore,

$$\frac{1}{rM}\frac{d}{dr}\left(r\frac{dM}{dr}\right)-\frac{1}{r^2}=C_0 \tag{3.22a}$$

$$\frac{1}{N}\frac{d^2N}{dz^2}+\omega^2\mu\varepsilon=C_0 \tag{3.22b}$$

We have thereby decoupled the *partial* differential equation in two variables (3.19b) into a pair of independent *ordinary* differential equations (3.22) of one variable each. The partial differential equation (3.19b) is thus said to be *separable*, and C_0 is the *separation constant*.

A fully rigorous solution to the above equations would involve finding all possible values of the separation constant that yield valid solutions, but we can shortcut that a bit with some educated guesses. First, since we have constrained the diameters of the coaxial line to be very small compared to a wavelength, we might guess that a solution to the r-dependent equation might be the same as the electrostatic case, the electrostatic field between two coaxial cylinders being inversely proportional to r, as in Section 1.2.3. Let us assume that $M\left(r\right)=C_1/r$ and try this in (3.22a),

$$\frac{1}{C_1}\frac{d}{dr}\left(r\frac{d}{dr}\frac{C_1}{r}\right)-\frac{1}{r^2}=\frac{1}{C_1}\frac{d}{dr}\left(-C_1r^{-1}\right)-\frac{1}{r^2}=\frac{1}{C_1}C_1r^{-2}-\frac{1}{r^2}=0 \tag{3.23a}$$

$$\therefore C_0=0 \tag{3.23b}$$

Thus, we find that the electrostatic field configuration is a solution for $M\left(r\right)$, so long as $C_0=0$. Putting this value into (3.22b), we have

$$\frac{d^2N}{dz^2}+\omega^2\mu\varepsilon N=0 \tag{3.24}$$

The solution to (3.24) is already known to us, as it is simply the wave equation in a single dimension. Therefore,

$$N(z) = N^+ e^{-\gamma z} + N^- e^{\gamma z} \tag{3.25}$$

where N^+ and N^- are the (possibly complex) phasor amplitudes of the forward- and backward-traveling waves, respectively, and γ is, as before, the propagation constant,

$$\gamma = j\beta = j\omega\sqrt{\mu\varepsilon} \tag{3.26}$$

Now combining our solutions for $M(r)$ and $N(z)$, we have for the total electric field

$$\mathbf{E} = E(r, z)\,\mathbf{r} = M(r)\,N(z)\,\mathbf{r} = \frac{C_1}{r}\left(N^+ e^{-\gamma z} + N^- e^{\gamma z}\right)\mathbf{r} \tag{3.27a}$$

$$= \frac{1}{r}\left(A^+ e^{-\gamma z} + A^- e^{\gamma z}\right)\mathbf{r} \tag{3.27b}$$

where the net phasor wave amplitudes are $A^+ = C_1 N^+$ and $A^- = C_1 N^-$.

The magnetic field may be found quite simply by Faraday's law,

$$\nabla \times \mathbf{E} = -j\omega\mathbf{B} \tag{3.28a}$$

$$\nabla \times \frac{1}{r}\left(A^+ e^{-\gamma z} + A^- e^{\gamma z}\right)\mathbf{r} = \frac{1}{r}\frac{\partial}{\partial z}\left(A^+ e^{-\gamma z} + A^- e^{\gamma z}\right)\boldsymbol{\theta} = -j\omega\mu\mathbf{H} \tag{3.28b}$$

$$\mathbf{H} = \frac{1}{r}\left(\frac{\gamma A^+}{j\omega\mu}e^{-\gamma z} - \frac{\gamma A^-}{j\omega\mu}e^{\gamma z}\right)\boldsymbol{\theta} = \frac{1}{r\eta}\left(A^+ e^{-\gamma z} - A^- e^{\gamma z}\right)\boldsymbol{\theta} \tag{3.28c}$$

where η, the wave impedance, was defined in (1.63).

3.2.2 Relating Terminal Parameters to Field Quantities

The foregoing field-based derivation for a transmission line having a coaxial cross-section yields qualitatively the same behavior as the previous derivation based on lumped elements, namely, guided wave propagation along the axis of the line, but expressed in different parameters. Recall that the voltage, v, and the current, i, are scalar substitutes in the quasistatic (lumped-element) case for the more general vector field quantities, such as $\mathbf{E}$, $\mathbf{H}$, and $\mathbf{J}$. These *terminal parameters* in Section 3.1 allowed us to define meaningful immittance quantities that describe the transmission line, such as the inductance or capacitance per unit length and, most

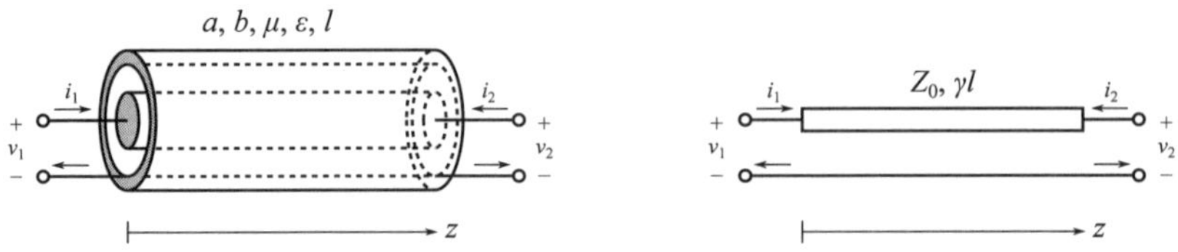

Figure 3.4 (a) Coaxial transmission line with port terminals attached. a and b are the inner and outer diameters of the coaxial line, respectively, while μ and ε describe the insulating medium. l is the physical length of the line. (b) Generalized schematic for a transmission line. γl is the electrical length, and Z_0 is the characteristic impedance.

importantly, the characteristic impedance. It is instructive to explore the relationship between the lumped and field-parameter systems at this time.

Because the cross-section of the coaxial line is assumed to be infinitesimally small, we may attach port terminals to the inner and outer conductors, as shown in Figure 3.4(a). The voltage from one terminal to the other will be the same no matter what the exact attachment point on the outer conductor, for example, and the current injected into either terminal will quickly distribute itself uniformly around the circle. The voltage at any point on the line may thus be found by application of (2.1) in the cross-section plane,

$$v(z) = \int_a^b \mathbf{E} \cdot d\mathbf{r} = \left(A^+ e^{-\gamma z} + A^- e^{\gamma z}\right) \int_a^b r^{-1} dr \tag{3.29a}$$

$$= \left(A^+ e^{-\gamma z} + A^- e^{\gamma z}\right) \ln\left(\tfrac{b}{a}\right) \tag{3.29b}$$

where the outer conductor has been used as the reference.

The current may be found by consideration of Ampère's law around a closed loop encircling the inner conductor,

$$i(z) = \int \mathbf{H} \cdot d\mathbf{l} = 2\pi a \frac{1}{a\eta} \left(A^+ e^{-\gamma z} - A^- e^{\gamma z}\right) \tag{3.30a}$$

$$= \frac{2\pi}{\eta} \left(A^+ e^{-\gamma z} - A^- e^{\gamma z}\right) \tag{3.30b}$$

The characteristic impedance may then be found by taking the ratio of the forward-traveling voltage and current phasor amplitudes,

$$Z_0 = \frac{v^+}{i^+} = \frac{A^+ \ln\left(\frac{b}{a}\right)}{A^+ 2\pi/\eta} = \frac{\eta}{2\pi}\ln\left(\frac{b}{a}\right) \tag{3.31}$$

Note that since we have used perfect conductors, the line is lossless and Z_0 is real-valued.

Alternatively, one may calculate the inductance and capacitance per unit length directly from the quasistatic fields in Section 1.2.3 and determine Z_0 using (3.13). From the general definition of capacitance in (2.21),

$$C' = \frac{Q}{v} = Q\left(\int_a^b \mathbf{E}\cdot\mathbf{r}dr\right)^{-1} = Q\left(\int_a^b \frac{Q}{2\pi\varepsilon r}dr\right)^{-1} = \frac{2\pi\varepsilon}{\ln\left(\frac{b}{a}\right)} \tag{3.32}$$

where C' denotes the capacitance per unit length, since Q in this case is not the total charge but rather charge per unit length. The inductance per unit length is easily found using the energetic definition in (2.55)

$$L' = \frac{2U_H}{I^2} = I^{-2}\iint \mathbf{B}\cdot\mathbf{H}dA = \mu I^{-2}\iint H^2 dA = 2\pi\mu I^{-2}\int_a^b H^2 r dr \tag{3.33a}$$

$$= 2\pi\mu I^{-2}\int_a^b \left(\frac{I}{2\pi r}\right)^2 r dr = \frac{\mu}{2\pi}\int_a^b r^{-1}dr = \frac{\mu}{2\pi}\ln\left(\frac{b}{a}\right) \tag{3.33b}$$

From (3.13), then,

$$Z_0 = \sqrt{\frac{R' + j\omega L'}{G' + j\omega C'}} = \sqrt{\frac{L'}{C'}} = \sqrt{\frac{\frac{\mu}{2\pi}\ln\left(\frac{b}{a}\right)}{2\pi\varepsilon/\ln\left(\frac{b}{a}\right)}} \tag{3.34a}$$

$$= \frac{\ln\left(\frac{b}{a}\right)}{2\pi}\sqrt{\frac{\mu}{\varepsilon}} = \frac{\eta}{2\pi}\ln\left(\frac{b}{a}\right) \tag{3.34b}$$

which is the same as (3.31).

We may also relate the power flow in a propagating wave in terms of either its terminal parameters or its field quantities. Since at any pair of terminals the complex power is given by $P = vi^*$, where v and i are the root-mean-square phasor amplitudes of the voltage and current, we may write the power flowing in a positive traveling wave as

$$P(z) = v(z)i^*(z) = \left(A^+e^{-\gamma z}\ln\left(\frac{b}{a}\right)\right)\left(\frac{2\pi}{\eta}A^+e^{-\gamma z}\right)^* \tag{3.35a}$$

$$= \frac{2\pi}{\eta}\left|A^+\right|^2\ln\left(\frac{b}{a}\right) \tag{3.35b}$$

Note that this is real-valued, indicating actual power flow rather than stored energy, and is independent of z.

For comparison, consider the cross-product $\mathbf{S} = \mathbf{E} \times \mathbf{H}^*$ of the fields for a forward-traveling wave, integrated over the cross-section of the coax line,

$$\iint \mathbf{S}\cdot\mathbf{dA} = \int_0^{2\pi}\int_a^b \mathbf{E}\times\mathbf{H}^*\cdot\mathbf{z}r\theta drd\theta \tag{3.36a}$$

$$= \int_0^{2\pi}\int_a^b \left(\frac{1}{r}A^+e^{-\gamma z}\mathbf{r}\right)\times\left(\frac{1}{r\eta}A^+e^{-\gamma z}\boldsymbol{\theta}\right)^*\cdot\mathbf{z}r\theta drd\theta \tag{3.36b}$$

$$= \frac{1}{\eta}\left|A^+\right|^2\int_0^{2\pi}\int_a^b \frac{1}{r}\theta drd\theta = \frac{2\pi}{\eta}\left|A^+\right|^2\ln\left(\frac{b}{a}\right) \tag{3.36c}$$

which is the same as the power flow predicted by terminal parameters in (3.35b). The cross-product $\mathbf{S}$ is known as the *Poynting vector* and is the mathematical expression of power flow density in an electromagnetic wave.

We see that both methods for describing the behavior of transmission lines (lumped-element and field) are thus consistent and equally valid, but which will be most useful in the general case? Not all transmission lines are coaxial, after all, but all transmission lines do exhibit qualitatively similar behavior, characterized by quasistatic terminal parameters with wave propagation in the longitudinal direction. It is not surprising that by convention, transmission lines are often described using the most convenient elements from both models: the characteristic impedance, Z_0 (a lumped-element/terminal parameter), and the propagation delay, γl (a wave propagation/field parameter), as shown in Figure 3.4(b). These two numbers are

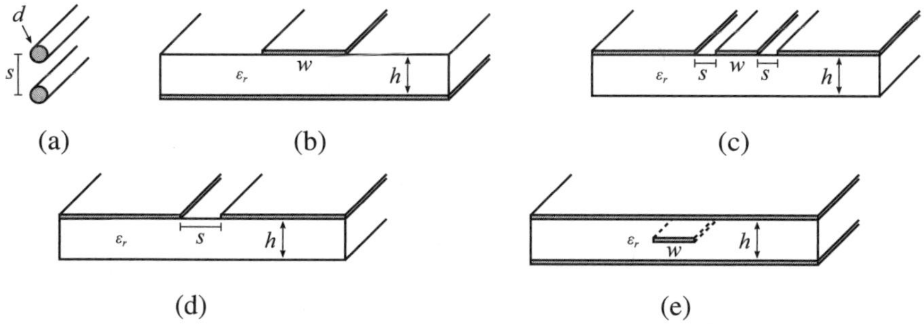

Figure 3.5 Gallery of some common transmission-line geometries besides coax. (a) Twin-wire. (b) Microstrip. (c) Coplanar waveguide (CPW). (d) Slotline. (e) Stripline.

sufficient to describe the port behavior of any transmission line, no matter what its physical geometry.

Examples of some other common transmission-line geometries are shown in Figure 3.5. Few practical structures admit closed-form solutions as readily as coax; instead the line parameters (Z_0 and γl) are calculated from the key dimensions (d, s, w, h) and material parameters (ε, σ) using empirical formulas [5] or numerical computer simulation. Formulas for microstrip will be given in Section 3.5.

3.3 TRANSMISSION LINES IN CIRCUITS

It is important to realize that the terminal parameters describe the cross-sectional behavior of a transmission line as though it were a lumped-element circuit, but lumped-element concepts do not extend to the longitudinal axis. Kirchhoff's voltage law, for example, may be applied anywhere within the cross-section of a transmission line, but it does not apply to any closed loop that extends a substantial distance along the axis of propagation. The z-dependent voltage described in (3.7a) and (3.29b) is not conservative. In this section, we will discuss how transmission lines fit into the framework of circuit structure alongside lumped elements.

3.3.1 The Terminated Transmission Line

Consider a transmission line terminated by a lumped impedance, Z_L, at $z = 0$, as shown in Figure 3.6(a). According to our solutions in Sections 3.1 and 3.2, we have both forward-traveling and backward-traveling waves on the transmission

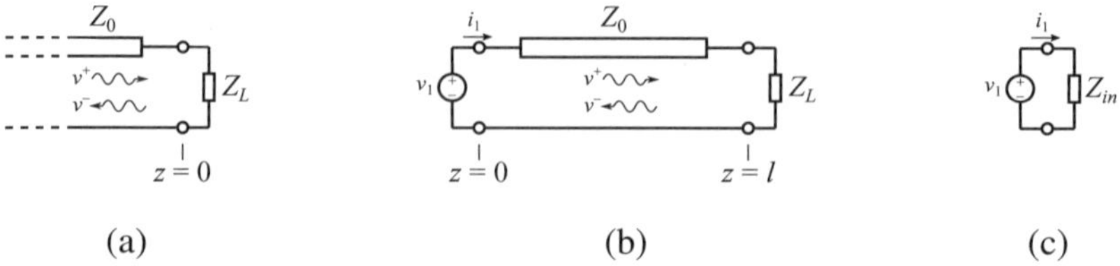

Figure 3.6 (a) Transmission line terminated with a load impedance, Z_L. (b) Transmission line driven with a voltage source at one end and a load impedance at the other end. (c) Lumped-element equivalent impedance as seen by the voltage source.

line. The forward-traveling wave is presumably coming from a signal source, while the backward-traveling wave arises as a reflection off of the lumped-element termination. These two waves must interfere with one another at the point of reflection in order to satisfy both the transmission line's and the lumped element's terminal parameters. Thus, the ratio of the total voltage and the total current at $z = 0$ must satisfy

$$\frac{v(0)}{i(0)} = \frac{v^+ + v^-}{i^+ + i^-} = \frac{v^+ + v^-}{v^+/Z_0 - v^-/Z_0} = \frac{v^+ + v^-}{v^+ - v^-} Z_0 = Z_L \tag{3.37a}$$

$$\therefore \frac{v^-}{v^+} = \frac{i^-}{i^+} = \frac{Z_L - Z_0}{Z_L + Z_0} = \Gamma \tag{3.37b}$$

The quantity Γ is known as the *reflection coefficient.*

If the load impedance, Z_L, is not at $z = 0$, but located some distance, l, away from the source, then the reflection that the source sees is delayed, as shown in Figure 3.6(b). This modifies the phase of the reflection as measured from the input end of the transmission line. This time, the ratio of the total voltage to the total current at port 2, or in other words at $z = l$, is constrained,

$$\frac{v_2}{i_2} = \frac{v(l)}{i(l)} = \frac{v^+ e^{-\gamma l} + v^- e^{\gamma l}}{i^+ e^{-\gamma l} - i^- e^{\gamma l}} = Z_0 \frac{v^+ e^{-\gamma l} + v^- e^{\gamma l}}{v^+ e^{-\gamma l} - v^- e^{\gamma l}} = Z_L \tag{3.38}$$

The ratio of the wave amplitudes is thus given by

$$\frac{v^-}{v^+} = \frac{i^-}{i^+} = \frac{Z_L e^{-\gamma l} - Z_0 e^{-\gamma l}}{Z_L e^{\gamma l} + Z_0 e^{\gamma l}} = \frac{Z_L - Z_0}{Z_L + Z_0} e^{-2\gamma l} = \Gamma e^{-2\gamma l} \tag{3.39}$$

We find, then, that the reflection coefficient from a distance is simply the reflection of the load directly, with a phase delay corresponding to twice the length of the line

— in other words, delayed by the round-trip time of the traveling wave from the source to the load termination and back.

To use such a line in a lumped-element circuit, we would like to find the net effective impedance, Z_{in}, seen by the voltage source at port 1. This is simply the ratio of the total voltage to the total current at $z = 0$,

$$Z_{in} = \frac{v_1}{i_1} = \frac{v(0)}{i(0)} = \frac{v^+ + v^-}{i^+ - i^-} = Z_0 \frac{i^+ + i^-}{i^+ - i^-} = Z_0 \frac{1 + (i^-/i^+)}{1 - (i^-/i^+)} \tag{3.40a}$$

$$= Z_0 \frac{1 + \frac{Z_L - Z_0}{Z_L + Z_0} e^{-2\gamma l}}{1 - \frac{Z_L - Z_0}{Z_L + Z_0} e^{-2\gamma l}} = Z_0 \frac{(Z_L + Z_0) e^{\gamma l} + (Z_L - Z_0) e^{-\gamma l}}{(Z_L + Z_0) e^{\gamma l} - (Z_L - Z_0) e^{-\gamma l}} \tag{3.40b}$$

$$= Z_0 \frac{Z_L \cosh(\gamma l) + Z_0 \sinh(\gamma l)}{Z_0 \cosh(\gamma l) + Z_L \sinh(\gamma l)} = Z_0 \frac{Z_L + Z_0 \tanh(\gamma l)}{Z_0 + Z_L \tanh(\gamma l)} \tag{3.40c}$$

If we further assume that the line is lossless, then we have $\gamma = j\beta$, and

$$Z_{in} = Z_0 \frac{Z_L + jZ_0 \tan(\beta l)}{Z_0 + jZ_L \tan(\beta l)} \tag{3.41}$$

Finally, the product $\beta l = \theta$, measured in radians, is termed the *electrical length* of the transmission line (implicitly a function of frequency), and we are left with

$$Z_{in} = Z_0 \frac{Z_L + jZ_0 \tan\theta}{Z_0 + jZ_L \tan\theta} \tag{3.42}$$

Thus, from the point of view of the signal generator, the terminated transmission line is indistinguishable from a lumped element having impedance Z_{in}, as shown in Figure 3.6(c). The impedance Z_L has effectively been transformed into Z_{in} according to (3.42). Had the circuit at port 1 taken on a more complicated form than a simple voltage source, the behavior of that circuit could be analyzed using lumped-element techniques with the impedance Z_{in} in place of the transmission-line terminals.

The picture that emerges is one in which the connection points of transmission lines lie within localized lumped-element circuits, coupling one such region to another while transforming their impedance over a distance, as indicated by example in the rather capriciously drawn Figure 3.7. One could solve this circuit by transforming the lumped-element impedances in subcircuits C and D into subcircuit B via transmission lines 2 and 3, and then from subcircuit B into subcircuit A using the characteristic parameters of transmission line 1. Such a structure is often

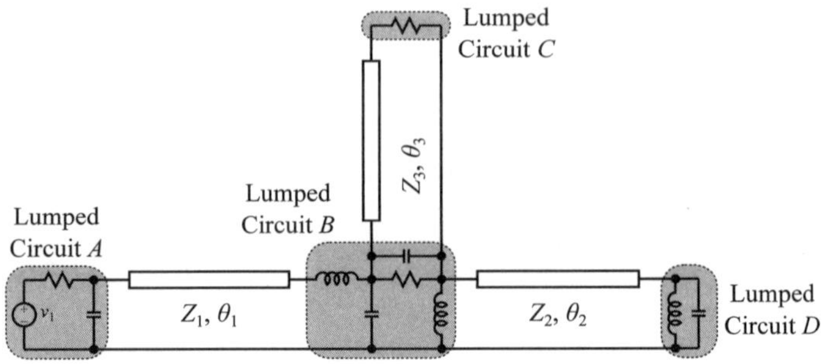

Figure 3.7 A transmission-line network comprising four lumped-element circuits (A, B, C, and D) interconnected by three transmission lines.

termed a network rather than a circuit, to distinguish it from purely lumped-element constructions. This convention is not followed universally, however; sometimes the terms are used interchangeably.

More complex networks, especially those having closed transmission-line loops, are analyzed more easily using other methods, which will be discussed in Chapter 4.

3.3.2 Stubs and Quarter-Wave Transformers

A common special case is one in which the transmission line is terminated with either an open-circuit or a short-circuit, for which $Z_L = \infty$ or $Z_L = 0$, respectively. Typically called a *stub*, the equivalent input impedance of such a transmission line is

$$Z_{in} = Z_0 \frac{Z_L + jZ_0 \tan\theta}{Z_0 + jZ_L \tan\theta} = \begin{cases} -jZ_0 \cot\theta & Z_L = \infty \text{ (open circuit)} \\ jZ_0 \tan\theta & Z_L = 0 \text{ (short circuit)} \end{cases} \tag{3.43}$$

Note that, just like lumped elements, stubs can be connected in series or in parallel with other elements. We can thus identify at least four different kinds of stubs depending on connection type and end termination, as illustrated in Figure 3.8(a–d). One should be careful not confuse a short-circuited, series connected stub, shown in Figure 3.8(d), with a transmission line connected in cascade, shown in Figure 3.8(e).

Another common case is when the transmission line is a quarter-wavelength long at the design frequency, or $\theta = \pi/2$. In that case, the input impedance is an

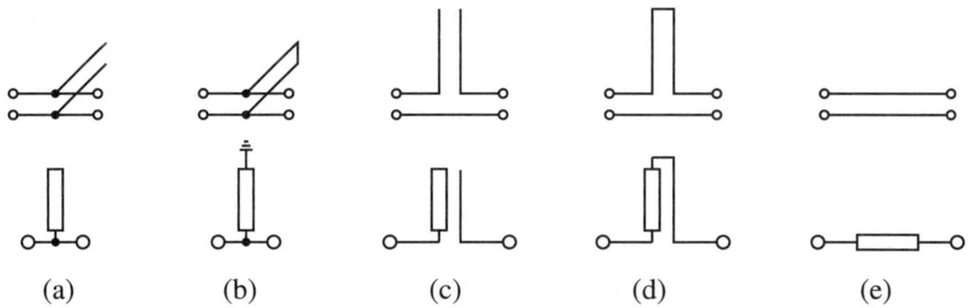

Figure 3.8 Transmission-line stubs drawn as differential lines (above) and with an implicit network ground (below). (a) Shunt- or parallel-connected, open-circuited stub. (b) Shunt- or parallel-connected, short-circuited stub. (c) Series-connected, open-circuited stub. (d) Series-connected, short-circuited stub. (e) Cascade transmission line.

inversion of the load impedance,

$$Z_{in}(\theta = \pi/2) = Z_0 \frac{Z_L + jZ_0 \tan\frac{\pi}{2}}{Z_0 + jZ_L \tan\frac{\pi}{2}} = \frac{Z_0^2}{Z_L} \tag{3.44}$$

Such a transmission line is often called a *quarter-wave transformer*. Note that in practice the electrical length is proportional to frequency, so the above inversion relationship is only strictly true at a single frequency point. It is therefore an approximation of a more idealized circuit element known as an *impedance inverter*, which will be discussed further in Section 4.7.2.

3.3.3 Infinite T-Lines and the Characteristic Impedance

Having determined what impedance is realized at one end of a length of transmission line as a function of the load impedance at the other end, it bears asking what impedance would be present at the input of an infinitely long transmission line. No real transmission line can be infinite, of course, but it is a useful thought experiment that provides insight into the real meaning of the characteristic impedance.

The scenario in question is drawn in Figure 3.9(a). A forward-traveling wave with phasor amplitude v^+ is illustrated propagating away from the voltage source. In this case, there can be no backward-traveling wave, since it would have taken an infinitely long time to arrive back at the input terminals. Hence, we know that

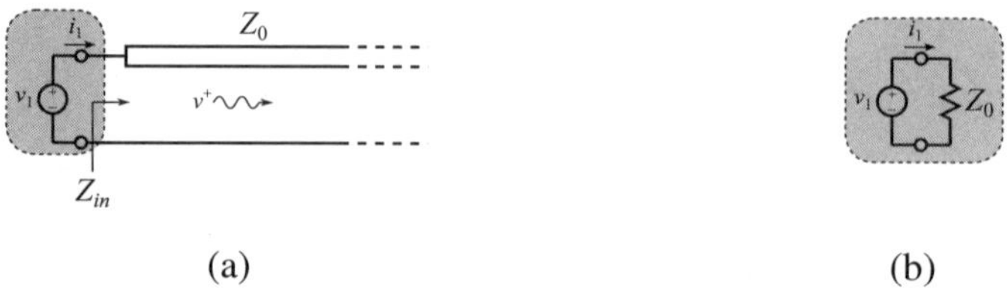

Figure 3.9 (a) A voltage source driving an infinitely long transmission line. (b) Equivalent lumped-element circuit.

$v^- = 0$. The input impedance is therefore found quite trivially,

$$Z_{in} = \frac{v_1}{i_1} = \frac{v(0)}{i(0)} = \frac{v^+}{i^+} = Z_0 \tag{3.45}$$

Clearly, the characteristic impedance of a transmission line is also the impedance that one sees at the input of a transmission line that is infinitely long.

Many young engineers become confused by how a transmission line that is lossless (assume it is made using perfect conductors) can have a real-valued impedance, Z_0, implying a dissipation of energy. What they are forgetting is that the terminal parameters (including Z_{in}) describe a lumped-element circuit which is local to the terminals of the transmission line (highlighted in gray in Figure 3.9). The signal energy that is dissipated in the equivalent resistor is nothing more than the energy propagated away from the terminals by the transmission line. It appears as dissipation in the equivalent lumped-element circuit, Figure 3.9(b), because the power really is extracted from the source by the transmission line, never to return in this scenario. It may not be absorbed in a global sense, but it is removed from the consideration of the local lumped elements.

A load resistor having value Z_0 is a perfectly good substitute for an infinitely long transmission line in most cases. Such a load is called a *matched* termination, because its impedance matches the characteristic impedance of the transmission line. Any waves incident upon it are absorbed without reflection.

If the transmission line is not infinite in length, and the termination not matched, then it is necessary to consider backward-traveling waves induced by reflection from that load impedance. Returning to (3.42), the exact condition for the input impedance from a lossless transmission line of finite length to have nonzero real part is

$$\text{Re}\{Z_{in}\} = \text{Re}\left\{Z_0 \frac{Z_L + jZ_0 \tan\theta}{Z_0 + jZ_L \tan\theta}\right\} > 0 \tag{3.46a}$$

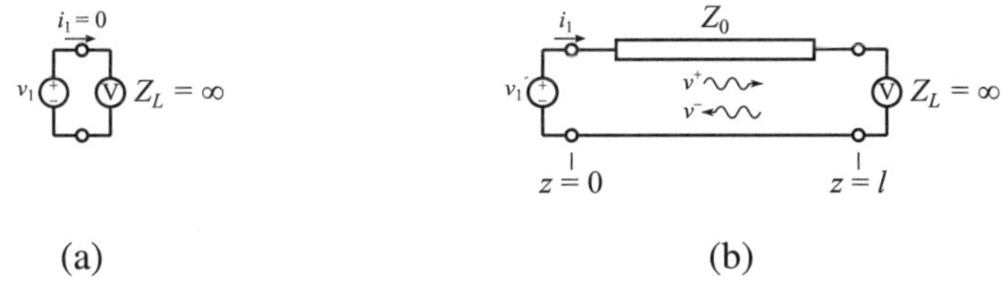

Figure 3.10 Voltage-mode data transfer (a) over a short distance, and (b) over a long distance.

$$\frac{\text{Re}\left\{Z_0\left(Z_L + Z_L^* \tan^2\theta\right) - j\left(|Z_L|^2 - Z_0^2\right)\tan\theta\right\}}{|Z_0 + jZ_L\tan\theta|^2} > 0 \tag{3.46b}$$

$$\text{Re}\left\{Z_L + Z_L^* \tan^2\theta\right\} > 0 \tag{3.46c}$$

$$\text{Re}\left\{Z_L\right\} > 0 \tag{3.46d}$$

Therefore, the input impedance has a real part only if the load impedance has a real part. Given lossless (and finite) transmission lines, the network dissipates energy only if the termination dissipates energy.

3.3.4 Signal Transmission with Transmission Lines

Perhaps the most classic application of transmission lines is that for which they are named: the transmission of information or power from one place to another over a distance. To understand why ordinary lumped-element circuits are ill-suited for this task, it is worth considering an example.

When data is to be transferred over a very short distance, say, from one transistor to another in an integrated circuit, it is convenient to encode that data as a voltage. If the information is binary, for example, a high voltage will typically represent a 1, while a low (or negative) voltage will typically represent a 0. Transistors at moderate frequencies tend to have very low output impedance but very high input impedance. Thus, one transistor driving another behaves much like a voltage source driving an open circuit (or a voltmeter with infinite internal impedance). The voltage information is therefore transmitted with near-perfect fidelity, and nominally without any dissipation of energy. This situation, which we will call *voltage-mode data transfer*, is depicted in Figure 3.10(a).

When the source and destination are far apart, and that distance is large compared to a wavelength, then intentionally or not, there is a transmission line separating them, as in Figure 3.10(b). Now the low output impedance and high input impedance work against us, as they tend to set up standing waves on the line between the source and load. In particular, the impedance seen by the transmitter voltage source is now given by (3.43),

$$Z_{in} = -jZ_0 \cot\theta \tag{3.47}$$

We therefore see that at certain frequencies (where $\theta = (2n+1)\,\pi/2$), the voltage source becomes shorted out. At these frequencies, it would take an infinite amount of power to drive the desired voltage on the line. In reality, no voltage source can have zero internal resistance, and no load can have truly infinite impedance, but the general conclusion that the transmission line with these kinds of terminations is blind to certain frequencies is a correct one.

One could try to arrange it such that the length of the interconnecting line (say, a coaxial cable) is the proper length to allow the desired signal frequencies through, but in practice this is very difficult. The bandwidth of the signal may be too wide to permit this over all required frequencies if the distance is too great (the transmission nulls are too close together) and it may be impractically sensitive to environmental conditions (e.g., thermal expansion or flexure of the cable).

A more robust solution is to design the transmitter and receiver at either end to be matched to the characteristic impedance of the transmission line. This ensures that a constant and predictable amplitude is delivered to the line by the source, and that signal pulses sent down the line do not return; there are no standing waves to corrupt the transmission. This arrangement has the disadvantage that it dissipates some power, but it is usually the only practical solution when the distance between source and destination is large.

3.3.5 Loss in Transmission Lines

Although relatively short transmission lines can sometimes be assumed to be lossless, if they are used to transmit signals over a distance then their losses can be quite significant. Recall that in the most general case, the propagation constant has both a real and imaginary part, so a forward traveling wave has the form,

$$v\,(l) = v^+ e^{-\gamma l} = v^+ e^{-(\alpha+j\beta)l} = v^+ e^{-\alpha l} e^{-j\beta l} \tag{3.48}$$

The second exponential above is the familiar phase term, but now it is preceded by a real factor that indicates a decay in amplitude exponentially dependent on length. That is, after a distance $1/\alpha$, the amplitude of the wave has been attenuated by a factor $1/e$. The total loss exponent, αl, is said to have units of *nepers* (Np). A more common unit for attenuation is the *decibel*[1] (dB) defined as 10 times the base-10 logarithm of the power ratio,

$$A_{\mathrm{dB}} = 10 \log\left(\frac{P_1}{P_2}\right) \tag{3.49}$$

Since the power in a traveling wave is proportional to the square of the voltage amplitude, the loss of a length of transmission line may be expressed in decibels as follows,

$$A_{\mathrm{dB}} = 10 \log\left(\frac{v^2(0)}{v^2(l)}\right) = 20 \log\left(e^{\alpha l}\right) = \alpha l \left(20 \log e\right) \approx 8.686 \alpha l \tag{3.50}$$

The conversion factor from nepers to decibels is thus about 8.686 dB/Np.

Consider the general propagation constant defined in (3.6). In most practical cases, the ohmic loss in the conductors will dominate over the dielectric loss in the insulator, so let us assume $G' = 0$,

$$\gamma = \sqrt{Z'Y'} = \sqrt{(R' + j\omega L')\, j\omega C'} = \sqrt{-\omega^2 L'C' + j\omega R'C'} \tag{3.51a}$$

$$= \sqrt{-\omega^2 L'C' \left(1 - j\frac{R'}{\omega L'}\right)} \approx \gamma_0 \left(1 - j\frac{R'}{2\omega L'}\right) = \alpha + j\beta \tag{3.51b}$$

where γ_0 is the (imaginary) propagation constant that would result if the line was lossless, and the approximation holds when the losses are small, or $R' \ll \omega L'$. The real part of the propagation constant is therefore given by

$$\alpha = -j\gamma_0 \frac{R'}{2\omega L'} = -j\left(j\omega\sqrt{L'C'}\right)\frac{R'}{2\omega L'} = \frac{R'}{2}\sqrt{\frac{C'}{L'}} \approx \frac{R'}{2Z_0} \tag{3.52}$$

This is known as the *loss constant* or *attenuation constant* of the line. Sometimes it is used with a subscript c, as in α_c, to indicate that it is the loss due to conductivity, as opposed to α_d, the loss due to dielectric leakage.

1 A decibel is one-tenth of a *bel*, named in honor of Alexander Graham Bell and based on power measurements associated with early telephony systems in the United States. The bel is no longer in common use.

3.3.6 The 50Ω Standard

For the convenience of interoperability, we must design our data-transmission components (transmitters, receivers, and interconnects) to each match a standard characteristic impedance. Otherwise, every system would need specialized components and cabling, custom-designed for the individual application. This would not be at all cost-effective.

By far the most common standard is 50Ω, with a second-most common standard of 75Ω, the latter used primarily for cable television (and the DOCSIS data communication specification that builds upon it [6]). There is some debate as to how exactly these standards came about, but a good place to start is to consider the optimization of losses in a coaxial cable. According to (3.52), to calculate the loss constant, we must first know both the characteristic impedance and the resistance per unit length, R'.

The characteristic impedance for a coaxial line was derived in Section 3.2.2. The resistance per unit length, R', is simply the sum of the resistances of the inner and outer conductors, which in turn are found by considering the skin effect sheet resistance (1.118) distributed over the entire circumference of each conductor,

$$R' = \frac{R_s}{2\pi a} + \frac{R_s}{2\pi b} = \frac{R_s}{2\pi}\left(\frac{1}{a} + \frac{1}{b}\right) = \frac{1}{2\pi}\left(\frac{1}{a} + \frac{1}{b}\right)\sqrt{\frac{\omega\mu'}{2\sigma}} \tag{3.53}$$

We have assumed, for simplicity, that the inner and outer conductors are made of the same metal, but one should note that this is not always the case.

Expressing this in terms of Z_0, and combining with (3.52), we have

$$\alpha = \frac{R'}{2Z_0} = \frac{1}{4\pi Z_0}\left(\frac{1}{a} + \frac{1}{b}\right)\sqrt{\frac{\omega\mu'}{2\sigma}} = \frac{1}{4\pi b Z_0}\left(\frac{b}{a} + 1\right)\sqrt{\frac{\omega\mu'}{2\sigma}} \tag{3.54a}$$

$$= \frac{1 + e^{2\pi Z_0/\eta}}{2b\eta\left(2\pi Z_0/\eta\right)}\sqrt{\frac{\omega\mu'}{2\sigma}} \tag{3.54b}$$

For a fixed outer diameter, $2b$, this expression is minimized when

$$\frac{2\pi Z_0}{\eta} \approx 1.2785 \tag{3.55a}$$

$$Z_0 \approx \frac{76.65\Omega}{\sqrt{\varepsilon_r}} \tag{3.55b}$$

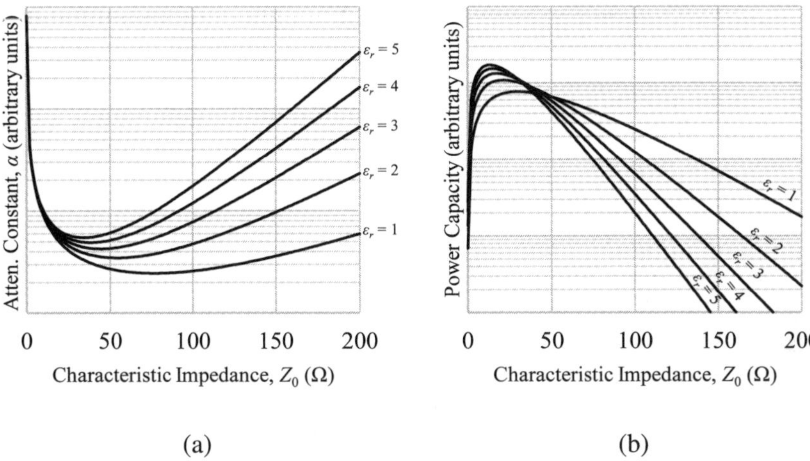

Figure 3.11 (a) Attenuation constant for coaxial transmission line as a function of characteristic impedance. The vertical scale will depend on the metal conductivity, frequency of operation, and line diameter, as given in (3.54b). (b) Power capacity of coaxial transmission line versus characteristic impedance. The vertical scale will depend on the critical field intensity and line diameter, as given in (3.57).

where we have assumed that $\mu_r = 1$, as is common for most simple dielectrics. (Further, the μ' in the square root obtained from the skin effect formula is for the conductor material, not the insulator between them, and is also typically $\mu' = \mu_0$ in most cases.) Note that while the loss itself is dependent upon conductivity, the outer diameter of the line, and the frequency, the optimum characteristic impedance that minimizes the loss is independent of all of these parameters. It depends only on the dielectric material filling the line. A plot of the loss of coaxial transmission line as a function of impedance and dielectric constant is shown in Figure 3.11(a).

Another important consideration is the power capacity of the transmission line, typically limited in the coaxial case by the voltage breakdown of the insulator. Assuming that the dielectric is uniform, the maximum field strength occurs at the inner conductor boundary, where

$$E_{\text{max}} = \frac{Q}{2\pi\varepsilon a} = \frac{C'V_0}{2\pi\varepsilon a} = \frac{C'\sqrt{PZ_0}}{2\pi\varepsilon a} = \frac{\sqrt{PZ_0}}{a\ln\left(\frac{b}{a}\right)} = \frac{\eta}{2\pi a}\sqrt{\frac{P}{Z_0}} \tag{3.56a}$$

$$= \frac{\eta}{2\pi b}\left(\frac{b}{a}\right)\sqrt{\frac{P}{Z_0}} = \frac{\eta}{2\pi b}e^{2\pi Z_0/\eta}\sqrt{\frac{P}{Z_0}} \tag{3.56b}$$

If we are to assume this field strength is at maximum for the given dielectric, then we may solve for the power capacity,

$$P_{\text{max}} = \left(\frac{2\pi b E_{\text{max}}}{\eta}\right)^2 Z_0 e^{-4\pi Z_0/\eta} \tag{3.57}$$

A plot of this function for several dielectric constants is shown in Figure 3.11(b). The power capacity is maximized when

$$Z_0 = \frac{\eta}{4\pi} \approx \frac{30\Omega}{\sqrt{\varepsilon_r}} \tag{3.58}$$

The value of the critical electric field, E_{max}, will depend on the dielectric material. In practice, it is wise to derate the theoretical value in (3.57) to account for a number of real factors, such as material imperfections, mechanical stress, and localized imperfections in the cross-sectional geometry which may tend to enhance breakdown at lower field intensities. Regardless of these practical concerns, however, the general statement about the optimal impedance in (3.58) applies.

Gilmour [7] suggested that the traditional selection of 50Ω as a standard impedance resulted from a compromise between these two optimizations for air-filled ($\varepsilon_r = 1$) coax lines used in early kilowatt radio transmitters of the 1930s — 50Ω being approximately the mean of 30Ω (maximum power capacity) and 77Ω (minimum loss). It is perhaps a coincidence that modern Teflon[2]-filled semirigid coaxial cables, having $\varepsilon_r = 2.2$, achieve almost exactly the minimum loss when designed for 50Ω ($\approx 77\Omega/\sqrt{2.2}$).

Note that both the loss and the power capacity may be improved by increasing the diameter, b, of the transmission line. As a general rule of thumb, more metal means less loss, and bigger means more power. However, in addition to the inconvenience of having to work with thicker, heavier, and probably very stiff coaxial cables (called *hardline* in the applications where it is used), there is an electromagnetic downside to increasing the size, which is that the cross-sectional geometry will no longer be electrically small at high frequencies, thereby ceasing to behave like a transmission line. The useful frequency range of coaxial lines is thus determined, in part, by their diameter. The precise nature of their failure at higher frequencies

2 Teflon is the Chemours/DuPont brand name for polytetrafluoroethylene, or PTFE, discovered accidentally by Roy Plunkett in 1938, after the common 50Ω standard was already established.

Table 3.2
Summary of Coaxial Lines

Parameter	Symbol	Expression
Characteristic impedance	Z_0	$\frac{\eta}{2\pi}\ln\left(\frac{b}{a}\right)$
Power flow	P	$\frac{2\pi}{\eta}\left\|A^+\right\|^2\ln\left(\frac{b}{a}\right)$
Attenuation constant	α	$\frac{1}{4\pi Z_0}\left(\frac{1}{a}+\frac{1}{b}\right)\sqrt{\frac{\omega\mu}{2\sigma}}$
Upper Frequency limit*	f_{max}	$\frac{1}{\pi(a+b)\sqrt{\mu\varepsilon}}$

*Derived in Section 7.6.

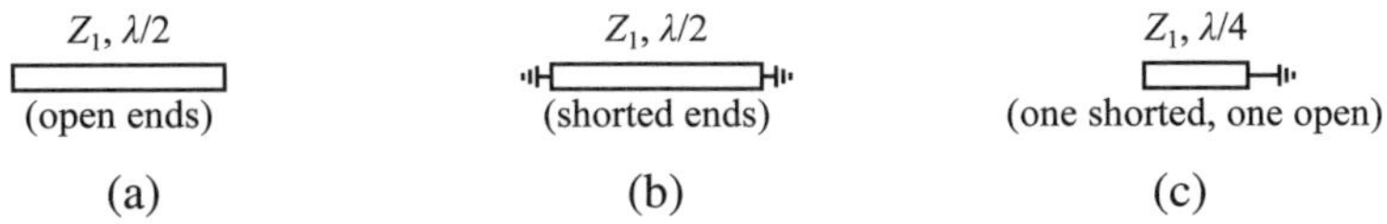

Figure 3.12 Three examples of transmission-line resonators. (a) Half-wave with open ends. (b) Half-wave with shorted ends. (c) Quarter-wave with one end open and one end shorted.

relates to the propagation of higher-order modes, which will be discussed in the context of waveguides in Chapter 7.

A summary of the results for coaxial line, including the upper-frequency limit which will be derived later, is given in Table 3.2.

3.4 TRANSMISSION-LINE RESONATORS

Resonators were discussed in the context of series or parallel lumped elements in Section 2.5. Transmission lines may also realize resonators, and share many of the same properties with their lumped-element counterparts. Typically, these fall into one of two categories, half-wave and quarter-wave resonators. Examples are shown schematically in Figure 3.12. Let us set aside for the moment the question of how one couples a signal into these resonators; assume instead that a standing wave is present and then consider its properties.

3.4.1 Half-Wave Resonators

Let us start with the resonator in Figure 3.12(a). Both ends are open circuited. The voltage and current on the line are given by the sum of forward and backward-traveling waves,

$$v(z) = v^+e^{-\gamma z} + v^-e^{\gamma z} \tag{3.59a}$$

$$i(z) = i^+e^{-\gamma z} - i^-e^{\gamma z} = \left(v^+e^{-\gamma z} - v^-e^{\gamma z}\right)/Z_1 \tag{3.59b}$$

The open-circuit boundary conditions at both ends require that

$$v^+ = v^- = v_0 \tag{3.60a}$$

$$v^-e^{\gamma l} = v^+e^{-\gamma l} \tag{3.60b}$$

$$e^{2\gamma l} = 1 \tag{3.60c}$$

$$\gamma l = (\alpha + j\beta)l = jn\pi \tag{3.60d}$$

$$\beta l = \theta = n\pi \tag{3.60e}$$

where n is a positive integer. Thus, a steady-state, time-harmonic solution is possible only if the line is lossless and a multiple of a half-wavelength long. For this analysis, we will assume initially that the lossless condition is met ($\alpha = 0$) and then consider the loss as a small perturbation of that solution. The total (phasor) voltage and current under these conditions is then given by

$$v(z) = 2v_0 \cos\left(\frac{n\pi z}{l}\right) \tag{3.61a}$$

$$i(z) = -j2\frac{v_0}{Z_1} \sin\left(\frac{n\pi z}{l}\right) \tag{3.61b}$$

Note that, unlike the series and parallel lumped-element resonators of Section 2.5, which have only one resonance, even simple transmission-line resonators have an infinite number of resonances at periodic frequencies. Just like the lumped-element resonators, transmission-line resonators operate by alternately storing electric and magnetic energy. Electric energy is stored as a voltage standing wave — the sum of forward and backward-traveling voltage waves — reaching its peaks at either end of the line and, if $n > 1$, at some nodes in between. Magnetic energy is stored during the alternate part of the cycle as a current standing wave — the sum of forward and backward-traveling current waves — reaching its peaks in the center of

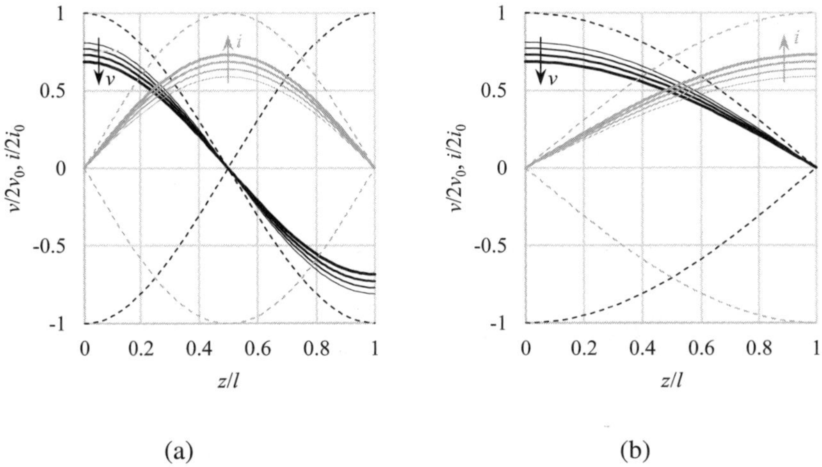

Figure 3.13 Voltage and current waveforms for the primary ($n = 1$) resonance of (a) a half-wave transmission-line resonator with open ends, and (b) a quarter-wave transmission-line resonator with one open and one shorted end. The waveforms evolve in time from the thin curves to the thick curves as indicated by the arrows. The dashed lines indicate the overall envelope within which each waveform oscillates.

the line and, if $n > 1$, at other nodes in between. A plot of the voltage and current waves over time are shown for $n = 1$ in Figure 3.13(a).

The average stored electric energy may be calculated by considering the voltage applied to the distributed capacitance along the line, and the average magnetic energy by considering the currents through the distributed inductance along the line as follows:

$$U_C = \frac{1}{4} C_1 \int_0^l |v(z)|^2 \, dz = \frac{1}{2} C_1 v_0^2 l \tag{3.62a}$$

$$U_L = \frac{1}{4} L_1 \int_0^l |i(z)|^2 \, dz = \frac{1}{2} L_1 \left(\frac{v_0}{Z_1} \right)^2 l \tag{3.62b}$$

where C_1 and L_1 are the capacitance and inductance per unit length of the transmission line, and

$$Z_1 = \sqrt{\frac{L_1}{C_1}} \tag{3.63a}$$

$$\beta = \omega_0 \sqrt{L_1 C_1} \tag{3.63b}$$

Therefore,

$$U_L = U_C = \frac{1}{2} C_1 v_0^2 l = \frac{v_0^2}{2\omega_0 Z_1} \beta l = \frac{n\pi v_0^2}{2\omega_0 Z_1} \tag{3.64}$$

Up to now we have assumed that the line is lossless, or nearly so. If there is any loss, then we may calculate the time-averaged power dissipation by considering the voltage across a small, distributed, shunt conductance, G_1, and the current through a small, distributed, series resistance, R_1,

$$P_l = \frac{1}{2} R_1 \int_0^l |i(z)|^2 \, dz + \frac{1}{2} G_1 \int_0^l |v(z)|^2 \, dz \tag{3.65a}$$

$$= \frac{v_0^2 l}{Z_1} \left(\frac{R_1}{Z_1} + Z_1 G_1 \right) = \frac{2 v_0^2 \alpha l}{Z_1} \tag{3.65b}$$

where the last step follows from (3.15d). The unloaded quality factor of the resonator, Q_0, then follows by definition,

$$Q_0 = \omega_0 \frac{U_L + U_C}{P_l} = \omega_0 \frac{n\pi v_0^2}{\omega_0 Z_1} \cdot \frac{Z_1}{2 v_0^2 \alpha l} = \frac{n\pi}{2\alpha l} = \frac{\beta}{2\alpha} \tag{3.66}$$

The half-wave resonator with shorted ends follows a derivation very much the same, but in this case the voltage wave assumes the sine form, and the current wave is the cosine. The total stored electric and magnetic energy is the same, as is the average power dissipation and quality factor.

3.4.2 Quarter-Wave Resonators

The resonator in Figure 3.12(c) has one open end and one shorted end. It behaves very much like the half-wave resonators described in Section 3.4.1 when n is odd and the line is cut in half. That is, let $n = 2m-1$, where m is the order of the quarter-wave resonance, and let the length of the new resonator be $l' = l/2$. The voltage and current solutions may then be applied directly to the new resonator. Since the time-averaged stored energy and power dissipation were distributed evenly in the two halves of the half-wave resonator, the quarter-wave resonator gets only half of each, and the quality factor remains the same. Therefore,

$$Q_0 = \frac{\beta}{2\alpha} = \frac{n\pi}{2\alpha l} = \frac{(2m-1)\pi}{4\alpha l'} \tag{3.67}$$

Table 3.3
Properties of Transmission-Line Resonators

Parameter	Symbol	Expression Half-wave	 Quarter-wave
End terminations		Open-open or short-short	Open-short
Phase constant at resonance	β	$\frac{n\pi}{l}$	$\frac{(n-\frac{1}{2})\pi}{l}$
Average total stored energy	U	$\frac{\beta l}{Z_1}\frac{v_0^2}{\omega_0}$	$\frac{\beta l}{2Z_1}\frac{v_0^2}{\omega_0}$
Average power dissipation	P_l	$\frac{2\alpha l}{Z_1}v_0^2$	$\frac{\alpha l}{Z_1}v_0^2$
Unloaded quality factor	Q_0	$\frac{\beta}{2\alpha}$	$\frac{\beta}{2\alpha}$

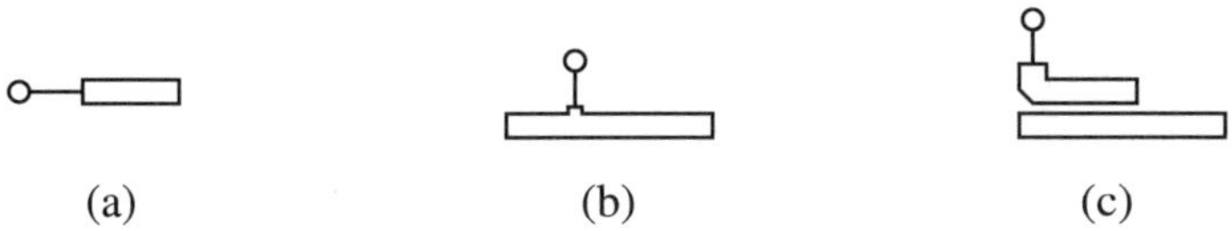

Figure 3.14 Three ways of coupling to a transmission-line resonator. (a) End-fed. (b) Tapped. (c) Edge-coupled.

Note that both types of resonators, half-wave and quarter-wave, if resonant at the same frequency, have the same Q.

A summary of the properties of transmission-line resonators is given in Table 3.3.

3.4.3 Coupling to a Transmission-Line Resonator

Coupling to a resonator inevitably involves loading the Q with an external load. In the case of a series LC resonator discussed in Section 2.5, that load appeared in series with the circuit elements and added to its resistance, whereas the load on a parallel LC resonator appeared in parallel with its components and added to its conductance.

The coupling geometry for a transmission-line resonator may take many forms, but the simplest is an end-fed structure, where the load is connected to a port at one end, as shown in Figure 3.14(a). The resonator then appears like a transmission-line stub, in this example, an open-circuited stub. At the quarter-wave resonance, the end to which the port is connected would formerly have been a short-circuit, thus the external load appears in series with the resonator's equivalent resistance at resonance. At the half-wave resonance, the port-connected end would formerly have been open, so the external load in this case appears in parallel with the resonator's equivalent conductance at resonance. If the stub was short-circuited, these features would be reversed. Either way, we see that an end-fed stub acts as both a quarter-wave and a half-wave resonator, and alternates between series-loaded and parallel-loaded configurations at harmonics of the quarter-wave resonance frequency.

In Figure 3.14(b), a half-wave resonator, open-circuited in this case, is coupled to a port through a tap somewhere along its length. The position of such a tap can alter the degree of coupling and external loading that the resonator experiences.

Figure 3.14(c) shows a half-wave resonator, also open-circuited for illustration, coupled to an adjacent transmission line over part of its length. We have not yet discussed coupled lines of this type, but will find that certain kinds of open-structure transmission-lines, especially the planar varieties discussed in the next section, induce waves propagating in neighboring lines simply by proximity, as a consequence of the intermingling of the electric and magnetic fields surrounding them.

3.5 PRINTED CIRCUIT TECHNOLOGIES

Although coaxial transmission lines are still very common as general-purpose interconnects and port interfaces for laboratory testing, their use as circuit elements has greatly diminished in favor of other transmission lines that can be printed on a flat substrate. Uniplanar designs such as microstrip, coplanar waveguide (CPW), and slotline (shown in Figures 3.5(b–d)) are the modern workhorse transmission lines for compact integrated circuits. Microstrip is especially common for its simplicity in manufacturing and the ease with which it integrates active devices (often fabricated directly within the substrate material) and transitions to coax and other long-distance interconnects.

3.5.1 Microstrip Design Equations

As stated earlier, planar printed-circuit geometries such as microstrip, though practically convenient, do not tend to permit straightforward, closed-form, mathematical

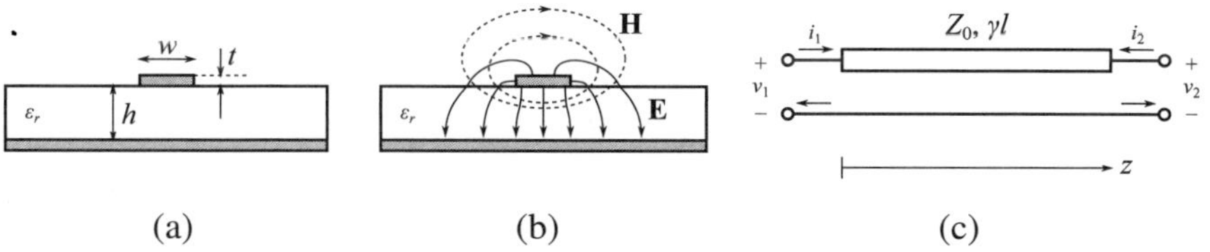

Figure 3.15 (a) Cross-sectional geometry of a microstrip line. (b) Field configuration. (c) Transmission-line model.

solutions. This is partly due to the fact that the electric and magnetic fields associated with the transmission line do not have a closed boundary and exist simultaneously in two dissimilar dielectric domains, namely the substrate and the air above it, as shown in Figure 3.15.

Despite these difficulties, the same, simple transmission-line model that was used for coax, comprising a characteristic impedance (Z_0) and a propagation factor (γl), applies here. The trace and substrate materials and thicknesses are typically fixed (or at least limited in selection) by the manufacturing technology, so the designer has only the trace width to control impedance of the line, and its length to control the propagation delay. In the absence of closed-form solutions, numerous empirical formulas have been derived for determining these parameters [5]. For example, one approximate form for the characteristic impedance of microstrip is

$$Z_0 \approx \begin{cases} \frac{60}{\sqrt{\varepsilon_e}} \ln\left(\frac{8h}{w} + \frac{w}{4h}\right) & \text{for } w \leq h \\ \frac{120\pi}{\sqrt{\varepsilon_e}\left(\frac{w}{h}+1.393+0.667\ln\left(\frac{w}{h}+1.444\right)\right)} & \text{for } w \geq h \end{cases} \tag{3.68}$$

where the *effective dielectric constant*, ε_e, is

$$\varepsilon_e \approx \frac{\varepsilon_r + 1}{2} + \frac{\varepsilon_r - 1}{2} \cdot \frac{1}{\sqrt{1 + 12\frac{h}{w}}} \tag{3.69}$$

The loss due to finite conductivity of the conductors is given by

$$\alpha_c \approx \frac{R_s}{2Z_0 w} \tag{3.70}$$

where R_s is the surface resistance of the conductors as determined by the skin-effect formula (1.118).

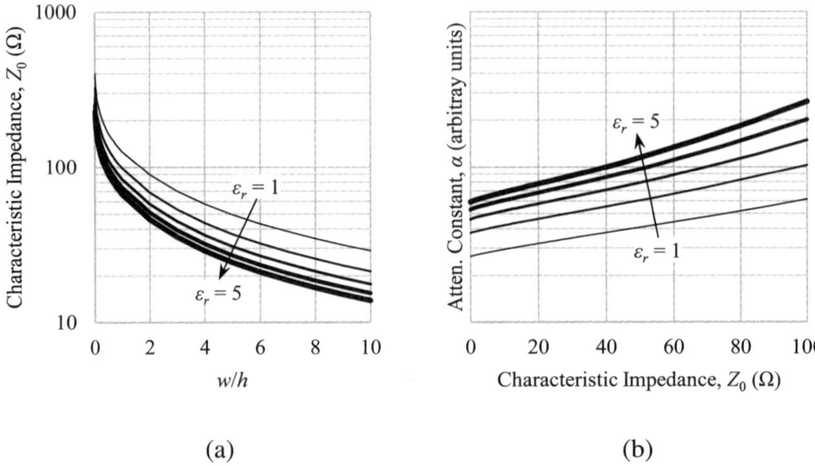

Figure 3.16 (a) Microstrip characteristic impedance as a function of trace width over substrate height, w/h. (b) Attenuation constant versus characteristic impedance. Vertical scale depends on the trace conductivity and other material parameters.

These relationships are plotted in Figure 3.16. Other similar formulas exist that solve the inverse problem of finding the key geometric parameter, w/h, given a desired characteristic impedance, Z_0 (since the formula in (3.68) is not easily invertible), and for capturing the subtle frequency dependence of the characteristic impedance. These in turn have been programmed into a number of readily available software tools allowing the user to simply enter what parameters they know and solve for those that remain.

In almost any electronic circuit involving transmission lines, lines of several different characteristic impedances will be needed. It is customary to choose a single system impedance (e.g., 50Ω) to which the input and output ports will be matched, while all other lines are scaled relative to it. Like coax, however, there are limits to the size parameters that microstrip can take on. If the substrate is too thick or the trace too wide, the cross-section may no longer be considered a quasistatic, lumped-element circuit, and it ceases to behave like a transmission line. Fabrication tolerance (e.g., lithography resolution) and practical metal conductivity also limit how narrow a trace can be used. One usually selects a substrate material and thickness that gives convenient dimensions for the overall characteristic impedance of the system at the center frequency of operation. A good rule of thumb then is

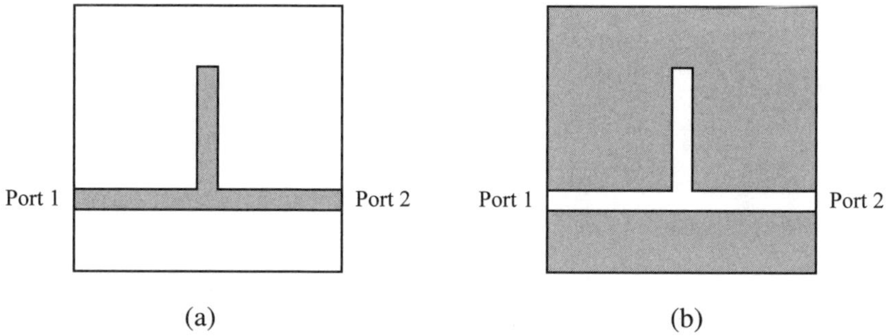

Figure 3.17 (a) Open-circuited, parallel-connected stub in microstrip. (b) Short-circuited, series-connected stub in slotline. The gray areas represent metal traces, while the white areas are the exposed substrate below.

that any transmission-line impedance within about a factor of 2 can be realized. Thus, in a 50Ω system, transmission lines having impedance between 25Ω and 100Ω are generally realizable, but not beyond, at least not without going to some extraordinary lengths.

3.5.2 Planar Junctions and Discontinuities

Certain practical features of planar circuit design warrant special attention. These typically involve the discontinuities in the line geometry, such as at the ends of stubs, between two transmission lines of different widths, and at the junction between three or more lines.

First, it must be noted that because of the common ground plane associated with microstrip, it is far more natural to have parallel connected transmission-line stubs than it is to have series-connected stubs, as shown in Figure 3.17(a). When series-stubs are used, it usually requires transition to another form of transmission-line. Slotline, in contrast, has the opposite constraint; series-connected stubs are trivial to make, while parallel-connected stubs are very difficult, as shown in Figure 3.17(b). In many ways, slotline and microstrip may be considered as duals of one another — loosely, circuits that are equivalent upon substitution of electric fields with magnetic fields, voltages with currents, and series with parallel and so forth. Duality will be discussed in greater detail in Section 5.1.7.

Confining the rest of this discussion to microstrip, let us examine the three forms of discontinuity in Figure 3.18. In the first case, Figure 3.18(a), there is an

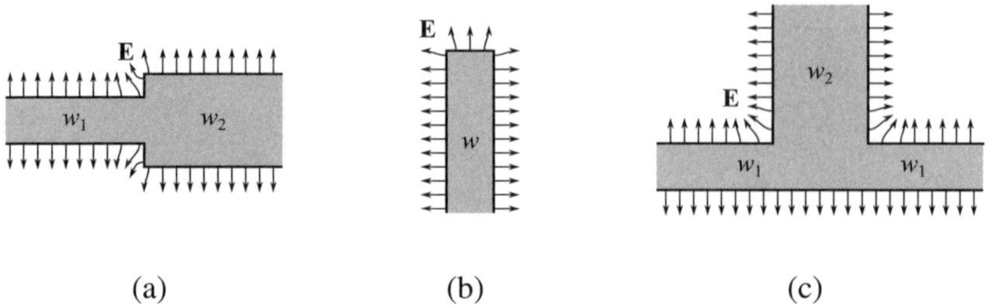

Figure 3.18 Microstrip discontinuities. (a) Step in width. (b) Open-ended stub. (c) Tee-junction.

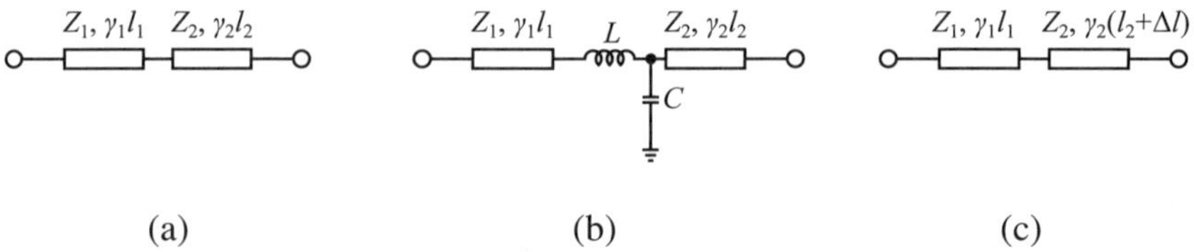

Figure 3.19 Three circuit models for a microstrip width discontinuity. (a) A simple transition in line parameters. (b) Local fields accounted for by lumped elements. (c) Local fields accounted for by extra line length.

abrupt change in trace width between two sections of transmission line. As we know, the width of the trace determines the characteristic impedance of waves propagating in each region, so this could be modeled as simply one transmission line having a certain characteristic in cascade with another line having a different characteristic impedance, as shown in Figure 3.19(a). (Since in microstrip, the common ground plane shared by all ports is always implicitly present beneath the circuit, it is convenient to draw only the top trace, and indicate the ground connection where needed with the appropriate symbol.) As a first-order model, this is correct, but it fails to account for the effect of the fringing electric and magnetic fields in the immediate vicinity of the width change.

To put it another way, the currents flowing in the outer edges of the microstrip trace have a short extra distance to travel outward from the narrow line into the wider line (or vice versa), roughly equal to the difference in width, which is not accounted for in the longitudinal length of either line. This could be modeled as a series inductance at the discontinuity. However, the accumulating charge in the corners of the wider line represent a localized excess capacitance to ground. Thus, virtual lumped elements could be added to the circuit model to account for the effect of the discontinuity, Figure 3.19(b), but it is also fairly accurate to simply add a small bit of extra effective length to the wider line, as shown in Figure 3.19(c). This last approach is often preferred because the discontinuity can be fully and directly compensated for with a physical reduction of the desired line length by an equal amount. As with other characteristics of microstrip, the amount of the excess line length that needs removing is found empirically [8],

$$\frac{\Delta l}{h} = 0.412\left(\frac{\varepsilon_e + 0.3}{\varepsilon_e - 0.258}\right)\left(\frac{w_2/h + 0.262}{w_2/h + 0.813}\right)\left(1 - \frac{w_1}{w_2}\right) \tag{3.71}$$

where $w_2 > w_1$, and ε_e is defined in (3.69).

Similarly, the fields beyond the end of an open-ended stub equate approximately to an extension of the line beyond its physical length, which can be corrected by shortening the stub during layout. The amount of the correction is

$$\frac{\Delta l}{h} = 0.412\left(\frac{\varepsilon_e + 0.3}{\varepsilon_e - 0.258}\right)\left(\frac{w_2/h + 0.262}{w_2/h + 0.813}\right) \tag{3.72}$$

Note that the correction for a step change in width is the same as the open-ended stub correction with an extra factor of $(1 - w_1/w_2)$ to account for that portion of the wider trace width which is overlapped by the narrower trace.

Finally, at the tee-junction between three transmission lines, the shunt arm is effectively shortened (rather than lengthened, in this case) by the currents shortcutting around the corners without traveling first to the center of the junction. In this case, the line must be lengthened in the layout to compensate, by an amount given approximately by

$$\frac{\Delta l}{h} = \frac{120\pi}{Z_1\sqrt{\varepsilon_{e1}}}\left(0.5 - 0.16\frac{Z_1}{Z_2}\left[1 - 2\ln\frac{Z_1}{Z_2}\right]\right) \tag{3.73}$$

where ε_{e1} is the effective dielectric constant of the through arm.

Of course, if a section of transmission line is associated with more than one of these discontinuities, say, an open-ended stub connected to the network at a tee-junction, then each appropriate length correction is applied in a cumulative fashion.

One commonly encountered discontinuity that is not easily accounted for by a length correction is a bend, as indicated in Figure 3.20(a). The larger area of a square corner creates an excess capacitance to ground. The most common way to compensate for this is to remove some of the metal in the corner by mitering,

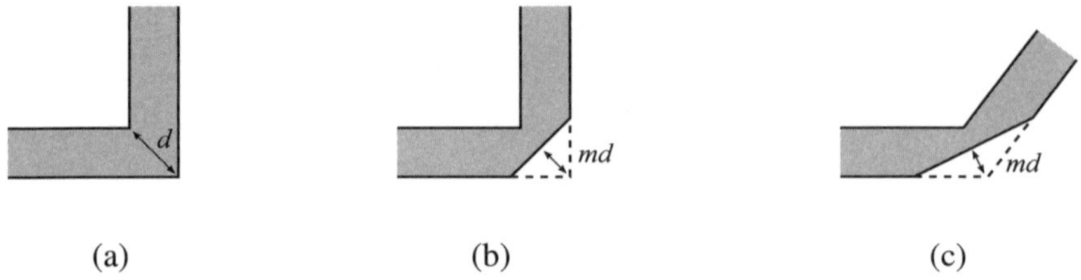

Figure 3.20 Compensation for a microstrip bend. (a) Uncompensated. (b) Compensation for a right angle bend. (c) Compensation for a bend of arbitrary angle.

as shown in Figures 3.20(b, c). The optimum *miter fraction*, m, defined as the relative distance along the diagonal from the outside corner to the edge of the miter, compared to the original distance from the outside to inside corner, is typically assumed to be about $m \approx 0.6$.

3.5.3 Radial Stubs

The quarter-wavelength, open-circuited, parallel stub is a common feature of many microwave circuits, providing an effective short-circuit to ground at high frequency while allowing lower frequencies to pass unimpeded. This is a critical element, for example, of a *bias tee*, a simple circuit through which DC power may be brought to an active device within the circuit while isolating the power supply from the primary signal path.

In these situations, we desire a stub which presents the lowest possible impedance in parallel to the main line over the widest possible bandwidth around its quarter-wave frequency. According to (3.43), then, it is best to use a stub with the lowest characteristic impedance, Z_0, available.

Of course, low-impedance microstrip lines are very wide, and wide parallel stubs would run into significant tee-junction effects as described in Section 3.5.2. While the stub-shortening effect of this discontinuity can be compensated as previously described, in the extreme case other effects come into play that are not so easily mitigated; the through arm, for example, will experience a substantial change in its characteristic impedance over the distance run from one edge of the wide stub to the other.

An inventive solution to these problems is the *radial stub* illustrated in Figure 3.21(a). It starts with a narrow tee junction at the main line, minimizing the parasitic effects of that discontinuity, and then widens as it gets further away.

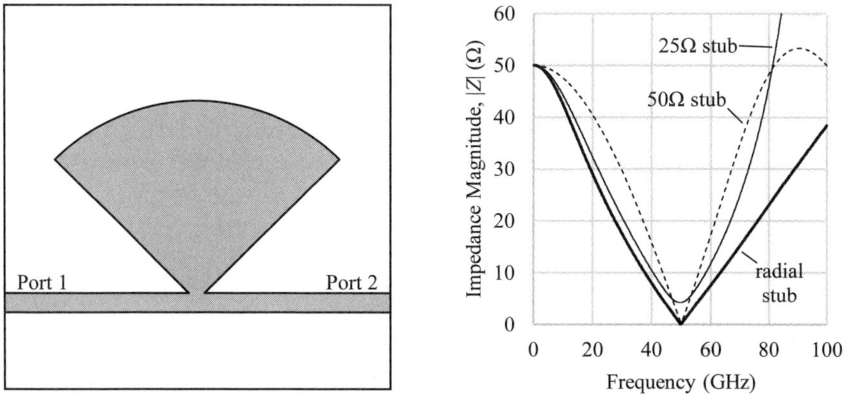

(a) (b)

Figure 3.21 (a) Radial stub on a microstrip transmission line. (b) Comparison of the impedance presented to the through line by a 50Ω shunt stub (dashed line), a 25Ω shunt stub (thin solid line), and a radial stub with 90° opening angle (thick solid line). For this model, we assume a 100-μm-thick GaAs substrate and a desired center frequency of 50 GHz.

The open end of the stub is traditionally drawn as a circular arc such that the whole stub has the appearance of a sector of a larger circle. There is significant fringing capacitance at the end of the stub, but this merely shifts the center frequency of operation for a given radius and is easily accounted for.

A comparison of the impedance magnitude presented by straight 50Ω and 25Ω shunt stubs and a radial stub is given in Figure 3.21(b). The 25Ω stub does present a lower impedance over a broader range of frequencies than the 50Ω stub, but it fails to ever reach a virtual short circuit at the center frequency of 50 GHz, and it rises rather sharply in the upper half of the band. The radial stub, in contrast, presents a broad and symmetric low-impedance response that does reach a null at 50 GHz.

3.5.4 Coupled Lines

Since the fields propagating along a microstrip line are not bounded the way they are in coax, there is the possibility that the fields of two different lines in the network may interlink with one another if the lines get too close. Physically, the accumulation of charges associated with voltage waves attracts opposite charges in the adjacent line, while the current waves in one line support magnetic fields that induce currents in its neighbor. In this way, waves on one transmission line may launch waves on another in close proximity. The two lines are said to be *coupled.*

While this coupling can be an unwanted side effect of compact circuits, it can also be used to the designer's advantage. The mathematical foundation of *coupled lines* as a design element shall wait until we have learned more about network parameters and even-/odd-mode analysis in Chapter 4.

Problems

3-1 What is the propagation constant of a transmission line having series resistance $R' = 1\,\Omega/\text{m}$, series inductance $L' = 7$ nH/m, shunt capacitance $C' = 7$ pF/m, and shunt resistance $R'' = 1$ kΩ·m at 10 GHz? What is its characteristic impedance?

3-2 What is the conductive attenuation constant of a transmission line having characteristic impedance $Z_0 = 100\Omega$ and a parasitic resistance $R' = 1$ Ω/m?

3-3 Calculate the capacitance per unit length of a coaxial line in which the dielectric constant changes from ε_{r1} to ε_{r2} at the mid-point between the inner and outer radii — that is, at $r = (a + b)/2$.

3-4 Suppose that an artificial transmission line were constructed using a large number of very small series capacitors and shunt inductors (as opposed to the series inductors and shunt capacitors of a normal, lossless transmission line). Derive expressions for the characteristic impedance and propagation constant of this line in terms of the series capacitance per unit length, C', and the shunt inductance per unit length, L'.

3-5 What ratio of outer to inner diameters is needed for air-filled coax to have a characteristic impedance of 50Ω?

3-6 Consider two transmission lines of different characteristic impedance, Z_1 and Z_2, but having the same length, l, and propagation constant, γ, connected in parallel at both ends. What is the impedance seen at one common port when a resistor, R, terminates the other common port?

3-7 What is the reflection coefficient presented by a load impedance of $Z_L = 40\Omega + j20\Omega$ seen from from a 50Ω source?

3-8 What is the load impedance that gives a reflection coefficient of $\Gamma_L = -0.5$ when seen from a 50Ω line?

3-9 If a particular load gives a reflection coefficient of $\Gamma_L = 0.3 + j0.3$, what is the reflection coefficient seen from a reference plane 45° removed from the load on a matched transmission line?

3-10 What is the input admittance measured at one end of a transmission line (having line parameters Z_1 and γl) that has identical shunt capacitors, C, connected at both ends? If the line is lossless, what does it resemble at frequencies near where $\gamma l = j\beta l = j\pi/2$?

3-11 How long (in degrees) would a short-circuited stub have to be to have the same input impedance as an open-circuited stub that is 25° long?

3-12 Typical atmospheric attenuation for plane-wave propagation at sea level is about 0.1 dB/km at 30 GHz. If we assume this is all due to dielectric losses (as opposed to scattering), what would the loss tangent of the air be?

3-13 Derive an expression for the loss of a coaxial line in which the inner conductor has conductivity σ_a and the outer conductor has conductivity σ_b. Ignore dielectric losses.

3-14 What is the ratio of the characteristic impedance for minimum loss of a coaxial line to the characteristic impedance for maximum power capacity of that line? Does it depend on the material filling the line?

3-15 What is the loaded quality factor of a half-wave resonator for which $\gamma = (0.01 + j2)\ \mathrm{m}^{-1}$, when the external $Q_e = 100$?

3-16 Suppose that a half-wave, transmission-line resonator is formed by having shorts at both ends, but the shorts are imperfect, each having a very small resistance, R. Calculate the unloaded Q accounting for this extra source of loss.

3-17 What is the characteristic impedance of a 5-mil-wide microstrip line on a 5-mil-thick alumina substrate ($\varepsilon_r = 9.8$)?

3-18 Which substrate supports a 50Ω microstrip line with less conductor loss — an alumina ($\varepsilon_r = 9.8$) substrate that is 125 μm thick, or a gallium arsenide ($\varepsilon_r = 12.9$) substrate which is 100 μm thick? Assume both use gold-plated metalization.

3-19 Calculate the amount that a microstrip open-circuit stub must be shortened to account for end capacitance if the line is 10 mils wide on a 10-mil-thick substrate of Teflon ($\varepsilon_r = 2.2$)?

3-20 A straight microstrip line on an alumina substrate 125 μm thick has a shunt, open-circuited stub connected at a tee-junction. Both the line and the stub are 50Ω. How long should the stub be to be a quarter-wavelength long at 10 GHz?

References

[1] D. M. Pozar, *Microwave Engineering*, 4th ed. New York: Wiley, 2011.

[2] G. Miano and A. Maffuci, *Transmission Lines and Lumped Circuits: Fundamentals and Applications*. London: Academic Press, 2001.

[3] S. Ramo, J. Whinnery, and T. Van Duzer, *Fields and Waves in Communication Electronics*. New York: Wiley, 1984.

[4] R. Haberman, *Elementary Applied Partial Differential Equations: with Fourier Series and Boundary Value Problems*, 2nd ed. Englewood Cliffs, NJ: Prentice Hall, 1987.

[5] B. Wadell, *Transmission Line Design Handbook*. Norwood, MA: Artech House, 1991.

[6] Wikipedia. (2017) DOCSIS. https://en.wikipedia.org/wiki/DOCSIS.

[7] A. S. Gilmour, *Microwave Tubes*. Norwood, MA: Artech House, 1986.

[8] E. Hammersted, "A microstrip handbook," University of Trondheim, Norway, Tech. Rep. STF 44 A74169, N7034, 1975.

Chapter 4

Network Parameters

Thus far, we have discussed two different classes of elements that reduce the full richness of Maxwell's equations to simpler mathematical forms for use as building blocks in electronic systems. These are the lumped elements, for which the expression of their behavior is algebraic in the complex frequency domain, and the transmission lines, for which the description has a trigonometric basis.

Methods for aggregating these elements into useful networks, however, have only been discussed in the most rudimentary terms: as series and parallel combinations, or transmission-line cascades yielding a single effective immittance quantity as measured at one port. To use these elements effectively, we must have a more general way of deriving and describing the behavior of larger networks with any number of input and output ports. This is the function of network parameters [1].

4.1 IMMITTANCE NETWORK PARAMETERS

Any linear electromagnetic circuit whatsoever having electrically small ports can be described in terms of its terminal voltages and currents. Such a network is shown conceptually in Figure 4.1(a). It is not necessary that the network itself be electrically small, or that the features inside necessarily conform to lumped-element or transmission-line descriptions, but the input and output ports must consist of terminal pairs that are electrically close to one another. This lone provision is necessary in order to properly define a conservative port voltage, and with it the immittance parameters that are key to the network description.

We assume that the ports are driven in such a way that the terminal currents are balanced, that is, the current entering the positive terminal of a port is matched

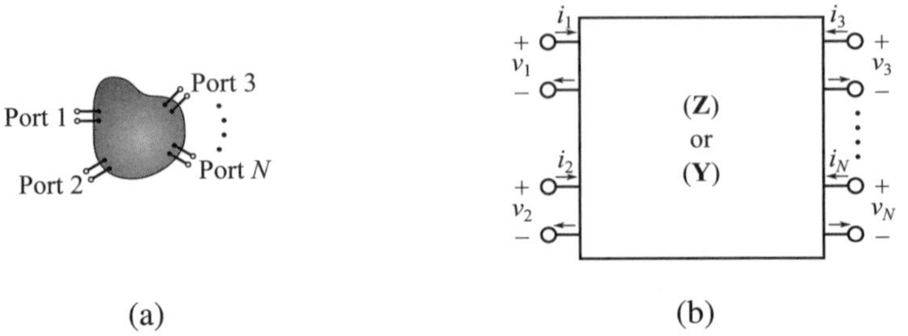

Figure 4.1 (a) Conceptual diagram of an arbitrary network with N terminal pairs, or ports. (b) Schematic of the network block illustrating voltage and current polarity, and the immittance-parameter ($\mathbf{Z}$ or $\mathbf{Y}$) description.

by an identical current leaving the negative terminal. This implies that any external generator or load may be connected between the terminals of one port only. If a connection is made between ports, then it must be included in the network description (along with any electrical length such a connection might entail). We adopt the *passive sign convention* for polarity of the voltage and current at each port, which is to say that positive current flows into the positive voltage terminal and out of the negative voltage terminal, as shown in Figure 4.1(b). This ensures that the product vi at any port, if positive, corresponds to the power delivered into the network through that port. If the product is negative, power is delivered out of the network.

4.1.1 Impedance Parameters

Let us now conduct a thought experiment wherein we test the network by attaching a current source to one of the ports, say, port number k, leaving the rest open-circuited. Clearly, a voltage will develop across the terminals of port k, and the ratio of that voltage to the source current must correspond to the effective impedance presented by the network to that port, which we will call z_{kk}. This is known as the *driving-point impedance*, since it is the impedance measured at the same point where the network is driven by the source.

The effect of the test source on our network does not end there, however. The flow of current through the internal elements of the network will similarly cause voltage potentials to appear at the terminals of the remaining ports as well. The ratio of the voltage measured at port l to the current incident at port k is denoted z_{lk} and is called a *transimpedance*.

Let us now assume that we have repeated this experiment N times with the test source connected to each of the ports in turn, the voltages at the remaining open ports measured, and the transimpedances and driving-point impedances calculated. What will be the effect of multiple independent current sources driving multiple ports simultaneously? Since the network is linear, the voltage at any port is simply the superposition of the voltages that would result from each of the current sources independently. That is,

$$v_l = \sum_{k=1}^{N} z_{lk} i_k \tag{4.1}$$

Note that for any port k which does not have a current source attached, $i_k = 0$. To generalize this result for all possible voltage outputs, we may write it in matrix form

$$\begin{pmatrix} v_1 \\ v_2 \\ \vdots \\ v_N \end{pmatrix} = \begin{pmatrix} z_{11} & z_{12} & \cdots & z_{1N} \\ z_{21} & z_{22} & \cdots & z_{2N} \\ \vdots & \vdots & \ddots & \vdots \\ z_{N1} & z_{N2} & \cdots & z_{NN} \end{pmatrix} \begin{pmatrix} i_1 \\ i_2 \\ \vdots \\ i_N \end{pmatrix} \tag{4.2}$$

or

$$\mathbf{v} = \mathbf{Z}\mathbf{i} \tag{4.3}$$

where $\mathbf{v}$ and $\mathbf{i}$ are column vectors containing the voltages and currents at every port, and $\mathbf{Z}$ is known as the *impedance matrix* for this network. The elements of the matrix, z_{lk}, are known as the *impedance parameters*, *z-parameters*, or *open-circuit parameters*, since the coefficients are calculated with all ports but the one having the test source left open-circuited (and the test source itself, as a current source, has infinite source impedance).

The impedance matrix is a complete description of the linear network in terms of its port behavior.[1] Though it may be easiest to apply when all inputs to the network are current sources and all outputs are voltages, the information contained within it is sufficient to calculate any and all outputs (voltage, or current, or both) in response to any number of inputs (again, voltage or current, or both). Importantly, the matrix does not provide sufficient information to uniquely describe the internal arrangement of elements that achieves this behavior. Indeed multiple networks may have the exact same impedance parameters, and thus would be indistinguishable

1 Assuming the matrix exists, that is. Lumped-element networks that have all floating components (none connected to ground) would have every z-parameter divergent to infinity. For such networks, the other network parameters described in this chapter may be more suitable.

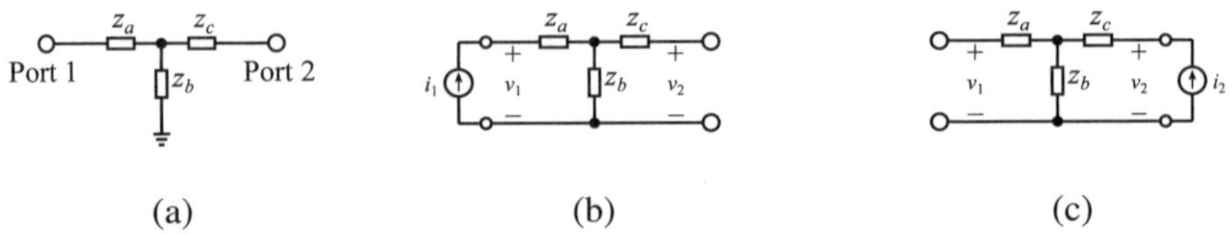

Figure 4.2 (a) Tee network of lumped elements. (b) Test current source at port 1. (c) Test current source at port 2.

from one another if placed inside a black box. Such networks are termed *identities*, as will be discussed in Section 5.2.

Let us work a simple example for two ports comprising the tee network of lumped elements in Figure 4.2(a). First, we attach a test current source to port 1, as in Figure 4.2(b). The port 1 driving-point impedance is found by considering that the source current passes through both z_a and z_b, which, for the purposes of this experiment, are in series. Thus,

$$z_{11} = \frac{v_1}{i_1} = z_a + z_b \tag{4.4}$$

To calculate the transimpedance from port 1 to port 2, consider that the test current flows through z_b, but none flows through z_c. The voltage at port 2 is thus the same as the voltage across z_b, or

$$z_{21} = \frac{v_2}{i_1} = z_b \tag{4.5}$$

Similarly, by attaching the test source to port 2, as in Figure 4.2(c), we find

$$z_{22} = \frac{v_2}{i_2} = z_b + z_c \tag{4.6a}$$

$$z_{12} = \frac{v_1}{i_2} = z_b \tag{4.6b}$$

The full impedance matrix for this network is then

$$\mathbf{Z} = \begin{pmatrix} z_{11} & z_{12} \\ z_{21} & z_{22} \end{pmatrix} = \begin{pmatrix} z_a + z_b & z_b \\ z_b & z_b + z_c \end{pmatrix} \tag{4.7}$$

Note that, in accordance with the reciprocity theorem of Section 2.7.1, the impedance matrix is symmetric, $\mathbf{Z} = \mathbf{Z}^T$.

Figure 4.3 (a) A two-port network comprising a simple transmission line. (b) Test current source at port 1.

Let us work another example, this time a simple transmission line, as in Figure 4.3. This time, the test current source launches voltage and current waves which propagate down the line and reflect off the open-circuit at the other end. If the line were lossless, the driving-point impedance would be given directly by (3.43) for an open-circuited stub. Let us be slightly more general, however, and allow for a lossy line. To do so, we may take the limit of (3.40c) as $Z_L \to \infty$,

$$z_{11} = Z_0 \coth(\gamma l) \tag{4.8}$$

To find the transimpedance, we must calculate the total voltage from the forward and backward-traveling waves at port 2.

$$v_2 = v^+ e^{-\gamma l} + v^- e^{\gamma l} = Z_0 \left(i^+ e^{-\gamma l} + i^- e^{\gamma l} \right) \tag{4.9}$$

The total current at port 1 is constrained by the test current source,

$$i_1 = i^+ - i^- \tag{4.10}$$

Therefore, the transimpedance is

$$z_{21} = \frac{v_2}{i_1} = Z_0 \frac{i^+ e^{-\gamma l} + i^- e^{\gamma l}}{i^+ - i^-} = Z_0 \frac{e^{-\gamma l} + (i^-/i^+)\, e^{\gamma l}}{1 - (i^-/i^+)} \tag{4.11a}$$

$$= Z_0 \frac{e^{-\gamma l} + \left(e^{-2\gamma l}\right) e^{\gamma l}}{1 - \left(e^{-2\gamma l}\right)} = Z_0 \frac{2}{e^{\gamma l} - e^{-\gamma l}} = Z_0 \operatorname{csch}(\gamma l) \tag{4.11b}$$

By symmetry, $z_{22} = z_{11}$ and $z_{12} = z_{21}$, so the full impedance matrix for the transmission line is

$$\mathbf{Z} = Z_0 \begin{pmatrix} \coth(\gamma l) & \operatorname{csch}(\gamma l) \\ \operatorname{csch}(\gamma l) & \coth(\gamma l) \end{pmatrix} \tag{4.12}$$

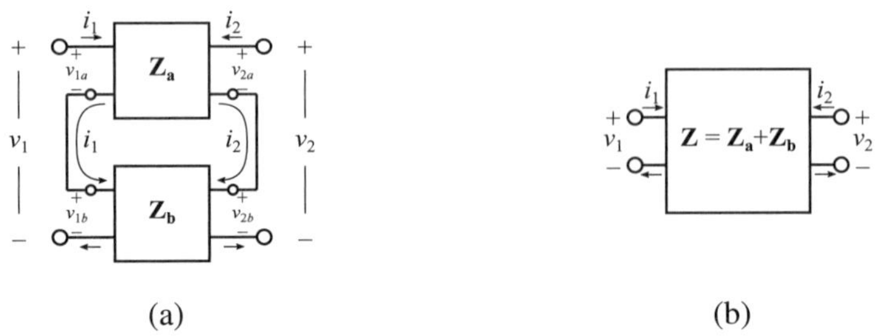

Figure 4.4 (a) Two networks combined in series. (b) Effective combined network.

or, if the line is lossless (such that $\gamma = j\beta$),

$$\mathbf{Z} = -jZ_0 \begin{pmatrix} \cot(\beta l) & \csc(\beta l) \\ \csc(\beta l) & \cot(\beta l) \end{pmatrix} \tag{4.13}$$

Although the foregoing examples of deriving z-parameters from the given networks are somewhat tedious, this representation has some properties that make it easier to form the matrix for more complex networks, once those of the basic building blocks are known. Consider the two subnetworks in Figure 4.4(a), described by the impedance parameter matrices $\mathbf{Z_a}$ and $\mathbf{Z_b}$. These have been combined port-by-port in series with one another; port 1 from the first network is stacked in series with port 1 from the second network, and so on.[2] Two-port networks are shown here for convenience, but conceptually we maintain that these could be arbitrary N-port networks for generality. The two networks are only required to have the same number of ports. Note that the same current flows in the corresponding ports of both networks, but their voltages add. Thus, using the notation indicated in the figure,

$$\begin{pmatrix} v_1 \\ v_2 \\ \vdots \\ v_N \end{pmatrix} = \begin{pmatrix} v_{1a} \\ v_{2a} \\ \vdots \\ v_{Na} \end{pmatrix} + \begin{pmatrix} v_{1b} \\ v_{2b} \\ \vdots \\ v_{Nb} \end{pmatrix} \tag{4.14a}$$

2 This is the rigorously correct usage of the term "series" when referring to networks. Many engineers casually describe networks that are connected port-to-port — that is, port 1 from the second network to port 2 of the preceding network — as being "in series" also, an arrangement more properly referred to as *cascade*. Network parameters more suitable for cascade connections will be described in more detail in Section 4.1.3.

$$= \begin{pmatrix} z_{11a} & \cdots & z_{1Na} \\ \vdots & \ddots & \vdots \\ z_{N1a} & \cdots & z_{NNa} \end{pmatrix} \begin{pmatrix} i_1 \\ \vdots \\ i_N \end{pmatrix} + \begin{pmatrix} z_{11b} & \cdots & z_{1Nb} \\ \vdots & \ddots & \vdots \\ z_{N1b} & \cdots & z_{NNb} \end{pmatrix} \begin{pmatrix} i_1 \\ \vdots \\ i_N \end{pmatrix} \tag{4.14b}$$

$$= \begin{pmatrix} z_{11a} + z_{11b} & z_{12a} + z_{12b} & \cdots & z_{1Na} + z_{1Nb} \\ z_{21a} + z_{21b} & z_{22a} + z_{22b} & \cdots & z_{2Na} + z_{2Nb} \\ \vdots & \vdots & \ddots & \vdots \\ z_{N1a} + z_{N1b} & z_{N2a} + z_{N2b} & \cdots & z_{NNa} + z_{NNb} \end{pmatrix} \begin{pmatrix} i_1 \\ i_2 \\ \vdots \\ i_N \end{pmatrix} \tag{4.14c}$$

or, as shown in Figure 4.4(b),

$$\mathbf{Z} = \mathbf{Z_a} + \mathbf{Z_b} \tag{4.15}$$

We may therefore construct the impedance parameters of complex networks by summing the contributions of more basic networks which are in series.

4.1.2 Admittance Parameters

Alternatively, one may consider characterizing an arbitrary network by positioning a test voltage source on each of the ports in turn and measuring the current that is produced in short-circuit terminations placed on all remaining ports. Denoting the ratio of the induced current at port l to the test voltage at port k as y_{lk}, superposition once again allows us to write a matrix representation for all output currents as a function of the applied voltages,

$$\begin{pmatrix} i_1 \\ i_2 \\ \vdots \\ i_N \end{pmatrix} = \begin{pmatrix} y_{11} & y_{12} & \cdots & y_{1N} \\ y_{21} & y_{22} & \cdots & y_{2N} \\ \vdots & \vdots & \ddots & \vdots \\ y_{N1} & y_{N2} & \vdots & y_{NN} \end{pmatrix} \begin{pmatrix} y_1 \\ v_2 \\ \vdots \\ v_N \end{pmatrix} \tag{4.16}$$

or

$$\mathbf{i} = \mathbf{Y}\mathbf{v} \tag{4.17}$$

where $\mathbf{Y}$ is known as the *admittance matrix* for this network. The elements of the matrix are called *admittance parameters*, *y-parameters*, or *short-circuit parameters*, since, in this case, the coefficients are calculated with short-circuit terminations on all ports but the one having the test voltage source (and the test voltage source has an internal impedance of zero).

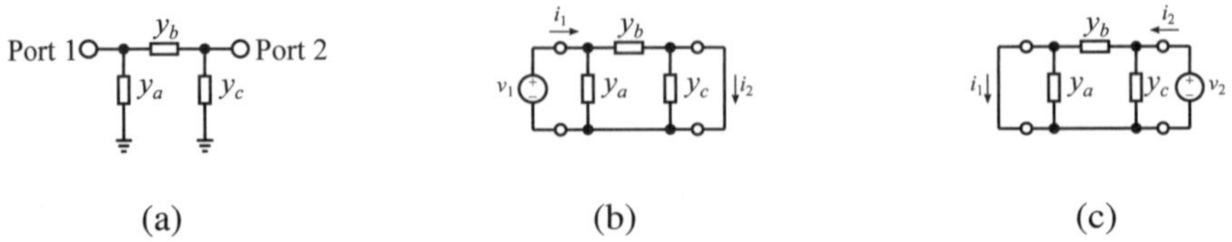

Figure 4.5 (a) Pi network of lumped elements. (b) Test voltage source at port 1. (c) Test voltage source at port 2.

Although the manner of determining the y-parameters of the network is conceptually different (in terms of the source type and terminations used) than that of the z-parameters, the current and voltage vectors in (4.3) and (4.17) are one in the same. It is quite apparent, therefore, that

$$\mathbf{Y} = \mathbf{Z}^{-1} \tag{4.18}$$

Keep in mind that this is the full matrix inverse; it does not imply that the individual elements of each matrix are reciprocals of one another (that is, $y_{jk} \neq 1/z_{jk}$).

Let us examine the y-parameters for some basic networks. First, consider the pi network of lumped elements shown in Figure 4.5(a). When a test voltage is applied to port 1 and a short-circuit termination to port 2, as in Figure 4.5(b), we find that y_a and y_b are in parallel while y_c is shorted out. The driving-point admittance at port 1 is therefore $y_{11} = y_a + y_b$.

The transadmittance from port 1 to port 2 is found by considering that the current flowing through port 2 is the same as that flowing through y_b. However, this current flows out of port 2, whereas the passive sign convention has currents flowing into the ports. This means that a negative sign is required, so $y_{21} = -y_b$.

Similar arguments apply to the situation where the test voltage is applied to port 2, Figure 4.5(c). The final y-parameters for this network are then given by

$$\mathbf{Y} = \begin{pmatrix} y_a + y_b & -y_b \\ -y_b & y_b + y_c \end{pmatrix} \tag{4.19}$$

The y-parameters for a simple transmission line could be found in much the same way as the z-parameters were found in Section 4.1.1. A much simpler method is to take advantage of the work that was already done to arrive at those z-parameters,

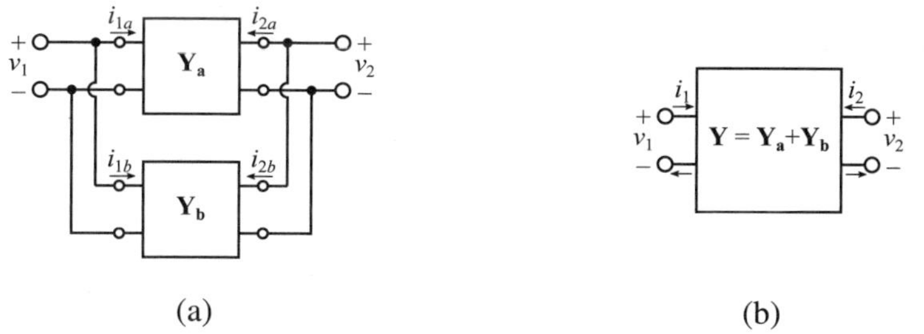

Figure 4.6 (a) Two networks combined in parallel. (b) Effective combined network.

and simply invert them according to (4.18),

$$\mathbf{Y} = \mathbf{Z}^{-1} = Z_0^{-1} \begin{pmatrix} \coth(\gamma l) & \operatorname{csch}(\gamma l) \\ \operatorname{csch}(\gamma l) & \coth(\gamma l) \end{pmatrix}^{-1} \tag{4.20a}$$

$$= Y_0 \begin{pmatrix} \coth(\gamma l) & -\operatorname{csch}(\gamma l) \\ -\operatorname{csch}(\gamma l) & \coth(\gamma l) \end{pmatrix} \tag{4.20b}$$

or, in the lossless case,

$$\mathbf{Y} = -jY_0 \begin{pmatrix} \cot(\beta l) & -\csc(\beta l) \\ -\csc(\beta l) & \cot(\beta l) \end{pmatrix} \tag{4.21}$$

As with the z-parameters, the y-parameters have properties that make it convenient to derive the network matrices for complex assemblies relatively simply from those of smaller building blocks. In this case, the relevant configuration is that of subnetworks combined in parallel, as in Figure 4.6(a). As one can see, the corresponding ports see the same voltage, but the currents are divided between them. Mathematically, we may derive these currents as follows,

$$\begin{pmatrix} i_1 \\ i_2 \\ \vdots \\ i_N \end{pmatrix} = \begin{pmatrix} i_{1a} \\ i_{2a} \\ \vdots \\ i_{Na} \end{pmatrix} + \begin{pmatrix} i_{1b} \\ i_{2b} \\ \vdots \\ i_{Nb} \end{pmatrix} \tag{4.22a}$$

$$= \begin{pmatrix} y_{11a} & \cdots & y_{1Na} \\ \vdots & \ddots & \vdots \\ y_{N1a} & \cdots & y_{NNa} \end{pmatrix} \begin{pmatrix} v_1 \\ \vdots \\ v_N \end{pmatrix} + \begin{pmatrix} y_{11b} & \cdots & y_{1Nb} \\ \vdots & \ddots & \vdots \\ y_{N1b} & \cdots & y_{NNb} \end{pmatrix} \begin{pmatrix} v_1 \\ \vdots \\ v_N \end{pmatrix} \tag{4.22b}$$

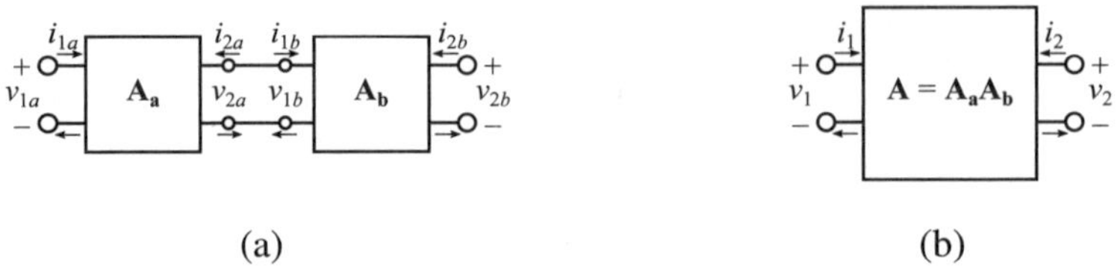

Figure 4.7 (a) Two networks combined in cascade. (b) Effective combined network.

$$= \begin{pmatrix} y_{11a} + y_{11b} & y_{12a} + y_{12b} & \cdots & y_{1Na} + y_{1Nb} \\ y_{21a} + y_{21b} & y_{22a} + y_{22b} & \cdots & y_{2Na} + y_{2Nb} \\ \vdots & \vdots & \ddots & \vdots \\ y_{N1a} + y_{N1b} & y_{N2a} + y_{N2b} & \cdots & y_{NNa} + y_{NNb} \end{pmatrix} \begin{pmatrix} v_1 \\ v_2 \\ \vdots \\ v_N \end{pmatrix} \tag{4.22c}$$

or, as shown in Figure 4.6(b),

$$\mathbf{Y} = \mathbf{Y_a} + \mathbf{Y_b} \tag{4.23}$$

4.1.3 ABCD-Parameters

Although the z- and y-parameters apply generally to networks with any number of ports, an important special case that covers many practical electromagnetic circuits is the two-port network. The parameters described in this section, $ABCD$-parameters, apply exclusively to two-ports, and were conceived specially for the purpose of analyzing cascades of individual subnetworks.

Consider the two networks shown in Figure 4.7(a). We note that at the connection point, the output voltage of the first network is equal to the input voltage of the second network, but the currents are reversed,

$$v_{2a} = v_{1b} \tag{4.24a}$$

$$i_{2a} = -i_{1b} \tag{4.24b}$$

Let us therefore define a matrix relating the input voltage and current to the output voltage and current for each of these subnetworks, but with a sign reversal of the output current,

$$\begin{pmatrix} v_{1a} \\ i_{1a} \end{pmatrix} = \begin{pmatrix} A_a & B_a \\ C_a & D_a \end{pmatrix} \begin{pmatrix} v_{2a} \\ -i_{2a} \end{pmatrix} \tag{4.25a}$$

$$\begin{pmatrix} v_{1b} \\ i_{1b} \end{pmatrix} = \begin{pmatrix} A_b & B_b \\ C_b & D_b \end{pmatrix} \begin{pmatrix} v_{2b} \\ -i_{2b} \end{pmatrix} \tag{4.25b}$$

We may now use (4.24) and (4.25) to express the input voltage and current to the output voltage and current of the combined network

$$\begin{pmatrix} v_{1a} \\ i_{1a} \end{pmatrix} = \begin{pmatrix} A_a & B_a \\ C_a & D_a \end{pmatrix} \begin{pmatrix} v_{2a} \\ -i_{2a} \end{pmatrix} = \begin{pmatrix} A_a & B_a \\ C_a & D_a \end{pmatrix} \begin{pmatrix} v_{1b} \\ i_{1b} \end{pmatrix} \tag{4.26a}$$

$$= \begin{pmatrix} A_a & B_a \\ C_a & D_a \end{pmatrix} \begin{pmatrix} A_b & B_b \\ C_b & D_b \end{pmatrix} \begin{pmatrix} v_{2b} \\ -i_{2b} \end{pmatrix} \tag{4.26b}$$

Thus, we can see that for the combined network, the $ABCD$-parameters multiply, or

$$\begin{pmatrix} A & B \\ C & D \end{pmatrix} = \begin{pmatrix} A_a & B_a \\ C_a & D_a \end{pmatrix} \begin{pmatrix} A_b & B_b \\ C_b & D_b \end{pmatrix} \tag{4.27}$$

as indicated in Figure 4.7(b). For this reason, $ABCD$-parameters are also sometimes called *chain parameters*. Note that matrix multiplication is not generally commutative ($\mathbf{A_a A_b} \neq \mathbf{A_b A_a}$). This is consistent with expectations, as the full port behavior of cascaded networks depends on the order of the cascade.

The somewhat awkward nature of the input and output vectors of the $ABCD$-matrix representation make conversions with z- and y-parameters relatively messy. Take the z-parameters, for instance. We must first write out the voltage and current relationships in long form,

$$v_1 = z_{11} i_1 + z_{12} i_2 \tag{4.28a}$$

$$v_2 = z_{21} i_1 + z_{22} i_2 \tag{4.28b}$$

Next, we must rearrange these equations algebraically so that they conform to the $ABCD$-matrix configuration. First, solve (4.28b) for i_1,

$$i_1 = \frac{1}{z_{21}} v_2 - \frac{z_{22}}{z_{21}} i_2 \tag{4.29}$$

Then substitute (4.29) into (4.28a),

$$v_1 = z_{11} \left(\frac{1}{z_{21}} v_2 - \frac{z_{22}}{z_{21}} i_2 \right) + z_{12} i_2 = \frac{z_{11}}{z_{21}} v_2 - \left(\frac{z_{11} z_{22} - z_{12} z_{21}}{z_{21}} \right) i_2 \tag{4.30a}$$

$$= \frac{z_{11}}{z_{21}} v_2 - \frac{\det(\mathbf{Z})}{z_{21}} i_2 \tag{4.30b}$$

where $\det(\mathbf{Z})$ is the determinant of the impedance matrix. Thus, we have

$$\begin{pmatrix} v_1 \\ i_1 \end{pmatrix} = \frac{1}{z_{21}} \begin{pmatrix} z_{11} & \det(\mathbf{Z}) \\ 1 & z_{22} \end{pmatrix} \begin{pmatrix} v_2 \\ -i_2 \end{pmatrix} \tag{4.31a}$$

$$\therefore \begin{pmatrix} A & B \\ C & D \end{pmatrix} = \frac{1}{z_{21}} \begin{pmatrix} z_{11} & \det(\mathbf{Z}) \\ 1 & z_{22} \end{pmatrix} \tag{4.31b}$$

The z-, y-, and $ABCD$-parameters are of such great value in calculating the response of complex networks that it is worth tabulating at this time some basic circuits for which the network parameters are especially simple. Such a listing is shown in Table 4.1. Typically, one would select whichever set of parameters will make the calculation easiest by breaking the full network down into series, parallel, and cascaded parts, then converting to the final desired set of network parameters at the end. (Often, the final target will be wave-amplitude network parameters, such as the scattering parameters, which will be discussed in Section 4.2.1.)

4.1.4 Hybrid and Inverse-Hybrid Parameters

A lesser-known set of two-port network parameters are the *hybrid* or *h-parameters* [2], defined by the equation,

$$\begin{pmatrix} v_1 \\ i_2 \end{pmatrix} = \begin{pmatrix} h_{11} & h_{12} \\ h_{21} & h_{22} \end{pmatrix} \begin{pmatrix} i_1 \\ v_2 \end{pmatrix} \tag{4.32}$$

As such, the diagonal elements, h_{11} and h_{22}, represent the input driving-point impedance and output admittance of the network, respectively (with the output terminated in a short, and the input with an open circuit), while the off-diagonal elements, h_{21} and h_{12}, represent the forward current gain and reverse voltage gain, respectively. For these characteristics, the hybrid parameters have found some applications in transistor modeling [3], but are otherwise fairly obscure. Their unique property as a general circuit analysis tool is that the hybrid matrices add when two networks are combined in series at the input and in parallel at the output, as shown in Figure 4.8. They are sometimes called *series-parallel parameters* for this reason.

The inverse of the hybrid parameter matrix contains the g- or *parallel-series parameters*,

$$\begin{pmatrix} i_1 \\ v_2 \end{pmatrix} = \begin{pmatrix} g_{11} & g_{12} \\ g_{21} & g_{22} \end{pmatrix} \begin{pmatrix} v_1 \\ i_2 \end{pmatrix} \tag{4.33}$$

Table 4.1

Some Basic Two-Port Circuits and Their Network Parameters

Circuit	Network Parameters
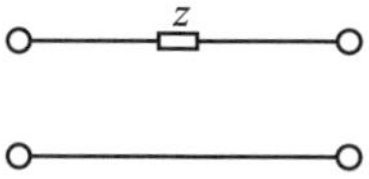	$ABCD = \begin{pmatrix} 1 & z \\ 0 & 1 \end{pmatrix}$
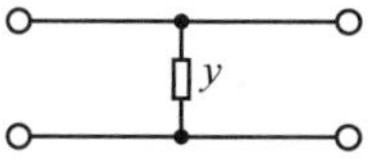	$ABCD = \begin{pmatrix} 1 & 0 \\ y & 1 \end{pmatrix}$
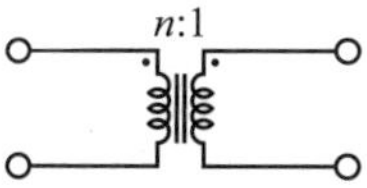	$ABCD = \begin{pmatrix} n & 0 \\ 0 & \frac{1}{n} \end{pmatrix}$
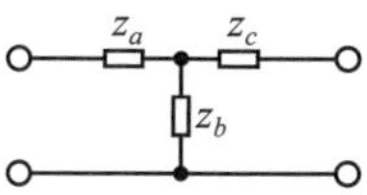	$\mathbf{Z} = \begin{pmatrix} z_a + z_b & z_b \\ z_b & z_b + z_c \end{pmatrix}$
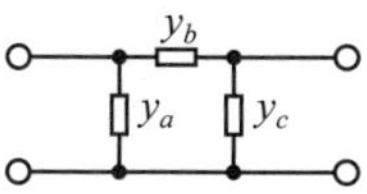	$\mathbf{Y} = \begin{pmatrix} y_a + y_b & -y_b \\ -y_b & y_b + y_c \end{pmatrix}$
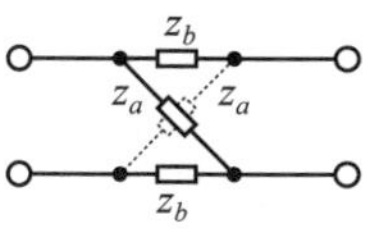	$\mathbf{Z} = \frac{1}{2} \begin{pmatrix} z_a + z_b & z_a - z_b \\ z_a - z_b & z_a + z_b \end{pmatrix}$
	$\mathbf{Z} = Z_0 \begin{pmatrix} \coth(\gamma l) & \operatorname{csch}(\gamma l) \\ \operatorname{csch}(\gamma l) & \coth(\gamma l) \end{pmatrix}$
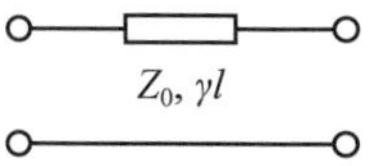	$\mathbf{Y} = Y_0 \begin{pmatrix} \coth(\gamma l) & -\operatorname{csch}(\gamma l) \\ -\operatorname{csch}(\gamma l) & \coth(\gamma l) \end{pmatrix}$
	$ABCD = \begin{pmatrix} \cosh(\gamma l) & Z_0 \sinh(\gamma l) \\ Y_0 \sinh(\gamma l) & \cosh(\gamma l) \end{pmatrix}$

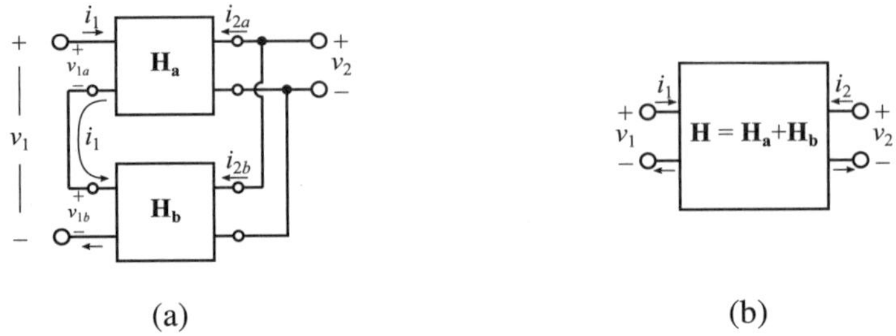

Figure 4.8 (a) Two networks combined in series at the input and in parallel at the output. (b) Effective combined network.

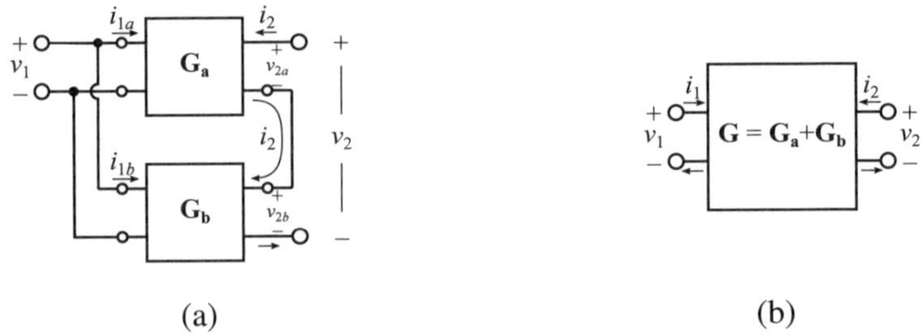

Figure 4.9 (a) Two networks combined in parallel at the input and in series at the output. (b) Effective combined network.

where

$$\begin{pmatrix} g_{11} & g_{12} \\ g_{21} & g_{22} \end{pmatrix} = \mathbf{G} = \mathbf{H}^{-1} \tag{4.34}$$

In this case, the diagonal elements contain the input admittance and output impedance, while the off-diagonal elements describe the forward voltage gain and reverse current gain. These parameters add when two networks are combined in parallel at the input and series at the output, as in Figure 4.9.

Table 4.2 shows all the conversions between the various immittance-based network parameters described in this section.

Table 4.2

Immittance Network Parameter Conversions

Z	Y	$ABCD$	H	G
$\mathbf{Z}$	$= \mathbf{Y}^{-1}$	$= \frac{1}{C}\begin{bmatrix} A & \Delta \\ 1 & D \end{bmatrix}$	$= \frac{1}{h_{22}}\begin{bmatrix} \Delta & h_{12} \\ -h_{21} & 1 \end{bmatrix}$	$= \frac{1}{g_{11}}\begin{bmatrix} 1 & -g_{12} \\ g_{21} & \Delta \end{bmatrix}$
$\mathbf{Z}^{-1}$	$= \mathbf{Y}$	$= \frac{1}{B}\begin{bmatrix} D & -\Delta \\ -1 & A \end{bmatrix}$	$= \frac{1}{h_{11}}\begin{bmatrix} 1 & -h_{12} \\ h_{21} & \Delta \end{bmatrix}$	$= \frac{1}{g_{22}}\begin{bmatrix} \Delta & g_{12} \\ -g_{21} & 1 \end{bmatrix}$
$\frac{1}{z_{21}}\begin{bmatrix} z_{11} & \Delta \\ 1 & z_{22} \end{bmatrix}$	$= \frac{-1}{y_{21}}\begin{bmatrix} y_{22} & 1 \\ \Delta & y_{11} \end{bmatrix}$	$= \begin{bmatrix} A & B \\ C & D \end{bmatrix}$	$= \frac{-1}{h_{21}}\begin{bmatrix} \Delta & h_{11} \\ h_{22} & 1 \end{bmatrix}$	$= \frac{1}{g_{21}}\begin{bmatrix} 1 & g_{22} \\ g_{11} & \Delta \end{bmatrix}$
$\frac{1}{z_{22}}\begin{bmatrix} \Delta & z_{12} \\ -z_{21} & 1 \end{bmatrix}$	$= \frac{1}{y_{11}}\begin{bmatrix} 1 & -y_{12} \\ y_{21} & \Delta \end{bmatrix}$	$= \frac{1}{D}\begin{bmatrix} B & \Delta \\ -1 & C \end{bmatrix}$	$= \mathbf{H}$	$= \mathbf{G}^{-1}$
$\frac{1}{z_{11}}\begin{bmatrix} 1 & -z_{12} \\ z_{21} & \Delta \end{bmatrix}$	$= \frac{1}{y_{22}}\begin{bmatrix} \Delta & y_{12} \\ -y_{21} & 1 \end{bmatrix}$	$= \frac{1}{A}\begin{bmatrix} C & -\Delta \\ 1 & B \end{bmatrix}$	$= \mathbf{H}^{-1}$	$= \mathbf{G}$

$\Delta =$ the determinant of the given matrix.

4.2 WAVE NETWORK PARAMETERS

Thus far, all of the network parameters that we have discussed pertain to immittance and trans-immittance related quantities: terminal voltages, currents, and their ratios. As we have seen, however, these same quantities may arise at the ports of a network as a consequence of waves incident upon it from external transmission lines, such as coaxial cables or traces on a printed circuit. It will be most convenient to define the network behavior in terms of its response to these waves entering and leaving the ports. In contrast to the foregoing immittance-based network parameters, we shall find that these wave-based network parameters are applicable even in situations where the ports are not electrically small (e.g., waveguides, which will be discussed in Chapter 7).

For now, however, let us relate the wave amplitudes to the known terminal quantities, so that the conversion between wave-based and immittance-based parameters for lumped-element and transmission-line networks will be readily apparent. Let us assume that the network we wish to analyze is embedded in a larger system connected by transmission lines having a standard characteristic impedance, Z_0 (often 50Ω, as was discussed in Section 3.3.6). In general, there will be forward and backward-traveling waves on the input and output transmission lines, as indicated

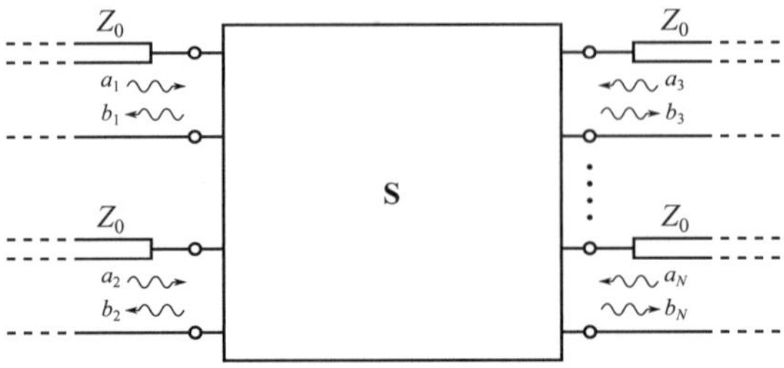

Figure 4.10 An N-port network embedded in a system with characteristic impedance Z_0, and with incident waves, a_k, and outgoing waves, b_k, at its ports. The network is fully described by its scattering-parameter matrix, $\mathbf{S}$.

in Figure 4.10. Denote the phasor amplitude of the wave entering the network at port k as a_k, and that of the wave leaving port k as b_k. The square of the wave amplitude will have units of power, so if the wave propagates along a transmission line with electrically small cross-section (as is the case here), the voltage and current amplitude will be related to the overall wave amplitude by a factor $\sqrt{Z_0}$. More specifically, the voltage and current phasors at port k due to the incident wave are

$$v_{ak} = a_k \sqrt{Z_0} \tag{4.35a}$$

$$i_{ak} = \frac{a_k}{\sqrt{Z_0}} \tag{4.35b}$$

and those due to the outgoing waves are

$$v_{bk} = b_k \sqrt{Z_0} \tag{4.36a}$$

$$i_{bk} = -\frac{b_k}{\sqrt{Z_0}} \tag{4.36b}$$

Note the negative sign in (4.36b) since the convention for current is that it is oriented in the direction of propagation, in this case away from the network, while the passive sign convention for network ports is that the current is directed into the positive terminal.

Superposition requires that the total voltage and current at any port is the sum of the voltages and currents at that port due to all present waves, both incident and outgoing, so

$$v_k = v_{ak} + v_{bk} = (a_k + b_k) \sqrt{Z_0} \tag{4.37a}$$

$$i_k = i_{ak} + i_{bk} = \frac{(a_k - b_k)}{\sqrt{Z_0}} \tag{4.37b}$$

or, in terms of the N-port wave-amplitude vectors,

$$\mathbf{v} = (\mathbf{a} + \mathbf{b})\sqrt{Z_0} \tag{4.38a}$$

$$\mathbf{i} = \frac{(\mathbf{a} - \mathbf{b})}{\sqrt{Z_0}} \tag{4.38b}$$

4.2.1 Scattering Parameters

The network will respond to the incident waves, a_k, by reflecting some of the energy back out as an outgoing wave, b_k. It also will transmit some energy into outgoing waves at other ports (the remainder being absorbed). This coupling of incident to outgoing waves is generally called *scattering*, and the parameters that describe this coupling are called *scattering parameters* or *s-parameters*,

$$\begin{pmatrix} b_1 \\ b_2 \\ \vdots \\ b_N \end{pmatrix} = \begin{pmatrix} s_{11} & s_{12} & \cdots & s_{1N} \\ s_{21} & s_{22} & \cdots & s_{2N} \\ \vdots & \vdots & \ddots & \vdots \\ s_{N1} & s_{N2} & \cdots & s_{NN} \end{pmatrix} \begin{pmatrix} a_1 \\ a_2 \\ \vdots \\ a_N \end{pmatrix} \tag{4.39}$$

or

$$\mathbf{b} = \mathbf{S}\mathbf{a} \tag{4.40}$$

The diagonal elements of the scattering matrix represent the reflection coefficients at each port, while the off-diagonal elements represent the transmission coefficient between ports.

The scattering parameters are of such value to high-frequency electronics that they will be the primary workhorse for network characterization throughout most of this book. The immittance parameters of Section 4.1 will take on a more subsidiary role as an intermediate tool for calculating the responses of certain networks, prior to the final conversion to s-parameters.

Conversion to and from the z- and y-parameters follows from the relationships in (4.38). Consider the z-parameters first,

$$\mathbf{v} = \mathbf{Z}\mathbf{i} \tag{4.41a}$$

$$(\mathbf{a} + \mathbf{b})\sqrt{Z_0} = \mathbf{Z}(\mathbf{a} - \mathbf{b})\frac{1}{\sqrt{Z_0}} \tag{4.41b}$$

$$(\mathbf{a} + \mathbf{Sa})\, Z_0 = \mathbf{Z}\,(\mathbf{a} - \mathbf{Sa}) \tag{4.41c}$$

$$Z_0\,(\mathbf{I} + \mathbf{S})\,\mathbf{a} = \mathbf{Z}\,(\mathbf{I} - \mathbf{S})\,\mathbf{a} \tag{4.41d}$$

where $\mathbf{I}$ is the identity matrix. Since the above must hold for all input vectors, $\mathbf{a}$, we may then write

$$Z_0\,(\mathbf{I} + \mathbf{S}) = \mathbf{Z}\,(\mathbf{I} - \mathbf{S}) \tag{4.42}$$

From this, it follows that

$$\mathbf{Z} = Z_0\,(\mathbf{I} + \mathbf{S})\,(\mathbf{I} - \mathbf{S})^{-1} \tag{4.43}$$

It is interesting to note that the terms in parentheses in (4.43) commute, unlike many pairs of square matrices. To prove this, let $\mathbf{A} = \mathbf{I} + \mathbf{S}$, and let $\mathbf{B} = \mathbf{I} - \mathbf{S}$, and consider their products in both orders,

$$\mathbf{AB} = (\mathbf{I} + \mathbf{S})\,(\mathbf{I} - \mathbf{S}) = (\mathbf{I} - \mathbf{S}) + \mathbf{S}\,(\mathbf{I} - \mathbf{S}) = \mathbf{I} - \mathbf{S} + \mathbf{S} - \mathbf{S}^2 = \mathbf{I} - \mathbf{S}^2 \tag{4.44a}$$

$$\mathbf{BA} = (\mathbf{I} - \mathbf{S})\,(\mathbf{I} + \mathbf{S}) = (\mathbf{I} + \mathbf{S}) - \mathbf{S}\,(\mathbf{I} + \mathbf{S}) = \mathbf{I} + \mathbf{S} - \mathbf{S} - \mathbf{S}^2 = \mathbf{I} - \mathbf{S}^2 \tag{4.44b}$$

$$\therefore \mathbf{AB} = \mathbf{BA} \tag{4.44c}$$

Thus, $\mathbf{A}$ and $\mathbf{B}$ commute. Further, if $\mathbf{A}$ and $\mathbf{B}$ commute, then $\mathbf{A}$ and $\mathbf{B}^{-1}$ must also commute, because

$$\mathbf{AB}^{-1} = \mathbf{B}^{-1}\mathbf{BAB}^{-1} = \mathbf{B}^{-1}\mathbf{ABB}^{-1} = \mathbf{B}^{-1}\mathbf{A} \tag{4.45}$$

This allows us to write the conversion formula from (4.43) with the product in either order,

$$\mathbf{Z} = Z_0\,(\mathbf{I} + \mathbf{S})\,(\mathbf{I} - \mathbf{S})^{-1} \tag{4.46a}$$

$$= Z_0\,(\mathbf{I} - \mathbf{S})^{-1}\,(\mathbf{I} + \mathbf{S}) \tag{4.46b}$$

Further, solving for $\mathbf{S}$,

$$\mathbf{S} = (\mathbf{Z} + Z_0\mathbf{I})^{-1}\,(\mathbf{Z} - Z_0\mathbf{I}) \tag{4.47a}$$

$$= (\mathbf{Z} - Z_0\mathbf{I})\,(\mathbf{Z} + Z_0\mathbf{I})^{-1} \tag{4.47b}$$

and, since $\mathbf{Y} = \mathbf{Z}^{-1}$,

$$\mathbf{Y} = Y_0\,(\mathbf{I} - \mathbf{S})\,(\mathbf{I} + \mathbf{S})^{-1} \tag{4.48a}$$

$$= Y_0\,(\mathbf{I} + \mathbf{S})^{-1}\,(\mathbf{I} - \mathbf{S}) \tag{4.48b}$$

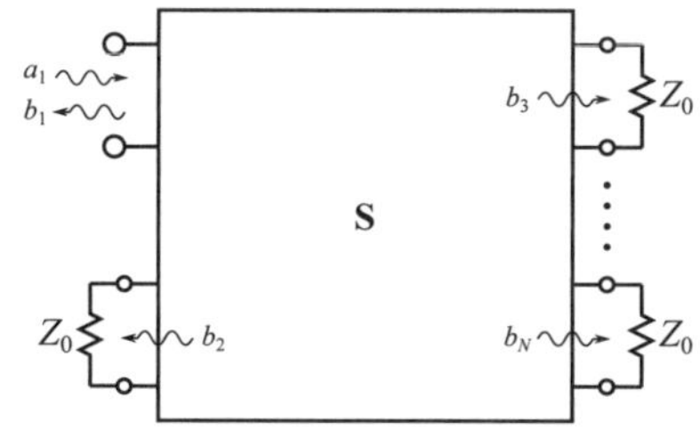

Figure 4.11 An N-port network with a wave source at port 1, and all other ports terminated with matched loads.

$$\mathbf{S} = (Y_0\mathbf{I} + \mathbf{Y})^{-1} (Y_0\mathbf{I} - \mathbf{Y}) \tag{4.48c}$$

$$= (Y_0\mathbf{I} - \mathbf{Y}) (Y_0\mathbf{I} + \mathbf{Y})^{-1} \tag{4.48d}$$

Unique among the network parameters described thus far, the scattering parameters are themselves dependent upon an assumed embedding system impedance,[3] Z_0. If, for example, you take a physical device that is normally connected to 50Ω cables, but remove those cables and replace them with 75Ω cables, the scattering of waves off of that device will change, because the relationship between voltage and current amplitudes within those waves will be different. The process of converting the s-parameters calculated for one characteristic impedance to those for another characteristic impedance is called *renormalization* and will be discussed in Section 4.6.2.

Whereas the z-parameters are found by considering a voltage source attached to each port with the remaining ports left open, and the y-parameters are found by considering current sources with short-circuit terminations (and the h- and g-parameters by an appropriate combination), the s-parameters are found by considering an incident wave, or *wave source* at one port, while all the remaining ports are terminated with matched loads, as in Figure 4.11. Recall that a matched load termination simulates the effect of an infinitely long transmission line, and guarantees that no waves will be incident upon those ports.

Let us consider a simple example, consisting of a two-port network with a series resistor, Z_0, followed by a shunt resistor, $2Z_0$, as in Figure 4.12(a). With a test wave incident at port 1, we have a matched termination resistor on port 2

3 For later wave-guiding structures that do not have a definable characteristic impedance, the scattering parameters will still assume and depend upon a particular input port geometry and propagating mode.

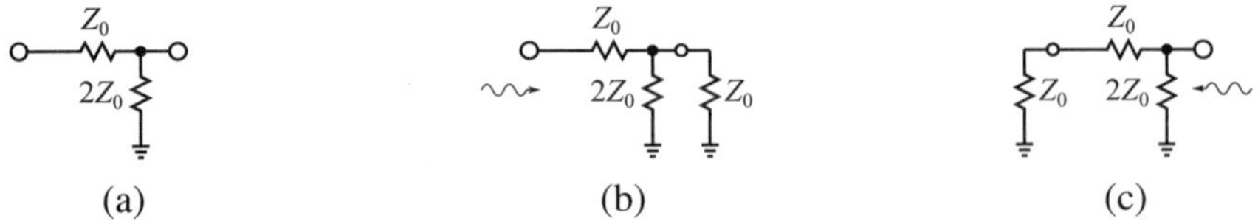

Figure 4.12 (a) A resistor network. (b) Wave source at port 1, with port 2 terminated. (c) Wave source at port 2, with port 1 terminated.

(Figure 4.12(b)). The total impedance seen from port 1 is found by considering simple series and parallel combinations of the resistors,

$$Z_{in,1} = Z_0 + \left((2Z_0)^{-1} + Z_0^{-1}\right)^{-1} = \tfrac{5}{3} Z_0 \tag{4.49}$$

The first s-parameter, s_{11}, is the ratio of the outgoing wave amplitude at port 1 to the incident wave amplitude. This in turn is equal to the ratio of the backward-traveling voltage-wave amplitude to the forward-traveling voltage-wave amplitude, which we recognize from (3.37b) as the reflection coefficient for this load impedance,

$$s_{11} = \frac{b_1}{a_1} = \frac{v^-}{v^+} = \Gamma = \frac{Z_{in,1} - Z_0}{Z_{in,1} + Z_0} = \frac{\frac{5}{2} Z_0 - Z_0}{\frac{5}{3} Z_0 + Z_0} = \frac{1}{4} \tag{4.50}$$

What about s_{21}, the coefficient of transmission from port 1 to port 2? First, from (4.37a) we know that the total voltage at port 1 is proportional to the sum of the incident and outgoing (reflected) wave amplitudes,

$$v_1 = (a_1 + b_1) \sqrt{Z_0} = a_1 (1 + s_{11}) \sqrt{Z_0} = \tfrac{5}{4} a_1 \sqrt{Z_0} \tag{4.51}$$

The total voltage at port 2 is determined by the voltage divider formed from the first series resistor and the parallel combination of the shunt resistor with the termination,

$$v_2 = \frac{\frac{2}{3} Z_0}{\frac{2}{3} Z_0 + Z_0} v_1 = \tfrac{2}{5} v_1 = \tfrac{2}{5} \cdot \tfrac{5}{4} a_1 \sqrt{Z_0} = \tfrac{1}{2} a_1 \sqrt{Z_0} \tag{4.52}$$

At port 2, there is no incident wave, so the port 2 voltage is just the amplitude of the outgoing wave at that port,

$$v_2 = b_2 \sqrt{Z_0} = \tfrac{1}{2} a_1 \sqrt{Z_0} \tag{4.53}$$

Finally, the transmission coefficient is just the ratio of b_2 to a_1,

$$s_{21} = \frac{b_2}{a_1} = \frac{1}{2} \tag{4.54}$$

This completes the first column of the 2-by-2 matrix. The second column is similarly found by stimulating the network with a wave at port 2, while port 1 is terminated, as in Figure 4.12(c). In this case, the input impedance is

$$Z_{in,2} = \left((2Z_0)^{-1} + (Z_0 + Z_0)^{-1}\right)^{-1} = Z_0 \tag{4.55}$$

which is a matched termination ($\Gamma = 0$). Therefore, the output reflection coefficient, s_{22}, is zero, and the voltage divider in this case directly divides the wave amplitude by the same amount,

$$s_{12} = \frac{b_1}{a_2} = \frac{v_1}{v_2} = \frac{1}{2} \tag{4.56}$$

Gathering our results, the final scattering matrix for this resistor network is

$$\mathbf{S} = \begin{pmatrix} \frac{1}{4} & \frac{1}{2} \\ \frac{1}{2} & 0 \end{pmatrix} \tag{4.57}$$

Let us consider another example, this time a simple transmission line, but not one matched to the characteristic impedance of the system. We could as before postulate the existence of forward-traveling and backward-traveling waves on the line, and solve the system of equations to determine the relationship between the various wave amplitudes. Since we already have the z-parameters for this network, however, let us simply apply the conversion rule from (4.47). Assuming the characteristic impedance of the line is Z_1 (we will reserve Z_0 for the port impedance, or the characteristic impedance of the system in which the transmission line is embedded), then,

$$\mathbf{S} = (\mathbf{Z} + Z_0\mathbf{I})^{-1} (\mathbf{Z} - Z_0\mathbf{I}) \tag{4.58a}$$

$$= \begin{pmatrix} Z_1 \coth \gamma l + Z_0 & Z_1 \operatorname{csch} \gamma l \\ Z_1 \operatorname{csch} \gamma l & Z_1 \coth \gamma l + Z_0 \end{pmatrix}^{-1} \cdot \begin{pmatrix} Z_1 \coth \gamma l - Z_0 & Z_1 \operatorname{csch} \gamma l \\ Z_1 \operatorname{csch} \gamma l & Z_1 \coth \gamma l - Z_0 \end{pmatrix} \tag{4.58b}$$

$$= \frac{\begin{pmatrix} z^2\cosh^2\gamma l - \sinh^2\gamma l - z^2 & 2z\sinh\gamma l \\ 2z\sinh\gamma l & z^2\cosh^2\gamma l - \sinh^2\gamma l - z^2 \end{pmatrix}}{2z\cosh\gamma l\sinh\gamma l + (z^2+1)\sinh^2\gamma l} \tag{4.58c}$$

$$= \frac{\begin{pmatrix} (z-y)\sinh\gamma l & 2 \\ 2 & (z-y)\sinh\gamma l \end{pmatrix}}{2\cosh\gamma l + (z+y)\sinh\gamma l} \tag{4.58d}$$

where $z = y^{-1} = Z_1/Z_0$.

In the special case where the line is lossless ($\gamma = j\beta$) and matched ($Z_1 = Z_0$),

$$\mathbf{S} = \frac{1}{\cos(\beta l) + j\sin(\beta l)}\begin{pmatrix} 0 & 1 \\ 1 & 0 \end{pmatrix} = \begin{pmatrix} 0 & e^{-j\beta l} \\ e^{-j\beta l} & 0 \end{pmatrix} \tag{4.59}$$

Thus, a matched transmission line introduces no reflections, but rather only delays the signals passing through it in either direction by a phase term, βl.

4.2.2 Scattering Transfer Parameters

For completeness, let us now briefly discuss a final set of network parameters closely related to the scattering parameters, but which perform a function for wave-based descriptions that is similar to what $ABCD$-matrices do for immittance. These are the *scattering transfer parameters*, *chain scattering parameters*, or *T-parameters* defined by the following equation [2],

$$\begin{pmatrix} a_1 \\ b_1 \end{pmatrix} = \begin{pmatrix} T_{11} & T_{12} \\ T_{21} & T_{22} \end{pmatrix}\begin{pmatrix} b_2 \\ a_2 \end{pmatrix} \tag{4.60}$$

Note that, like the $ABCD$-parameters, the definition for T-parameters applies exclusively to two-port networks.

The utility of this network description is illustrated in Figure 4.13(a), where two networks have been combined in cascade. As the figure makes clear, the outgoing wave from the first network at port 2 is equal to the incident wave to the second network at port 1 ($b_{2a} = a_{1b}$) and the incident wave to the first network at port 2 is equal to the outgoing wave from the second network at port 1 ($a_{2a} = b_{1b}$). Thus, we may write for the combined network

$$\begin{pmatrix} a_{1a} \\ b_{1a} \end{pmatrix} = \begin{pmatrix} T_{11a} & T_{12a} \\ T_{21a} & T_{22a} \end{pmatrix}\begin{pmatrix} b_{2a} \\ a_{2a} \end{pmatrix} = \begin{pmatrix} T_{11a} & T_{12a} \\ T_{21a} & T_{22a} \end{pmatrix}\begin{pmatrix} a_{1b} \\ b_{1b} \end{pmatrix} \tag{4.61a}$$

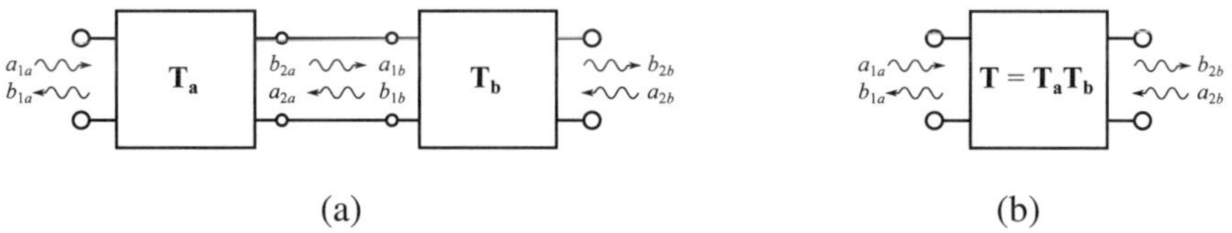

Figure 4.13 (a) Two networks combined in cascade. (b) Effective combined network.

$$= \begin{pmatrix} T_{11a} & T_{12a} \\ T_{21a} & T_{22a} \end{pmatrix} \begin{pmatrix} T_{11b} & T_{12b} \\ T_{21b} & T_{22b} \end{pmatrix} \begin{pmatrix} b_{2b} \\ a_{2b} \end{pmatrix} \tag{4.61b}$$

In other words, the T-matrix of the cascade is equal to the product of the individual T-matrices of the subnetworks,

$$\mathbf{T} = \mathbf{T_a}\mathbf{T_b} \tag{4.62}$$

A cautionary note is warranted that some references define the T-parameters with the incident and outgoing waves in the opposite order, that is,

$$\begin{pmatrix} b_1 \\ a_1 \end{pmatrix} = \begin{pmatrix} T'_{11} & T'_{12} \\ T'_{21} & T'_{22} \end{pmatrix} \begin{pmatrix} a_2 \\ b_2 \end{pmatrix} \tag{4.63}$$

The elements of the matrix are the same as those in (4.60), but exchanged across diagonals (e.g., $T'_{11} = T_{22}$, $T'_{12} = T_{21}$). The cascade-multiplication property in (4.62) remains unchanged. For all cases in this book, the convention of (4.60) will be used.

Conversion equations between T-parameters and scattering parameters, as well as between the scattering parameters and the most important immittance parameters, are summarized in Table 4.3.

4.3 PROPERTIES OF NETWORKS

In addition to the special properties that some network parameters have (e.g., the z-parameters add when networks are combined in series), there are certain properties that are intrinsic to the network itself. In this section, we will explore some of those properties and explain how to identify them in the various network-parameter representations.

Table 4.3
Scattering Parameter Conversions

Conversions to S-parameters	Conversions from S-parameters
$\mathbf{S} = (\mathbf{Z} + Z_0\mathbf{I})^{-1}(\mathbf{Z} - Z_0\mathbf{I}) = (\mathbf{Z} - Z_0\mathbf{I})(\mathbf{Z} + Z_0\mathbf{I})^{-1}$	$\mathbf{Z} = Z_0(\mathbf{I} + \mathbf{S})(\mathbf{I} - \mathbf{S})^{-1} = Z_0(\mathbf{I} - \mathbf{S})^{-1}(\mathbf{I} + \mathbf{S})$
$\mathbf{S} = (Y_0\mathbf{I} + \mathbf{Y})^{-1}(Y_0\mathbf{I} - \mathbf{Y}) = (Y_0\mathbf{I} - \mathbf{Y})(Y_0\mathbf{I} + \mathbf{Y})^{-1}$	$\mathbf{Y} = Y_0(\mathbf{I} - \mathbf{S})(\mathbf{I} + \mathbf{S})^{-1} = Y_0(\mathbf{I} + \mathbf{S})^{-1}(\mathbf{I} - \mathbf{S})$
$\mathbf{S} = \frac{1}{T_{22}}\begin{pmatrix} T_{12} & \det\mathbf{T} \\ 1 & -T_{21}\end{pmatrix}$	$\mathbf{T} = \frac{1}{s_{21}}\begin{pmatrix} -\det\mathbf{S} & s_{11} \\ -s_{22} & 1\end{pmatrix}$
$\mathbf{S} = \begin{pmatrix} \frac{A+BY_0-CZ_0-D}{A+BY_0+CZ_0+D} & \frac{2(AD-BC)}{A+BY_0+CZ_0+D} \\ \frac{2}{A+BY_0+CZ_0+D} & \frac{-A+BY_0-CZ_0+D}{A+BY_0+CZ_0+D}\end{pmatrix}$	$\begin{pmatrix} A & B \\ C & D\end{pmatrix} = \begin{pmatrix} \frac{(1+s_{11})(1-s_{22})+s_{12}s_{21}}{2s_{21}} & \frac{(1+s_{11})(1+s_{22})-s_{12}s_{21}}{2s_{21}Y_0} \\ \frac{(1-s_{11})(1-s_{22})-s_{12}s_{21}}{2s_{21}Z_0} & \frac{(1-s_{11})(1+s_{22})+s_{12}s_{21}}{2s_{21}}\end{pmatrix}$

4.3.1 Reciprocity

Reciprocity is one such property that we have discussed twice already — first in Section 1.7 as it pertains generally to electromagnetic fields, and then again in Section 2.7.1 as it pertains to terminal quantities. To review in loose terms, these theorems stated that certain quantities, when regarded as inputs and outputs, could be exchanged in a fixed system without altering the transfer function between them. However, in both of those contexts, the inputs and outputs being exchanged were of dissimilar types, such as $\mathbf{E}$ for $\mathbf{J}$, and v for i. We shall now see that the transimmittance reciprocity proven for a certain class of networks in Section 2.7.1 also implies wave-scattering reciprocity. In other words, if a network is *reciprocal*, then not only is a voltmeter (or ammeter) measurement unaltered by exchanging its port position with a current source (or voltage source), but the transmission coefficient of waves from port k to port i is also equivalent to the transmission coefficient of waves from port i to port k.

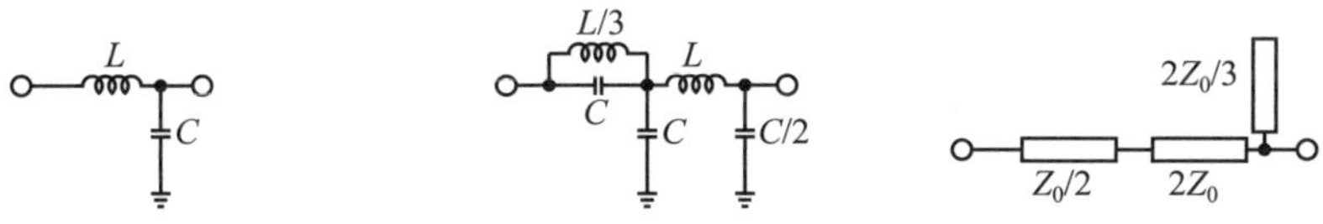

(a) (b) (c)

Figure 4.14 (a) A network which is reciprocal but not symmetric. (b, c) Networks that are electrically symmetric but not physically symmetric. All transmission lines are assumed to be equal in length.

Transimpedance reciprocity implies that the impedance matrix is symmetric, ($z_{ik} = z_{ki}$ or $\mathbf{Z} = \mathbf{Z}^T$). Converting this to scattering parameters, we find

$$\mathbf{S}^T = \left((\mathbf{Z} + Z_0\mathbf{I})^{-1} (\mathbf{Z} - Z_0\mathbf{I}) \right)^T = (\mathbf{Z} - Z_0\mathbf{I})^T (\mathbf{Z} + Z_0\mathbf{I})^{-T} \quad (4.64a)$$

$$= \left(\mathbf{Z}^T - Z_0\mathbf{I}\right) \left(\mathbf{Z}^T + Z_0\mathbf{I}\right)^{-1} = (\mathbf{Z} - Z_0\mathbf{I}) (\mathbf{Z} + Z_0\mathbf{I})^{-1} = \mathbf{S} \quad (4.64b)$$

Therefore, $s_{ik} = s_{ki}$. This proves that transimpedance reciprocity is also wave-scattering reciprocity.

For two-port networks, we can see from Tables 4.2 and 4.3 that, for the $ABCD$-matrix, reciprocity further implies $\det(\mathbf{A}) = AD - BC = 1$.

4.3.2 Symmetry

We shall define electrical symmetry as the property that a two-port network may have, wherein the driving-point characteristics are the same from either port, given equivalent terminations. Although z_{11} and z_{22} are not the input impedances themselves (unless the loads are open circuits), their equality is sufficient to guarantee that the input impedances will be the same given equivalent terminations, even if the network is not reciprocal. So symmetry is established by $z_{11} = z_{22}$, $y_{11} = y_{22}$, or $s_{11} = s_{22}$, all of which are equivalent conditions. Additionally, Table 4.2 illustrates that electrical symmetry is apparent from the $ABCD$-matrix if $A = D$.

Note that reciprocity itself does not typically imply symmetry, nor does electrical symmetry always imply physical symmetry. See Figure 4.14 for some examples. The z-parameters for the circuit in Figure 4.14(a) are easily found using Table 4.1, specifically using the tee-network configuration with $z_a = j\omega L$,

$z_b = (j\omega C)^{-1}$, and $z_c = 0$,

$$\mathbf{Z} = \begin{pmatrix} j\omega L + (j\omega C)^{-1} & (j\omega C)^{-1} \\ (j\omega C)^{-1} & (j\omega C)^{-1} \end{pmatrix} \tag{4.65}$$

Obviously, the network is not symmetric, physically or electrically, but since $z_{12} = z_{21}$, this network is reciprocal (Figure 4.12(a) is another example).

Let us look at the network in Figure 4.14(b). We may calculate its $ABCD$-parameters by breaking it down into four sections (a series impedance, a shunt admittance, another series impedance, and another shunt admittance) and then multiplying,

$$\mathbf{A} = \begin{pmatrix} 1 & \frac{1}{j\omega C + \left(j\frac{1}{3}\omega L\right)^{-1}} \\ 0 & 1 \end{pmatrix} \begin{pmatrix} 1 & 0 \\ j\omega C & 1 \end{pmatrix} \begin{pmatrix} 1 & j\omega L \\ 0 & 1 \end{pmatrix} \begin{pmatrix} 1 & 0 \\ j\frac{1}{2}\omega C & 1 \end{pmatrix} \tag{4.66a}$$

$$= \begin{pmatrix} 1 - \omega^2 LC & j\omega L\left(\frac{4-2\omega^2 LC}{3-\omega^2 LC}\right) \\ j\frac{1}{2}\omega C\left(3 - \omega^2 LC\right) & 1 - \omega^2 LC \end{pmatrix} \tag{4.66b}$$

Since $A = D$, this network is electrically symmetric. Further,

$$AD - BC = \left(1 - \omega^2 LC\right)^2 - j\omega L\left(\frac{4-2\omega^2 LC}{3-\omega^2 LC}\right) j\frac{1}{2}\omega C\left(3 - \omega^2 LC\right) \tag{4.67a}$$

$$= \left(1 - \omega^2 LC\right)^2 + \omega^2 LC\left(2 - \omega^2 LC\right) = 1 \tag{4.67b}$$

proving that this network is also reciprocal. In other words, despite having an asymmetrical construction, this network behaves the same forwards as backwards in all respects.

Finally, we consider the network in Figure 4.14(c). (We shall adopt a convention, here and elsewhere in this book, that unless otherwise noted, all transmission lines in a network are of equal electrical length, γl. These are known as *commensurate-line networks*.) Once again, we may calculate the $ABCD$-parameters by multiplication,

$$\mathbf{A} = \begin{pmatrix} \cosh\gamma l & \frac{1}{2}Z_0\sinh\gamma l \\ 2Y_0\sinh\gamma l & \cosh\gamma l \end{pmatrix} \begin{pmatrix} \cosh\gamma l & 2Z_0\sinh\gamma l \\ \frac{1}{2}Y_0\sinh\gamma l & \cosh\gamma l \end{pmatrix} \begin{pmatrix} 1 & 0 \\ y & 1 \end{pmatrix} \tag{4.68}$$

where y is the input admittance of the stub, given by

$$y = \tfrac{3}{2}Y_0\tanh\left(\gamma l\right) \tag{4.69}$$

Thus,

$$\mathbf{A} = \begin{pmatrix} \cosh\gamma l & \frac{1}{2}Z_0\sinh\gamma l \\ 2Y_0\sinh\gamma l & \cosh\gamma l \end{pmatrix} \begin{pmatrix} \cosh\gamma l & 2Z_0\sinh\gamma l \\ \frac{1}{2}Y_0\sinh\gamma l & \cosh\gamma l \end{pmatrix} \cdot \begin{pmatrix} 1 & 0 \\ \frac{3}{2}Y_0\tanh\gamma l & 1 \end{pmatrix} \quad (4.70a)$$

$$= \begin{pmatrix} \cosh^2\gamma l + 4\sinh^2\gamma l & \frac{5}{2}Z_0\sinh\gamma l\cosh\gamma l \\ 4Y_0\sinh\gamma l\cosh\gamma l + 6Y_0\sinh^2\gamma l\tanh\gamma l & \cosh^2\gamma l + 4\sinh^2\gamma l \end{pmatrix} \quad (4.70b)$$

As before, since $A = D$, this network is electrically symmetric without being physically symmetric.

4.3.3 Antimetry

Antimetry is the property of a two-port network that the input immittances from both ports are inverses of one another (with respect to a given characteristic impedance) when the terminations are also inverses, or

$$\frac{Z_{in,1}}{Z_0} = \frac{Z_0}{Z_{in,2}} \quad (4.71)$$

Typically, the terminations will be matched to the characteristic impedance itself, which makes them normalized inverses automatically. The conditions for antimetry are

$$s_{11} = -s_{22} \quad (4.72a)$$

$$\det(\mathbf{Z}) = Z_0^2 \quad (4.72b)$$

$$\det(\mathbf{Y}) = Y_0^2 \quad (4.72c)$$

and for $ABCD$-matrices,

$$\frac{B}{C} = Z_0^2 \quad (4.73)$$

Remarkably, symmetry and antimetry are not mutually exclusive, nor are the networks that achieve them both simultaneously (leading to the condition $s_{11} = s_{22} = 0$) limited solely to trivial degenerate cases. Rather, they are the basis of a rich class of networks the author has termed *reflectionless* which may have surprisingly complex behaviors in spite of these constraints [4]. Figures 4.15(a, b) are two such networks. These will be more easily analyzed when we talk about even and odd-mode analysis in Section 4.5.

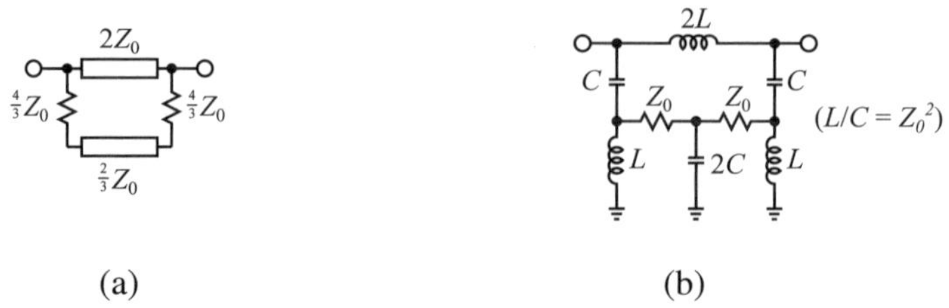

Figure 4.15 (a, b) Networks that are simultaneously symmetric and antimetric.

4.3.4 Losslessness

If a network is lossless, then it dissipates no power internally. Equivalently, the power delivered to the network at all ports is equal to the power emitted from the network at all ports,

$$\sum_{k=1}^{N} |a_k|^2 = \sum_{k=1}^{N} |b_k|^2 \tag{4.74}$$

Putting this in matrix form,

$$\mathbf{a}^*\mathbf{a} = \mathbf{b}^*\mathbf{b} = (\mathbf{Sa})^*\,\mathbf{Sa} = (\mathbf{a}^*\mathbf{S}^*)\,\mathbf{Sa} = \mathbf{a}^*\mathbf{S}^*\mathbf{Sa} \tag{4.75}$$

where the superscript asterisk denotes the conjugate transpose. Since this equation must hold for all input vectors, $\mathbf{a}$, the matrix $\mathbf{S}$ is required to be unitary ($\mathbf{S}^*\mathbf{S} = \mathbf{I}$, or $\mathbf{S}^* = \mathbf{S}^{-1}$). Further, if $\mathbf{S}$ is unitary, then

$$\mathbf{Z}^* = Z_0\,(\mathbf{I}-\mathbf{S}^*)^{-1}\,(\mathbf{I}+\mathbf{S}^*) = Z_0\left(\mathbf{I}-\mathbf{S}^{-1}\right)^{-1}\left(\mathbf{I}+\mathbf{S}^{-1}\right) \tag{4.76a}$$

$$= Z_0\left(\mathbf{I}-\mathbf{S}^{-1}\right)^{-1}\mathbf{S}^{-1}\mathbf{S}\left(\mathbf{I}+\mathbf{S}^{-1}\right) = Z_0\,(\mathbf{S}-\mathbf{I})^{-1}\,(\mathbf{S}+\mathbf{I}) \tag{4.76b}$$

$$= -Z_0\,(\mathbf{I}-\mathbf{S})^{-1}\,(\mathbf{I}+\mathbf{S}) = -\mathbf{Z} \tag{4.76c}$$

This, then, is a necessary condition for $\mathbf{S}$ to be unitary. Sufficiency is proven by substituting this condition ($\mathbf{Z}^* = -\mathbf{Z}$) back into $\mathbf{S}^*\mathbf{S}$,

$$\mathbf{S}^*\mathbf{S} = (\mathbf{Z}^*-Z_0\mathbf{I})\,(\mathbf{Z}^*+Z_0\mathbf{I})^{-1}\,(\mathbf{Z}-Z_0\mathbf{I})\,(\mathbf{Z}+Z_0\mathbf{I})^{-1} \tag{4.77a}$$

$$= (-\mathbf{Z}-Z_0\mathbf{I})\,(-\mathbf{Z}+Z_0\mathbf{I})^{-1}\,(\mathbf{Z}-Z_0\mathbf{I})\,(\mathbf{Z}+Z_0\mathbf{I})^{-1} \tag{4.77b}$$

$$= (\mathbf{Z} + Z_0\mathbf{I})(\mathbf{Z} - Z_0\mathbf{I})^{-1}(\mathbf{Z} - Z_0\mathbf{I})(\mathbf{Z} + Z_0\mathbf{I})^{-1} = \mathbf{I} \tag{4.77c}$$

confirming that $\mathbf{S}$ is unitary. Note that if the network is both lossless ($\mathbf{Z}^* = -\mathbf{Z}$) and reciprocal ($\mathbf{Z} = \mathbf{Z}^T$), then we can further say that the z-parameters are all imaginary.

Losslessness, in the most general case, is not easy to evaluate based on some parameters, such as the $ABCD$-parameters. If the network is both reciprocal and lossless, then it can be shown that the diagonal elements (A and D) are real, while the off-diagonal elements (B and C) are imaginary.

4.3.5 Passivity

Closely related to losslessness is the condition of passivity, the property that a network cannot generate power for any inputs, though it may dissipate power. This may be verified by ensuring that the power input to the network is always greater than or equal to the power emitted from the network, or in wave-vector form,

$$\mathbf{a}^*\mathbf{a} - \mathbf{b}^*\mathbf{b} \geq 0 \tag{4.78a}$$

$$\mathbf{a}^*\mathbf{a} - \mathbf{a}^*\mathbf{S}^*\mathbf{S}\mathbf{a} \geq 0 \tag{4.78b}$$

$$\mathbf{a}^*(\mathbf{I} - \mathbf{S}^*\mathbf{S})\,\mathbf{a} \geq 0 \tag{4.78c}$$

which shall be true for all input vectors $\mathbf{a}$ if the bracketed quantity is a *positive semi-definite matrix* (p.s.d.), meaning all of its eigenvalues are non-negative (if there were a negative eigenvalue, and $\mathbf{a}$ was the corresponding eigenvector, then the product above would be negative, violating the inequality). That is,

$$\lambda_k\{\mathbf{I} - \mathbf{S}^*\mathbf{S}\} \geq 0 \tag{4.79}$$

for all k. Note that, passive or not, the quantity in braces is always *Hermitian* (equal to its own conjugate transpose) guaranteeing that its eigenvalues are real.

In terms of the immittance parameters, the passivity condition is simply that the corresponding matrix ($\mathbf{Z}$ or $\mathbf{Y}$) is *positive-real* (p.r.) [5, 6].

A summary of the network properties discussed in this section and how they manifest in the various network parameters is given in Table 4.4.

4.3.6 A Constraint on Three-Port Networks

For most kinds of circuits, it is desirable that wave signals pass through them without reflecting from any port, or in terms of the scattering parameters, $s_{kk} = 0$

Table 4.4

Properties of Networks and Network Parameters

Matrix	Reciprocity	Symmetry	Antimetry	Losslessness	Passivity
$\mathbf{Z}$	$\mathbf{Z} = \mathbf{Z}^T$	$z_{11} = z_{22}$	$\det \mathbf{Z} = Z_0^2$	$\mathbf{Z}^* = -\mathbf{Z}$	$\mathbf{Z}$ is p.r.
$\mathbf{Y}$	$\mathbf{Y} = \mathbf{Y}^T$	$y_{11} = y_{22}$	$\det \mathbf{Y} = Y_0^2$	$\mathbf{Y}^* = -\mathbf{Y}$	$\mathbf{Y}$ is p.r.
$ABCD$	$AD - BC = 1$	$A = D$	$B/C = Z_0^2$	...	...
$\mathbf{H}$	$h_{21} = -h_{12}$	$\det \mathbf{H} = 1$	$h_{11}/h_{22} = Z_0^2$	...	...
$\mathbf{G}$	$g_{21} = -g_{12}$	$\det \mathbf{G} = 1$	$g_{11}/g_{22} = Y_0^2$	...	...
$\mathbf{S}$	$\mathbf{S} = \mathbf{S}^T$	$s_{11} = s_{22}$	$s_{11} = -s_{22}$	$\mathbf{S}^*\mathbf{S} = \mathbf{I}$	$\mathbf{I} - \mathbf{S}^*\mathbf{S}$ is p.s.d.
$\mathbf{T}$	$\det \mathbf{T} = 1$	$T_{21} = -T_{12}$	$T_{21} = T_{12}$	...	...

for all k. Since this requires each port to present a matched impedance of Z_0 when the other ports are likewise terminated, such a network is called a *matched network*. Impedance matching will be covered in more detail in Chapter 6.

Unfortunately, there are some fundamental constraints on how well certain kinds of networks can be matched, and under what conditions. For a well-known example, it is impossible for a three-port network (say, a two-way splitter) to be reciprocal, lossless, and matched at the same time. To see this, one may write the general scattering parameters with $s_{kk} = 0$ (to enforce matching) and $s_{ik} = s_{ki}$ (to enforce reciprocity),

$$\mathbf{S} = \begin{pmatrix} 0 & s_{21} & s_{31} \\ s_{21} & 0 & s_{32} \\ s_{31} & s_{32} & 0 \end{pmatrix} \tag{4.80}$$

and substitute this into the condition for losslessness,

$$\mathbf{S}^*\mathbf{S} = \mathbf{I} \tag{4.81a}$$

$$\begin{pmatrix} 0 & s_{21}^* & s_{31}^* \\ s_{21}^* & 0 & s_{32}^* \\ s_{31}^* & s_{32}^* & 0 \end{pmatrix} \begin{pmatrix} 0 & s_{21} & s_{31} \\ s_{21} & 0 & s_{32} \\ s_{31} & s_{32} & 0 \end{pmatrix} = \begin{pmatrix} 1 & 0 & 0 \\ 0 & 1 & 0 \\ 0 & 0 & 1 \end{pmatrix} \tag{4.81b}$$

$$\begin{pmatrix} |s_{21}|^2 + |s_{31}|^2 & s_{31}^* s_{32} & s_{21}^* s_{32} \\ s_{31} s_{32}^* & |s_{21}|^2 + |s_{32}|^2 & s_{21}^* s_{31} \\ s_{21} s_{32}^* & s_{21} s_{31}^* & |s_{31}|^2 + |s_{32}|^2 \end{pmatrix} = \begin{pmatrix} 1 & 0 & 0 \\ 0 & 1 & 0 \\ 0 & 0 & 1 \end{pmatrix} \quad (4.81c)$$

Note that there are only three remaining independent variables (s_{21}, s_{31}, and s_{32}). To satisfy the off-diagonal components in (4.81c), at least two of these must be zero, but that would preclude satisfying at least one of the diagonal elements. Therefore, there is no set of numbers that can satisfy this equation.

4.4 SIGNAL-FLOW GRAPHS

Because of the intuitive way that scattering parameters elucidate the reflection and transmission of waves incident upon the ports of a network, they will be the preferred description for high-frequency electromagnetic circuits throughout this book. However, they lack the easy combination rules that some of the other network parameters have. It will be useful, then, to come up with a formalism for calculating the response of two networks interacting with one another through their scattering parameters. This will be the function of *signal-flow graphs*.

4.4.1 Principles of Signal-Flow Graphs

Let us start with a simple, one-port network. Such a network is described entirely by its driving-point impedance, or in terms of scattering parameters, by its reflection coefficient, s_{11}. We may draw a directed graph where the nodes represent the incident and outgoing wave amplitudes, a_1 and b_1, and the edges represent the gain between these two quantities. The signal-flow graph for our one-port network is thus shown in Figure 4.16(a). Note that the value of the wave amplitude, b_1, may be found by multiplying the value of the antecedent wave amplitude, a_1, by the gain of the edge that connects them, or $b_1 = s_{11}a_1$.

If we next consider a two-port network, the graph becomes moderately more complicated, as shown in Figure 4.16(b). We now have four wave amplitudes to consider — the incident waves a_1 and a_2, and the outgoing waves b_1 and b_2 — and four scattering parameters that couple them through both reflections and transmissions. We may now write

$$b_1 = s_{11}a_1 + s_{12}a_2 \quad (4.82a)$$

$$b_2 = s_{21}a_1 + s_{22}a_2 \quad (4.82b)$$

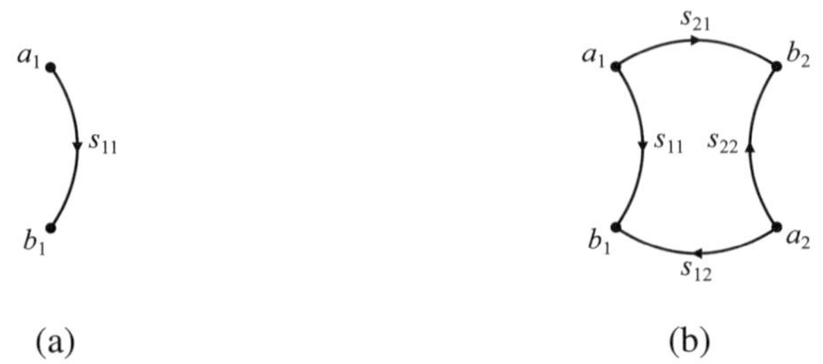

Figure 4.16 (a) Signal-flow graph for a one-port network. (b) Signal-flow graph for a two-port network.

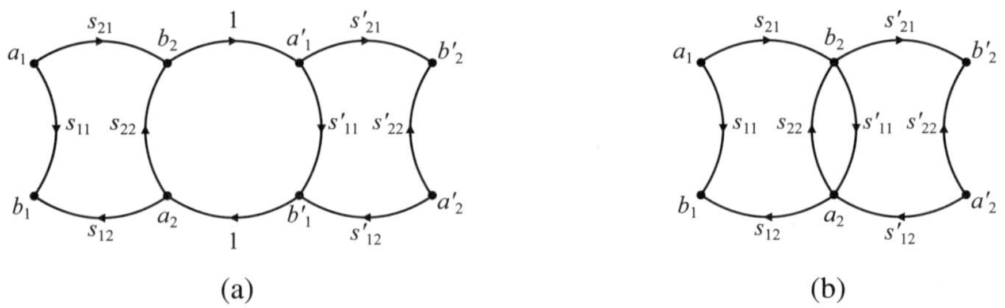

Figure 4.17 (a) Signal-flow graph of two networks in cascade. (b) Simplified signal-flow graph of two networks in cascade.

in accordance with the scattering matrix definition in (4.40) where $N = 2$.

Now let us see what happens when we combine two such networks in cascade, as in Figure 4.17(a). We'll use primed parameters for the second network. Since the outgoing wave at port 2 of the first network becomes the incident wave at port 1 of the second network, we may write $a'_1 = b_2$. In the graph, these may be connected with a gain of unity. Equivalently, one may simply draw a single node for the wave, using either label, and connect the edges accordingly, as in Figure 4.17(b).

We wish to write the scattering parameters of the combined network, as follows

$$\begin{pmatrix} b_1 \\ b'_2 \end{pmatrix} = \begin{pmatrix} s''_{11} & s''_{12} \\ s''_{12} & s''_{22} \end{pmatrix} \begin{pmatrix} a_1 \\ a'_2 \end{pmatrix} \tag{4.83}$$

The new (double-primed) scattering parameters will simply be the net effective gain from the corresponding source node, a_k, to the output node, b_k. The gain of a particular path is the product of the edge gains along that path, and the total net gain from source to output is, by superposition, the sum of the gains of the various paths. For s''_{11}, then,

$$s''_{11} = \frac{b_1}{a_1} = s_{11} + s_{21}s'_{11}s_{12} + s_{21}s'_{11}s_{22}s'_{11}s_{12} + \cdots \tag{4.84a}$$

$$= s_{11} + s_{21}s'_{11}s_{12}\sum_{k=0}^{\infty}\left(s_{22}s'_{11}\right)^k \tag{4.84b}$$

Note that the loop, $s_{22}s'_{11}$, has caused us to have an infinite series. The series is geometric, so it will be convergent if the loop gain (which is complex) has a magnitude less than unity. (If the loop gain is greater than unity, which would require nonpassive networks, then the combination may be unstable.) The solution of an infinite geometric series is well known:

$$s''_{11} = s_{11} + s_{21}s'_{11}s_{12}\sum_{k=0}^{\infty} = s_{11} + \frac{s_{21}s'_{11}s_{12}}{1 - s_{22}s'_{11}} \tag{4.85}$$

Similarly, the transmission coefficient may be found,

$$s''_{21} = s_{21}s'_{21}\sum_{k=0}^{\infty}\left(s_{22}s'_{11}\right)^k = \frac{s_{21}s'_{21}}{1 - s_{22}s'_{11}} \tag{4.86}$$

The remaining two scattering parameters may be found in a similar fashion, or, if the two networks are reciprocal, they will be equivalent to the above.

4.4.2 Decomposition Rules

Depending on a particular engineer's inclination, more complex signal-flow graphs may be solved graphically, or mathematically by formula. For the graphical approach, it is useful to identify rules for reducing large signal-flow graphs to smaller ones. These graph reduction or decomposition rules are laid out in Figure 4.18.

The first rule, Figure 4.18(a), illustrates that the gains of sequential edges in the graph are multiplied. The second rule, Figure 4.18(b), shows that the gains of parallel paths add. The third rule, Figure 4.18(c), concerns branching paths, and shows that the branch point can be shifted one node back with duplicate edge gains. The final rule, Figure 4.18(d), quantifies the infinite series associated with closed loops.

To see how these rules apply, let us redraw the original cascaded graph connected to the a_1 input (that is, delete a'_2 and the edges that emanate from it).

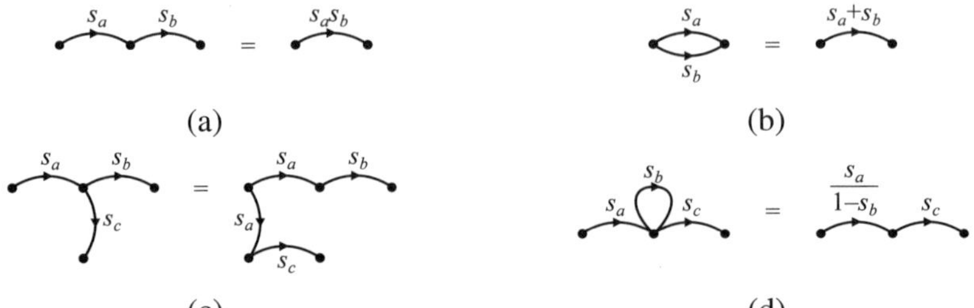

Figure 4.18 Decomposition rules for signal flow graphs. (a) Sequential edges. (b) Parallel edges. (c) Branch paths. (d) Loops.

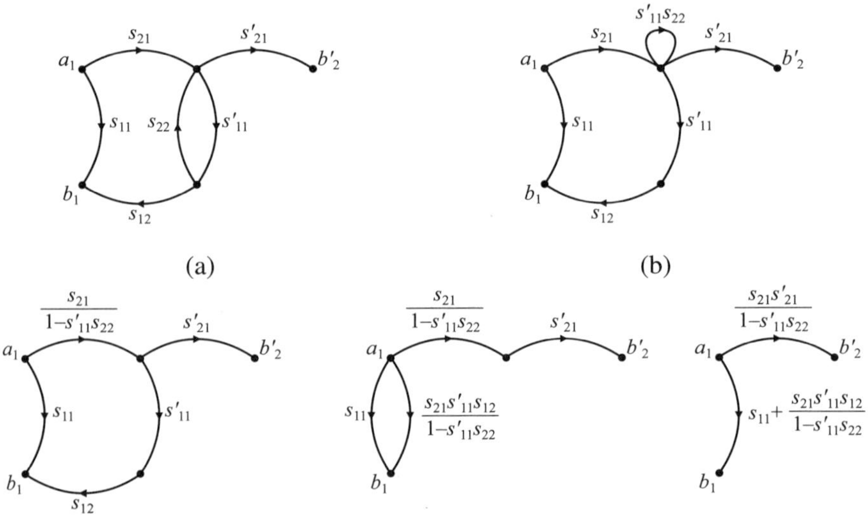

Figure 4.19 Application of decomposition rules to a signal-flow graph. (a) Original network without unused input, a'_2. (b) Application of branch-point and product rules to edge s_{22}. (c) Application of loop rule. (d) Product and branch rules applied to lower path. (e) Sum and product rules applied to lower and upper path, respectively.

This is shown in Figure 4.19(a). In Figure 4.19(b), we have used the branch and product rules to isolate a closed loop. That closed loop is removed with the loop rule in Figure 4.19(c). In Figure 4.19(d), the branch point between the upper and lower paths is moved back to the input with the branch rule, and then simplified according to the product rule. Finally, in Figure 4.19(e), the sum rule is applied to the lower path, while the product rule is applied to the upper path, reducing the graph to a simple three-node graph with the forward scattering coefficients (from a_1) directly labeled.

4.4.3 Mason's Rule

For those who tire quickly of drawing graphs, or are more mathematically inclined, the calculations behind the decomposition rules can be summarized in a fairly compact, universal formula, known as *Mason's rule*,

$$G = \frac{1}{\Delta} \sum_{k} G_k \Delta_k \tag{4.87}$$

where G is the net path gain (the scattering parameter) to be calculated, G_k is the nonlooping gain of forward path k from input to output, Δ is the determinant of the graph, and Δ_k is the determinant of the subgraph not touching (or not sharing any nodes with) path k. The determinants are calculated as

$$\Delta = 1 - \sum L_i + \sum L_i L_j - \sum L_i L_j L_k + \cdots \tag{4.88}$$

where L_i is the total gain of each closed loop, $L_i L_j$ is the product of each non-touching pair of loops, $L_i L_j L_k$ is the product of each set of three non-touching loops, and so on. Individual loops are sometimes referred to as *first-order* loops, while the non-touching loop pairs are called *second-order* loops, the non-touching sets of three *third-order* loops, etc.

To see how Mason's rule applies, it is useful to consider a slightly more complex graph, having multiple, interlinked loops, such as that in Figure 4.20(a). There is only one direct (non-looping) forward path from node 1 to node 2, having gain $G_1 = ab$. The sub-graph disconnected from this path has no loops, so $\Delta_1 = 1$. The determinant of the whole graph, however, requires consideration of multiple loops. First-order loops are shown in Figure 4.20(b-d). Therefore,

$$\sum L_i = cd + ef + bcfg \tag{4.89}$$

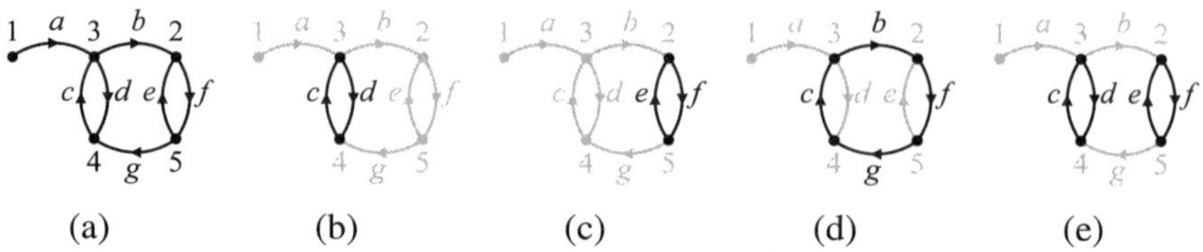

Figure 4.20 (a) A signal-flow graph with multiple, interlinked loops. (b–d) First-order loops. (e) A second-order loop.

There is only one second-order loop (a pair of loops without any common nodes), shown in Figure 4.20(e), having product gain

$$\sum L_i L_j = cdef \tag{4.90}$$

There are no loops of order three or higher. Therefore, the effective gain from node 1 to node 2 is

$$G = \frac{1}{\Delta}\sum_k G_k \Delta_k = \frac{G_1 \Delta_1}{1 - \sum L_i + \sum L_i L_j} \tag{4.91a}$$

$$= \frac{ab}{1 - (cd + ef + bcfg) + cdef} \tag{4.91b}$$

Note that adding even a single branch can alter the gain of the graph significantly. Take Figure 4.21, for example. We have added a single branch, h, to the graph, but doing so adds two extra first-order loops, adh and $abfgh$, and an extra second-order loop, $adefh$, which all appear in the determinant of Mason's formula,

$$G' = \frac{ab}{1 - (cd + ef + bcfg + adh + abfgh) + (cdef + adefh)} \tag{4.92}$$

4.4.4 Equivalent Matrix Form

One may also solve signal flow graphs using an equivalent matrix form of Mason's rule [7]. Imagine initializing each node k with a particular value, $v_k(0)$, and that hypothetically, each branch propagates those values to the connected nodes with the appropriate gain after a 1-second delay. Therefore, the node values at time $t = 1$

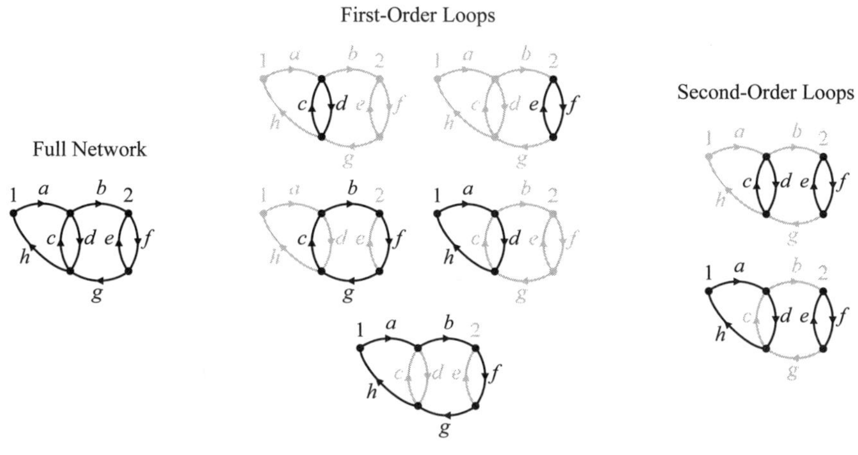

Figure 4.21 (a) The same signal flow graph from Figure 4.20(a), but with a single extra branch added, with gain h. (b) First-order loops. (c) Second-order loops.

are given by

$$\begin{pmatrix} v_1(1) \\ v_2(1) \\ \vdots \\ v_N(1) \end{pmatrix} = \begin{pmatrix} h_{11} & h_{12} & \cdots & h_{1N} \\ h_{21} & h_{22} & \cdots & h_{2N} \\ \vdots & \vdots & \ddots & \vdots \\ h_{N1} & h_{N2} & \cdots & h_{NN} \end{pmatrix} \begin{pmatrix} v_1(0) \\ v_2(0) \\ \vdots \\ v_N(0) \end{pmatrix} \tag{4.93}$$

or, for any time step t,

$$\mathbf{V}_t = \mathbf{H}\mathbf{V}_{t-1} \tag{4.94}$$

where h_{jk} is the sum of the direct branch gains to node j from node k. The equilibrium solution of the graph is found in the limit as $t \to \infty$,

$$\mathbf{V}_\infty = \mathbf{V}_0 + \mathbf{H}\mathbf{V}_0 + \mathbf{H}^2\mathbf{V}_0 + \mathbf{H}^3\mathbf{V}_0 + \cdots \tag{4.95a}$$

$$= \left(\mathbf{I} + \mathbf{H} + \mathbf{H}^2 + \mathbf{H}^3 + \cdots\right)\mathbf{V}_0 = \mathbf{G}\mathbf{V}_0 \tag{4.95b}$$

$$\therefore \mathbf{G} = (\mathbf{I} - \mathbf{H})^{-1} \tag{4.95c}$$

and the elements g_{jk} represent the net effective gain to node j from node k. Citing Figure 4.20 as an example, we may write

$$\mathbf{H} = \begin{pmatrix} 0 & 0 & 0 & 0 & 0 \\ 0 & 0 & b & 0 & e \\ a & 0 & 0 & c & 0 \\ 0 & 0 & d & 0 & g \\ 0 & f & 0 & 0 & 0 \end{pmatrix} \tag{4.96}$$

Substituting this into (4.95c), we have

$$\mathbf{G} = (\mathbf{I} - \mathbf{H})^{-1} = \begin{pmatrix} 1 & 0 & 0 & 0 & 0 \\ 0 & 1 & -b & 0 & -e \\ -a & 0 & 1 & -c & 0 \\ 0 & 0 & -d & 1 & -g \\ 0 & -f & 0 & 0 & 1 \end{pmatrix}^{-1} \tag{4.97a}$$

$$= \frac{\begin{pmatrix} 1 & 0 & 0 & 0 & 0 \\ ab & 1-cd & b & bc & e+bcg-cde \\ a-aef & cfg & 1-ef & 1-cef & cg \\ ad+abfg-adef & fg & d+bfg-def & 1-ef & g \\ abf & f-cdf & bf & bcf & 1-cd \end{pmatrix}}{1-(cd+ef+bcfg)+cdef} \tag{4.97b}$$

Note that the element g_{21} is the same expression that we derived in (4.91b). For very large graphs, this method is most easily applied by computer.

4.5 EVEN AND ODD-MODE ANALYSIS

When a reciprocal network is physically symmetric as well as electrically symmetric, a useful technique for characterizing it is *even-/odd-mode analysis*. The method begins with an arbitrary two-port network in a box, as shown in Figure 4.22(a). We imagine cutting the box open and pulling the two halves apart, as shown in Figure 4.22(b), exposing wires where nodes lie along the symmetry line. Then we test the network by applying two special excitations to its ports.

The first excitation, Figure 4.22(c), is known as the *even mode*, and consists of an arbitrary signal applied simultaneously to both ports. Due to the symmetry, there can be no current flowing in the exposed wires, for the direction of the current would be indeterminate. We may therefore calculate the response of the network

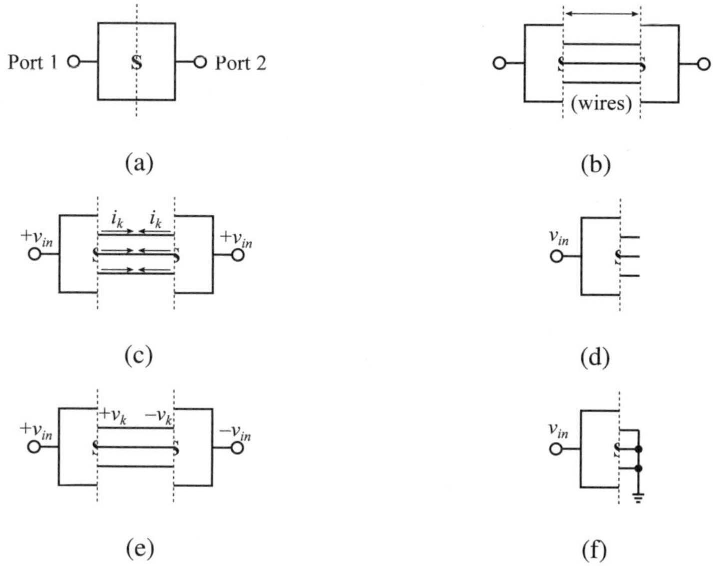

Figure 4.22 (a) A physically symmetric two-port network. (b) The network split in half to expose wires on the symmetry line. (c) Even-mode excitation. (d) Even-mode equivalent circuit. (e) Odd-mode excitation. (f) Odd-mode equivalent circuit.

to this excitation by considering only the elements in the left half of the box, with an open-circuit boundary condition applied to the nodes along the symmetry line. This is the *even-mode equivalent circuit*. In terms of the scattering parameters of the original two-port network, the reflection response of this equivalent half-circuit may be written as

$$\Gamma_{\text{even}} = \frac{b_1}{a_1} = \frac{s_{11}a_1 + s_{12}a_2}{a_1} = \frac{s_{11}a_1 + s_{12}a_1}{a_1} = s_{11} + s_{22} \tag{4.98}$$

since, for the even-mode excitation, $a_1 = a_2$.

The second excitation, Figure 4.22(d), is known as the *odd mode*, and consists of an arbitrary signal again applied to both ports, but this time with a sign reversal. In the frequency domain, this corresponds to a 180° phase difference between the ports. In this case, there can be no voltage on the symmetry nodes with respect to ground, for the sign of that voltage would be indeterminate. We may therefore calculate the response of this equivalent half-circuit by placing a short-circuit boundary condition on the symmetry nodes. This is the *odd-mode equivalent circuit*. In terms of the original s-parameters, its reflection coefficient is

$$\Gamma_{\text{odd}} = \frac{b_1}{a_1} = \frac{s_{11}a_1 + s_{12}a_2}{a_1} = \frac{s_{11}a_1 - s_{12}a_1}{a_1} = s_{11} - s_{22} \tag{4.99}$$

since, for the odd-mode excitation, $a_1 = -a_2$.

Putting these two results together and solving for the scattering parameters of the original network

$$s_{11} = s_{22} = \tfrac{1}{2}\left(\Gamma_{\text{even}} + \Gamma_{\text{odd}}\right) \tag{4.100a}$$

$$s_{12} = s_{21} = \tfrac{1}{2}\left(\Gamma_{\text{even}} - \Gamma_{\text{odd}}\right) \tag{4.100b}$$

We may therefore calculate the full network parameters of a symmetric, reciprocal network by considering two simpler networks employing only half the elements. This may greatly simplify the analysis of certain kinds of networks.

Even-/odd-mode analysis applies directly to immittance parameters as well. Take the z-parameters, for instance. In this case, the even and odd-mode stimuli comprise simultaneous in-phase and out-of-phase current excitations rather than wave inputs, and the outputs are the superpositions of the voltage responses rather than scattered waves, but the same basic premise and boundary conditions apply. We may therefore write

$$Z_{11} = Z_{22} = \tfrac{1}{2}\left(Z_{\text{in,even}} + Z_{\text{in,odd}}\right) \tag{4.101a}$$

$$Z_{12} = Z_{21} = \tfrac{1}{2}\left(Z_{\text{in,even}} - Z_{\text{in,odd}}\right) \tag{4.101b}$$

4.5.1 Two-Port Example

Take, for example, the circuit shown Figure 4.15(b), which was claimed to be both symmetric and antimetric at the same time (requiring $s_{11} = s_{22} = 0$). We will now prove that this is the case.

First, we must redraw the circuit such that only nodes/wires lie along the symmetry line. This requires splitting the top inductor and the bottom capacitor into two, as illustrated in Figure 4.23(a). The even-mode equivalent circuit is shown in Figure 4.23(b). The top inductor appears in gray because it is open-circuited in this equivalent circuit, thus having no effect. The input impedance of the equivalent circuit may be found by simple series and parallel combination of the elements,

$$Z_{\text{even}} = (j\omega C)^{-1} + \left((j\omega L)^{-1} + \left(Z_0 + (j\omega C)^{-1}\right)^{-1}\right)^{-1} \tag{4.102}$$

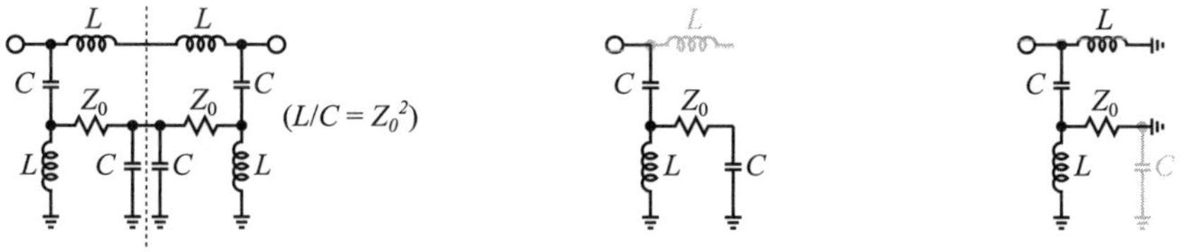

Figure 4.23 (a) A symmetric two-port network divided in two. (b) Even-mode equivalent circuit. (c) Odd-mode equivalent circuit.

Since we have a constraint on the relative values of L, C, and Z_0, it will be useful to make a simplifying substitution,

$$x = \frac{j\omega L}{Z_0} = j\omega C Z_0 \tag{4.103}$$

which clearly satisfies the constraint, $L/C = Z_0^2$. Thus, Z_{even} becomes

$$Z_{\text{even}} = Z_0 \left(x^{-1} + \left(x^{-1} + \left(1 + x^{-1} \right)^{-1} \right)^{-1} \right) \tag{4.104a}$$

$$= Z_0 \left(\frac{x^3 + 2x^2 + x + 1}{x^3 + x^2 + x} \right) \tag{4.104b}$$

The even-mode reflection coefficient is given by

$$\Gamma_{\text{even}} = \frac{Z_{\text{even}} - Z_0}{Z_{\text{even}} + Z_0} = \frac{x^2 + 1}{2x^3 + 3x^2 + 2x + 1} \tag{4.105}$$

Similarly, the odd-mode input impedance is

$$Z_{\text{odd}} = \left((j\omega L)^{-1} + \left((j\omega C)^{-1} + \left(Z_0^{-1} + (j\omega L)^{-1} \right)^{-1} \right)^{-1} \right)^{-1} \tag{4.106a}$$

$$= Z_0 \left(x^{-1} + \left(x^{-1} + \left(1 + x^{-1} \right)^{-1} \right)^{-1} \right)^{-1} \tag{4.106b}$$

$$= Z_0 \left(\frac{x^3 + x^2 + x}{x^3 + 2x^2 + x + 1} \right) \tag{4.106c}$$

and the odd-mode reflection coefficient is

$$\Gamma_{\text{odd}} = \frac{Z_{\text{odd}} - Z_0}{Z_{\text{odd}} + Z_0} = \frac{-x^2 - 1}{2x^3 + 3x^2 + 2x + 1} \tag{4.107}$$

Note that $\Gamma_{\text{even}} = -\Gamma_{\text{odd}}$, so the scattering parameters are

$$s_{11} = s_{22} = \tfrac{1}{2}\left(\Gamma_{\text{even}} + \Gamma_{\text{odd}}\right) = 0 \tag{4.108a}$$

$$s_{12} = s_{21} = \tfrac{1}{2}\left(\Gamma_{\text{even}} - \Gamma_{\text{odd}}\right) = \Gamma_{\text{even}} = \frac{x^2 + 1}{2x^3 + 3x^2 + 2x + 1} \tag{4.108b}$$

$$= \frac{1 - (\omega C Z_0)^2}{1 - 3(\omega C Z_0)^2 + j2(\omega C Z_0) - j2(\omega C Z_0)^3} \tag{4.108c}$$

As claimed, the network is both symmetric ($s_{11} = s_{22}$) and antimetric ($s_{11} = -s_{22}$). The frequency response in (4.108c), incidentally, is that of a third-order Chebyshev Type-II low-pass filter. Filters will be discussed in more detail in Chapter 11.

4.5.2 Coupled Lines

Even-/odd-mode analysis allows us now to characterize coupled transmission lines, which were alluded to at the end of Chapter 3 but not mathematically defined. When a pair of open-structure transmission lines, such as microstrip, are brought into close proximity, their fields become linked, and the behavior of the effective four-port device may be broken down into even- and odd-mode excitations.

For the even mode, shown in cross-section in Figure 4.24(a), the symmetry plane between the two lines becomes an open-circuit, or magnetic-wall boundary condition (the magnetic field must be everywhere perpendicular to this boundary). The resulting field geometry gives each element of the pair a particular characteristic impedance, Z_e, and propagation constant, γ_e.

Under odd-mode excitation, shown in Figure 4.24(b), the symmetry plane becomes a short-circuit, or electric-wall boundary condition (the electric field must be everywhere perpendicular to this boundary). This modifies the field geometry so that the pair of lines has a different characteristic impedance and propagation constant, Z_o and γ_o. In particular, the vertical boundary walls acts like an additional ground plane, which will generally make the characteristic impedance lower ($Z_o < Z_e$). Further, if the transmission lines are microstrip, the fields will exist more in the air and less in the substrate than they do for the even mode, causing the effective

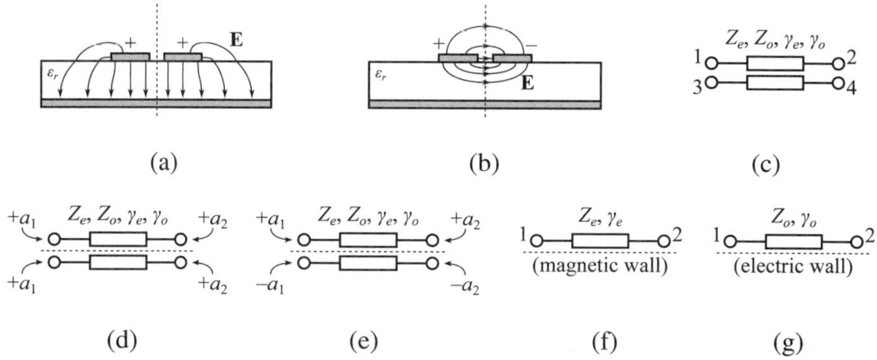

Figure 4.24 Coupled microstrip line geometry with (a) even-mode and (b) odd-mode excitation. (c) Schematic symbol for coupled lines with (d) even-mode and (e) odd-mode excitation. (f) Even-mode equivalent circuit. (g) Odd-mode equivalent circuit.

dielectric constant to be lower and the propagation constant to be smaller ($\gamma_o < \gamma_e$). In geometries like stripline, where there is only one dielectric medium, the propagation constants will typically be equivalent. In general, however, it takes four parameters to characterize a coupled-line section, as indicated in Figure 4.24(c).

As with two-port devices, the full scattering parameters of this four-port structure may be expressed in terms of the scattering parameters of the corresponding even- and odd-mode equivalent circuits. The even mode, for example, may be thought of as the simultaneous excitation of ports 1 and 3 with one signal (a_1), and of ports 2 and 4 with another signal (a_2), as indicated in Figure 4.24(d). So the even-mode circuit may be considered a two-port with reflection coefficient that is the superposition of the port 1 reflection added to the reverse coupling from port 3,

$$s_{11,e} = s_{11} + s_{13} \tag{4.109}$$

and its transmission coefficient is the superposition of the transmission from port 1 added to the forward coupling from port 3,

$$s_{21,e} = s_{21} + s_{23} \tag{4.110}$$

Under the odd-mode excitation, the couplings from port 3 have a negative sign, so

$$s_{11,o} = s_{11} - s_{13} \tag{4.111a}$$

$$s_{21,o} = s_{21} - s_{23} \tag{4.111b}$$

These relations, along with reciprocity and the obvious symmetry of the coupled lines, allow us to solve for the complete set of four-port scattering parameters,

$$s_{11} = s_{22} = s_{33} = s_{44} = \tfrac{1}{2}\left(s_{11,e} + s_{11,o}\right) \tag{4.112a}$$

$$s_{12} = s_{21} = s_{34} = s_{43} = \tfrac{1}{2}\left(s_{21,e} + s_{21,o}\right) \tag{4.112b}$$

$$s_{13} = s_{31} = s_{24} = s_{42} = \tfrac{1}{2}\left(s_{11,e} - s_{11,o}\right) \tag{4.112c}$$

$$s_{14} = s_{41} = s_{23} = s_{32} = \tfrac{1}{2}\left(s_{21,e} - s_{21,o}\right) \tag{4.112d}$$

Since the even- and odd-mode equivalent circuits are simple transmission lines, their s-parameters are already known, from (4.58d). Therefore,

$$s_{11} = s_{22} = s_{33} = s_{44} = \frac{\frac{1}{2}(z_e - y_e)\sinh(\gamma_e l)}{\Delta_e} + \frac{\frac{1}{2}(z_o - y_o)\sinh(\gamma_o l)}{\Delta_o} \tag{4.113a}$$

$$s_{12} = s_{21} = s_{34} = s_{43} = \frac{1}{\Delta_e} + \frac{1}{\Delta_o} \tag{4.113b}$$

$$s_{13} = s_{31} = s_{24} = s_{42} = \frac{\frac{1}{2}(z_e - y_e)\sinh(\gamma_e l)}{\Delta_e} - \frac{\frac{1}{2}(z_o - y_o)\sinh(\gamma_o l)}{\Delta_o} \tag{4.113c}$$

$$s_{14} = s_{41} = s_{23} = s_{32} = \frac{1}{\Delta_e} - \frac{1}{\Delta_o} \tag{4.113d}$$

where

$$z_e = y_e^{-1} = \frac{Z_e}{Z_0} \tag{4.114a}$$

$$z_o = y_o^{-1} = \frac{Z_o}{Z_0} \tag{4.114b}$$

$$\Delta_e = 2\cosh\left(\gamma_e l\right) + \left(z_e + y_e\right)\sinh\left(\gamma_e l\right) \tag{4.114c}$$

$$\Delta_o = 2\cosh\left(\gamma_o l\right) + \left(z_o + y_o\right)\sinh\left(\gamma_o l\right) \tag{4.114d}$$

The z-parameters of the coupled lines are similarly found by superposing the z-parameters of the even- and odd-mode equivalent transmission lines,

$$z_{11} = z_{22} = z_{33} = z_{44} = \tfrac{1}{2}\left(Z_e \coth\left(\gamma_e l\right) + Z_o \coth\left(\gamma_o l\right)\right) \tag{4.115a}$$

$$z_{12} = z_{21} = z_{43} = z_{34} = \tfrac{1}{2}\left(Z_e \operatorname{csch}\left(\gamma_e l\right) + Z_o \operatorname{csch}\left(\gamma_o l\right)\right) \tag{4.115b}$$

$$z_{13} = z_{31} = z_{24} = z_{42} = \tfrac{1}{2}\left(Z_e \coth\left(\gamma_e l\right) - Z_o \coth\left(\gamma_o l\right)\right) \tag{4.115c}$$

$$z_{14} = z_{41} = z_{23} = z_{32} = \frac{1}{2}\left(Z_e \operatorname{csch}(\gamma_e l) - Z_o \operatorname{csch}(\gamma_o l)\right) \tag{4.115d}$$

It should be emphasized that the even and odd-mode characteristic impedances are defined as the effective impedances seen by each individual trace when both traces are excited simultaneously, but from separate sources, in-phase for the even mode and out-of-phase for the odd mode. When the two lines are driven in parallel as a unit from a common source, the effective characteristic impedance seen by that source is known as the *common-mode impedance*, Z_c. Since the common mode will involve the same voltage as the even mode but twice the current, the common-mode impedance is one-half the even-mode impedance ($Z_c = Z_e/2$). Similarly, when the pair is driven out-of-phase by a common source, the characteristic impedance that source sees is called the *differential-mode impedance*, Z_d. In this case, the voltage amplitude is twice that of the odd mode, while the current is the same, thus the differential-mode impedance is twice the odd-mode impedance ($Z_d = 2Z_o$). Therefore, whereas for coupled lines the even-mode impedance is greater than the odd-mode impedance, the common-mode impedance is usually less than the differential-mode impedance.

Even-/odd-mode analysis is a powerful technique for characterizing symmetric structures. We will encounter it again in Chapter 10 when we talk about symmetric devices such as splitters, couplers, and hybrids.

4.6 NORMALIZATION OF THE SCATTERING PARAMETERS

In all of our discussion of scattering parameters so far, we have assumed that all the ports are terminated with transmission lines of the same characteristic impedance, Z_0, and this impedance in many practical cases is 50Ω. Some situations arise, however, where it is beneficial to attach transmission lines of differing impedances at different ports, such as the transition from one wave-propagating structure to another, say, from coaxial cable to slotline.

4.6.1 Generalized Scattering Parameters

In these situations, it is useful to define *generalized scattering parameters*. Like before, these relate the amplitudes of incident and outgoing waves, denoted a_k and b_k, respectively, but their ratios are no longer simply equal to the voltage and current wave ratios. That is, for the standard scattering parameters,

$$s_{ik} = \frac{b_i}{a_k} = \frac{v_{bi}}{v_{ak}} = \frac{i_{bi}}{i_{ak}} \tag{4.116}$$

but for the generalized scattering parameters,

$$s_{ik} = \frac{b_i}{a_k} = \frac{v_{bi}}{v_{ak}}\sqrt{\frac{Z_{0k}}{Z_{0i}}} = \frac{i_{bi}}{i_{ak}}\sqrt{\frac{Z_{0i}}{Z_{0k}}} \tag{4.117}$$

where Z_{0k} is the characteristic impedance associated with port k.

Let us see how the generalized scattering parameters relate to the impedance parameters of a network, which do not depend on characteristic impedances at all. The total voltage and current at a given port, k, is the sum of contributions from the incident and outgoing waves at that port,

$$v_k = (a_k + b_k)\sqrt{Z_{0k}} \tag{4.118a}$$

$$i_k = (a_k - b_k)\sqrt{Y_{0k}} \tag{4.118b}$$

where $Y_{0k} = Z_{0k}^{-1}$. As always, the negative sign in (4.118b) is due to the convention that positive current flows in the direction of propagation, which, for an outgoing wave, is flowing out of the port. Expressed in vector form,

$$\mathbf{v} = \sqrt{\mathbf{z}}\,(\mathbf{a} + \mathbf{b}) \tag{4.119a}$$

$$\mathbf{i} = \sqrt{\mathbf{y}}\,(\mathbf{a} - \mathbf{b}) \tag{4.119b}$$

where $\sqrt{\mathbf{z}}$ is a real-valued diagonal matrix with elements corresponding to the square roots of the individual port characteristic impedances, and $\sqrt{\mathbf{y}}$ is its inverse. These equations are analogous to (4.38), only now the immittance terms are matrices instead of scalars. Following a very similar chain of reasoning, we may write

$$\mathbf{v} = \mathbf{Z}\mathbf{i} \tag{4.120a}$$

$$\sqrt{\mathbf{z}}\,(\mathbf{a} + \mathbf{b}) = \mathbf{Z}\sqrt{\mathbf{y}}\,(\mathbf{a} - \mathbf{b}) \tag{4.120b}$$

$$\sqrt{\mathbf{z}}\,(\mathbf{a} + \mathbf{S}\mathbf{a}) = \mathbf{Z}\sqrt{\mathbf{y}}\,(\mathbf{a} - \mathbf{S}\mathbf{a}) \tag{4.120c}$$

$$\sqrt{\mathbf{z}}\,(\mathbf{I} + \mathbf{S})\,\mathbf{a} = \mathbf{Z}\sqrt{\mathbf{y}}\,(\mathbf{I} - \mathbf{S})\,\mathbf{a} \tag{4.120d}$$

Since the above must hold for all input vectors, $\mathbf{a}$, we have

$$\sqrt{\mathbf{z}}\,(\mathbf{I} + \mathbf{S}) = \mathbf{Z}\sqrt{\mathbf{y}}\,(\mathbf{I} - \mathbf{S}) \tag{4.121}$$

This may then be solved for the z-parameters,

$$\mathbf{Z} = \sqrt{\mathbf{z}}\,(\mathbf{I} + \mathbf{S})\,(\mathbf{I} - \mathbf{S})^{-1}\sqrt{\mathbf{z}} \tag{4.122a}$$

$$= \sqrt{\mathbf{z}}\,(\mathbf{I} - \mathbf{S})^{-1}\,(\mathbf{I} + \mathbf{S})\sqrt{\mathbf{z}} \tag{4.122b}$$

or for the generalized scattering parameters,

$$\mathbf{S} = \left(\sqrt{\mathbf{y}}\mathbf{Z}\sqrt{\mathbf{y}} + \mathbf{I}\right)^{-1} \left(\sqrt{\mathbf{y}}\mathbf{Z}\sqrt{\mathbf{y}} - \mathbf{I}\right) \tag{4.123a}$$

$$= \left(\sqrt{\mathbf{y}}\mathbf{Z}\sqrt{\mathbf{y}} - \mathbf{I}\right) \left(\sqrt{\mathbf{y}}\mathbf{Z}\sqrt{\mathbf{y}} + \mathbf{I}\right)^{-1} \tag{4.123b}$$

Note that, as before, the matrix products in these equations are commutable [8]. Similarly, for the y-parameters,

$$\mathbf{Y} = \sqrt{\mathbf{y}}\left(\mathbf{I} - \mathbf{S}\right)\left(\mathbf{I} + \mathbf{S}\right)^{-1}\sqrt{\mathbf{y}} \tag{4.124a}$$

$$= \sqrt{\mathbf{y}}\left(\mathbf{I} + \mathbf{S}\right)^{-1}\left(\mathbf{I} - \mathbf{S}\right)\sqrt{\mathbf{y}} \tag{4.124b}$$

$$\mathbf{S} = \left(\mathbf{I} - \sqrt{\mathbf{z}}\mathbf{Y}\sqrt{\mathbf{z}}\right)\left(\mathbf{I} + \sqrt{\mathbf{z}}\mathbf{Y}\sqrt{\mathbf{z}}\right)^{-1} \tag{4.124c}$$

$$= \left(\mathbf{I} + \sqrt{\mathbf{z}}\mathbf{Y}\sqrt{\mathbf{z}}\right)^{-1}\left(\mathbf{I} - \sqrt{\mathbf{z}}\mathbf{Y}\sqrt{\mathbf{z}}\right) \tag{4.124d}$$

4.6.2 Renormalization

Now that we know how to apply scattering parameters for a network having different characteristic impedances at its ports, we should consider the conversion from scattering parameters measured with one set of characteristic impedances to scattering parameters with another set of characteristic impedances. This may be required, for example, when the measurements or simulations are not performed in the same impedance environment as the intended usage, or when the available test equipment was designed for a characteristic impedance other than the that of the device under test.

Since the impedance parameters are not dependent on characteristic impedance, we may do the calculation by first converting the original s-parameters into z-parameters using (4.122a), and then back to s-parameters using (4.123a), but with a different set of normalizing impedances. Let $\mathbf{S}$, $\mathbf{z}$, and $\mathbf{y}$ be the set of original s-parameters and the associated characteristic immittances, and let $\mathbf{S}'$, $\mathbf{z}'$, and $\mathbf{y}'$ be the set to which we wish to convert. We then have

$$\mathbf{S}' = \left(\sqrt{\mathbf{y}'}\mathbf{Z}\sqrt{\mathbf{y}'} - \mathbf{I}\right)\left(\sqrt{\mathbf{y}'}\mathbf{Z}\sqrt{\mathbf{y}'} + \mathbf{I}\right)^{-1} \tag{4.125a}$$

$$= \left(\sqrt{\mathbf{y}'}\sqrt{\mathbf{z}}\,(\mathbf{I}+\mathbf{S})\,(\mathbf{I}-\mathbf{S})^{-1}\sqrt{\mathbf{z}}\sqrt{\mathbf{y}'}-\mathbf{I}\right)$$
$$\cdot\left(\sqrt{\mathbf{y}'}\sqrt{\mathbf{z}}\,(\mathbf{I}+\mathbf{S})\,(\mathbf{I}-\mathbf{S})^{-1}\sqrt{\mathbf{z}}\sqrt{\mathbf{y}'}+\mathbf{I}\right)^{-1} \tag{4.125b}$$

$$= \left(\sqrt{\mathbf{y}'}\sqrt{\mathbf{z}}\,(\mathbf{I}+\mathbf{S}) - \sqrt{\mathbf{z}'}\sqrt{\mathbf{y}}\,(\mathbf{I}-\mathbf{S})\right)$$
$$\cdot\left(\sqrt{\mathbf{y}'}\sqrt{\mathbf{z}}\,(\mathbf{I}+\mathbf{S}) + \sqrt{\mathbf{z}'}\sqrt{\mathbf{y}}\,(\mathbf{I}-\mathbf{S})\right)^{-1} \tag{4.125c}$$

$$= \sqrt{\mathbf{y}'}\sqrt{\mathbf{y}}\,(\mathbf{z}\,(\mathbf{I}+\mathbf{S}) - \mathbf{z}'\,(\mathbf{I}-\mathbf{S}))$$
$$\cdot\,(\mathbf{z}\,(\mathbf{I}+\mathbf{S}) + \mathbf{z}'\,(\mathbf{I}-\mathbf{S}))^{-1}\sqrt{\mathbf{z}}\sqrt{\mathbf{z}'} \tag{4.125d}$$

$$= \sqrt{\mathbf{y}'}\sqrt{\mathbf{y}}\,((\mathbf{z}-\mathbf{z}') + (\mathbf{z}+\mathbf{z}')\,\mathbf{S})\,((\mathbf{z}+\mathbf{z}') + (\mathbf{z}-\mathbf{z}')\,\mathbf{S})^{-1}\sqrt{\mathbf{z}}\sqrt{\mathbf{z}'} \tag{4.125e}$$

$$= \sqrt{\mathbf{y}'}\sqrt{\mathbf{y}}\,(\mathbf{z}+\mathbf{z}')\left((\mathbf{z}+\mathbf{z}')^{-1}(\mathbf{z}-\mathbf{z}')+\mathbf{S}\right)$$
$$\cdot\left(\mathbf{I}+(\mathbf{z}+\mathbf{z}')^{-1}(\mathbf{z}-\mathbf{z}')\,\mathbf{S}\right)^{-1}(\mathbf{z}+\mathbf{z}')^{-1}\sqrt{\mathbf{z}}\sqrt{\mathbf{z}'} \tag{4.125f}$$

In the derivation above, we have used the fact that products of diagonal matrices (e.g., $\sqrt{\mathbf{z}}$ and $\sqrt{\mathbf{y}}$) always commute. To simplify the expression, let

$$\mathbf{\Gamma} = (\mathbf{z}'+\mathbf{z})^{-1}(\mathbf{z}'-\mathbf{z}) \tag{4.126a}$$
$$= (\mathbf{z}'-\mathbf{z})\,(\mathbf{z}'+\mathbf{z})^{-1} \tag{4.126b}$$

$$\mathbf{X} = (\mathbf{z}'+\mathbf{z})^{-1}\sqrt{\mathbf{z}}\sqrt{\mathbf{z}'} \tag{4.126c}$$

Note that $\mathbf{X}$ and $\mathbf{\Gamma}$ are both diagonal matrices. The matrix $\mathbf{\Gamma}$ in particular is a diagonal matrix of reflection coefficients as seen by the original transmission lines when terminated with the new characteristic impedances, or

$$\Gamma_{kk} = \frac{Z'_{0k} - Z_{0k}}{Z'_{0k} + Z_{0k}} \tag{4.127}$$

Substituting these into (4.125f), we find

$$\mathbf{S}' = \mathbf{X}^{-1}(\mathbf{S}-\mathbf{\Gamma})\,(\mathbf{I}-\mathbf{\Gamma}\mathbf{S})^{-1}\,\mathbf{X} \tag{4.128}$$

4.7 PARAMETER-DEFINED NETWORKS

Having described now a wide variety of network parameters that one can use to characterize general, linear, multiport networks, and then deriving the parameter matrices for the known lumped and transmission-line elements, we now allow ourselves to define new networks from the network parameters directly, without immediately constraining ourselves to known elemental forms. Although such networks do not always lend themselves easily to physical realization, we shall find that they are nevertheless useful mathematical artifices for analysis, and that they do approximate the behavior of some useful electromagnetic structures, at least over a limited frequency range.

4.7.1 Gyrators

In all of our discussion of reciprocity and how to verify that our network parameters describe reciprocal networks, we have failed to mention the fact that none of the circuit elements defined so far, whether lumped or transmission-line, are capable of producing a network that is not reciprocal. Typically, nonreciprocal devices require materials such as ferrites that have anisotropic constitutive parameters (ε or μ). Without going into the electromagnetic details of those constructions, we may define a kind of canonical, nonreciprocal, two-port network directly in terms of its scattering parameters,

$$\mathbf{S} = \begin{pmatrix} 0 & -1 \\ 1 & 0 \end{pmatrix} \tag{4.129}$$

Known as a *gyrator*, this matched (reflectionless) network simply transmits the signal incident at either port to the other port, but with a 180° phase shift in the reverse direction. According to Table 4.4, it is clearly lossless (since $\mathbf{S}^*\mathbf{S} = \mathbf{I}$) but nonreciprocal ($\mathbf{S} \neq \mathbf{S}^T$).

The gyrator was first proposed by Bernard Tellegen (of Tellegen's theorem) as a new circuit element [9], along with a circuit symbol, Figure 4.25(a), and some suggested implementations. Gyrators in principle are passive devices,[4] but in modern times they are most often synthesized at low frequencies using active transistor-feedback circuits.

4 In fact, some claim that Tellegen introduced the gyrator as a direct counterexample to the claim that all passive circuits must be reciprocal.

Figure 4.25 (a) Tellegen's schematic symbol for a gyrator. (b) Single-wire schematic symbol for a gyrator.

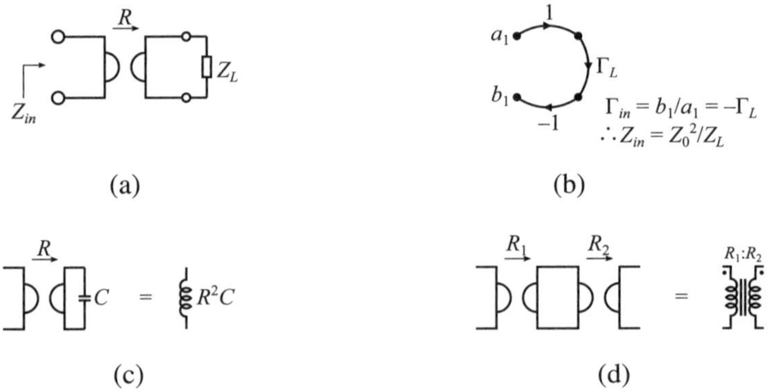

Figure 4.26 (a) Gyrator with an impedance-element termination. (b) Signal flow graph for the terminated gyrator. (c) Conversion of a capacitor into an inductor. (d) A transformer constructed with two gyrators in cascade.

The impedance parameters are traditionally written in terms of the *gyration resistance*, R, where $R = Z_0$,

$$\mathbf{Z} = Z_0\left(\mathbf{I}+\mathbf{S}\right)\left(\mathbf{I}-\mathbf{S}\right)^{-1} = R\begin{pmatrix}1 & -1\\ 1 & 1\end{pmatrix}\begin{pmatrix}1 & 1\\ -1 & 1\end{pmatrix}^{-1} \tag{4.130a}$$

$$= \tfrac{1}{2}R\begin{pmatrix}1 & -1\\ 1 & 1\end{pmatrix}\begin{pmatrix}1 & -1\\ 1 & 1\end{pmatrix} = \begin{pmatrix}0 & -R\\ R & 0\end{pmatrix} \tag{4.130b}$$

The arrow labeled R in the schematic symbol of Figure 4.25(a) represents the directionality of the positive phase shift. In another common symbol, Figure 4.25(b), the arrow is labeled instead with the characteristic phase shift of π radians.

When one port of the gyrator is terminated with a load impedance, Z_L, as in Figure 4.26(a), the effective impedance presented to the other port is

$$Z_{in} = \frac{Z_0^2}{Z_L} = \frac{R^2}{Z_L} \tag{4.131}$$

This may be proven by considering the signal flow graph for the network, Figure 4.26(b). The gyrator imparts a negative sign to the reflection coefficient from the original load termination, effectively inverting its value relative to the characteristic impedance, Z_0. This effectively converts an inductor into a capacitor, a capacitor into an inductor, a series LC resonator into a parallel LC resonator, and so on.

Taken to an extreme, the gyrator converts an open-circuit to a short-circuit and vice versa. This makes the gyrator an ideal current-sensing element, for one may insert one branch of the gyrator into the current path and attach a voltmeter (an open-circuit) to the other branch.

Let us now consider two gyrators in cascade. First, we convert the z-parameters from (4.130) to $ABCD$-parameters using the formula in Table 4.2

$$\mathbf{A} = \frac{1}{z_{21}} \begin{pmatrix} z_{11} & \det \mathbf{Z} \\ 1 & z_{22} \end{pmatrix} = \frac{1}{R} \begin{pmatrix} 0 & R \\ 1 & 0 \end{pmatrix} = \begin{pmatrix} 0 & R \\ R^{-1} & 0 \end{pmatrix} \tag{4.132}$$

If the first gyrator has gyration resistance R_1, and the second has gyration resistance R_2, then the $ABCD$-matrix of the combination is

$$\mathbf{A}' = \begin{pmatrix} 0 & R_1 \\ R_1^{-1} & 0 \end{pmatrix} \begin{pmatrix} 0 & R_2 \\ R_2^{-1} & 0 \end{pmatrix} = \begin{pmatrix} \frac{R_1}{R_2} & 0 \\ 0 & \frac{R_2}{R_1} \end{pmatrix} \tag{4.133}$$

which, according to Table 4.1, is a transformer having a turns ratio $n = R_1/R_2$.

It is a matter of some academic note, then, that all lumped-element networks can be realized with just three fundamental pieces: resistors, gyrators, and one kind of reactive element, say, capacitors. The role of an inductor may then be filled with a gyrator and a capacitor, as in Figure 4.26(c), and that of the transformer may be filled with two gyrators in cascade, as in Figure 4.26(d). The gyrator is critical to this set, as there is no way to create a nonreciprocal element with a combination of the remaining reciprocal elements.

The shared ability of gyrators and quarter-wave transformers to invert the magnitude of a load impedance — compare (4.131) to (3.44) — has led some to call the quarter-wave transformer a type of gyrator. This is erroneous, however, as the gyrator is by design a nonreciprocal device, and transmission lines, as we already know, are quite reciprocal. To identify one with the other on the basis of one shared, single-port behavior, when the gyrator is defined explicitly by its two-port nonreciprocity, is misleading at best — to say nothing of the fact that transmission lines exhibit impedance inversion at only a single frequency point, and that there are numerous other (reciprocal) networks which accomplish the same thing.

4.7.2 Impedance Inverters

In terms of the s-parameters, assuming the ports are matched, the only feature that is required for impedance inversion is that the total phase shift from one port to the other and back is 180°, or π radians. To do that with a reciprocal s-matrix simply requires distributing this phase shift evenly in both directions,

$$\mathbf{S} = \begin{pmatrix} 0 & -j \\ -j & 0 \end{pmatrix} \tag{4.134}$$

Note that either $+j$ or $-j$ could have been used for the transmission coefficients, but since causality would generally require a delay in phase over wide-ranging frequencies, the lagging instead of leading phase term has been selected here. It is interesting to consider the immittance parameters that result from this formulation,

$$\mathbf{Z} = Z_0 \left(\mathbf{I} + \mathbf{S}\right) \left(\mathbf{I} - \mathbf{S}\right)^{-1} = Z_0 \begin{pmatrix} 1 & -j \\ -j & 1 \end{pmatrix} \begin{pmatrix} 1 & j \\ j & 1 \end{pmatrix}^{-1} \tag{4.135a}$$

$$= \tfrac{1}{2} Z_0 \begin{pmatrix} 1 & -j \\ -j & 1 \end{pmatrix} \begin{pmatrix} 1 & -j \\ -j & 1 \end{pmatrix} = \begin{pmatrix} 0 & -jZ_0 \\ -jZ_0 & 0 \end{pmatrix} \tag{4.135b}$$

$$\therefore \mathbf{Y} = \mathbf{Z}^{-1} = Y_0 \begin{pmatrix} 0 & -j \\ -j & 0 \end{pmatrix}^{-1} = \begin{pmatrix} 0 & jY_0 \\ jY_0 & 0 \end{pmatrix} \tag{4.135c}$$

and the $ABCD$-parameters derived from the conversion formula,

$$\begin{pmatrix} A & B \\ C & D \end{pmatrix} = \frac{-1}{Y_{21}} \begin{pmatrix} Y_{22} & 1 \\ \Delta & Y_{11} \end{pmatrix} = \frac{j}{Y_0} \begin{pmatrix} 0 & 1 \\ Y_0^2 & 0 \end{pmatrix} = j \begin{pmatrix} 0 & Z_0 \\ Y_0 & 0 \end{pmatrix} \tag{4.136}$$

Although mathematically identical, it is common in some circles to call impedance inverters defined by the normalizing impedance *Z-inverters* or *impedance inverters*, whereas those defined by the normalizing admittance are called *Y-inverters* or *admittance inverters*. In still other texts, the normalizing impedance is labeled K, or the normalized admittance is labeled J, in which case they are referred to as *K-inverters* or *J-inverters*, respectively.

Whatever the network is called, a comparison of these parameter matrices with the forms in Table 4.1 suggests numerous possible network structures for such

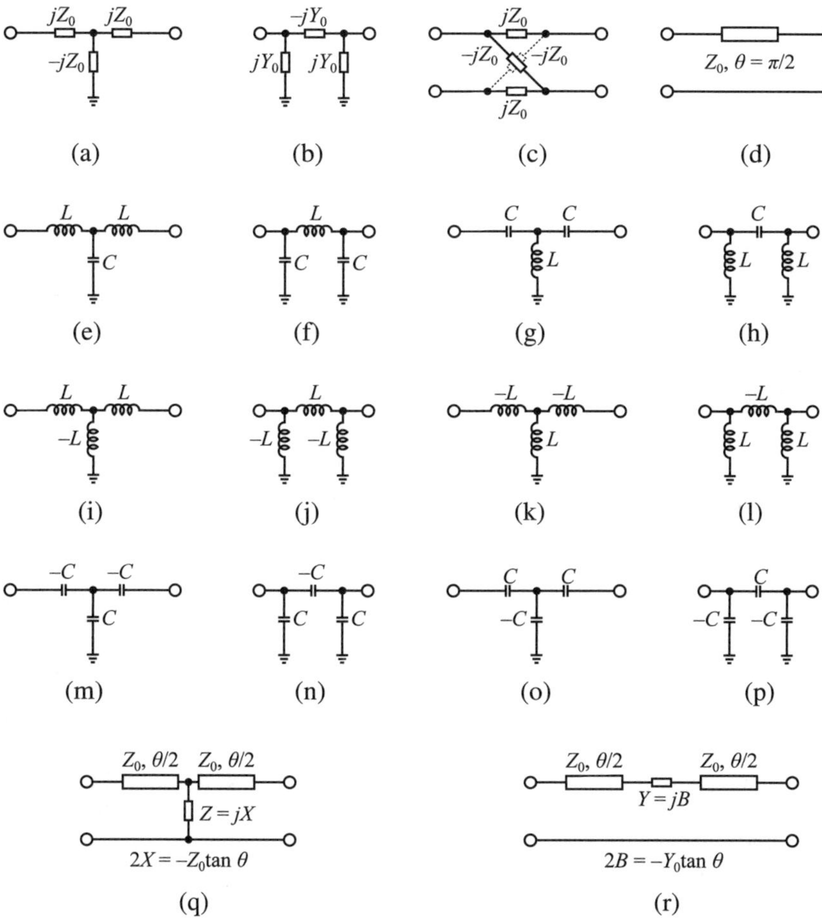

Figure 4.27 Impedance inverters. (a) Ideal tee network. (b) Ideal pi network. (c) Ideal lattice network. (d) 90° transmission line. (e) Low-pass tee network. (f) Low-pass pi network. (g) High-pass tee network. (h) High-pass pi network. (i–l) All-inductor forms. (m–p) All-capacitor forms. (q) Shunt reactance with port extensions. (r) Series susceptance with port extensions.

an inverter, which are illustrated in Figure 4.27. The first three, Figures 4.27(a–c), are tee, pi, and lattice networks formed with constant (frequency-independent) reactance and susceptance elements. In principle, at least, these implement the impedance-inverting function represented by (4.134) exactly. Figure 4.27(d) shows a transmission line with $\pi/2$ radians of electrical length. (Note that this result was anticipated in Section 3.3.2, when it was called a quarter-wave transformer.) Were the transmission line able to maintain this electrical length across a broad range of frequencies, then it too would implement the impedance-inverting function exactly.

In reality, lumped reactances are typically not frequency independent, and transmission lines do have frequency-dependent electrical lengths. Approximations of the impedance inverter are possible, however, using realistic circuit elements. Figures 4.27(e, f) show the tee- and pi-network configurations implemented with capacitors in place of the negative reactances and inductors in place of the negative susceptances. These networks implement a perfect impedance inverter (with real-valued normalizing impedance) at only a single frequency, at which the elements are in resonance. Still, they are good enough to approximate the desired behavior over a narrow bandwidth. Note also that these structures tend toward an open-circuit or short-circuit at high frequency. Thus, they may be described as low-pass inverters.

Recall, also, that we had a choice in (4.134) about which sign to use in the transmission coefficients. Had we made the opposite choice then, the immittances in Figures 4.27(a, b) would be negated, and the approximate inverter networks we arrived at here would instead have taken on high-pass forms, as shown in Figures 4.27(g, h). Again, these replicate the desired impedance inversion exactly at only a single frequency point.

Alternatively, one may approximate the inverter using only one kind of element: inductors in Figures 4.27(i–l) and capacitors in Figures 4.27(m–p). While this requires some of the elements in every case to take on negative values, which is nonphysical, these structures still find extensive use in filter synthesis where the negative elements can sometimes be absorbed into neighboring elements having larger values. Compared to the mixed element-type networks, these have the advantage that they do implement an impedance inversion at all frequencies, although the real-valued normalizing impedance does change with frequency.

Finally, two inverter forms that are useful when designing filters involving reactive discontinuities along a transmission line are shown in Figures 4.27(q, r). The inverting behavior of the first network having a shunt reactance can be verified by examining the $ABCD$-parameters of the cascade,

$$\begin{pmatrix} A & B \\ C & D \end{pmatrix} = \begin{pmatrix} \cos(\theta/2) & jZ_0\sin(\theta/2) \\ jY_0\sin(\theta/2) & \cos(\theta/2) \end{pmatrix} \cdot \begin{pmatrix} 1 & 0 \\ (jX)^{-1} & 1 \end{pmatrix} \begin{pmatrix} \cos(\theta/2) & jZ_0\sin(\theta/2) \\ jY_0\sin(\theta/2) & \cos(\theta/2) \end{pmatrix} \tag{4.137a}$$

$$= j\begin{pmatrix} 0 & Z_0\tan(\theta/2) \\ Y\cot(\theta/2) & 0 \end{pmatrix} \tag{4.137b}$$

which is the same as an impedance inverter with normalizing impedance $K = Z_0\tan(\theta/2)$. A similar analysis shows that the network in Figure 4.27(r) behaves as an admittance inverter with normalizing admittance $J = -Y_0\tan(\theta/2)$.

Problems

4-1 What are the z-parameters of a three-port network having series 1Ω resistors connecting each port to a common node, and a shunt 1Ω resistor from the common node to ground?

4-2 What are the z-parameters of a two-port, two-element network comprising a series 50Ω resistor between the two ports and a shunt 100Ω resistor on port 2? What are its y-parameters?

4-3 Ports 1 and 2 of network A, a three-port network, are combined in parallel with ports 1 and 2 of network B, a two-port network. What are the y-parameters of the combined network in terms of the y-parameters of the original networks?

4-4 Given the $ABCD$-parameters of a two-port network, write down the $ABCD$-parameters for that same network in reverse (port 1 becomes port 2 and vice versa).

4-5 Derive the $ABCD$-parameters for the two-port networks shown in the first three rows of Table 4.1.

4-6 Convert the z-parameters for the tee network shown in the fourth row of Table 4.1 into $ABCD$-parameters.

4-7 Both ends of a two-wire transmission line with characteristic impedance Z_0 and electrical length θ are joined together in parallel. What is the input impedance of the structure if the ends are joined in like polarity (positive-to-positive, and negative-to-negative)? What if the line is twisted, that is, the ends are connected in opposite polarity (positive-to-negative and negative-to-positive)?

4-8 Write the z-, y-, $ABCD$-, and s-parameters of a lossless transmission line that has characteristic impedance Z_0 and is a quarter wavelength long.

4-9 Convert the z-parameters for the tee-network given in Table 4.1 to scattering parameters.

4-10 Write the transfer scattering parameters for a transmission line.

4-11 Prove that the antimetry condition stated in (4.71) is, for s-parameters, that $s_{11} = -s_{22}$, and for $ABCD$-parameters that $B/C = Z_0^2$.

4-12 Write down a 2×2 scattering parameter matrix that is reciprocal and nonpassive, despite satisfying the condition that $|s_{11}|^2 + |s_{21}|^2 = 1$ and $|s_{22}|^2 + |s_{21}|^2 = 1$. Give an example of an input wave vector, $\mathbf{a}$, for which the output has more total power than the input.

4-13 Use signal-flow graph decomposition or Mason's rule to derive the reflection coefficient at port 1 of a three-port network, where port 2 is terminated with a reflection, Γ_2, and port 3 is terminated with a reflection, Γ_3.

4-14 Derive the impedance matrix of a tee-network having series arms z_a and shunt arm z_b using even-/odd-mode analysis.

4-15 Calculate the even and odd-mode impedances that a pair of coupled lines must have when a quarter-wavelength long to couple half the signal power input at port 1 to port 3, while keeping the input reflectionless ($s_{11} = 0$). Assume that the even and odd-mode propagation constants are the same and the lines are lossless ($\gamma_e = \gamma_o = j\beta$).

4-16 What are the four-port admittance parameters of a general coupled-line section?

4-17 A coaxial cable is found to exhibit a reflection coefficient of $s_{11} = 5/13$ and a transmission coefficient of $s_{21} = -j12/13$ at a particular frequency when embedded in a system with characteristic impedance $Z_0 = 50\Omega$. Calculate what its scattering parameters would be in a 75Ω system.

4-18 Derive expressions for the normalizing impedance of the impedance inverters in Figures 4.27(e), (i), and (m).

References

[1] D. M. Pozar, *Microwave Engineering*, 4th ed. New York: Wiley, 2011.

[2] G. Gonzalez, *Microwave Transistor Amplifiers: Analysis and Design*, 2nd ed. Upper Saddle River, NJ: Prentice Hall, 1996.

[3] Wikipedia. (2017) Two-port network. https://en.wikipedia.org/wiki/Two-port_network.

[4] M. A. Morgan, *Reflectionless Filters*. Norwood, MA: Artech House, 2017.

[5] S. Boyd and L. Chua, "On the passivity criterion for LTI N-ports," *International Journal of Circuit Theory and Applications*, vol. 10, no. 4, pp. 323–333, October 1982.

[6] L. Weinberg and P. Slepian, "Positive real matrices," *Journal of Mathematics and Mechanics*, vol. 9, no. 1, pp. 71–83, 1960.

[7] Wikipedia. (2018) Mason's gain formula. https://en.wikipedia.org/wiki/Mason's_gain_formula.

[8] P. Russer, *Electromagnetics, Microwave Circuit and Antenna Design for Communications Engineering*, 2nd ed. Norwood, MA: Artech House, 2006.

[9] B. D. H. Tellegen, "The gyrator, a new electric network element," *Phillips Research Report*, vol. 3, pp. 81–101, 1948.

Chapter 5

Transformations and Identities

In this chapter, we study two kinds of equivalence between networks. First, we have transformations, which relate the behavior of one network to the behavior of another given a change in independent variables (frequency or impedance), and second, we have identities, which are pairs of networks which have different topological forms but are otherwise indistinguishable from one another in terms of their port behavior.

5.1 TRANSFORMATIONS

In some cases, having gone through the effort of designing and characterizing a network over a given frequency range and for a particular port impedance, it may be useful to convert the elements of that design in such a way that the same essential result is achieved, but for a different port impedance or over a different frequency regime. This is the purpose of transformations.

The classic example is a filter, which may be first optimized to provide the best selectivity (or slope) in a low-pass configuration, given impedance, ripple, and rejection requirements, but which one then wishes to implement in a high-pass or band-pass configuration without losing the desirable features of the original prototype. Filters will not be studied in great detail until Chapter 11, but the transformations will be studied here because they apply to many kinds of networks, not just filters. In Chapter 6, for example, we will use some of these transformations to generate eight distinct solutions to impedance-matching problems, after having meticulously derived only one.

5.1.1 Impedance Scaling

Section 4.6.2 showed us how to renormalize the scattering parameters of a network to see how it would behave with different port impedances. Suppose, instead, we wanted to modify the network so that the elements and normalizing impedances change, but the scattering parameters stay the same.

Since the behavior of a network is completely described by its network parameters, and those parameters may be constructed cumulatively using operations on more basic building blocks (e.g., $\mathbf{Z}$ adding in series, or $\mathbf{Y}$ adding in parallel), most transformations may be applied on an element-by-element basis. Preserve the behavior of each single element, and the aggregate behavior of the network is preserved as well.

Consider an inductor that is part of a larger network. In its most basic form, the inductor itself may be considered a single-port network (embedded in a larger one) whose impedance matrix is a scalar having value $z_{11} = j\omega L$. When terminated with a normalizing impedance of Z_0, it has a single scattering parameter,

$$\mathbf{S} = (s_{11}) = (\mathbf{Z} + Z_0\mathbf{I})^{-1}(\mathbf{Z} - Z_0\mathbf{I}) = \frac{z_{11} - Z_0}{z_{11} + Z_0} = \frac{j\omega L - Z_0}{j\omega L + Z_0} \tag{5.1}$$

We now wish to scale the port impedance from Z_0 to Z_0', but to preserve the scattering parameter as it is. The answer may seem obvious, but it will be instructive for later sections if we go through the math:

$$s_{11} = \frac{j\omega L - Z_0}{j\omega L + Z_0} = \frac{j\omega L' - Z_0'}{j\omega L' + Z_0'} \tag{5.2a}$$

$$\frac{L}{Z_0} = \frac{L'}{Z_0'} \tag{5.2b}$$

$$\therefore L' = \frac{Z_0'}{Z_0} L \tag{5.2c}$$

A similar exercise finds, for the resistors and capacitors,

$$R' = \frac{Z_0'}{Z_0} R \tag{5.3a}$$

$$C' = \frac{Y_0'}{Y_0} C \tag{5.3b}$$

Hence, to preserve the behavior of a network containing resistors, inductors, and capacitors while modifying the port impedance, we must scale the resistors and inductors by the ratio of the port impedances, and the capacitors by the ratio of the port admittances.

If the network contains transmission lines, we should be slightly more careful. It could be a mistake to reduce the transmission line to a single-port building block — a stub — for not all transmission lines in a network are stubs. Nonetheless, by equating the full two-port s-parameters for a transmission line (4.58d) for the different port impedances,

$$\frac{\begin{pmatrix}(z-y)\sinh(\gamma l) & 2 \\ 2 & (z-y)\sinh(\gamma l)\end{pmatrix}}{2\cosh(\gamma l)+(z+y)\sinh(\gamma l)} = \frac{\begin{pmatrix}(z'-y')\sinh(\gamma' l') & 2 \\ 2 & (z'-y')\sinh(\gamma' l')\end{pmatrix}}{2\cosh(\gamma' l')+(z'+y')\sinh(\gamma' l')} \tag{5.4}$$

we obtain a very simple scaling law,

$$z = z' \tag{5.5a}$$

$$\gamma l = \gamma' l' \tag{5.5b}$$

Recall that z in this formula is the normalized characteristic impedance, so (5.5a) becomes

$$\frac{Z_1}{Z_0} = \frac{Z_1'}{Z_0'} \tag{5.6a}$$

$$\therefore Z_1' = Z_1 \frac{Z_0'}{Z_0} \tag{5.6b}$$

Therefore, the characteristic impedance scales just like the resistor or inductor values in the network. It is interesting to note that (5.5b) does allow for the propagation constant and line length to both change under this transformation, so long as they compensate one another. Traditionally, however, both are assumed to remain constant.

5.1.2 Frequency Scaling

To most, the impedance scaling laws of the previous section will seem rather obvious, but it was deemed useful to go through the exercise to obtain familiarity with a rigorous mathematical approach for use in later transformations. In this section, we consider the case where not the impedance, but the frequency range is to be scaled by a constant factor. For example, let us assume that we have a network that exhibits the desired behavior at one fixed frequency, ω_0, but we wish to modify it so that the same behavior now appears at a new fixed frequency, ω_0'. Equating the impedance of the inductor before and after the change of frequency

$$j\omega_0 L = j\omega_0' L' \tag{5.7a}$$

$$L' = \frac{\omega_0}{\omega_0'} L \tag{5.7b}$$

and for resistors and capacitors,

$$R' = R \tag{5.8a}$$

$$C' = \frac{\omega_0}{\omega_0'} C \tag{5.8b}$$

For transmission lines, it is useful to expand the propagation factor, γl, in terms of frequency, ω_0. We will limit ourselves to the case where the line is lossless (or near lossless), so

$$\gamma l = j\beta l = \frac{j\omega_0 l}{v_p} \tag{5.9}$$

where v_p is the phase velocity of the line (see (3.11c)). Equating the network parameters of the transmission line before and after the frequency scaling, we have

$$\frac{j\omega_0 l}{v_p} = \frac{j\omega_0' l'}{v_p'} \tag{5.10}$$

In principle, we could engineer the transmission lines to have different phase velocities, but usually it is easier just to modify the length,

$$l' = \frac{\omega_0}{\omega_0'} l \tag{5.11}$$

5.1.3 Frequency Inversion (High-Pass Transformation)

In some cases, it is useful to invert the frequency response of a network — to change it so that its behavior at low frequencies is exchanged with its behavior at high frequencies and vice versa. In filter design, this is often used to convert a low-pass prototype to a high-pass configuration.

We would therefore like to make the following substitution while preserving the impedance parameters,

$$\frac{\omega}{\omega_0} = -\frac{\omega_0'}{\omega'} \tag{5.12}$$

A few comments about this equation are warranted. First, the negative sign may seem superfluous (and indeed is absent in many other books [1]). The rationale for its presence will soon become clear. For now, it is sufficient to recognize that any real, physical network must have a response that is symmetric about zero frequency anyway, owing to the properties of the Fourier transform. Consequently, the presence of the negative sign has no effect on the intended transformation.

Second, we note in this case that the free variables, ω and ω', are inversely related, with two fixed frequency points, ω_0 and ω_0', acting as normalizing constants. If $\omega_0 = \omega_0'$, then this becomes the pivot point about which the entire frequency scale is inverted. If $\omega_0 \neq \omega_0'$, then the transformation combines inversion simultaneously with frequency scaling, as described in Section 5.1.2. (Equivalently, one may consider it a simple inversion about a fixed point that is the geometric mean of ω_0 and ω_0'.)

Evaluating the impedance of an inductor, we have

$$j\omega L = j\left(\tfrac{\omega}{\omega_0}\right)\omega_0 L = j\left(-\tfrac{\omega_0'}{\omega'}\right)\omega_0 L = \frac{1}{j\omega'\left(\omega_0'\omega_0 L\right)^{-1}} = \frac{1}{j\omega' C'} \tag{5.13}$$

Thus, we find that the impedance of the inductor becomes that of a capacitor having value $C' = (\omega_0'\omega_0 L)^{-1}$. Likewise, the impedance of a capacitor is

$$\frac{1}{j\omega C} = \left(j\tfrac{\omega}{\omega_0}\omega_0 C\right)^{-1} = \left(-j\tfrac{\omega_0'}{\omega'}\omega_0 C\right)^{-1} = j\omega'\left(\omega_0'\omega_0 C\right)^{-1} = j\omega' L' \tag{5.14}$$

which is equivalent to an inductor having value $L' = (\omega_0'\omega_0 C)^{-1}$. Resistors remain unchanged under this transformation. Note that, had we not used the negative sign in (5.12), the elements corresponding to the transformed network would have been more difficult to recognize.

Consider, as an example, the network from Figure 4.15(b) which was analyzed using even and odd-mode analysis in Section 4.5.1. We claimed at the time that it implemented a third-order Chebyshev Type-II filter, and further proved that it was reflectionless ($s_{11} = s_{22} = 0$) at all frequencies. We would like now to design a network that preserves these desirable properties, but in a high-pass instead of low-pass form, without having to start over from scratch.

The original network has been redrawn in Figure 5.1(a), with element values

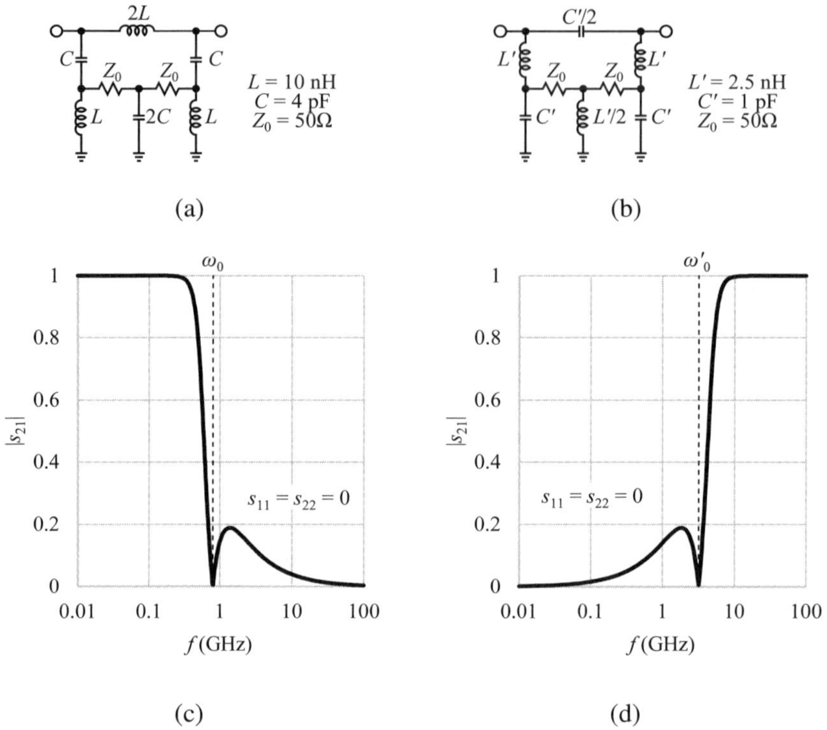

Figure 5.1 (a) Original low-pass network with transmission null at $\omega_0 = 5 \cdot 10^9$ rad/s ($\approx$ 0.8 GHz). (b) Modified high-pass network with transmission zero at $\omega_0' = 20 \cdot 10^9$ rad/s ($\approx$ 3.2 GHz). (c) Frequency response of the original network. (d) Frequency response of the modified network.

assigned such that it has a transmission null at $\omega_0 = 5 \cdot 10^9$ rad/s ($\approx$ 0.8 GHz). The full frequency response is plotted in Figure 5.1(c). We wish to invert this spectrum, while moving the fixed point (the transmission null in this case) to $\omega_0' = 20 \cdot 10^9$ rad/s ($\approx$ 3.2 GHz). We therefore substitute capacitors for the inductors,

$$C' = (\omega_0 \omega_0' L)^{-1} = \frac{1}{(5 \cdot 10^9 \text{ rad/s})\,(20 \cdot 10^9 \text{ rad/s})\,(10 \text{ nH})} = 1 \text{ pF} \qquad (5.15)$$

and inductors for the capacitors

$$L' = (\omega_0 \omega_0' C)^{-1} = \frac{1}{(5 \cdot 10^9 \text{ rad/s})\,(20 \cdot 10^9 \text{ rad/s})\,(4 \text{ pF})} = 2.5 \text{ nH} \qquad (5.16)$$

The modified network is thus shown in Figure 5.1(b). Note that the factors of 2 for some elements have converted to factors of one-half in the modified network. The new frequency response is shown in Figure 5.1(d), and is noted to be a shifted, mirror image of the original, as was our intent. More importantly, all other aspects of the network's behavior, such as the transition slope, peak ripple level, and the reflectionless condition, remain intact.

5.1.4 Band-Pass Transformation

Suppose instead we wished to localize the passband of the filter, currently covering 0 to ω_0, around a particular range of frequencies, ω_1' to ω_2'. We could then employ the following transformation,

$$\frac{\omega}{\omega_0} = \frac{1}{\Delta}\left(\frac{\omega'}{\omega_0'} - \frac{\omega_0'}{\omega'}\right) \tag{5.17a}$$

$$\omega_0' = \sqrt{\omega_1'\omega_2'} \tag{5.17b}$$

$$\Delta = \frac{\omega_2' - \omega_1'}{\omega_0'} \tag{5.17c}$$

where ω_0' is the center (geometric mean) of the target frequency range, and Δ is the relative bandwidth. It may be useful to recognize this as the combination (or the sum in frequency space) of a simple frequency-scaling transformation with scaling factor $1/\Delta$ and the high-pass transformation of (5.12), similarly scaled. The impedance of any inductors, then, is given by

$$Z_L(\omega) = j\omega L = j\frac{\omega_0}{\Delta}\left(\frac{\omega'}{\omega_0'} - \frac{\omega_0'}{\omega'}\right)L \tag{5.18a}$$

$$= j\omega'\left(\frac{\omega_0 L}{\omega_0'\Delta}\right) + \left(j\omega'\frac{\Delta}{\omega_0\omega_0' L}\right)^{-1} \tag{5.18b}$$

$$= j\omega' L_1' + \left(j\omega' C_1'\right)^{-1} = Z_{L_1'}(\omega') + Z_{C_1'}(\omega') \tag{5.18c}$$

which may be recognized as an inductor and capacitor in series, where

$$L_1' = \frac{\omega_0 L}{\omega_0'\Delta} \tag{5.19a}$$

$$C_1' = \frac{\Delta}{\omega_0 \omega_0' L} \tag{5.19b}$$

Similarly, the admittance of any capacitor in the network would become

$$Y_C(\omega) = j\omega C = j\frac{\omega_0}{\Delta}\left(\frac{\omega'}{\omega_0'} - \frac{\omega_0'}{\omega'}\right) C \tag{5.20a}$$

$$= j\omega'\left(\frac{\omega_0 C}{\omega_0' \Delta}\right) + \left(j\omega' \frac{\Delta}{\omega_0 \omega_0' C}\right)^{-1} \tag{5.20b}$$

$$= j\omega' C_2' + (j\omega' L_2')^{-1} = Y_{C_2'}(\omega') + Y_{L_2'}(\omega') \tag{5.20c}$$

which has the form of a capacitor in parallel with an inductor, where

$$C_2' = \frac{\omega_0 C}{\omega_0' \Delta} \tag{5.21a}$$

$$L_2' = \frac{\Delta}{\omega_0 \omega_0' C} \tag{5.21b}$$

Let us apply these results to our example network from Figure 5.1(a). Suppose we wish to locate the transmission nulls at $\omega_1' = 1 \cdot 10^9$ rad/s and $\omega_2' = 9 \cdot 10^9$ rad/s. We can then calculate

$$\omega_0' = \sqrt{\omega_1' \omega_2'} = \sqrt{(1 \cdot 10^9 \text{ rad/s})(9 \cdot 10^9 \text{ rad/s})} = 3 \cdot 10^9 \text{ rad/s} \tag{5.22a}$$

$$\Delta = \frac{\omega_2' - \omega_1'}{\omega_0'} = \frac{(9 \cdot 10^9 \text{ rad/s}) - (1 \cdot 10^9 \text{ rad/s})}{(3 \cdot 10^9 \text{ rad/s})} = \frac{8}{3} \tag{5.22b}$$

$$L_1' = \frac{\omega_0 L}{\omega_0' \Delta} = \frac{(5 \cdot 10^9 \text{ rad/s})(10 \text{ nH})}{(3 \cdot 10^9 \text{ rad/s})\left(\frac{8}{3}\right)} = 6.25 \text{ nH} \tag{5.22c}$$

$$C_1' = \frac{\Delta}{\omega_0 \omega_0' L} = \frac{\left(\frac{8}{3}\right)}{(5 \cdot 10^9 \text{ rad/s})(3 \cdot 10^9 \text{ rad/s})(10 \text{ nH})} = 17.8 \text{ pF} \tag{5.22d}$$

$$C_2' = \frac{\omega_0 C}{\omega_0' \Delta} = \frac{(5 \cdot 10^9 \text{ rad/s})(4 \text{ pF})}{(3 \cdot 10^9 \text{ rad/s})\left(\frac{8}{3}\right)} = 2.5 \text{ pF} \tag{5.22e}$$

$$L_2' = \frac{\Delta}{\omega_0 \omega_0' C} = \frac{\left(\frac{8}{3}\right)}{(5 \cdot 10^9 \text{ rad/s})(3 \cdot 10^9 \text{ rad/s})(4 \text{ pF})} = 44.4 \text{ nH} \tag{5.22f}$$

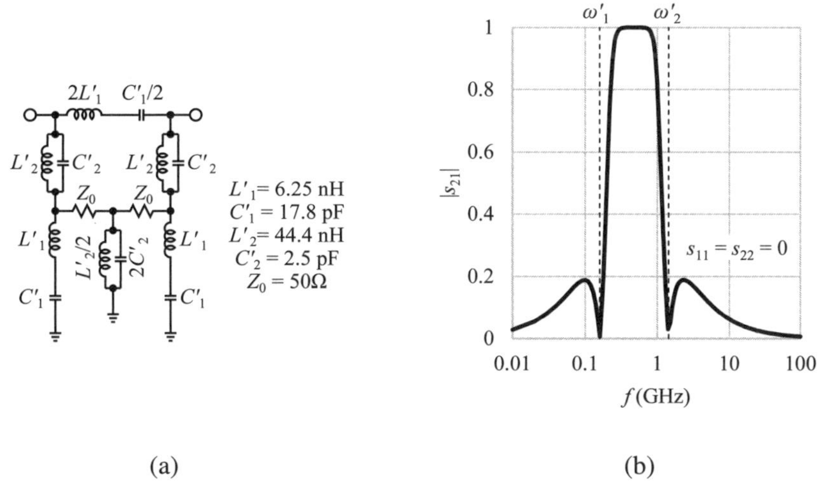

Figure 5.2 (a) Modified network with $\omega_1' = 1 \cdot 10^9$ rad/s ($\approx$159 MHz) and $\omega_2' = 9 \cdot 10^9$ rad/s ($\approx$1.43 GHz). (b) Frequency response.

The resulting network and its frequency response are shown in Figure 5.2.

It is interesting to consider the transformation under the limit as ω_2' approaches infinity,

$$\frac{\omega}{\omega_0} = \lim_{\omega_2' \to \infty} \frac{1}{\Delta}\left(\frac{\omega'}{\omega_0'} - \frac{\omega_0'}{\omega'}\right) = \lim_{\omega_2' \to \infty} \frac{\sqrt{\omega_1'\omega_2'}}{\omega_2' - \omega_1'}\left(\frac{\omega'}{\sqrt{\omega_1'\omega_2'}} - \frac{\sqrt{\omega_1'\omega_2'}}{\omega'}\right) \tag{5.23a}$$

$$= \lim_{\omega_2' \to \infty}\left(\frac{\omega'}{\omega_2' - \omega_1'} - \frac{\omega_1'\omega_2'}{(\omega_2' - \omega_1')\,\omega'}\right) = 0 - \frac{\omega_1'}{\omega'} = -\frac{\omega_1'}{\omega'} \tag{5.23b}$$

Note that it has reduced to the frequency-inversion (high-pass) transformation of Section 5.1.3. This consistency further underscores the reasoning behind having the negative sign in that transformation.

5.1.5 Band-Stop Transformation

Finally, if we wanted to confine the stopband of the filter, currently ω_0 to infinity, to a finite range between ω_1' and ω_2', we could apply a transformation that is the multiplicative inverse of the band-pass transformation (alternatively, a band-pass and high-pass transformation combined),

$$\frac{\omega}{\omega_0} = \Delta\left(\frac{\omega_0'}{\omega'} - \frac{\omega'}{\omega_0'}\right)^{-1} \tag{5.24a}$$

$$Y_L(\omega) = \frac{1}{j\omega L} = \frac{1}{j\omega_0 L\Delta}\left(\frac{\omega_0'}{\omega'} - \frac{\omega'}{\omega_0'}\right) \tag{5.24b}$$

$$= \frac{1}{j\omega'\left(\frac{\omega_0 L\Delta}{\omega_0'}\right)} + j\omega'\left(\frac{1}{\omega_0\omega_0' L\Delta}\right) \tag{5.24c}$$

$$= Y_{L_1'}(\omega') + Y_{C_1'}(\omega') \tag{5.24d}$$

$$Z_C(\omega) = \frac{1}{j\omega C} = \frac{1}{j\omega_0 C\Delta}\left(\frac{\omega_0'}{\omega'} - \frac{\omega'}{\omega_0'}\right) \tag{5.24e}$$

$$= \frac{1}{j\omega'\left(\frac{\omega_0 C\Delta}{\omega_0'}\right)} + j\omega'\left(\frac{1}{\omega_0\omega_0' C\Delta}\right) \tag{5.24f}$$

$$= Z_{C_2'}(\omega') + Z_{L_2'}(\omega') \tag{5.24g}$$

where the band's center frequency, ω_0', and relative bandwidth, Δ, are defined as before in (5.17b) and (5.17c), respectively. In this case, each inductor transforms to a scaled inductor and capacitor in parallel, where each capacitor transforms to a scaled inductor and capacitor in series.

A summary of the transformations discussed so far in this chapter is given in Table 5.1. These transformations will be revisited when we talk more about filters in Chapter 11.

5.1.6 Richard's Transformation

Let us now consider a new transformation where the normal, linear frequency scale acquires a trigonometric dependence,

$$\frac{\omega}{\omega_0} = \tan\left(\frac{\omega' l}{v_p}\right) \tag{5.25}$$

Applying this transformation to the impedance of an inductor yields

$$j\omega L = j\omega_0 L \tan\left(\frac{\omega' l}{v_p}\right) = j\omega_0 L \tan\theta \tag{5.26}$$

Table 5.1
Network Transformations

Original / Low-Pass	Impedance Scaling	Frequency Scaling	Frequency Inversion / High-Pass	Band-Pass	Band-Stop
$\frac{\omega}{\omega_0} =$	$\frac{\omega}{\omega_0}$	$\frac{\omega'}{\omega'_0}$	$-\frac{\omega'_0}{\omega'}$	$\frac{1}{\Delta}\left(\frac{\omega'}{\omega'_0} - \frac{\omega'_0}{\omega'}\right)$	$\Delta\left(\frac{\omega'_0}{\omega'} - \frac{\omega'}{\omega'_0}\right)^{-1}$
fixed points:					
$\omega = 0$	$\omega = 0$	$\omega' = 0$	$\omega' = \infty$	$\omega' = \omega'_0$	$\omega' = 0, \infty$
$\omega = \omega_0$	$\omega = \omega_0$	$\omega' = \omega'_0$	$\omega' = \omega_0$	$\omega' = \omega'_1, \omega'_2$	$\omega' = \omega'_1, \omega'_2$
$\omega = \infty$	$\omega = \infty$	$\omega' = \infty$	$\omega' = 0$	$\omega' = 0, \infty$	$\omega' = \omega'_0$
L	$\frac{Z'_0}{Z_0}L$	$\frac{\omega_0}{\omega'_0}L$	$(\omega_0\omega'_0 L)^{-1}$	$\frac{\omega_0 L}{\omega'_0\Delta}, \frac{\Delta}{\omega_0\omega'_0 L}$	$\frac{\omega_0 L\Delta}{\omega'_0} // \frac{1}{\omega_0\omega'_0 L\Delta}$
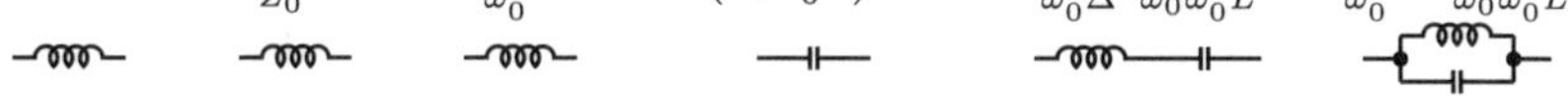					
C	$\frac{Y'_0}{Y_0}C$	$\frac{\omega_0}{\omega'_0}C$	$(\omega_0\omega'_0 C)^{-1}$	$\frac{\Delta}{\omega_0\omega'_0 C} // \frac{\omega_0 C}{\omega'_0\Delta}$	$\frac{1}{\omega_0\omega'_0 C\Delta}, \frac{\omega_0 C\Delta}{\omega'_0}$
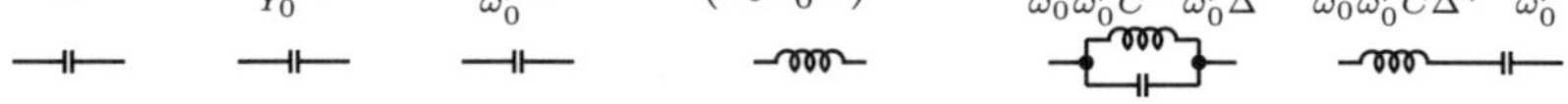					
R	$\frac{Z'_0}{Z_0}R$	R	R	R	R

$Z_1, \gamma l$	$\frac{Z'_0}{Z_0}Z_1, \gamma l$	$Z_1, \frac{\omega_0}{\omega'_0}\gamma l$			
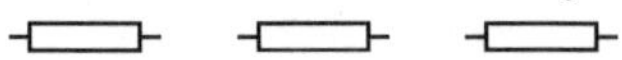					

ω'_0, ω'_1, ω'_2, and Δ are defined in the text.

which, by design, is the input impedance of a short-circuited transmission-line stub having characteristic impedance $Z_0 = \omega_0 L$. Likewise, transforming a capacitor yields

$$j\omega C = j\omega_0 C \tan\theta \tag{5.27}$$

which is the input admittance of an open-circuited transmission-line stub having characteristic impedance $Z_0 = (\omega_0 C)^{-1}$. In both cases, the line is one-eighth of a wavelength long ($\theta = \frac{\pi}{4}$) at the fixed frequency point (transformed from $\omega = \omega_0$). Resistors, having no frequency dependence, are unchanged by this transformation.

Called *Richard's transformation*, this frequency substitution provides a mathematical link between networks designed using lumped elements and those designed using transmission lines. To illustrate, let us apply it to the example

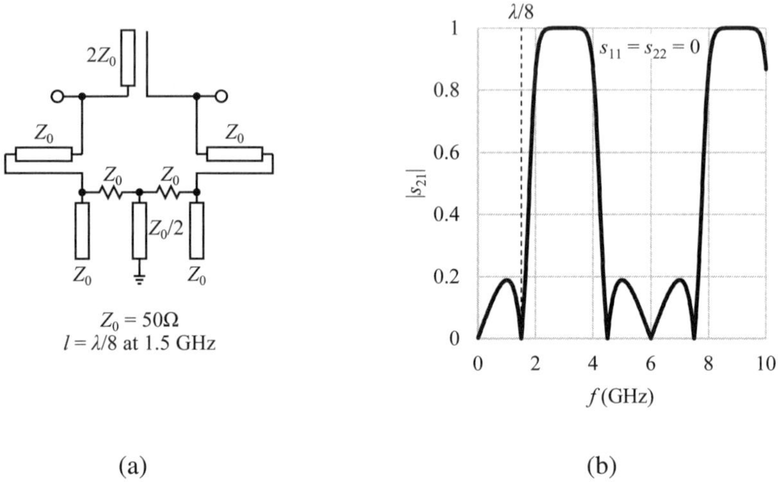

Figure 5.3 (a) Reflectionless network converted to the transmission-line form using Richard's transformation. (b) Frequency response.

in Figure 5.1(b). The capacitor between the two ports becomes an open-circuit transmission-line stub having characteristic impedance

$$Z_{C/2} = \frac{1}{\frac{1}{2}\omega_0 C} = \frac{1}{\frac{1}{2}\left(20 \cdot 10^9 \text{ rad/s}\right)(1 \text{ pF})} = 100\Omega = 2Z_0 \tag{5.28}$$

Since the original capacitor is floating (ungrounded), the transmission-line stub must be connected in series, as shown in Figure 5.3(a). Similarly, the inductors branching off from the ports have characteristic impedance

$$Z_L = \omega_0 L = \left(20 \cdot 10^9 \text{ rad/s}\right)(2.5 \text{ nH}) = 50\Omega = Z_0 \tag{5.29}$$

and are likewise connected in series. The reactive elements along the bottom are grounded, so the corresponding transmission-line stubs are connected in parallel.

Let us choose a fixed frequency point for the transformed network (corresponding to the transmission null from the original network) as 1.5 GHz. Each transmission line is thus assumed to have length $l = \lambda/8$ at that frequency. The resulting frequency response is given in Figure 5.3(b).

Several features of this transformation are worth noting. First, every transmission line will always have the same length if this transformation is used.

Transmission-line networks wherein all lines are the same length or an integral multiple of that length are called *commensurate-line networks* [2]. Since the response of such elements has a trigonometric dependence on frequency, the network parameters will always be periodic, as the figure shows. Lumped-element networks that are high-pass, such as the example from which we started, become band-pass in the transformed network, with the center of the passband at the frequency where the lines are a quarter-wavelength long. Low-pass networks then become band-stop after Richard's transformation.

Another important feature of this transformation is that the transmission lines it produces are only stubs, open-circuited or short-circuited. No cascade transmission lines are produced directly. This presents a challenge for layout in many cases since all the stubs in such a network must emanate from an electrically small, lumped-element cluster. The more elements in the original network, the more transmission lines extend outward from it and it soon becomes too difficult to connect them all without unacceptably high parasitic coupling between them.

Additionally, some of the transmission-line stubs are almost invariably series-connected, while others are parallel-connected. This is a problem for many transmission-line geometries since, as stated previously, most are better suited to having one kind of connection than the other (e.g., microstrip is good for parallel connections, but ill-suited to series connections). For these reasons, the network in Figure 5.3(a), while mathematically valid, is not practical in most cases.

The solution to these issues relies on the use of transmission-line identities, which will be discussed in Section 5.2.

5.1.7 Immittance Inversion (Duality Transformation)

Suppose now we wish to create a network from a preexisting prototype that acts on voltages the way the original network acts on current — effectively replacing impedance expressions with admittance, inductors with capacitors, and so forth. This is possible thanks to the principle of *duality*. Unlike the preceding frequency transformations, this cannot be accomplished fully with direct element-for-element substitutions. A change in topology is required. This is analogous to the duality of the electric and magnetic fields discussed in Section 1.1.2, where the boundary conditions were inverted so that the original solution for **E** now applied to **H**.

It is tempting to claim (as some have) that a dual network is one whose impedance-parameter matrix is the same as the admittance-parameter matrix of the original network and vice versa. This is not quite true. Recall that the duality of **E** and **H** in Maxwell's equations also involved a sign change. Something similar

Figure 5.4 A two-port model for the behavior of dual networks. (a) The prototype network, and (b) its dual.

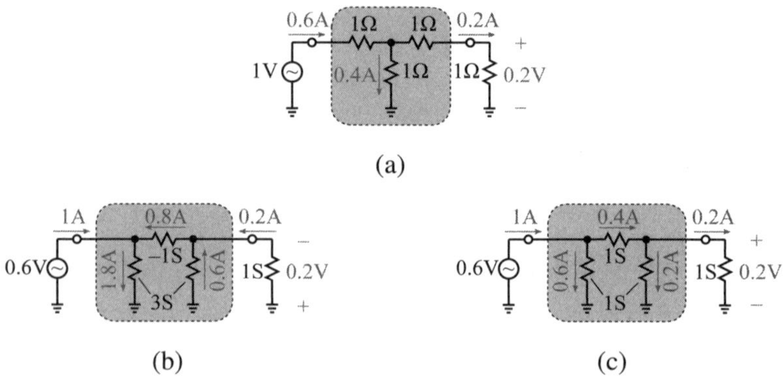

Figure 5.5 (a) A resistive network. (b) "False" dual (S is for siemens, or inverse-ohms). (c) True dual.

happens with electrical networks. To see why, consider the two-port model in Figure 5.4(a). A signal generator at port 1 excites the network, driving a known voltage and current through the termination resistor at port 2. For the dual network in Figure 5.4(b), we would like to see the roles of current and voltage reversed. As the figure shows, the current and voltage labels have been swapped, appropriately scaled by a factor $Z_0 = 1/Y_0$. Importantly, the polarities of the voltage and current have not changed.

Now let us work a simple example, where the two-port network comprises three unit-valued resistors, shown in Figure 5.5(a). We will use $Z_0 = 1\Omega$ in this example for simplicity. As determined from Table 4.1, the impedance matrix for this network is

$$\mathbf{Z} = \begin{pmatrix} 2 & 1 \\ 1 & 2 \end{pmatrix} \Omega \tag{5.30}$$

If we assume that the admittance matrix of the dual network is just the same as this matrix after scaling units, then we have

$$\mathbf{Y} = Y_0^2 \mathbf{Z} = \begin{pmatrix} 2 & 1 \\ 1 & 2 \end{pmatrix} \text{S} \tag{5.31}$$

where the unit for conduction, S, is siemens, or inverse-ohms. An equivalent circuit for this admittance matrix is shown in Figure 5.5(b). The negative conductor in this network is perhaps suspicious, but in itself is not proof that our dual is incorrect. The net conductance between any two ports is still positive, after all, and we have not yet proven that there is not another topology that may realize the same admittance matrix without containing negative elements (in fact, there is one, as we will see later). Instead, the problem with this proposed dual is that, while indeed swapping the voltage and current values at both ports, it has also reversed their polarity at the output port. This is not at all what we wanted.

The true dual network is shown in Figure 5.5(c). A sign change has appeared in the transadmittance parameters, but the port voltage and currents have been swapped, as desired, without altering their polarity. Note that in comparison with the original network, the 1Ω resistors have become 1S conductors, the series elements have become shunt elements, and the shunt element has become a series element.

In terms of the scattering parameters, the dual of a network negates the reflection coefficients ($s'_{ii} = -s_{ii}$) while the transmission coefficients remain unchanged ($s'_{ik} = s_{ik}$ when $i \neq k$). Reciprocal, antimetric networks are their own duals drawn in reverse.

For simple ladder networks, such as the tee and pi networks used in the preceding example, one may form the dual simply by exchanging series elements with shunt elements, inductors with capacitors, and resistors with the appropriate conductors. Each impedance element scales inversely proportional to Z_0^2, or

$$L' = Z_0^2 C \tag{5.32a}$$

$$C' = Y_0^2 L \tag{5.32b}$$

$$R' = Z_0^2 G \tag{5.32c}$$

$$G' = Y_0^2 R \tag{5.32d}$$

Similarly, open-circuited transmission-line stubs are exchanged with short-circuited stubs, and parallel-connected stubs with series-connected stubs, each with characteristic impedance scaled like a resistor in (5.32c). Cascaded transmission lines are

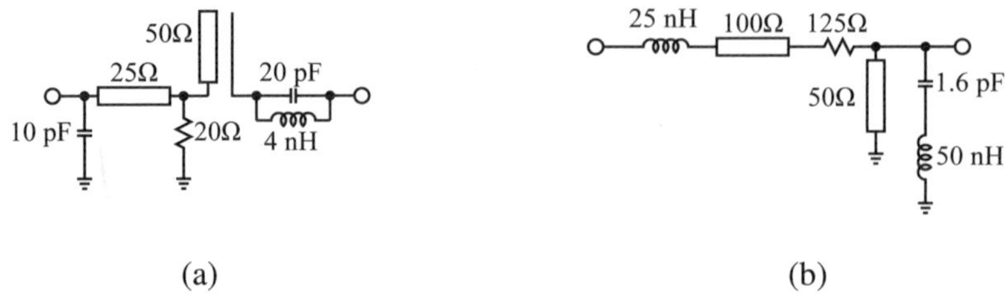

Figure 5.6 (a) A ladder-network comprising lumped elements and transmission lines. (b) The dual of that network in a 50Ω system, found by exchanging impedances with the corresponding admittances, series elements with shunt elements, and cascade transmission lines with others having the dual characteristic impedance.

exchanged for new cascaded lines having the dual characteristic impedance. An example is shown in Figure 5.6.

For more complex lumped-element networks, one may use a graphical procedure, outlined in Figure 5.7. Let us use, as an example, the reflectionless circuit analyzed by even- and odd-mode analysis in Section 4.5.1. This circuit is redrawn in Figure 5.7(a). First, any ports must be loaded (at least notionally) with sources, as indicated in Figure 5.7(b); otherwise, one might mistake the port for an open-circuit and replace it with its dual impedance, the short. We must also draw explicitly the ground node so that the complete current path is represented.

In the next step, Figure 5.7(c), the new nodes for the dual circuit are identified in the center of each mesh (simple loop) in the circuit. An additional node is placed outside the circuit.

In Figure 5.7(d), dual elements are drawn in, crossing perpendicularly over each original element. In other words, each element is replaced with its dual (e.g., a capacitor for an inductor) connecting the new nodes separated by the original element. To be rigorous, the sources should be replaced by their dual counterparts (e.g., a Thévenin voltage source for a Norton current source), but since the sources here are only notional anyway, we can ignore this detail. Their orientation would also, in a strict sense, be connected to the reference current direction for the meshes involved, but that too can be disregarded in most cases.

In Figure 5.7(e), the elements have been labeled with their new values, scaled according to the relationships in (5.32). (Note, in this case, we have used the fact that $L/C = Z0^2$, as given by the original network.) We have also identified a new ground node as that which is common to both sources.

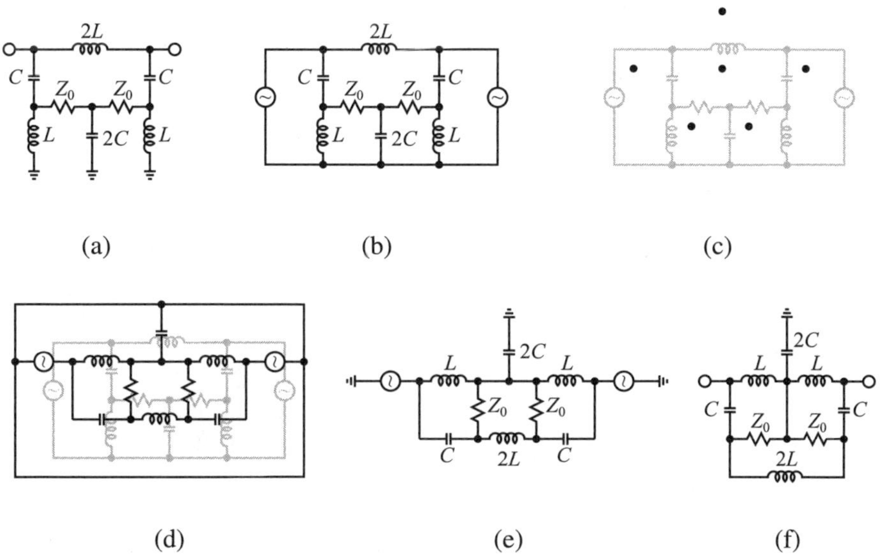

Figure 5.7 Formation of a dual network of lumped elements. (a) Original network, where $L/C = Z_0^2$. (b) Redrawn with explicit sources and ground node. (c) Dual nodes identified. (d) Dual elements drawn in. (e) Ground nodes identified and element values determined. (f) Final dual network.

Finally, in Figure 5.7(f), the notional sources have been replaced again with port nodes and the circuit drawing reorganized for clarity. As a dual network, this has exactly the same transmission coefficients (s_{ik}) as the original, but the reflection coefficients (s_{kk}) pick up a 180° phase shift. In this case, since $s_{kk} = 0$ (as Section 4.5.1 showed), the reflection coefficients are also unchanged. This circuit is thus not only a dual for the original network, but also an *identity*, which will be discussed in the Section 5.2.

One of the weaknesses of this approach is that it only applies to lumped-element networks, which are planar, meaning that they can be drawn on a flat piece of paper without crossovers. If the circuit cannot be redrawn to eliminate crossovers, then identification of the mesh loops that become the dual-node locations is ambiguous. Several methods that can be used in the nonplanar case are given in [3].

Figure 5.8 (a) A delta network of three elements. (b) A wye network of three elements.

5.2 IDENTITIES

Most of us learn in our early courses on trigonometry that there are frequently multiple expressions that have the same value for all possible arguments, for example the cosine double-angle formula, $\cos(2x) = \cos^2 x - \sin^2 x$ (and many others in Appendix A). Known as trigonometric identities, these are often used to simplify the solutions of many kinds of problems, and we spend no small amount of time memorizing a wide variety of them. Similarly, many of us are aware of numerous vector identities that relate equivalent expressions involving the products and differentials of vector quantities (a fairly extensive list can be found in Appendix C).

We have an analogous entity in the world of electromagnetic circuits, where two or more networks may have identical behaviors for all possible stimuli. The Thévenin and Norton equivalent sources studied in Section 2.6, for example, may be thought of as a pair embodying one such circuit identity, since the two cannot be distinguished from one another solely from their external port characteristics.

When two circuits are both known to be linear, it is sufficient to show that they have the same network parameters to establish that they are identities for one another. We will explore a number of different common and useful circuit identities in this section.

5.2.1 Delta-Wye Identity

Perhaps the most well-known is the *delta-wye identity*. (Although listed in most books as the *delta-wye transform* or *transformation*, by the definitions used in this book it is more properly called an identity.) The general form of the delta-wye identity is shown in Figure 5.8. On the left is a ring of three lumped admittance elements — a configuration known as a *delta* network — and on the right is a trio of elements emanating from a common node — known as a *wye* ("Y") network. The equivalence between these two representations depends on a specific relationship between the element values, which we derive by equating their network parameters. Note that the element subscripts in the delta network identify the ports connected by the corresponding element, whereas the wye-network element subscripts denote the single port to which the central node is connected by that element.

Since there is no connection to ground in either circuit, the test current that flows when any port is stimulated and the others left open must always be zero. This is an example, then, of a case where the impedance parameters would all be undefined (infinite). Instead, therefore, we shall use the admittance parameters. The admittance matrix for the delta network in Figure 5.8(a) is

$$\mathbf{Y}_{\Delta} = \begin{pmatrix} y_{12} + y_{13} & -y_{12} & -y_{13} \\ -y_{12} & y_{12} + y_{23} & -y_{23} \\ -y_{13} & -y_{23} & y_{13} + y_{23} \end{pmatrix} \tag{5.33}$$

For the wye network in Figure 5.8(b), the admittance matrix is

$$\mathbf{Y}_{Y} = \frac{1}{y_1 + y_2 + y_3} \begin{pmatrix} y_1 (y_2 + y_3) & -y_1 y_2 & -y_1 y_3 \\ -y_1 y_2 & y_2 (y_1 + y_3) & -y_2 y_3 \\ -y_1 y_3 & -y_2 y_3 & y_3 (y_1 + y_2) \end{pmatrix} \tag{5.34}$$

These two admittance matrices are equivalent when

$$y_{12} = \frac{y_1 y_2}{y_1 + y_2 + y_3} \tag{5.35a}$$

$$y_{13} = \frac{y_1 y_3}{y_1 + y_2 + y_3} \tag{5.35b}$$

$$y_{23} = \frac{y_2 y_3}{y_1 + y_2 + y_3} \tag{5.35c}$$

or, in terms of the elemental impedances ($z_k = y_k^{-1}$ and $z_{ik} = y_{ik}^{-1}$),

$$z_{12} = z_1 z_2 (y_1 + y_2 + y_3) = z_1 z_2 \sum_{k=1}^{3} y_k \tag{5.36a}$$

$$z_{13} = z_1 z_3 (y_1 + y_2 + y_3) = z_1 z_3 \sum_{k=1}^{3} y_k \tag{5.36b}$$

Figure 5.9 (a) A three-element wye network and (b) its delta equivalent.

$$z_{23} = z_2 z_3 \left(y_1 + y_2 + y_3\right) = z_2 z_3 \sum_{k=1}^{3} y_k \tag{5.36c}$$

Alternatively, one may solve for the wye elements in terms of the delta elements,

$$y_1 = y_{12} y_{13} \left(z_{12} + z_{13} + z_{23}\right) = y_{12} y_{13} \sum_{k \neq i} z_{ik} \tag{5.37a}$$

$$y_2 = y_{12} y_{23} \left(z_{12} + z_{13} + z_{23}\right) = y_{12} y_{23} \sum_{k \neq i} z_{ik} \tag{5.37b}$$

$$y_3 = y_{13} y_{23} \left(z_{12} + z_{13} + z_{23}\right) = y_{13} y_{23} \sum_{k \neq i} z_{ik} \tag{5.37c}$$

Take note of the similarity between (5.36) and (5.37), essentially a consequence of duality since impedance is the dual of admittance and the delta topology is the dual of the wye topology.

Mathematically, the identity formulas above are valid for any kind of immittances, even if the individual branches are not elemental but combinations of elements in series and/or parallel. There is no guarantee, however, that the resulting immittance will be realizable using the traditional lumped elements in every case. In fact, the only conditions under which the identity equivalent of three basic elements is realizable is when all three of the original elements are of the same type (resistors, inductors, or capacitors), or when at least two of them are resistors [4]. In the first case, the resulting alternate network has only single elements in each branch, of the same type as the original. In the latter case, the branches of the alternate network will generally have twice as many elements as the original. Take, for example, the wye network in Figure 5.9(a), comprising a capacitor and two resistors. The corre-

Figure 5.10 (a) A star network of $N = 4$ elements and (b) its equivalent mesh network.

sponding branches of the equivalent delta network are given by

$$z_{12} = z_1 z_2 \sum_{k=1}^{3} y_k = \frac{R_2}{j\omega C_1}\left(j\omega C_1 + R_2^{-1} + R_3^{-1}\right) = R_2 + \frac{1 + R_2 R_3^{-1}}{j\omega C_1} \tag{5.38a}$$

$$z_{13} = z_1 z_3 \sum_{k=1}^{3} y_k = \frac{R_3}{j\omega C_1}\left(j\omega C_1 + R_2^{-1} + R_3^{-1}\right) = R_3 + \frac{1 + R_2^{-1} R_3}{j\omega C_1} \tag{5.38b}$$

$$z_{23} = z_2 z_3 \sum_{k=1}^{3} y_k = R_2 R_3 \left(j\omega C_1 + R_2^{-1} + R_3^{-1}\right) \tag{5.38c}$$

$$= j\omega C_1 R_2 R_3 + R_2 + R_3 \tag{5.38d}$$

The first two branch impedances above may be recognized as a resistor in series with a capacitor, whereas the last branch impedance is a resistor in series with an inductor. The final delta network is thus shown in Figure 5.9(b).

In cases where there is only one resistor and two reactive elements, or all reactive elements of mixed types, the resulting branch immittances have quadratic frequency dependence. More specifically, they are not positive-real, and thus cannot be realized using the conventional set of lumped elements (this does not, of course, preclude the identity from being used as a valid mathematical equivalent). A complete listing of realizable three-element delta-wye identities is given in Appendix E.

5.2.2 Star-Mesh Identity

The results of Section 5.2.1 can be generalized (in part) to arbitrarily large numbers of ports. The relevant identity in this case is shown in Figure 5.10. A star network, comprising N elements emanating from a common central node, equates to a mesh

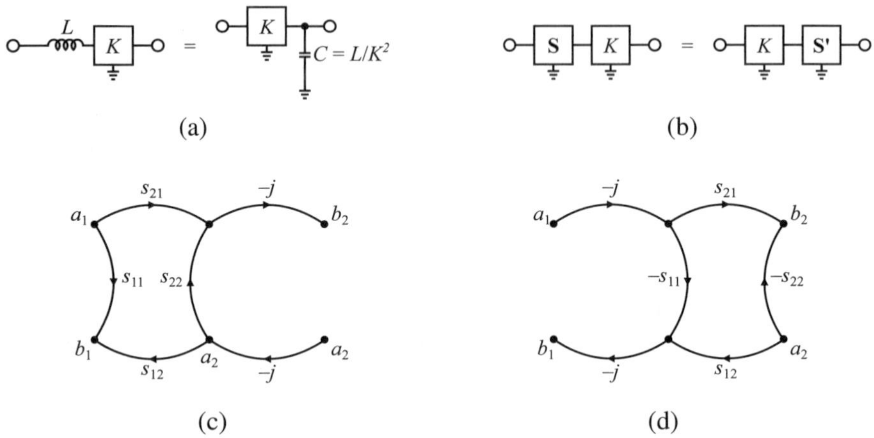

Figure 5.11 Illustration of the inverter-dual identity. (a) A series inductor is replaced with a shunt capacitor on the opposite side of an impedance inverter. (b) A general network is replaced with its dual on the opposite side. (c) Signal-flow diagram when the inverter is on the output side of the original network. (d) Signal-flow diagram when the inverter is on the input side of the dual network.

of $\frac{1}{2}N(N-1)$ elements linking every pair of remaining nodes. Analogously from (5.36), the element values are given by

$$z_{ik} = z_i z_k \sum_{l=1}^{N} y_l \tag{5.39}$$

Note that when a star network is converted to a mesh, the central node is removed, while for $N > 3$ the number of elements increases. Consequently, the general identity is one-directional (from star to mesh) when $N > 3$. The reverse identity does not apply to large, complex meshes without added constraints.

5.2.3 Inverter-Dual Identity

Another identity concerns the equivalence of dual networks on either side of an ideal impedance inverter. Consider first the series inductor on one side of a K-inverter, as shown in Figure 5.11(a). As a consequence of this identity, the reactive element may be moved to the opposite side of the inverter, and is replaced with its dual capacitor, where the normalizing impedance is $Z_0 = K$. A diagram very much like this is often seen in books as an example of one of *Kuroda's identities*, which will be discussed later in Section 5.2.6. In Kuroda's case, however, the inductor and capacitor are merely symbolic substitutes for short-circuited and open-circuited transmission line stubs, respectively, while the inverter represents a cascade quarter-wave transformer. (In that identity, however, the normalizing impedance, K — really the characteristic impedance of the quarter-wave transformer — would also change.)

In the present case, however, a far more general statement can be made. Imagine that on one side of the inverter is a ladder network of elements. These elements may be shifted from one side to the other, replacing them as we go with their duals, until the entire ladder appears on the opposite side. In this way, the ladder network is replaced with its topological dual (also a ladder). If the original network on one side of the inverter was not a ladder network, then it may not be possible to apply this identity one element at a time. Nevertheless, the entire network can be moved from one side to the other, and in so doing, is transformed into its dual, as indicated in Figure 5.11(b).

The simplest proof of this is to consider the signal flow diagrams, first with the inverter coming after the original network (Figure 5.11(c)) and then with the inverter preceding the dual (Figure 5.11(d)). We have used (4.134) for the s-parameters of the inverter, and negated the reflection coefficients for the dual network. Almost by inspection, one may verify that all four scattering parameters in both cases are exactly the same (as they would be if the transmission coefficients of the inverter were $+j$ instead of $-j$).

5.2.4 Equivalent Impedance Groups

We already know that simple two-terminal immittances formed by complex groupings of a single type of lumped element (e.g., resistors, inductors, or capacitors) can ultimately be reduced to a single equivalent element of the same type using series and parallel combinations as well as the star-mesh identities. When the types of elements are mixed, reduction may not always be possible; however, there are certain equivalent forms that apply. One such example is given in Figure 5.12(a). Circles and squares have been used in place of the usual schematic symbols to emphasize that the types of the elements are arbitrary (the labels correspond to normalized impedance values). An example, where the squares are taken to be inductors and the circles are taken to be capacitors is given in Figure 5.12(b).

As more elements (and more types of elements) are added, the number of similar equivalent-impedance identities that one may find is really quite large [5, 6], and they will not be enumerated here in detail. Instead, a minimal sampling of such

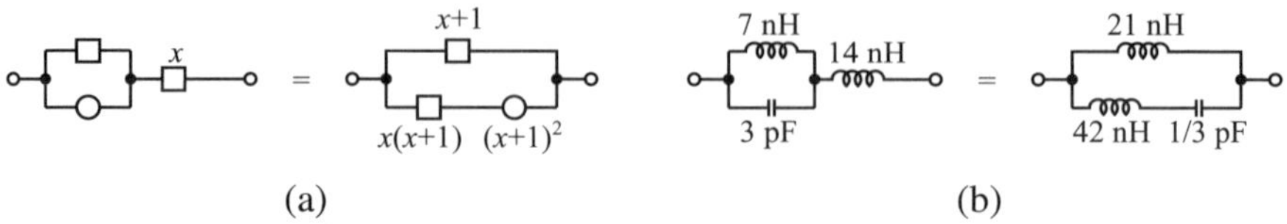

Figure 5.12 An equivalent-impedance identity. (a) General form for two types of elements. Labels correspond to normalized impedance values, where the unlabelled elements have unit value. (b) Example application using inductors and capacitors.

identities is shown in Figure 5.13 to give the reader an idea of what is possible [6]. The intermediate parameters for all cases shown are given by

$$a = 1 + x + y \tag{5.40a}$$

$$b = \sqrt{a^2 - 4xy} \tag{5.40b}$$

$$c = \frac{(1+x)(1+y)}{(x-y)^2} \tag{5.40c}$$

$$d = \frac{b - a + 2y}{2b} \tag{5.40d}$$

$$e = \frac{b + a - 2y}{2b} \tag{5.40e}$$

5.2.5 Transformer Identities

We showed in Section 2.3.4 that the impedance of a termination resistor as seen through a transformer is scaled according to the square of the turns ratio, n. This was implicitly used in Section 2.8.3 to move parasitic elements from one side of the transformer to another. In Figure 5.14(a), another illustration of this property is given where a series inductor and shunt capacitor are each moved from one side to the other with the necessary scaling. In this section, we are prepared to make the general statement that a two-port network, regardless of its topology or element type (lumped or transmission-line) can be moved from one side of a transformer to another with the appropriate scaling of its impedance matrix, as shown in Figure 5.14(b).

Further, if one happens to have a transformer adjacent to an *L-network* (a network comprising a series and a shunt element) with like-kind elements, then one

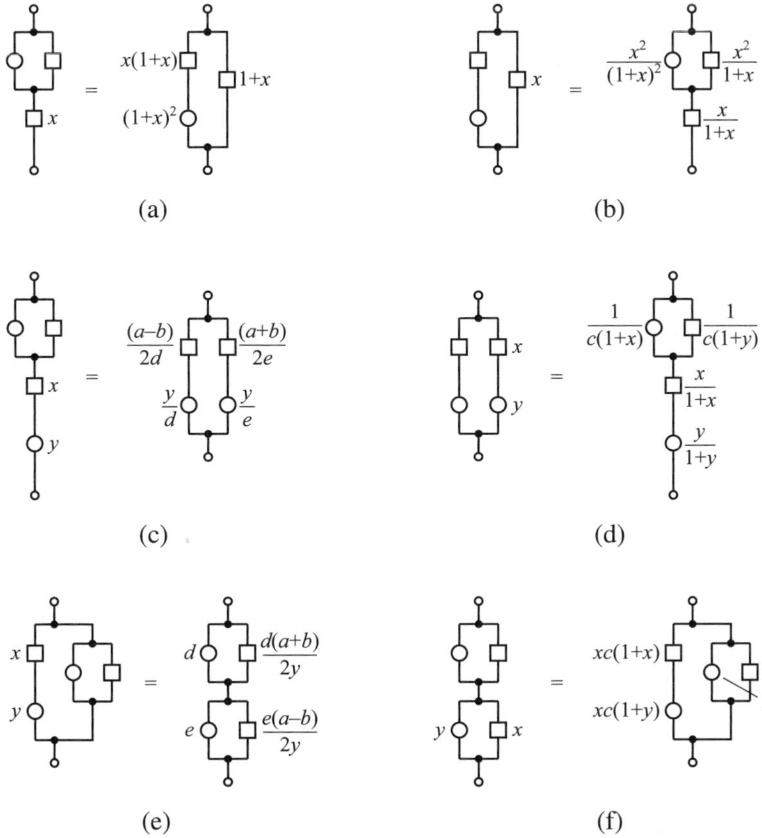

Figure 5.13 A sample of lumped-element topologies that can be made to have equivalent immittances given arbitrary element types. Formulas for the intermediate parameters a, b, c, d, and e are given in the text.

may eliminate the transformer entirely by reversing the L-network and scaling its elements as shown in Figures 5.14(c, d).

Finally, a transformer wired for reversing the polarity of current flow can be used to synthesize passive networks equivalent to tee- and pi-configurations where the innermost elements are negative, as in Figure 5.15. This can be quite useful for implementing circuits that might otherwise seem unrealizable [7].

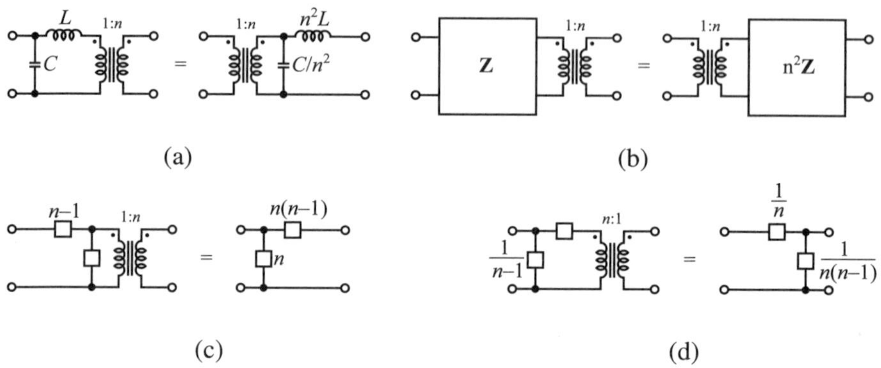

Figure 5.14 Transformer identities. (a) Movement of elements from one side of a transformer to the other with impedance scaling. (b) Movement of a network from one side to the other with scaling of the impedance matrix. (c) Elimination of a step-up transformer. (d) Elimination of a step-down transformer.

5.2.6 Kuroda's Identities

At this time, we shall turn our attention to identities that apply specifically to transmission-line networks. These are especially important, for as explained previously, certain transmission-line structures are difficult to implement, due to the electrically-long nature of the transmission line. Transmission-line identities are often used to transform such networks from an unrealizable (or at least impractical) topology to a more suitable one.

The most notable of these are Kuroda's identities [8, 9], summarized in Figure 5.16. Much of the early literature on Kuroda's identities draws them with the open-circuited and short-circuited stubs replaced by lumped capacitors and inductors respectively, as if anticipating the use of Richard's transformation, while the cascade transmission line is drawn as an ideal impedance inverter. Many modern books therefore follow suit [1]. This at first may appear to be justified by Figure 5.11(a), which looks quite similar and is also true. Presenting them in this way is a dangerous practice, as students tend to assume, then, that a cascade transmission line and an ideal impedance inverter are proper counterparts under Richard's transformation, and that the identity applies in both domains, but this is not the case. Consider Figure 5.17, which shows how this analogy breaks down. Series lumped elements equate to shunt elements on opposite sides of an impedance inverter, whether those lumped elements are inductors or capacitors (or, for that

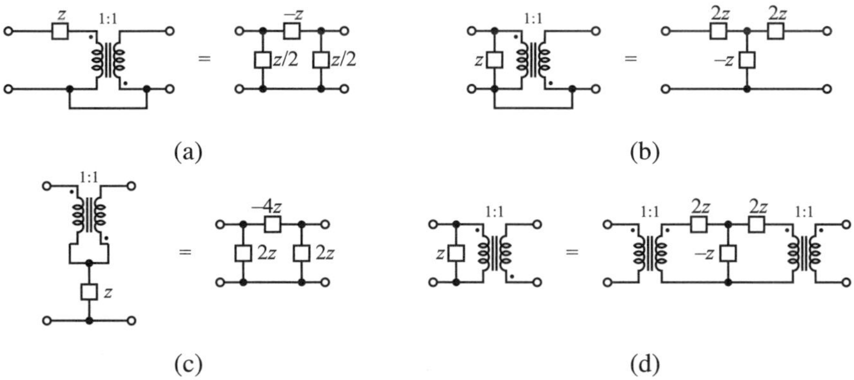

Figure 5.15 Current-reversing transformer identities that have equivalent (a) pi- and (b) tee-networks with negative elements. (c) Alternate form having an equivalent pi-network with negative elements. (d) Fully differential form with an equivalent tee-network having negative elements.

matter, resistors), but the only series-connected stub that equates to a parallel-connected stub after a cascade transmission line is the short-circuited series stub. The last case in Figure 5.17, where the series-open stub is presumed identical to a parallel-short stub, can be true at the single frequency point where the lines are a quarter-wavelength long, but unlike the others it does not hold at all frequencies and thus is not an identity.

The first of Kuroda's identities in Figure 5.16(a) may be the most widely used transmission-line identity of all, because it provides for the elimination of a very inconvenient structure: the series-connected stub. Recall that some of the most popular transmission line geometries (e.g., microstrip) do not permit series-connected stubs, at least not easily. Let us take Figure 5.18 as an example. We start with a low-pass ladder-network filter, Figure 5.18(a). The specific values of the elements are not important at this stage; we are only concerned with the topology. Assuming that we wish to implement this filter using a transmission-line technology, say, microstrip, we first convert it using Richard's transformation, as in Figure 5.18(b). This topology immediately presents us with two problems. The first is that we have series-connected stubs, which microstrip prohibits. The second problem is that all of the stubs must now emanate from an electrically small region. In practice, the proximity of the stubs would undoubtedly lead to undesired coupling, which would upset the filter response.

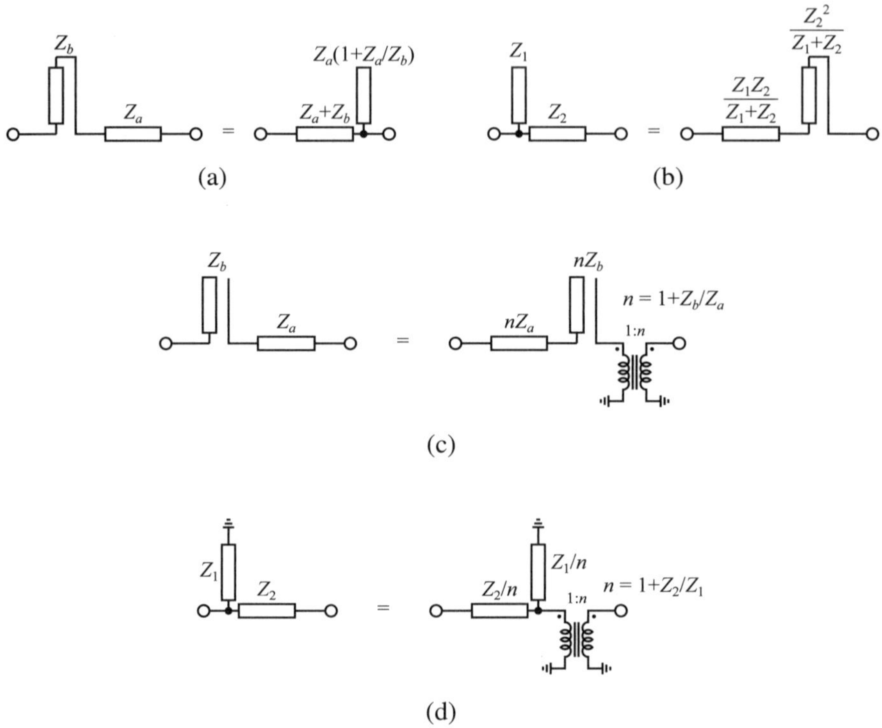

Figure 5.16 Kuroda's identities.

The solution is to first add a port extension of two matched transmission lines to one of the ports, shown in Figure 5.18(c). Since these lines are matched to the characteristic impedance of the system, they do not alter the transfer coefficient, s_{21}, in magnitude, only in phase. Such alteration is deemed acceptable since it is usually the magnitude of s_{21} we are most interested in. Now, having cascade lines to work with, we may apply Kuroda's identity as many times as needed to arrive at a topology with only shunt stubs and with all stubs spaced apart by cascade transmission lines, arriving at Figure 5.18(d).

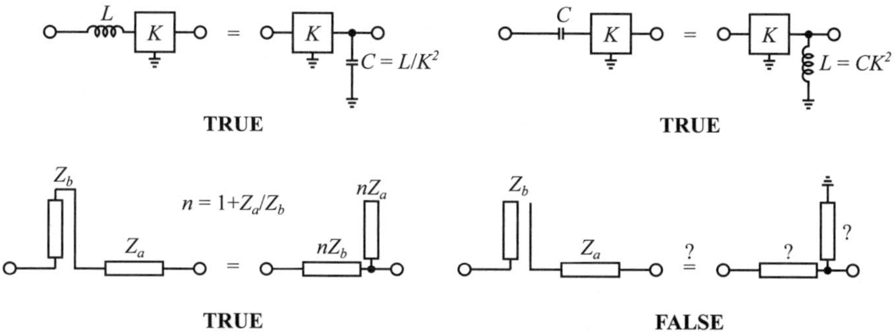

Figure 5.17 The fallacy of drawing an analogy between the lumped-element inverter-dual identity of Section 5.2.3 and Kuroda's identities for transmission lines.

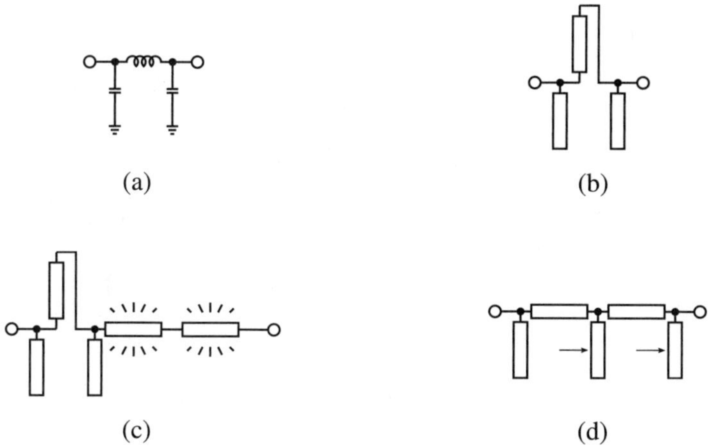

Figure 5.18 Example utilization of Kuroda's identity. (a) Prototype lumped-element network. (b) Application of Richard's transformation. (c) Addition of port extension. (d) Application of Kuroda's identity.

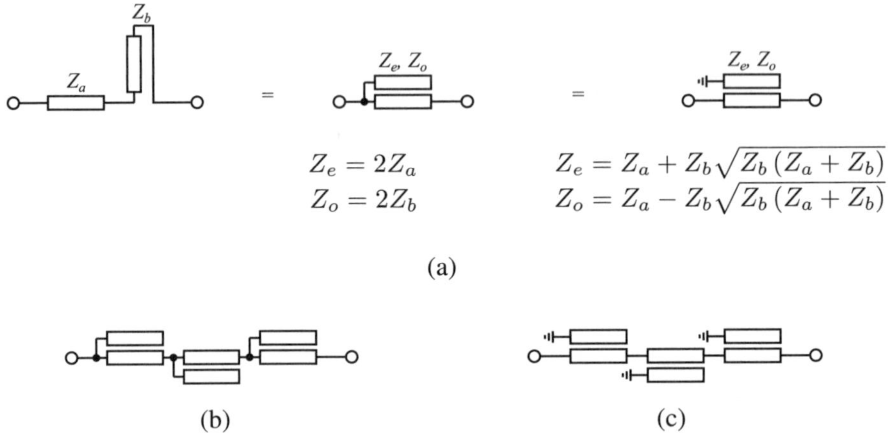

Figure 5.19 (a) Coupled-line identities that are useful for eliminating series-connected short-circuited stubs. (b, c) Alternate transmission-line topologies for the low-pass filter originating from a lumped-element ladder.

5.2.7 Coupled-Line Identities

It is worth noting that the series-to-shunt stub Kuroda's identity is not the only option available to the circuit designer. Other identities exist that would lead to different topologies having the same response. Consider the two identities involving coupled lines shown in Figure 5.19(a). These suggest two possible alternate topologies for the filter from the previous example, now shown in Figures 5.19(b, c). Coupled lines also provide a solution for eliminating series-connected open-circuited stubs, as in Figure 5.20(a). Still other coupled-line forms take the place of groupings of series- and parallel-connected stubs simultaneously, as shown in Figures 5.20(b, c). In the latter two cases, the formulas for the characteristic impedances that guarantee equality are

$$Z_e = Z_1 = Z_a + 2Z_b \tag{5.41a}$$

$$Z_o = \frac{Z_1 Z_2}{2Z_1 + Z_2} = Z_a \tag{5.41b}$$

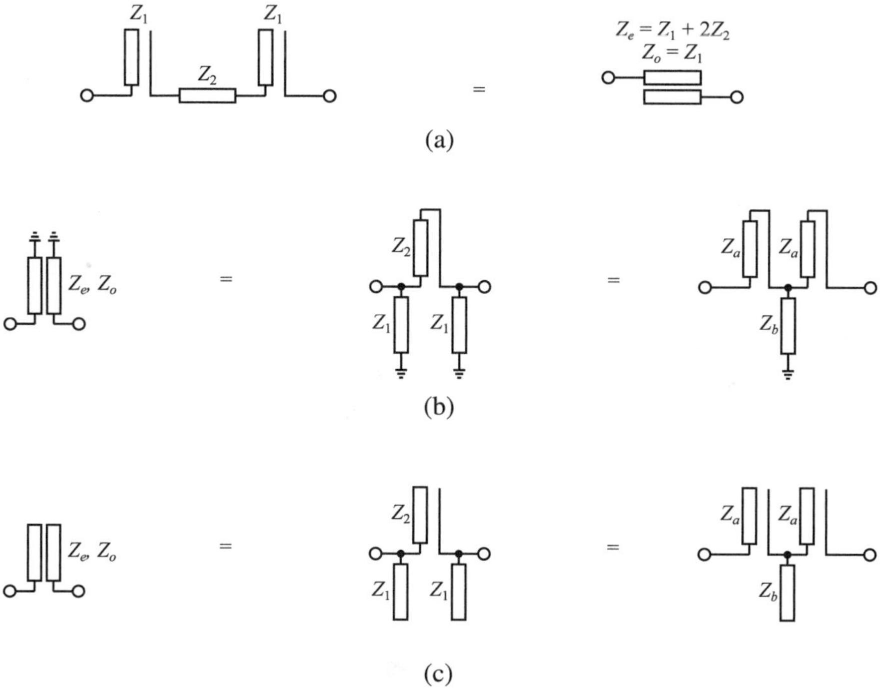

Figure 5.20 (a) A coupled-line identity for eliminating series-open stubs. (b) A coupled-line identity for replacing groups of short-circuited stubs. (c) A coupled-line identity for replacing groups of open-circuited stubs.

5.2.8 A Three-Port Transmission-Line Identity

So far, all of the transmission-line identities we have discussed have been for two-port networks. There are, of course, three-port identities as well. One that has special importance is shown in Figure 5.21. The element values for this identity are given by

$$Z_e = Z_1 (1 + n) \tag{5.42a}$$

$$Z_o = Z_1 (1 - n) \tag{5.42b}$$

$$n^2 = \frac{Z_1}{Z_1 + Z_2} \tag{5.42c}$$

Figure 5.21 A three-port transmission-line identity.

Because the significance of this identity was not realized until fairly recently [4], it is not found in other texts. Its proof is therefore a good one for us to work through explicitly.

As described at the beginning of Section 5.2, an identity is proven by showing that any one set of network parameters is identical in both cases. For this example, let us use the z-parameters. Since the z-parameters are found by assuming open-circuit terminations on all unstimulated ports, the z-parameters for the coupled lines on the left side of Figure 5.21 are the same as those for the generic four-port coupled lines given in (4.115), only with the parameters involving the last port omitted,

$$Z_{11} = Z_{22} = Z_{33} = \tfrac{1}{2}\left(Z_e + Z_o\right)\coth\left(\gamma l\right) \tag{5.43a}$$

$$Z_{12} = Z_{21} = \tfrac{1}{2}\left(Z_e + Z_o\right)\operatorname{csch}\left(\gamma l\right) \tag{5.43b}$$

$$Z_{13} = Z_{31} = \tfrac{1}{2}\left(Z_e - Z_o\right)\operatorname{csch}\left(\gamma l\right) \tag{5.43c}$$

$$Z_{23} = Z_{32} = \tfrac{1}{2}\left(Z_e - Z_o\right)\coth\left(\gamma l\right) \tag{5.43d}$$

Note that we have assumed for this identity that the phase velocities of the even and odd modes are equal, $\gamma_e = \gamma_o = \gamma$. Substituting (5.42a) into these parameters, we find

$$Z_{11} = Z_{22} = Z_{33} = Z_1 \coth\left(\gamma l\right) \tag{5.44a}$$

$$Z_{12} = Z_{21} = Z_1 \operatorname{csch}\left(\gamma l\right) \tag{5.44b}$$

$$Z_{13} = Z_{31} = nZ_1 \operatorname{csch}\left(\gamma l\right) \tag{5.44c}$$

$$Z_{23} = Z_{32} = nZ_1 \coth\left(\gamma l\right) \tag{5.44d}$$

Now let us derive the z-parameters of the network on the right side of Figure 5.21. By leaving port 3 open, the transformer and series stub are rendered inactive, giving us the first four z-parameters as simply those of a single transmission line,

$$Z'_{11} = Z'_{22} = Z_1 \coth\left(\gamma l\right) \tag{5.45a}$$

$$Z'_{12} = Z'_{21} = Z_1 \operatorname{csch}(\gamma l) \tag{5.45b}$$

which matches the first part of (5.44a). The last part of (5.44a) is verified by considering port 3 with ports 1 and 2 left open. The impedance seen looking into port 3 is simply the series connection of two open-circuited stubs, modified by the transformer turns ratio,

$$Z'_{33} = n^2 (Z_1 + Z_2) \coth(\gamma l) = Z_1 \coth(\gamma l) \tag{5.46}$$

To verify the remaining parameters, we must consider the voltage developed at ports 1 and 2 when a current is injected through the transformer at port 3. That transformer steps up the current by a factor n. Further, since the open-circuited stub is in series, that current flows through it unmodified. We are thus once again left with a simple transmission line, only with a current stimulus n-times larger than it otherwise would be,

$$Z'_{23} = Z'_{32} = nZ_1 \coth(\gamma l) \tag{5.47a}$$

$$Z'_{13} = Z'_{31} = nZ_1 \operatorname{csch}(\gamma l) \tag{5.47b}$$

Having now obtained all the same z-parameters from both networks in Figure 5.21, the identity may be considered proven.

The significance of this identity lies in its ability to make non-ladder, reflectionless filter topologies similar to that of Figure 5.3 realizable. This usage will be illustrated in more detail during our discussion of filter design in Chapter 11.

The transmission-line identities featured in this chapter are considered to be the most broadly useful; however, a much larger array of identities has been identified. For a more extensive list, the reader is referred to Appendix F.

Problems

5-1 It was stated that application of Richard's transformation to a high-pass lumped-element network yields a commensurate-line network having a band-pass response. Assuming that the passband of the high-pass network starts at ω_0, what would be the fractional bandwidth of the transformed transmission-line network? Rewrite the transformation scaling law, (5.25), so that the result has a fractional bandwidth of Δ.

5-2 Draw the dual of the network shown in Figure 5.2(a).

5-3 Calculate the equivalent delta network for a wye network comprising a resistor, R, an inductor, L, and a capacitor, C. Are all of the branches of the delta network realizable? Which ones?

5-4 Validate the inverter-dual identity by cascading an arbitrary tee network with the ideal inverter in Figure 4.27(a), and then manipulating the combined network using delta-wye identities until the dual pi-network is obtained at the other side of another ideal inverter.

5-5 Prove the equivalent-impedance identity shown in Figure 5.12(a).

5-6 Prove the transformer identity shown in Figure 5.14(c).

5-7 Use Kuroda's identity to prove that the commensurate-line network in Figure 4.14(c) is electrically symmetric.

5-8 Prove the coupled-line identities illustrated in Figure 5.20(c). Assume for the coupled lines that $\gamma_e = \gamma_o$.

5-9 Prove that the false identity in the lower right of Figure 5.17 cannot be satisfied for any finite values of the equivalent network parameters.

References

[1] D. M. Pozar, *Microwave Engineering*, 4th ed. New York: Wiley, 2011.

[2] P. Richard, "Resistor-transmission line circuits," *Proceedings of the IRE*, vol. 36, no. 2, pp. 210–220, February 1948.

[3] A. Bloch, "On methods for the construction of networks dual to non-planar networks," *Proceedings of the Physical Society*, vol. 58, no. 6, pp. 677–694, November 1946.

[4] M. A. Morgan, *Reflectionless Filters*. Norwood, MA: Artech House, 2017.

[5] R. M. Foster and G. A. Campbell, "Maximum output networks for telephone substation and repeater circuits," *Transactions of the American Institute of Electrical Engineers*, vol. 39, no. 1, pp. 230–290, January 1920.

[6] Wikipedia. (2017) Equivalent impedance transforms. https://en.wikipedia.org/wiki/Equivalent_impedance_transforms.

[7] M. A. Morgan, "Deep rejection reflectionless filters," U.S. Patent Application 62/652,731, April 4, 2018.

[8] H. Ozaki and J. Ishii, "Synthesis of a class of strip-line filters," *IEEE Transactions on Circuit Theory*, vol. 5, pp. 104–109, June 1958.

[9] K. Kuroda, "Derivation methods of distributed constant filters from lumped constant filters," *Joint meeting of Kansai branch of IECE*, p. 32, October 1952.

Chapter 6

Impedance Matching

Perhaps no other task defines the daily work of a microwave engineer more directly than impedance matching. This is the process by which we start with a device — maybe an antenna, a transistor, or a laser diode — and add elements around it so as to make its input or output impedance equal to that of its source or load (or to some other target impedance that may be required). We saw, for example, in Section 2.6, that, when the source had an internal resistance of 50Ω, the power delivered to a load resistor was maximized when it also had a value of 50Ω. We also saw, in Section 3.3.4, that when the termination of a transmission line is not equal to the characteristic impedance of that line, the result is reflection, standing waves, and ultimately at least partial cancellation of the desired waveform.

Impedance matching allows us to optimize the performance of our system by changing the apparent impedance of one device as seen by another to maximize the power transfer between them, the efficiency, the bandwidth, or immunity to noise. We shall explore in this chapter both the necessity for and the methodologies behind matching various kinds of impedances.

6.1 SINGLE FREQUENCY MATCHING

Let us consider an example, shown in Figure 6.1(a). We have a generator, in this case a Thévenin-equivalent voltage source, with 5-V rms voltage amplitude and 50Ω internal impedance. We also have a load, which in this case is modeled by a 20Ω resistor in parallel with a 10-pF capacitor. We assume that both the generator and the load are closed boxes to us, meaning we can model their internal workings using

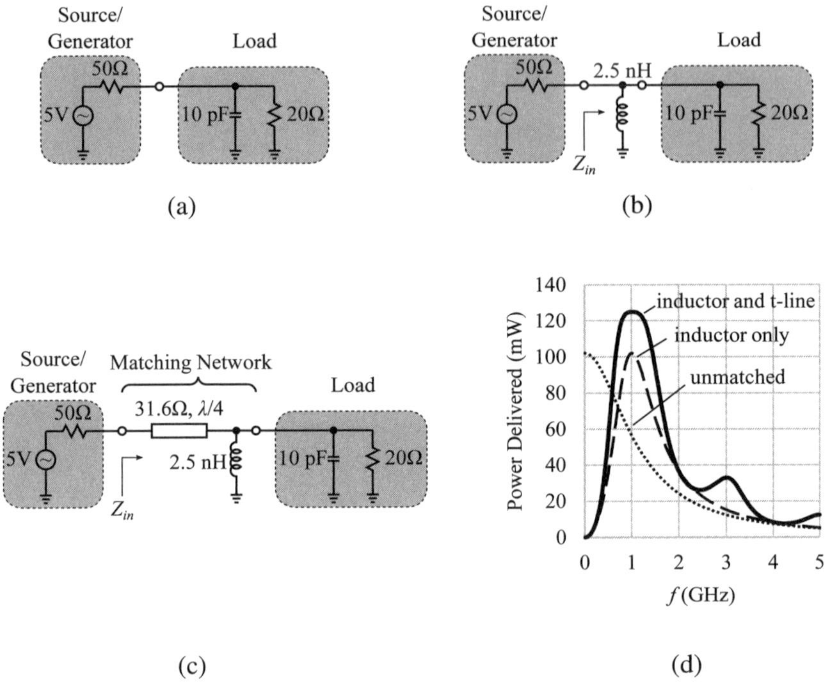

Figure 6.1 Conjugate matching example. (a) Original (unmatched) source and load. (b) With shunt inductive element. (c) With inductance and quarter-wave transformer. The transmission line is a quarter wavelength long at 1 GHz. (d) Power delivered to the load when unmatched (dotted line), with series inductor (dashed line), and with inductor and quarter-wave transformer (solid line).

these elements, but we cannot change them; we can only add elements externally. The power delivered to the load is plotted in Figure 6.1(d) with a dotted line.

Suppose we wish to maximize the power delivered to the load at 1 GHz. At present, we are only getting about 56 mW into the load at that frequency, but we know from our experience in Section 2.6 that we should be able to get more than twice that much. First, we notice that the power delivered is a decreasing function of frequency. This is a consequence of the 10-pF capacitor tending to short out the load resistor at high frequency. Even at 1 GHz, though the load resistor is not entirely shorted out, the imaginary admittance of the capacitor shunts some of the current provided by the source that could otherwise flow through the load. That admittance is

$$Y_C = j\omega C = j\,(2\pi \cdot 1 \text{ GHz})\,(10 \text{ pF}) \approx j0.063\text{S} \tag{6.1}$$

One way to deal with this is to add an element having the same but negative admittance to cancel out the effect of this capacitor. We are of course talking about an inductor and may calculate its value as follows

$$Y_L = \frac{1}{j\omega L} = -Y_C \tag{6.2a}$$

$$\therefore L = -\frac{1}{j\omega Y_C} = -\frac{1}{j\,(2\pi \cdot 1 \text{ GHz})\,(j0.063\text{S})} \approx 2.5 \text{ nH} \tag{6.2b}$$

This is shown in Figure 6.1(b). The impedance seen from the source at this point (at 1 GHz) should be $Z_{in} = 20\Omega$. The power delivered to the load is now plotted in Figure 6.1(d) using a dashed line.

The power now peaks at 1 GHz, which is good, but it is still less than the 125 mW we got in Section 2.6. To achieve that level, we must raise the real part of the input impedance, Z_{in}, to 50Ω. We could add a 30Ω resistor in series, but that resistor too would steal some of the power we are trying to couple into the load. When matching impedance for maximum power transfer, it is best to stick with lossless components.

Fortunately, we may remember from Section 3.3.2 that a lossless quarter-wavelength transmission line is capable of changing the real part of the impedance. We need only determine the characteristic impedance of the line. From (3.44), we determine that it must be the geometric mean of the current load impedance (30Ω) and the impedance we wish to match to (50Ω). Therefore,

$$Z_1 = \sqrt{(30\Omega \cdot 50\Omega)} = 31.6\Omega \tag{6.3}$$

This is shown in Figure 6.1(c), and the corresponding power delivered is plotted in Figure 6.1(d) with a solid line. The peak has now been raised to 125 mW, the same as what we found with a simple 50Ω resistor in Section 2.6. The group of elements that we added between the source and the load to achieve this result is termed an *impedance-matching network*, or simply a *matching network*.

It is instructive to examine the impedance at 1 GHz as seen looking into and from the source, Figure 6.2(a), and looking into and from the load, Figure 6.2(b). At the source, we have

$$Z_{s1} = 50\Omega \tag{6.4a}$$

$$Z_{s2} = (31.6\Omega)^2 \left(-j0.063\text{S} + j0.063\text{S} + (20\Omega)^{-1}\right) = 50\Omega \tag{6.4b}$$

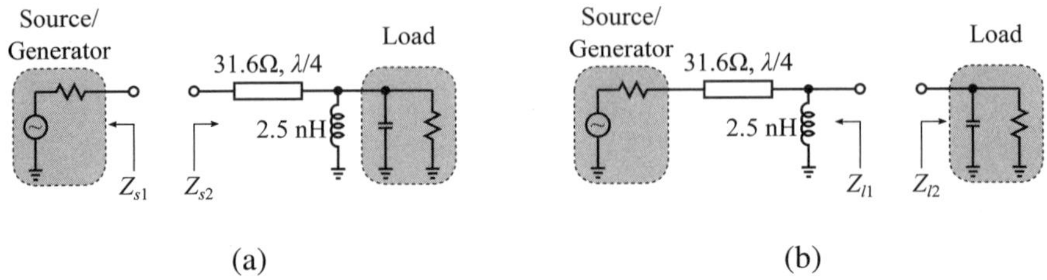

Figure 6.2 Conjugate impedance match seen from (a) the source and (b) the load.

As expected, the load with matching network as seen from the source is equal to the source impedance of 50Ω. This was by design. Consider now the impedances at the load,

$$Z_{l1} = \left(\frac{50\Omega}{(31.6\Omega)^2} - j0.063\text{S} \right)^{-1} = 7.73 + j9.74\Omega \tag{6.5a}$$

$$Z_{l2} = \left(j0.063\text{S} + (20\Omega)^{-1} \right)^{-1} = 7.73 - j9.74\Omega \tag{6.5b}$$

We note that in both cases, the impedance seen looking in one direction is the conjugate of the impedance seen looking in the other direction. In fact, if we had calculated the forward-looking and backward-looking impedances at a point between the transmission line and the inductor, or between the capacitor and resistor inside the load, or at any point along the transmission line whatsoever, we would find that one impedance is always the conjugate of the other. This is called a *conjugate impedance match* and is the precise condition for maximizing the gain of a linear network (the gain, in this case, manifesting as the power delivered from a fixed-amplitude source).

When the impedance is real, then gain is maximized by matching that impedance exactly. It is useful and customary, then, to match the impedance of every component to the characteristic impedance of the system, and then connect those components with cables having that characteristic impedance. Then the precise length of those cables, which may be difficult to control, is rendered unimportant.

6.1.1 Voltage Standing-Wave Ratio (VSWR)

Maximization of gain is but one reason to match the impedance of devices in a system. For another, recall in Section 3.3.4 we found that reflections on a long transmission line may cause cancellation of the signal at specific frequencies and at

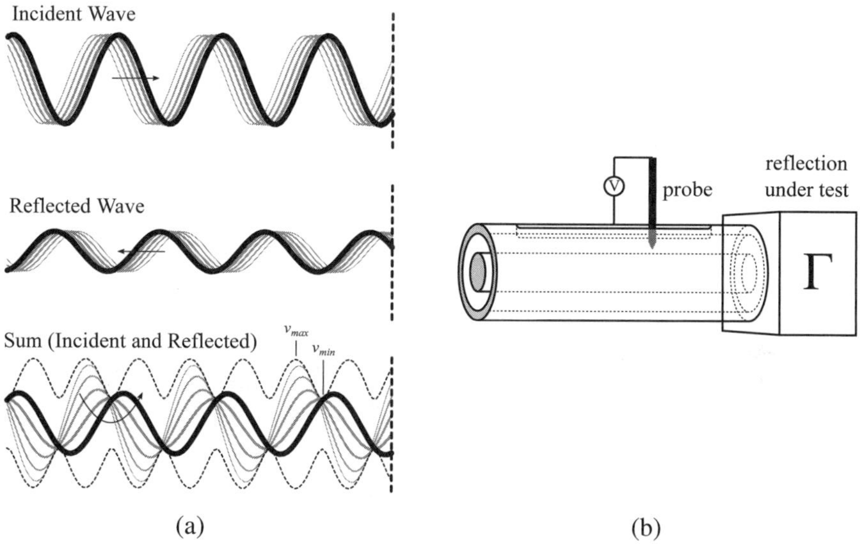

Figure 6.3 (a) An incident wave, a reflected wave, and the sum of the two. Line shading indicates motion — a thick dark line for the most recent instant in time, and thin gray lines for previous instants. (b) Slotted line for measuring VSWR.

specific locations along the line. Even partial cancellation due to reflections less than unity in magnitude may cause unacceptable amplitude variations across frequency and sensitivity to cable length.

Consider the plot in Figure 6.3(a). An incident wave traveling rightward impinges upon a partially reflecting boundary in the top panel. The reflected wave traveling leftward in the center panel is reduced in amplitude and shifted in phase. The sum of the two waves, shown in the bottom panel, appears to travel rightward, while undulating in amplitude following a fixed-in-space envelope defined by

$$v_{env}(v) = \left|1 + \Gamma e^{j2\beta l}\right| \tag{6.6}$$

The ratio of the peak to minimum of the amplitude envelope above is known as the voltage standing-wave ratio (VSWR),

$$\text{VSWR} = \frac{1 + |\Gamma|}{1 - |\Gamma|} \tag{6.7}$$

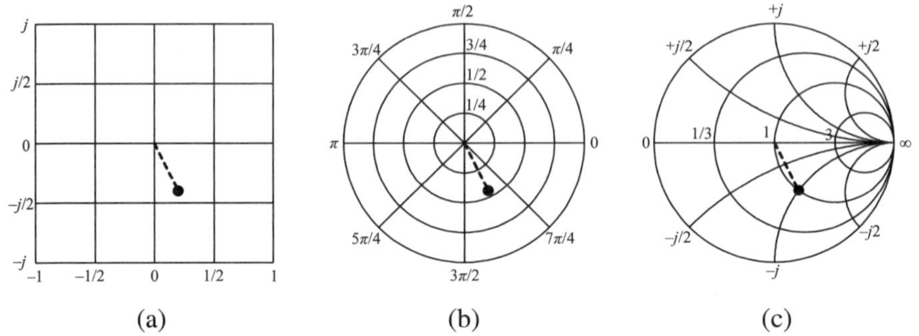

Figure 6.4 Three ways to plot a reflection coefficient. (a) Cartesian coordinates. (b) Polar coordinates. (c) A Smith chart, where the lines correspond to the real and imaginary parts of impedance.

Note that if the load impedance is matched ($\Gamma = 0$), then VSWR $= 1$. However, if the the load is totally reflecting, ($|\Gamma| = 1$), then perfect nulls appear at some points along the line, and VSWR $= \infty$.

In the early days of microwave engineering, VSWR was measured using a device called a *slotted line*, shown diagrammatically in Figure 6.3(b). It consisted of a coaxial line (or waveguide) with a slot in the outer shield through which an electric field probe was inserted. The probe could slide forward and backward on the line, providing a relative measurement of the wave amplitude at different points while disturbing the fields inside the line as little as possible. By measuring the amplitude ratio, or VSWR, one could then determine the magnitude of the reflection coefficient, $|\Gamma|$. Further, by noting the location of the minima, and the distance between them, one could deduce the phase of the reflection coefficient and the precise wavelength of the test signal, respectively.

Reflections are more easily measured now with computer-controlled *vector network analyzers (VNAs)*, but VSWR is still a useful way to express reflection magnitude as it highlights the total amplitude variation that one can expect from a poorly matched load.

6.1.2 The Smith Chart

There is a graphical tool which is a powerful aid for impedance-matching problems, known as a *Smith chart* — after Phillip H. Smith who invented it [1, 2]. To understand it, we first plot in the complex plane a reflection coefficient, say, $\Gamma = 0.2 - j0.4$, as in Figure 6.4(a). In this case, we have used Cartesian axes for the real and imaginary parts of the reflection coefficient. We may also highlight its magnitude and phase by using polar coordinates, as in Figure 6.4(b). Note that the position of the point plotted in the complex plane has not changed, only our axis lines are different. A Smith chart is nothing more than another plot of the reflection coefficient in the complex plane, but where the axis lines correspond to the real and imaginary parts of the normalized impedance (relative to the system characteristic impedance, typically 50Ω) of the termination that produces that reflection. This is shown in Figure 6.4(c). Note that the Smith chart indicates a normalized impedance of $z = 1 - j$. Applying this to (3.37b), we find

$$\Gamma = \frac{z-1}{z+1} = \frac{-j}{2-j} = 0.2 - j0.4 \tag{6.8}$$

which is indeed the reflection coefficient we originally plotted.

The full circles on the Smith chart indicate contours of constant resistance (the real part of impedance) while the arcs extending leftward from the right-most point indicate contours of constant reactance (the imaginary part). These are highlighted in Figures 6.5(a, b), respectively. The outer boundary of the chart, for zero resistance, has a radius of 1, indicating that the largest possible reflection coefficient from a passive impedance has unit magnitude (100% reflection). Larger reflection coefficients would require negative impedance values. The right-most point ($\Gamma = 1$) corresponds to infinite impedance (either the real or imaginary part), otherwise known as an open-circuit, whereas the left-most point is a short-circuit ($z = 0$ or $\Gamma = -1$). The bottom half of the chart (for which reactance is negative) corresponds to capacitive loads, while the top half (positive reactance) is for inductive loads.

Sometimes, it is useful to draw axes for the real and imaginary parts of normalized admittance instead of impedance. Constant conductance and susceptance contours are thus shown in Figures 6.5(c, d), respectively. The Smith chart in this case appears very similar, but rotated 180°, as shown in Figure 6.5(e). (Thus, to convert a known impedance to its equivalent admittance graphically, one needs only to reflect that impedance point through the origin.) In still other cases, it may be useful to show both sets of axes, as in Figure 6.5(f), but some may consider this view too cluttered.

6.1.3 Movements on the Smith Chart

The value of the Smith chart lies in knowing how a given impedance point (a reflection coefficient) moves around the chart in response to different kinds of

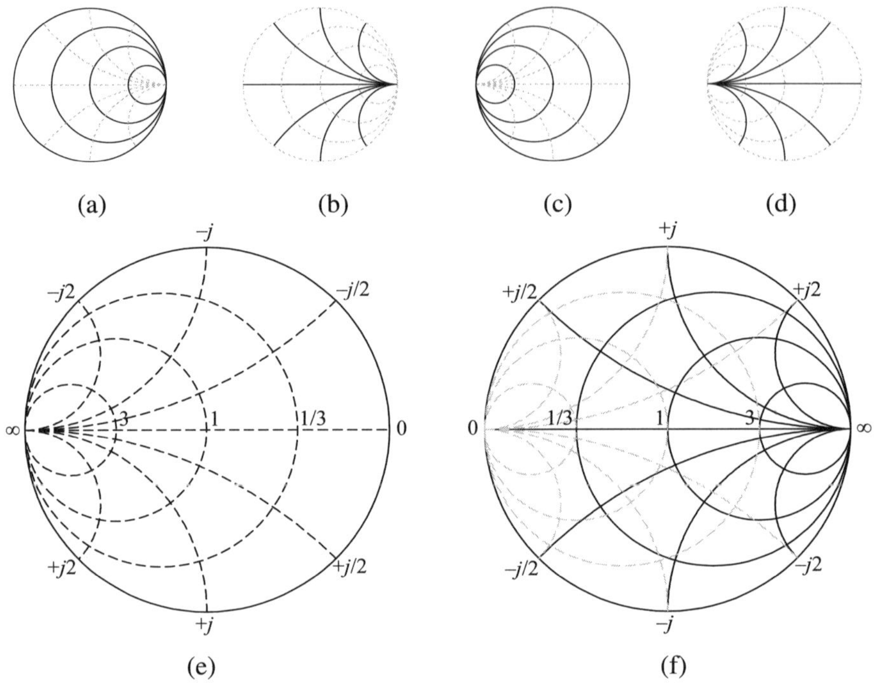

Figure 6.5 Smith charts highlighting (a) constant resistance contours, (b) constant reactance contours, (c) constant conductance contours, and (d) constant susceptance contours. (e) An admittance Smith chart. (f) A Smith chart with both admittance and impedance axes.

elements in the matching network. We will consider the effects of each type of element in turn.

First, we consider series lumped elements — resistors, capacitors, or inductors — as shown in Figure 6.6. Since the lines on the Smith chart correspond to the real and imaginary parts of impedance, series impedance elements simply move the point of the reflection coefficient along these lines. The resistor increases the real part of impedance, moving the point along the circular arcs towards the open-circuit at the right side of the plot. The distance moved is found by considering the value of the resistor (normalized to the characteristic impedance, Z_0). For example, if the starting point is $z = 1 - j$, as shown, and a $Z_0/2$ resistor is added in series, the point moves along the $-j$ contour to its intersection with the circle corresponding to a real part $r = 1 + \frac{1}{2} = \frac{3}{2}$.

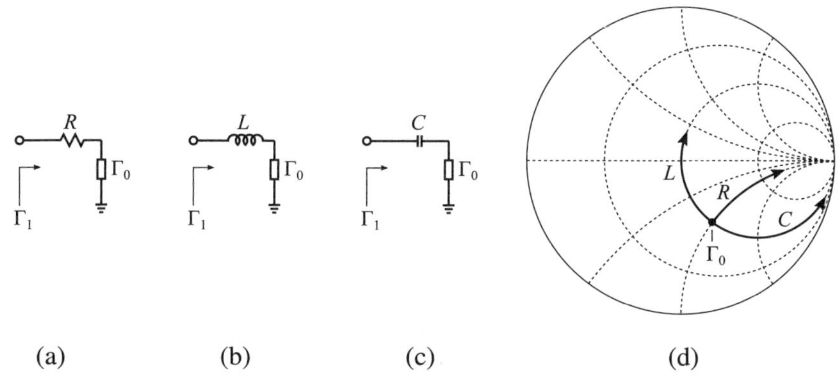

Figure 6.6 Series (a) R, (b) L, and (c) C elements, and (d) their effect on the reflection coefficient plotted on an impedance Smith chart.

A series inductor, on the other hand, adds to the reactance an amount given by $x = j\omega L$. This corresponds to a clockwise movement around the circle denoting constant resistance, as shown. The capacitor instead subtracts from the reactance an amount equal to $1/j\omega C$, causing the reflection coefficient to move counterclockwise around the circle for constant resistance.

Note that series-connected transmission-line stubs behave exactly as series lumped elements (at least at a given frequency point), where the reactance added is calculated using (3.43). Typically, short-circuited stubs less than a quarter-wavelength long are used to add positive reactance (like an inductor), whereas open-circuited stubs less than a quarter wavelength long are used to add negative reactance (like a capacitor), but longer stubs may be used to add reactances of the opposite sign. The shorter stubs are preferred, as they will usually offer greater bandwidth around the central frequency point. Broadband impedance matching will be discussed in Section 6.2.

Next, let us look at lumped elements added in parallel, as illustrated in Figure 6.7. For this case, it is useful to draw the Smith chart with admittance contours (experienced microwave engineers may become adept at visualizing these contours without them actually being drawn in). The parallel conductor (or resistor) moves the reflection coefficient along the arc contours of constant susceptance, whereas the inductor and capacitor move the point along the circular contours of constant conductance. This time, the inductor moves the point in a counterclockwise direction, while the capacitor moves it in a clockwise direction. Parallel-connected

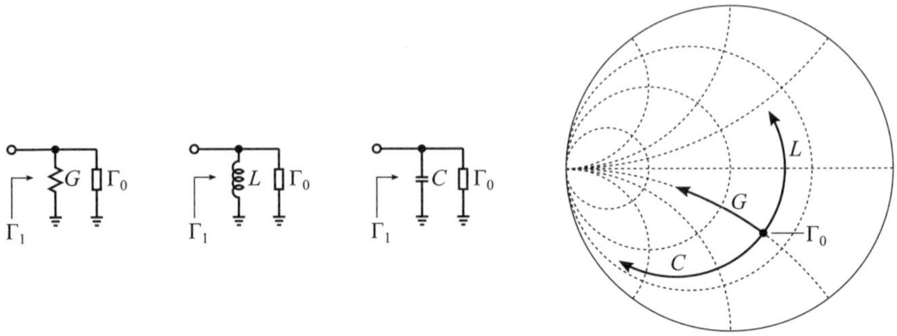

(a) (b) (c) (d)

Figure 6.7 Parallel (a) G, (b) L, and (c) C elements, and (d) their effect on the reflection coefficient plotted on an admittance Smith chart.

transmission-line stubs act in much the same way, where the susceptance added is the inverse of the reactance given in (3.43).

Note that series elements of all kinds move the reflection coefficient toward the open circuit, but via different paths. Likewise, the parallel elements move the reflection coefficient toward the short circuit via different paths. Inductors, whether series or parallel, always move the reflection coefficient toward or through the upper-half of the Smith chart, whereas capacitors always move the reflection coefficient toward or through the lower half of the Smith chart.

Consider now a matched transmission line in cascade. We saw in Section 3.3.1 that the transmission line extension simply adds a phase delay to the reflection coefficient corresponding to the round-trip time from the input terminals of the line to the load and back. Therefore, adding a transmission line matched to the characteristic impedance of the system (or, more accurately, to the normalizing impedance of the Smith chart) simply causes the reflection coefficient to orbit clockwise around the origin, as shown in Figure 6.8. Moreover, since it is the round-trip delay that adds to the reflection, the angle subtended by the path is twice the electrical length of the transmission line, θ.

If the cascade transmission line is not matched to the normalizing impedance of the Smith chart, the reflection coefficient still orbits clockwise along a circular path, but that path is no longer centered at the origin — nor is it precisely centered on a point corresponding to the characteristic impedance of the transmission line. If the transmission line has normalized characteristic impedance z_1, then the circular

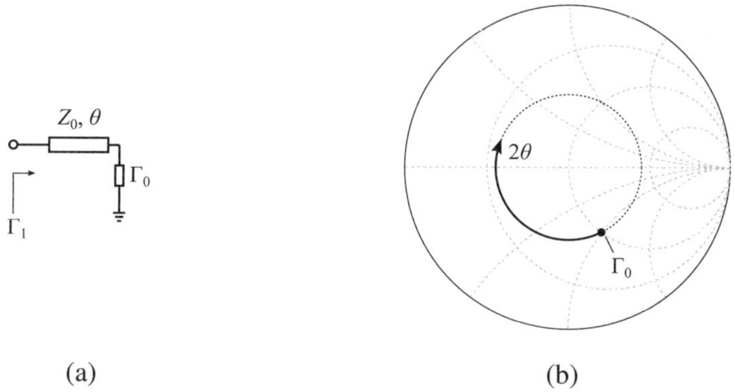

Figure 6.8 (a) Matched cascade transmission line and (b) its effect on the reflection coefficient plotted on an impedance Smith chart.

path the reflection coefficient takes around the Smith chart intersects the horizontal axis at two points corresponding $z_1 x$ and z_1/x where x is a constant determined by the starting point, Γ_0. Graphically, one may draw a circle corresponding to the path they would like for the reflection coefficient to follow, look at the intercept points, z_a and z_b, and then calculate the necessary impedance of the transmission line, $z_1 = \sqrt{z_a z_b}$. This is illustrated in Figure 6.9.

In Smith's time, his charts were used as a mathematical aid, and decades later they are still available as graph paper with finely detailed immittance contours and graduated scales allowing students to solve impedance-matching problems using nothing more than a compass. These calculations are now more readily done by computer, but the Smith chart remains an invaluable guide for designing matching network topologies, presenting data, and understanding the underlying principles and causes of impedance mismatch in electronic systems.

Let us now revisit the matching problem given at the beginning of this chapter, shown in Figure 6.1. The solution found at the time, comprising a shunt inductance and a cascade transmission line, is illustrated as Smith chart movements on the admittance chart in Figure 6.10(a). Other possible solutions to this same matching problem are shown in Figures 6.10(b–d). Although each accomplishes the same thing as far as impedance matching at a single frequency, one or another may be preferred in specific applications. For example, the designer may need a DC block, in which case the series capacitor in the solution of Figure 6.10(c) may be an

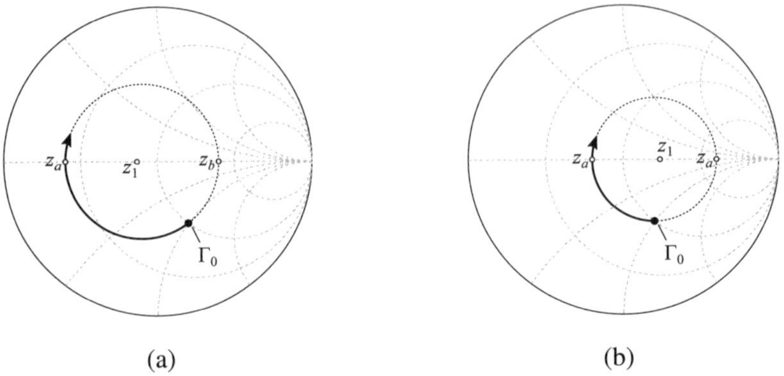

Figure 6.9 The effect on reflection coefficient of a cascade transmission line with normalized characteristic impedance z_1, where (a) $z_1 < 1$ and (b) $z_1 > 1$. The horizontal intercepts are related by $z_1^2 = z_a z_b$.

advantage, or the specific geometry may not permit easy access to ground, in which case the transmission-line solution of Figure 6.10(d) may be preferred.

6.1.4 Two-Parameter Matching Networks

Practicality will often be a critical factor in selecting a matching network for a particular application, but it is useful to have some canonical scenarios at one's fingertips. Let us take an arbitrary, normalized load impedance, $z = r + jx$, and solve for the element values in a given matching network that matches that load to the normalizing impedance at $\omega = 1$. Since impedance is a two-dimensional quantity — it has both magnitude and phase, or real and imaginary parts — two lumped elements are required to achieve the match in the general case.

We thus begin by considering all possible two-element matching networks. First, let us examine the case in Figure 6.11(a) where we have a series inductor, l, preceded by a shunt capacitor, c. The element values are determined by equating the net effective normalized impedance to unity,

$$(r + jx + jl)^{-1} + jc = 1 \tag{6.9a}$$

$$1 + jcr - cx - cl = r + jx + jl \tag{6.9b}$$

$$(1 - cx - cl - r) + j\,(cr - x - l) = 0 \tag{6.9c}$$

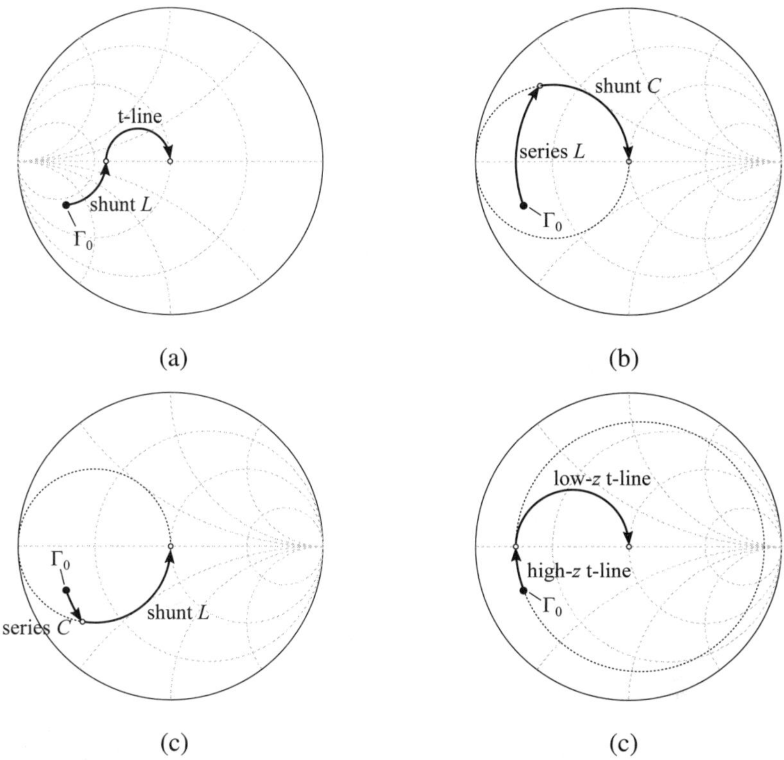

Figure 6.10 Four possible solutions to the impedance-matching problem presented at the beginning of this chapter. (a) Shunt (parallel) inductor and quarter-wave transformer. (b) Series inductor and shunt capacitor. (c) Series capacitor and shunt inductor. (d) High-impedance transmission line (less than a quarter-wavelength) and quarter-wavelength, low-impedance transmission line.

Equating the real and imaginary parts to 0, we may solve for l and c simultaneously,

$$c = \sqrt{\frac{1-r}{r}} \tag{6.10a}$$

$$l = \sqrt{r\,(1-r)} - x \tag{6.10b}$$

Note that the above equations are valid when $r < 1$ and $\sqrt{r(1-r)} > x$. Outside of these ranges, the required elements would be negative or imaginary-valued, and the proposed matching network is not suitable. The latter condition is met if $x < 0$,

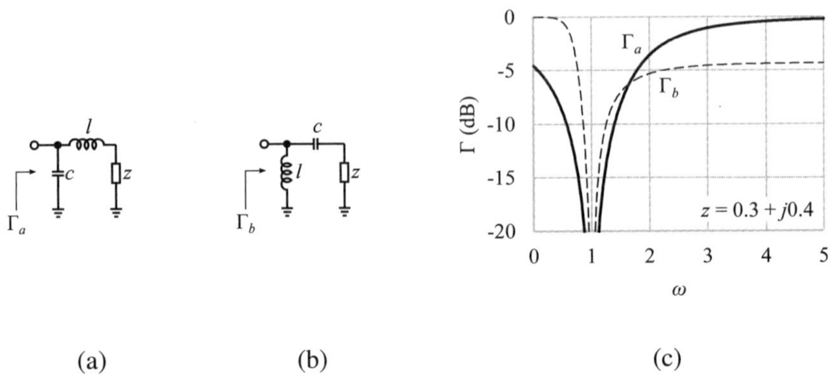

Figure 6.11 Matching networks consisting of (a) a series inductor followed by a shunt capacitor, and (b) a series capacitor followed by a shunt inductor. (c) Comparison of the reflection coefficients with both networks for a normalized load impedance given by $z = 0.3 + j0.4$.

or if

$$r(1 - r) > x^2 \tag{6.11a}$$

$$r - r^2 > x^2 \tag{6.11b}$$

$$\frac{r}{r^2 + x^2} > 1 \tag{6.11c}$$

$$g > 1 \tag{6.11d}$$

where $(r + jx)^{-1} = g + jb$ and g is the normalized conductance. The design equations and validity conditions for this matching network are thus summarized in the first row of Table 6.1.

Alternatively, the same load may be matched using the circuit in Figure 6.11(b), where a series capacitor is preceded by a shunt inductor. The derivation of matching elements is greatly simplified by recognizing that this is merely a high-pass-transformation of the network in Figure 6.11(a). One may then utilize the substitutions from the frequency inversion column of Table 5.1, taking care to include the original load impedance within that transformation ($r + jx$ becomes $r - jx$ due to the negation of the frequency variable). We thus find for this network,

$$l = \sqrt{\frac{r}{1 - r}} \tag{6.12a}$$

Table 6.1
Two-Parameter Matching Networks
$$z = r + jx = (g + jb)^{-1}$$

Network	First Parameter	Second Parameter	Valid for
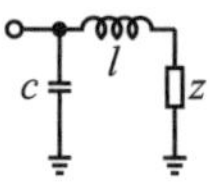	$l = \sqrt{r(1-r)} - x$	$c = \sqrt{\frac{1-r}{r}}$	$r < 1$ and ($x < 0$ or $g > 1$)
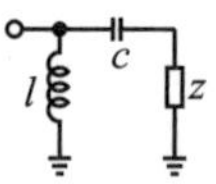	$c = \frac{1}{\sqrt{r(1-r)}+x}$	$l = \sqrt{\frac{r}{1-r}}$	$r < 1$ and ($x > 0$ or $g > 1$)
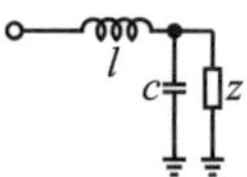	$c = \sqrt{g(1-g)} - b$	$l = \sqrt{\frac{1-g}{g}}$	$g < 1$ and ($b < 0$ or $r > 1$)
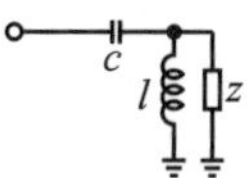	$l = \frac{1}{\sqrt{g(1-g)}+b}$	$c = \sqrt{\frac{g}{1-g}}$	$g < 1$ and ($b > 0$ or $r > 1$)
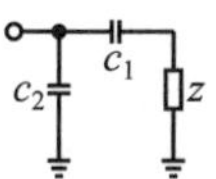	$c_1 = \frac{1}{x-\sqrt{r(1-r)}}$	$c_2 = \sqrt{\frac{1-r}{r}}$	$r < 1$ and $g < 1$ and $x > 0$
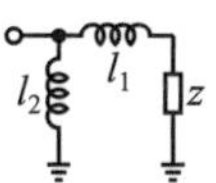	$l_1 = -x - \sqrt{r(1-r)}$	$l_2 = \sqrt{\frac{r}{1-r}}$	$r < 1$ and $g < 1$ and $x < 0$
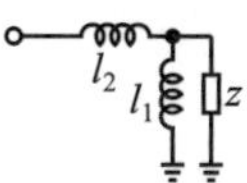	$l_1 = \frac{1}{b-\sqrt{g(1-g)}}$	$l_2 = \sqrt{\frac{1-g}{g}}$	$r < 1$ and $g < 1$ and $b > 0$
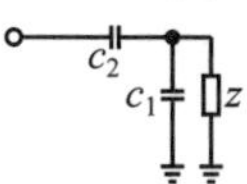	$c_1 = -b - \sqrt{g(1-g)}$	$c_2 = \sqrt{\frac{g}{1-g}}$	$r < 1$ and $g < 1$ and $b < 0$
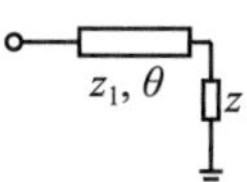	$\tan^2\theta = \left(\frac{r(1-r)}{x^2} - 1\right)(1-r)$	$z_1 = \frac{x}{1-r}\tan\theta$	$r > 1$ or $g > 1$

$$c = \frac{1}{\sqrt{r(1-r)} + x} \tag{6.12b}$$

as indicated in the second row of Table 6.1.

Additional matching networks may be derived by using the duality transformation from Section 5.1.7. This requires exchanging the l's and c's, the series elements and shunt elements, and the load impedance, $r + jx$, with a load admittance, $g + jb = (r + jx)^{-1}$. This completes the first four rows of Table 6.1.

What about networks where both elements are of the same type? Say, for example, we want to use a series capacitor preceded by a shunt capacitor. Since we are only concerned with a single frequency here, we may take the network from Figure 6.11(a) and replace the series inductor with a series capacitor having the same impedance at $\omega = 1$, or $c' = -1/l$. This approach, along with the aforementioned high-pass and duality transformations, gives us the next four rows in the table.

Note that we have only had to go through a meticulous derivation for one matching network so far, in the first row of Table 6.1. Clever application of transformations and substitutions have then multiplied the fruits of our labor by eight.

If we use transmission lines instead of lumped elements, a single line already has two free parameters — namely the characteristic impedance, z_1, and the electrical length, θ. A single transmission line, then, could match a general load impedance, at least in principle. Lone stubs do not work in the general case, as they produce only a series or parallel reactance at a single frequency, although that reactance can be achieved with multiple combinations of z_1 and θ. Instead we use a cascade transmission line, as shown in the final row of Table 6.1. Applying (3.42) in normalized quantities and setting the result equal to 1, we have

$$z_1 \frac{r + jx + jz_1 \tan\theta}{z_1 + j(r + jx)\tan\theta} = 1 \tag{6.13a}$$

$$rz_1 + jxz_1 + jz_1^2 \tan\theta = z_1 + jr\tan\theta - x\tan\theta \tag{6.13b}$$

$$rz_1 - z_1 + x\tan\theta = 0 \tag{6.13c}$$

$$z_1 = \frac{x}{1-r}\tan\theta \tag{6.13d}$$

$$\tan^2\theta = \left(\frac{r(1-r)}{x^2} - 1\right)(1-r) \tag{6.13e}$$

$$\therefore \tan\theta = \pm\sqrt{\left(\frac{r(1-r)}{x^2} - 1\right)(1-r)} \tag{6.13f}$$

Care must be exercised in the application of (6.13f), since only one of the roots (positive or negative) will be correct in any given problem, and the arctangent must be taken such that θ is positive.

For convenience, we have worked exclusively with normalized elements in deriving the results of this section (suitable, say, for a system characteristic impedance of $Z_0 = 1\Omega$ and at an operating frequency of $\omega = 1$ rad/s). The reader is referred to the scaling laws introduced in Sections 5.1.1 and 5.1.2 to convert these into real element values.

6.1.5 Two Transmission-Line Matching Networks

The set of matching solutions presented in Table 6.1 are considered minimal in the sense that the fewest possible elements have been used. Other matching-network topologies are possible, and may have practical advantages in certain situations.

Of particular note is that only one transmission-line network has been given, and the range of load impedances to which it can be applied is restricted to a particular region of the Smith chart (where either $r > 1$ or $g > 1$). Access to the remainder of the Smith chart requires a second transmission line.

Obviously, any one of the lumped-element solutions presented in the previous section could be converted almost trivially to a double-stub transmission-line solution using Richard's transformation. Unfortunately, this will inevitably result in a combination of series-connected and parallel-connected stubs, which, as explained in Sections 3.5.2 and 5.2.6, is most often inconvenient in practice.

Instead, let us consider two-line matching networks in which at least one is a cascade transmission line. There are many possibilities. First, let us examine matching networks comprising a parallel-connected stub, either open-circuited or short-circuited, preceded by a cascade transmission line, as shown in Figure 6.12. Since parallel-connected stubs add susceptance to the load, it will be useful to draw the Smith chart with admittance contours shown. If the load immittance starts in the upper half of the Smith chart, it will be most useful to use an open-circuited stub to add some capacitive susceptance to it, as shown in Figure 6.12(a). If the load immittance starts in the lower half of the chart, a short-circuited stub may be used to add inductive susceptance. Either way, there is some choice as to how far the the susceptance should be modified before inserting the cascade transmission line (the intermediate admittance point must, however, lie within the valid region for cascade-line matching identified in the previous section, namely $r > 1$ or $g > 1$).

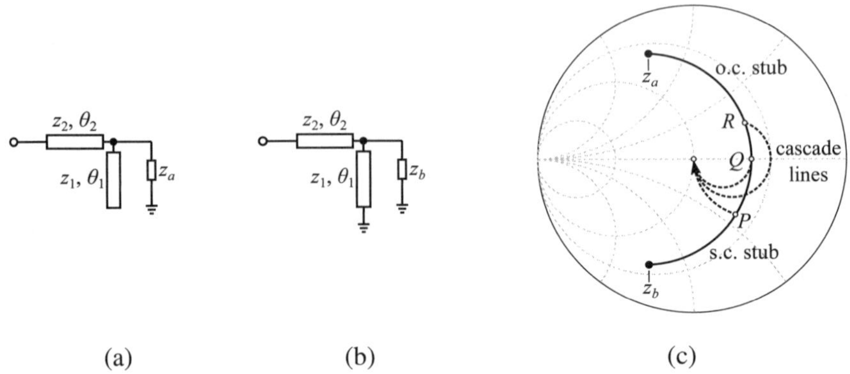

Figure 6.12 Impedance matching using parallel-connected (a) open-circuit and (b) short-circuit stubs, preceded by a cascade transmission line. (c) Admittance movements caused by stubs are drawn with solid lines, while those introduced by cascade lines are drawn with dotted lines.

Three such intermediate admittances are shown in Figure 6.12(c), labeled P, Q, and R. If the stub brings the admittance to point P, only a short (less than a quarter-wavelength) cascade line is then needed to bring it back to the origin, thus completing the match. If instead the intermediate admittance is point Q, which lies on the horizontal contour of the Smith chart, then a cascade line that is exactly a quarter-wavelength long (a quarter-wave transformer) is needed. This may be simplest mathematically, since the susceptance of the stub is simply the imaginary part of original load admittance, and the characteristic impedance of the quarter-wave transformer may be derived from the real part using (3.44). Finally, if one brings the admittance to point R, a fairly long cascade line is needed to bring the admittance to a match — while the stub, either open-circuited or short-circuited, is comparatively shorter than if it had been brought to points P or Q. Note that regardless of which intermediate point is used, the cascade lines move the admittance point clockwise toward the origin.

Alternatively, a cascade line attached directly to the load impedance may be preceded by a stub, as shown in Figure 6.13. Once again, the Smith chart is drawn with admittance contours, and there are many possible trajectories one might use. Depending on the characteristic impedance and length of the cascade line, the admittance may be brought to intermediate points P, Q, or R (or any other that lies on the $g = 1$ contour of the Smith chart). Then, depending on the sign of the susceptance at that point, either an open-circuited or short-circuited stub may bring

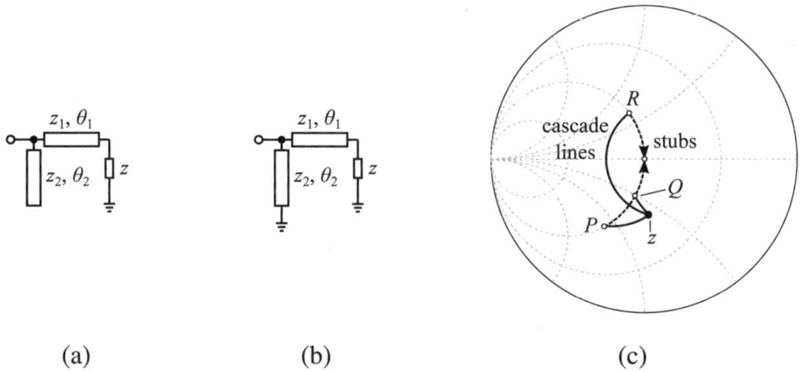

Figure 6.13 Impedance matching using a cascade transmission line preceded by (a) an open-circuited stub, and (b) a short-circuited stub. (c) Admittance movements caused by cascade transmission lines are drawn with sold lines, while those due to stubs are drawn with dotted lines.

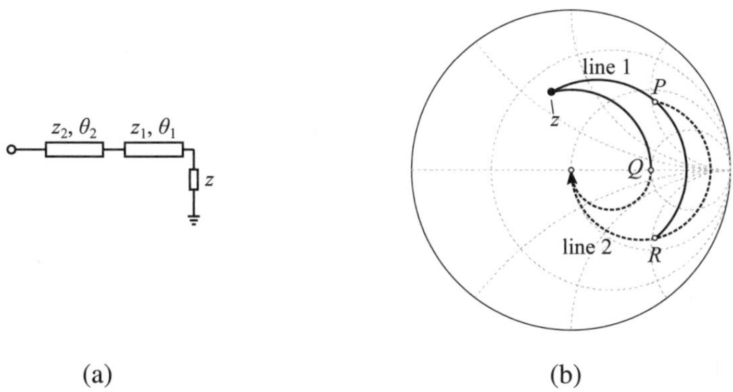

Figure 6.14 Impedance matching using two cascade transmission lines of differing impedance. (a) Schematic. (b) Immittance movements on a Smith chart.

it back to the origin. Again, the cascade line permits only clockwise movements, while the stubs may add susceptance in either direction (positive or negative).

A final two transmission-line alternative utilizes two cascade lines of differing impedance (if they were the same impedance, then it would be the same as having only one cascade line with the sum of the lengths, a scenario that was already covered in Section 6.1.4). This scenario is shown in Figure 6.14. As in every other two transmission-line case, multiple trajectories are possible to arrive at the origin. The only constraint is that the intermediate point lies within the valid region for single cascade-line matching given in Table 6.1.

6.2 BROADBAND IMPEDANCE MATCHING

Thus far, we have talked about impedance matching at a single frequency point. In many situations, however, we would like for the impedance as seen through our matching network to remain constant for at least some moderate bandwidth around the center frequency, if for nothing else than to guard against manufacturing tolerances which might tend to detune the match. For example, the reflection coefficient null shown in Figure 6.11(c) is wider when network (a) is used than when network (b) is used, at least for the load impedance given in that example. The latter network, then, is more prone to detuning from manufacturing errors than the former. We would like to gain some insight into why one network is better suited than the other and how to design networks that maximize the matched bandwidth for a given load.

6.2.1 Comparison of Two-Parameter Matching Networks

A good way to start is by examining how the minimal matching networks of Table 6.1 compare to one another in performance over a small range around the center frequency. Not surprisingly, this will depend on the load impedance they are trying to match. A 55Ω load should be easier to match to 50Ω than a $5\text{k}\Omega$ load, for example, and a slightly inductive load should be compensated rather more easily with a single capacitor than a short-circuited stub followed by a transmission line.

To make this quantitative, we shall consider all possible impedances that one might wish to match (i.e., the entire Smith chart) and then look at how quickly the reflection coefficient diverges from zero for a given type of matching network. It is somewhat artificial to assume that the load impedance, $z = r + jx$, is constant for all frequencies (inevitably, it will have some frequency dependence, possibly a lot), but in the absence of other assumptions, this is a good place to start.

First, let us write down the admittance, $y(\omega)$, seen looking into the load through the matching network in the first row of Table 6.1,

$$y(\omega) = (r + jx + j\omega l)^{-1} + j\omega c \tag{6.14}$$

The bandwidth may be characterized by the rate of change of the reflection coefficient around the match point, $\omega = 1$,

$$\left|\frac{d\Gamma}{d\omega}\right|^2_{\omega=1} = \left|\frac{d}{d\omega}\frac{1-y(\omega)}{1+y(\omega)}\right|^2_{\omega=1} = \left|\frac{-(1+y)-(1-y)}{(1+y)^2}\frac{dy}{d\omega}\right|^2_{y=1} \tag{6.15a}$$

$$= \frac{1}{4}\left|\frac{dy}{d\omega}\right|^2_{\omega=1} = \frac{1}{4}\left|\frac{d}{d\omega}\left((r+jx+j\omega l)^{-1}+j\omega c\right)\right|^2_{\omega=1} \tag{6.15b}$$

$$= \frac{1}{4}\left|-jl\left(r+j(x+l)\right)^{-2}+jc\right|^2 = \frac{1}{4}\left|\left(r+j(x+l)\right)^{-2}l-c\right|^2 \tag{6.15c}$$

where l and c are given in the table. Similar expressions may be derived for the next three rows in the table by using the same transformations as before (high-pass and duality). The two-capacitor network in the fifth row was derived from the first by exchanging the series inductor for its equivalent negative capacitor. One should be cautious about that substitution here, as we had only convinced ourselves that the value of the reactance would be the same for both elements; nothing at all was said about the rate of change of that reactance. Nonetheless, it turns out that a simple substitution of $c_1 = -1/l$ gives the correct answer,

$$\left|\frac{d\Gamma}{d\omega}\right|^2_{\omega=1} = \frac{1}{4}\left|\left(r+j\left(x-c_1^{-1}\right)\right)^{-2}c_1^{-1}-c_2\right|^2 \tag{6.16}$$

Again, the corresponding expressions for the next three rows follow simply from high-pass and duality transformations.

We may therefore gather these results for the eight minimal matching networks comprising only lumped elements and see which is best for each region of the Smith chart — where "best" is measured by which gives the slowest rate of change of the reflection coefficient around the match point. The outcome of this experiment is summarized in Figure 6.15. Each of these networks has an optimal place in the Smith chart where they produce the best (most broadband) results. These regions are bounded by the horizontal and vertical axes as well as the $r = 1$ and $g = 1$ contours.

A similar calculation can be performed for the last minimal matching network comprising a single cascade transmission line,

$$\left|\frac{d\Gamma}{d\omega}\right|^2_{\omega=1} = \frac{1}{4}\left|\frac{d}{d\omega}z_1\frac{r+jx+jz_1\tan\theta}{z_1+j(r+jx)\tan\theta}\right|^2_{\omega=1} \tag{6.17a}$$

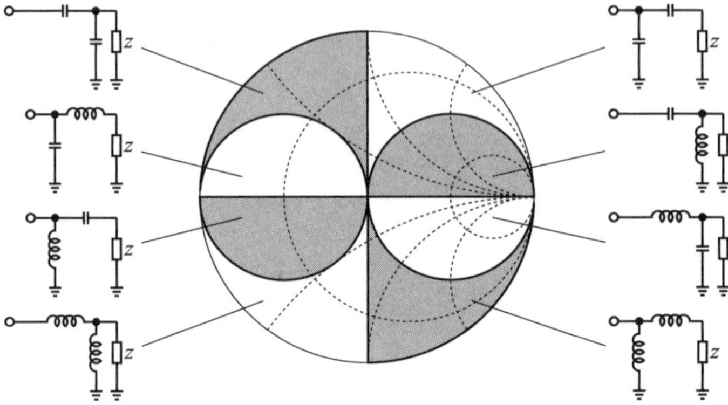

Figure 6.15 Regions of the Smith chart for which various kinds of minimal, lumped-element matching networks produce the most broadband impedance match.

$$= \frac{z_1^2\theta^2}{4}\left|\frac{(z_1+j(r+jx)\tan\theta)\big(jz_1\sec^2\theta\big)-(r+jx+jz_1\tan\theta)\big(j(r+jx)\sec^2\theta\big)}{(z_1+j(r+jx)\tan\theta)^2}\right|^2 \qquad (6.17b)$$

$$= \frac{z_1^2\theta^2}{4\cos^2\theta}\left|\frac{(z_1+j(r+jx)\tan\theta)jz_1-(r+jx+jz_1\tan\theta)j(r+jx)}{(z_1+j(r+jx)\tan\theta)^2}\right|^2 \qquad (6.17c)$$

where the nominal values of z_1 and θ are given in the table. (Note that although the expression given for z_1 depends on the nominal value of θ at the center frequency, it is considered a constant versus frequency in this calculation.) If this result is included, the map of optimal matching networks across the Smith chart is somewhat changed, as shown in Figure 6.16. The cascade-line matching network has eclipsed the series-inductor/shunt-capacitor networks entirely, offering better performance in their entire optimal regions. It also supersedes the series-capacitor/shunt-inductor networks over a small slice of their optimal regions just beyond the horizontal, creating a now curved horizontal boundary across the middle.

The exact placement of these boundaries should be taken loosely, as we have ignored for now any variability in the load impedance with frequency. The trajectory of this changing impedance can be oriented in any direction on the Smith chart whatsoever, inevitably helping certain matching networks while working against others to create a more or less broadband match, effecting a shift in the boundary line between one best matching network and another. We have also ignored any practical constraint on the realizable values of certain parameters, such as the characteristic impedance of the transmission lines. These charts are thus not meant

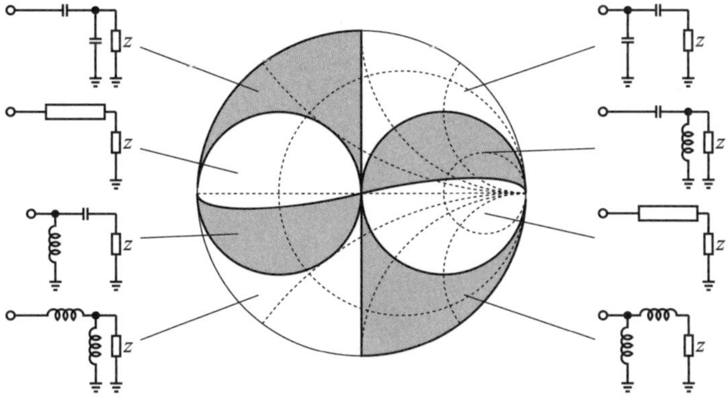

Figure 6.16 Regions of the Smith chart for which various kinds of minimal, lumped-element or transmission-line matching networks produce the most broadband impedance match.

to be entirely rigorous, but only starting points to guide engineers in which matching networks might work best for new impedance-matching problems.

6.2.2 Carter Chart

While the minimal, two-parameter networks of Section 6.1.4 offer the most direct route to matching a load immittance, more complex networks may permit wider-bandwidth performance. A good rule of thumb is to minimize the loaded Q of each tuning element; recall from Section 2.5 that Q is inversely proportional to bandwidth around a null. For a lumped matching element in series, we may apply (2.80) with normalized quantities

$$Q = \frac{x}{r} \tag{6.18}$$

whereas for parallel tuning elements, we apply (2.85),

$$Q = \frac{b}{g} \tag{6.19}$$

For a given load, these two definitions are the same (except for a sign change, which merely indicates which type of element, inductor or capacitor, is to be added to

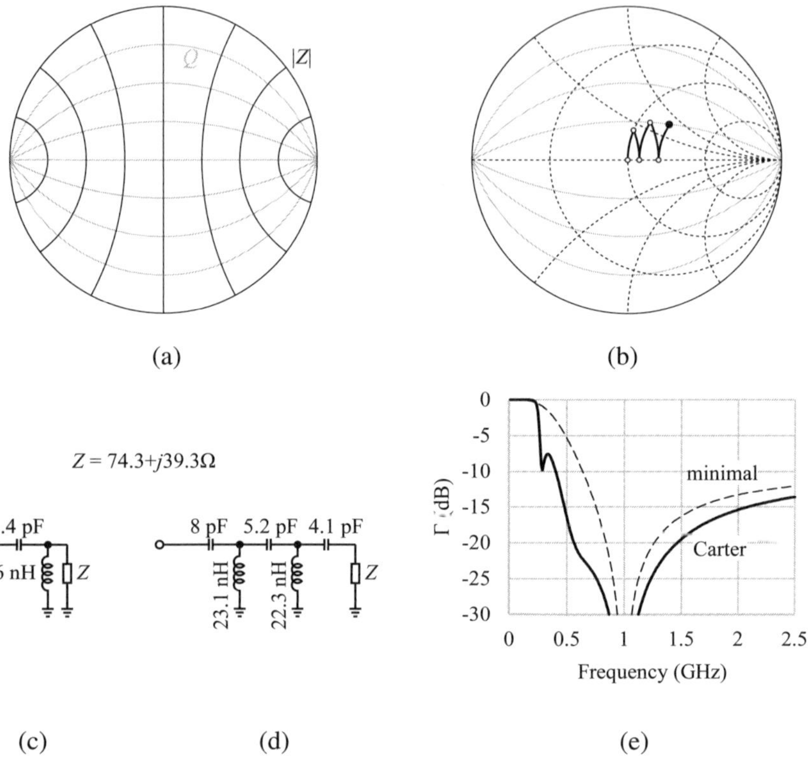

Figure 6.17 (a) Carter chart showing contours of impedance magnitude and Q. (b) Multi-element matching example using constant-Q contours. (c) Best minimal matching network. (d) Multi-element matching network represented by the Smith/Carter chart in (b). (e) Comparison of reflection coefficient achieved by the two networks.

complete the resonator),

$$Q = \frac{b}{g} = \frac{\frac{-x}{r^2+x^2}}{\frac{r}{r^2+x^2}} = -\frac{x}{r} \tag{6.20}$$

We may thus visualize the problem by drawing contours of constant Q on the Smith chart. Conventionally, when combined with contours of impedance magnitude, as in Figure 6.17(a), this is called a Carter chart [3]. For impedance matching work, it is easiest to simply overlay the constant-Q contours on a conventional Smith chart, as shown in Figure 6.17(b). The goal, then, is to add elements which move the match point closer to the origin, but staying within the Q contour on which the original load immittance lies. This prevents any one tuning element from increasing the Q of the match. Once the match point intersects $r = 1$ or $g = 1$, the final element may be used to bring it back to the origin.

For this example, an arbitrary load of $Z = 74.3 + j39.3\Omega$ has been used. The best two-element matching network for this load is given in Figure 6.17(c). This is contrasted with a multi-element matching network derived graphically from the Smith/Carter chart. The initial load point in this example is generally inductive with a high resistance. We therefore add a sequence of small series capacitors and shunt inductors which move the impedance point in small increments between the zero reactance and constant-Q contours, as indicated by the trajectory drawn in Figure 6.17(b). The final network is given in Figure 6.17(d). The reflection coefficients derived from both matching networks are plotted in Figure 6.17(e). While both match the immittance perfectly at 1 GHz, the multi-element network derived here has a wider match bandwidth, approaching 3:1 at the 20-dB level.

6.2.3 Immittance Folding

To make further progress in developing broadband matching techniques, we must not only take into account the contributions of our matching elements, but also the trajectory of frequency variation in the load immittance itself. It is important to recognize that some movements on the Smith chart scale in proportion to frequency, while others scale inversely to frequency. Specifically:

- Cascade transmission lines move immittance points in a clockwise circle on the Smith chart by an amount that increases with increasing frequency. In other words, higher frequencies move further.
- Series inductors or series-connected, short-circuit transmission-line stubs move immittance points clockwise along constant-resistance contours by an amount that increases with increasing frequency. Again, higher frequencies move further.
- Shunt capacitors or parallel-connected, open-circuit transmission-line stubs move immittance points clockwise along constant-conductance contours by an amount that increases with increasing frequency. Higher frequencies move further.

- Series capacitors or series-connected, open-circuit transmission-line stubs move immittance points counterclockwise along constant-resistance contours by an amount that increases with decreasing frequency. In this case, lower frequencies move further.
- Shunt inductors or parallel-connected, short-circuit transmission-line stubs move immittance points counterclockwise along constant-conductance contours by an amount that increases with decreasing frequency. Lower frequencies move further.

Note the pattern: In clockwise movements, higher frequencies move further, while in counterclockwise movements it is the lower frequencies that move further. This information is summarized in Table 6.2.

Each of these movements, then, acting on a locus of immittance points representing a frequency-dependent load, alters the shape of that locus as it moves. If the locus is already oriented with higher frequencies in the direction of greater movement, then the effect is to stretch out the locus of immittance points. However, if the locus is oriented opposite the direction of greatest movement, the locus of immittance points may be made to collapse or fold back upon itself as it moves. This latter effect is the key to achieving a broadband impedance match.

Let us work an example qualitatively. We start with a given load impedance modeled as a small-value resistor with some series capacitance (typical for, say, the input of a field-effect transistor). This load is shown in Figure 6.18(a). We wish to match this load at around 1 GHz to a reflection coefficient of 20 dB or more with as much bandwidth as possible. The matching network we will use is given in Figure 6.18(b). To see why this network is superior to the minimal, single frequency networks in Table 6.1, we will examine the movements of the immittance locus one step at a time.

First, in Figure 6.18(c) we show the original immittance locus for the unmodified load over a range of frequencies, 0.85–1.2 GHz. The first matching element, a 4.6-nH shunt inductor, moves the immittance in a counterclockwise direction along contours of constant conductance, as shown in Figure 6.18(d). Note that we have switched the Smith chart to admittance axes to aid in the visualization of this step. The locus has moved close to the center of the chart, but since the low-frequency end moves further, the locus of immittances also stretches out as shown.

In Figure 6.18(e), we have added a 1.6-pF capacitor. This moves the immittance locus in a counterclockwise direction again, this time along constant-resistance contours (notice that our Smith chart is once again drawn with impedance axes). Importantly, the low-frequency end moves further, but since it was further

Table 6.2
Immittance Movements

Matching Elements	Smith Chart	Frequency Dependence
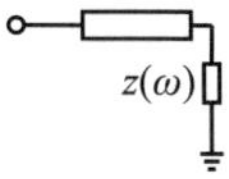	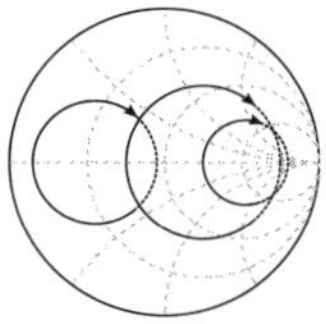	*Higher* frequencies move further.
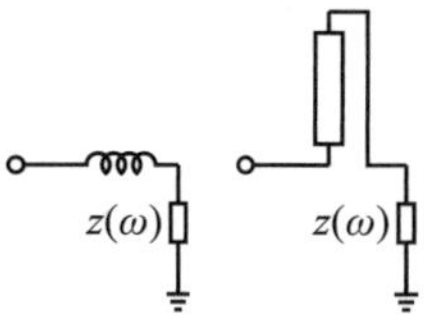	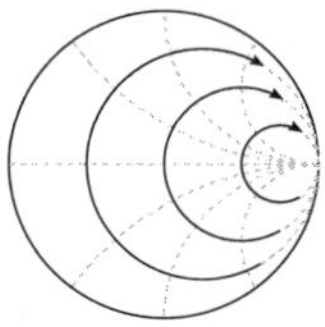	*Higher* frequencies move further.
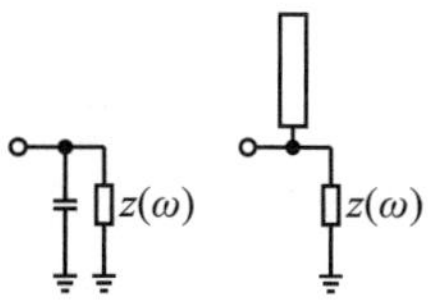	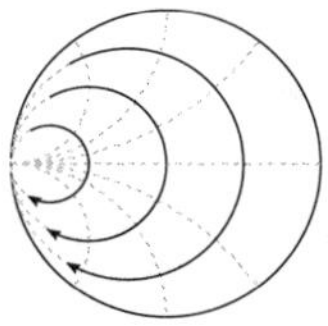	*Higher* frequencies move further.
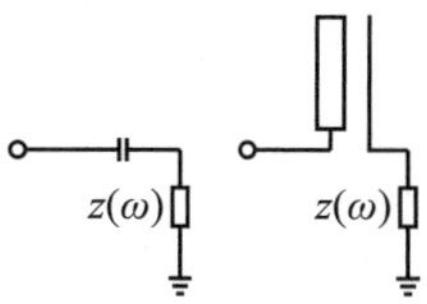	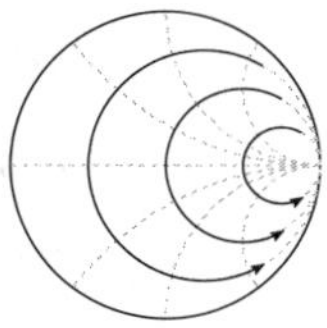	*Lower* frequencies move further.
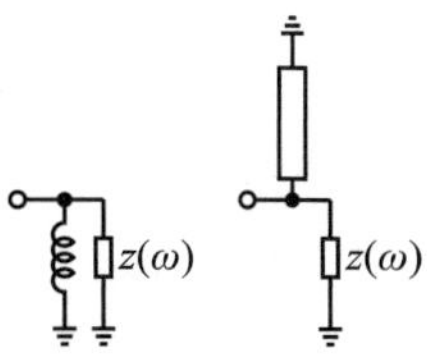	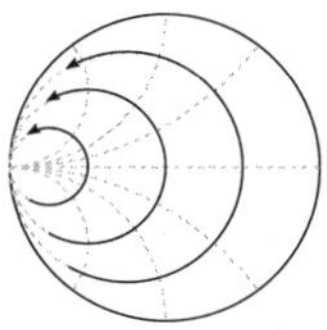	*Lower* frequencies move further.

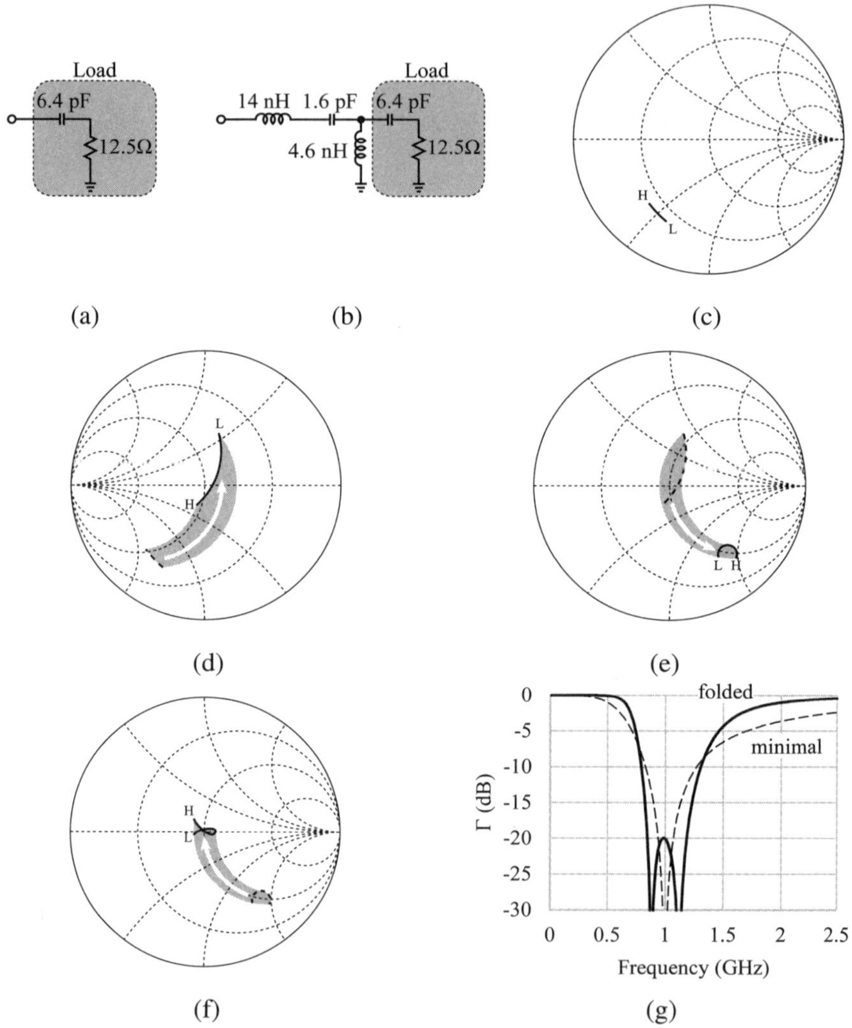

Figure 6.18 (a) A complex load with frequency dependence. (b) Lumped-element, immittance-folding matching network. (c) Load immittance ("L" and "H" correspond to the low-frequency and high-frequency ends, respectively). (d) Locus movement for the 4.6-nH shunt inductor, (e) the 1.6-pF series capacitor, and (f) the 14-nH series inductor. (g) Comparison of reflection coefficient using a minimal, two-parameter matching network (dashed line) and the immittance-folding network (solid line).

behind the $r = 1$ contour to begin with, it has the effect of folding the immittance locus back upon itself as it moves.

In the last step, Figure 6.18(f), a series 14-nH inductor is added, moving the locus back along the $r = 1$ contour toward the origin. In this case, the high-frequency end moves further, but in this direction the high-frequency end was lagging so the net effect is once again to fold the immittance curve even tighter as it moves. The end result is a tiny knot of immittance points clustered around the origin of the Smith chart.

Observe that the final two steps, comprising a series capacitor and a series inductor, nearly cancel each other out in terms of the overall movement on the Smith chart, but effectively pinch the immittance locus at the origin as a consequence of the greater impact each one has at the low-frequency or high-frequency end. If the locus prior to these steps had been oriented in the opposite direction (with the high-frequency end vertically above the low-frequency end), then a parallel LC resonator would have accomplished the same thing.

Figure 6.18(g) shows the reflection coefficient in decibels as seen through this matching network with a solid line. For comparison, we also show with a dotted line the reflection coefficient that would be obtained for the same load using the network in row 7 of Table 6.1 (Figure 6.16 indicates that this would be optimal for those networks). While the latter network matches the load perfectly at 1 GHz, the more complex immittance folding network matches it almost perfectly at two frequencies, and keeps the reflection coefficient better than 20 dB over a wider bandwidth.

Immittance folding works with transmission lines as well. Consider the same load and an alternate immittance folding network shown in Figures 6.19(a, b). The load immittance plot, Figure 6.19(c), is the same, but this time in Figure 6.19(d) we use a short cascade transmission line to move the locus into the $g = 1$ circle. The locus stretches out a bit, but not by much since the line is only 11° long. Next, in Figure 6.19(e), a shunt, short-circuited stub adds inductive susceptance, moving the locus around a circle of constant conductance, folding the low-frequency end around faster than the high-frequency end. The immittance locus is already getting more compact. In the final step, Figure 6.19(f), another cascade transmission line moves the locus back toward the origin. Since the high-frequency end moves fastest, but started from farther away along the relevant contour, the immittance locus tightens up even further, resulting once again in a compact knot at the origin.

Figure 6.19(g) shows, as before, a comparison of the resultant reflection coefficient (solid line) with that of the minimal two-parameter matching network from Table 6.1.

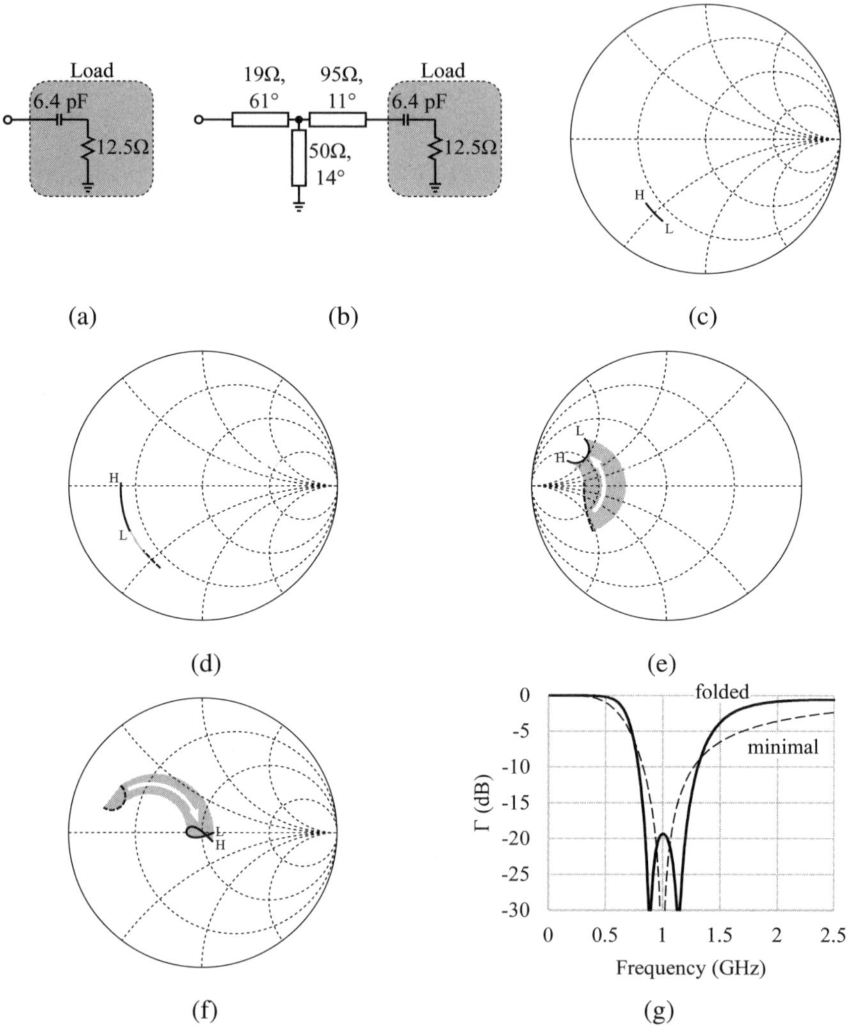

Figure 6.19 (a) A complex load with frequency dependence. (b) Transmission-line, immittance-folding matching network. (c) Load immittance ("L" and "H" correspond to the low-frequency and high-frequency ends, respectively). (d) Locus movement for the 95Ω cascade line, (e) the 50Ω shunt stub, and (f) the 19Ω cascade line. (g) Comparison of reflection coefficient using a minimal, two-parameter matching network (dashed line) and the immittance-folding network (solid line).

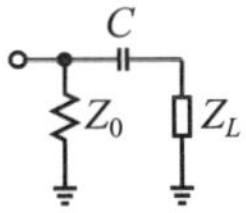

Figure 6.20 A lossy matching network that may provide arbitrarily broad matching to any load, Z_L, at the cost of coupling efficiency to that load.

6.2.4 The Bode-Fano Criteria

It is interesting to note that both the lumped-element and transmission-line folded solutions have two reflection nulls and give about the same bandwidth for a given load. More complex folding networks could have resulted in more nulls and slightly wider bandwidth, but the consistency of our results strongly suggests that a fundamental limit on matching bandwidth may be at work. It seems logical to ask, what is that limit, and how does it depend on the properties of the load?

A complete answer to that question is fairly complex, and requires delving into the mathematical requirements of the network parameters that may be realized using our limited repertoire of tuning elements (lumped and transmission-line). If there were no limit to the elements that could be realized, nor any concern over physical absolutes such as causality and passivity, then perfect, lossless, broadband impedance matching would be possible; one could simply synthesize an impedance element that exactly negates the reactive part of the load at all frequencies, then scale the residual real-part using an ideal transformer. Sadly, even if we cast aside the practical issues of nonideality, the required impedance element would almost certainly be noncausal; it would have to respond to stimuli before they occurred.

Further, if lossy elements (e.g., resistors) are allowed, then in principle there is no limit to how broadly a given immittance may be matched. For a trivial example, consider the RC matching network shown in Figure 6.20. Since the resistor is matched to Z_0, the input reflection coefficient may be made arbitrarily small at all frequencies simply by decreasing the value of the coupling capacitor, C. This is hardly of any use since doing so effectively decouples the load from the input port. Although this is a fairly pathological example, the point that lossy elements broaden the impedance match only at the cost of coupling from the source to the load is a general feature of such networks.

We constrain the problem, then, to matching networks which have no lossy elements. All real elements have some incidental losses anyway, of course, but the ideal, lossless case provides us with useful theoretical limits. The first such limit

Table 6.3

The Bode-Fano Matching Constraints

Network	Bode-Fano Limit
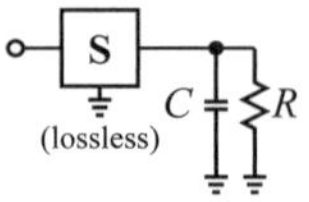	$\int_0^\infty \ln \frac{1}{\lvert\Gamma(\omega)\rvert} d\omega \leq \frac{\pi}{RC}$
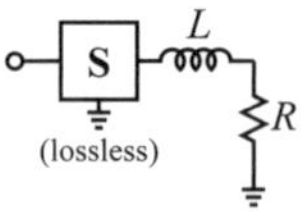	$\int_0^\infty \ln \frac{1}{\lvert\Gamma(\omega)\rvert} d\omega \leq \frac{\pi R}{L}$
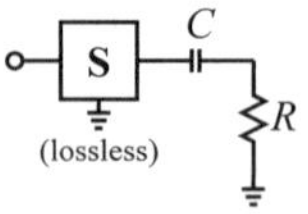	$\int_0^\infty \frac{1}{\omega^2} \ln \frac{1}{\lvert\Gamma(\omega)\rvert} d\omega \leq \pi RC$
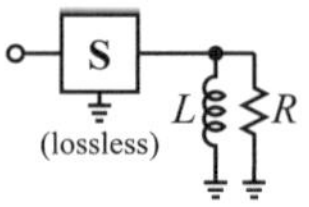	$\int_0^\infty \frac{1}{\omega^2} \ln \frac{1}{\lvert\Gamma(\omega)\rvert} d\omega \leq \frac{\pi L}{R}$

was derived by Hendrik Bode [4] for the specific case of a load resistor, R, in parallel with a capacitor, C. He found, for an arbitrarily complex, lossless matching network, comprising only lumped elements, the following integral constraint,

$$\int_0^\infty \ln \frac{1}{\lvert\Gamma(\omega)\rvert} d\omega \leq \frac{\pi}{RC} \tag{6.21}$$

Robert Fano later extended this work to more general cases [5, 6]. The results, which came to be known as the *Bode-Fano criteria*,[1] are summarized for the four most basic loads in Table 6.3. Real-world loads are more complex than this, but usually not in a way that is helpful to the goal of broadband impedance matching. It is typically sufficient, then, to approximate the real load impedance with one of these simple models around the frequencies of interest to obtain a reasonable estimate of the matched bandwidth that can be achieved. For a rigorous and thorough treatment

1 Most people write the Bode-Fano "criterion," despite the fact there is clearly more than one.

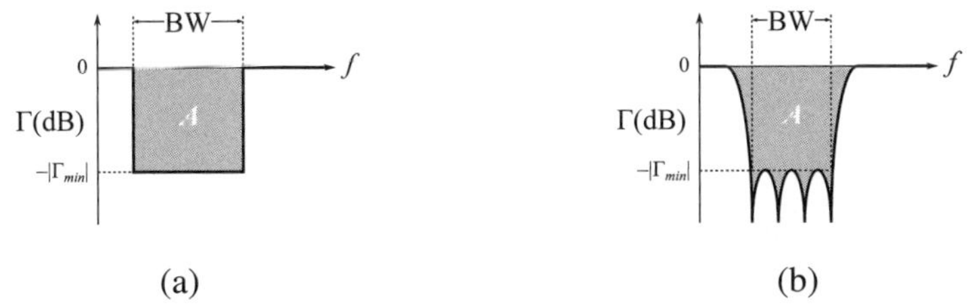

Figure 6.21 (a) Ideal return loss curve to maximize bandwidth given a minimum return loss specification. (b) Causal approximation to the ideal return loss curve. The shaded regions have the same area, A.

of the theoretical limit in the more general cases, one is best advised to consult the original source [4–8].

Some comments are warranted on the form and implications of the Bode-Fano criteria. First, notice that the integrand of (6.21) is proportional to the return loss expressed in decibels. The constraint may thus be interpreted as a limitation on the allowable area under the return loss curve when plotted in decibels from DC to infinity. Ideally, given a minimum acceptable return loss, $|\Gamma_{min}|$, we would like to maximize the bandwidth by spreading this area out in a flat-bottom return loss curve as shown in Figure 6.21(a). As discussed previously, such frequency characteristics are noncausal. The time-domain, impulse response of this function (found by inverse Fourier transform) would be a sync function, having outputs at all times from minus infinity to plus infinity. No phase-frequency characteristic possible would delay this response enough to make it causal.

The best, physically realizable (i.e., causal) responses that most closely approximate such a curve are based on functions like Chebyshev polynomials, and result in return loss peaks and nulls as shown in Figure 6.21(b). The number of nulls and steepness of the transitions relates to the order of the response function, which in turn derives from the number of elements used.

6.3 RESISTANCE MATCHING

An important special case in matching problems is when one wishes to match one purely real impedance (i.e., a resistance) to another. This occurs, for example, when one wishes to couple one transmission-line system using an atypical characteristic

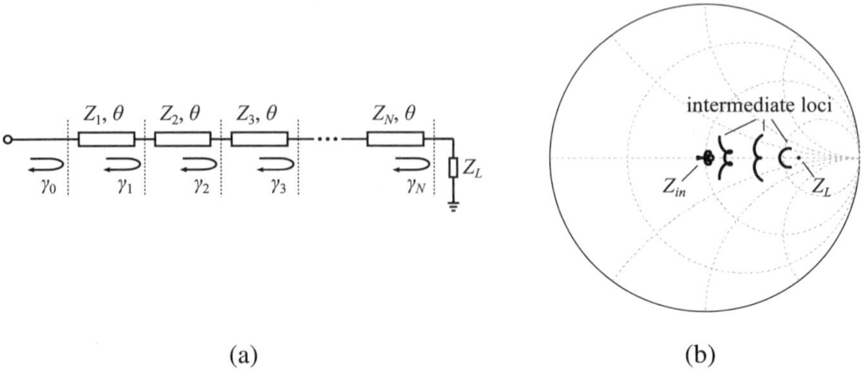

Figure 6.22 (a) Multisection matching transformer illustrating the theory of small reflections. (b) Smith chart showing a resistance match utilizing a multisection transformer. Each individual loci from right to left correspond to the net impedance looking into the cascaded transformer sections.

impedance, say, 100Ω, to another using 50Ω. In another example, antennas commonly have very high radiation resistance that must be matched to 50Ω transceivers.

6.3.1 Multisection Transformers

A very simple way to do this is with a quarter-wave transformer. If the ratio of the impedances to be matched is very large, however, the resulting bandwidth of the match may be insufficient. The bandwidth may be improved by using multiple transformers in cascade corresponding to intermediate impedances, as shown in Figure 6.22(a). This may be understood as an immittance folding technique, as illustrated in Figure 6.22(b). The first transformer stretches out the load, Z_L, a high impedance in this example, into a circular arc with the high-frequency end at the top. Subsequent transformers fold that locus back upon itself, leaving multiple looped loci after each step, centered on the real impedance axis, and again with the high-frequency end at the top. The final transformer then centers the resulting knot on the origin of the Smith chart. (The loci shown in the figure are for an equiripple, or Chebyshev, matching transformer, which will be discussed in Section 6.3.3.)

We can quantify this behavior by considering the final reflection coefficient presented by the array of characteristic impedance values, Z_k, for $k = 1 \ldots N$. This could be done rigorously by first calculating the input impedance with repeated, compound applications of the transmission-line impedance formula (3.42) followed by (3.37b). A simpler solution may be found if we utilize a technique known as the *theory of small reflections* [9]. We assume that the impedance ratio between adjacent sections is small, so that the intermediate reflection coefficients, γ_k, are also small,

$$|\gamma_k| \leq \epsilon \ll 1 \text{ for all } k \tag{6.22}$$

The lowercase γ is used here to emphasize that these coefficients represent the direct, localized reflections resulting from the immediate interface to the adjoining line segment, and do not account for the net effect of multiple reflections with other interfaces. The aggregation of these individual reflections into a net effective reflection coefficient, Γ, will be dealt with next.

A wave entering the port of this network will thus experience a very small reflection, γ_0, at the interface to the first section, Z_1, while most of the energy will continue on propagating into that section. At the next interface, another small reflection will occur, γ_1, while most of the energy continues on to the next section, and so forth. These will be known as the *principal reflections*.

To rigorously account for the steady-state of this network, we would also have to consider the subsequent reflections of the return waves traveling backward through each segment (*secondary reflections*), plus the reflections of those waves now traveling forward (*tertiary reflections*), and so on, in an infinite sequence. Recall that such an infinite series was calculated for the cascade of two networks in Section 4.4.1. Since we have assumed that the local reflections are small, however, the principal reflections dominate. The rest are smaller by at least a factor of ϵ. We may thus write approximately for the total reflection coefficient of the network,

$$\Gamma \approx \gamma_0 + \gamma_1 e^{-j2\theta} + \gamma_2 e^{-j4\theta} + \cdots + \gamma_N e^{-j2N\theta} = \sum_{k=0}^{N} \gamma_k e^{-j2k\theta} \tag{6.23}$$

where we have accounted for the round-trip phase delay for each principal reflection (θ is the electrical length of each segment). We may recognize this as a finite-length Fourier series for a periodic function in θ. It is customary (but not strictly required) to convert this to a Fourier cosine series by constraining the array of principal reflections to be symmetric about the center ($\gamma_k = \gamma_{N-k}$),

$$\Gamma \approx \sum_{k=0}^{\lfloor N/2 \rfloor} c_k \gamma_k \left(e^{-j2k\theta} + e^{-j2(N-k)\theta} \right) \tag{6.24a}$$

$$= 2e^{-jN\theta} \sum_{k=0}^{\lfloor N/2 \rfloor} c_k \gamma_k \cos\left((N-2k)\theta\right) \tag{6.24b}$$

where

$$c_k = \begin{cases} \frac{1}{2} & k = N/2 \\ 1 & k \neq N/2 \end{cases} \tag{6.25}$$

The coefficient c_k is simply required to ensure that the middle term of (6.23) is not double-counted when N is even.

Thus, it is possible to approximately synthesize any reflection profile that is desired by first equating the principal reflections to its cosine series coefficients. Before applying this method to some canonical cases, however, a few words regarding its limitations is warranted.

First, by assuming that the principal reflection at any given interface is very small, we have neglected the attenuation of the residual wave traveling forward to the next interface. While this may represent a very small error at any given interface, the cumulative effect is to substantially overestimate the signal power reaching the final load impedance. We should not be surprised, then, to find that conservation of energy is not strictly adhered to in our predicted results, especially outside the matched bandwidth.

Second, the approximations made by this theory are not entirely compatible with the usual formula, namely (3.37b), for relating the reflection coefficient to the line impedances,

$$\gamma_k \stackrel{?}{=} \frac{Z_{k+1} - Z_k}{Z_{k+1} + Z_k} \tag{6.26}$$

While theoretically true, applying this formula sequentially to recover the line impedances from the array of coefficients derived in the approximate manner described above would not be internally consistent; starting from Z_0, one would ultimately calculate a value for Z_{k+1} that is not equal to Z_L. This discrepancy is related to the nonconservation of energy explained in the previous paragraph. The Z_{k+1} that is obtained is what the load impedance would need to be in order to compensate for the excess signal amplitude reaching it.

To get around these issues, we use an alternate formula to recover the line impedances from γ_k, based on the following approximation,

$$\ln x \approx 2\frac{x-1}{x+1} \text{ for } x \approx 1 \tag{6.27}$$

Applied to (6.26), this yields

$$\gamma_k \approx \frac{1}{2} \ln \frac{Z_{k+1}}{Z_k} \tag{6.28}$$

This formula, it turns out [10], yields line impedances that are consistent with the approximations made previously.

6.3.2 Maximally Flat (Binomial) Matching Transformers

Let us now apply the theory of small reflections to derive some optimum resistance matching networks. In Section 6.2.1, we loosely quantified the bandwidth of an impedance match by considering its derivative at the match point in the center of the band. In this section, we prescribe the first and higher derivatives to be identically zero at that point, limited by the available degrees of freedom, to ensure that the reflection coefficient stays as low as possible for as long as possible as you move away from the center frequency. This condition we describe as *maximally flat*. (The reflection profile would indeed appear very flat near center on a linear scale, but the description is not so apt on a logarithmic, decibel scale, the way reflection coefficients are more often plotted.)

Let us postulate, then, a net reflection coefficient of the following form,

$$\Gamma = \frac{\Gamma_0}{2^N}\left(1 + e^{-j2\theta}\right)^N \tag{6.29}$$

where Γ_0 is the reflection coefficient at DC ($\theta = 0$). The magnitude of this formula is given by

$$|\Gamma| = \left|\frac{\Gamma_0}{2^N}\right| \left|1 + e^{-j2\theta}\right|^N = \left|\frac{\Gamma_0}{2^N}\right| \left|e^{-j\theta}\left(e^{j\theta} + e^{-j\theta}\right)\right|^N = |\Gamma_0|\,|\cos\theta|^N \tag{6.30}$$

The kth derivative of this reflection coefficient magnitude therefore has the following form,

$$\frac{d^k}{d\theta^k}|\Gamma| = |\Gamma_0|\,|\cos\theta|^{N-k}(\cdots) \tag{6.31}$$

Regardless of the trailing expression in parentheses, then, the value of this derivative at the middle frequency ($\theta = \pi/2$) is identically zero for $0 \le k \le N-1$. This, therefore, represents a maximally flat reflection response.

Recall now the binomial theorem from basic algebra,

$$(x+y)^N = \sum_{k=0}^{N} \binom{N}{k} x^{N-k} y^k \tag{6.32}$$

where

$$\binom{N}{k} = \frac{N!}{(N-k)!k!} \tag{6.33}$$

Applying this to (6.29) we have

$$\Gamma = \frac{\Gamma_0}{2^N}\left(1 + e^{-j2\theta}\right)^N = \frac{\Gamma_0}{2^N}\sum_{k=0}^{N}\binom{N}{k}e^{-j2k\theta} \tag{6.34}$$

Comparison of this expression with (6.23), then, yields

$$\gamma_k = \frac{\Gamma_0}{2^N}\binom{N}{k} \text{ for } k = 0 \ldots N \tag{6.35}$$

For consistency, we determine Γ_0 using the approximate formula (6.28) with the source and load impedances,

$$\Gamma_0 = \frac{1}{2}\ln\frac{Z_L}{Z_0} \tag{6.36}$$

and similarly for the line impedances,

$$\frac{1}{2}\ln\frac{Z_{k+1}}{Z_k} = \gamma_k = \frac{\Gamma_0}{2^N}\binom{N}{k} = \frac{1}{2^N}\binom{N}{k}\frac{1}{2}\ln\frac{Z_L}{Z_0} \tag{6.37a}$$

$$\therefore \ln\frac{Z_{k+1}}{Z_k} = 2^{-N}\binom{N}{k}\ln\frac{Z_L}{Z_0} \tag{6.37b}$$

$$Z_{k+1} = Z_k\left(\frac{Z_L}{Z_0}\right)^{2^{-N}\binom{N}{k}} \tag{6.37c}$$

As an example, a five-section binomial transformer designed to match a 10:1 impedance ratio is shown in Figure 6.23(a). The reflection coefficient obtained with these values from (6.23) is shown in Figure 6.23(b) with a dashed line. The function is periodic, as expected for a commensurate-line network. Note also the inaccurate prediction of gain at the off-match frequencies, in violation of the conservation of energy. For comparison, the exact result that would be achieved with this network is plotted with a solid line.

More exact line impedances that account for multiple reflections in each segment can be obtained via numerical optimization. Some useful cases are tabulated in Table 6.4. In each row, the load and transformer section impedances are each normalized to the impedance of the input that is being matched, and assume that $z_L > 1$. If $z_L < 1$, then the same tables may be used with the roles of the source and load reversed. Alternatively, one may consider the values in the z_L and z_k columns as the normalized admittances instead of impedances.

Table 6.4

Maximally Flat (Binomial) Matching Transformers

N	z_L	z_1	z_2	z_3	z_4	z_5	z_5
2	1.5	1.1067	1.3554				
2	2.0	1.1892	1.6818				
2	3.0	1.3161	2.2795				
2	4.0	1.4142	2.8285				
2	6.0	1.5651	3.8336				
2	8.0	1.6818	4.7568				
2	10.0	1.7783	5.6233				
3	1.5	1.0520	1.2247	1.4259			
3	2.0	1.0907	1.4142	1.8337			
3	3.0	1.1479	1.7321	2.6135			
3	4.0	1.1907	2.0000	3.3594			
3	6.0	1.2544	2.4495	4.7832			
3	8.0	1.3022	2.8284	6.1434			
3	10.0	1.3409	3.1623	7.4577			
4	1.5	1.0257	1.1351	1.3215	1.4624		
4	2.0	1.0444	1.2421	1.6102	1.9150		
4	3.0	1.0718	1.4105	2.1269	2.7990		
4	4.0	1.0919	1.5442	2.5903	3.6633		
4	6.0	1.1215	1.7553	3.4182	5.3500		
4	8.0	1.1436	1.9232	4.1597	6.9955		
4	10.0	1.1613	2.0651	4.8424	8.6110		
5	1.5	1.0128	1.0790	1.2247	1.3902	1.4810	
5	2.0	1.0220	1.1391	1.4142	1.7558	1.9569	
5	3.0	1.0354	1.2300	1.7321	2.4390	2.8974	
5	4.0	1.0452	1.2995	2.0000	3.0781	3.8270	
5	6.0	1.0596	1.4055	2.4495	4.2689	5.6625	
5	8.0	1.0703	1.4870	2.8284	5.3800	7.4745	
5	10.0	1.0789	1.5541	3.1623	6.4346	9.2687	
6	1.5	1.0064	1.0454	1.1496	1.3048	1.4349	1.4905
6	2.0	1.0110	1.0790	1.2693	1.5757	1.8536	1.9782
6	3.0	1.0176	1.1288	1.4599	2.0549	2.6577	2.9481
6	4.0	1.0225	1.1661	1.6129	2.4800	3.4302	3.9120
6	6.0	1.0296	1.2219	1.8573	3.2305	4.9104	5.8275
6	8.0	1.0349	1.2640	2.0539	3.8950	6.3291	7.7302
6	10.0	1.0392	1.2982	2.2215	4.5015	7.7030	9.6228

From: D. M. Pozar, 2011 [9].

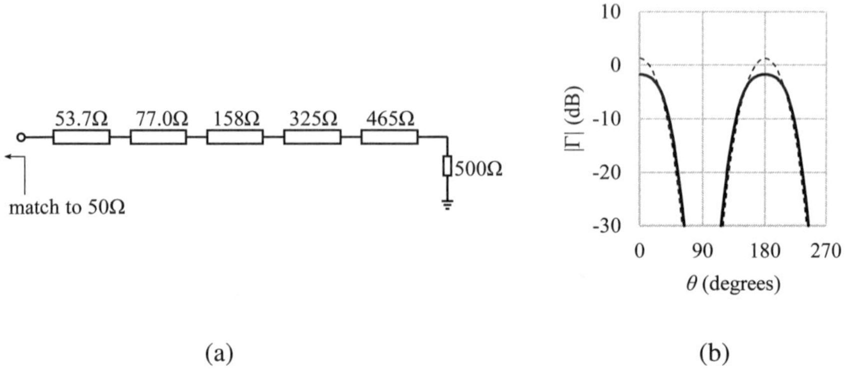

Figure 6.23 (a) Binomial matching transformer derived from the theory of small reflections for a 10:1 impedance ratio. (b) Plot of the reflection coefficient obtained from the small-reflection theory (dashed line) compared to the exact result (solid line).

6.3.3 Equiripple (Chebyshev) Matching Transformers

Another useful way to optimize a multisection transformer is by maximizing the slope outside the band for a given minimum return-loss constraint in-band. For finite-order transformers, this results in return-loss ripples that peak at the constrained value, Γ_{min}, similar to the area-optimizing curve in Figure 6.21(b).

Among the class of realizable reflection functions, those that maximize the bandwidth in this way are the Chebyshev polynomials. Discussed in more detail in Appendix G, these are normalized such that the ripples have unit amplitude over a range $-1 \leq x \leq 1$. For example, the fifth-order Chebyshev polynomial is

$$T_5(x) = 16x^5 - 20x^3 + 5x \tag{6.38}$$

and its magnitude is plotted in Figure 6.24(a). As a basic polynomial, this would most directly describe the immittance function of a lumped-element circuit versus frequency (with some appropriate amplitude scaling). Since we are using transmission-line circuits in this case, we must convert it to a function with trigonometric frequency dependence. We thus substitute for the argument of the polynomial,

$$x = B \cos \theta \tag{6.39}$$

where B is a scaling factor which indirectly determines the bandwidth. The resulting amplitude function, $|T_5(B \cos \theta)|$, is plotted in Figure 6.24(b) (in which

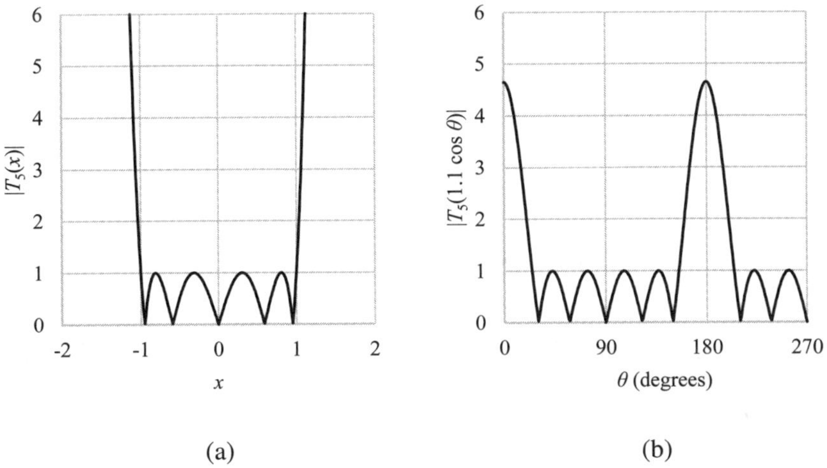

Figure 6.24 (a) Magnitude of the fifth-order Chebyshev polynomial. (b) Magnitude of the same polynomial where $x = 1.1 \cos\theta$.

$B = 1.1$ for the purpose of illustration). Note that the function is now limited in magnitude. To make it consistent with a realizable reflection coefficient, we must add the necessary phase for a five-section transformer ($e^{-j5\theta}$, as indicated in (6.24b)), and scale it to align with the expected return loss at DC. Thus, let

$$\Gamma(\theta) = \frac{\Gamma_0}{T_5(B)} e^{-j5\theta} T_5(B\cos\theta) \tag{6.40}$$

The constant Γ_0 uses (6.28), the approximation for small reflections,

$$\Gamma_0 = \frac{1}{2} \ln \frac{Z_L}{Z_0} \tag{6.41}$$

and the constant B may be determined by considering the fractional bandwidth, w, which causes the argument of the Chebyshev polynomial to equal ± 1,

$$B\cos\left(\tfrac{\pi}{2}\left(1 \pm \tfrac{w}{2}\right)\right) = \pm B \sin\left(\tfrac{w\pi}{4}\right) = \pm 1 \tag{6.42a}$$

$$\therefore B = \frac{1}{\sin\left(\frac{w\pi}{4}\right)} = \csc\left(\tfrac{w\pi}{4}\right) \tag{6.42b}$$

Note that since the ripple amplitude of the original Chebyshev polynomial was unity, the new scaling factor is also the peak reflection coefficient in the matched bandwidth,

$$\Gamma_{min} = \frac{\Gamma_0}{T_5(B)} = \frac{1}{2T_5(\csc\left(\frac{w\pi}{4}\right))} \ln \frac{Z_L}{Z_0} \tag{6.43}$$

To determine the principal reflections, γ_k, of the multisection transformer, we must expand the expression in (6.40),

$$\Gamma(\theta) = \frac{\Gamma_0}{T_5(B)} e^{-j5\theta} T_5(B \cos\theta) \tag{6.44a}$$

$$= \frac{\Gamma_0}{T_5(B)} e^{-j5\theta} \left(16B^5 \cos^5\theta - 20B^3 \cos^3\theta + 5B\cos\theta\right) \tag{6.44b}$$

$$= \frac{\Gamma_0}{T_5(B)} e^{-j5\theta} \left(B^5 \cos(5\theta) + 5B^3 \left(B^2 - 1\right) \cos(3\theta) + 5B\left(2B^4 - 3B^2 + 1\right)\cos\theta\right) \tag{6.44c}$$

and compare it with the cosine-series expression in (6.24b) to arrive at

$$\gamma_0 = \gamma_5 = \tfrac{1}{2} \frac{\Gamma_0}{T_5(B)} B^5 \tag{6.45a}$$

$$\gamma_1 = \gamma_4 = \tfrac{5}{2} \frac{\Gamma_0}{T_5(B)} B^3 \left(B^2 - 1\right) \tag{6.45b}$$

$$\gamma_2 = \gamma_3 = \tfrac{5}{2} \frac{\Gamma_0}{T_5(B)} B \left(2B^4 - 3B^2 + 1\right) \tag{6.45c}$$

As before, we use the logarithmic approximation for the principal reflection coefficients to derive the line impedances sequentially,

$$\gamma_k = \frac{1}{2} \ln \frac{Z_{k+1}}{Z_k} \tag{6.46a}$$

$$\therefore Z_{k+1} = Z_k e^{2\Gamma_k} \tag{6.46b}$$

An example five-section transformer designed using the calculations in this section is given in Figure 6.25(a). Once again, the small-reflection theory makes imperfect predictions, most notably some signal gain at the off-match frequencies. Improved line impedances derived from more sophisticated calculations that account for multiple reflections are tabulated for convenience in Table 6.5.

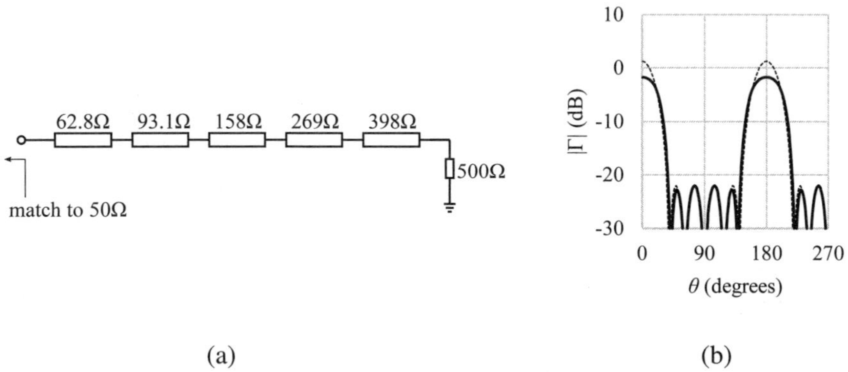

Figure 6.25 (a) Chebyshev matching transformer derived from the theory of small reflections for a 10:1 impedance ratio and fractional bandwidth $w = 1.2$. (b) Plot of the reflection coefficient obtained from the small-reflection theory (dashed line) compared to the exact result (solid line).

6.3.4 Tapered Transformers

The theory of small reflections works best when the number of sections, N, is large, thus minimizing the reflection at each individual interface. We may also observe that the bandwidth increases with N and is centered on a frequency where the individual section length is $\theta = \pi/2$. One may speculate, then, that as the number of sections is allowed to become arbitrarily large while keeping the total electrical length fixed, that bandwidth too may become arbitrarily large with a center frequency that diverges to infinity. In the limit, we obtain a continuously tapered transformer, illustrated in Figure 6.26(a).

Here we specify the total length of the transformer, L, and the characteristic impedance as a function of position on that line, $Z(x)$, where $0 \leq x \leq L$. The line may be modeled as infinitely many short transmission-line sections with incremental contributions to the overall reflection coefficient, as shown in Figure 6.26(b).

In accordance with the theory of small reflections, then, we may rewrite the summation of (6.23) as an integral over x with $\gamma_k \rightarrow d\gamma(x)$ and $k\theta \rightarrow \beta x$ where β is the phase constant of the line. Thus,

$$\Gamma = \int_{x=0}^{x=L} d\gamma(x) e^{-j2\beta x} \tag{6.47}$$

Table 6.5
Equiripple (Chebyshev) Matching Transformers

		$\Gamma_{min} = 0.05$				$\Gamma_{min} = 0.20$			
N	z_L	z_1	z_2	z_3	z_4	z_1	z_2	z_3	z_4
2	1.5	1.1347	1.3219			1.2247	1.2247		
2	2.0	1.2193	1.6402			1.3161	1.5197		
2	3.0	1.3494	2.2232			1.4565	2.0598		
2	4.0	1.4500	2.7585			1.5651	2.5558		
2	6.0	1.6047	3.7389			1.7321	3.4641		
2	8.0	1.7244	4.6393			1.8612	4.2983		
2	10	1.8233	5.4845			1.9680	5.0813		
3	1.5	1.1029	1.2247	1.3601		1.2247	1.2247	1.2247	
3	2.0	1.1475	1.4142	1.7429		1.2855	1.4142	1.5558	
3	3.0	1.2171	1.7321	2.4649		1.3743	1.7321	2.1829	
3	4.0	1.2662	2.0000	3.1591		1.4333	2.0000	2.7908	
3	6.0	1.3383	2.4495	4.4833		1.5193	2.4495	3.9492	
3	8.0	1.3944	2.8284	5.7372		1.5766	2.8284	5.0742	
3	10	1.4385	3.1623	6.9517		1.6415	3.1623	6.0920	
4	1.5	1.0892	1.1742	1.2775	1.3772	1.2247	1.2247	1.2247	1.2247
4	2.0	1.1201	1.2979	1.5409	1.7855	1.2727	1.3634	1.4669	1.5715
4	3.0	1.1586	1.4876	2.0167	2.5893	1.4879	1.5819	1.8965	2.0163
4	4.0	1.1906	1.6414	2.4369	3.3597	1.3692	1.7490	2.2870	2.9214
4	6.0	1.2290	1.8773	3.1961	4.8820	1.4415	2.0231	2.9657	4.1623
4	8.0	1.2583	2.0657	3.8728	6.3578	1.4914	2.2428	3.5670	5.3641
4	10	1.2832	2.2268	4.4907	7.7930	1.5163	2.4210	4.1305	6.5950

From: D. M. Pozar, 2011 [9].

where

$$d\gamma(x) = \frac{Z(x+dx) - Z(x)}{Z(x+dx) + Z(x)} \approx \frac{dZ(x)}{2Z(x)} = \frac{1}{2}\frac{1}{Z(x)/Z_0}d\left(Z(x)/Z_0\right) \tag{6.48a}$$

$$= \frac{1}{2}d\left(\ln\frac{Z(x)}{Z_0}\right) \tag{6.48b}$$

Therefore,

$$\Gamma = \frac{1}{2}\int_0^L \frac{d}{dx}\left(\ln\frac{Z(x)}{Z_0}\right)e^{-j2\beta x}dx \tag{6.49}$$

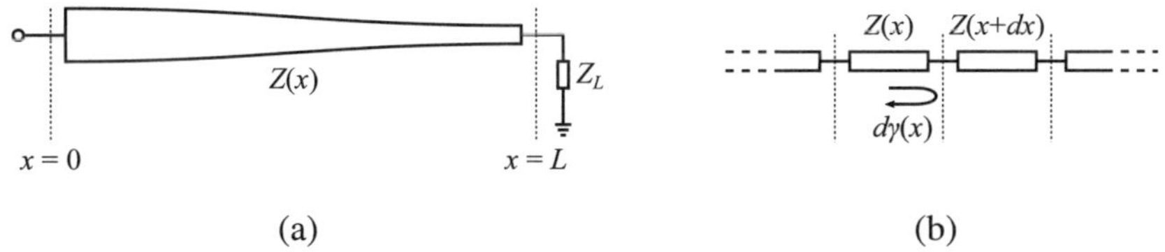

Figure 6.26 (a) Continuously tapered transmission-line transformer. (b) Differential model of infinitesimal impedance steps.

We are thus equipped to calculate the reflection response of any given impedance taper, $Z(x)$. The inverse problem of synthesizing such a taper — that is, prescribing the desired reflection coefficient, Γ, and then deriving the necessary impedance curve, $Z(x)$ — can be quite challenging. Instead, we will simply analyze the properties of a few canonical cases.

6.3.5 Exponential Taper

Given the form (6.49), one of the easiest tapers to analyze is known as an *exponential taper*, where

$$Z(x) = Z_0 e^{ax} \tag{6.50}$$

and, to ensure that $Z(L) = Z_L$, we let

$$a = \frac{1}{L} \ln \frac{Z_L}{Z_0} \tag{6.51}$$

Putting this expression into (6.49), we find

$$\Gamma = \frac{1}{2} \int_0^L \frac{d}{dx} \left(\ln e^{ax}\right) e^{-j2\beta x} dx = \frac{1}{2} \int_0^L \frac{d}{dx} (ax) e^{-j2\beta x} dx \tag{6.52a}$$

$$= \frac{a}{2} \int_0^L e^{-j2\beta x} dx = \frac{1}{2L} \ln \frac{Z_L}{Z_0} \left[\frac{1}{-j2\beta} e^{-j2\beta x} \right]_0^L \tag{6.52b}$$

$$= \frac{1}{2\beta L} \ln \frac{Z_L}{Z_0} e^{-j\beta L} \sin \beta L = \frac{1}{2} \ln \frac{Z_L}{Z_0} e^{-j\beta L} \operatorname{sinc} \beta L \tag{6.52c}$$

This is plotted in Figure 6.27. As with the multisection transformers, the approx-

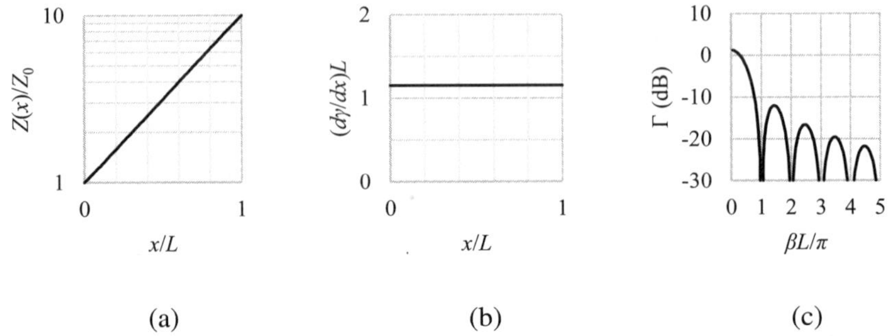

Figure 6.27 (a) Impedance profile of an exponential taper with $Z_L/Z_0 = 10$. (Note logarithmic scale.) (b) Differential reflection coefficient as a function of position. (c) Reflection coefficient as a function of electrical length.

imations of the theory of small reflections lead to a nonphysically large reflection coefficient outside the passband (in this case, DC), but the general characteristics are evident. Most notably, the reflection peaks diminish as the electrical length (and hence the frequency) increases. The first null appears at the point where the transformer is half a wavelength long, or $\beta L = \pi$.

Although the exponential taper is not an optimal design, the key point of this exercise is to show that tapered lines provide a broad high-pass matching response, in contrast to the band-pass responses of the multisection transformers, and the longer the taper, the lower in frequency the matched band will reach. This will generally be true of any smooth taper, regardless of its functional form.

These points enable the microwave engineer to be sometimes quite lazy in matching resistances, if given enough space to draw a long, smooth taper, and if the resulting losses can be tolerated. For those cases where loss must be minimized, and/or space is at a premium, an optimal taper has been derived which minimizes the passband corner and peak reflection coefficient for a given length. Essentially a limiting case of the Chebyshev multisection transformer, this is known as a *Klopfenstein taper* [11, 12].

6.3.6 Klopfenstein Taper

Klopfenstein showed that the optimum equiripple taper is given by

$$\ln \frac{Z(x)}{Z_0} = \frac{1}{2} \ln \left(\frac{Z_L}{Z_0} \right) + \frac{\Gamma_0}{\cosh A} A^2 \phi(2x/L - 1, A) \text{ for } 0 \le x \le L \tag{6.53}$$

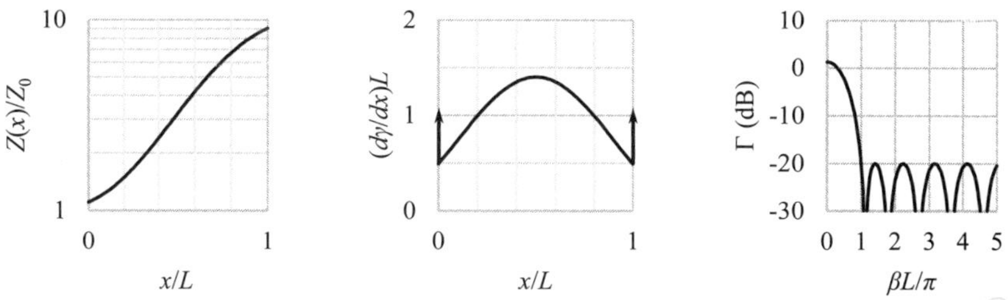

(a) (b) (c)

Figure 6.28 (a) Impedance profile of a Klopfenstein taper with $Z_L/Z_0 = 10$ and $\Gamma_{min} = 0.1$. (Note logarithmic scale.) (b) Differential reflection coefficient as a function of position. (c) Reflection coefficient as a function of electrical length.

where

$$\phi(u, A) = \int_0^u \frac{I_1\left(A\sqrt{1-y^2}\right)}{A\sqrt{1-y^2}} dy \text{ for } |u| \leq 1 \tag{6.54}$$

and I_1 is the first-order modified Bessel function of the first kind. (Bessel functions are discussed in more detail in Appendix G.) Γ_0 is the reflection coefficient at DC as evaluated through the usual small-reflection approximation,

$$\Gamma_0 = \frac{1}{2} \ln \frac{Z_L}{Z_0} \tag{6.55}$$

and the constant A is related to the peak reflection coefficient,

$$\Gamma_{min} = \frac{\Gamma_0}{\cosh A} \tag{6.56}$$

Grossberg [13] showed that the function $\phi(u, A)$ can be evaluated efficiently by computer as follows,

$$\phi(u, A) = \sum_{k=0}^{\infty} a_k b_k \tag{6.57}$$

where a_k and b_k are given by the recurrence relations

$$a_0 = 1 \tag{6.58a}$$

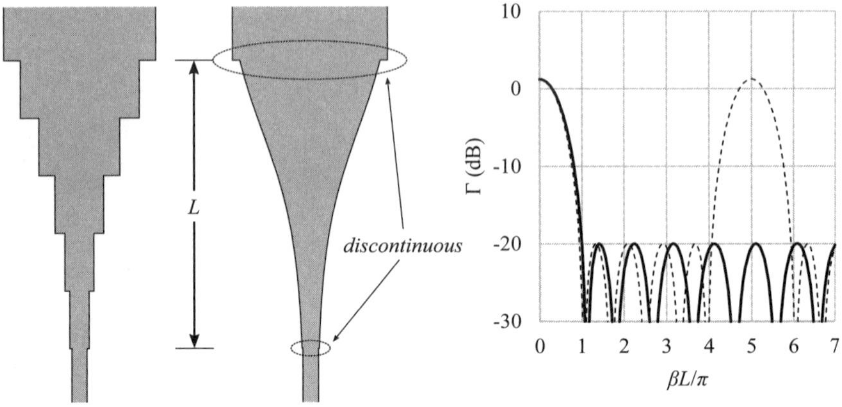

(a) (b)

Figure 6.29 (a) Optimum Chebyshev five-section transformer and Klopfenstein taper for $Z_L/Z_0 = 10$ and $\Gamma_{min} = 0.1$. (b) Performance of the Chebyshev (dashed line) and Klopfenstein (solid line) transformers. As in the previous examples, the slight positive gain exhibited by the Chebyshev transformer at $\beta L = 5\pi$ and by both transformers at $\beta L = 0$ is due to the imperfect conservation of energy in the small-reflection theory used to analyze them.

$$a_k = \frac{A^2}{4k(k+1)} a_{k-1} \tag{6.58b}$$

$$b_0 = \frac{u}{2} \tag{6.58c}$$

$$b_k = \frac{\frac{1}{2}u\left(1-u^2\right)^k + 2kb_{k-1}}{2k+1} \tag{6.58d}$$

Finally, the reflection coefficient achieved by the Klopfenstein taper is

$$\Gamma = \Gamma_{min} e^{-j\beta L} \cosh \sqrt{A^2 - (\beta L)^2} \tag{6.59}$$

Design and performance curves for a Klopfenstein taper with $Z_L/Z_0 = 10$ are shown in Figure 6.28. Note that the taper impedance does not quite reach Z_0 and Z_L at the end points, and consequently $d\gamma/dx$ has impulse peaks at both ends. This can be understood as a consequence of the undiminishing reflection peaks in the high-frequency limit. As the frequency increases, the taper becomes infinitely smooth electrically, such that no reflection occurs inside the taper. Only the impedance discontinuities at the ends of the taper contribute to reflection in this regime, and are needed to maintain the equiripple characteristic. Eliminating the discontinuities would suppress this ringing, but at the cost of increasing the low-frequency passband corner.

It is useful to compare the performance of the optimum taper (Klopfenstein) to an optimum multisection transformer (Chebyshev) of the same total length. Such a comparison is shown in Figure 6.29. Once again, note that the Klopfenstein taper does have a discontinuity at each end. The performance is, not surprisingly, quite similar, except for the periodicity of the multisection design. The low-frequency passband corner is also very slightly lower for the multisection design, as a consequence of the occasional larger reflection ripple. The total matched bandwidth of the taper, however, is substantially larger. In practice, the high-frequency performance will be limited by the particular implementation of the transmission line, as eventually it will no longer be electrically small in its cross-section and will cease to behave according to the mathematics we have used thus far. In that regime, the lines are said to become overmoded, and we must consider the electromagnetic structures we create as waveguides, the subject of the next chapter.

Problems

6-1 What is the VSWR of a load with reflection coefficient $\Gamma_L = 0.3 - j0.2$?

6-2 What is the range of possible reflection coefficients having a VSWR of 10?

6-3 Consider a source with real impedance Z_0. Calculate the average power delivered to an arbitrary load, relative to the maximum power that could be delivered to an optimal load (called the *available power*), as a function of (a) the load impedance, (b) the load admittance, (c) the load reflection coefficient, and (d) the load VSWR.

6-4 Name the contours (e.g., "constant-conductance") traversed by a point on the Smith chart when adding a series resistor, a series inductor, a series capacitor, a shunt capacitor, and a shunt resistor.

6-5 Prove that the range of validity for the cascade transmission line matching network at the bottom of Table 6.1 is $r > 1$ or $g > 1$.

6-6 Which of the matching networks in Table 6.1 are capable of matching a load of $Z_L = 30\Omega + j40\Omega$ to 50Ω?

6-7 Derive analytical solutions for matching an arbitrary load impedance at a single frequency using a double-stub tuner, a matching network comprising two parallel-connected stubs set a fixed distance apart on a 50Ω transmission line. Assume that the first stub is connected immediately adjacent to the load, and the stub lengths are variable, but the separation, d, is fixed and given. Provide solutions for cases where the stubs are either open-circuited or short-circuited. What range of impedances can be matched?

6-8 Derive solutions for a double-slug tuner, a matching network comprising two sliding quarter-wave transformers (slugs) that can be moved independently along a 50Ω transmission line. Assume that each slug has normalized characteristic impedance, $Z_1/Z_0 = z_1$, and is a quarter-wavelength long at the desired matching frequency. The lengths of the 50Ω lines between the slugs and from the first slug to the load are free variables.

6-9 Sketch out the paths on the Smith chart that a load impedance would take for each of the matching networks in Figure 6.16 from a point within the optimal regions of the those networks to the origin.

6-10 Derive expressions for the center and radius of the constant Q and constant $|Z|$ contours on the Carter chart.

6-11 For Carter chart matching using lumped elements as illustrated in Figure 6.17, from which regions of the Smith/Carter chart does the resulting network have a low-pass topology (i.e., where the source and load are connected at DC and not grounded)? From which regions is it high-pass (connected at high frequency and not grounded)? From which regions is it neither?

6-12 Consider the matching problem explored in Figure 6.18. What is the widest bandwidth centered at 1 GHz over which the load could be matched to 50Ω with better than 20-dB return loss according to the Bode-Fano criteria? How close do the minimal and immittance-folding matching networks presented in Figure 6.18(g) come to meeting that standard (a visual estimate from the plot is acceptable)?

6-13 Assume that a load comprising a 5-nH inductor and 25Ω resistor is matched to 50Ω with the smallest constant reflection coefficient magnitude permitted by the Bode-Fano criterion from 1 GHz to 2 GHz. What is the magnitude of that reflection coefficient if the inductor and resistor are in series? If they are in parallel?

6-14 How many sections would a Chebyshev transformer need to have in order to match a $10Z_0$ load to Z_0 with better than 20-dB return loss over a 7:1 bandwidth? How many sections would a binomial transformer need to meet the same specification?

6-15 What are the impedances of a fourth-order Chebyshev matching transformer which matches a load resistance of 400Ω to 50Ω to a peak reflection coefficient of 26 dB?

6-16 Derive the reflection coefficient curve of a taper having impedance function $Z(x) = Z_0 e^{a(1-\cos(\pi x/L))/2}$ from $x = 0$ to $x = L$ using the theory of small reflections.

6-17 Calculate the lower passband corner (the lowest frequency at which the match is equal to the ripple level) for the multisection Chebyshev transformer and the Klopfenstein taper in Figure 6.29. Assume the center frequency of the Chebyshev transformer is 1 GHz and both have the same total length.

References

[1] P. H. Smith, "Transmission line calculator," *Electronics*, vol. 12, no. 1, pp. 29–31, January 1939.

[2] P. H. Smith, "Transmission line calculator," *Electronics*, vol. 17, no. 1, p. 130, January 1944.

[3] Engineering and Technology History Wiki. (2015) History of broadband impedance matching. http://ethw.org/History_of_Broadband_Impedance_Matching.

[4] H. W. Bode, *Network Analysis and Feedback Amplifier Design.* New York: D. Van Nostrand Company, 1945.

[5] R. M. Fano, "Theoretical limitations on the broadband matching of arbitrary impedances," *Journal of the Franklin Institute*, vol. 249, no. 1, pp. 57–83, January 1950.

[6] R. M. Fano, "Theoretical limitations on the broadband matching of arbitrary impedances, cont'd," *Journal of the Franklin Institute*, vol. 249, no. 2, pp. 139–154, February 1950.

[7] H. J. Carlin and P. J. Crepeau, "Comments on 'theoretical limitations on the broadband matching of arbitrary impedances'," *IRE Transactions on Circuit Theory*, vol. 8, no. 2, p. 165, June 1961.

[8] D. Nie and B. Hochwald, "Improved Bode-Fano broadband matching bound," *IEEE International Symposium on Antennas and Propagation*, pp. 179–180, June 2016.

[9] D. M. Pozar, *Microwave Engineering*, 4th ed. New York: Wiley, 2011.

[10] R. E. Collin, *Foundations for Microwave Engineering*, 2nd ed. New York: IEEE Press, 2001.

[11] R. E. Collin, "The optimum tapered transmission line matching section," *Proceedings of the IRE*, vol. 44, pp. 539–548, April 1956.

[12] R. W. Klopfenstein, "A transmission line taper of improved design," *Proceedings of the IRE*, vol. 44, pp. 31–55, January 1956.

[13] M. A. Grossberg, "Extremely rapid computation of the Klopfenstein taper," *Proceedings of the IEEE*, vol. 56, pp. 1629–1630, January 1968.

Chapter 7

Waveguides

When electromagnetic structures become too large in their cross-section to be accurately modeled as transmission lines, we must begin treating them as *waveguides*. Waveguides retain some of the properties of transmission lines, such as wave propagation along the longitudinal axis, but the manner and specific properties (e.g., characteristic impedance) of that propagation will not depend exclusively on the relative dimensions of the cross-section, as was the case with transmission lines, but on their absolute physical size as well. Most notably, more than one mode of propagation shall be possible simultaneously for a given structure.

Recall from Chapter 4 that the definition of our network parameters was dependent upon the assumption of electrically small ports. This was the only way that certain scalar quantities — like voltage — could be defined unambiguously. Since waveguides are too large for these definitions to hold, many of the network parameters that we used in that chapter will not apply. The scattering parameters are the exception and are the formalism of choice for describing the interaction of waves going into and coming out of a waveguide network, so long as the different propagating modes of a waveguide input or output are considered separate ports, despite occupying the same physical space within a waveguide aperture. Further, the scattering parameters at waveguide junctions will depend not just on the propagation characteristics of the lines that are joined, but on the exact physical geometry of the junction as well.

7.1 WAVEGUIDE MODES

A gallery of possible waveguide geometries is shown in Figure 7.1. The first, Fig-

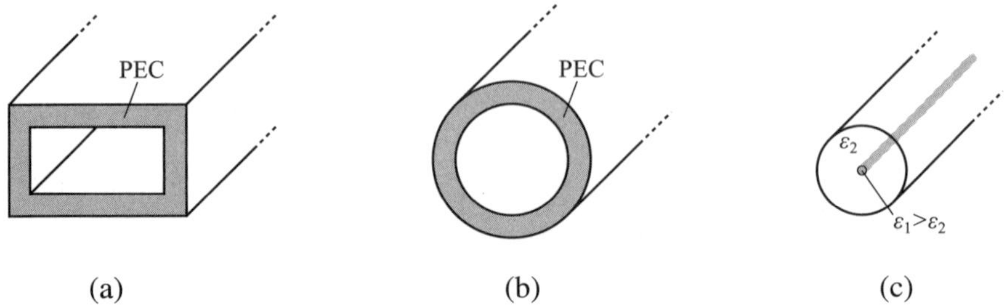

Figure 7.1 Several waveguide geometries. (a) Rectangular waveguide. (b) Circular waveguide. (c) Optical fiber.

ure 7.1(a), is known as rectangular waveguide and is comprised of a closed metal pipe with rectangular cross-section. This is a workhorse for many high-frequency electronic systems and will be described in some detail in Section 7.2. The second, Figure 7.1(b), is similar but has a circular cross-section and is used primarily in antenna feed systems where dual polarization operation is required. Finally, we show an optical fiber in Figure 7.1(c), which comprises two concentric dielectrics (usually glass, but sometimes transparent plastic), having different dielectric constants. Fiber, as is well known, is designed to carry information modulated on light-wave (actually, infrared) carriers, and is the modern standard for high data rate transmission over great distances.

Before focusing on any particular waveguide, however, let us derive some general formulas for the propagation of *guided waves*, waves that are supported by fixed cross-sectional boundaries rather than propagating through free space. We will assume without loss of generality that the guide will support waves propagating in the positive-z direction. By definition, then, the only dependence the fields will have on the z coordinate in the time-harmonic case is a complex exponential phase term,

$$\mathbf{E}(x, y, z) = \mathbf{E}'(x, y)e^{-\gamma z} \tag{7.1a}$$

$$\mathbf{H}(x, y, z) = \mathbf{H}'(x, y)e^{-\gamma z} \tag{7.1b}$$

where $\gamma = \alpha + j\beta$. $\mathbf{E}'$ and $\mathbf{H}'$ are thus the phasor amplitudes of the electric and magnetic field vectors as a function of x and y coordinates at $z = 0$.

7.1.1 Cutoff Frequencies

Substituting (7.1a) into the Helmholtz equation (1.73), we have

$$\left(\nabla^2 + k^2\right)\mathbf{E}'e^{-\gamma z} = 0 \tag{7.2a}$$

$$\left(\frac{\partial^2}{\partial x^2} + \frac{\partial^2}{\partial y^2} + \frac{\partial^2}{\partial z^2} + k^2\right)\mathbf{E}'e^{-\gamma z} = 0 \tag{7.2b}$$

$$\left(\frac{\partial^2}{\partial x^2} + \frac{\partial^2}{\partial y^2} + \gamma^2 + k^2\right)\mathbf{E}'e^{-\gamma z} = 0 \tag{7.2c}$$

$$\left(\nabla_t^2 + k_c^2\right)\mathbf{E}'e^{-\gamma z} = 0 \tag{7.2d}$$

where we have substituted

$$\nabla_t^2 = \frac{\partial^2}{\partial x^2} + \frac{\partial^2}{\partial y^2} \tag{7.3a}$$

$$k_c^2 = k^2 + \gamma^2 \tag{7.3b}$$

The subscript t on the Laplacian operator indicates that it applies only to the coordinates in the plane that is *transverse* to the direction of propagation, and k_c is known as the *cutoff wavenumber*. When $k_c = 0$, we must have

$$\gamma = jk = j\omega\sqrt{\mu\varepsilon} = j\beta \tag{7.4}$$

as was the case in (1.85) for an unguided plane wave. For guided waves, we will find that k_c is often finite, in which case the propagation constant is given by

$$\gamma = \sqrt{k_c^2 - k^2} \tag{7.5}$$

Note that if $k > k_c$, then γ is purely imaginary ($\gamma = j\beta$), which is normal for the propagation constant of a propagating mode. However, if $k < k_c$, then γ is real-valued ($\gamma = \alpha$) indicating a wave that decays exponentially (in the $+z$ or $-z$ direction, depending on the sign of α), without any change in phase. Such a wave is not propagating, and is said to be *cutoff*. Since k is proportional to frequency ($k = \omega\sqrt{\mu\varepsilon}$), k_c may also be associated with a *cutoff frequency*, below which a given mode cannot propagate. The cutoff frequency may be calculated from the cutoff wavenumber by

$$f_c = \frac{\omega_c}{2\pi} = \frac{k_c}{2\pi\sqrt{\mu\varepsilon}} \tag{7.6}$$

Keep in mind that here the material parameters, μ and ε, refer to the insulating medium inside the waveguide (usually air), not the walls of the waveguide itself.

In general, we shall find that a waveguide supports an infinite number of modes, with each having its own cutoff frequency. The mode that has the lowest cutoff frequency is called the *dominant mode*. Since there is a range of frequencies for which this is the only mode that can propagate, it is common for waveguide networks to operate within this regime, and only the dominant mode needs to be considered. Beyond that frequency range, the waveguide is said to be *overmoded* and is typically associated with a breakdown of the orderly operation of the waveguide network.

Note that it is possible for two modes (including the dominant modes) to have the same cutoff frequency. Such modes are called *degenerate*. A common example of degeneracy is the pair of orthogonal polarizations which may propagate in a waveguide with 90° rotational symmetry, such as square or circular. It is important to recognize that if two modes are degenerate, then by superposition they may combine in an infinite number of ways to create new modes.[1] This tells us that the list of modes we come up with to describe the behavior of a particular waveguide is not always unique; other systems of modes are also possible.

7.1.2 Transverse Fields

We now turn our attention to deriving the field equations that describe these waveguide modes. We start from (7.2d) and its corresponding expression for **H**,

$$\left(\nabla_t^2 + k_c^2\right) \mathbf{E} = 0 \tag{7.7a}$$

$$\left(\nabla_t^2 + k_c^2\right) \mathbf{H} = 0 \tag{7.7b}$$

where we have dropped the z-dependent term (since the z-dependence is already known, it is sufficient to solve these equations at $z = 0$). Both of these expressions represent a system of three equations — one for each of the x , y, and z components of the field vectors. The z components of these equations are of special importance, as they are directed along the axis of propagation. It is useful to write these, then, as separate scalar equations,

$$\left(\nabla_t^2 + k_c^2\right) E_z = 0 \tag{7.8a}$$

1 Nondegenerate modes combine by superposition also, but their different cutoff frequencies imply different propagation velocities, such that the combined field configuration does not remain intact as the waves travel down the length of the waveguide. It is not, therefore, reasonable to call such a combination a stable mode.

$$\left(\nabla_t^2 + k_c^2\right) H_z = 0 \tag{7.8b}$$

It will be easiest in many cases to solve these equations for the longitudinal components first based on the boundary conditions of the waveguide, and then derive the transverse fields from them using (1.71), the differential form of Maxwell's equations in the spectral domain.

To facilitate that calculation, we must isolate the transverse field components from Maxwell's equations, and solve for them in terms of the longitudinal components. Thus, Faraday's law (1.71c) becomes

$$\nabla \times \mathbf{E} = -j\omega\mu\mathbf{H} \tag{7.9a}$$

$$\nabla \times (\mathbf{E}_t + E_z\mathbf{z}) = -j\omega\mu\left(\mathbf{H}_t + H_z\mathbf{z}\right) \tag{7.9b}$$

$$\nabla \times \mathbf{E}_t + \nabla E_z \times \mathbf{z} = -j\omega\mu\left(\mathbf{H}_t + H_z\mathbf{z}\right) \tag{7.9c}$$

where E_t and H_t refer to the transverse field components and, in the last step, we have used the vector identity (C.7). Note that because of the known z-dependence, we may use the following substitution for the del operator,

$$\nabla = \nabla_t + \frac{\partial}{\partial z}\mathbf{z} = \nabla_t - \gamma\mathbf{z} \tag{7.10}$$

Thus,

$$\nabla_t \times \mathbf{E}_t - \gamma\mathbf{z} \times \mathbf{E}_t + \nabla_t E_z \times \mathbf{z} = -j\omega\mu\left(\mathbf{H}_t + H_z\mathbf{z}\right) \tag{7.11}$$

We may now pick out the transverse components from this equation,

$$-\gamma\mathbf{z} \times \mathbf{E}_t + \nabla_t E_z \times \mathbf{z} = -j\omega\mu\mathbf{H}_t \tag{7.12}$$

Similarly, from Ampère's law (1.71d), we obtain

$$-\gamma\mathbf{z} \times \mathbf{H}_t + \nabla_t H_z \times \mathbf{z} = j\omega\varepsilon\mathbf{E}_t \tag{7.13}$$

We may now solve for the transverse electric field components, $\mathbf{E}_t$, by eliminating $\mathbf{H}_t$ from the above pair of equations. First, we cross multiply $\gamma\mathbf{z}$ with (7.12),

$$\gamma^2\mathbf{E}_t + \gamma\nabla_t E_z = -j\gamma\omega\mu\mathbf{z} \times \mathbf{H}_t \tag{7.14}$$

and multiply (7.13) by $j\omega\mu$

$$-j\gamma\omega\mu\mathbf{z} \times \mathbf{H}_t + j\omega\mu\nabla_t H_z \times \mathbf{z} = -k^2\mathbf{E}_t \tag{7.15}$$

then add the two together, cancelling the common terms,

$$\left(\gamma^2 + k^2\right) \mathbf{E}_t + \gamma \nabla_t E_z + j\omega\mu \nabla_t H_z \times \mathbf{z} = 0 \tag{7.16a}$$

$$\therefore \mathbf{E}_t = \frac{\gamma}{k_c^2} \left(\frac{j\omega\mu}{\gamma} \mathbf{z} \times \nabla_t H_z - \nabla_t E_z \right) \tag{7.16b}$$

Similarly, by multiplying (7.12) by $j\omega\varepsilon$ and adding it to (7.13) crossed with $\gamma\mathbf{z}$, we find

$$\mathbf{H}_t = \frac{\gamma}{k_c^2} \left(\frac{j\omega\varepsilon}{\gamma} \nabla_t E_z \times \mathbf{z} - \nabla_t H_z \right) \tag{7.17}$$

We now have expressions for the transverse fields, both electric and magnetic, in terms of the longitudinal field components. Note that these expressions are quite general; we had to assume almost nothing about the waveguide geometry itself in order to derive them. The only constraints are that the z axis is oriented in the direction of propagation, and the medium filling the waveguide aperture is linear, isotropic, source-free, and *homogeneous*, meaning that the constituent parameters ε and μ are constant throughout the space where the fields exist. (The optical fiber in Figure 7.1(c) is one example where this assumption does not work.)

Note that there is no direct requirement in these equations for both longitudinal components, E_z and H_z, to exist. Unless both are required to satisfy the boundary conditions of the specific waveguide geometry, one may solve for the transverse fields due to E_z and H_z independently, assuming the other is zero, and then combine them using superposition. If these two solutions have degenerate cutoff frequencies, as described earlier, then that combination itself may properly be called a mode.

It is customary, then, to classify the modes in a waveguide according to the absence or presence of its longitudinal components, as follows,

- *Transverse electric and magnetic (TEM) modes* are characterized by the electric and magnetic fields having only transverse components, that is, $E_z = H_z = 0$. The dominant mode of coaxial transmission line, derived in Section 3.2.1, is an example. TEM modes typically have zero cutoff frequency.

- *Transverse electric (TE) modes* are characterized by the electric field having only transverse components, while the magnetic field has a longitudinal component, or $E_z = 0$ but $H_z \neq 0$. Due to the presence of H_z, these are also sometimes called *H-modes*.

- *Transverse magnetic (TM) modes* are characterized by the magnetic field having only transverse components, while the electric field has a longitudinal component, or $H_z = 0$ but $E_z \neq 0$. Due to the presence of E_z, these are also sometimes called *E-modes*.
- *Hybrid modes* have both electric and magnetic longitudinal components. These are most often associated with coupling between TE and TM modes due to finite conductivity at the waveguide boundary.

7.1.3 Backward-Traveling Waves

Consider the effect on the above solutions if we negate the propagation constant, γ, and the longitudinal electric field, E_z. That is,

$$\gamma = -\gamma' \tag{7.18a}$$

$$E_z = -E_z' \tag{7.18b}$$

Substituting this into (7.16b) and (7.17), we have,

$$\mathbf{E}_t' = \frac{-\gamma'}{k_c^2}\left(\frac{j\omega\mu}{-\gamma'}\mathbf{z} \times \nabla_t H_z + \nabla_t E_z'\right) \tag{7.19a}$$

$$= \frac{\gamma'}{k_c^2}\left(\frac{j\omega\mu}{\gamma'}\mathbf{z} \times \nabla_t H_z - \nabla_t E_z'\right) \tag{7.19b}$$

$$\mathbf{H}_t' = \frac{-\gamma'}{k_c^2}\left(-\frac{j\omega\varepsilon}{-\gamma'}\nabla_t E_z' \times \mathbf{z} - \nabla_t H_z\right) \tag{7.19c}$$

$$= -\frac{\gamma'}{k_c^2}\left(\frac{j\omega\varepsilon}{\gamma'}\nabla_t E_z' \times \mathbf{z} - \nabla_t H_z\right) \tag{7.19d}$$

The transverse electric field has the same form as the original, but the transverse magnetic field is negated. If we had negated H_z instead of E_z, then it would have been the other way around. This means that the solutions we derive for forward-traveling waves may be reused for backward-traveling waves if we negate the transverse components of one field and the longitudinal component of the other. This fact will be useful in scenarios where the fields of reflected and incident guided waves combine.

7.1.4 Power Flow in Waveguide Modes

The longitudinal power, P, flowing through a traveling waveguide mode is given by the Poynting vector, $\mathbf{S}$, integrated over the aperture of the guide,

$$P = \iint \mathbf{S} \cdot d\mathbf{A} = \iint \left(\mathbf{E} \times \mathbf{H}^*\right) \cdot d\mathbf{A} \tag{7.20a}$$

$$= \iint \left[(\mathbf{E}_t + E_z\mathbf{z}) \times (\mathbf{H}_t + H_z\mathbf{z})^*\right] \cdot d\mathbf{A} \tag{7.20b}$$

$$= \iint \left[\mathbf{E}_t \times \mathbf{H}_t^* + E_z\left(\mathbf{z} \times \mathbf{H}_t^*\right) + H_z^*\left(\mathbf{E}_t \times \mathbf{z}\right)\right] \cdot d\mathbf{A} \tag{7.20c}$$

$$= \iint \left(\mathbf{E}_t \times \mathbf{H}_t^*\right) \cdot d\mathbf{A} \tag{7.20d}$$

where $d\mathbf{A} = \mathbf{z}dA$. Note that by isolating the z-directed component of the Poynting vector, we have neglected power terms that flow laterally. In principle, for a lossless waveguide, these would be reactive and represent energy that is locally stored in electric or magnetic fields and then released again as the wave passes. If real, these components would necessarily indicate loss, either from resistive or dielectric dissipation, or leakage from the waveguide into the surrounding medium (making the waveguide also a kind of antenna). In almost all cases, it is the power flowing longitudinally, derived above, that we are interested in.

Note also that the energy flow is oriented in the direction of propagation. Recall that for backward-traveling waves, one of the transverse fields will be negated, inverting the direction of the Poynting vector.

7.1.5 TE and TM Modes

As TEM modes are most often associated with the dominant mode of structures used as transmission lines (like coax), and hybrid modes are rarely encountered in metal-enclosed waveguide geometries, the TE and TM modes are by far the most commonly encountered in the practice of high-frequency circuits. It is useful, then, to provide simplified equations for these special cases.

For TE modes, we note that $E_z = 0$, thus (7.16b) and (7.17) become

$$\mathbf{H}_t = -\frac{\gamma}{k_c^2}\nabla_t H_z \tag{7.21a}$$

$$\mathbf{E}_t = Z_{TE}\left(\mathbf{H}_t \times \mathbf{z}\right) \tag{7.21b}$$

where

$$Z_{TE} = Y_{TE}^{-1} = \frac{j\omega\mu}{\gamma} \tag{7.22}$$

For TM modes, we note that $H_z = 0$ and have

$$\mathbf{E}_t = -\frac{\gamma}{k_c^2}\nabla_t E_z \tag{7.23a}$$

$$\mathbf{H}_t = Y_{TM}\left(\mathbf{z} \times \mathbf{E}_t\right) \tag{7.23b}$$

where

$$Y_{TM} = Z_{TM}^{-1} = \frac{j\omega\varepsilon}{\gamma} \tag{7.24}$$

In either case, we have

$$\mathbf{E}_t = Z_{TE/TM}\left(\mathbf{H}_t \times \mathbf{z}\right) \tag{7.25a}$$

$$\mathbf{H}_t = Y_{TE/TM}\left(\mathbf{z} \times \mathbf{E}_t\right) \tag{7.25b}$$

illustrating that the transverse fields are everywhere orthogonal to one another.

The ratios between transverse field components, Z_{TE} and Z_{TM}, have units of impedance and are constant throughout the aperture of the waveguide. They are known as the *wave impedances*. Note that in the high-frequency limit where $\gamma \to jk = j\omega\sqrt{\mu\varepsilon}$ we may write

$$Z_{TE/TM} \to \sqrt{\frac{\mu}{\varepsilon}} = \eta \tag{7.26}$$

which is the free-space wave impedance given in (1.63).

A summary of the general equations for homogeneously filled waveguides is given in Table 7.1.

7.2 RECTANGULAR WAVEGUIDE

Consider now the rectangular metal waveguide shown in Figure 7.2, where the long and short sides of the rectangle have inner dimensions a and b, aligned with the x and y coordinate axes, respectively. This is known as *rectangular waveguide*. For simplicity, we assume the walls of the waveguide comprise a perfect electric conductor (PEC) and need only consider the fields inside the aperture (those outside are isolated and independent of the waveguide).

Table 7.1

Equations for Arbitrary Homogeneous Waveguides

Parameter	Symbol	Expression
Wavenumber (unguided)	k	$\omega\sqrt{\mu\varepsilon}$
Wave impedance (unguided)	η	$\sqrt{\frac{\mu}{\varepsilon}}$
Propagation constant	$\gamma = \alpha + j\beta$	$\sqrt{k_c^2 - k^2}$
Cutoff wavelength	λ_c	$\frac{2\pi}{k_c}$
Guided wavelength	λ_g	$\frac{2\pi}{\beta}$
Phase velocity	v_p	$\frac{\omega}{\beta}$
Group velocity	v_g	$\frac{d\omega}{d\beta}$
Transverse E-field (general)*	$\mathbf{E}_t$	$\frac{\gamma}{k_c^2}\left(\frac{j\omega\mu}{\gamma}\mathbf{z} \times \nabla_t H_z - \nabla_t E_z\right)$
Transverse H-field (general)*	$\mathbf{H}_t$	$\frac{\gamma}{k_c^2}\left(\frac{j\omega\varepsilon}{\gamma}\nabla_t E_z \times \mathbf{z} - \nabla_t H_z\right)$
Transverse E-field (TE modes)*	$\mathbf{E}_t$	$Z_{TE}\left(\mathbf{H}_t \times \mathbf{z}\right)$
Transverse H-field (TE modes)*	$\mathbf{H}_t$	$-\frac{\gamma}{k_c^2}\nabla_t H_z$
Transverse E-field (TM modes)*	$\mathbf{E}_t$	$-\frac{\gamma}{k_c^2}\nabla_t E_z$
Transverse H-field (TM modes)*	$\mathbf{H}_t$	$Y_{TM}\left(\mathbf{z} \times \mathbf{E}_t\right)$
TE wave impedance	Z_{TE}	$\frac{j\omega\mu}{\gamma} = \frac{jk\eta}{\gamma}$
TM wave admittance	Y_{TM}	$\frac{j\omega\varepsilon}{\gamma} = \frac{jk}{\eta\gamma}$
Power flow	P	$\iint\left(\mathbf{E}_t \times \mathbf{H}_t^*\right) \cdot d\mathbf{A}$

*For waves traveling in the positive z direction.

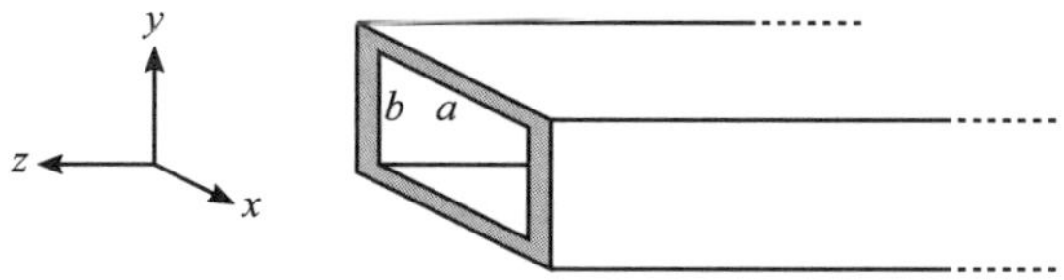

Figure 7.2 Geometry of a rectangular waveguide.

7.2.1 TM Modes

To solve for the TM modes in a rectangular waveguide, we start with (7.8a). A form of the Helmholtz equation in two dimensions, the solutions are given by complex exponentials,

$$E_z(x, y, z) = \left(E_1 e^{jk_x x} + E_2 e^{-jk_x x}\right)\left(E_3 e^{jk_y y} + E_4 e^{-jk_y y}\right) e^{-\gamma z} \tag{7.27}$$

where

$$k_x^2 + k_y^2 = k_c^2 \tag{7.28}$$

The unknown constants in this equation are determined by applying the boundary conditions for a perfect electric conductor (PEC) at the walls of the waveguide. According to (1.9b), the tangential electric field must vanish at these boundaries. Therefore, for the short walls at $x = 0$ and $x = a$,

$$E_z(0, y, z) = (E_1 + E_2)\left(E_3 e^{jk_y y} + E_4 e^{-jk_y y}\right) e^{-\gamma z} = 0 \tag{7.29a}$$

$$\therefore E_2 = -E_1 \tag{7.29b}$$

and

$$E_z(a, y, z) = E_1\left(e^{jk_x a} - e^{-jk_x a}\right)\left(E_3 e^{jk_y y} + E_4 e^{-jk_y y}\right) e^{-\gamma z} = 0 \tag{7.30a}$$

$$\therefore \left(e^{jk_x a} - e^{-jk_x a}\right) = j2\sin(k_x a) = 0 \tag{7.30b}$$

$$k_x = \frac{m\pi}{a} \tag{7.30c}$$

for any integer m. Likewise, for the long walls at $y = 0$ and $y = b$, we find that

$$E_4 = -E_3 \tag{7.31a}$$

$$k_y = \frac{n\pi}{b} \tag{7.31b}$$

for any integer n. Putting these results together, we have

$$E_z(x, y, z) = E_1 E_3 \sin\left(\frac{m\pi x}{a}\right) \sin\left(\frac{n\pi y}{b}\right) e^{-\gamma z} \tag{7.32a}$$

$$k_c^2 = \left(\frac{m\pi}{a}\right)^2 + \left(\frac{n\pi}{b}\right)^2 \tag{7.32b}$$

Note that if either index (m or n) is zero, then the electric field vanishes. Further, if either is negative, the solution is the same but for a scaling factor. Thus, for all possible, nontrivial TM solutions, we may consider the range of the indices to be 1 to infinity. Any combination of these indices within that range is a valid solution to wave propagation within the waveguide. Transverse magnetic modes will be labeled TM_{mn} where the subscripts correspond to these indices.

The prefactor $E_1 E_3$ is of course arbitrary. Some simplification of the final expressions will result if we let $E_1 E_3 = E_0 k_c/\gamma$; therefore,

$$E_z = E_0 \frac{k_c}{\gamma} \sin\left(\frac{m\pi x}{a}\right) \sin\left(\frac{n\pi y}{b}\right) e^{-\gamma z} \tag{7.33}$$

The transverse electric fields for these modes are derived using (7.23a),

$$\mathbf{E}_t = -\frac{\gamma}{k_c^2} \nabla_t E_z \tag{7.34a}$$

$$= -\frac{E_0}{k_c} \left(\frac{\partial}{\partial x}\mathbf{x} + \frac{\partial}{\partial y}\mathbf{y}\right) \sin\left(\frac{m\pi x}{a}\right) \sin\left(\frac{n\pi y}{b}\right) e^{-\gamma z} \tag{7.34b}$$

$$= -\frac{E_0}{k_c} \Big(\tfrac{m\pi}{a} \cos\left(\tfrac{m\pi x}{a}\right) \sin\left(\tfrac{n\pi y}{b}\right) \mathbf{x} + \tfrac{n\pi}{b} \sin\left(\tfrac{m\pi x}{a}\right) \cos\left(\tfrac{n\pi y}{b}\right) \mathbf{y}\Big) e^{-\gamma z} \tag{7.34c}$$

The transverse magnetic fields are then found using (7.23b),

$$\mathbf{H}_t = Y_{TM} \left(\mathbf{z} \times \mathbf{E}_t\right) \tag{7.35a}$$

$$= -\frac{H_0}{k_c} \Big(\tfrac{m\pi}{a} \cos\left(\tfrac{m\pi x}{a}\right) \sin\left(\tfrac{n\pi y}{b}\right) \mathbf{y} - \tfrac{n\pi}{b} \sin\left(\tfrac{m\pi x}{a}\right) \cos\left(\tfrac{n\pi y}{b}\right) \mathbf{x}\Big) e^{-\gamma z} \tag{7.35b}$$

where $H_0 = Y_{TM}E_0$. To summarize in a compact form, the fields associated with the TM mode are,

$$\mathbf{E} = E_0 e^{-\gamma z} \begin{pmatrix} \mathbf{x} & \mathbf{y} & \mathbf{z} \end{pmatrix} \begin{pmatrix} -\frac{1}{k_c}\frac{m\pi}{a}\cos\left(\frac{m\pi x}{a}\right)\sin\left(\frac{n\pi y}{b}\right) \\ -\frac{1}{k_c}\frac{n\pi}{b}\sin\left(\frac{m\pi x}{a}\right)\cos\left(\frac{n\pi y}{b}\right) \\ \frac{k_c}{\gamma}\sin\left(\frac{m\pi x}{a}\right)\sin\left(\frac{n\pi y}{b}\right) \end{pmatrix} \tag{7.36a}$$

$$\mathbf{H} = H_0 e^{-\gamma z} \begin{pmatrix} \mathbf{x} & \mathbf{y} & \mathbf{z} \end{pmatrix} \begin{pmatrix} \frac{1}{k_c}\frac{n\pi}{b}\sin\left(\frac{m\pi x}{a}\right)\cos\left(\frac{n\pi y}{b}\right) \\ -\frac{1}{k_c}\frac{m\pi}{a}\cos\left(\frac{m\pi x}{a}\right)\sin\left(\frac{n\pi y}{b}\right) \\ 0 \end{pmatrix} \tag{7.36b}$$

The power flowing in the TM modes are then given by (7.20d),

$$P = \iint (\mathbf{E}_t \times \mathbf{H}_t^*) \cdot d\mathbf{A} = \int_0^a \int_0^b (E_x H_y^* - E_y H_x^*)\, dx dy \tag{7.37a}$$

$$= \frac{E_0 H_0^*}{k_c^2} \int_0^a \int_0^b \left(\frac{m\pi}{a}\right)^2 \cos^2\left(\frac{m\pi x}{a}\right) \sin^2\left(\frac{n\pi y}{b}\right) dx dy$$
$$+ \frac{E_0 H_0^*}{k_c^2} \int_0^a \int_0^b \left(\frac{n\pi}{b}\right)^2 \sin^2\left(\frac{m\pi x}{a}\right) \cos^2\left(\frac{n\pi y}{b}\right) dx dy \tag{7.37b}$$

$$= \frac{E_0 H_0^*}{k_c^2} \left(\frac{m\pi}{a}\right)^2 \frac{ab}{4} + \frac{E_0 H_0^*}{k_c^2} \left(\frac{n\pi}{b}\right)^2 \frac{ab}{4} = E_0 H_0^* \frac{ab}{4} \tag{7.37c}$$

where the indices, m and n, are both known to be greater than zero. Observe that with the normalization we have chosen, the power flow is a constant, having no dependence on frequency. The longitudinal component, E_z, however, is inversely proportional to the propagation constant, γ, which does depend on frequency according to (7.5) and (1.74). As frequency increases above the cutoff, the transverse field components stay fixed in amplitude while the longitudinal component diminishes, becoming TEM-like in the limit.

7.2.2 TE Modes

To solve for the TE modes in a rectangular waveguide, we start with (7.8b). As with the TM modes, the solutions to this equation take the form of complex exponentials,

$$H_z(x,y,z) = \left(H_1 e^{jk_x x} + H_2 e^{-jk_x x}\right)\left(H_3 e^{jk_y y} + H_4 e^{-jk_y y}\right) e^{-\gamma z} \tag{7.38}$$

where, again,

$$k_x^2 + k_y^2 = k_c^2 \tag{7.39}$$

In this case, we apply the differential boundary condition (1.24c). At the short walls

$$\frac{\partial H}{\partial n} = \pm\frac{\partial H}{\partial x} = 0 \tag{7.40a}$$

$$\therefore H_2 = H_1 \tag{7.40b}$$

$$k_x = \frac{m\pi}{a} \tag{7.40c}$$

Likewise, at the long walls,

$$\frac{\partial H}{\partial n} = \pm\frac{\partial H}{\partial y} = 0 \tag{7.41a}$$

$$\therefore H_4 = H_3 \tag{7.41b}$$

$$k_y = \frac{n\pi}{b} \tag{7.41c}$$

Thus, letting $H_1 H_3 = H_0 k_c/\gamma$, we have

$$H_z = H_0 \frac{k_c}{\gamma} \cos\left(\frac{m\pi x}{a}\right) \cos\left(\frac{n\pi y}{b}\right) e^{-\gamma z} \tag{7.42a}$$

$$k_c^2 = \left(\frac{m\pi}{a}\right)^2 + \left(\frac{n\pi}{b}\right)^2 \tag{7.42b}$$

The transverse magnetic field may then be found from (7.21a),

$$\mathbf{H}_t = -\frac{\gamma}{k_c^2}\nabla_t H_z \tag{7.43a}$$

$$= -\frac{H_0}{k_c}\left(\frac{\partial}{\partial x}\mathbf{x} + \frac{\partial}{\partial y}\mathbf{y}\right) \cos\left(\frac{m\pi x}{a}\right) \cos\left(\frac{n\pi y}{b}\right) e^{-\gamma z} \tag{7.43b}$$

$$= \frac{H_0}{k_c}\left(\tfrac{m\pi}{a}\sin\left(\tfrac{m\pi x}{a}\right)\cos\left(\tfrac{n\pi y}{b}\right)\mathbf{x} + \tfrac{n\pi}{b}\cos\left(\tfrac{m\pi x}{a}\right)\sin\left(\tfrac{n\pi y}{b}\right)\mathbf{y}\right)e^{-\gamma z} \quad (7.43c)$$

and the transverse electric field from (7.21b),

$$\mathbf{E}_t = Z_{TE}\left(\mathbf{H}_t \times \mathbf{z}\right) \quad (7.44a)$$

$$= \frac{E_0}{k_c}\left(\tfrac{n\pi}{b}\cos\left(\tfrac{m\pi x}{a}\right)\sin\left(\tfrac{n\pi y}{b}\right)\mathbf{x} - \tfrac{m\pi}{a}\sin\left(\tfrac{m\pi x}{a}\right)\cos\left(\tfrac{n\pi y}{b}\right)\mathbf{y}\right)e^{-\gamma z} \quad (7.44b)$$

where $E_0 = Z_{TE}H_0$. In compact form,

$$\mathbf{E} = E_0 e^{-\gamma z}\begin{pmatrix}\mathbf{x} & \mathbf{y} & \mathbf{z}\end{pmatrix}\begin{pmatrix} \frac{1}{k_c}\frac{n\pi}{b}\cos\left(\frac{m\pi x}{a}\right)\sin\left(\frac{n\pi y}{b}\right) \\ -\frac{1}{k_c}\frac{m\pi}{a}\sin\left(\frac{m\pi x}{a}\right)\cos\left(\frac{n\pi y}{b}\right) \\ 0 \end{pmatrix} \quad (7.45a)$$

$$\mathbf{H} = H_0 e^{-\gamma z}\begin{pmatrix}\mathbf{x} & \mathbf{y} & \mathbf{z}\end{pmatrix}\begin{pmatrix} \frac{1}{k_c}\frac{m\pi}{a}\sin\left(\frac{m\pi x}{a}\right)\cos\left(\frac{n\pi y}{b}\right) \\ \frac{1}{k_c}\frac{n\pi}{b}\cos\left(\frac{m\pi x}{a}\right)\sin\left(\frac{n\pi y}{b}\right) \\ \frac{k_c}{\gamma}\cos\left(\frac{m\pi x}{a}\right)\cos\left(\frac{n\pi y}{b}\right) \end{pmatrix} \quad (7.45b)$$

Note that, in this case, either m or n may be zero, but not both. (If both were zero, all fields would vanish, including H_z since k_c would also become zero in that case.) As before with the TM modes, the TE modes will be labeled by these indices as TE_{mn}.

Power flow in this case is given by

$$P = \iint \left(\mathbf{E}_t \times \mathbf{H}_t^*\right)\cdot d\mathbf{A} = \int_0^a\int_0^b \left(E_x H_y^* - E_y H_x^*\right)dx\,dy \quad (7.46a)$$

$$= E_0 H_0^* \frac{ab}{4}\begin{cases} 2 & m \neq 0, n = 0 \\ 2 & m = 0, n \neq 0 \\ 1 & m \neq 0, n \neq 0 \end{cases} = E_0 H_0^* \frac{ab}{4}\left(1+\delta_m\right)\left(1+\delta_n\right) \quad (7.46b)$$

where δ_i is the *Kronecker delta function* (equal to 1 when $i = 0$, and 0 otherwise). The possibility that either m or n might be zero in this case changes the result of the integral. As before, however, the normalization used ensures that the transverse fields and the power in any mode remain constant, while the longitudinal field component, H_z, decreases with increasing frequency.

7.2.3 Charge and Current Distributions

Electromagnetic waves can propagate through free space without any local sources (charges or currents) to support them, but in a hollow metallic waveguide, such as rectangular, the confined electric field vectors must terminate on accumulated charges, and the confined magnetic fields must be enclosed by currents. In this way, a propagating mode in waveguide is just as much a charge and current wave as it is an electromagnetic field wave.

The distribution of these charges and currents may be found using the boundary conditions derived in Section 1.1.4. For example, the charge density is given by (1.7b) applied at the walls. By symmetry, the charge densities on opposite walls must be the same, except possibly for a difference in phase. Therefore, for simplicity, we will concentrate on the charge density at $x = 0$ and $y = 0$. Thus,

$$\rho_s = \varepsilon(\mathbf{n} \cdot \mathbf{E}) - \varepsilon(\mathbf{n} \cdot \mathbf{E}_t)\big|_{x=0 \text{ or } y=0} \tag{7.47a}$$

$$= \varepsilon \frac{E_0}{k_c} e^{-\gamma z} \begin{cases} \frac{n\pi}{b} \sin\left(\frac{n\pi y}{b}\right) & \text{TE, } x = 0 \\ -\frac{m\pi}{a} \sin\left(\frac{m\pi x}{a}\right) & \text{TE, } y = 0 \\ -\frac{m\pi}{a} \sin\left(\frac{n\pi y}{b}\right) & \text{TM, } x = 0 \\ -\frac{n\pi}{b} \sin\left(\frac{m\pi x}{a}\right) & \text{TM, } y = 0 \end{cases} \tag{7.47b}$$

The currents are given by (1.10b),

$$\mathbf{J}_s = \mathbf{n} \times \mathbf{H} = \begin{cases} \mathbf{x} \times (H_y\mathbf{y} + H_z\mathbf{z}) & x = 0 \\ \mathbf{y} \times (H_x\mathbf{x} + H_z\mathbf{z}) & y = 0 \end{cases} = \begin{cases} H_y\mathbf{z} - H_z\mathbf{y} & x = 0 \\ H_z\mathbf{x} - H_x\mathbf{z} & y = 0 \end{cases} \tag{7.48a}$$

$$= H_0 e^{-\gamma z} \begin{cases} \frac{1}{k_c}\frac{n\pi}{b} \sin\left(\frac{n\pi y}{b}\right)\mathbf{z} - \frac{k_c}{\gamma} \cos\left(\frac{n\pi y}{b}\right)\mathbf{y} & \text{TE, } x = 0 \\ \frac{k_c}{\gamma} \cos\left(\frac{m\pi x}{a}\right)\mathbf{x} - \frac{1}{k_c}\frac{m\pi}{a} \sin\left(\frac{m\pi x}{a}\right)\mathbf{z} & \text{TE, } y = 0 \\ -\frac{1}{k_c}\frac{m\pi}{a} \sin\left(\frac{n\pi y}{b}\right)\mathbf{z} & \text{TM, } x = 0 \\ -\frac{1}{k_c}\frac{n\pi}{b} \sin\left(\frac{m\pi x}{a}\right)\mathbf{z} & \text{TM, } y = 0 \end{cases} \tag{7.48b}$$

7.2.4 Mode Plots

It is a useful exercise in visualization to plot the fields and sources associated with a waveguide mode. Let us start with the dominant TE_{10} mode, whose fields are

$$\mathbf{E}_{10} = E_0 e^{-\gamma z} \begin{pmatrix} \mathbf{x} & \mathbf{y} & \mathbf{z} \end{pmatrix} \begin{pmatrix} 0 \\ -\frac{1}{k_c}\frac{\pi}{a}\sin\left(\frac{\pi x}{a}\right) \\ 0 \end{pmatrix} \tag{7.49a}$$

$$\mathbf{H}_{10} = H_0 e^{-\gamma z} \begin{pmatrix} \mathbf{x} & \mathbf{y} & \mathbf{z} \end{pmatrix} \begin{pmatrix} \frac{1}{k_c}\frac{\pi}{a}\sin\left(\frac{\pi x}{a}\right) \\ 0 \\ \frac{k_c}{\gamma}\cos\left(\frac{\pi x}{a}\right) \end{pmatrix} \tag{7.49b}$$

and whose charge and current distributions are given by

$$\rho_{s,10} = \varepsilon \frac{E_0}{k_c} e^{-\gamma z} \begin{cases} 0 & x = 0 \\ -\frac{\pi}{a}\sin\left(\frac{\pi x}{a}\right) & y = 0 \end{cases} \tag{7.50a}$$

$$\mathbf{J}_{s,10} = H_0 e^{-\gamma z} \begin{cases} -\frac{k_c}{\gamma}\mathbf{y} & x = 0 \\ \frac{k_c}{\gamma}\cos\left(\frac{\pi x}{a}\right)\mathbf{x} - \frac{1}{k_c}\frac{\pi}{a}\sin\left(\frac{\pi x}{a}\right)\mathbf{z} & y = 0 \end{cases} \tag{7.50b}$$

These quantities are plotted in Figure 7.3. In the first column, Figure 7.3(a), we plot the transverse $\mathbf{E}$ and $\mathbf{H}$ fields in the aperture plane of the waveguide. Note that the electric field lines are straight and vertically polarized, while tapering in intensity toward the side walls. The transverse magnetic field is oriented at a right angle to the electric field. In Figure 7.3(b), we show these same fields at a longitudinal plane at the mid-point of the broad wall. This is known as the E-plane, since the electric field of the dominant mode lies parallel within it, while the magnetic field is perpendicular to it. Both fields oscillate sinusoidally in the direction of propagation. In Figure 7.3(c), we show these fields in the orthogonal longitudinal plane, known as the H-plane. In this case, the magnetic field of the dominant mode lies within the plane, while the electric field is everywhere perpendicular to it. Note that the electric field has no longitudinal component, but the magnetic field loops in alternating clockwise and counterclockwise directions along the axis of propagation.

In Figures 7.3(d, e), we show the surface charge and current distributions on the conductive walls associated with this mode of propagation. Noteworthy features

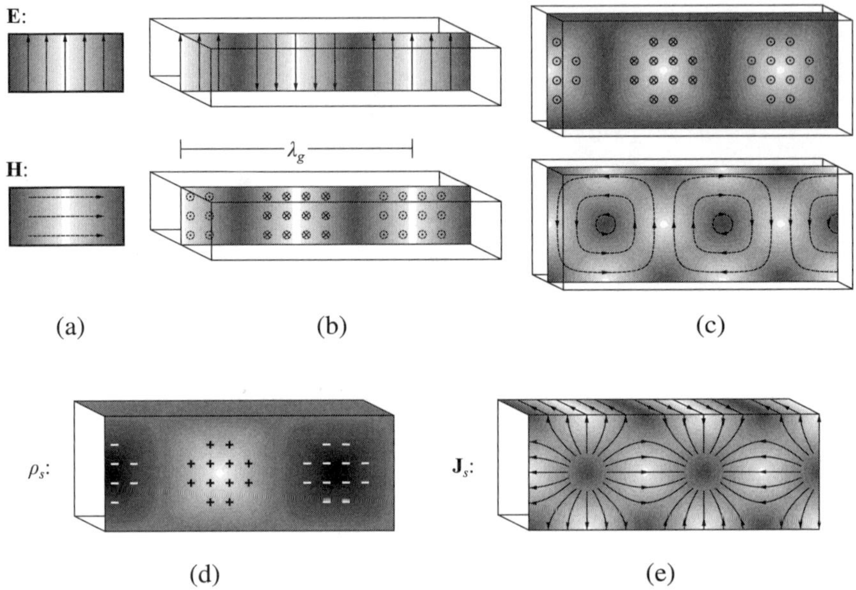

Figure 7.3 Fields and source distributions for the dominant, TE_{10} mode in rectangular waveguide. (a) Aperture view. (b) E-plane view. (c) H-plane view. Electric fields are plotted with solid lines and magnetic fields are plotted with dashed lines, while the shading represents the relative intensity of the field. (d) Surface charge distribution. (e) Surface current distribution.

include the fact that there is no net charge on the narrow side walls of the guide, that the current is purely circumferential on these walls, and the current on the broad top and bottom walls has a quadrupolar pattern. Of particular note is the fact that the current along the mid-point of the broad well is entirely longitudinal — the symmetric H-plane is not crossed by any current. This will be of great importance in Chapter 8 when we talk about split-block construction of waveguide components.

Let us now consider the lowest-order TM mode, TM_{11}, shown in Figure 7.4. According to (7.36), the fields are given by

$$\mathbf{E} = E_0 e^{-\gamma z} \begin{pmatrix} \mathbf{x} & \mathbf{y} & \mathbf{z} \end{pmatrix} \begin{pmatrix} -\frac{1}{k_c}\frac{\pi}{a}\cos\left(\frac{\pi x}{a}\right)\sin\left(\frac{\pi y}{b}\right) \\ -\frac{1}{k_c}\frac{\pi}{b}\sin\left(\frac{\pi x}{a}\right)\cos\left(\frac{\pi y}{b}\right) \\ \frac{k_c}{\gamma}\sin\left(\frac{\pi x}{a}\right)\sin\left(\frac{\pi y}{b}\right) \end{pmatrix} \tag{7.51a}$$

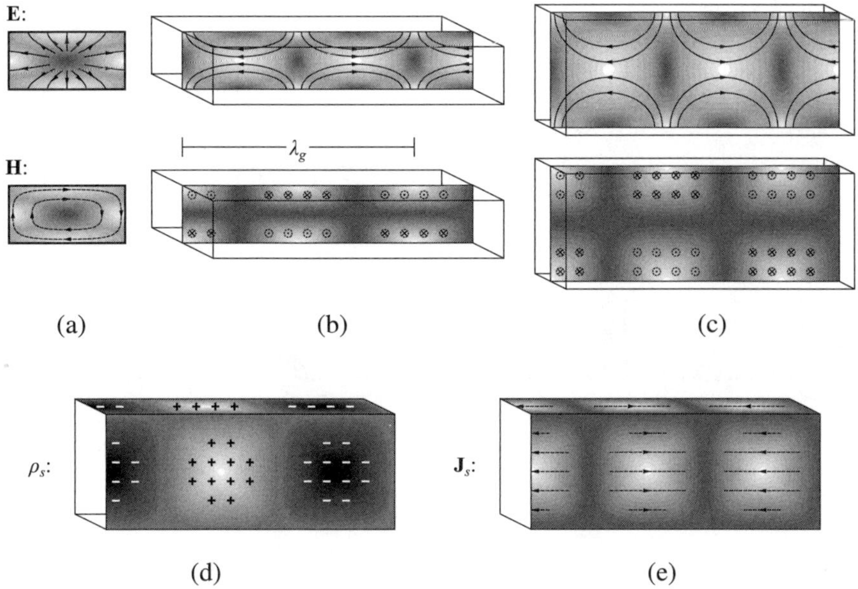

Figure 7.4 Fields and source distributions for the TM_{11} mode in rectangular waveguide. (a) Aperture view. (b) E-plane view. (c) H-plane view. Electric fields are plotted with solid lines and magnetic fields are plotted with dashed lines, while the shading represents the relative intensity of the field. (d) Surface charge distribution. (e) Surface current distribution.

$$\mathbf{H} = H_0 e^{-\gamma z} \begin{pmatrix} \mathbf{x} & \mathbf{y} & \mathbf{z} \end{pmatrix} \begin{pmatrix} \frac{1}{k_c}\frac{\pi}{b}\sin\left(\frac{\pi x}{a}\right)\cos\left(\frac{\pi y}{b}\right) \\ -\frac{1}{k_c}\frac{\pi}{a}\cos\left(\frac{\pi x}{a}\right)\sin\left(\frac{\pi y}{b}\right) \\ 0 \end{pmatrix} \tag{7.51b}$$

and the charge and current distributions are given by

$$\rho_s = \varepsilon\frac{E_0}{k_c}e^{-\gamma z}\begin{cases} -\frac{\pi}{a}\sin\left(\frac{\pi y}{b}\right) & x=0 \\ -\frac{\pi}{b}\sin\left(\frac{\pi x}{a}\right) & y=0 \end{cases} \tag{7.52a}$$

$$\mathbf{J}_s = H_0 e^{-\gamma z}\begin{cases} -\frac{1}{k_c}\frac{\pi}{a}\sin\left(\frac{\pi y}{b}\right)\mathbf{z} & x=0 \\ -\frac{1}{k_c}\frac{\pi}{b}\sin\left(\frac{\pi x}{a}\right)\mathbf{z} & y=0 \end{cases} \tag{7.52b}$$

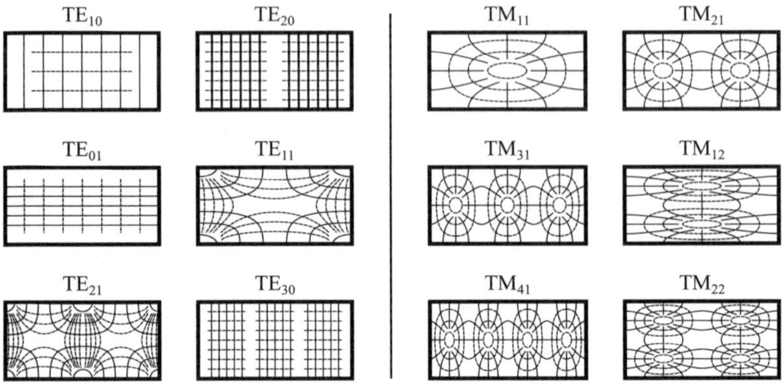

Figure 7.5 First few TE modes (left side) and TM modes (right side) for rectangular waveguide. Electric fields are shown with solid lines, and magnetic fields with dashed lines.

In this case, the magnetic field lies in the transverse plane, looping in alternating directions as the phase progresses. The electric field emanates from the circumference of the waveguide, before threading longitudinally and then back to the walls again. This time, all four walls (top, bottom, and sides) have charge accumulations as well as currents. The currents in this case are fully longitudinal, reversing direction as the wave propagates.

Higher-order modes may essentially be considered harmonics of these field patterns across the width and height of the aperture. The aperture patterns for a variety of these modes is illustrated in Figure 7.5. These are the eigenfunctions upon which one may perform a decomposition of any aperture field pattern whatsoever as an infinite series, in the same way that the Fourier sine and cosine series are used to represent arbitrary waveforms. That field pattern, incident upon the waveguide, will launch a wavefront down the longitudinal axis which evolves according to the characteristics of the individual modes in the series, some of which are cutoff, others propagating, but most having different propagation velocities. These propagation characteristics will be studied in the following section.

7.2.5 Dispersion Diagrams

Although the cutoff frequencies are formulaically identical for both TE and TM modes — compare (7.32b) and (7.42b) — the valid range of the indices is different.

For TM modes, neither subscript can be zero. For TE modes, either can be zero, but not both. This makes the TE_{10} mode (having lower cutoff than TE_{01} since $a \geq b$) the dominant mode of the waveguide. Note that the cutoff wavelength in this case is

$$\lambda_c = \frac{2\pi}{k_c} = \frac{2\pi}{\pi/a} = 2a \tag{7.53}$$

Thus, a rectangular waveguide must be at least a half a wavelength wide to support any propagating modes at all.

In addition to the field patterns, it is instructive to map out the relative cutoff frequencies and propagation velocities of the modes of a given waveguide on what is known as a *dispersion diagram*, also called $k - \beta$ *diagram* or *Brillouin diagram*, such as that shown for rectangular waveguide in Figure 7.6.

There are many common variations in dispersion diagrams, with some engineers choosing to plot against frequency instead of wavenumber, and sometimes the vertical and horizontal axes are reversed, but the general premise is the same: to show how the guided wave propagation characteristics differ from those of waves in free (or unguided) space. My preference is to plot propagation constant as a function of wavenumber, as these two will have the same units, with wavenumber on the horizontal axis as the independent variable. It should be kept in mind always that the wavenumber, k, is directly proportional to frequency.

It is also customary to plot the imaginary part of the propagation constant, β, on the positive y axis, and the real part, α, on the negative y axis. In this way, one may look at both the propagation velocity and the attenuation of a given mode simultaneously. In Figure 7.6, we have normalized both axes by the width of the guide, a. Thus, βa and ka may be thought of as radians per guide width in the waveguide and in free space (or the unguided medium defined by ε and μ) respectively. For cutoff modes, in turn, αa represents the attenuation in Nepers per guide width.

Note that the phase constants, β, are hyperbolic in the upper half of the plot, asymptotically approaching the line of unguided waves at high frequency. The attenuation constants, α, are quarter circles in the lower half of the plot with perpendicular intercepts on the horizontal and vertical axis.

The horizontal intercepts correspond to the cutoff wavenumbers for each mode. As TE_{10} is dominant, it has the left-most intercept. The next higher-order modes are TE_{20} and TE_{01}, which are degenerate in this case, a factor of 2 higher. Thus the rectangular waveguide with 2:1 aspect ratio ($b = a/2$) has a single-mode bandwidth of 1 octave. This is the largest bandwidth that a rectangular waveguide can have; the TE_{01} cutoff moves lower as the aspect ratio decreases (becoming a

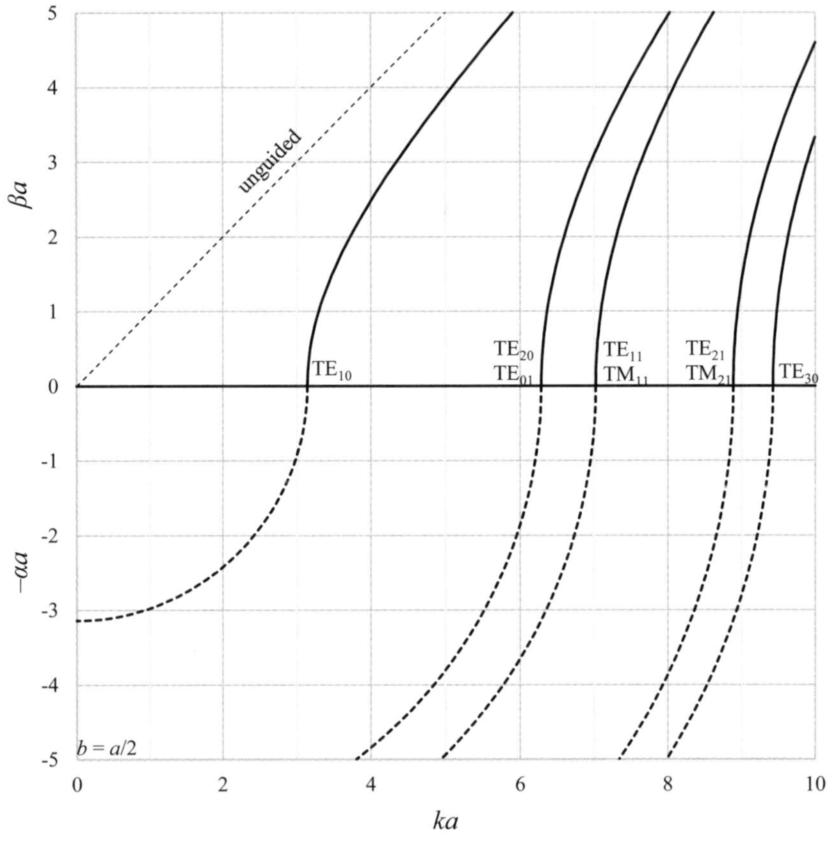

Figure 7.6 Dispersion diagram for rectangular waveguide having $b = a/2$. The phase constants, β, are shown with solid lines in the upper half, while the attenuation constants, α, are shown in the lower half with dashed lines.

degenerate dominant mode when the waveguide is square, $b = a$). Waveguides with higher aspect ratios (where $b < a/2$) still exhibit 2:1 single-mode bandwidths (limited by the TE_{20} mode), but have higher loss. For this reason, and since single-mode operation is most often desired, the 2:1 aspect ratio is a common standard for most rectangular waveguides.

7.2.6 Phase Velocity and Group Velocity

It would be a lost opportunity not to point out that modes in a closed metal waveguide such as this have a rather peculiar phase velocity, as determined by (3.11c),

$$v_p = \frac{\omega}{\beta} = \frac{k/\sqrt{\mu\varepsilon}}{\sqrt{k^2 - k_c^2}} \tag{7.54}$$

which, for an air-filled waveguide (where $\mu = \mu_0$ and $\varepsilon = \varepsilon_0$), is equal to

$$v_p = \frac{k/\sqrt{\mu_0\varepsilon_0}}{\sqrt{k^2 - k_c^2}} = \frac{c}{\sqrt{1 - \left(\frac{k_c}{k}\right)^2}} \tag{7.55}$$

In the propagating regime, then, where $k > k_c$, we have $v_p > c$. In other words, the phase fronts inside a waveguide move faster than the speed of light. This is a characteristic of any waveguide mode for which the propagation curves in the dispersion diagram lie below the diagonal for unguided waves.

How is this possible? The answer lies in understanding the true nature of the phase velocity and how it is constructed from more basic wave elements inside a waveguide. Consider the interference pattern of two plane waves in free space, as shown in Figure 7.7(a). The shading indicates wave amplitude. Where the two waves intersect, they interfere with one another. The portion of that interference pattern that lies between the dashed horizontal lines is exactly the same as that shown for the TE_{10} mode in Figure 7.3. In fact, we may consider the TE_{10} mode — and all other modes in rectangular waveguide, for that matter — as the superposition of plane waves reflecting at oblique angles off the walls. The phase velocity of the modal field pattern is nothing more than the speed at which the maximum of the interference pattern moves rightward as the intersecting waves pass by one another. This is given simply by the vector projection onto the individual wave velocities from the horizontal axis, as shown.

Does this imply that energy travels down the waveguide faster than the speed of light? Clearly not. Just as a crowd of people performing "the wave" in a stadium may create a ripple effect that propagates faster than any individual could run, the interference effect here travels faster than the energy embodied in the individual plane waves. Note that in Figure 7.7(a) the rightward-moving interference pattern vanishes once it reaches the side edges of the individual beams. In the same way, the modal field pattern inside the waveguide may only extend into the waveguide as far as the individual plane waves have had time to reach. Since these plane waves are traveling at an angle relative to the longitudinal axis, the leading edge of a transient

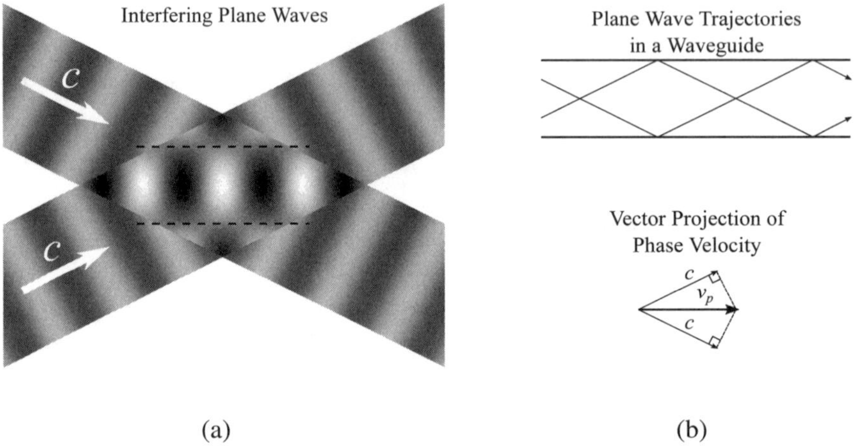

Figure 7.7 Construction of TE_{10} mode in rectangular waveguide as the interference of two plane waves. (a) Plane waves in free space. (b) Ricochet trajectories of plane waves in a waveguide and vector projection of their phase velocities.

waveform must actually propagate down the length of the guide more slowly than the unguided velocity of light in the medium.

That speed, the speed at which the envelope of a transient or modulated waveform propagates, is known as *group velocity* and is given by

$$v_g = \frac{d\omega}{d\beta} \tag{7.56}$$

(Note the similarity with the equation for phase velocity, (3.11c).) The relationship between phase and group velocity for a Gaussian wave packet is thus illustrated in Figure 7.8(a). This may be considered a plot of the electric field on a line running longitudinally down the middle of the waveguide. As time evolves, the wave packet contained within the Gaussian envelope moves to the right at the group velocity, less than c, while the phase fronts of the underlying carrier wave move through it at the phase velocity, greater than c. On a dispersion diagram, Figure 7.8(b), the group velocity is inversely proportional to the slope of the propagation curve at the point defined by its carrier frequency, while the phase velocity is inversely proportional to the slope of the line connecting that point to the origin.

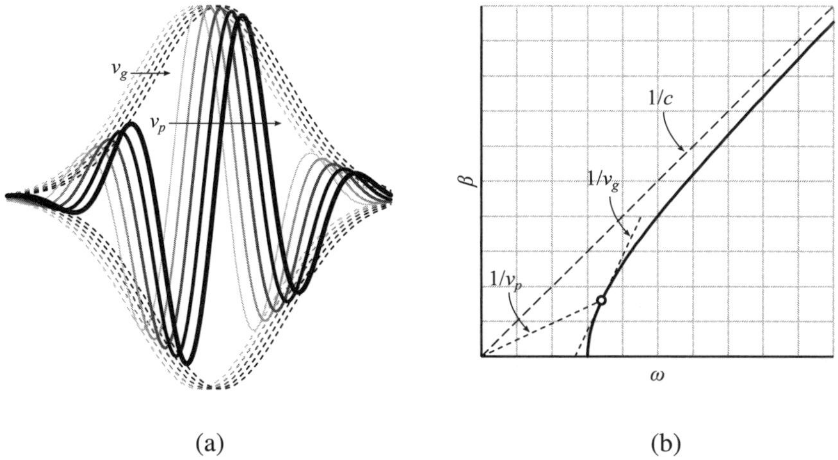

Figure 7.8 (a) Graphical illustration of the phase and group velocity in a Gaussian wave packet. (b) Slopes associated with phase and group velocity on a dispersion diagram.

7.2.7 Losses

Thus far, we have assumed the walls of the waveguide are perfect electric conductors. In the case where the walls have finite but large conductivity, we may approximate the losses by assuming the field solutions already derived are still good and determine the amount of power dissipated in the surface currents on those walls per unit length. This idea of assuming that the losses are small enough for the overall field solutions using PEC boundary conditions to remain valid is known as the *perturbation method.*

First, we note that since the power flow given by the Poynting vector is a product of **E** and **H**, and the individual field terms both decay in the lossy case by a factor $e^{-\alpha_c z}$, the power flow as a function of unit length may be written

$$P(z) = P_0 e^{-2\alpha_c z} \tag{7.57}$$

The attenuation constant may therefore be determined by considering the rate of change at $z = 0$,

$$\frac{dP}{dz} = -2\alpha_c P_0 \tag{7.58a}$$

$$\therefore \alpha_c = \frac{1}{2P_0}\left(-\frac{dP}{dz}\right) \tag{7.58b}$$

The term in parentheses is simply the power loss per unit length, P_l, thus

$$\alpha_c = \frac{P_l}{2P_0} \tag{7.59}$$

P_0 was given in (7.46b) — derived in that section for TE modes, but resulting in a form that is valid for both types. P_l may be derived from Ohm's law by integrating the square magnitude of the surface current from (7.48b) times the skin effect sheet resistance of the walls (1.118) around the full circumference of the waveguide,

$$P_l = \oint \mathbf{J}_s \cdot (\mathbf{J}_s R_s)^* \, dl = R_s \oint |\mathbf{J}_s|^2 \, dl = \sqrt{\frac{\omega\mu'}{2\sigma}} \oint |\mathbf{J}_s|^2 \, dl \tag{7.60}$$

The conjugated term in parentheses above, $\mathbf{J}_s R_s$, may be thought of as a small perturbation of the electric field, a component that is tangential to the walls, resulting from the finite conductivity of the waveguide. Also, keep in mind that the permeability, μ', under the radical in the expression for skin depth resistance is the constituent parameter of the metal walls, not the insulating fill material inside the waveguide (usually air). Evaluating this expression for the dominant TE_{10} mode, we have

$$P_l = 2\sqrt{\frac{\omega\mu'}{2\sigma}}\left(\int_0^a |\mathbf{J}_s(x,0)|^2 \, dx + \int_0^b |\mathbf{J}_s(0,y)|^2 \, dy\right) \tag{7.61a}$$

$$= 2|H_0|^2 \sqrt{\frac{\omega\mu'}{2\sigma}}\left(\left(\frac{k_c}{\beta}\right)^2 \frac{a}{2} + \left(\frac{\pi}{k_c a}\right)^2 \frac{a}{2} + \left(\frac{k_c}{\beta}\right)^2 b\right) \tag{7.61b}$$

$$= \frac{|H_0|^2}{a^2\beta^2}\sqrt{\frac{\omega\mu'}{2\sigma}}\left(\pi^2 a + a^3\beta^2 + 2b\pi^2\right) \tag{7.61c}$$

Therefore, the attenuation constant is

$$\alpha_c = \frac{P_l}{2P_0} = \frac{|H_0|^2\left(\pi^2 a + a^3\beta^2 + 2b\pi^2\right)}{a^2\beta^2 E_0 H_0^* ab}\sqrt{\frac{\omega\mu'}{2\sigma}} \tag{7.62a}$$

$$= \frac{\pi^2 a + a^3\beta^2 + 2b\pi^2}{a^3 b\beta^2 Z_{TE}}\sqrt{\frac{\omega\mu'}{2\sigma}} = \frac{\pi^2 a + a^3\beta^2 + 2b\pi^2}{a^3 b\beta\omega\mu}\sqrt{\frac{\omega\mu'}{2\sigma}} \tag{7.62b}$$

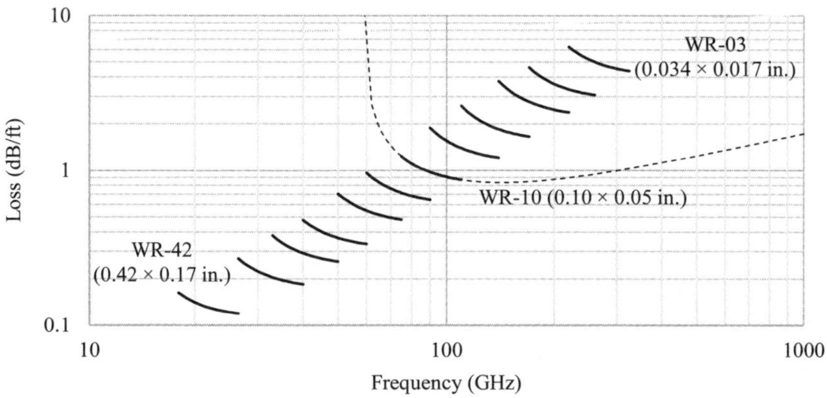

Figure 7.9 Losses in the dominant (TE_{10}) mode of rectangular waveguide of 12 different standard sizes operating in the millimeter-wave frequency range. Standard operating frequencies are shown with solid lines, while the full frequency dependence for a particular waveguide (WR-10) is shown with the dashed line. We have assumed a conductivity of $\sigma = 4/1 \cdot 10^7 S/m$, consistent with gold-plated waveguide.

$$= \frac{a^3\left(k_c^2+\beta^2\right)+2b\pi^2}{a^3 b\beta k\eta}\sqrt{\frac{\omega\mu'}{2\sigma}} = \frac{a^3k^2+2b\pi^2}{a^3 b\beta k\eta}\sqrt{\frac{\omega\mu'}{2\sigma}} \tag{7.62c}$$

The frequency dependence of loss is not easily read from this complex expression, so an illustration is shown in Figure 7.9. This plot captures the loss of the TE_{10} mode for 12 different standard waveguide sizes, designated WR-xx (where "xx" roughly corresponds to the waveguide width, a, in hundredths of an inch). The loss of one guide in particular, WR-10 (with dimensions $a = 0.1$ in. and $b = 0.05$ in.), is shown with a dashed line from its cutoff frequency of about 59 GHz up to 1 THz. Note that there is a minimum for this standard configuration near 142 GHz. Unfortunately, the waveguide is overmoded at this point, so the typical useful range is lower, in this case 75–110 GHz, shown with a solid line.

Waveguide is thus atypical for having a negative loss slope across its operating range; contrast that with the loss for coax transmission line, which is monotonically increasing with frequency. However, note that the loss does go up as the waveguides get smaller, as they must in order to support operation at progressively higher frequencies.

(a) (b)

Figure 7.10 Waveguide step-changes in (a) height and (b) width.

Bear in mind that the losses calculated here are based on the perturbation method, which assumes from the outset that the loss is small. The results are thus liable to be inaccurate near the cutoff where attenuation is very high. Further, actual losses in real waveguides will be larger than is determined by this theoretical calculation (by as much as a factor of 2) due to roughness of the interior surface, metal purity, and other factors.

A summary of the results obtained for rectangular waveguide thus far is given in Table 7.2.

7.2.8 Modal Discontinuity Analysis

Waveguides may seem a lot like transmission lines, in that both are used to propagate (or transmit) waves along one axis while confining the electromagnetic fields along the other two. The theory of ideal transmission lines rested on the ability to clearly define lumped-element port parameters, such as voltage and current. In contrast, rectangular waveguides must be at least a half-wavelength wide even to support a propagating mode. This is not some parasitic effect brought on by imperfect manufacturing, it is fundamental to their operation. Therefore, lumped port descriptions and the associated network parameters are not applicable.

This distinction is made most clear when one considers the behavior of the waveguide at a junction or discontinuity, such as the height step indicated in Figure 7.10(a). One might be tempted to use the wave impedance associated with the dominant mode, Z_{TE}, in the transmission-line formula for reflection coefficient, that is, (3.37b). The result would be immediately suspect, however, since both waveguide sections have the same width, the same propagation constant, the same wave impedance, and thus (we would erroneously conclude) no reflection. It is a mistake to equate the wave impedance of waveguides (which relates the transverse electric and magnetic field amplitudes) to the characteristic or *line impedance* of transmission lines (which relates the voltage and current). The line impedance is a property of the transmission-line guiding structure, whereas the wave impedance

Table 7.2
Equations for Rectangular Waveguide

Parameter	Symbol	Expression
Cutoff wavenumber	k_c	$\sqrt{\left(\frac{m\pi}{a}\right)^2 + \left(\frac{n\pi}{b}\right)^2}$
TE mode E-field	$\begin{pmatrix} E_x \\ E_y \\ E_z \end{pmatrix}$	$E_0 e^{-\gamma z} \begin{pmatrix} \frac{1}{k_c}\frac{n\pi}{b}\cos\left(\frac{m\pi x}{a}\right)\sin\left(\frac{n\pi y}{b}\right) \\ -\frac{1}{k_c}\frac{m\pi}{a}\sin\left(\frac{m\pi x}{a}\right)\cos\left(\frac{n\pi y}{b}\right) \\ 0 \end{pmatrix}$
TE mode H-field	$\begin{pmatrix} H_x \\ H_y \\ H_z \end{pmatrix}$	$H_0 e^{-\gamma z} \begin{pmatrix} \frac{1}{k_c}\frac{m\pi}{a}\sin\left(\frac{m\pi x}{a}\right)\cos\left(\frac{n\pi y}{b}\right) \\ \frac{1}{k_c}\frac{n\pi}{b}\cos\left(\frac{m\pi x}{a}\right)\sin\left(\frac{n\pi y}{b}\right) \\ \frac{k_c}{\gamma}\cos\left(\frac{m\pi x}{a}\right)\cos\left(\frac{n\pi y}{b}\right) \end{pmatrix}$
TE mode wave impedance	Z_{TE}	$\frac{E_0}{H_0} = \frac{j\omega\mu}{\gamma} = \frac{jk\eta}{\gamma}$
TM mode E-field	$\begin{pmatrix} E_x \\ E_y \\ E_z \end{pmatrix}$	$E_0 e^{-\gamma z} \begin{pmatrix} -\frac{1}{k_c}\frac{m\pi}{a}\cos\left(\frac{m\pi x}{a}\right)\sin\left(\frac{n\pi y}{b}\right) \\ -\frac{1}{k_c}\frac{n\pi}{b}\sin\left(\frac{m\pi x}{a}\right)\cos\left(\frac{n\pi y}{b}\right) \\ \frac{k_c}{\gamma}\sin\left(\frac{m\pi x}{a}\right)\sin\left(\frac{n\pi y}{b}\right) \end{pmatrix}$
TM mode H-field	$\begin{pmatrix} H_x \\ H_y \\ H_z \end{pmatrix}$	$H_0 e^{-\gamma z} \begin{pmatrix} \frac{1}{k_c}\frac{n\pi}{b}\sin\left(\frac{m\pi x}{a}\right)\cos\left(\frac{n\pi y}{b}\right) \\ -\frac{1}{k_c}\frac{m\pi}{a}\cos\left(\frac{m\pi x}{a}\right)\sin\left(\frac{n\pi y}{b}\right) \\ 0 \end{pmatrix}$
TM mode wave admittance	Y_{TM}	$\frac{H_0}{E_0} = \frac{j\omega\varepsilon}{\gamma} = \frac{jk}{\eta\gamma}$
Power	P	$E_0 H_0^* \frac{ab}{4}(1+\delta_m)(1+\delta_n)$
Attenuation constant (ohmic, dominant mode)	α_c	$\frac{a^3k^2+2b\pi^2}{a^3b\beta k\eta}\sqrt{\frac{\omega\mu'}{2\sigma}}$

is a property of the waveguide mode, and modes are not strictly isolated from one another at a discontinuity, even if all but one of them are cutoff. In the limit of electrically small cross-sections enjoyed by transmission lines, the line and wave impedances are one and the same, but that is not generally the case for waveguides.

Fortunately, it is possible to define scattering parameters for a waveguide network, but the mechanism from which they arise is much more complex than the simple impedance ratios of transmission lines. Instead, they depend on an intricate interaction of all possible modes, whether propagating or not, and each mode, for the purposes of the scattering parameter matrix, must be considered a separate potential port despite occupying the same physical space as all the others.

Let us analyze the width step in Figure 7.10(b) as an example. We assume the discontinuity is excited by a wave propagating in the dominant TE_{10} mode toward the junction from the left side. In principle, there are an infinite number of reflected modes propagating away (or decaying evanescently) from the junction in this segment also, as well as an infinite number of transmitted modes propagating (decaying) in the right section, which has width $a' < a$. Typically, many of these modes will be cutoff, but still contribute to the final result as the exponentially decaying fields that they produce will store electromagnetic energy at the junction and impart reactive contributions to the scattering coefficients.

We may begin by writing down equations for the electric and magnetic fields, expanded on each side of the boundary in terms of an infinite series of modes, and keeping in mind that the transverse electric field and longitudinal magnetic field on the left side must vanish for $x > a'$ in accordance with the boundary conditions for a PEC,

$$\mathbf{E} = a_1\mathbf{E}_1 + \sum_{l=1}^{\infty} b_l\mathbf{E}_l = \sum_{p=1}^{\infty} c_p\mathbf{E}'_p \text{ for } 0 \leq x \leq a' \tag{7.63a}$$

$$\mathbf{H} = a_1\mathbf{H}_1 - \sum_{l=1}^{\infty} b_l\mathbf{H}_l = \sum_{p=1}^{\infty} c_p\mathbf{H}'_p \text{ for } 0 \leq x \leq a' \tag{7.63b}$$

$$E_x = E_y = H_z = 0 \text{ for } a' < x \leq a \tag{7.63c}$$

where a_1 is the phasor amplitude of the incident TE_{10} wave, b_l is the phasor amplitude of the reflected wave in mode l, and c_p is the phasor amplitude of the transmitted wave in mode p. The prime notation on the right side indicates that these modes are for a waveguide of width a' instead of width a.

Note that the magnetic field has been negated for the backward-traveling waves. In practice, this choice was arbitrary; one of $\mathbf{E}$ or $\mathbf{H}$ had to change sign in order to reverse the direction of the Poynting vector, but either would have been acceptable.

The geometry of the problem does permit some simplifications to be made in this case. First, since there is no y variation in either the geometry or the field pattern of the incident mode, we may conclude that no TE_{mn} or TM_{mn} mode is excited in either guide for which $n \neq 0$ (which excludes TM modes entirely). Thus, mode l on the left side is the TE_{l0} mode, and mode p on the right side is the TE'_{p0}

mode, where[2]

$$k_{c,l0} = \frac{l\pi}{a} \tag{7.64a}$$

$$k'_{c,p0} = \frac{p\pi}{a'} \tag{7.64b}$$

$$Z_l = \frac{E_0}{H_{0,l}} = \frac{j\omega\mu}{\gamma_{l0}} \tag{7.64c}$$

$$Z'_p = \frac{E_0}{H'_{0,p}} = \frac{j\omega\mu}{\gamma'_{p0}} \tag{7.64d}$$

Further, there is some redundancy in (7.63), since the field components are not all independent. Let us write just the E_y and H_x components (at $z = 0$),

$$a_1 E_0 \sin\left(\tfrac{\pi x}{a}\right) + \sum_{l=1}^{\infty} b_l E_0 \sin\left(\tfrac{l\pi x}{a}\right) = \begin{cases} \sum c_p E_0 \sin\left(\tfrac{p\pi x}{a'}\right) & 0 \leq x \leq a' \\ 0 & a' \leq x \leq a \end{cases} \tag{7.65a}$$

$$a_1 H_{0,1} \sin\left(\tfrac{\pi x}{a}\right) - \sum_{l=1}^{\infty} b_l H_{0,l} \sin\left(\tfrac{l\pi x}{a}\right) = \sum_{p=1}^{\infty} c_p H'_{0,p} \sin\left(\tfrac{p\pi x}{a'}\right) \quad 0 \leq x \leq a' \tag{7.65b}$$

$$\therefore a_1 \frac{E_0}{Z_1} \sin\left(\tfrac{\pi x}{a}\right) - \sum_{l=1}^{\infty} b_l \frac{E_0}{Z_l} \sin\left(\tfrac{l\pi x}{a}\right) = \sum_{p=1}^{\infty} c_p \frac{E_0}{Z'_p} \sin\left(\tfrac{p\pi x}{a'}\right) \quad 0 \leq x \leq a' \tag{7.65c}$$

To remove the x dependence from (7.65a), we multiply both sides by $\sin(m\pi x/a)$ where $m \geq 1$ and then integrate from $x = 0$ to a. The orthogonality relationships

2 Note that for simplicity we have chosen to normalize amplitudes such that E_0 is the same for all modes. However, due to the differing wave impedances, H_0 cannot be the same for each and must depend on the mode index.

in Section A.4.1 may be useful here. After dropping the common terms, we have

$$a_1 \frac{a}{2} \delta_{m1} + b_m \frac{a}{2} = \sum_{p=1}^{\infty} c_p I_{mp} = \sum_{k=1}^{\infty} c_k I_{mk} \tag{7.66}$$

where

$$I_{mp} = \int_0^{a'} \sin\left(\frac{m\pi x}{a}\right) \sin\left(\frac{p\pi x}{a'}\right) dx \tag{7.67a}$$

$$= \frac{1}{2} \int_0^{a'} \left(\cos\left(\left(\frac{m}{a} - \frac{p}{a'}\right)\pi x\right) - \cos\left(\left(\frac{m}{a} + \frac{p}{a'}\right)\pi x\right)\right) dx \tag{7.67b}$$

$$= \frac{a'}{2} \operatorname{sinc}\left(\left(\frac{m}{a} - \frac{p}{a'}\right)\pi a'\right) - \frac{a'}{2} \operatorname{sinc}\left(\left(\frac{m}{a} + \frac{p}{a'}\right)\pi a'\right) \tag{7.67c}$$

Similarly, if we multiply both sides of (7.65c) by $\sin(k\pi x/a')$ and integrate from $x = 0$ to a', we obtain

$$a_1 \frac{I_{1k}}{Z_1} - \sum_{l=1}^{\infty} b_l \frac{I_{lk}}{Z_l} = c_k \frac{a'}{2Z'_k} \tag{7.68a}$$

$$\therefore c_k = a_1 \frac{2I_{1k} Z'_k}{a' Z_1} - \sum_{l=1}^{\infty} b_l \frac{2I_{lk} Z'_k}{a' Z_l} \tag{7.68b}$$

Substituting this expression for c_k back into (7.66), we have

$$a_1 \frac{a}{2} \delta_{m1} + b_m \frac{a}{2} = \sum_{k=1}^{\infty} \left(a_1 \frac{2I_{1k} Z'_k}{a' Z_1} - \sum_{l=1}^{\infty} b_l \frac{2I_{lk} Z'_k}{a' Z_l} \right) I_{mk} \tag{7.69a}$$

$$\frac{b_m}{a_1} + \sum_{k=1}^{\infty} \sum_{l=1}^{\infty} \frac{b_l}{a_1} \frac{4I_{lk} I_{mk} Z'_k}{a' a Z_l} = \sum_{k=1}^{\infty} \frac{4I_{1k} I_{mk} Z'_k}{a' a Z_1} - \delta_{m1} \tag{7.69b}$$

Although computer codes may be written to solve these kinds of problems while taking into account a very large number of modes, we still must truncate the list at some point so that the summations above may be evaluated. Let us say that we will consider the first N modes on each side of the junction. The result above then becomes

$$\frac{b_m}{a_1} + \sum_{k=1}^{N}\sum_{l=1}^{N} \frac{b_l}{a_1}\frac{4I_{lk}I_{mk}Z'_k}{a'aZ_l} = \sum_{k=1}^{N} \frac{4I_{1k}I_{mk}Z'_k}{a'aZ_1} - \delta_{m1} \tag{7.70}$$

or, in matrix form,

$$\mathbf{Q\Gamma} = \mathbf{P} \tag{7.71}$$

where $\mathbf{\Gamma}$ is the $N \times 1$ column vector of reflection and mode-conversion coefficients in the left-hand waveguide, whose elements are

$$\Gamma_l = \frac{b_l}{a_1} \tag{7.72}$$

and $\mathbf{Q}$ is the $N \times N$ matrix and $\mathbf{P}$ the $N \times 1$ column vector whose elements are given by

$$Q_{ml} = \delta_{ml} + \sum_{k=1}^{N} \frac{4I_{lk}I_{mk}Z'_k}{a'aZ_l} \tag{7.73a}$$

$$P_m = \sum_{k=1}^{N} \frac{4I_{1k}I_{mk}Z'_k}{a'aZ_1} - \delta_{m1} \tag{7.73b}$$

We may then solve for $\mathbf{\Gamma}$ by inverting $\mathbf{Q}$,

$$\mathbf{\Gamma} = \mathbf{Q}^{-1}\mathbf{P} \tag{7.74}$$

The first element of this vector is the reflection coefficient in the incident mode (the rest describe conversion of energy from the incident mode into higher-order modes, many of which may be cut off).

Let us plot this reflection coefficient as a function of frequency in Figure 7.11. For this example, we use a WR-10 waveguide with internal dimensions of 100×50 mils stepping down to a waveguide having dimensions 80×50 mils. The magnitude and phase of the dominant-mode reflection coefficient is plotted from 60 to 110 GHz, a range over which the incident waveguide is single-moded. Note the abrupt discontinuity at 73.77 GHz (this is correct in theory for perfect conductors, but in practice, the abruptness of this transition would be smoothed out by conductive losses in the waveguide walls). This transition marks the point at which the first mode in the secondary waveguide turns on. Below this frequency, there is no other mode to propagate energy away from the junction, so the reflection coefficient has

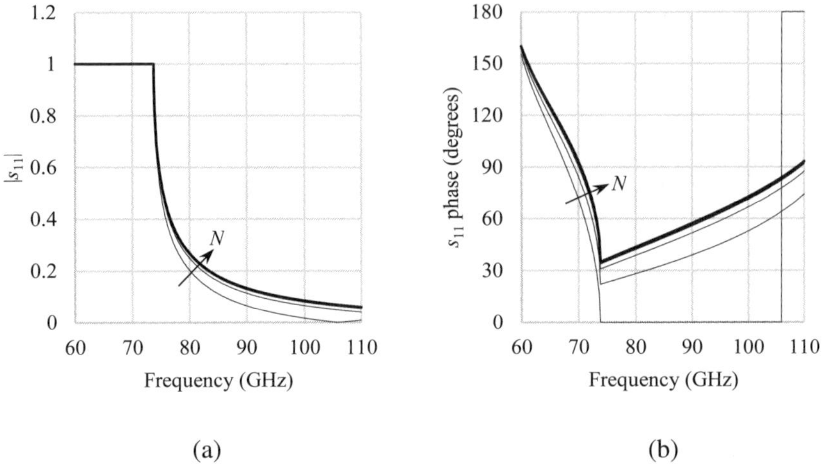

Figure 7.11 Reflection coefficient (a) magnitude and (b) phase for an H-plane step in WR-10 rectangular waveguide derived by modal analysis considering $N = 1 \ldots 10$ modes. For this example, $a' = 0.8a$.

unit magnitude. At higher frequencies, energy couples from the input waveguide into the output waveguide, so that the reflection coefficient rapidly becomes smaller.

It is interesting to plot this reflection coefficient on the Smith chart, as shown in Figure 7.12(a). The use of a Smith chart here is slightly incongruous since we have not identified a consistent characteristic impedance for the waveguide to which it can be normalized. Nevertheless, the meaning of the scattering parameter is clear, and it is common to draw an analogy with the immittance-related parameters of transmission lines in this context. We see that the reflection coefficient traces a clockwise path around the upper part of the Smith chart in the frequency range between the cutoffs of the two waveguides. In the transmission-line analogy, then, the discontinuity appears as an inductive termination. We thus draw the approximate equivalent circuit in Figure 7.12(b) comprising a transmission line in place of our waveguide terminated with a lumped inductance.

It is important not to take this analogy too far, however, The phase slope of the reflection coefficient is much too fast in this example to be represented by a constant, lumped inductance. Thus, we show the inductor as having frequency dependence. Further, the characteristic impedance itself is arbitrary. We may use any value we want — say, 50Ω — so long as the inductor is scaled accordingly. Some researchers choose the guide's dominant wave impedance as Z_0, but the fact

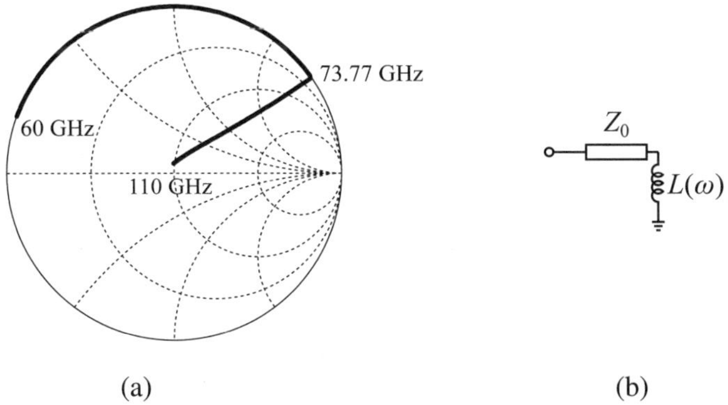

Figure 7.12 (a) Reflection coefficient of the H-plane step plotted on a Smith chart. (b) Transmission-line model for low-band response.

that it too depends on frequency complicates the analysis, and one must remember that the usual transmission line formulas for step changes in impedance cannot be applied to these values.

7.2.9 Waveguide Discontinuity Equivalent Circuits

With these caveats in mind, it is useful to catalog analogous transmission-line equivalents for some of the most common waveguide discontinuities, shown in Table 7.3. In the first three rows, we show thin-wall irises extending inward from the walls of the waveguide, which appear as lumped reactive components in parallel with the transmission line. Next, we show E- and H-plane steps, much like the one just analyzed, with similar equivalent circuits. Finally, we have E- and H-plane junctions between three waveguides which have the equivalent circuits of series- and parallel-connected transmission lines, respectively, with associated reactive parasitics.

As before, the reactive elements in these equivalent circuits must have frequency-dependent values if they are to be valid over more than a narrow bandwidth. As the previous exercise shows, derivation of these equivalents is computationally intensive, and is best done now using readily available computer-aided design programs. Marcuvitz's *Waveguide Handbook* [1] is the definitive reference containing empirical formulas for these and even more sophisticated forms, which remain useful at least in the early planning and design stages.

Table 7.3

Common Rectangular Waveguide Discontinuities and Junctions

Discontinuity	T-Line Equivalent Circuit
Inductive Iris 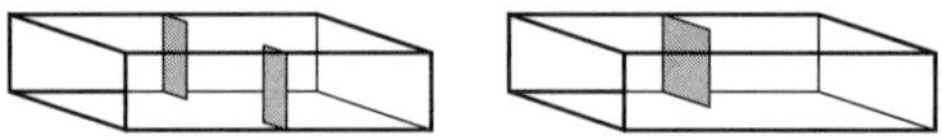	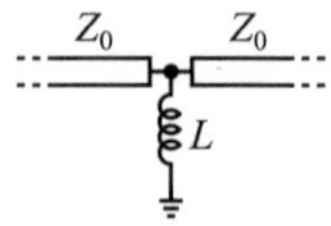
Capacitive Iris 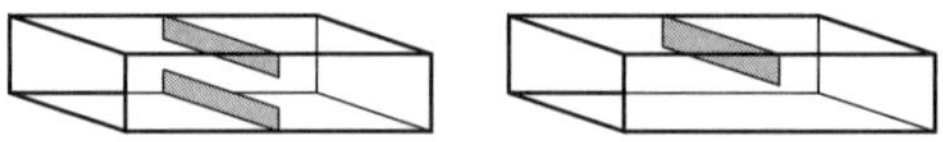	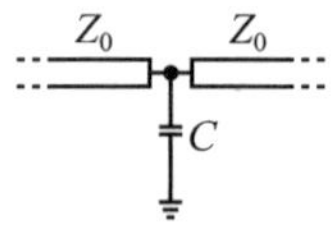
Resonant Iris 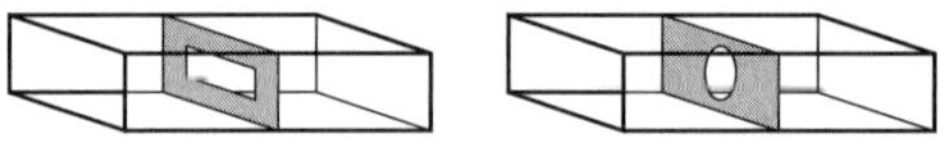	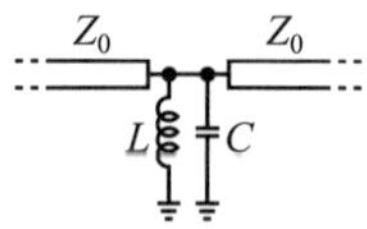
E-Plane Step 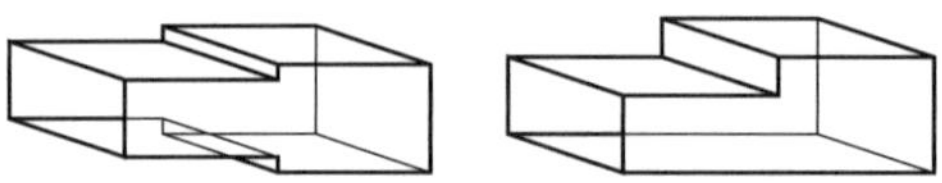	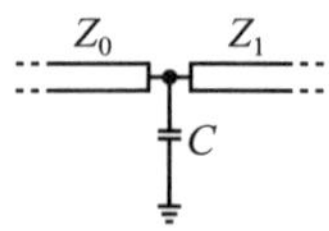
H-Plane Step 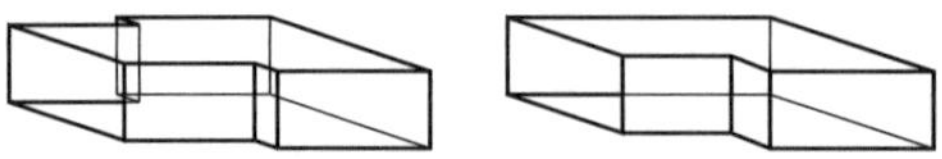	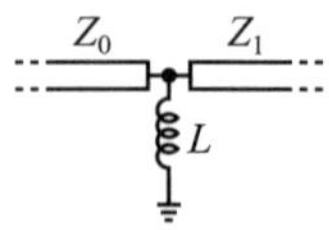
E-Plane Tee Junction 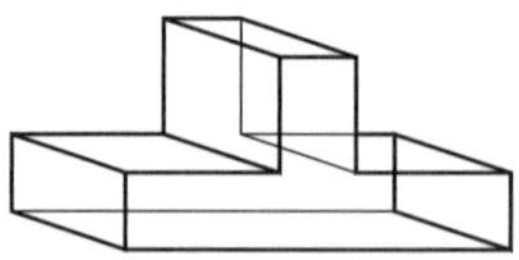	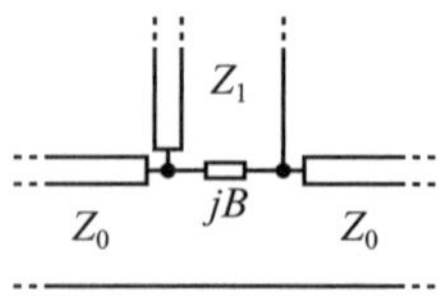
H-Plane Tee Junction 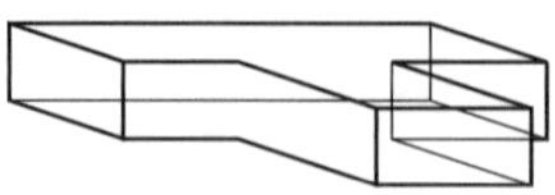	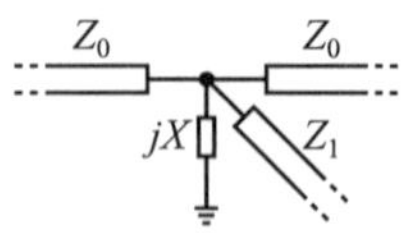

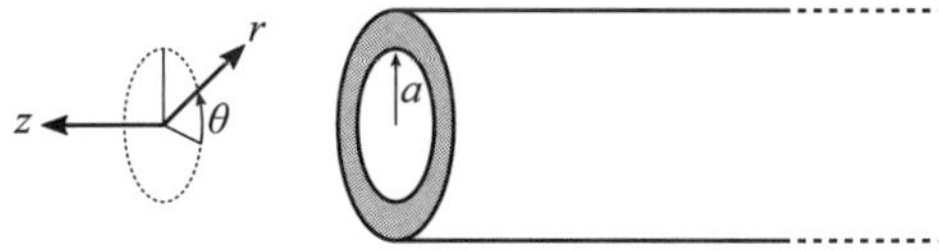

Figure 7.13 Geometry of a circular waveguide.

7.3 CIRCULAR WAVEGUIDE

While rectangular waveguide is by far the most common and convenient for single-mode operation, there are times when it is desirable to support two degenerate dominant modes over short distances, often associated with the horizontal and vertical polarizations received from an antenna. This could be achieved with a square waveguide — a special case of rectangular waveguide where $a = b$ — but another choice is circular waveguide like that shown in Figure 7.13.

7.3.1 TM Modes

As with rectangular waveguide, we may derive the TM modes for circular waveguide by solving the Helmholtz equation, (7.8a), in two dimensions, only this time it is most convenient to work with cylindrical coordinates,

$$\left(\nabla_t^2 + k_c^2\right) E_z = \left\{ \frac{1}{r}\frac{\partial}{\partial r}\left(r\frac{\partial}{\partial r}\right) + \frac{1}{r^2}\frac{\partial^2}{\partial \theta^2} + k_c^2 \right\} E_z = 0 \tag{7.75}$$

Let us solve this by separation of variables. We therefore postulate a solution comprising the product of three functions, one dependent on r, another dependent upon θ, and an exponential for the prescribed propagation term in z,

$$E_z(r, \theta, z) = R(r)\Theta(\theta)e^{-\gamma z} \tag{7.76}$$

Substituting this into (7.75), we have

$$\left\{ \frac{1}{r}\frac{\partial}{\partial r}\left(r\frac{\partial}{\partial r}\right) + \frac{1}{r^2}\frac{\partial^2}{\partial \theta^2} + k_c^2 \right\} R(r)\Theta(\theta)e^{-\gamma z} = 0 \tag{7.77a}$$

$$\therefore \frac{1}{r}\frac{\partial}{\partial r}\left(r\frac{\partial}{\partial r}R(r)\Theta(\theta)\right) + \frac{1}{r^2}\frac{\partial^2}{\partial \theta^2}R(r)\Theta(\theta) + k_c^2 R(r)\Theta(\theta) = 0 \tag{7.77b}$$

$$\therefore \frac{r}{R}\frac{\partial}{\partial r}\left(r\frac{\partial R}{\partial r}\right) + k_c^2 r^2 = -\frac{1}{\Theta}\frac{\partial^2 \Theta}{\partial \theta^2} \tag{7.77c}$$

The left side depends only on r and the right side depends only on θ, so we may assume that both sides are equal to a constant, which we will call k_θ^2. Let us start with the right side,

$$-\frac{1}{\Theta}\frac{\partial^2 \Theta}{\partial \theta^2} = k_\theta^2 \tag{7.78a}$$

$$\therefore \frac{\partial^2 \Theta}{\partial \theta^2} + \Theta k_\theta^2 = 0 \tag{7.78b}$$

which is simply the wave equation in one dimension, whose general solution can be written in terms of sines and cosines,

$$\Theta(\theta) = E_1 \sin(k_\theta \theta) + E_2 \cos(k_\theta \theta) \tag{7.79}$$

or, alternatively (and equivalently),

$$\Theta(\theta) = E_3 \cos(k_\theta(\theta + \phi)) \tag{7.80}$$

where the polarization angle ϕ is arbitrary. The separation constant is constrained to be an integer, since the solution must be periodic in the angle θ. We therefore let $k_\theta = n$,

$$\Theta(\theta) = E_3 \cos(n(\theta + \phi)) \tag{7.81}$$

Note that by choosing the cosine form in (7.80) instead of sine, we have avoided the possibility of the solution vanishing when $n = 0$.

Now we use this separation constant in the equation for $R(r)$,

$$\frac{r}{R}\frac{\partial}{\partial r}\left(r\frac{\partial R}{\partial r}\right) + k_c^2 r^2 = n^2 \tag{7.82a}$$

$$\therefore r^2 \frac{\partial^2 R}{\partial r^2} + r\frac{\partial R}{\partial r} + \left(k_c^2 r^2 - n^2\right) R = 0 \tag{7.82b}$$

This is Bessel's equation with argument $x = k_c r$ and order n, and as such has for its solutions the Bessel functions of the first and second kinds,

$$R(r) = E_4 J_n(k_c r) + E_5 Y_n(k_c r) \tag{7.83}$$

Since the Bessel functions of the second kind become infinite at the origin, we may conclude that $E_5 = 0$. Thus, we have for the longitudinal component of the electric field,

$$E_z(r,\theta,z) = R(r)\Theta(\theta)e^{-\gamma z} = E_3E_4J_n(k_cr)\cos(n(\theta+\phi))e^{-\gamma z} \tag{7.84}$$

Once again the prefactors are arbitrary, and we assign for convenience $E_3E_4 = E_0k_c/\gamma$,

$$E_z(r,\theta,z) = E_0\frac{k_c}{\gamma}J_n(k_cr)\cos(n(\theta+\phi))e^{-\gamma z} \tag{7.85}$$

To determine the cutoff wavenumber, k_c, we must enforce the electric field boundary condition at $r = a$,

$$E_z(a,\theta,z) = E_0\frac{k_c}{\gamma}J_n(k_ca)\cos(n(\theta+\phi))e^{-\gamma z} = 0 \tag{7.86a}$$

$$\therefore J_n(k_ca) = 0 \tag{7.86b}$$

The zeros of the Bessel functions of the first kind are given in Appendix G (Table G.1) and shall be denoted z_{nm} (the mth root of the nth-order Bessel function). Thus,

$$k_c = \frac{z_{nm}}{a} \tag{7.87}$$

where $n \geq 0$ and $m \geq 1$. The first TM mode to propagate, then, is TM_{01} with a cutoff wavenumber $2.4048/a$.

The transverse electric and magnetic fields are found using the results from Table 7.1,

$$\mathbf{E}_t = -\frac{\gamma}{k_c^2}\nabla_t E_z \tag{7.88a}$$

$$= -\frac{E_0}{k_c}\left\{\frac{\partial}{\partial r}\mathbf{r} + \frac{1}{r}\frac{\partial}{\partial\theta}\boldsymbol{\theta}\right\}J_n(k_cr)\cos(n(\theta+\phi))e^{-\gamma z} \tag{7.88b}$$

$$= -\frac{E_0}{k_c}e^{-\gamma z}\begin{pmatrix} k_cJ_n'(k_cr)\cos(n(\theta+\phi))\mathbf{r} \\ -\frac{n}{r}J_n(k_cr)\sin(n(\theta+\phi))\boldsymbol{\theta}\end{pmatrix} \tag{7.88c}$$

and

$$\mathbf{H}_t = Y_{TM}(\mathbf{z}\times\mathbf{E}_t) \tag{7.89a}$$

$$= -\frac{H_0}{k_c}e^{-\gamma z}\begin{pmatrix} \frac{n}{r}J_n(k_cr)\sin(n(\theta+\phi))\mathbf{r} \\ k_cJ_n'(k_cr)\cos(n(\theta+\phi))\boldsymbol{\theta}\end{pmatrix} \tag{7.89b}$$

where again $H_0 = Y_{TM} E_0$.

In summary, for the TM_{nm} mode in a circular waveguide,

$$\mathbf{E} = E_0 e^{-\gamma z} \begin{pmatrix} \mathbf{r} & \boldsymbol{\theta} & \mathbf{z} \end{pmatrix} \begin{pmatrix} -J_n'(k_c r)\cos(n(\theta+\phi)) \\ \frac{n}{k_c r} J_n(k_c r)\sin(n(\theta+\phi)) \\ \frac{k_c}{\gamma} J_n(k_c r)\cos(n(\theta+\phi)) \end{pmatrix} \tag{7.90a}$$

$$\mathbf{H} = H_0 e^{-\gamma z} \begin{pmatrix} \mathbf{r} & \boldsymbol{\theta} & \mathbf{z} \end{pmatrix} \begin{pmatrix} -\frac{n}{k_c r} J_n(k_c r)\sin(n(\theta+\phi)) \\ -J_n'(k_c r)\cos(n(\theta+\phi)) \\ 0 \end{pmatrix} \tag{7.90b}$$

$$k_c = \frac{z_{nm}}{a} \tag{7.90c}$$

$$\frac{H_0}{E_0} = Y_{TM} = \frac{j\omega\varepsilon}{\gamma} \tag{7.90d}$$

Note that we have an arbitrary constant, ϕ, the polarization angle, which is not constrained by the boundary conditions. In practice, this can take on any value whatsoever, but the solutions for one reference angle, ϕ_a, will be orthogonal to those for another angle, $\phi_b = \phi_a + \pi/(2n)$, and the fields for all other values of ϕ will be superpositions of these two. It is thus seen that all TM modes (all modes, in fact), except for those having the rotational index $n = 0$, come in degenerate pairs.

7.3.2 TE Modes

The derivation of TE modes proceeds in much the same way, but with H_z in place of E_z, up to the point that the electric wall boundary conditions are applied. Thus, we find that

$$H_z(r,\theta,z) = H_0 \frac{k_c}{\gamma} J_n(k_c r)\cos(n(\theta+\phi))e^{-\gamma z} \tag{7.91}$$

According to Table 1.1, the derivative of H_z in the direction of the surface normal ($\mathbf{r}$) must be zero at $r = a$. Therefore,

$$H_0 \frac{k_c^2}{\gamma} J_n'(k_c a)\cos(n(\theta+\phi))e^{-\gamma z} = 0 \tag{7.92a}$$

$$\therefore J_n'(k_c a) = 0 \tag{7.92b}$$

The cutoff wavenumber is thus found by locating the extrema of the Bessel functions, which are also given in Appendix G (Table G.2). Therefore,

$$k_c = \frac{e_{nm}}{a} \tag{7.93}$$

where e_{nm} is the mth positive extremum of the nth-order Bessel function.[3] The first TE mode, then, is TE_{11}, with cutoff wavenumber $k_c = 1.8412/a$. With a lower cutoff than TM_{01}, this is also the dominant mode of the waveguide.

Once again, we derive the transverse fields using the results from Table 7.1,

$$\mathbf{H}_t = -\frac{\gamma}{k_c^2}\nabla_t H_z \tag{7.94a}$$

$$= -\frac{H_0}{k_c}\left\{\frac{\partial}{\partial r}\mathbf{r} + \frac{1}{r}\frac{\partial}{\partial \theta}\boldsymbol{\theta}\right\} J_n(k_c r)\cos(n(\theta+\phi))e^{-\gamma z} \tag{7.94b}$$

$$= -\frac{H_0}{k_c}e^{-\gamma z}\begin{pmatrix} k_c J_n'(k_c r)\cos(n(\theta+\phi))\mathbf{r} \\ -\frac{n}{r}J_n(k_c r)\sin(n(\theta+\phi))\boldsymbol{\theta}\end{pmatrix} \tag{7.94c}$$

and

$$\mathbf{E}_t = Z_{TE}(\mathbf{H}_t \times \mathbf{z}) \tag{7.95a}$$

$$= \frac{E_0}{k_c}e^{-\gamma z}\begin{pmatrix} \frac{n}{r}J_n(k_c r)\sin(n(\theta+\phi))\mathbf{r} \\ k_c J_n'(k_c r)\cos(n(\theta+\phi))\boldsymbol{\theta}\end{pmatrix} \tag{7.95b}$$

where $E_0 = Z_{TE}H_0$.

Thus, the complete field expressions for a TE_{nm} mode in circular waveguide are

$$\mathbf{E} = E_0 e^{-\gamma z}\begin{pmatrix}\mathbf{r} & \boldsymbol{\theta} & \mathbf{z}\end{pmatrix}\begin{pmatrix} \frac{n}{k_c r}J_n(k_c r)\sin(n(\theta+\phi)) \\ J_n'(k_c r)\cos(n(\theta+\phi)) \\ 0\end{pmatrix} \tag{7.96a}$$

3 By convention, the extremum at the origin for the first-order Bessel function is excluded from this list. This is not a valid solution as H_z would vanish, and all other field components along with it.

$$\mathbf{H} = H_0 e^{-\gamma z} \begin{pmatrix} \mathbf{r} & \boldsymbol{\theta} & \mathbf{z} \end{pmatrix} \begin{pmatrix} -J_n'(k_c r)\cos(n(\theta+\phi)) \\ \frac{n}{k_c r} J_n(k_c r)\sin(n(\theta+\phi)) \\ \frac{k_c}{\gamma} J_n(k_c r)\cos(n(\theta+\phi)) \end{pmatrix} \tag{7.96b}$$

$$k_c = \frac{e_{nm}}{a} \tag{7.96c}$$

$$\frac{E_0}{H_0} = Z_{TE} = \frac{j\omega\mu}{\gamma} \tag{7.96d}$$

It is worth pointing out that the transverse electric fields for TE modes as formulated above are orthogonal to those of the TM modes having the same rotational index, n, and polarization angle, ϕ — for example, the radial component depends on $\sin(n\theta)$ for TE modes, but on $\cos(n\theta)$ for TM modes. In mode-coupling problems, then, it is useful to rotate one or the other set by $\phi = \pi/(2n)$, as this more naturally represents a pairing of TE and TM modes that are most likely to couple given the symmetries of common junctions.

7.3.3 Modes, Dispersion, and Shielding

A dispersion diagram showing the relative cutoff wavenumbers of the first few TE and TM modes in circular waveguide is given in Figure 7.14. Field patterns for these modes are illustrated in Figure 7.15. Keep in mind that every mode for which $n \neq 0$ has a degenerate counterpart which is rotated by $\pi/(2n)$ radians.

An interesting feature of waveguides below cutoff is that the attenuation constant, α, approaches a constant value at very low frequencies, rather than diverging to infinity. In Figure 7.14(a), for example, the attenuation constant for the dominant mode approaches a low-frequency limit of $\alpha \to k_c \approx 1.8412/a$. This has implications for the effectiveness of electrical shielding, and specifically of penetrations through that shielding.

Imagine an enclosure for an electronic system through which we need to pass something nonmetallic, say, air (for ventilation) or an optical fiber. Some kind of hole or opening will thus be required in the metal walls. In order to block internal emissions from leaking outside, or to improve the isolation of the interior system from external interference, we could choose to minimize the diameter of the hole, or else to lengthen its passage into a tube, as illustrated in Figure 7.16. For this example, we will assume the pinhole diameter is equal to the wall thickness, while the tube stretches out to a length four times the diameter. If we assume the diameter of either opening is much smaller than a wavelength, then we may derive an upper

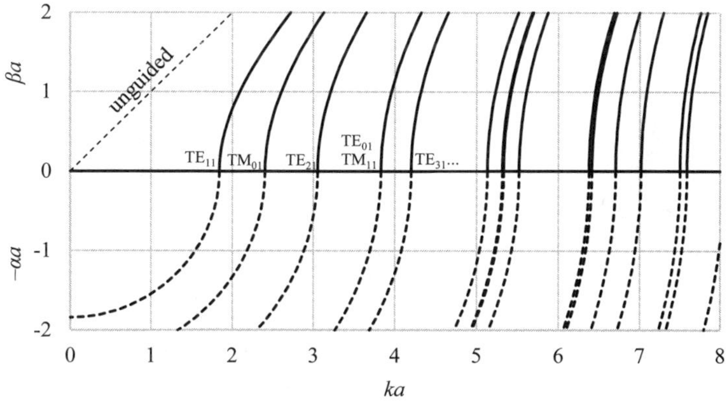

Figure 7.14 Dispersion diagram showing the lowest-order modes in circular waveguide with inner radius a.

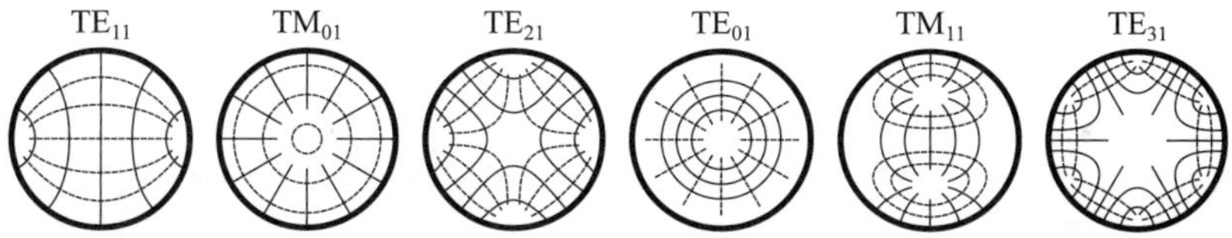

Figure 7.15 First few modes in circular waveguide, excluding the degenerate, rotated counterparts. Electric fields are shown with solid lines, and magnetic fields are shown with dashed lines.

bound for the leakage through these holes by modeling them as circular waveguides, and evaluate the attenuation of the dominant mode. For a cutoff waveguide with diameter $2a$ and length L, the minimum, low-frequency attenuation, A, in decibels is given by

$$A = -20 \log \left(e^{-\alpha L}\right) \to k_c L \cdot 20 \log e \approx \left(\frac{L}{2a}\right) 1.8412 \cdot 40 \log e \tag{7.97a}$$

$$\approx 32 \text{ dB}/\square \tag{7.97b}$$

or "32 dB per square." In other words, the long tube will provide $4 \cdot 32 = 128$ dB of attenuation, whereas the pinhole provides only 32 dB, despite being much smaller.

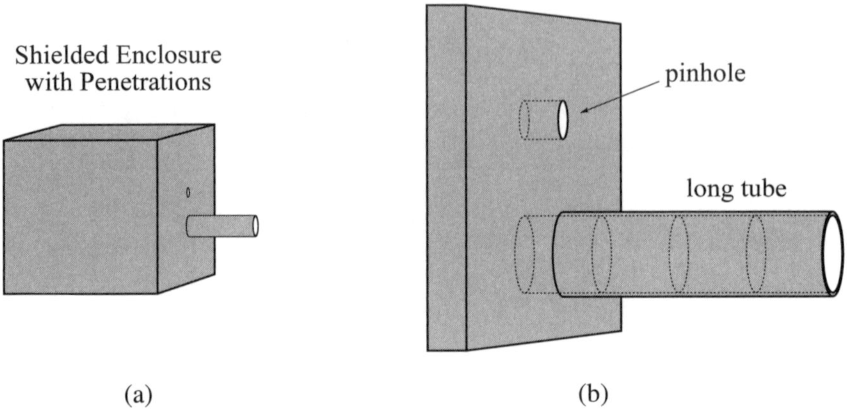

Figure 7.16 (a) Shielded enclosure with two kinds of penetrations: a pinhole, and a long tube. (b) Detail of wall penetrations. The pinhole has a 1:1 aspect ratio, whereas the tube has a 4:1 aspect ratio.

Note that this does not imply that a smaller diameter hole would not block emissions at even lower frequencies; only that for a given frequency, once a hole is small enough, there is no point in making it smaller unless the aspect ratio also changes. The same general conclusion applies to holes of any shape, but the above rule of thumb is worth remembering since circular holes are generally the easiest to manufacture.

7.3.4 Losses

For propagating modes, losses due to conductivity may be calculated as before using the perturbation method. First, let us determine the power propagating in the TE_{nm} mode,

$$P_0 = \iint (\mathbf{E}_t \times \mathbf{H}_t^*) \cdot d\mathbf{A} = \int_0^a \int_0^{2\pi} (E_r H_\theta^* - E_\theta H_r^*)\, r d\theta dr \tag{7.98a}$$

$$= E_0 H_0^* \int_0^a \int_0^{2\pi} \left(\frac{n^2}{k_c^2 r^2} J_n^2 (k_c r) \sin^2(n\theta) + J_n'^2 (k_c r) \cos^2(n\theta) \right) r d\theta dr \tag{7.98b}$$

$$= (1+\delta_n)\,\pi E_0 H_0^* \int_0^a \left(\frac{n^2}{k_c^2 r^2} J_n^2(k_c r) + J_n'^2(k_c r)\right) r dr \tag{7.98c}$$

$$= (1+\delta_n)\,\frac{\pi E_0 H_0^*}{2k_c^2}\left(e_{nm}^2 - n^2\right) J_n^2(e_{nm}) \tag{7.98d}$$

where for simplicity we have assumed $\phi = 0$ and the integral evaluated in the last step is given in Appendix G.

The power loss per unit length is determined by the surface currents and skin effect sheet resistance, where the currents in turn derive from the magnetic fields at the boundary ($r = a$),

$$P_l = R_s \oint |\mathbf{J}_s|^2\, dl = \sqrt{\frac{\omega\mu'}{2\sigma}} \oint |\mathbf{n}\times\mathbf{H}|^2\, dl \tag{7.99a}$$

$$= a\sqrt{\frac{\omega\mu'}{2\sigma}} \int_0^{2\pi} \left(|H_\theta|^2 + |H_z|^2\right) d\theta \tag{7.99b}$$

$$= a\,|H_0|^2 J_n^2(e_{nm}) \sqrt{\frac{\omega\mu'}{2\sigma}} \int_0^{2\pi} \left(\frac{n^2}{e_{nm}^2}\sin^2(n\theta) + \frac{k_c^2}{\beta^2}\cos^2(n\theta)\right) d\theta \tag{7.99c}$$

$$= (1+\delta_n)\,\pi a\,|H_0|^2 J_n^2(e_{nm}) \sqrt{\frac{\omega\mu'}{2\sigma}} \left(\frac{n^2}{e_{nm}^2} + \frac{k_c^2}{\beta^2}\right) \tag{7.99d}$$

Thus, the attenuation constant is

$$\alpha_c = \frac{P_l}{2P_0} = \frac{k_c^2 a H_0 \left(\frac{n^2}{e_{nm}^2} + \frac{k_c^2}{\beta^2}\right)}{E_0\left(e_{nm}^2 - n^2\right)}\sqrt{\frac{\omega\mu'}{2\sigma}} = \frac{k_c^2 a\beta \left(\frac{n^2}{e_{nm}^2} + \frac{k_c^2}{\beta^2}\right)}{\omega\mu\left(e_{nm}^2 - n^2\right)}\sqrt{\frac{\omega\mu'}{2\sigma}} \tag{7.100a}$$

$$= \frac{\beta^2 n^2 + e_{nm}^2 k_c^2}{ak\eta\beta\left(e_{nm}^2 - n^2\right)}\sqrt{\frac{\omega\mu'}{2\sigma}} = \frac{k}{a\eta\beta}\left(\frac{k_c^2}{k^2} + \frac{n^2}{e_{nm}^2 - n^2}\right)\sqrt{\frac{\omega\mu'}{2\sigma}} \tag{7.100b}$$

Be reminded that the terms under the square root sign describe the skin effect resistance, so μ' is the permeability of the conductive metal, not the dielectric material filling the waveguide.

For TM modes, we may calculate the power as

$$P_0 = \iint \left(\mathbf{E}_t \times \mathbf{H}_t^*\right)\cdot d\mathbf{A} = \int_0^a \int_0^{2\pi} \left(E_r H_\theta^* - E_\theta H_r^*\right) r d\theta dr \tag{7.101a}$$

$$= E_0 H_0^* \int_0^a \int_0^{2\pi} \left(J_n'^2 (k_c r) \cos^2(n\theta) + \frac{n^2}{k_c^2 r^2} J^2(k_c r) \sin^2(n\theta) \right) r d\theta dr \quad (7.101b)$$

$$= (1 + \delta_n) \pi E_0 H_0^* \int_0^a \left(J_n'^2 (k_c r) + \frac{n^2}{k_c^2 r^2} J^2(k_c r) \right) r dr \quad (7.101c)$$

$$= (1 + \delta_n) \frac{\pi E_0 H_0^*}{2k_c^2} z_{nm}^2 J_n'^2 (z_{nm}) \quad (7.101d)$$

and the power loss per unit length as

$$P_l = R_s \oint |\mathbf{J}_s|^2 dl = \sqrt{\frac{\omega\mu'}{2\sigma}} \oint |\mathbf{n} \times \mathbf{H}|^2 dl \quad (7.102a)$$

$$= a\sqrt{\frac{\omega\mu'}{2\sigma}} \int_0^{2\pi} |H_\theta|^2 d\theta = a |H_0|^2 J'^2(z_{nm}) \sqrt{\frac{\omega\mu'}{2\sigma}} \int_0^{2\pi} \cos^2(n\theta) d\theta \quad (7.102b)$$

$$= (1 + \delta_n) \pi a |H_0|^2 J'^2(z_{nm}) \sqrt{\frac{\omega\mu'}{2\sigma}} \quad (7.102c)$$

so that the attenuation constant is

$$\alpha_c = \frac{P_l}{2P_0} = \frac{k_c^2 a H_0}{E_0 z_{nm}^2} \sqrt{\frac{\omega\mu'}{2\sigma}} = \frac{\omega\varepsilon}{a\beta} \sqrt{\frac{\omega\mu'}{2\sigma}} = \frac{k}{a\eta\beta} \sqrt{\frac{\omega\mu'}{2\sigma}} \quad (7.103)$$

The losses for a number of modes in a copper waveguide where $a = 1$ in. are plotted in Figure 7.17. The TE_{0m} modes have the rather extraordinary property that their losses asymptotically approach zero at high frequency. The fields retreat from the walls into the center of the waveguide [2], propagating along the axis with almost no current in the walls to confine them. However, this fragile state of affairs is difficult to maintain, as any deviation from a straight line or imperfection in the circular geometry will tend to convert energy into other propagating modes. Nevertheless, there was a time when the TE_{01} mode in circular waveguide was thought to be the best hope for long-distance wired communication [3]. Waveguides were made using insulated wire helices — designed to support the circumferential currents of the desired mode while suppressing other modes having longitudinal currents — along with couplers and other components needed to launch and receive the signal in the desired mode. These schemes were ultimately abandoned in favor of fiber optics for data transmission, but are still of some interest in high power applications where heating due to ohmic losses is a concern [4].

The key results for circular waveguide are summarized in Table 7.4.

Table 7.4
Equations for Circular Waveguide

Parameter	Symbol	Expression
	TE Modes	
Cutoff wavenumber	k_c	$\frac{e_{nm}}{a}$
Electric field	$\begin{pmatrix} E_r \\ E_\theta \\ E_z \end{pmatrix}$	$E_0 e^{-\gamma z} \begin{pmatrix} \frac{n}{k_c r} J_n(k_c r) \sin(n(\theta+\phi)) \\ J_n'(k_c r) \cos(n(\theta+\phi)) \\ 0 \end{pmatrix}$
Magnetic field	$\begin{pmatrix} H_r \\ H_\theta \\ H_z \end{pmatrix}$	$H_0 e^{-\gamma z} \begin{pmatrix} -J_n'(k_c r) \cos(n(\theta+\phi)) \\ \frac{n}{k_c r} J_n(k_c r) \sin(n(\theta+\phi)) \\ \frac{k_c}{\gamma} J_n(k_c r) \cos(n(\theta+\phi)) \end{pmatrix}$
Wave impedance	Z_{TE}	$\frac{E_0}{H_0} = \frac{j\omega\mu}{\gamma} = \frac{jk\eta}{\gamma}$
Power	P	$(1+\delta_n) \frac{\pi E_0 H_0^*}{2k_c^2} \left(e_{nm}^2 - n^2\right) J_n^2(e_{nm})$
Attenuation constant (ohmic)	α_c	$\frac{k}{a\eta\beta} \left(\frac{k_c^2}{k^2} + \frac{n^2}{e_{nm}^2 - n^2}\right) \sqrt{\frac{\omega\mu'}{2\sigma}}$
	TM Modes	
Cutoff wavenumber	k_c	$\frac{z_{nm}}{a}$
Electric field	$\begin{pmatrix} E_r \\ E_\theta \\ E_z \end{pmatrix}$	$E_0 e^{-\gamma z} \begin{pmatrix} -J_n'(k_c r) \cos(n(\theta+\phi)) \\ \frac{n}{k_c r} J_n(k_c r) \sin(n(\theta+\phi)) \\ \frac{k_c}{\gamma} J_n(k_c r) \cos(n(\theta+\phi)) \end{pmatrix}$
Magnetic field	$\begin{pmatrix} H_r \\ H_\theta \\ H_z \end{pmatrix}$	$H_0 e^{-\gamma z} \begin{pmatrix} -\frac{n}{k_c r} J_n(k_c r) \sin(n(\theta+\phi)) \\ -J_n'(k_c r) \cos(n(\theta+\phi)) \\ 0 \end{pmatrix}$
Wave admittance	Y_{TM}	$\frac{H_0}{E_0} = \frac{j\omega\varepsilon}{\gamma} = \frac{jk}{\eta\gamma}$
Power	P	$(1+\delta_n) \frac{\pi E_0 H_0^*}{2k_c^2} z_{nm}^2 J_n'^2(z_{nm})$
Attenuation constant (ohmic)	α_c	$\frac{k}{a\eta\beta} \sqrt{\frac{\omega\mu'}{2\sigma}}$

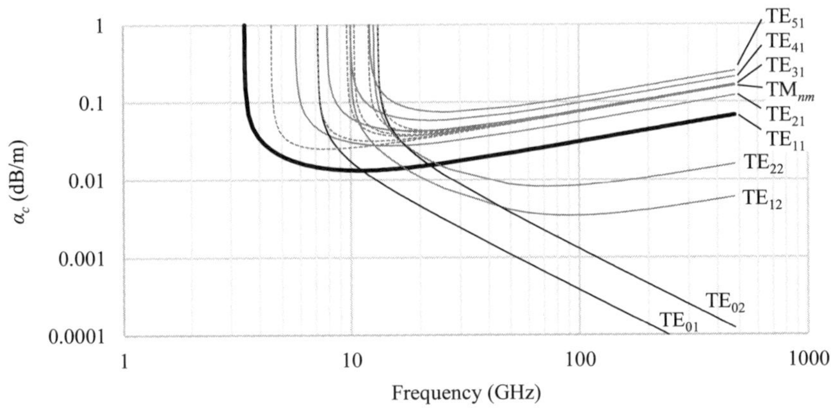

Figure 7.17 Attenuation constants for the first 15 modes of a circular waveguide with copper walls and radius $a = 1$ in.

7.4 WAVEGUIDE CAVITY RESONATORS

We have talked in previous chapters about resonators formed using both lumped elements and transmission lines. It should be no surprise, then, that we can make resonators out of closed, metal waveguide sections, or *cavities*, as well; closed metal cavities have the highest quality factors of any type of resonator in common use.

7.4.1 Rectangular Cavity Resonators

The simplest way to make a cavity from waveguide is to terminate the ends with flat, metal walls, much like a short-circuit termination on a section of transmission line.[4] Also like transmission lines, the resonant modes of such a waveguide cavity take the form of standing waves in each of the previously derived waveguide modes.

Consider a rectangular waveguide with cross-sectional dimensions a and b, and length L, as shown in Figure 7.18. The mode solutions already given in Table 7.2 still apply, leaving us simply to constrain the forward-traveling and

4 Recall that transmission-line resonators were formed by terminating the ends of a half- or quarter-wavelength section with either open-circuits or short-circuits. Since waveguide cross-sections are not electrically small, open-ended waveguides radiate their energy into free space surprisingly well, giving open-ended waveguide resonators unacceptably low Q.

Figure 7.18 (a) Rectangular and (b) cylindrical waveguide cavity resonators.

backward-traveling wave amplitudes to satisfy the boundary conditions at the end-points. Before we do that, let us simplify our notation with the following substitutions,

$$k_x = \frac{m\pi}{a} \tag{7.104a}$$

$$k_y = \frac{n\pi}{b} \tag{7.104b}$$

Then, we may write the standing wave solution for TE waves as follows,

$$\mathbf{E} = \begin{pmatrix} E_x \\ E_y \\ E_z \end{pmatrix} = \frac{E_0^+}{k_c} e^{-j\beta z} \begin{pmatrix} k_y \cos(k_x x) \sin(k_y y) \\ -k_x \sin(k_x x) \cos(k_y y) \\ 0 \end{pmatrix} + \frac{E_0^-}{k_c} e^{j\beta z} \begin{pmatrix} k_y \cos(k_x x) \sin(k_y y) \\ -k_x \sin(k_x x) \cos(k_y y) \\ 0 \end{pmatrix} \tag{7.105a}$$

$$\mathbf{H} = \begin{pmatrix} H_x \\ H_y \\ H_z \end{pmatrix} = \frac{H_0^+}{k_c \beta} e^{-j\beta z} \begin{pmatrix} \beta k_x \sin(k_x x) \cos(k_y y) \\ \beta k_y \cos(k_x x) \sin(k_y y) \\ -j k_c^2 \cos(k_x x) \cos(k_y y) \end{pmatrix} + \frac{H_0^-}{k_c \beta} e^{j\beta z} \begin{pmatrix} -\beta k_x \sin(k_x x) \cos(k_y y) \\ -\beta k_y \cos(k_x x) \sin(k_y y) \\ -j k_c^2 \cos(k_x x) \cos(k_y y) \end{pmatrix} \tag{7.105b}$$

Note that we have inverted the transverse components of $\mathbf{H}$ for the backward-traveling wave, in accord with the conclusions of Section 7.1.3. To satisfy the electric-wall boundary condition at $z = 0$, we let $E_0^- = -E_0^+ = E_0/2$. Therefore,

$$\mathbf{E} = \frac{jE_0}{k_c} \begin{pmatrix} k_y \cos(k_x x) \sin(k_y y) \sin(\beta z) \\ -k_x \sin(k_x x) \cos(k_y y) \sin(\beta z) \\ 0 \end{pmatrix} \tag{7.106}$$

This also ensures that $H_0^- = -H_0^+ = H_0/2$, where $H_0 = Y_{TE}E_0$; therefore,

$$\mathbf{H} = \frac{H_0}{k_c\beta}\begin{pmatrix} -\beta k_x \sin(k_x x)\cos(k_y y)\cos(\beta z) \\ -\beta k_y \cos(k_x x)\sin(k_y y)\cos(\beta z) \\ k_c^2 \cos(k_x x)\cos(k_y y)\sin(\beta z) \end{pmatrix} \tag{7.107}$$

To satisfy the transverse-electric boundary condition at $z = L$, we must have

$$\sin(\beta L) = 0 \tag{7.108a}$$

$$\therefore \beta L = L\sqrt{k^2 - k_c^2} = l\pi \tag{7.108b}$$

where l is a non-negative integer. This may then be called the TE_{mnl} mode of the resonator. The resonant frequency is determined by solving (7.108b),

$$\omega_0 = \frac{k}{\sqrt{\mu\varepsilon}} = \frac{1}{\sqrt{\mu\varepsilon}}\sqrt{k_c^2 + \left(\frac{l\pi}{L}\right)^2} \tag{7.109a}$$

$$= \frac{1}{\sqrt{\mu\varepsilon}}\sqrt{\left(\frac{m\pi}{a}\right)^2 + \left(\frac{n\pi}{b}\right)^2 + \left(\frac{l\pi}{L}\right)^2} \tag{7.109b}$$

or, by substituting $k_z = \beta = l\pi/L$, we find the resonant wavenumber

$$k_0 = \sqrt{k_x^2 + k_y^2 + k_z^2} \tag{7.110}$$

The complete field solutions for TE_{mnl} modes are then given by

$$\mathbf{E} = \frac{jE_0}{k_c}\begin{pmatrix} k_y \cos(k_x x)\sin(k_y y)\sin(k_z z) \\ -k_x \sin(k_x x)\cos(k_y y)\sin(k_z z) \\ 0 \end{pmatrix} \tag{7.111a}$$

$$\mathbf{H} = \frac{H_0}{k_c k_z}\begin{pmatrix} -k_x k_z \sin(k_x x)\cos(k_y y)\cos(k_z z) \\ -k_y k_z \cos(k_x x)\sin(k_y y)\cos(k_z z) \\ k_c^2 \cos(k_x x)\cos(k_y y)\sin(k_z z) \end{pmatrix} \tag{7.111b}$$

Not surprisingly, the derivation of TM_{mnl} modes is quite similar and need not be repeated in as much detail. The resonant frequency is still given by (7.109b), and the field solutions are given by superposition,

$$\mathbf{E} = \begin{pmatrix} E_x \\ E_y \\ E_z \end{pmatrix} = \frac{E_0^+}{k_c k_z} e^{-jk_z z} \begin{pmatrix} -k_x k_z \cos(k_x x) \sin(k_y y) \\ -k_y k_z \sin(k_x x) \cos(k_y y) \\ -jk_c^2 \sin(k_x x) \sin(k_y y) \end{pmatrix} + \frac{E_0^-}{k_c k_z} e^{jk_z z} \begin{pmatrix} -k_x k_z \cos(k_x x) \sin(k_y y) \\ -k_y k_z \sin(k_x x) \cos(k_y y) \\ jk_c^2 \sin(k_x x) \sin(k_y y) \end{pmatrix} \quad (7.112a)$$

$$= \frac{jE_0}{k_c k_z} \begin{pmatrix} -k_x k_z \cos(k_x x) \sin(k_y y) \sin(k_z z) \\ -k_y k_z \sin(k_x x) \cos(k_y y) \sin(k_z z) \\ k_c^2 \sin(k_x x) \sin(k_y y) \cos(k_z z) \end{pmatrix} \quad (7.112b)$$

$$\mathbf{H} = \begin{pmatrix} H_x \\ H_y \\ H_z \end{pmatrix} = \frac{H_0^+}{k_c} e^{-jk_z z} \begin{pmatrix} k_y \sin(k_x x) \cos(k_y y) \\ -k_x \cos(k_x x) \sin(k_y y) \\ 0 \end{pmatrix} + \frac{H_0^-}{k_c} e^{jk_z z} \begin{pmatrix} -k_y \sin(k_x x) \cos(k_y y) \\ k_x \cos(k_x x) \sin(k_y y) \\ 0 \end{pmatrix} \quad (7.112c)$$

$$= \frac{H_0}{k_c} \begin{pmatrix} -k_y \sin(k_x x) \cos(k_y y) \cos(k_z z) \\ k_x \cos(k_x x) \sin(k_y y) \cos(k_z z) \\ 0 \end{pmatrix} \quad (7.112d)$$

The electric field for TE modes vanishes if $l = 0$ or if both $m = 0$ and $n = 0$, thus the dominant TE mode is TE_{101} (assuming $a > b$). For TM modes, however, we can have $l = 0$, and the dominant TM mode is TM_{110}. The fields for such a mode have no z-dependence, and thus it has the peculiar property that the resonant frequency is independent of L, and L may be arbitrarily small.

The field solutions for rectangular cavity resonators are summarized in Table 7.5.

7.4.2 Cylindrical Cavity Resonators

The modes of cylindrical resonators can be derived in much the same way, as standing waves in a truncated (shorted) length of circular waveguide using the

Table 7.5
Field Solutions for Rectangular Cavity Resonators

Quantity	Symbol	Expression
Resonant wavenumber	k_0	$\sqrt{\left(\frac{m\pi}{a}\right)^2+\left(\frac{n\pi}{b}\right)^2+\left(\frac{l\pi}{L}\right)^2} = \sqrt{k_x^2+k_y^2+k_z^2} = \sqrt{k_c^2+k_z^2}$
Electric field, TE modes	$\begin{pmatrix} E_x \\ E_y \\ E_z \end{pmatrix}$	$\frac{jE_0}{k_c}\begin{pmatrix} k_y\cos(k_x x)\sin(k_y y)\sin(k_z z) \\ -k_x\sin(k_x x)\cos(k_y y)\sin(k_z z) \\ 0 \end{pmatrix}$
Magnetic field, TE modes	$\begin{pmatrix} H_x \\ H_y \\ H_z \end{pmatrix}$	$\frac{H_0}{k_c k_z}\begin{pmatrix} -k_x k_z\sin(k_x x)\cos(k_y y)\cos(k_z z) \\ -k_y k_z\cos(k_x x)\sin(k_y y)\cos(k_z z) \\ k_c^2\cos(k_x x)\cos(k_y y)\sin(k_z z) \end{pmatrix}$
Electric field, TM modes	$\begin{pmatrix} E_x \\ E_y \\ E_z \end{pmatrix}$	$\frac{jE_0}{k_c k_z}\begin{pmatrix} k_x k_z\cos(k_x x)\sin(k_y y)\sin(k_z z) \\ -k_y k_z\sin(k_x x)\cos(k_y y)\sin(k_z z) \\ k_c^2\sin(k_x x)\sin(k_y y)\cos(k_z z) \end{pmatrix}$
Magnetic field, TM modes	$\begin{pmatrix} H_x \\ H_y \\ H_z \end{pmatrix}$	$\frac{H_0}{k_c}\begin{pmatrix} -k_y\sin(k_x x)\cos(k_y y)\cos(k_z z) \\ k_x\cos(k_x x)\sin(k_y y)\cos(k_z z) \\ 0 \end{pmatrix}$

propagating mode solutions from Table 7.4. For TE modes,

$$\mathbf{E} = \begin{pmatrix} E_r \\ E_\theta \\ E_z \end{pmatrix} = \frac{E_0^+}{k_c r}e^{-j\beta z}\begin{pmatrix} nJ_n(k_c r)\sin(n(\theta+\phi)) \\ k_c r J_n'(k_c r)\cos(n(\theta+\phi)) \\ 0 \end{pmatrix} + \frac{E_0^-}{k_c r}e^{j\beta z}\begin{pmatrix} nJ_n(k_c r)\sin(n(\theta+\phi)) \\ k_c r J_n'(k_c r)\cos(n(\theta+\phi)) \\ 0 \end{pmatrix} \quad (7.113a)$$

$$\mathbf{H} = \frac{H_0^+}{j\beta k_c r} e^{-j\beta z} \begin{pmatrix} -j\beta k_c r J_n' (k_c r) \cos (n(\theta + \phi)) \\ j\beta n J_n (k_c r) \sin (n(\theta + \phi)) \\ k_c^2 r J_n (k_c r) \cos (n(\theta + \phi)) \end{pmatrix} + \frac{H_0^-}{j\beta k_c r} e^{j\beta z} \begin{pmatrix} j\beta k_c r J_n' (k_c r) \cos (n(\theta + \phi)) \\ -j\beta n J_n (k_c r) \sin (n(\theta + \phi)) \\ k_c^2 r J_n (k_c r) \cos (n(\theta + \phi)) \end{pmatrix} \tag{7.113b}$$

The boundary conditions on the transverse fields are the same in this case as they were for the rectangular cavity, so once again, we let

$$E_0^- = -E_0^+ = E_0/2 \tag{7.114a}$$

$$H_0^- = -H_0^+ = H_0/2 \tag{7.114b}$$

where $H_0 = Y_{TE} E_0$, and

$$\beta = k_z = l\pi/L \tag{7.115}$$

We thus have for the resonant wavenumber,

$$k_0 = \sqrt{k_c^2 + k_z^2} = \sqrt{\left(\frac{e_{nm}}{a}\right)^2 + \left(\frac{l\pi}{L}\right)^2} \tag{7.116}$$

and for the final field solutions,

$$\mathbf{E} = \frac{jE_0}{k_c r} \begin{pmatrix} n J_n (k_c r) \sin (n(\theta + \phi)) \sin (k_z z) \\ k_c r J_n' (k_c r) \cos (n(\theta + \phi)) \sin (k_z z) \\ 0 \end{pmatrix} \tag{7.117a}$$

$$\mathbf{H} = \frac{H_0}{k_c k_z r} \begin{pmatrix} k_c k_z r J_n' (k_c r) \cos (n(\theta + \phi)) \cos (k_z z) \\ -k_z n J_n (k_c r) \sin (n(\theta + \phi)) \cos (k_z z) \\ k_c^2 r J_n (k_c r) \cos (n(\theta + \phi)) \sin (k_z z) \end{pmatrix} \tag{7.117b}$$

For TM modes, we have

$$k_0 = \sqrt{k_c^2 + k_z^2} = \sqrt{\left(\frac{z_{nm}}{a}\right)^2 + \left(\frac{l\pi}{L}\right)^2} \tag{7.118a}$$

$$\mathbf{E} = \frac{E_0}{jk_c k_z r} \begin{pmatrix} k_c k_z r J_n' (k_c r) \cos (n(\theta + \phi)) \sin (k_z z) \\ -k_z n J_n (k_c r) \sin (n(\theta + \phi)) \sin (k_z z) \\ k_c^2 r J_n (k_c r) \cos (n(\theta + \phi)) \cos (k_z z) \end{pmatrix} \tag{7.118b}$$

Table 7.6
Cylindrical Cavity Resonators

Quantity	Symbol	Expression
Resonant wavenumber	k_0	$\sqrt{k_c^2+k_z^2}$ $=\sqrt{\left(\frac{e_{nm}}{a}\right)^2+\left(\frac{l\pi}{L}\right)^2}$ (TE modes) $=\sqrt{\left(\frac{z_{nm}}{a}\right)^2+\left(\frac{l\pi}{L}\right)^2}$ (TM modes)
Electric field, TE modes	$\begin{pmatrix}E_r\\E_\theta\\E_z\end{pmatrix}$	$\frac{jE_0}{k_cr}\begin{pmatrix}nJ_n(k_cr)\sin(n(\theta+\phi))\sin(k_zz)\\k_crJ_n'(k_cr)\cos(n(\theta+\phi))\sin(k_zz)\\0\end{pmatrix}$
Magnetic field, TE modes	$\begin{pmatrix}H_r\\H_\theta\\H_z\end{pmatrix}$	$\frac{H_0}{k_ck_zr}\begin{pmatrix}k_ck_zrJ_n'(k_cr)\cos(n(\theta+\phi))\cos(k_zz)\\-nk_zJ_n(k_cr)\sin(n(\theta+\phi))\cos(k_zz)\\k_c^2rJ_n(k_cr)\cos(n(\theta+\phi))\sin(k_zz)\end{pmatrix}$
Electric field, TM modes	$\begin{pmatrix}E_r\\E_\theta\\E_z\end{pmatrix}$	$\frac{E_0}{jk_ck_zr}\begin{pmatrix}k_ck_zrJ_n'(k_cr)\cos(n(\theta+\phi))\sin(k_zz)\\-nk_zJ_n(k_cr)\sin(n(\theta+\phi))\sin(k_zz)\\k_c^2rJ_n(k_cr)\cos(n(\theta+\phi))\cos(k_zz)\end{pmatrix}$
Magnetic field, TM modes	$\begin{pmatrix}H_r\\H_\theta\\H_z\end{pmatrix}$	$\frac{H_0}{k_cr}\begin{pmatrix}-nJ_n(k_cr)\sin(n(\theta+\phi))\cos(k_zz)\\-k_crJ_n'(k_cr)\cos(n(\theta+\phi))\cos(k_zz)\\0\end{pmatrix}$

$$\mathbf{H}=\frac{H_0}{k_cr}\begin{pmatrix}-nJ_n(k_cr)\sin(n(\theta+\phi))\cos(k_zz)\\-k_crJ_n'(k_cr)\cos(n(\theta+\phi))\cos(k_zz)\\0\end{pmatrix} \tag{7.118c}$$

These field solutions are summarized in Table 7.6.

7.4.3 Stored Energy in a Cavity

The averaged stored electric and magnetic energy in a cavity at resonance is given by (2.45) and (2.54) in phasor form,

$$U_E=\frac{1}{2}\iiint\mathbf{E}\cdot\mathbf{D}^*dV \tag{7.119a}$$

$$U_H = \frac{1}{2} \iiint \mathbf{B} \cdot \mathbf{H}^* dV \tag{7.119b}$$

where the field quantities represent rms phasor amplitudes. Substituting (7.111) for the TE_{mnl} mode of a rectangular waveguide, we find the stored electric energy,

$$U_E = \frac{\varepsilon}{2} \frac{|E_0|^2}{k_c^2} \frac{abL}{8} \cdot \left[k_y^2 (1 + \delta_m)(1 - \delta_n) + k_x^2 (1 - \delta_m)(1 + \delta_n)\right] (1 - \delta_l) \tag{7.120}$$

where we have used the trigonometric integral formulas in Section A.4.1. Recall that $l > 0$ for all TE cavity modes, so we may conclude that $\delta_l = 0$. Also, the expression in square brackets may be simplified by noting that m and n cannot both be zero, and

$$k_y^2 (1 + \delta_m)(1 - \delta_n) + k_x^2 (1 - \delta_m)(1 + \delta_n) = \begin{cases} 2k_c^2 & m = 0, n \neq 0 \\ 2k_c^2 & m \neq 0, n = 0 \\ k_c^2 & m \neq 0, n \neq 0 \end{cases} \tag{7.121a}$$

$$= k_c^2 (1 + \delta_m)(1 + \delta_n) \tag{7.121b}$$

Thus, we may write

$$U_E = \frac{abL\varepsilon}{16} |E_0|^2 (1 + \delta_m)(1 + \delta_n) \tag{7.122}$$

A similar calculation shows that the stored magnetic energy is

$$U_H = \frac{\mu}{2} \frac{|H_0|^2}{k_c^2 k_z^2} \frac{abL}{8} \left(k_z^2 k_c^2 + k_c^4\right)(1 + \delta_m)(1 + \delta_n) \tag{7.123a}$$

$$= \frac{abL\mu}{16} \frac{|H_0|^2}{k_z^2} k^2 (1 + \delta_m)(1 + \delta_n) \tag{7.123b}$$

$$= \frac{abL\varepsilon}{16} |E_0|^2 (1 + \delta_m)(1 + \delta_n) = U_E \tag{7.123c}$$

As with all other resonators, then, we see that the average stored electric and magnetic energies in a cavity at resonance are equal.

7.4.4 Quality Factors of Cavity Resonators

The quality factor due to dielectric losses in a cavity resonator take on a surprisingly simple form. Recall that AC dielectric loss is indistinguishable from conductivity in Maxwell's equations, with a total effective conductivity that is proportional to the loss tangent,

$$\sigma_e = \sigma + \omega\varepsilon'' = \omega\varepsilon \tan\delta \tag{7.124}$$

We may thus calculate the loss in the interior using Ohm's law applied to this effective conductivity,

$$P_l = \frac{1}{\sigma_e} \iiint |\mathbf{J}|^2 \, dV = \frac{1}{\sigma_e} \iiint |\sigma_e \mathbf{E}|^2 \, dV = \sigma_e \iiint \mathbf{E} \cdot \mathbf{E}^* dV \tag{7.125a}$$

$$= \omega\varepsilon \tan\delta \iiint \mathbf{E} \cdot \mathbf{E}^* dV = \omega \tan\delta \iiint \mathbf{E} \cdot \mathbf{D}^* dV = 2\omega U_E \tan\delta \tag{7.125b}$$

The quality factor due to dielectric loss is thus given simply by

$$Q_d = \omega \frac{U_E + U_H}{P_l} = \omega \frac{2U_E}{P_l} = \frac{1}{\tan\delta} \tag{7.126}$$

independent of the cavity's geometry.

Nevertheless, in most cavity resonators, the loss due to conductivity in the walls is dominant. These losses may be found by the integral of the currents times the surface resistance,

$$P_l = R_s \oint\!\!\oint |\mathbf{J}_s|^2 \, dS = \sqrt{\frac{\omega\mu'}{2\sigma}} \oint\!\!\oint |\mathbf{H}|^2 \, dS \tag{7.127}$$

The resulting, geometry-dependent expressions are not as simple as that for the dielectric quality factor and will not be derived here, but given a calculation of the two separately, the conductive and dielectric quality factors combine in the same manner as internal and external Q,

$$\frac{1}{Q} = \frac{1}{Q_c} + \frac{1}{Q_d} \tag{7.128}$$

From the expressions above it is clear that stored energy is proportional to the field intensity and the volume over which they extend, whereas conductive losses (which are usually dominant) are proportional to field intensity and the surface area of the conductors. A general rule of thumb tells us that the highest Q is achieved by the resonator that maximizes volume per unit conductive surface area — more metal, less loss, higher Q. Cavity resonators are necessarily much bigger than transmission-line and lumped-element resonators of the same frequency, and have the largest Qs of any resonator in common use.

7.4.5 Other Cavity and Resonator Types

Of course, any closed metal cavity may act as a resonator, whether or not it conforms to a recognizable waveguide geometry along any axis. A spherical cavity, for example, would certainly have resonant modes and frequencies that may be calculated using the Helmholtz equation in spherical coordinates, and one would expect it to have quality factors similar to those of rectangular and cylindrical cavities of similar size. One might also reason that the three-axis symmetry of such a cavity would lead to multiple degenerate, resonant modes.

Similarly, dielectric resonators exist comprising slabs or "pucks" of high-permittivity dielectric materials. These are commonly mounted on planar circuits to which they couple as a means of manufacturing filters. The resonant frequencies and modes of such structures will not be calculated here, but the general principles and features of resonators described in this section still apply.

7.5 RIDGED WAVEGUIDE

Because the dominant modes of closed metal waveguides have finite cutoff frequencies, they have limited relative bandwidth over which they can achieve single-mode operation. The most that can be realized with rectangular waveguides is 1 octave, as described in Section 7.2.5. Circular waveguides have no single-mode regime at all, operating instead with degenerate dominant modes over a 26% fractional bandwidth (corresponding to the dual TE_{11} modes with orthogonal polarization angles). If we wish to design waveguide circuits with significantly broader bandwidth, then we must turn to ridged waveguide structures.

7.5.1 Double-Ridged Waveguide

One example known as double-ridged waveguide is given in Figure 7.19. The dominant effect of the ridges in an otherwise rectangular waveguide is to lower the cutoff frequency of the dominant TE_{10} mode without lowering that of the higher-order modes. The single-mode bandwidth is therefore broadened.

Figure 7.19 (a) Geometry of double-ridged waveguide. (b) Critical dimensions and dominant-mode field patterns (E-fields with solid lines, H-fields with dashed lines).

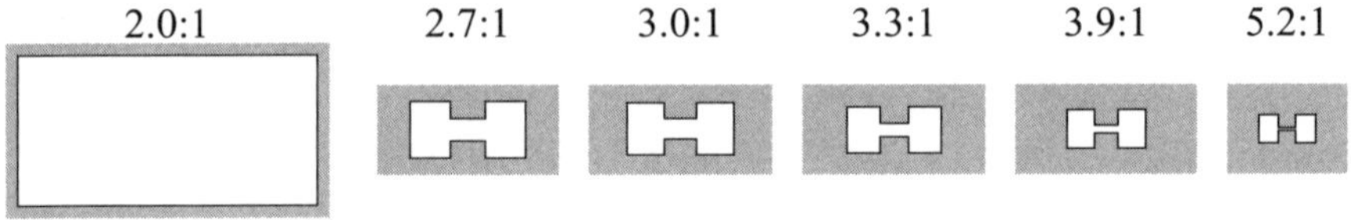

Figure 7.20 Several waveguide cross-sections having the same dominant cutoff frequency but different single-mode bandwidths, labeled above.

Alternatively, the overall width and height of a ridged waveguide is comparatively much smaller than that of a simple rectangular waveguide having the same dominant-mode cutoff frequency. A gallery of rectangular and double-ridged waveguides all having the same cutoff frequency but widely ranging single-mode bandwidths is shown in Figure 7.20. Note that each increase in bandwidth is associated with tighter and tighter gaps between the ridges. One might think that the relatively short current path from the end of one ridge around the perimeter to the opposite ridge across the gap would tend to short out any field potential that exists between them (where the dominant mode's fields are strongest); however, the gap creates a large distributed capacitance there that tends to resonate out the distributed inductance of the current path around the perimeter, allowing the potential difference to be maintained.

Ridged waveguides do not lend themselves to closed-form solutions like rectangular or circular waveguide, so instead we must rely on computer-aided numerical solutions. A plot of the cutoff frequencies for the first 10 modes in double-ridged waveguide as a function of gap size is shown in Figure 7.21. Note that as the gap gets smaller, to the left side of the plot, all of the higher-order modes' cutoff frequencies diverge exponentially to infinity, increasing the operating

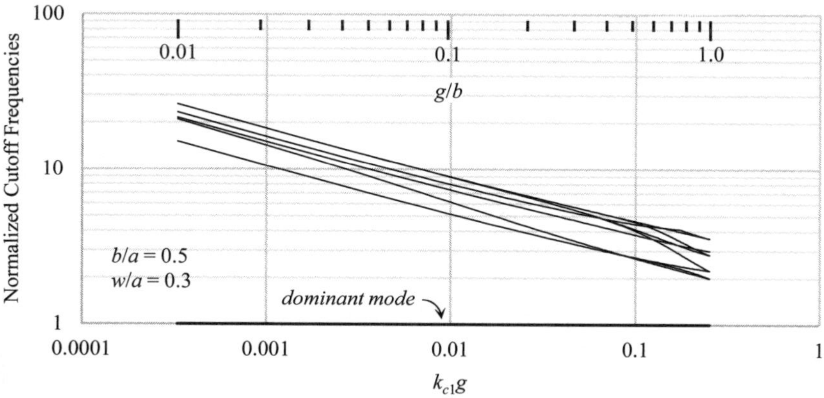

Figure 7.21 Normalized cutoff frequency versus gap size for the first 10 modes in a double-ridged waveguide. The overall aspect ratio is fixed at $b/a = 0.5$, and the relative ridge width $w/a = 0.3$. The gap-to-height ratio, g/b, is given by the scale at the top of the figure. When $g/b = 1$, the waveguide reverts to a simple rectangular geometry.

bandwidth of the waveguide. At the right side of the plot, where $g = b$, the waveguide is just an open rectangle with 2:1 aspect ratio and therefore has 1 octave of single-mode bandwidth.

Figure 7.21 can be used to design ridged waveguides optimized for a particular bandwidth with the largest possible gap (since gap size is typically the limiting factor in fabrication). For example, say that a particular application requires operation from 1 GHz to 4 GHz. To allow for some margin, let us design a waveguide having only one propagating mode from 0.9 GHz to 4.5 GHz, or 5:1 bandwidth. According to Figure 7.21, this bandwidth is achieved when $g/b \approx 0.1$, and $k_{c1}g \approx 0.01$. We may thus determine the waveguide dimensions as follows,

$$g = \frac{k_{c1}g}{k_{c1}} = \frac{0.01}{(2\pi \cdot 0.9 \text{ GHz})/(299792458 \text{ m/s})} = 21 \text{ mils} \tag{7.129a}$$

$$b = \frac{g}{g/b} = \frac{21 \text{ mils}}{0.1} = 210 \text{ mils} \tag{7.129b}$$

$$a = \frac{b}{b/a} = \frac{210 \text{ mils}}{0.5} = 420 \text{ mils} \tag{7.129c}$$

$$w = a \cdot w/a = (420\text{ mils}) \cdot 0.3 = 126\text{ mils} \quad (7.129d)$$

where a *mil* is one-thousandth of an inch.

Ridged waveguides represent a compromise between more open waveguides, like rectangular, which have lower loss and more power-handling capacity, and transmission lines like coax or microstrip which are much smaller. In the limit as the gap size approaches zero, the dominant mode becomes very TEM-like, making the structure behave in its single-mode regime very much like a transmission line, and transmission-line network design techniques can be applied successfully.

7.5.2 Quad-Ridged and Triple-Ridged Waveguides

If dual-polarization operation is desired, but with more than the meager 26% relative bandwidth afforded by circular waveguide, one may use waveguides having more ridges. A common choice is some form of quad-ridge waveguide, such as that shown in Figure 7.22(a). The two dominant modes are shown in the upper row, labeled mode A and B. Mode A corresponds to what we might call *vertical polarization* with opposite charges accumulating in the north and south ridges, while Mode B corresponds to *horizontal polarization* in which the opposing charges accumulate in the east and west ridges. One could also draw diagonal polarizations, such as those shown in the second row, but while these are also perfectly valid mode solutions, we do not count them as additional independent modes since they can be formed by superposition of the previous two.

There is a catch with quad-ridge waveguide, namely, that there is a third mode whose cutoff is never very far away from the dominant modes. This is Mode C, having a quadrupolar field pattern and shown in the last row of Figure 7.22(a). A plot of the normalized cutoff frequencies in Figure 7.23(a) illustrates the problem. While the rest of the higher-order modes' normalized cutoff frequencies diverge to infinity as the gaps get smaller, much like the double-ridged waveguide of Section 7.5.1, this third mode becomes asymptotically degenerate with the dominant modes. Electromagnetic components which make use of quad-ridge waveguide, such as some *orthomode transducers* [5], must find some way of suppressing this mode, usually by absorption.

A lesser-known alternative is triple-ridged waveguide, as shown in Figure 7.22(b). Although at first glance it may appear to have three modes, each rotated by 120°, any one can be written as the superposition of the other two, thus there are only two dominant, independent modes for this structure [6]. The plot of normalized cutoffs in Figure 7.23(b) further highlights that only the two dominant modes remain for very small gaps.

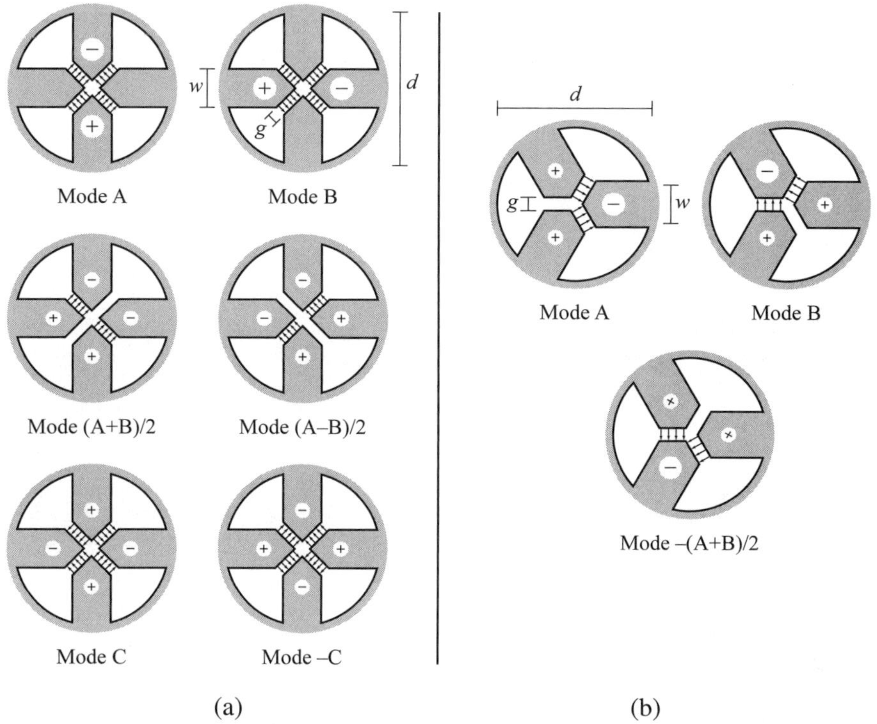

Figure 7.22 Modes in (a) quad-ridge and (b) triple-ridge waveguides.

7.6 COAXIAL WAVEGUIDE

Coaxial cable was discussed in Section 3.2 in the context of transmission lines, where the cross-section radii, a and b, were assumed to be very small compared to a wavelength. The field patterns that we developed in that limit may be considered more generally as the dominant mode of a coaxial waveguide, in this case a TEM mode, with cutoff frequency at zero.[5] It is worth briefly thinking about the higher-order modes in a coaxial waveguide where the diameter is not electrically small, not because we would ever use it in such a regime, but so that we can put a rigorous

5 Transmission lines of all kinds may be thought of as a "skinny" type of waveguide, whereas waveguides in general cannot all be considered transmission lines (hollow cylindrical waveguides, such as rectangular or circular, certainly are not).

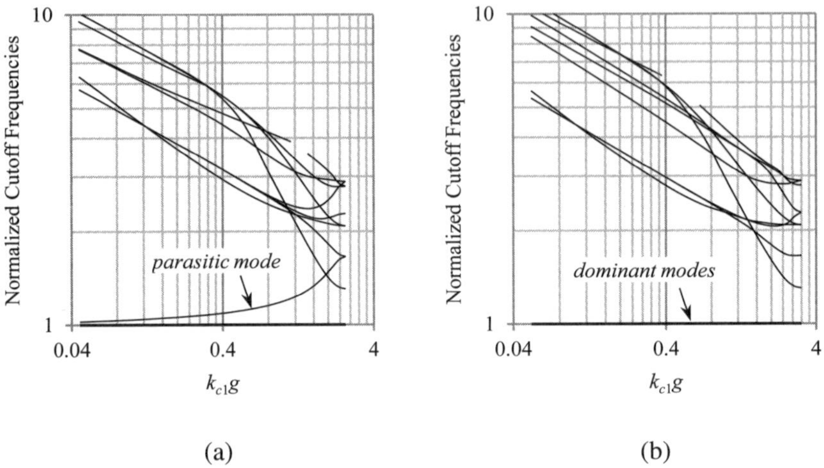

Figure 7.23 Plots of the normalized cutoff frequencies versus gap size for (a) quad-ridged and (b) triple-ridged waveguide. The terminus of the curves at the far right of the plots correspond to circular waveguide.

upper limit on the frequency range over which a coax line of given size will behave as a single-moded transmission line.

The first higher-order mode in a coaxial line after the dominant TEM mode is a TE mode. As with circular waveguide, we must solve the Helmholtz equation for the H_z field in cylindrical coordinates. The derivation is the same until boundary conditions are applied; thus,

$$H_z = \frac{k_c}{\gamma}\left(H_1 J_n\left(k_c r\right) + H_2 Y_n\left(k_c r\right)\right)\cos\left(n(\theta + \phi)\right) \tag{7.130}$$

This time, we cannot exclude the Bessel function of the second kind since the fields do not extend to the origin where it becomes infinite. Instead, we must constrain the sum of both terms to meet the boundary condition ($\partial H_z / \partial r = 0$) at $r = a$ and $r = b$,

$$H_1 J_n'\left(k_c a\right) + H_2 Y_n'\left(k_c a\right) = 0 \tag{7.131a}$$

$$H_1 J_n'\left(k_c b\right) + H_2 Y_n'\left(k_c b\right) = 0 \tag{7.131b}$$

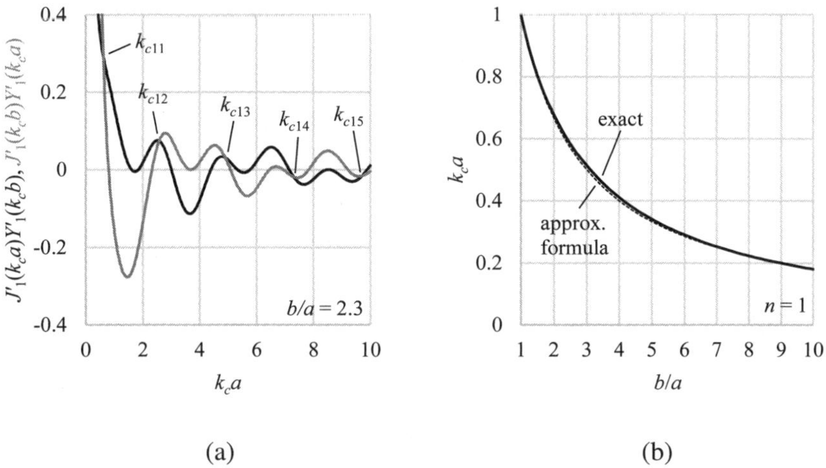

Figure 7.24 Solution for cutoff wavenumbers of TE modes in coaxial waveguide. (a) Left and right sides of (7.133b), showing TE_{1m} wavenumbers for $b/a = 2/3$. (b) TE_{11} cutoff wavenumber as a function of b/a.

or, in matrix form,

$$\begin{pmatrix} J'_n(k_ca) & Y'_n(k_ca) \\ J'_n(k_cb) & Y'_n(k_cb) \end{pmatrix} \begin{pmatrix} H_1 \\ H_2 \end{pmatrix} = \begin{pmatrix} 0 \\ 0 \end{pmatrix} \tag{7.132}$$

This equation is trivially satisfied if $H_1 = H_2 = 0$, but nontrivial solutions require that the determinant of the square matrix is zero,

$$\det \begin{pmatrix} J'_n(k_ca) & Y'_n(k_ca) \\ J'_n(k_cb) & Y'_n(k_cb) \end{pmatrix} = 0 \tag{7.133a}$$

$$\therefore J'_n(k_ca)\, Y'_n(k_cb) = J'_n(k_cb)\, Y'_n(k_ca) \tag{7.133b}$$

The solutions to this equation will determine the cutoff wavenumbers, $k_{c,nm}$, for each TE_{nm} mode. Closed-form solutions are not generally available, but can be approximated using numerical methods. A plot showing the left and right sides of (7.133b) for $n = 1$ and $b/a = 2.3$ is given in Figure 7.24(a). The abscissa of the first intersection of these curves corresponds to the cutoff wavenumber of the TE_{11} mode, which is the first higher-order mode after the dominant TEM mode. The

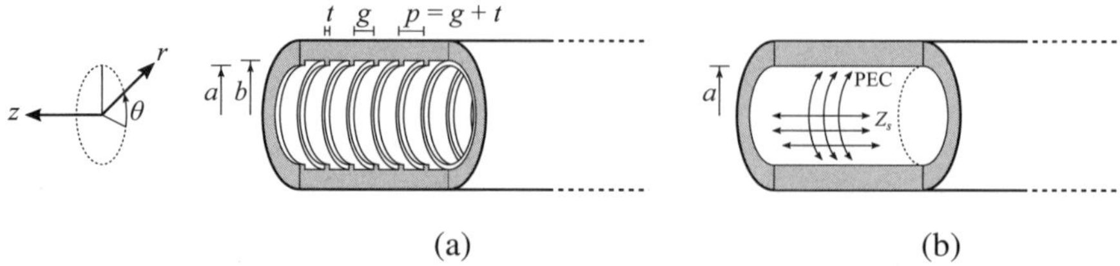

Figure 7.25 (a) Geometry of a corrugated circular waveguide. The sections having smaller internal radius a will be called *teeth* and have thickness t, whereas the sections having larger internal radius b will be called *grooves* and have thickness g. (b) Approximate model using PEC boundary conditions in the azimuthal direction and a reactive surface impedance, Z_s, in the longitudinal direction.

solution for this cutoff as a function of b/a is given in Figure 7.24(b), along with a very good but simple approximation [7] given by

$$k_c = \frac{2}{a+b} \tag{7.134}$$

In other words, coax is single-moded up to the point where the circumference around the mid-point between the two conductors is approximately a full wavelength inside the dielectric material. In terms of an upper frequency-limit, then

$$f_{max} = \frac{v_p}{\lambda} = \frac{v_p k_c}{2\pi} = \frac{v_p}{\pi(a+b)} = \frac{1}{\pi(a+b)\sqrt{\mu\varepsilon}} \tag{7.135}$$

as first reported in Table 3.2.

7.7 PERIODIC WAVEGUIDE AND FLOQUET MODES

Thus far, all of the transmission-line and waveguide structures we have considered have had smooth-wall geometries, that is, the cross-section is the same everywhere along the longitudinal axis of propagation. There are scenarios in which it is useful to model wave propagation along a structure with a periodic cross-section profile. Such structures, especially those with circular cross-sections like that shown in Figure 7.25, are sometimes referred to as *corrugated waveguides*. Conventionally for this particular geometry, the smaller-diameter sections are called *teeth*, and the larger-diameter sections are called *grooves*.

In a smooth-wall waveguide, the defining condition of a forward propagating mode may be written

$$\mathbf{E}(z + \Delta z) = \mathbf{E}e^{-\gamma\Delta z} \tag{7.136}$$

or a similar expression in $\mathbf{H}$, where γ is the propagation constant and Δz is any finite distance in the longitudinal direction. This condition cannot be strictly met in a corrugated waveguide, since the boundary conditions in different regions will necessarily alter the field patterns beyond a simple propagation delay. However, the above condition can be met if Δz is constrained to be an integer multiple of the corrugation period, or $\Delta z = np$. In other words, a field solution may be considered a propagating mode in corrugated waveguide if the fields repeat their configuration with a consistent propagation delay every corrugation period. This is known as the *Floquet condition* [8]. Note that

$$e^{-\gamma pn} = e^{-\alpha pn}e^{-j\beta pn} = e^{-\alpha pn}e^{-j(\beta p+2\pi)n} \tag{7.137}$$

for any integer n, proving that the true dispersion curve for a resulting Floquet mode must be periodic in β with period $2\pi/p$.

In principle, the Floquet modes are mixtures, or superpositions, of the normal (smooth-wall) modes in each section, tooth or groove. Ordinarily, the normal modes' differing velocities would make it impossible for the unique, combined field pattern of the Floquet mode to remain whole as it propagated, but in a periodic waveguide, the relative phases of the normal modes are essentially reset at each corrugation boundary, giving the combined field pattern enough cohesion to propagate down the guide intact.

7.7.1 Surface Impedance Approximation

An exact solution for the Floquet modes in a periodic waveguide requires either a numerical simulation or mode-matching analysis similar to that described in Section 7.2.8. Fortunately, electromagnetic simulation software packages are now readily available that can be programmed with the periodic boundary conditions needed to solve for these modes, and efficient mode-matching algorithms exist that allow the full spectrum of modes to be calculated with very little computational effort [9, 10]. Nevertheless, it is useful to have at least an approximate analytical theory from which one can gain insight into the different kinds of modes which can propagate in a structure such as this.

If the corrugation period, p, is electrically small, then we may approximate the effect of the corrugations by defining a surface impedance, $Z_s = jX_s$, which applies to currents flowing in the longitudinal direction, while the PEC boundary conditions remain applicable to circumferentially directed currents, as shown in Figure 7.25(b). In this model, the corrugation slots or grooves act almost like series-connected, short-circuited transmission-line stubs. Shallow grooves, like short stubs, present an inductive impedance to the surface, while quarter-wavelength deep slots present an open-circuit to the surface, and deeper slots become capacitive [11].

To satisfy these combined boundary conditions in general, we need longitudinal components of both the **E** and **H** fields. The solutions we derive will therefore be categorized as hybrid modes. Let us start with the Helmholtz equations for E_z and H_z,

$$\left(\nabla_t^2 + K^2\right) E_z = 0 \tag{7.138a}$$

$$\left(\nabla_t^2 + K^2\right) H_z = 0 \tag{7.138b}$$

where

$$K^2 = \gamma^2 + k^2 \tag{7.139}$$

In all our previous examples, we used the cutoff wavenumber, k_c in place of K above, but for corrugated waveguide we shall find that K in this equation is frequency-dependent, so a different nomenclature is deemed appropriate.

The solutions to these equations in cylindrical coordinates are well known to us now,

$$E_z = E_1 \frac{K}{\gamma} J_n(Kr)\cos(n\theta)e^{-\gamma z} \tag{7.140a}$$

$$H_z = H_1 \frac{K}{\gamma} J_n(Kr)\sin(n\theta)e^{-\gamma z} \tag{7.140b}$$

For simplicity, we have chosen the cosine term in θ for E_z, and the sine term for H_z, neglecting the free parameter, ϕ which, as before, leads to identical degenerate modes with rotated polarization. The azimuthal vector component of the electric field, E_θ, is found using the general formula in Table 7.1 for the transverse components,

$$\mathbf{E}_t = \frac{\gamma}{K^2}\left(\frac{j\omega\mu}{\gamma}\mathbf{z} \times \nabla_t H_z - \nabla_t E_z\right) \tag{7.141a}$$

$$\therefore E_\theta = \frac{\gamma}{K^2}\left(\frac{j\omega\mu}{\gamma}\frac{\partial}{\partial r}H_z - \frac{1}{r}\frac{\partial}{\partial \theta}E_z\right) \tag{7.141b}$$

$$= E_1\left(\frac{jk}{\gamma}\Lambda J_n'(Kr) + \frac{n}{Kr}J_n(Kr)\right)\sin(n\theta)e^{-\gamma z} \tag{7.141c}$$

$$= E_1 \frac{jk}{\gamma} \frac{J_n(Kr)}{Kr} \left(\Lambda F_n(Kr) + \frac{n\gamma}{jk} \right) \sin(n\theta) e^{-\gamma z} \tag{7.141d}$$

where $\Lambda = \eta H_1 / E_1$ is known as the normalized hybrid factor, a critical parameter in defining the nature of the hybrid modes, and

$$F_n(x) = x \frac{J_n'(x)}{J_n(x)} \tag{7.142}$$

Similarly, the azimuthally directed magnetic field is given by

$$\mathbf{H}_t = \frac{\gamma}{K^2} \left(\frac{j\omega\varepsilon}{\gamma} \nabla_t E_z \times \mathbf{z} - \nabla_t H_z \right) \tag{7.143a}$$

$$\therefore H_\theta = \frac{\gamma}{K^2} \left(-\frac{j\omega\varepsilon}{\gamma} \frac{\partial}{\partial r} E_z - \frac{1}{r} \frac{\partial}{\partial \theta} H_z \right) \tag{7.143b}$$

$$= -E_1 \left(\frac{jk}{\eta\gamma} J_n'(Kr) + \frac{n}{Kr} \frac{\Lambda}{\eta} J_n(Kr) \right) \cos(n\theta) e^{-\gamma z} \tag{7.143c}$$

$$= -E_1 \frac{jk}{\gamma\eta} \frac{J_n(Kr)}{Kr} \left(F_n(Kr) + \frac{n\gamma}{jk} \Lambda \right) \cos(n\theta) e^{-\gamma z} \tag{7.143d}$$

To determine the cutoff frequency, we must now apply the boundary conditions. The circumferential PEC boundary condition requires that $E_\theta = 0$ at $r = a$, so

$$E_1 \frac{jk}{\gamma} \frac{J_n(Ka)}{Ka} \left(\Lambda F_n(Ka) + \frac{n\gamma}{jk} \right) \sin(n\theta) e^{-\gamma z} = 0 \tag{7.144a}$$

$$\therefore \Lambda = -\frac{n\gamma}{jk} \frac{1}{F_n(Ka)} \tag{7.144b}$$

In addition, the longitudinal boundary condition requires that $H_\theta / E_z = Z_s^{-1}$ at $r = a$ (because H_θ is associated with longitudinal current in the walls, and E_z is associated with the localized voltage differential). Therefore,

$$\frac{H_\theta}{E_z} = -\frac{jk}{K^2 a \eta} \left(F_n(Ka) + \frac{n\gamma}{jk} \Lambda \right) = Z_s^{-1} = -jX_s^{-1} \tag{7.145a}$$

$$\therefore \left(\frac{n\gamma}{jk} \right)^2 = F_n(Ka) \left(F_n(Ka) - \frac{K^2 a}{k} \frac{\eta}{X_s} \right) \tag{7.145b}$$

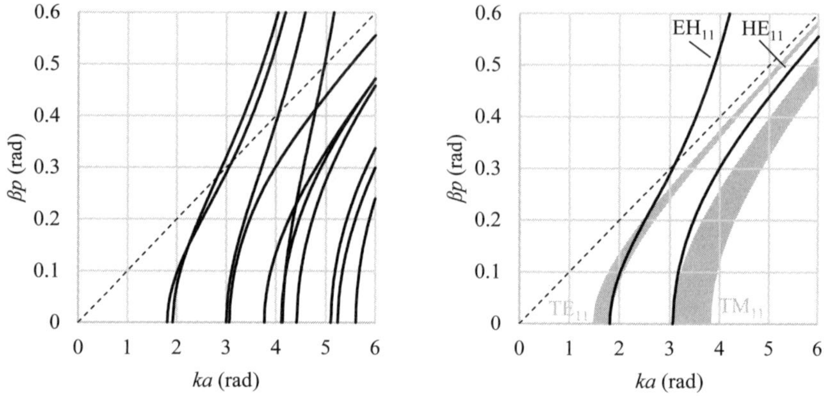

(a) (b)

Figure 7.26 (a) Propagation constant, β, for the lowest order Floquet modes in a circular corrugated waveguide with $a/b = 0.8$, $p = a/10$ and $t = p/100$. (b) Detail of EH_{11} and HE_{11} modes. The shaded regions represent the range of possible dispersion curves for the TE_{11} and TM_{11} modes in a smooth-walled circular waveguide with radii between a and b.

Equation (7.145b) is a characteristic equation from which we can calculate the propagation constant, γ, given the longitudinal surface reactance, X_s. We have not defined an exact frequency dependence for the surface impedance yet, but nonetheless we can use this relationship to identify the features of some special points on the dispersion curve.

7.7.2 Hybrid Modes

Consider the dispersion diagram in Figure 7.26(a), derived using mode-matching analysis for a corrugated waveguide with $a/b = 0.8$, $p = a/10$, and $t = p/100$. The simple hyperbolic dispersion curves common to all smooth-walled waveguides take on more complex forms in periodic waveguides. Some modes turn on at a particular frequency and then asymptotically approach the dashed free-space light curve at high frequency, as do the TE and TM modes in a smooth-walled waveguide, while others first turn on and then approach a vertical asymptote where they turn off again.

Let us examine the characteristics of the initial cutoff frequencies, at which $\gamma = 0$ and $K = k = k_c$. Substituting this into (7.145b), we are given two possibilities,

$$F_n(k_c a) = 0 \tag{7.146}$$

or

$$F_n(k_c a) = k_c a \frac{\eta}{X_s} \tag{7.147}$$

In the first case, nontrivial solutions require that $\Lambda \to \infty$, indicating that the mode exhibits a transverse-electric (TE) field pattern in the limit. Moreover, (7.146) is satisfied when

$$J'_n(k_c a) = 0 \tag{7.148a}$$

$$\therefore k_c = \frac{e_{nm}}{a} \tag{7.148b}$$

which is the cutoff of a TE mode in smooth, circular waveguide of radius a. As the propagation constant increases from zero, the normalized hybrid factor, Λ, takes on finite values and the mode changes from transverse electric to one having hybrid characteristics. This is known as an EH hybrid mode. The EH_{11} mode is plotted in Figure 7.26(b), emanating from a low-frequency cutoff equivalent to the upper end of the TE_{11} range.

In the latter case, we have from (7.147) that $F_n(k_c a)$ is finite, and thus from (7.144b) that $\Lambda = 0$. Thus, the mode at cutoff in this case is transverse-magnetic, acquiring hybrid characteristics as the propagation constant increases. This is known as an HE mode, and Figure 7.26(b) shows that its cutoff is the same as that of a TM mode in smooth, circular waveguide with radius b.

Note that as the corrugations become deeper, the range of cutoffs for the TE and TM modes — the shaded regions in Figure 7.26(b) — expand to the point of overlapping, at which time the cutoffs of the EH_{nm} and HE_{nm} modes coincide. Given their relationship to the TE_{nm} and TM_{nm} cutoffs, we may solve for this condition as

$$k_c = \frac{e_{nm}}{a} = \frac{z_{nm}}{b} \tag{7.149a}$$

$$\therefore \frac{a}{b} = \frac{e_{nm}}{z_{nm}} \tag{7.149b}$$

For example, the EH_{11} and HE_{11} modes have the same lower cutoff frequency when

$$\frac{a}{b} \approx \frac{1.8412}{3.8317} \approx 0.4805 \tag{7.150}$$

It is important to realize that these cutoffs are an approximation resulting from the surface-impedance model, and are most accurate when both p and t/p are small. Full-wave mode-matching analyses will generally confirm the nature of the hybrid modes described above, but the actual cutoff wavenumbers may differ slightly.

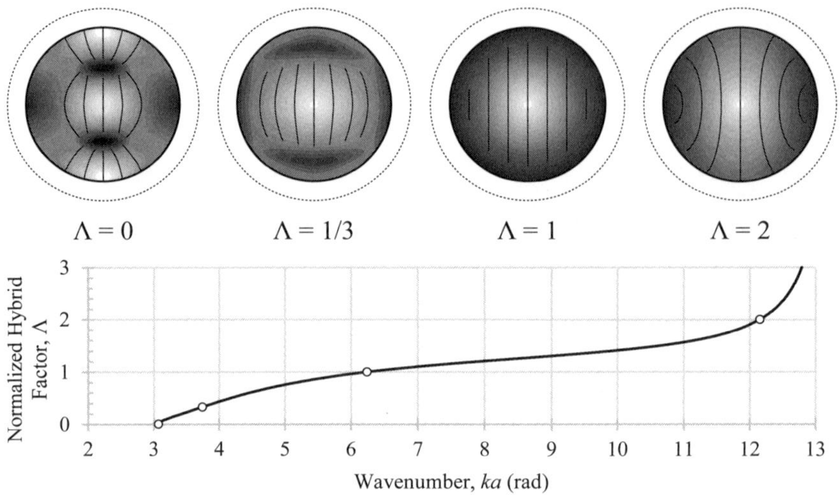

Figure 7.27 Evolution of the HE_{11} mode from cutoff ($\gamma = 0$) to the balanced hybrid condition ($\Lambda = +1$) and beyond. The transverse electric field lines and intensity are shown above, while the normalized hybrid factor is plotted below.

7.7.3 The Balanced Hybrid Condition

In contrast with smooth-walled waveguide modes, whose transverse fields are constant for all frequencies, the field patterns of the Floquet modes evolve as the frequency increases. Consider the HE_{11} mode in Figure 7.27. At the initial cutoff, $\Lambda = 0$ and it resembles the TM_{11} mode in smooth circular waveguide. As the wavenumber increases, so does the hybrid factor until the wave impedance becomes equal to the free-space wave impedance, η. This is known as the *balanced hybrid condition* and is characterized by $\Lambda = +1$ for HE modes (and by $\Lambda = -1$ for EH modes [11]). As the wavenumber increases beyond the balanced hybrid condition, the HE_{11} mode takes on more transverse electric characteristics and more closely resembles the TE_{11} mode in smooth circular waveguide.

In terms of the surface impedance model, the balanced-hybrid condition is met when the longitudinal currents are zero, or the longitudinal surface impedance becomes infinite. This can be seen by substituting $X_s \to \infty$ into the hybrid characteristic equation (7.145b); thus,

$$\left(\frac{n\gamma}{jk}\right)^2 = F_n^2(Ka) \tag{7.151}$$

Then, according to (7.144b), we must have $|\Lambda|^2 = 1$. This will asymptotically be the case when the corrugations are a quarter-wavelength deep — as though the slots themselves are radial transmission-line stubs, a quarter-wavelength long and short-circuited by the outer wall.

The balanced hybrid condition in the HE_{11} mode is considered an important operating point for corrugated waveguides due to a number of unique properties. As Figure 7.27 shows, the transverse fields become perfectly linearly polarized, and have a circularly symmetric field intensity. This is especially useful in the aperture of an antenna, especially horns that illuminate a parabolic dish, as it tends to radiate a beam that is optimally polarized and circular. Horn antennas will be discussed in more detail in Section 9.9.4.

7.7.4 Fast, Slow, Forward, Backward, and Complex Waves

Floquet modes in corrugated waveguides can exhibit a wide variety of unusual behaviors. Consider the dispersion diagrams in Figure 7.28 for corrugated circular waveguide with $a/b = 0.9$, 0.7, and 0.5. For simplicity, only modes having an angular index of 1, such as EH_{1m} and HE_{1m}, are shown. Note that when $a/b = 0.9$ the dispersion curves closely resemble that of smooth-walled circular waveguide, with only minor perturbation. Nonetheless, some unusual features can be identified. Recall that in smooth-walled waveguides, all TE and TM modes have imaginary propagation constants, $\gamma = j\beta$, which fall below the dashed, diagonal line, indicating that these modes have phase velocity which is faster than the speed of light in the medium filling the waveguide. These are called *fast waves* for obvious reasons. The EH_{11} mode in Figure 7.28(a) crosses over this diagonal line at about $ka \approx 4.5$ into the upper-left triangle of the plot, indicating that this wave has phase velocity slower than the speed of light. These are called *slow waves* in this regime, and the crossing point marks the fast-wave to slow-wave transition for the given mode.

As the ratio a/b decreases, we see another unusual feature emerge. While most modes have propagation constants that are either totally real (if the mode is cutoff) or totally imaginary (if the mode is propagating), we now see modes emerge which have both real and imaginary components, α and β, simultaneously. Two

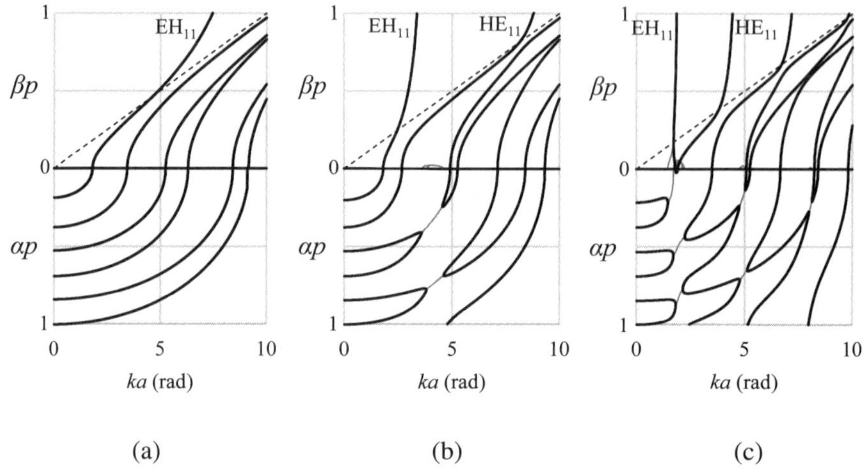

Figure 7.28 Dispersion diagrams for corrugated circular waveguides with (a) $a/b = 0.9$, (b) $a/b = 0.7$, and (c) $a/b = 0.5$. In all cases, $p = a/10$ and $t = p/100$. Only modes with angular index $n = 1$ are shown. Light gray segments indicate complex modes, having propagation constants with both real and imaginary parts.

such modes appearing at about $ka = 4$ in Figure 7.28(b) are shown with thin gray lines. These are called *complex waves*.

Finally, in Figure 7.28(c), the corrugations have become so deep that the EH_{11} mode has tilted past vertical, and actually runs slightly backwards in frequency from the cutoff near the horizontal axis. This indicates that the group velocity (associated with the slope of this curve) is in the opposite direction of the phase velocity — in other words, as a wave packet propagates forward along the axis of the waveguide, the phase fronts underlying that wave packet are threading through it in the opposite direction. These are called *backward waves* in this regime.[6]

To summarize, in addition to the transverse electric, magnetic, and hybrid mode classifications, modes in any waveguide may be further categorized by the following criteria,

- *Fast waves* ($\beta < k$) — Waves in which the phase velocity is faster than the speed of light (in the corresponding medium). As such, the wavelength

6 These should not be confused with backward-traveling waves, such as reflected waves, which are simply waves propagating in a direction opposite to some forward reference direction, but having codirectional phase and group velocities.

is longer (larger) than it would be in free (unguided) space for the same frequency. The phase constant, β, is therefore smaller than it would be in free space. This is the usual type of mode in a uniform, hollow, metallic waveguide.

- *Slow waves ($\beta > k$)* — Waves in which the phase velocity is slower than the speed of light. As such, the wavelength is shorter (smaller) than it would be in free space for the same frequency. The phase constant, β, is therefore larger than it would be in free space. Some modes in periodic structures may be of this type.
- *Forward waves ($v_p v_g > 0$)* — Waves in which the phase fronts move in the same direction as the energy flow. Characterized by the phase velocity ($v_p = k/\beta$) having the same sign as the group velocity ($v_g = dk/d\beta$). This is again standard for modes in a uniform, hollow, metallic waveguide.
- *Backward waves ($v_p v_g < 0$)* — Waves in which the phase fronts move in the direction opposite to the energy flow. Characterized by the phase velocity ($v_p = k/\beta$) having the opposite sign as the group velocity ($v_g = dk/d\beta$). Some modes in periodic structures may be of this type.
- *Complex waves (α and β both nonzero)* — Waves in which the propagation constant has both real and imaginary parts, corresponding to both amplitude decay and phase evolution. Some modes in periodic structures may be of this type, even if the walls are lossless.

7.8 DIELECTRIC WAVEGUIDES

In addition to the hollow metallic waveguides discussed thus far in this chapter, waveguides may also be formed entirely by dielectric materials, or by dielectrics combined with conductors. Some examples are given in Figure 7.29.

Substrate-integrated optical waveguide, shown in Figure 7.29(a), is typically formed by printing or etching a layer of dielectric material in slabs or strips on top of a substrate having lower dielectric constant. It is a convenient way to form complex circuits along with active solid-state devices such as lasers and photodiodes. Imageline and ribline, Figures 7.29(b, c), comprise dielectric slabs patterned on top of a metallic ground plane.

By far the most well-known type of dielectric waveguide is optical fiber, shown in Figure 7.29(d), and this is the structure to which the remainder of this

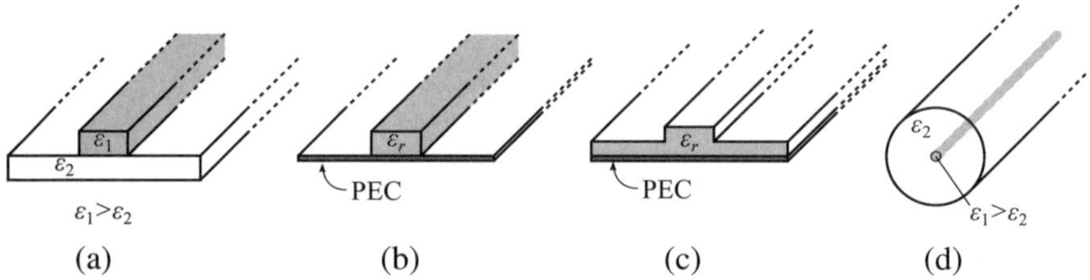

Figure 7.29 Several types of dielectric waveguides. (a) Integrated optical waveguide (raised strip). (b) Imageline. (c) Ribline. (d) Optical fiber.

chapter will be dedicated. As the name implies, fiber is most often used as a waveguide at near-optical wavelengths — typically in the infrared spectrum with wavelengths between 850 and 1,550 nm (between 190 and 350 THz) — but the same electromagnetic principles still apply.

7.8.1 Confinement by Total Internal Reflection

One of the most intuitive ways to understand the wave guiding properties of optical fiber is through the principle of total internal reflection (TIR), described in Section 1.5.2. Recall that a plane wave incident upon a dielectric interface has perfect reflection if the dielectric constant of the internal medium is greater than that of the surrounding medium and the angle of incidence is sufficiently large (grazing).

We thus have a situation very similar to that discussed in Section 7.2.6, where the modes in a hollow, metallic waveguide were constructed as a superposition of plane waves bouncing off the metal walls at oblique angles. This time, it is the TIR of the dielectric media that provide confinement of the mode, rather than the metal boundary conditions, but the basic principle is the same, as indicated in Figure 7.30.

There are two subtle differences, however. First, while the fields in a metallic waveguide are completely contained within the interior, the dielectric waveguide does allow evanescent fields to exist close to the interface in the surrounding medium. These fields are bound to the interface, so their presence does not represent leakage from the waveguide, but it does require that the surrounding medium is sufficiently large to isolate the fields from external influence, and its properties do contribute to the loss and other characteristics of the propagating mode, just as the internal medium does. The second difference is that the angle of incidence of the bouncing waves in the metallic waveguide is unlimited, while the angle of incidence in the dielectric waveguide must be large enough for TIR to occur in accordance

(a) (b)

Figure 7.30 Waveguide modes formed by a superposition of reflected waves in (a) a hollow metallic waveguide due to conductive boundary conditions, and (b) a dielectric waveguide due to total internal reflection.

with Snell's law. At steeper angles, TIR does not occur, and the waves are no longer guided. That critical angle was given in (1.105),

$$\theta_c = \sin^{-1}\left(\frac{n_2}{n_1}\right) \tag{7.152}$$

7.8.2 Fiber Cladding

In principle, a plain strand of low-loss glass could function as a dielectric waveguide, since the dielectric constant of the surrounding air would definitely be lower than that of the glass. In practice, however, such a fiber would be very sensitive to conditions at the surface; small defects or contaminants would scatter light away from the waveguide. Additionally, as stated above, any protective covering would interfere with the evanescent fields which extend beyond the glass itself.

Practical fibers, therefore, comprise two different kinds of glass with different dielectric constants: the *core*, where most of the guided energy is concentrated, and the *cladding* which surrounds it concentrically, as shown in Figure 7.29(d). The cladding provides a high-quality, low-loss medium for the evanescent fields, while being large enough that defects at its outer surface do not affect the modes concentrated at the core. (In real optical fibers, the cladding itself is also covered by an external coating and an opaque protective layer, known as the *buffer*, to guard against physical damage.)

Regarding the choice of dielectric constants, wave-guiding operation can be assured so long as the dielectric constant of the cladding is less than that of the core, but the greater the contrast between the two, the more difficult it is to achieve single-mode operation, and the greater the modal dispersion that will be exhibited in multimode operation [12]. Most fibers for long-distance communications will therefore use fiber with very small dielectric contrasts. For example, the very common SMF-28 fiber from Corning has a core with $n_1 = 1.45205$ and cladding with $n_2 = 1.44681$. Such a fiber is called *weakly guided*. It has the additional benefit of enlarging the area occupied by the mode (beyond the core), which helps to avoid nonlinear effects in the glass at higher power levels, but it also makes it sensitive to bending losses and reduces the angle of acceptance of light as described above, which makes it more difficult to launch a wave into the fiber.

7.8.3 LP Modes in Optical Fiber

If the dielectric contrast is low (i.e., the fiber is weakly guided), then the modes take on a relatively simple form wherein the electric and magnetic fields are almost entirely transverse and linearly polarized. These are called *LP* modes. (Strongly guided fibers will have TE, TM, HE, and EH modes, which will usually have to be derived by numerical approximation.)

For LP modes in weakly guided fibers, then, the Helmholtz equation reduces to a scalar form,

$$\left(\nabla_t^2 + k^2 + \gamma^2\right) E = 0 \tag{7.153}$$

Until now, all of the waveguides we have studied have been homogeneously filled — ε and μ were continuous throughout — and as a consequence k was a constant over the cross-section proportional to frequency. In this case, k is a function of position,

$$k(r) = k_0 \begin{cases} n_1 & r \le a \\ n_2 & r > a \end{cases} \tag{7.154}$$

where k_0 is the vacuum wavenumber and a is the radius of the core. For simplicity, let us assume that the wave is propagating ($\gamma = j\beta$) and define the following two quantities,

$$u = \sqrt{k^2 + \gamma^2} = \sqrt{k_0^2 n_1^2 - \beta^2} \quad \text{when } r \le a \tag{7.155a}$$

$$w = \sqrt{-k^2 - \gamma^2} = \sqrt{\beta^2 - k_0^2 n_2^2} \quad \text{when } r > a \tag{7.155b}$$

Note that these are defined in such a way that both are real when the phase velocity falls somewhere between that of unguided waves in the core material and in the cladding material, that is, $k_0 n_2 \le \beta \le k_0 n_1$.

As we know, the Helmholtz equation in cylindrical coordinates is separable, with solutions that are trigonometric in the angular coordinate and that are Bessel functions in the radial coordinate. The discontinuity at $r = a$ suggests that we should apply these solutions in piecewise fashion. In the core, the Bessel function of the first kind is appropriate, as the field must be finite at the origin. In the cladding, the modified Bessel function of the second kind, K_n, is more suitable as it decays to zero at infinity (as the cladding is typically very large, it is customary for the purposes of this calculation to assume that it is infinite). Thus,

$$E(r, \theta, z) = E_0 e^{-\gamma z} \cos(n(\theta + \phi)) \begin{cases} J_n(ur) & r \leq a \\ C_0 K_n(wr) & r > a \end{cases} \tag{7.156}$$

The coefficient C_0 and propagation constant β must now be chosen to satisfy the boundary conditions at the core-cladding interface (be reminded that u and w are both functions of β).

According to Table 1.1, the tangential component of the electric field must be continuous across the dielectric boundary, and so must the derivative of the normal component along the normal direction. Since we have assumed that the field is linearly polarized, it inevitably becomes entirely tangential to the boundary at some points around the circle, and entirely normal at others. The radial function for E must therefore be both continuous and smooth at $r = a$. Consequently,

$$C_0 = \frac{J_n(ua)}{K_n(wa)} \tag{7.157}$$

and

$$u \frac{J_n'(ua)}{J_n(ua)} = w \frac{K_n'(wa)}{K_n(wa)} \tag{7.158}$$

As a transcendental equation, this must usually be solved by numerical methods. For each order n at a given frequency (a given k_0), only a finite number of phase constants, β, will satisfy it. These will be denoted β_{nm} and each represents a propagating LP_{nm} mode at that frequency. Several of these modes for a multimode optical fiber are plotted in Figure 7.31. The vector direction of the electric (or magnetic) field is not shown in these plots, because it is arbitrary.

Note that the LP_{13} and LP_{32} modes in particular have significant field intensity outside the core. This is characteristic of modes with higher m indices, for which the propagation constant is very near to that of unguided waves in the cladding ($\beta_{nm} \approx k_0 n_2$). These modes are only barely guided by the fiber; waves

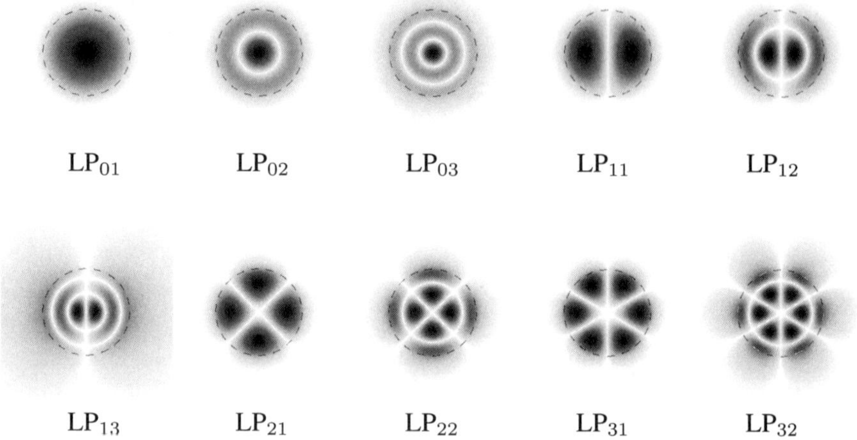

Figure 7.31 Field intensity for several LP modes of a multimode optical fiber. The polarizations of the electric and magnetic fields are linear, transverse, and orthogonal to each other, but otherwise arbitrary. The dashed circles represent the extent of the core.

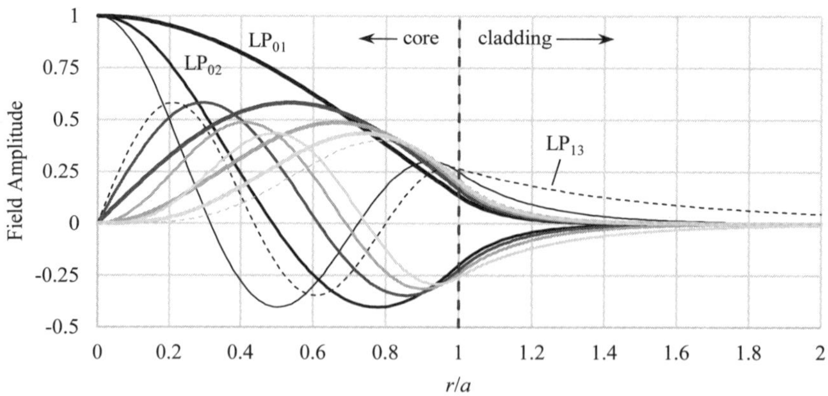

Figure 7.32 Radial profiles for several LP modes in a multimode fiber. LP_{13} in this case is very near cutoff and decays very slowly in the cladding.

of a slightly longer wavelength would escape entirely. The resulting long tail of the radial profile extends well into the cladding, as illustrated in Figure 7.32. Such modes are typically most sensitive to bending loss.

Since practical fibers are designed to be weakly guiding, the indices of refraction in the core and in the cladding are very similar, and the range of possible propagation constants for the modes is quite narrow. This makes it possible to use even multimode fibers having a very large number of modes for short-range data transmission. If information is modulated as pulses of light, then those pulses remain intact, even if many different modes are involved, as long as the pulses are not too short (i.e., the bit rate is not too fast) and the distance traveled is not too great.

For faster data rates and/or longer distances, even the slightest modal dispersion can smear out the pulses to an unacceptable degree. For these applications, single-mode fiber is used, in which only the LP_{01} mode propagates over the range of wavelengths needed. These typically have very tiny cores, about 8.2 μm in diameter in the case of SMF-28 mentioned previously. Even single-mode fiber is not immune to dispersion, for the LP_{01} mode (like any mode in a waveguide) is still subject to velocity variation as a result of the wave-guiding action as well as the slightly wavelength-dependent dielectric constant of the glass. This chromatic dispersion ultimately puts a limit on the distance-bandwidth product that a single-mode fiber-optic waveguide can practically support.

The carrier wavelength, or color, is chosen in many cases to optimize for the longest transmission with the greatest signal fidelity. For example, 1,550 nm is the wavelength of lowest dielectric loss for most standard glass materials used in fibers, while 1,310 nm is the wavelength corresponding to the minimum chromatic dispersion. More sophisticated fibers use graded refractive-index profiles (rather than the step-indexed profiles analyzed here) to shift the dispersion curve so that minimum dispersion is coincident with minimum loss.

Still other fibers may be designed with asymmetric shapes, asymmetric index profiles, or may even incorporate stress rods in the cladding to induce birefringence in the glass in such as way as to isolate the two polarizations from one another. These polarization-maintaining fibers are most useful when interfacing with active devices that may be sensitive to the polarization state of the light passing through them.

Note that the cladding in any fiber will also support modes in itself, which are parasitic to the waveguide; however, these modes are usually much more lossy (due to the effects of the outer coatings and the buffer), and one is usually careful not to excite these modes.

Problems

7-1 A hollow (air-filled) metallic waveguide with triangular cross-section has its dominant TE-mode cutoff frequency at 1 GHz. What is the wave impedance at 1.3 GHz?

7-2 What width must an air-filled rectangular waveguide have to be cut off at 10 GHz? At 50 GHz?

7-3 What aspect ratio, a/b, should a rectangular waveguide have in order to maximize the fractional TM bandwidth (over which only a single TM mode propagates)? What TM bandwidth does it achieve?

7-4 What is the location and direction of the peak current density on the walls of a rectangular waveguide in the dominant mode? Does it depend on frequency?

7-5 At what unguided wavenumber, k, in a 2:1 rectangular waveguide do the first parasitic modes decay in amplitude by a factor $1/e$ over the same distance as one wavelength in the dominant mode?

7-6 Describe the limiting behavior of the phase velocity, group velocity, and guided wavelength of modes in a hollow, metallic waveguide as the mode approaches cutoff. How do they trend as the frequency diverges to infinity?

7-7 Write an expression for the group velocity of a mode in a hollow metallic waveguide as a function of the ratio f/f_c.

7-8 What are the relative dimensions of a rectangular waveguide that provides minimum, in-band loss in the dominant mode while remaining single-moded over a 1.5:1 bandwidth?

7-9 Write an expression for the frequency of minimum conductive loss for a rectangular waveguide in the TE_{10} mode. Verify the result by calculating the minimum loss frequency for WR-10 ($a = 2b = 0.1$ in.) plotted in Figure 7.9.

7-10 Draw an approximate transmission-line equivalent circuit for a junction between a waveguide with dimensions a and b and another waveguide with dimensions $a' = 1.1a$ and $b' = 1.1b$.

7-11 Estimate the leakage at 10 MHz through a square hole in a metallic wall that is 1 cm wide and 5 cm long.

7-12 Write expressions for the surface currents on circular waveguide in a TE_{0m} mode. How does this explain the asymptotic loss for such modes at high frequency?

7-13 Two metallic cavities are constructed having identical geometries, but one is filled with an unknown dielectric, while the other is filled with a known dielectric having a similar dielectric constant and loss tangent $\tan \delta_0 = 10^{-4}$. The cavity having the known dielectric is measured to have a Q of 5,000, while the cavity with the unknown dielectric has a better Q of 8,000. Estimate the loss tangent of the unknown dielectric.

7-14 Give approximate dimensions for a double-ridged waveguide having cutoff frequency at 1 GHz and 10:1 single-mode bandwidth.

7-15 What is the upper frequency limit of a Teflon-filled ($\varepsilon_r = 2.1$) coaxial cable designed for 75Ω impedance with outer diameter of 2 mm?

7-16 Under the surface impedance model, what is the cutoff frequency of the HE_{10} mode in an air-filled square waveguide having side length of 1 cm, with fine-pitch, 3-mm corrugations in all four walls. (The mode subscripts refer to harmonics in the x and y dimensions, in the same manner as the TE_{mn} and TM_{mn} mode subscripts for smooth-wall rectangular waveguide.)

References

[1] N. Marcuvitz, *Waveguide Handbook.* Stevenage, UK: Peter Peregrinus Ltd., 1986.

[2] R. Gadhavi, A. Patel, R. Butani, and S. K. Hadia, "Analysis of Ohmic loss in oversized smooth walled circular waveguide using FEM," *International Journal of Rengineering Research & Technology*, vol. 3, no. 3, pp. 1263–1265, March 2014.

[3] A. E. Karbowiak and L. Solymar, "Characteristics of waveguides for long-distance transmission," *Journal of Research of the National Bureau of Standards — D. Radio Propagation*, vol. 65D, no. 1, pp. 75–88, January 1961.

[4] W. A. Huting, J. W. Warren, and K. Jerry A, "Recent progress in circular high-power overmoded waveguide," *Johns Hopkins APL Technical Digest*, vol. 12, no. 1, pp. 60–74, 1991.

[5] G. M. Coutts, "Wideband diagonal quadruple-ridge orthomode transducer for circular polarization detection," *IEEE Transactions on Antennas and Propagation*, vol. 59, no. 6, pp. 1902–1909, June 2011.

[6] M. A. Morgan, "Dual-mode propagation in triangular and triple-ridged waveguides," *NRAO Electronics Division Technical Note*, no. 218, February 2011.

[7] D. M. Pozar, *Microwave Engineering*, 4th ed. New York: Wiley, 2011.

[8] R. E. Collin, *Field Theory of Guided Waves*, 1st ed. New York: McGraw-Hill, 1960.

[9] S. Amari, J. Bornemann, and R. Vahldieck, "Accurate analysis of scattering from multiple waveguide discontinuities using the coupled-integral equations technique," *Journal of Electromagnetic Waves and Applications*, vol. 10, pp. 1623–1644, December 1996.

[10] S. Amari, R. Vahldieck, J. Bornemann, and P. Leuchtmann, "Spectrum of corrugated and periodically loaded waveguides from classical matrix eigenvalues," *IEEE Transactions on Microwave Theory and Techniques*, vol. 48, no. 3, pp. 453–460, March 2000.

[11] P. J. B. Clarricoats and A. D. Olver, *Corrugated Horns for Microwave Antennas*. London: Peter Peregrinus Ltd., 1984.

[12] R. Paschotta. (2018) Encyclopedia of laser physics and technology. https://www.rp-photonics.com/encyclopedia.html.

Chapter 8

Launchers and Transitions

In the first part of this book, we have developed a wide variety of structures for propagating and manipulating electromagnetic signals, from lumped elements to transmission lines and waveguides of many different kinds. Rarely does an electronic system comprise exclusively just one of these structures. It is therefore necessary to consider transitional elements designed to bridge between these systems. Such components are commonly referred to as *launchers* and/or *transitions*.

These words are used somewhat interchangeably, as there is no universally accepted definition for what constitutes a launcher instead of a transition, or even a probe. Nonetheless, the author offers the following suggested guidelines for the sake of clarity,

- *A transition* is the bridge between two different media of the same general category (e.g., from transmission line to transmission line, or from waveguide to waveguide). Thus, a connection between microstrip and stripline would be a transition, as would a junction between rectangular waveguide and circular waveguide.

- *A launcher* connects one form of media to a physically larger or less lossy one for the purpose of transmission over greater distances. Thus, connections between microstrip and waveguide, or lumped elements and optical fibers, could be considered launchers.

- *A probe* is simply a descriptive term for certain features of these interconnects that do not generally fit into the formalism of either media. Many microstrip to waveguide launchers incorporate a probe, for example, as we shall soon

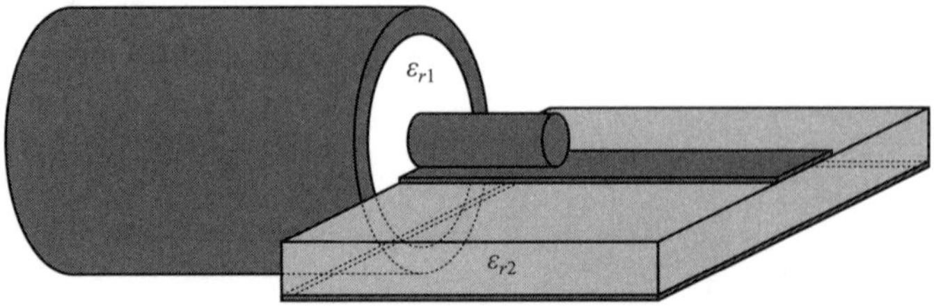

Figure 8.1 Conceptual diagram of a microstrip-to-coax end-launch transition.

see. That probe is not strictly a microstrip element or a waveguide element, but has features of both.

Note that these definitions allow some components to fall into more than one category. A junction between microstrip and coax is a transition (since both are transmission lines) as well as a launcher (since coax is typically larger and less lossy than microstrip and is often used to carry the signal a significant distance away from the microstrip circuit). It is with this example that we will start this chapter.

8.1 MICROSTRIP-TO-COAX AND CPW-TO-COAX TRANSITIONS

Microstrip is one of the most common media for high-frequency, compact integrated circuits today, and coax is the standard for physical interconnection between modular circuits and systems, at least below 50 GHz or so. Usually, when one talks about a coax launcher, the transition from microstrip is implied. Numerous methods of transitioning between these two have been devised, but they generally fall into one of two categories: end-launch, and right-angle (or perpendicular).

8.1.1 End-Launch Transition

The basic geometry of an end-launch coaxial transition is shown in Figure 8.1. The premise is that the center conductor of the coaxial line extends outward past the outer conductor and makes contact with the trace on the microstrip substrate, while the outer conductor is electrically connected to the microstrip's ground plane. As both are TEM (or quasi-TEM) transmission lines, the impedance match and transmission from one to the other are usually quite good if both are designed with the same characteristic impedance, and so long as the cross-sections remain electrically small. At higher frequencies, there is some local shunt capacitance associated with the termination of the outer conductor around much of its periphery, and series inductance associated with the ground current path from the bottom of the microstrip substrate into and around the outer conductor before fully coupling into the coax mode. These effects combine to render the transition low-pass.

A detailed equivalent circuit for this junction will depend on the relative dimensions of the coaxial line and microstrip substrate as well as the dielectric materials on both sides. Impedance-matching features may be included beyond the junction in either the coax line or the microstrip circuit or both. With the aid of numerical simulations, good-performing transitions can be designed well into the millimeter-wave frequency range [1]. It is helpful to minimize the parasitic effects by sizing the coax line such that the distance between the inner and outer conductors is similar to the thickness of the microstrip substrate, and so that the width of the microstrip trace is not too much different from the diameter of the inner conductor (very often the coax line will get larger by steps away from the transition).

In practice, the electrical connection between the center pin and microstrip trace, and between the outer conductor and the substrate ground plane must be secured. The pin is most often connected by soldering, though tiny, horizontal spring contacts designed to be surface mounted on microstrip substrates do exist. The ground connection can be soldered as well, but coaxial launchers frequently incorporate threaded fasteners or clamping structures for additional ruggedness.

Many end-launch connector systems are actually designed to transition from grounded coplanar waveguide (GCPW) instead of microstrip, having top-side ground planes to the left and right of the trace as shown in Figure 3.5, in addition to the ground plane on the bottom. A secondary transition from microstrip to GCPW is often then required in the immediate vicinity of the coax launcher.

8.1.2 Right-Angle Transition

In some cases, it is more useful to transition to a coaxial line out of the plane of the microstrip circuit, as shown in Figure 8.2. The coaxial line may run vertically upward from the microstrip line (in which case connections to the ground plane are typically made through plated via holes in the substrate) or beneath the ground plane, as shown in the figure. In either case, a soldered or press-fit connection to the center pin is usually required.

In the case where the coax line extends beneath the substrate (as shown), broadband matching may be improved by shaping the aperture in the ground metal [2]. Normally, if the aperture were coincident with the outer conductor of

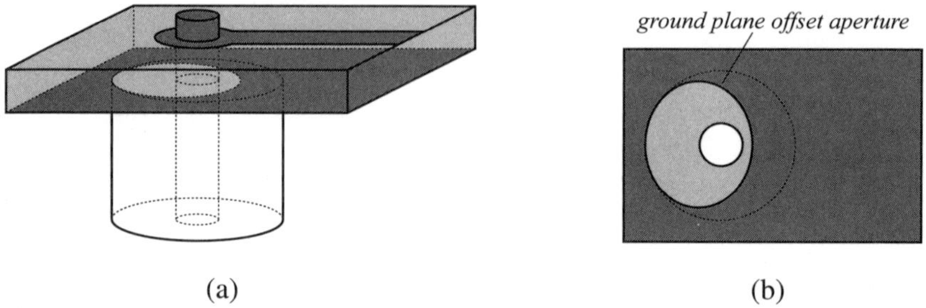

Figure 8.2 (a) Perpendicular microstrip-to-coax transition. (b) Detail of ground plane with offset aperture on microstrip substrate.

the coax, ground currents incident upon the junction from the microstrip would be concentrated on the side nearest the trace, and the long path it must take around the outer conductor to complete the coaxial mode would add significant parasitic inductance.

By shrinking the diameter of the aperture in the ground plane, a shortcut is provided for ground currents from beneath the trace to the opposite side. Further, by offsetting this aperture toward the back of the transition and shaping it elliptically as shown, the lengths of the current paths from the microstrip side to points all around the periphery of the coaxial outer conductor are more balanced.

To illustrate this effect, we compare the modeled performance of a junction with circular and elliptical ground apertures in Figure 8.3. In the first case, Figure 8.3(a), the coax line opens up fully into the ground plane of the microstrip substrate with an aperture matched in size to the coaxial outer conductor. The simulated performance shows that it achieves good impedance match only at the lowest frequencies. In Figure 8.3(b), the aperture has been reduced in size, but is still concentric with the coaxial line. Ground currents have a shorter path to get from the microstrip side to the opposite side of the outer circle. This improves the return loss at moderately higher frequencies. Finally, in Figure 8.3(c), we shape the aperture in the ground plane, optimizing its performance over a much wider band, as illustrated by the plot in Figure 8.3(d). For these examples, the microstrip substrate is 4-mil-thick alumina (Al_2O_3, $\varepsilon_r \approx 9.9$) with a 3.4-mil-wide trace, and the coax line has glass dielectric ($\varepsilon_r \approx 4.8$) with inner and outer diameters of 12 mils and 74 mils, respectively. The concentric aperture in Figure 8.3(b) has a diameter of 42 mils, and the elliptical aperture in Figure 8.3(c) has minor and major axes of 48 and 64 mils, respectively.

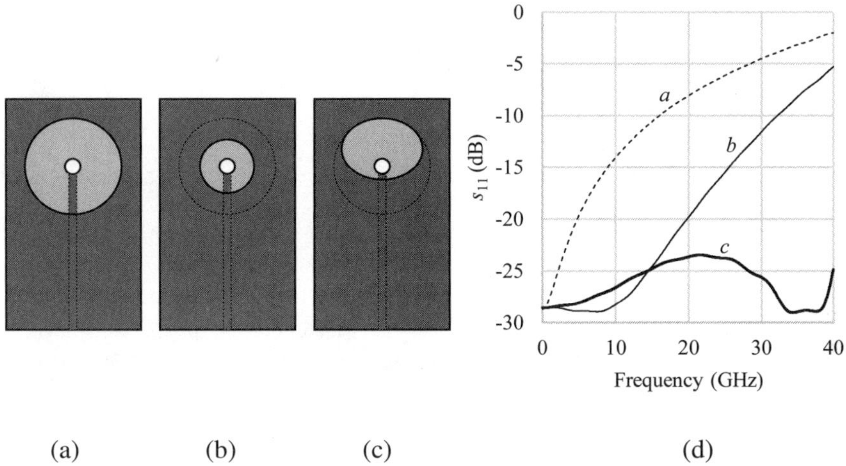

Figure 8.3 Comparison of right-angle coax launcher substrates with various ground plane apertures. (a) Fully open aperture matching the coax outer conductor's diameter. (b) Concentric, circular aperture reduced in size. The dashed circle represents the coaxial outer conductor diameter. (c) Offset elliptical aperture. (d) Comparison of s_{11} achieved with each of these apertures.

8.2 ALL-PLANAR TRANSITIONS

The wide variety of planar transmission lines that have been devised for integrated circuits has led also to the proliferation of techniques for transitioning between them. Sometimes two or even three different kinds of transmission-line structures will be mixed and used in a single circuit element (see, for example, the CPW-slotline rat-race hybrid in Section 10.7.1). Transitions are thus required to be compact, broadband, and easy to fabricate. A variety of such transitions will be explored in this section.

8.2.1 Microstrip to CPW and GCPW

Although microstrip is most common for the internal network of an integrated circuit, it is not unusual to see transitions to CPW at the ports, to facilitate coaxial launchers as described in Section 8.1.1, or to enable testing with wafer probes which require a top-side ground contact pad. A fairly straightforward transition from microstrip to CPW is shown in Figure 8.4(a), where the bottom ground plane is transitioned to the top-side ground planes with via holes.

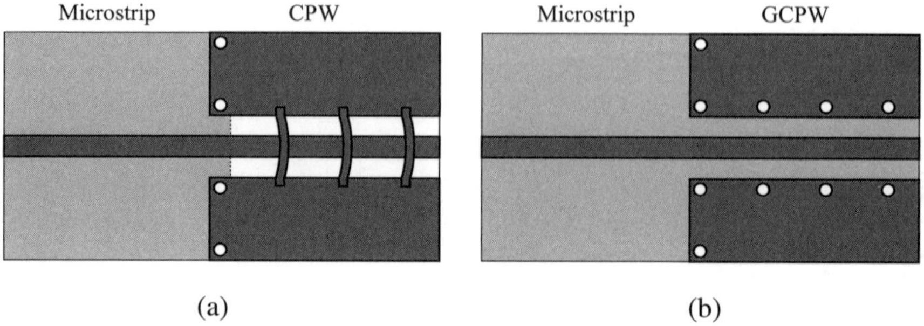

Figure 8.4 (a) Top-view of microstrip-to-CPW transition with air bridges. (b) Microstrip-to-GCPW transition with via fence. The bottom metalization is shown in light gray, while the top metal is shown in dark gray.

True CPW has no ground plane on the bottom, and the top-side ground planes are electrically isolated. Strict symmetry must be maintained in order to keep the two top-side ground planes at the same electrical potential.[1] In practice, since perfect symmetry can never be maintained, periodic air bridges are required, as shown in the figure, to short the two ground planes together. One must ensure that the air bridges are much less than a quarter-wavelength apart to avoid introducing resonances in the ground path.

In many cases, GCPW will actually be more practical, as mounting the substrate in an enclosure is easier if a bottom-side ground plane is present. In these cases, a fence of vias along the sides of the transmission line may replace the air bridges, as shown in Figure 8.4(b). Once again, the vias must be spaced much less than a quarter-wavelength apart to avoid resonances. When this is not possible, a combination of vias and air bridges can be used.

Like most of the transitions illustrated in this chapter, the figures are meant to illustrate the general principles and critical features, but not the exact layout. Good-performing transitions require careful modeling of the parasitics, which depend on the specific dimensions of the substrates and transmission lines, the dielectric materials, and the frequency range of operation. The abrupt junctions in Figure 8.4 might in some cases be modified to incorporate tapered trace and gap widths or other impedance-matching features to enhance their performance.

1 In principle, any long structure having a cross-section with N isolated conductors will support $N-1$ independent TEM or quasi-TEM modes. CPW, having $N = 3$ isolated conductors will support two quasi-TEM modes: the symmetric CPW mode in which both ground planes are at the same potential, and a parasitic coupled-slot mode where the center trace is at an average potential between the two.

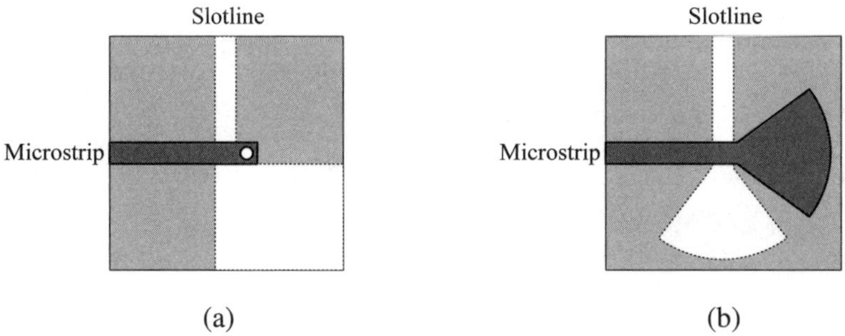

Figure 8.5 Microstrip-to-slotline transitions using (a) a via connection, and (b) radial stubs.

8.2.2 Microstrip to Slotline

The most direct way to transition from microstrip to slotline is to open up the slot within the microstrip's ground plane, as illustrated in Figure 8.5. One simply needs to connect the top trace of the microstrip to the opposite side of the slot. In Figure 8.5(a), this is accomplished using a via hole. The end of the slotline is then left open at the junction by leaving the two halves of the ground plane separated.

A common alternative replaces the via connection with a radial stub (effectively shorting the trace to the ground plane at the stem of the radial stub), as shown in Figure 8.5(b). A radial slot, the dual of a radial stub, is then used at the end of the slotline to effect the open-circuit condition. This transition has a band-pass response, in contrast to the transition in Figure 8.5(a) which is low-pass.

8.2.3 CPW to Slotline

CPW-to-slotline transitions usually take advantage of the fact that CPW itself already has two slots in an extended ground plane. One needs only to combine the energy from the two slots into one without introducing reflections. Two simple examples are shown in Figure 8.6. The first, Figure 8.6(a), uses a radial slot to open-circuit the ground current on one side of the CPW line, while an air-bridge provides a path for it to combine with ground current on the other side. Another variation, Figure 8.6(b), introduces a 180° phase shift in one of the slots before combining it with the other slot [3].

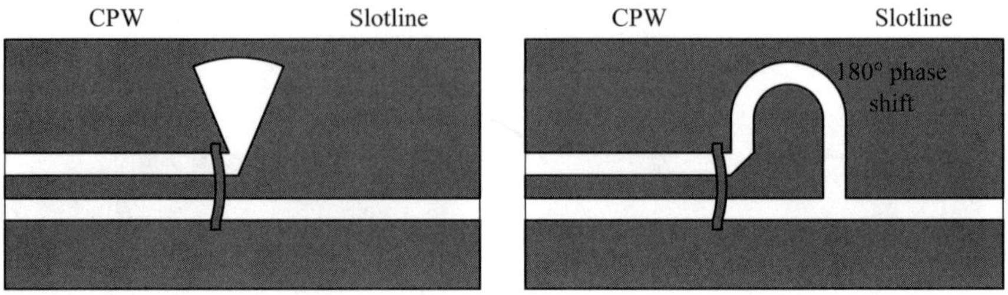

Figure 8.6 Two variants of a CPW-to-slotline transition, using (a) a radial slot, and (b) a 180° phase delay.

8.3 BALANCED TO UNBALANCED

All transmission lines generally fall into one of two classes:

- *Unbalanced*, also called *single-ended*, characterized by having signal and ground terminals. Usually, the ground conductor is larger than and/or encompasses the signal conductor, and is typically contiguous throughout the circuit (i.e., series elements are only inserted in the path of the signal conductor, not the ground). Examples include coaxial cable, microstrip, stripline, and CPW.
- *Balanced*, also called *differential*, in which the conductors attached to the terminals may be called positive and negative for reference, but are nevertheless symmetric in both size and form. There is no definable ground terminal, only a virtual ground, situated symmetrically between the two conductors. Examples include twin wire, slotline, and coplanar strips (CPS).

A transition that connects a balanced transmission line to an unbalanced transmission line is called a *balun*, a portmanteau for "balanced-to-unbalanced." As viewed from the unbalanced end, a balun should divide the input signal into two equal parts with 180° phase shift. It may also be necessary in certain applications to isolate the output pins from ground at DC. In addition to the usual figures of merit such as loss and reflection coefficient, key performance metrics for baluns include the amplitude balance (how well matched the gain is to the two output pins), phase balance (how close the phase differential is to 180°), and common-mode rejection (the ability to isolate in-phase excitations at the balanced ports from the single-ended port, or conversely the leakage of a signal input at the single-ended port to in-phase components at the terminals of the balanced port).

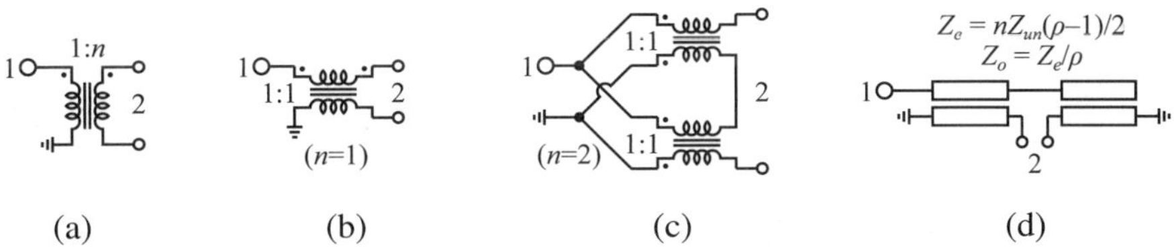

Figure 8.7 (a) Balun transformer. (b) Current transformer or Guanella 1:1 balun. (c) Guanella 1:4 balun. (d) Marchand balun. In all cases $n^2 = Z_{bal}/Z_{un}$.

Baluns are especially common when interfacing to antennas, which frequently have balanced terminals. As their driving point impedances are also not conveniently 50Ω in every case, it is not uncommon to integrate an impedance transformation in the balun as well. Antennas will be discussed in more detail in Chapter 9.

In the lumped-element domain, baluns typically involve transformers. In fact, the simplest balun of all is a single transformer with the single-ended input coil grounded, and the balanced output coil floating, as shown in Figure 8.7(a). The turns ratio, n, provides a convenient means of matching the balanced impedance, Z_{bal}, to the unbalanced impedance Z_{un}, where $n^2 = Z_{bal}/Z_{un}$.

The transformer configuration in Figure 8.7(a), with the coils connected in shunt, is sometimes referred to as a *voltage transformer*. It can also be used with the coils connected in series, as shown in Figure 8.7(b), in which case it is referred to as a *current transformer*. As a balun, it is called a *Guanella 1:1 balun*, and is only suitable when the balanced and unbalanced ports have the same characteristic impedance. The more complex network in Figure 8.7(c) is known as a *Guanella 1:4 balun*, and is applicable when $Z_{bal}/Z_{un} = 4$, or $n = 2$.

A very popular transmission-line balun, known as the *Marchand balun* [4], is shown in Figure 8.7(d). It uses two pairs of coupled transmission lines where

$$Z_e = \tfrac{1}{2} n Z_{un}(\rho - 1) \tag{8.1a}$$

$$Z_o = \frac{Z_e}{\rho} \tag{8.1b}$$

Under the above conditions, the balun is perfectly matched at the center frequency for all coupling factors, ρ, which may be left as a free parameter affecting the bandwidth. A good choice sets $\rho = 3 + 2\sqrt{2} \approx 5.83$, as this results in a three-pole return loss response for large Z_{bal}.

A typical, planar layout for a Marchand balun and its simulated response are thus given in Figure 8.8.

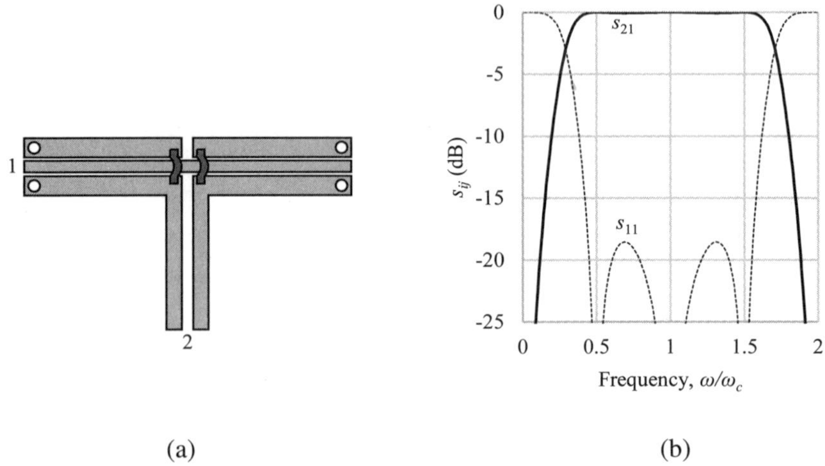

Figure 8.8 Marchand balun (a) layout and (b) simulated performance.

8.4 COAX TO WAVEGUIDE

Transitions from coaxial cable to waveguide are common adapters in microwave development labs, used to mate low-loss waveguide apparatus to test equipment that more often has coaxial ports. The geometric form of the transition may depend on a number of factors, including especially the bandwidth over which the waveguide, and hence the transition, are required to operate.

8.4.1 Coax to Rectangular Waveguide

For applications where less than a 2:1 bandwidth is acceptable, simple rectangular waveguide is usually the preferred solution as it minimizes the loss. The most common form of transition in these cases is shown in Figure 8.9. The outer conductor of the coaxial line merges with the waveguide walls, while the inner conductor extends out as a probe from the broad wall. The end of the waveguide section is short-circuited approximately one quarter-wavelength away from the probe, a feature known as the *backshort*, presenting an open-circuit in parallel with the probe from that direction. The precise shape of the probe varies from one particular design to the next and may be optimized for impedance match, power

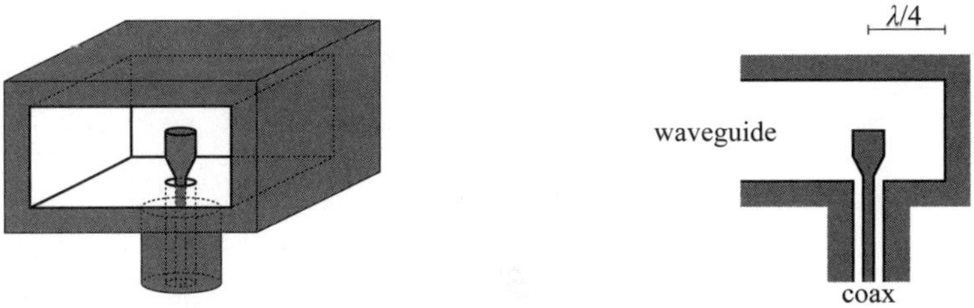

(a) (b)

Figure 8.9 (a) Coax-to-rectangular-waveguide transition. (b) Cutaway side-view.

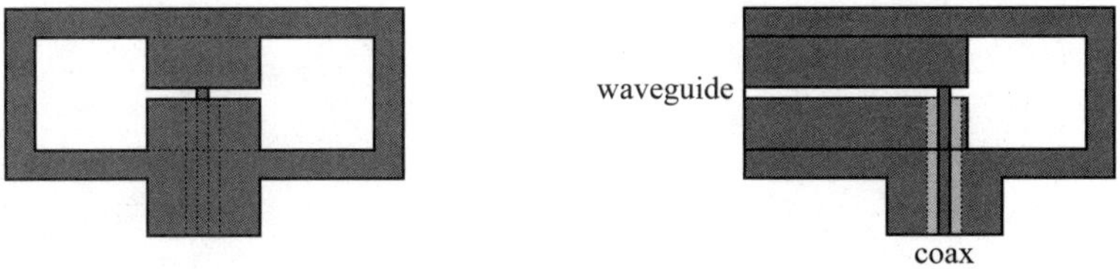

(a) (b)

Figure 8.10 Coax-to-double-ridged-waveguide launcher. (a) Front view. (b) Side view.

handling, losses, and/or bandwidth, but it is generally resonant, like a transmission-line stub, at the center frequency of the transition.

8.4.2 Coax to Double-Ridged Waveguide

For applications where wider than 2:1 bandwidth is required, double-ridged waveguide can be used. Naturally, if interfacing to coax is desired, an equally wideband transition is needed. One such example is given in Figure 8.10. In this case, the center conductor of the coaxial line extends across the gap and makes electrical contact with the opposite side. This is necessary for the transition to work over wide bandwidths, since a resonant probe is inherently band-limited. The backshort is also modified in this case by removing the ridges. No longer easily associated with a quarter-wavelength, this geometric feature helps to provide a near open-circuit across the gap region from the back side of the launcher over a wide range of frequencies.

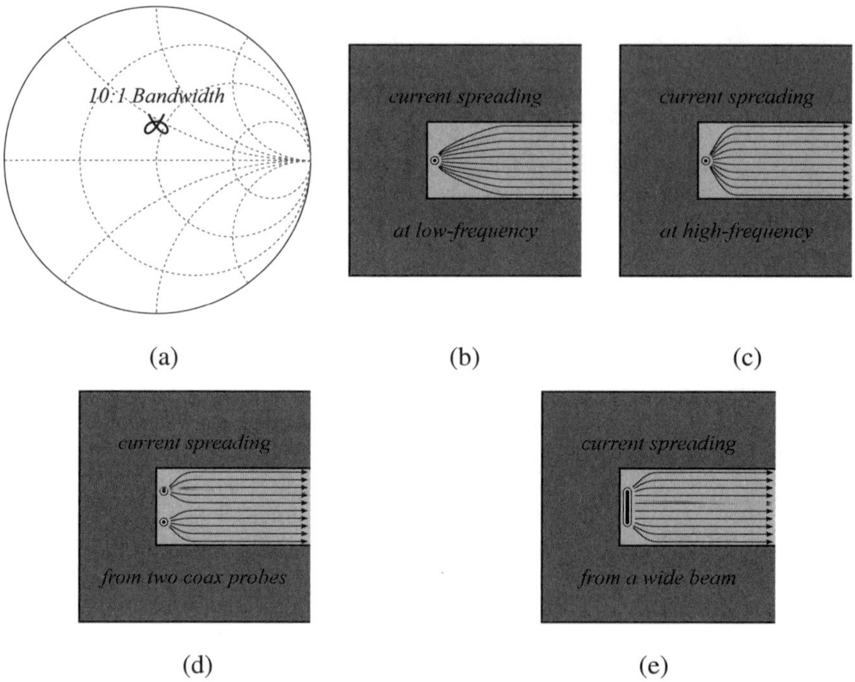

Figure 8.11 Illustration of spreading inductance effects in double-ridged waveguide launchers. (a) Typical impedance locus for a single probe, having nearly constant positive reactance. (b) Illustration of the current-spreading at low frequency, having a long path length. (c) Current-spreading at high frequency, having a shorter path length. (d) Reduction of current-spreading effects with multiple coaxial probes, and (e) with a wide beam probe.

In extreme cases approaching a decade (10:1) of bandwidth, the *spreading inductance* associated with currents diverging from the coaxial inner and outer conductors to span the width of the ridge can be a subtle but important parasitic. Upon optimizing the dimensions of the ridged waveguide, the probe position, the backshort, and other features of the transition, one often ends up with an impedance that looks like Figure 8.11(a), a tight locus near the origin but consistently inductive at all frequencies. One might think this could be matched with a small amount of extra capacitance, but that is not the case. Despite the compactness of the grouping on the Smith chart, one must keep in mind that this is plotted over a very wide bandwidth, say, 10:1. Any capacitor, be it lumped or transmission-line based, will have roughly 10 times as much reactance at one end of the band as it does at the other. This will inevitably stretch the locus of points out along a capacitive arc, and could at best improve the impedance match at one end or the other.

This fact only emphasizes the extraordinary nature of the mismatch in the first place. The transition is not suffering simply from an excess inductance, but from an excess positive reactance which is constant with frequency. This ought, at first, to seem puzzling, as it would imply a current path which is shorter at high frequency than it is at low frequency, just so to compensate the usual reactive proportionality. In fact, this is exactly what we have, as illustrated in Figures 8.11(b, c). Working under the assumption that the coax probe diameter is small compared to the ridge width (usually the case), the currents from the center and outer conductors of the coax must diverge from the launch point into a uniform sheet that spans the width of the ridge. Numerical simulations have shown [5] that the region over which this spreading occurs is large for low-frequency oscillations, and smaller for high-frequency oscillations. This modifies the effective path length versus frequency in an inversely proportional fashion.

A broadband solution to this problem is to first divide the feed lines into several, then launching into the ridges across an array, as indicating in Figure 8.11(d), or to widen the coax line itself into a wide fin or beam lead, as in Figure 8.11(e). Although this does not eliminate the spreading inductance entirely, it can reduce the inductance to such a degree that it becomes negligible.

8.5 MICROSTRIP TO WAVEGUIDE

Active electronics are now most often fabricated as planar circuits on semiconductor wafers. They are too small and too fragile to be handled directly, creating the need for a housing with a more rugged external interface. Coaxial connectors are commonly used up to about 50 GHz or so, but above that, waveguide is the preferred interface. This makes the transition from planar microstrip to waveguide a critical component of modern, high-frequency electronics.

8.5.1 Longitudinal Probes

The most common microstrip-to-waveguide launchers are essentially variations of the coaxial-probe geometry which was shown in Figure 8.9. One example is the longitudinal probe in Figure 8.12(a). The probe itself is a flattened paddle printed on the microstrip substrate extending roughly halfway into the waveguide. It is

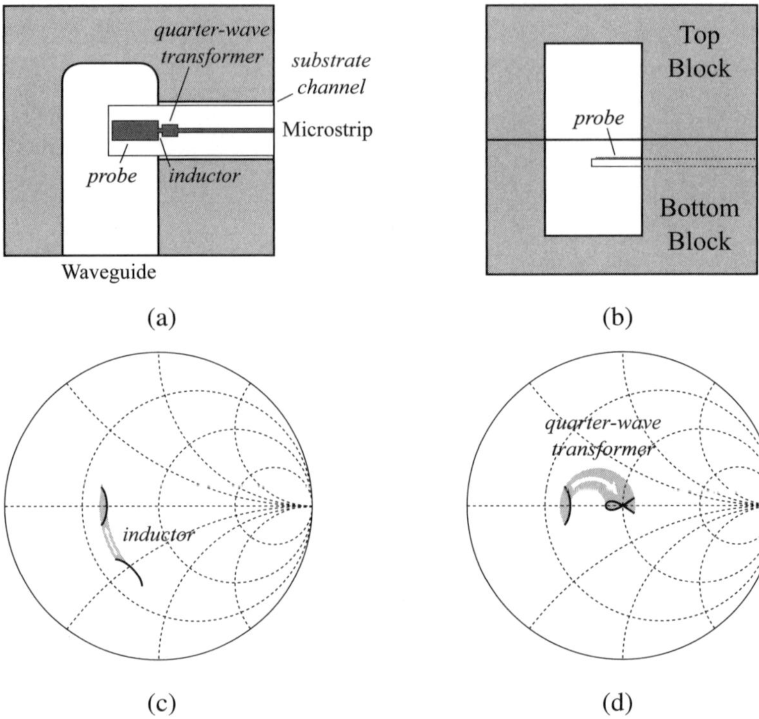

Figure 8.12 Microstrip-to-waveguide transition using a longitudinal electric-field probe. (a) View of substrate mounted in waveguide bottom-half block. (b) View from waveguide port with top and bottom blocks. (c) Smith chart of impedance locus before and after the addition of the matching inductor section. (d) Impedance locus before and after introduction of quarter-wave transformer.

called *longitudinal* because the flat plane of the probe is parallel to the direction of propagation in the waveguide. This is especially convenient because it allows the housing to be split along the mid-point of the broad wall of the waveguide, giving access to the interior for insertion of the active circuitry, a technique known as split-block construction. Recall from Section 7.2.4 that there is no current crossing this symmetry plane in the dominant, TE_{10} mode. Thus, the need for superior electrical contact between the two halves of the housing is relaxed. Split blocks that cut the waveguide along a different plane typically suffer from much greater losses [6].

The probe substrate itself, along with the rest of the active components, typically sits within a shallow channel cut into one-half of the split block. The enclosed channel must be narrow enough that it does not by itself support a parasitic waveguide mode which passes over the microstrip line.

One could arrange for the metallic surface of the probe to be symmetrically located in the waveguide, but this is unnecessary, as a small to moderate offset, such as that indicated in Figure 8.12(b), can easily be accounted for in its design. Since the waveguide is most often used in its single-mode range, the lack of perfect symmetry is of little concern. Also, since the waveguide will typically be cut from a metal housing using a milling machine, the circular cutting tool will inevitably leave rounded corners in the backshort, as shown. This has a minor effect on the electrical length of the backshort from the axis of the probe, but once again this is easily accounted for in its design.

Generally, the combination of the probe paddle and the backshort lead to an impedance curve, as seen from the microstrip side, which is partially capacitive, plotted in the lower curve on the Smith chart of Figure 8.12(c). Matching elements may be included on the microstrip substrate to correct for this impedance. For example, one may include a short section of narrow (high-impedance) transmission line to bring the impedance up to the horizontal axis on the Smith chart. We will call this short piece an inductor, since it effectively cancels out the capacitance of the probe. The next section is a low-impedance quarter-wave transformer, with the result shown in Figure 8.12(d). The immittance curve is thus folded in to the center of the Smith chart as desired. With this design, microstrip-to-waveguide transitions operating with better than 20-dB return loss over the full single-mode bandwidth of the waveguide can be achieved [7].

8.5.2 Vertical Probes

Despite the performance advantages of split-block construction where the waveguide sections are cut along their broad symmetry plane, some applications require a flatter package in which the waveguide port extends vertically from the plane in which the active circuits are mounted. In such cases, a vertical microstrip launcher such as that shown in Figure 8.13(a) may be suitable. Based on the same general principles as the longitudinal microstrip probe, and of the coax-to-waveguide launcher before that, it simply aligns the plane of the probe perpendicular to the direction of propagation in the waveguide. The probe paddle is generally wider than it is in the longitudinal probe for good impedance matching. The entrance into the probe may be slightly tapered to reduce edge capacitance across this extra wide feature, but other additional matching features are typically unnecessary while

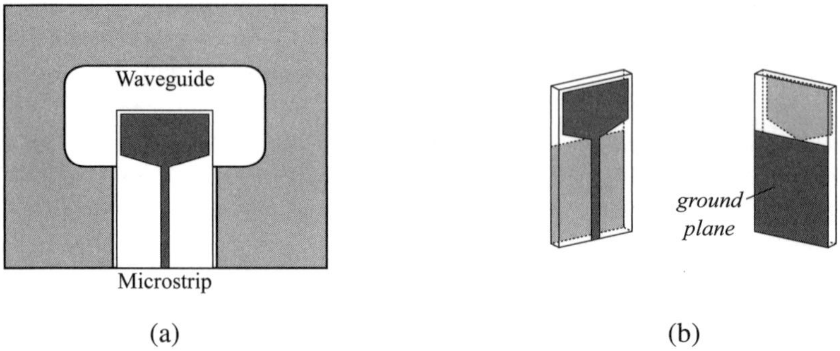

Figure 8.13 A vertical microstrip-to-waveguide launcher. (a) Top view. (b) Oblique views.

still achieving better than 20-dB return loss over the full, single-mode waveguide bandwidth.

The probe may be designed such that the waveguide flange is either above it (in what would be the lid covering the active circuits) or below it. In the former case, the backshort is a machined pocket over which the end of the substrate hangs, as shown in the figure. Its depth will be roughly a quarter-wavelength at the center frequency of the waveguide. If, instead, the waveguide flange is to be underneath the active circuitry, the waveguide aperture is machined clear through, and a backshort cavity is cut into the lid. Either way, the cutting tool will usually leave rounded corners in both the backshort and the waveguide aperture, minor features that can easily be accommodated in the design.

It is important to note, for both the longitudinal and vertical microstrip launchers, that the ground plane of the microstrip is not to extend into the waveguide beneath the probe. Thus, if the ground plane is not just formed by the housing but rather is integral to the substrate (as is most often the case), then that ground plane must be etched back beneath the paddle, as indicated for the vertical launcher's substrate in Figure 8.13(b). Backside patterning is a common feature of many passive printed-circuit technologies (e.g., circuit board and thin-film), but may not be available from standard semiconductor wafer fabrication.

8.6 WAVEGUIDE TAPERS

When transitioning from one type of waveguide to another, one must be mindful of the fact that any geometric discontinuity whatsoever will be prone to the excitation of higher-order modes. Even if these modes are cutoff, they will contribute reactive impedance terms to the junction, causing reflections in the dominant mode. This is essentially a consequence of that fact that waveguide cross-sections in general are not electrically small, so no fundamental change in shape or aspect ratio is negligible. It is thus usually best to transition from one waveguide form to another gradually, as in a multisection transformer or taper, much like was done for transmission lines in Section 6.3.

There is a caveat to this, however. In the case of transmission lines, the matching problem was strictly a matter of impedance, not mode content. When all but one mode is cutoff throughout the waveguide transition, the same general principles usually apply. However, if high-order modes can propagate anywhere within the structure, then one must be extremely careful. It is usually not possible to ensure that such modes never get exited — real-world limitations on manufacturing symmetry and precision are more than sufficient to guarantee their presence — so instead we must work to prevent such modes from becoming trapped and resonant. These are the circumstance under which vanishingly small imperfections can have disastrous consequences [8].

8.6.1 Trapped Mode Resonances

Take, for example, the benign-looking taper from circular to rectangular waveguide in Figure 8.14(a), formed from the intersection of a long, narrow cone and a four-sided pyramid. Such transitions are common in wideband antenna-feed structures, for example. It is smooth, monotonic, and gradual enough to ensure that any impedance-related reflections are minimized. All of our prior analysis on transmission-line tapers suggest that this should be a nearly ideal transition. Even most forms of computer-aided electromagnetic modeling of the complete structure, if not very cautiously configured, would give us exactly the result we expect, a smooth, low-loss transmission coefficient shown in the dashed line in Figure 8.14(b). If the computer says it works, after all, then must it not be true?

Not so. If one built such a transition and measured it, one is very likely to get a result instead like that shown with the solid line in Figure 8.14(b). Though smooth and low loss over most of the band, there appears somewhere in the middle a sharp null, known as a *suckout*.

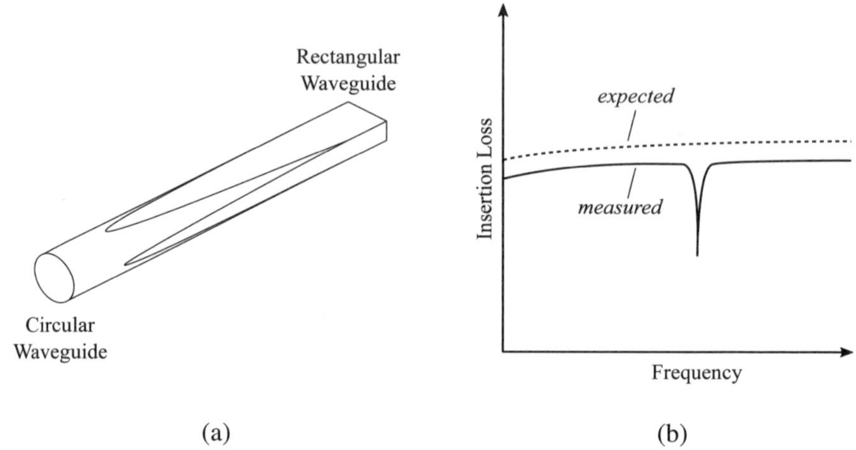

Figure 8.14 (a) Smooth taper between circular and rectangular waveguide. (b) Comparison of expected, smooth insertion loss versus likely measured insertion loss revealing a suckout due to trapped mode resonance.

So why did our computer-aided, electromagnetic simulation fail us? There are at least two possibilities. First, many forms of finite-element models solve Maxwell's equations independently at discrete frequency points, interpolating in between. When we expect a smooth result, we tend not to request a solution at too many frequencies, and issues such as these are so narrowband as to be very easily skipped over if the sampling is not dense enough.

Second, even if we had used very dense frequency sampling, or a transient, time-domain solver that cannot pass over a sharp feature no matter how narrow, it is common practice to draw a symmetric structure such as this with perfect symmetry, and even to enforce that symmetry condition in the model to reduce its size and save compute time. Perfect symmetry reduces to exactly zero any coupling between some modes, but in practice no manufacturing process is capable of producing mathematically perfect symmetry. In some sense, the harder we try, the worse it gets, for we shall soon see that what we have inadvertently created is an almost perfect cavity resonator which resonates in a high-order mode. The closer to perfection in symmetry that we get, the weaker the external coupling to that resonator, which only increases its loaded Q. Any decrease in coupling thus makes the resonance narrower, and sharper, but not shallower.

To see how this has occurred, let us look closer at the mode content along the length of the taper. At the circular end, we have the TE_{nm} and TM_{nm} modes in circular waveguide derived in Section 7.3, and at the rectangular end we have the TE_{mn} and TM_{mn} modes of rectangular waveguide derived in Section 7.2. Obviously, small perturbations in the circular or rectangular geometry should only introduce small perturbations in the modal field solutions, so one would expect the modal fields to gradually change from the circular modes to the rectangular modes in a taper such as this.

Keep in mind that the subscript indices given to the modes in our derivations related to spatial harmonics along the coordinate axes of whatever coordinate systems we used, cylindrical coordinates in the case of the circular waveguide, and Cartesian coordinates in the case of rectangular waveguide. In the middle of the taper, no such closed form solutions are possible, and the particular indices used at one end of the taper have nothing to do with those for essentially the same mode at the other end of the taper. Nonetheless, the modes (both their fields and their cutoff frequencies) vary smoothly from one end of the tapered waveguide to the other, and the names given to the modes at one end can be mapped, or cross-referenced, to the names given to the equivalent modes at the other.

This is illustrated in Figure 8.15(a), where it can be seen that the dominant TE_{11} mode in circular waveguide maps to the dominant TE_{10} mode in the rectangular waveguide. Similarly, the TE_{21} mode in circular waveguide maps to the TE_{11} mode in rectangular waveguide. Recall also that mode degeneracy leads to multiple ways of defining certain modes. Sometimes a mode in one waveguide maps not to a single mode in the other, but rather to a combination of modes that may be degenerate when both waveguides possess the same symmetry properties (e.g., circular and square). When that symmetry is broken, as it is here when we transition to rectangular waveguide (instead of square), the combined degenerate modes which were superposed to create the original mode split into separate modes.

Despite these complexities, we can plot the cutoff frequencies of all the modes in a waveguide transition as a function of position along the taper, as is done in Figure 8.15(b). We start with circular waveguide at the left edge of the plot. Note that many degenerate pairs split into separate modes as we move into the taper and the circular symmetry is lost. The cutoff of the dominant TE mode gets lower in cutoff frequency as most of the others diverge to higher frequencies. One mode in particular, a remnant of the circular TE_{21} mode, trends slightly lower at first, then higher as you move through the taper. This is quite significant, as it indicates that there are some frequencies (one of which is indicated by the horizontal dashed line)

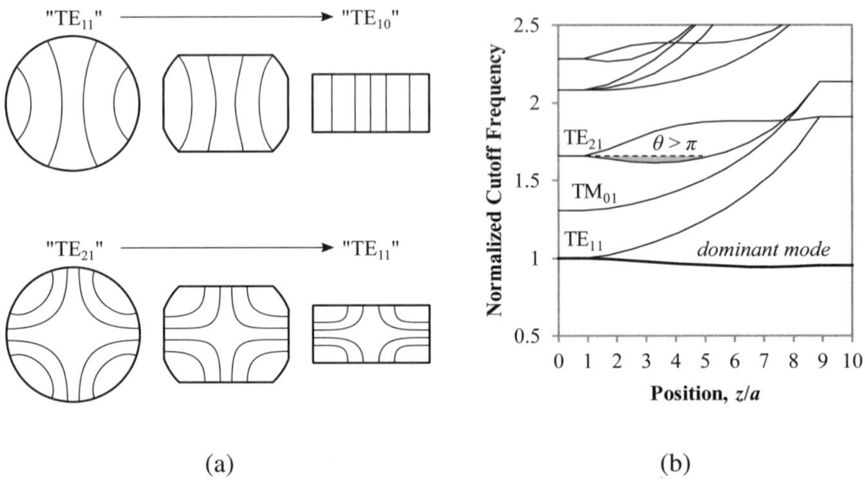

Figure 8.15 (a) Example mapping from circular waveguide modes to rectangular waveguide modes. (b) Plot of cutoff frequencies versus position along the taper. The circular waveguide is at the left edge of the plot ($z = 0$) while rectangular is at the right. The labels correspond to the circular waveguide nomenclature. The shaded region below the dashed line indicates a range of frequencies for which one of the modes is trapped. Because the total phase is greater than 180° (π radians), the trapped mode has become resonant.

at which the given mode can propagate in the center of the taper, but not at either end. Such a mode is said to be trapped at that frequency.

A trapped mode in itself does not always lead to undesirable effects. The danger is that such a trapped mode accumulates enough phase to become resonant. Although a rigorous analysis would require full mode-matching and a determination of the reactive impedance presented by cutoff at each end for the given mode, a very good approximation is that the mode is resonant when the trap is half a wavelength long, analogous to the half-wave transmission-line resonators discussed in Section 3.4.1 or the waveguide cavities in Section 7.4.

When the mode is trapped in a cavity formed by one or more discrete, uniform waveguide sections, the total phase may be calculated by adding the contributions of each individual section. When the cavity is part of a waveguide taper, as is the case here, we must integrate the phase contributed by a continuously variable propagation constant as follows,

$$\theta(f) = \int_{z_1}^{z_2} \beta(z)dz = \int_{z_1}^{z_2} \sqrt{k^2 - k_c^2(z)}dz = 2\pi\sqrt{\mu\varepsilon}\int_{z_1}^{z_2} \sqrt{f^2 - f_c^2(z)}dz \qquad (8.2)$$

where z_1 and z_2 are the boundaries of the trap along the z axis, and f is the highest frequency at which the mode being considered is still trapped. If $\theta \geq n\pi$, then there are potentially n resonant frequencies between the minimum of the cutoff frequency curve and f. For the circular-to-rectangular taper with cutoff frequencies plotted in Figure 8.14(b), we have $\theta \approx 1.5\pi$, indicating a single resonance.

There are at least three ways around this problem. The simplest method is to shorten the taper so that the phase integral in (8.2) is less than π. The mode will still be trapped, but not at any frequency which may become resonant. If the total length of the waveguide cannot be shortened due to external mechanical factors, then the length subtracted from the taper can be added back as a uniform length of circular or rectangular waveguide at one end.

However, if the shortened taper is too abrupt, it may unacceptably degrade the reflection coefficient of the transition. If that is the case, then another approach is to change the intermediate shape of the taper, while maintaining its total length. This may be accomplished by altering the profile of the circular and rectangular elements along the direction of propagation. For instance, instead of the linear taper in a and b of the rectangular waveguide, one dimension or the other could follow a quadratic profile, forming a kind of wedge instead of a four-side pyramid to intersect with the circular cone to create the taper. There is a risk in this approach, however. Although one could render monotonic the cutoff profile of a particular mode, one might find that in the process a basin has been inadvertently created in the profile of another mode, which could then become resonant at a different frequency.

There is a better approach. Simple rules of thumb for common, linear tapers were articulated in [8] that have monotonic cutoff profiles in at least the first 10 or so modes (modes higher than tenth-order would tend to produce resonances only well above the operating band and thus are of little consequence). These rules of thumb are given in Figure 8.16. The first, Figure 8.16(a), states that for circular-to-rectangular tapers where the rectangular waveguide has a 2:1 aspect ratio, monotonic cutoff frequencies can be assumed as long as the diameter of the circular waveguide is more than 1.25 times the width of the rectangular waveguide. The second, Figure 8.16(b), gives a similar inequality for the relative dimensions of circular and square waveguides (i.e., rectangular waveguides with unity aspect ratio).

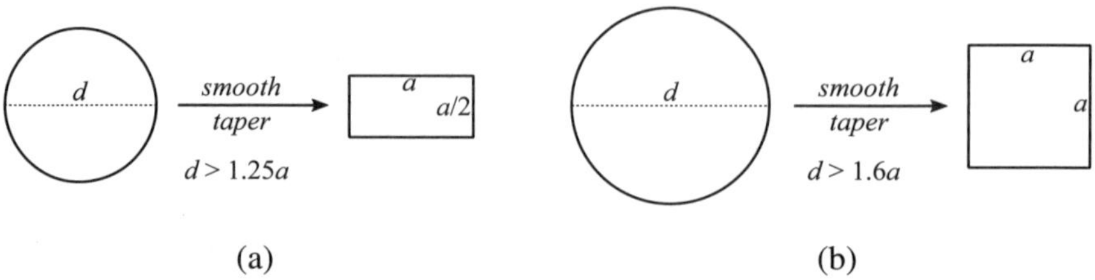

Figure 8.16 rules of thumb for avoiding trapped modes in linear (a) circular-to-rectangular and (b) circular-to-square waveguide tapers. The monotonicity of the first few waveguide modes is assured if the given inequalities are satisfied.

If these inequalities are already met by the required dimensions of the input and output waveguide, then one may employ simple linear taper as previously described without risk. If not, then it must be the case that the circular waveguide is too small, or the rectangular waveguide is too big. One possible solution, then, is to shrink the rectangular waveguide in a linear taper, scaling both the width and the height by the same factor, until the appropriate condition from Figure 8.16 is met. Then a linear taper from circular to square completes the transition. The geometry of such a transition is shown in Figure 8.17(a). The cutoff-frequency curves for such a transition are shown in Figure 8.17(b). The modes are monotonic on either side of the midpoint, and the cutoffs at the midpoint are higher than at either end. This ensures that no trapped mode can exist, and one is free to make this taper as long as is needed to achieve good return loss.

8.7 HYBRID-MODE LAUNCHER

Another important transition we must consider is that which launches a hybrid or Floquet mode in corrugated waveguide from the dominant mode in smooth-walled waveguide. Most often, the goal will be to launch the HE_{11} mode at the balanced hybrid condition, which we recall occurs for large waveguides when the corrugations are a quarter-wavelength deep. Instinctively, we might be tempted to use a kind of taper where the corrugation depth slowly increases from zero, as is shown for the circularly symmetric case in Figure 8.18(a), but empirical studies have shown that such a transition does not yield good results [9].

A far better solution is to start with an initial corrugation a half-wavelength deep at the center of the desired band for the launcher, as shown in Figure 8.18(b).

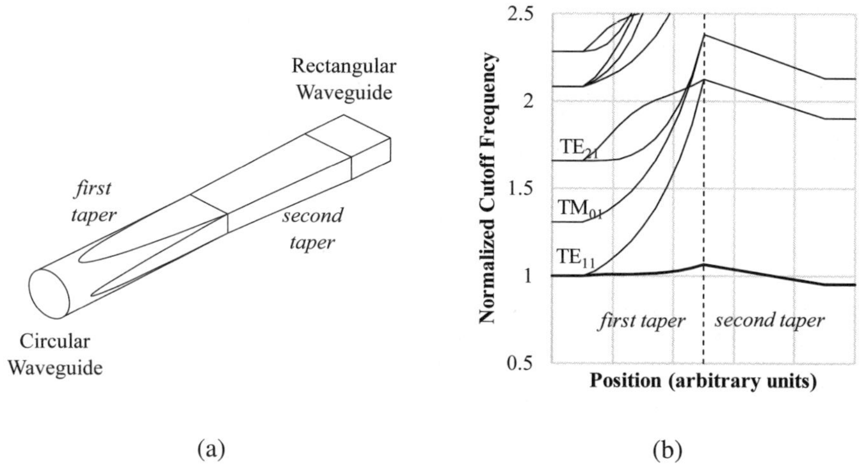

Figure 8.17 (a) Geometry of a two-part circular-to-rectangular waveguide transition. (b) Plot of the normalized cutoff frequencies as a function of position.

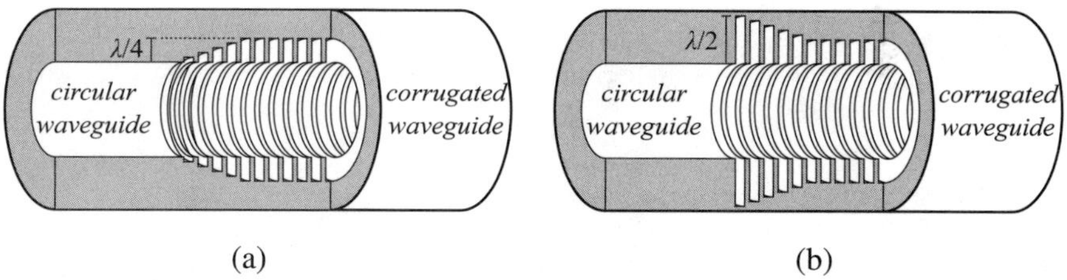

Figure 8.18 Two approaches to launching a hybrid mode in circular corrugated waveguide. (a) Slot depth increasing from zero. (b) Slot depth decreasing from $\lambda/2$.

Thought of as a half-wavelength, short-circuited stub, one can see that this presents a near short-circuit at the tips of the corrugations, essentially matching the surface impedance of the metal in the smooth-wall waveguide. The corrugation depths may then slowly be reduced until they are a quarter-wavelength deep, as shown. This results in a much more broadband matched response.

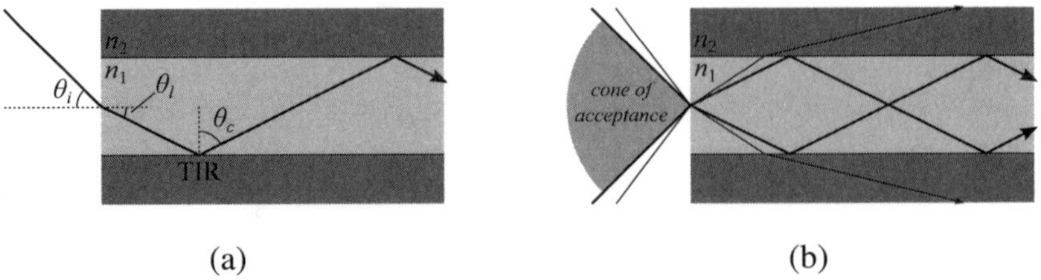

Figure 8.19 Derivation of numerical aperture. (a) Calculation of the maximum incidence angle based on the critical angle for total internal reflection (we assume that outside the fiber is air, $n_0 = 1$). (b) Illustration of the cone of acceptance. The thick lines represent rays that are captured by the fiber, while the thin lines, greater than the maximum angle of incidence, escape into the cladding.

8.8 BEAM COUPLING TO OPTICAL FIBERS

Finally, we consider the problem of launching a guided wave into optical fiber. Due to the way optical fibers are manufactured — by drawing glass down into a thin filament — the active circuitry, typically lasers and photodiodes, are not integral to the structure. Joints can be made by mechanical force (pressing two fibers together end to end) or by splicing (fusing or melting the glass from two fibers into one), but the connection to an active component or to an integrated circuit is typically accomplished by free-space optical coupling into the open end of the fiber. The ease with which this is accomplished is largely determined by a measure known as the *numerical aperture*.

8.8.1 Numerical Aperture

It was stated in Section 7.8.1 that the angle at which plane waves of light would be confined by total internal reflection was limited; beyond that, the wave would escape without being guided. In practice, this means that light shining on the end of an optical fiber must be sufficiently close to axial in order for the light to be captured by the fiber. Let us examine that limit now.

Consider the diagram in Figure 8.19. The rays inside the fiber are presumed to be reflecting from the walls at the critical angle for the fiber, given by (1.105). Assuming the end of the fiber where the light enters is flat and perpendicular to the axis, we have that its launch angle from that interface inside the glass is

$$\theta_l = \frac{\pi}{2} - \theta_c = \frac{\pi}{2} - \sin^{-1}\left(\frac{n_2}{n_1}\right) \tag{8.3}$$

The incidence angle, θ_i, from outside the fiber is then given by Snell's law, where we assume the external medium is air, having a refractive index of $n_0 = 1$,

$$\sin\theta_i = \frac{n_1}{n_0}\sin\theta_l = n_1 \sin\left(\frac{\pi}{2} - \sin^{-1}\left(\frac{n_2}{n_1}\right)\right) \tag{8.4a}$$

$$= n_1 \cos\left(\sin^{-1}\left(\frac{n_2}{n_1}\right)\right) = n_1 \frac{\sqrt{n_1^2 - n_2^2}}{n_1} = \sqrt{n_1^2 - n_2^2} \tag{8.4b}$$

This quantity is known as the numerical aperture, denoted NA. Note that the radial scale factors, u and w, in the LP_{nm} mode solutions given by (7.155) satisfy

$$u^2 + w^2 = k_0^2\left(n_1^2 - n_2^2\right) = (k_0\text{NA})^2 \tag{8.5}$$

8.9 BEAM GENERATION

Implied in the foregoing discussion of numerical aperture is the assumption that we can generate a propagating beam in free space with controlled characteristics, or that we can capture one with definable receptivity. The type of structure that accomplishes this, an antenna, may be viewed as a special kind of transition, from guided to unguided waves, but the wide variety of forms and the elegant, mathematical theory behind them is too rich not to be given their own chapter. Antennas will therefore be discussed along with other optical and quasi-optical components in Chapter 9.

Problems

8-1 Compare the loss per unit length at 10 GHz of a 50Ω microstrip line on 125-μm-thick alumina ($\varepsilon_r = 9.8$) having gold-plated traces ($\sigma = 4.1 \times 10^7$ S/m), and a 50Ω coaxial line having copper conductors ($\sigma = 5.96 \times 10^7$ S/m) and Teflon insulation ($\varepsilon_r = 2.1$) where the inner conductor is 0.28 mm in diameter. Consider conductive losses only, neglecting any surface roughness. Assuming that a transition between the two has 0.5 dB of insertion loss, what is the maximum distance one could cover with the microstrip before it becomes worthwhile to transition to coax and back?

8-2 For the purposes of designing an end-launch microstrip-to-coax transition, write down an expression for the dielectric constant of the microstrip substrate, ε_m, in terms of the dielectric constant of the coax insulator, ε_c, such that the distance between the center conductor and outer conductor is equal to the substrate height, and the diameter of the center conductor is equal to the width of the microstrip trace. Assume both lines are 50Ω, and that the lowest dielectric constant available is 2.1 (for Teflon). If the coax dielectric is Teflon, what must the dielectric constant of the substrate be?

8-3 Two identical, low-loss, and broad bandwidth coax-to-waveguide launchers are measured back-to-back from the waveguide ports with a long cable between them. The return loss shows a periodic response with 12-dB peaks in between nulls. What is the approximate magnitude of the return loss of a single launcher?

8-4 Is a rectangular waveguide balanced or unbalanced?

8-5 Show that the Marchand balun given in Figure 8.7(d) is perfectly matched at the center frequency (where the lines are a quarter-wavelength long) regardless of the values of Z_{bal}, Z_{un}, and ρ. Hint: some of the identities in Appendix F might be useful.

8-6 What are the coupled-line even- and odd-mode impedances needed for a Marchand balun matching $Z_{bal} = 100\Omega$ and $Z_{un} = 50\Omega$, as a function of ρ?

8-7 Nonplanar Marchand baluns often use isolated transmission lines (e.g., coax) in place of the coupled lines. Assume that such a transmission line may be modeled as as a coupled line where both Z_e and ρ diverge to infinity, while Z_o remains constant. Write an expression for Z_o, which may be considered the isolated transmission-line impedance in this limit, in terms of Z_{bal} and Z_{un}.

8-8 Consider the microstrip-to-waveguide launcher shown in Figure 8.12. Assuming that the dielectric material of the substrate can be neglected, how narrow must the channel be to allow operation up to 110 GHz?

8-9 Two square waveguides with side length a are connected by an intervening section of circular waveguide with diameter d having the same dominant mode cutoff frequency. How long can the circular waveguide section be without introducing a trapped mode resonance within the operating bandwidth of the square waveguide (defined by the onset of its first nondominant mode)?

8-10 An "optical fiber" is formed by submerging a rod of quartz ($n = 1.46$) in water ($n = 1.333$). If the end of the rod is brought to the surface of the water, what is the angle of the cone of light which is accepted by the fiber?

8-11 What is the spot size that an open-ended fiber (having $n_1 = 1.46$ and $n_2 = 1.45$) would project onto a perpendicular surface 10 mm away?

References

[1] Southwest Microwave. (2018) End launch connectors DC to 110 GHz. http://mpd.southwestmicrowave.com/showImage.php?image=1171.

[2] M. A. Morgan and S. Weinreb, "A millimeter-wave perpendicular coax-to-microstrip transition," *IEEE MTT-S International Microwave Symposium Digest*, vol. 2, June 2002, pp. 817–820.

[3] K.-P. Ma, Y. Qian, and T. Itoh, "Analysis and applications of a new CPW-slotline transition," *IEEE Transactions on Microwave Theory and Techniques*, vol. 47, no. 4, pp. 426–432, April 1999.

[4] N. Marchand, "Transmission line conversion transformers," *Electronics*, vol. 17, pp. 142–145, December 1944.

[5] M. A. Morgan and T. A. Boyd, "A 10–100 GHz double-ridged horn antenna and coax launcher," *IEEE Transactions on Antennas and Propagation*, vol. 63, no. 8, pp. 3417–3422, August 2015.

[6] M. A. Morgan, "Millimeter-wave MMICs and applications," Ph.D. dissertation, California Institute of Technology, Pasadena, CA, May 2003.

[7] Y. C. Leong and S. Weinreb, "Full-band waveguide-to-microstrip probe transitions," *IEEE MTT-S International Microwave Symposium Digest*, pp. 1435–1438, 1999.

[8] M. A. Morgan and S.-K. Pan, "Graphical prediction of trapped mode resonances in sub-mm and THz waveguide networks," *IEEE Transactions on Terahertz Science and Technology*, vol. 3, no. 1, pp. 72–80, January 2013.

[9] P. J. B. Clarricoats and A. D. Olver, *Corrugated Horns for Microwave Antennas*. London: Peter Peregrinus Ltd., 1984.

Chapter 9

Antennas and Quasioptics

Just as the rules governing lumped elements derived from an approximation of Maxwell's equations in which the physical structures defining the boundary conditions are assumed to be very small compared to a wavelength, the rules governing geometric optics (colloquially "ray tracing") can be derived from the opposite assumption, that the physical structures are very large compared to a wavelength. The term *quasioptics* is typically reserved for a somewhat intermediate case where an unguided electromagnetic wave is reasonably well collimated — meaning that it does not diverge too quickly from the axis of propagation — and has cross-sectional dimensions that are moderate in terms of wavelengths.

9.1 PARALLEL-PLATE WAVEGUIDE

Let us preface our study of quasioptical systems by first considering a region bounded by two perfectly conducting planes of infinite extent, as shown in Figure 9.1. Although commonly referred to as *parallel-plate waveguide* and introduced in most books alongside other waveguides such as rectangular and circular, the lack of confinement of the modes in such a structure along the cross-section plane would seem to disqualify them from that definition. The number of dominant modes in such a waveguide at any frequency is infinite,[1] and the proper excitation of any one

1 We will shortly derive what seems to be a single, dominant TEM mode and other higher-order modes; however, these are based on the assumption of infinite uniformity in the lateral direction and on a chosen axis of propagation. Any true excitation between two parallel plates will necessarily excite an infinite number of these modes, all propagating independently in different directions and with different phases.

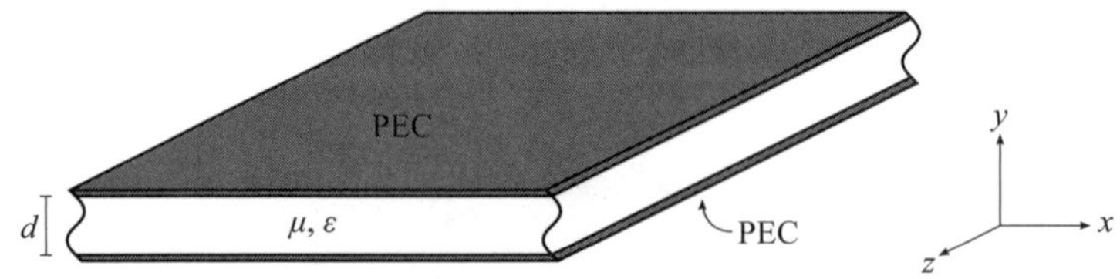

Figure 9.1 Geometry of parallel-plate waveguide.

of those modes requires the incidence of a plane wave along an infinitely wide strip. To my way of thinking, this puts the parallel-plate waveguide firmly in the realm of quasioptical components.

9.1.1 TEM Modes

Nonetheless, it is possible to define something like discrete modes for this waveguide if we first assume that the electric and magnetic fields are uniform in configuration along an infinitely wide strip perpendicular to the axis of propagation. That is, if we let $+z$ be the direction of propagation and the y axis be perpendicular to the guiding planes, then we write

$$\mathbf{E} = \mathbf{E}'(y)e^{-\gamma z} \tag{9.1a}$$

$$\mathbf{H} = \mathbf{H}'(y)e^{-\gamma z} \tag{9.1b}$$

Since there are two conductors, a TEM mode is supported and may be found as though in a transmission line using an electrostatic solution for the transverse fields. We therefore start with Laplace's equation (1.34) for the electrostatic potential between the two plates,

$$\nabla^2 \varphi = 0 \tag{9.2}$$

In the electrostatic case, the uniformity of the boundary conditions in x and y is sufficient alone to ensure that $\varphi = \varphi(y)$. Thus,

$$\nabla^2 \varphi = \frac{d^2}{dy^2}\varphi = 0 \tag{9.3a}$$

$$\therefore \varphi = C_1 y + C_2 \tag{9.3b}$$

and the electrostatic field is

$$\mathbf{E}'(y) = -\nabla\varphi(y) = -\frac{d}{dy}\varphi(y)\mathbf{y} = -C_1\mathbf{y} = E_0\mathbf{y} \tag{9.4}$$

The propagating electric field is then

$$\mathbf{E} = \mathbf{E}'(y)e^{-\gamma z} = E_0 e^{-\gamma z}\mathbf{y} \tag{9.5}$$

and the magnetic field is given by Faraday's law (1.71c)

$$\mathbf{H} = -\frac{1}{j\omega\mu}\nabla\times\mathbf{E} = \frac{E_0}{j\omega\mu}\frac{d}{dz}e^{-\gamma z}\mathbf{x} = -\frac{E_0\gamma}{j\omega\mu}e^{-\gamma z}\mathbf{x} \tag{9.6}$$

For a TEM mode, we have $\gamma = jk = j\omega\sqrt{\mu\varepsilon}$,

$$\mathbf{H} = -\frac{E_0 j\omega\sqrt{\mu\varepsilon}}{j\omega\mu}e^{-jkz}\mathbf{x} = -\frac{E_0}{\eta}e^{-jkz}\mathbf{x} \tag{9.7}$$

where we have used the standard intrinsic impedance definition, $\eta = \sqrt{\mu/\varepsilon}$. Note that this is exactly the same as a plane wave in free space, only bounded by the metal plates in the y direction.

Since the mode is infinite in extent, the total power propagating in the mode is also infinite, underscoring the view that parallel plates can only loosely be considered a waveguide. Nonetheless, a power density per unit width can be defined for the wave, given by integrating the Poynting vector across the gap in the y direction,

$$P' = \int_0^d (\mathbf{E}_t \times \mathbf{H}_t^*)\cdot\mathbf{z}dy = E_0 H_0^* d \tag{9.8}$$

Similarly, the power loss per unit width and per unit distance propagated may be calculated by considering both conductive and dielectric losses,

$$P_c' = R_s\left|\mathbf{J}_s(0)\right|^2 + R_s\left|\mathbf{J}_s(d)\right|^2 = 2\sqrt{\frac{\omega\mu'}{2\sigma}}\left|H_0\right|^2 \tag{9.9a}$$

$$P_d' = \omega\varepsilon\tan\delta\int_0^d |\mathbf{E}|^2 dy = \omega\varepsilon\tan\delta\left|E_0\right|^2 d \tag{9.9b}$$

The conductive and dielectric attenuation constants are thus found quite simply,

$$\alpha_c = \frac{P'_c}{2P'} = \sqrt{\frac{\omega\mu'}{2\sigma}}\frac{|H_0|^2}{E_0 H_0^* d} = \sqrt{\frac{\omega\mu'}{2\sigma}}\frac{1}{\eta d} \tag{9.10a}$$

$$\alpha_d = \frac{P'_d}{2P'} = \frac{\omega\varepsilon\tan\delta\,|E_0|^2\,d}{2E_0 H_0^* d} = \frac{1}{2}k\tan\delta \tag{9.10b}$$

9.1.2 TE Modes

We may also derive TE modes by solving for $H_z(y)$ in the Helmholtz equation,

$$\left(\nabla^2 + k^2\right) H'_z e^{-\gamma z} = 0 \tag{9.11a}$$

$$\left(\frac{d^2}{dy^2} + \gamma^2 + k^2\right) H'_z e^{-\gamma z} = 0 \tag{9.11b}$$

$$\left(\frac{d^2}{dy^2} + k_c^2\right) H'_z = 0 \tag{9.11c}$$

The solutions to this now-familiar equation are sines and and cosines in $k_c y$. Choosing the cosine form so that the differential boundary condition is met at $y = 0$, we have

$$H'_z = H_0 \frac{k_c}{\gamma} \cos(k_c y) \tag{9.12}$$

where the usual prefactor, k_c/γ, has been added to simplify the subsequent expressions. If the separation between planes is d, then we must also have

$$k_c = \frac{n\pi}{d} \tag{9.13}$$

where $n \geq 1$ and

$$\gamma = \sqrt{k_c^2 - k^2} = \sqrt{\left(\frac{n\pi}{d}\right)^2 - k^2} \tag{9.14}$$

The transverse fields may then be found using the general relations in Table 7.1,

$$\mathbf{H}'_t = -\frac{\gamma}{k_c^2}\nabla_t H'_z = -\frac{\gamma}{k_c^2}\frac{d}{dy} H_0 \frac{k_c}{\gamma}\cos(k_c y)\,\mathbf{y} = H_0 \sin(k_c y)\,\mathbf{y} \tag{9.15a}$$

$$\mathbf{E}_t' = Z_{TE}\left(\mathbf{H}_t' \times \mathbf{z}\right) = E_0 \sin\left(k_c y\right)\mathbf{x} \tag{9.15b}$$

where $E_0 = Z_{TE}H_0$. Putting these results together, we have for the TE$_n$ mode,

$$\mathbf{E} = E_0 e^{-\gamma z}\sin\left(k_c y\right)\mathbf{x} \tag{9.16a}$$

$$\mathbf{H} = H_0 e^{-\gamma z}\left(\sin\left(k_c y\right)\mathbf{y} + \frac{k_c}{\gamma}\cos\left(k_c y\right)\mathbf{z}\right) \tag{9.16b}$$

The power density is calculated once again from the Poynting vector,

$$P' = \int_0^d \left(\mathbf{E}_t \times \mathbf{H}_t^*\right) \cdot \mathbf{z} dy = E_0 H_0^* \int_0^d \sin^2\left(k_c y\right) dy = \frac{d}{2} E_0 H_0^* \tag{9.17}$$

and the losses due to conductivity and dielectric loss tangent are

$$P_c' = R_s \left|\mathbf{J}_s(0)\right|^2 + R_s \left|\mathbf{J}_s(d)\right|^2 = 2\sqrt{\frac{\omega\mu'}{2\sigma}} \left|H_0\right|^2 \frac{k_c^2}{|\gamma|^2} \tag{9.18a}$$

$$P_d' = \omega\varepsilon \tan\delta \int_0^d |\mathbf{E}|^2 dy = \omega\varepsilon \tan\delta \left|E_0\right|^2 \frac{d}{2} \tag{9.18b}$$

Thus, in the case that the wave is propagating ($\gamma = j\beta$), we may write the attenuation constants as

$$\alpha_c = \frac{P_c'}{2P'} = \sqrt{\frac{\omega\mu'}{2\sigma}} \frac{2k_c^2}{k\beta\eta d} \tag{9.19a}$$

$$\alpha_d = \frac{P_d'}{2P'} = \frac{k^2}{2\beta} \tan\delta \tag{9.19b}$$

9.1.3 TM Modes

The TM modes may be derived in a very similar fashion. This time we let

$$E_z' = E_0 \frac{k_c}{\gamma} \sin\left(k_c y\right) \tag{9.20}$$

which solves the Helmholtz equation for E'_z as well as the tangential electric field boundary conditions when

$$k_c = \frac{n\pi}{d} \tag{9.21a}$$

$$\gamma = \sqrt{k_c^2 - k^2} = \sqrt{\left(\frac{n\pi}{d}\right)^2 - k^2} \tag{9.21b}$$

and $n \geq 1$ as before (when $n = 0$, the solution is the same as the TEM mode already found). The rest of the fields for the TM_n mode follow naturally from the equations in Table 7.1,

$$\mathbf{E} = E_0 e^{-\gamma z} \left(-\cos(k_c y)\,\mathbf{y} + \frac{k_c}{\gamma} \sin(k_c y)\,\mathbf{z} \right) \tag{9.22a}$$

$$\mathbf{H} = H_0 e^{-\gamma z} \cos(k_c y)\,\mathbf{x} \tag{9.22b}$$

Power density and loss are calculated as before

$$P' = \int_0^d (\mathbf{E}_t \times \mathbf{H}_t^*) \cdot \mathbf{z} dy = E_0 H_0^* \int_0^d \cos^2(k_c y)\, dy = \frac{d}{2} E_0 H_0^* \tag{9.23a}$$

$$P'_c = R_s \left|\mathbf{J}_s(0)\right|^2 + R_s \left|\mathbf{J}_s(d)\right|^2 = 2\sqrt{\frac{\omega\mu'}{2\sigma}} \left|H_0\right|^2 \tag{9.23b}$$

$$P'_d = \omega\varepsilon \tan\delta \int_0^d |\mathbf{E}|^2 dy \tag{9.23c}$$

$$= \omega\varepsilon \tan\delta \left|E_0\right|^2 \int_0^d \left(\cos^2(k_c y) + \frac{k_c^2}{\beta^2} \sin^2(k_c y) \right) dy \tag{9.23d}$$

$$= \omega\varepsilon \tan\delta \left|E_0\right|^2 \frac{d}{2} \left(1 + \frac{k_c^2}{\beta^2} \right) = \omega\varepsilon \tan\delta \left|E_0\right|^2 \frac{d}{2} \frac{k^2}{\beta^2} \tag{9.23e}$$

$$= \frac{k^3 d}{2\beta^2 \eta} \tan\delta \left|E_0\right|^2 \tag{9.23f}$$

Therefore,

$$\alpha_c = \frac{P'_c}{2P'} = \sqrt{\frac{\omega\mu'}{2\sigma}} \frac{2k}{\beta\eta d} \tag{9.24a}$$

Table 9.1
Equations for Parallel-Plate Waveguide

Parameter	Symbol	Expression
	TEM Modes	
Electric field	$\mathbf{E}$	$E_0 e^{-jkz}\mathbf{y}$
Magnetic field	$\mathbf{H}$	$-H_0 e^{-jkz}\mathbf{x}$
Wave impedance	Z_{TEM}	$\frac{E_0}{H_0} = \eta = \sqrt{\frac{\mu}{\varepsilon}}$
Power density per unit width	P'	$E_0 H_0^* d$
Attenuation constant (conduction)	α_c	$\sqrt{\frac{\omega\mu'}{2\sigma}}\frac{1}{\eta d}$
Attenuation constant (dielectric)	α_d	$\frac{1}{2}k\tan\delta$
	TE Modes	
Electric field	$\mathbf{E}$	$E_0 e^{-\gamma z}\sin(k_c y)\,\mathbf{x}$
Magnetic field	$\mathbf{H}$	$H_0 e^{-\gamma z}\left(\sin(k_c y)\,\mathbf{y} + \frac{k_c}{\gamma}\cos(k_c y)\,\mathbf{z}\right)$
Wave impedance	Z_{TE}	$\frac{E_0}{H_0} = \frac{j\omega\mu}{\gamma} = \frac{jk\eta}{\gamma}$
Cutoff wavenumber	k_c	$\frac{n\pi}{d}$
Power density per unit width	P'	$\frac{d}{2}E_0 H_0^*$
Attenuation constant (conduction)	α_c	$\sqrt{\frac{\omega\mu'}{2\sigma}}\frac{2k_c^2}{k\beta\eta d}$
Attenuation constant (dielectric)	α_d	$\frac{k^2}{2\beta}\tan\delta$
	TM Modes	
Electric field	$\mathbf{E}$	$E_0 e^{-\gamma z}\left(-\cos(k_c y)\,\mathbf{y} + \frac{k_c}{\gamma}\sin(k_c y)\,\mathbf{z}\right)$
Magnetic field	$\mathbf{H}$	$H_0 e^{-\gamma z}\cos(k_c y)\,\mathbf{x}$
Wave admittance	Y_{TM}	$\frac{H_0}{E_0} = \frac{j\omega\varepsilon}{\gamma} = \frac{jk}{\eta\gamma}$
Cutoff wavenumber	k_c	$\frac{n\pi}{d}$
Power density per unit width	P'	$\frac{d}{2}E_0 H_0^*$
Attenuation constant (conduction)	α_c	$\sqrt{\frac{\omega\mu'}{2\sigma}}\frac{2k}{\beta\eta d}$
Attenuation constant (dielectric)	α_d	$\frac{k^2}{2\beta}\tan\delta$

$$\alpha_d = \frac{P'_d}{2P} = \frac{k^2}{2\beta} \tan \delta \tag{9.24b}$$

A summary of the TEM, TE, and TM mode solutions for parallel-plate waveguide is given in Table 9.1.

9.2 FOURIER OPTICS

Although the parallel-plate mode solutions derived in Section 9.1 are calculable and physically valid, they are nonetheless somewhat artificial in that none can exist in isolation; one cannot excite a uniform wavefront across an infinitely wide strip, so any real excitation will, by necessity, excite an infinite number of such modes, all perhaps having the same mode identity (such as TEM, TE_1, or TM_2) but with differing phases and directions of propagation. Said another way, the wave is guided by the plates on the y axis, but propagates freely (unguided) in the two-dimensional xz plane, as though in a two-dimensional analog of free-space. This will serve as our introduction to quasioptical techniques.

9.2.1 Planar Fourier Optics

Let us consider, for example, a parallel-plate structure where the spacing between the plates is small enough to support only the TEM mode at a given frequency. Imagine linear wavefronts in the TEM mode (the two-dimensional analog of plane waves) propagating at different angles through a boundary line on the x axis, as shown in Figure 9.2. A snapshot of the aperture field distribution (at the bottom of the figure) shows that each plane wave produces a sinusoidal pattern in the aperture with a spatial frequency that depends on the angle of incidence. In particular, we may write the total wavenumber, k (which is fixed for a given frequency), as a combination of the axial (z-directed) and lateral (x-directed) wavenumbers using nothing more than the Pythagorean theorem,

$$k^2 = k_x^2 + k_z^2 \tag{9.25}$$

The aperture pattern is then simply given by writing down the wave solution for arbitrary k_x and k_z

$$E(x, z) = E_0 e^{-j(k_x x + k_z z)} \tag{9.26}$$

and substituting $z = 0$,

$$E(x, 0) = E_0 e^{-jk_x x} \tag{9.27}$$

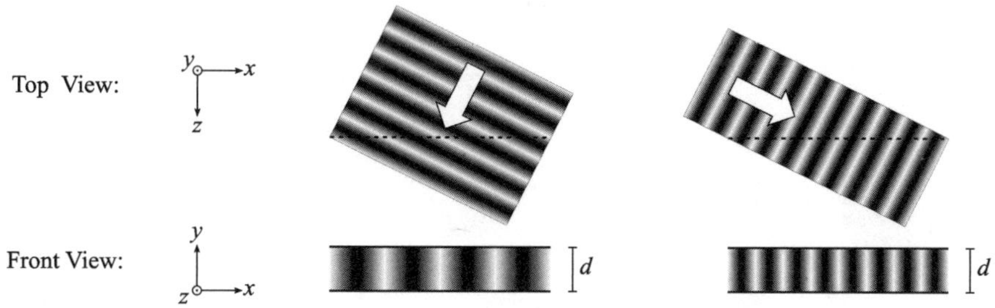

Figure 9.2 Linear TEM wavefronts propagating at oblique angles through an aperture in parallel-plate waveguide. The shading represents field amplitude as seen from above looking down on the parallel plates (top view) and in the aperture plane (front view).

Let us now postulate a more complex excitation of the parallel-plate waveguide comprising an arbitrary field pattern along the x-axis, oscillating in time at a single frequency for which the waveguide is consider single-moded. Elements of the field pattern that are not uniform in the y direction do not conform to the TEM mode and are thus cutoff, but that still leaves an arbitrary intensity distribution in the x direction. We may write this arbitrary intensity profile as a Fourier transform of spatial frequency components,

$$E(x, 0) = \mathfrak{F}\{\mathcal{E}\}(x) = \int_{-\infty}^{\infty} \mathcal{E}(\omega_x)\, e^{-j\omega_x x} d\omega_x \tag{9.28}$$

We see by comparing (9.28) with (9.27) that the Fourier spectrum of the field distribution, $\mathcal{E}(\omega_x)$, is a decomposition of the aperture fields into a continuum of linear TEM-wave solutions [1], where the spatial frequency, ω_x, is equal to k_x, the lateral wavenumber. We may therefore calculate the field pattern at any point in the xz plane resulting from this excitation by tracking the propagation of these modes. This is the basis of *Fourier optics* [2].

Note that when $k_x > k$, we find from (9.25) that k_z is imaginary, representing an evanescent mode in the z direction. Thus, spatial components in the excitation that are smaller than a wavelength decay exponentially and do not contribute to the fields far away from the source. This is known as the *diffraction limit* in optics, and the distant region in which the evanescent fields have decayed sufficiently to be considered negligible is called the *far field*.

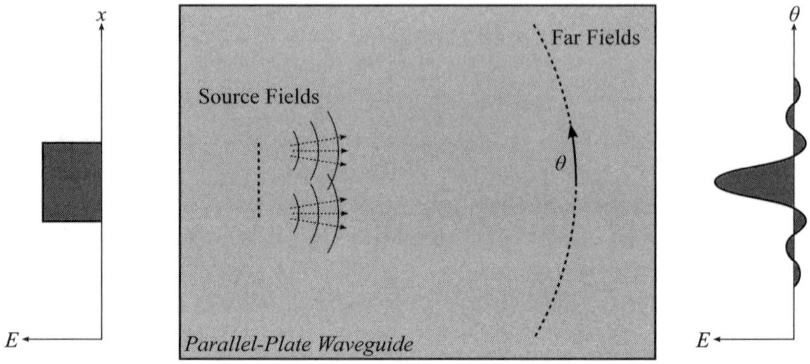

Figure 9.3 Evolution of a local source field distribution into its far-field radiation pattern in parallel-plate waveguide.

Assuming that the source is compact, or finite in extent,[2] we may calculate the far-field radiation pattern by taking the limit as the distance from the source diverges to infinity, as illustrated in Figure 9.3. Instinctively, we might expect that only the linear TEM-wave component which is directed along a particular axis will contribute to the far-field radiation pattern in that direction. This is, in fact, the case, but we can prove it mathematically. First, we write the total field at coordinates (x,z) as the integral over the contributions from all linear TEM-wave components,

$$E(x,z) = \int_{-\infty}^{\infty} \mathcal{E}\left(\omega_x\right) e^{-j\left(k_x(\omega_x)x + k_z(\omega_x)z\right)} d\omega_x \tag{9.29}$$

where

$$k_x\left(\omega_x\right) = \omega_x \tag{9.30a}$$

$$k_z\left(\omega_x\right) = \sqrt{k^2 - \omega_x^2} \tag{9.30b}$$

Therefore,

$$E(x,z) = \int_{-\infty}^{\infty} \mathcal{E}\left(\omega_x\right) e^{-j\left(\omega_x x + z\sqrt{k^2 - \omega_x^2}\right)} d\omega_x \tag{9.31a}$$

2 If the source has infinite extent, as was the case for the fundamental-mode solutions in Section 9.1, then it is not possible to get far enough away from it to be in the far field.

$$= \int_{-\infty}^{\infty} \mathcal{E}(\omega_x)\, e^{-jr\left(\omega_x \sin\theta + \cos\theta\sqrt{k^2-\omega_x^2}\right)} d\omega_x \tag{9.31b}$$

where we have substituted $x = r\sin\theta$ and $z = r\cos\theta$ (thus $\theta = 0$ is in the direction perpendicular to the excitation aperture). Note that when $\omega_x^2 > k^2$, we have $k_z = \pm j\sqrt{\omega_x^2 - k^2}$. Only the negative sign makes sense, as otherwise the exponential term (having a positive real part) would diverge to infinity for large values of r. However, those terms having the negative real part represent evanescent fields, and since we have assumed r is large, the integrand would be vanishingly small. We may thus tighten the limits on the integral to $\pm k$,

$$E(r,\theta) = \int_{-k}^{k} \mathcal{E}(\omega_x)\, e^{-jr\left(\omega_x \sin\theta + \cos\theta\sqrt{k^2-\omega_x^2}\right)} d\omega_x \tag{9.32}$$

Let us now substitute $\omega_x = k\sin\phi$, where ϕ is the propagation direction in the xz plane of the linear TEM wave associated with ω_x,

$$E(r,\theta) = \int_{-\pi/2}^{\pi/2} \mathcal{E}(k\sin\phi)\, e^{-jkr(\sin\phi\sin\theta + \cos\phi\cos\theta)} (k\cos\phi) d\phi \tag{9.33a}$$

$$= \int_{-\pi/2}^{\pi/2} \mathcal{E}(k\sin\phi)\, e^{-jkr\cos(\theta-\phi)} (k\cos\phi) d\phi \tag{9.33b}$$

Since $kr \gg 1$, the exponential inside the integral undergoes rapid oscillations for all arguments except when $\theta = \phi$. It is only near that point that contributions to the integral are significant. Evaluating the integral using *stationary phase approximation*[3] [3], we obtain

$$E(r,\theta) \approx (k\cos\theta)\mathcal{E}(k\sin\theta) \int_{-\pi/2}^{\pi/2} e^{-jkr\cos(\theta-\phi)} d\phi \tag{9.34a}$$

3 The stationary phase approximation is essentially the result of taking the limit of the integral (9.33b) as $kr \to \infty$. Increasingly rapid oscillations of the exponential term effectively wash out all contributions to the end result except where θ and ϕ are the same; at that value of θ alone, the phase is stationary under this limit, and the value of the integral converges. This allows terms containing the variable of integration, ϕ, to be moved outside the integral with the substitution, $\theta = \phi$.

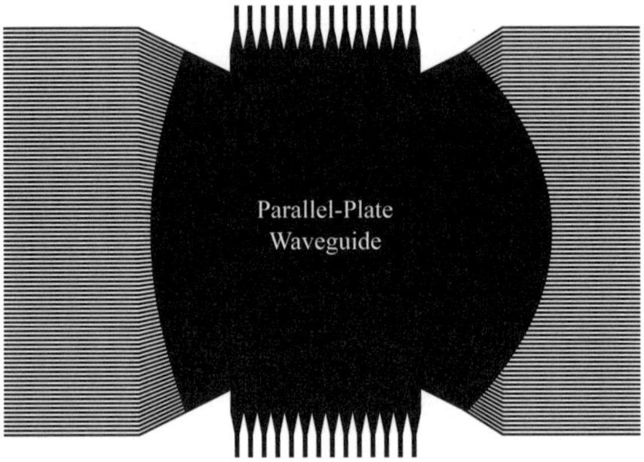

Figure 9.4 A Rotman lens. The feeder lines at the left launch a patterned wavefront into the parallel-plate waveguide section in the middle, and the pickups at the right contain the Fourier transform of that pattern. The dummy ports or baffles along the top and bottom edge prevent stray wave components outside the capture window from reflecting back into the device and corrupting the results.

$$= (k \cos\theta)\sqrt{\frac{2\pi}{kr}}e^{j\pi/4}e^{-jkr}\mathcal{E}\,(k\sin\theta) \tag{9.34b}$$

The $\cos\theta$ term is simply the projected angle of the aperture, and is nearly constant close to $\theta = 0$. The exponential term describes the expected phase rotation of a radially propagating wave, and the $\sqrt{r}$ in the denominator simply provides for conservation of energy in a cylindrically expanding wavefront (since the magnetic field, H, will include a similar term). Within a scaling factor, we are left to conclude that the angular radiation pattern is given by the Fourier transform of the field distribution in the excitation aperture.

9.2.2 Rotman Lens

This result is the foundation of an elegant device known as a *Rotman lens* [4]. Details vary, but an approximate typical layout is shown in Figure 9.4. Typically used for beamforming applications in radars, the Rotman lens works by feeding the inputs from multiple antennas into a parallel-plate waveguide, allowing the combined wavefront to propagate across a large distance, then reading out the Fourier transform of that input pattern at the far end. Note that this device is required to be electrically large, since the pickup antennas must be in the far field of the feeder lines.

9.2.3 Fourier Optics in Free Space

In Section 9.2.1, we derived some of the key properties of Fourier optics in a planar region bounded by parallel-plate waveguide, but the same general conclusions apply to unguided waves in free space. Consider a local field distribution in a two-dimensional aperture normal to the z axis. We may express this field distribution as a two-dimensional Fourier transform of spatial frequencies in x and y,

$$E(x, y, 0) = \mathfrak{F}\{\mathcal{E}\}(x, y) \tag{9.35a}$$

$$= \iint \mathcal{E}\left(k_x, k_y\right) e^{-j(k_x x + k_y y)} dk_x dk_y \tag{9.35b}$$

Each component of the Fourier spectrum, $\mathcal{E}\left(k_x, k_y\right)$, corresponds to a plane wave with wavenumber, k, where

$$k^2 = k_x^2 + k_y^2 + k_z^2 \tag{9.36}$$

As before, we may deduce that since the source field distribution is compact, only that plane-wave component that is oriented in a particular direction will survive to contribute to the far-field radiation pattern in that direction. The general relation in this case [2] is

$$E(r, \theta, \phi) = j2\pi(k\cos\theta)\frac{e^{-jkr}}{r}\mathcal{E}\left(k_x, k_y\right) \tag{9.37}$$

where

$$k_x = k \sin\theta \cos\phi \tag{9.38a}$$

$$k_y = k \sin\theta \sin\phi \tag{9.38b}$$

9.3 GAUSSIAN BEAMS

The problem with Fourier optics is that it only tells us how the local fields affect radiated fields very far away from them; it does not give us any information about the fields in between. Despite the lack of a wave-guiding structure, we would like to derive some form of mode solution that shows how a wavefront evolves as it propagates away from the source. To do this, we must do away with our simple plane waves that have infinite lateral extent and instead define a compact cross-section profile which confines the propagating waves within a localized beam.

It would furthermore simplify analysis if the profile that localizes the wavefront is unchanged (in form, if not in scale) between the near and far fields. Based on Fourier optics, then, we must choose a profile that is its own Fourier transform, the unique property of a Gaussian distribution.

9.3.1 The Paraxial Helmholtz Equation

Let us start once again with the Helmholtz equation,

$$\left(\nabla^2 + k^2\right) E = 0 \tag{9.39}$$

where E is assumed to be transverse and linearly polarized. For propagation primarily in the $+z$ direction, we may write

$$E(x, y, z) = u(x, y, z)e^{-jkz} \tag{9.40}$$

The exponential is the usual phase term for propagating waves, and $u(x, y, z)$ is the envelope within which the wavefronts move. In contrast with fully guided waves, this complex amplitude can have a dependence on z in addition to x and y, but that longitudinal dependence is expected to be smooth compared to the normal phase progression. Putting this back into the Helmholtz equation, we have

$$\left(\nabla^2 + k^2\right) E = \left(\nabla_t^2 + \frac{\partial^2}{\partial z^2} + k^2\right) ue^{-jkz} = 0 \tag{9.41}$$

where, as before, ∇_t is the Laplacian operator in transverse coordinates. Note that

$$\frac{\partial^2}{\partial z^2}\left(ue^{-jkz}\right) = \frac{\partial}{\partial z}\left(\frac{\partial u}{\partial z}e^{-jkz} - jkue^{-jkz}\right) \tag{9.42a}$$

$$= \left(\frac{\partial^2 u}{\partial z^2} - j2k\frac{\partial u}{\partial z} - k^2 u\right) e^{-jkz} \tag{9.42b}$$

Therefore, we have

$$\left(\nabla_t^2 u + \frac{\partial^2 u}{\partial z^2} - j2k\frac{\partial u}{\partial z}\right) e^{-jkz} = 0 \tag{9.43}$$

or, more simply,

$$\nabla_t^2 u + \frac{\partial^2 u}{\partial z^2} - j2k\frac{\partial u}{\partial z} = 0 \tag{9.44}$$

This is sometimes referred to as the *reduced wave equation* [5].

To define a propagating mode in the absence of any guiding structure, we have to make some assumptions about the geometry of our solution. First, we assume that the wave is not rapidly divergent from the z axis. This means that the transverse variation of the envelope will be more rapid than its longitudinal variation,

$$\left|\frac{\partial^2 u}{\partial z^2}\right| \ll \left|\nabla_t^2 u\right| \tag{9.45}$$

Further, we may conclude that its amplitude decay in the longitudinal direction over a wavelength distance will be slow, allowing us to write

$$\left|\frac{\partial^2 u}{\partial z^2}\right| \ll \left|2k\frac{\partial u}{\partial z}\right| \tag{9.46}$$

These assumptions collectively define the *paraxial approximation*. Applying these to (9.44), we have

$$\nabla_t^2 u - j2k\frac{\partial u}{\partial z} = 0 \tag{9.47}$$

which is known as the *paraxial Helmholtz equation*. If we further cast this equation in cylindrical coordinates,

$$\frac{1}{r}\frac{\partial}{\partial r}\left(r\frac{\partial u}{\partial r}\right) + \frac{1}{r^2}\frac{\partial^2 u}{\partial \theta^2} - j2k\frac{\partial u}{\partial z} = 0 \tag{9.48}$$

and assume that it is rotationally constant ($\partial u/\partial\theta = 0$), then we have

$$\frac{1}{r}\frac{\partial}{\partial r}\left(r\frac{\partial u}{\partial r}\right) - j2k\frac{\partial u}{\partial z} = 0 \tag{9.49a}$$

$$\frac{\partial^2 u}{\partial r^2} + \frac{1}{r}\frac{\partial u}{\partial r} - j2k\frac{\partial u}{\partial z} = 0 \tag{9.49b}$$

which is the *axially symmetric paraxial Helmholtz equation*.

9.3.2 The Gaussian Mode Solution

Some progress has been made, but there are still infinitely many solutions to this equation. To narrow down the choices, we may now apply our prescribed Gaussian profile to the aperture fields at $z = 0$. Recall that a Gaussian profile was selected since it has the unique property of being its own Fourier transform, suggesting through Fourier optics that it shall remain intact, like a mode, from the near field into the far field. A Gaussian profile is simply one that has an e^{-r^2} dependence. Adding some scale factors to this form, we may write

$$u(r,0) = u(0,0)e^{-\frac{r^2}{w_0^2}} \tag{9.50}$$

The parameter w_0 shall be called the *beam waist radius* and is the point at which the field amplitude in the aperture is reduced by a factor of $1/e$ relative to its value on-axis.

Let us then postulate a z-dependent solution having the form

$$u(r,z) = A(z)e^{-\frac{r^2}{b(z)}} \tag{9.51}$$

where

$$b(0) = w_0^2 \tag{9.52}$$

Substituting this into (9.49), we obtain

$$A(z)\frac{1}{r}\frac{\partial}{\partial r}\left(r\frac{\partial}{\partial r}e^{-\frac{r^2}{b(z)}}\right) - j2k\frac{\partial}{\partial z}\left(A(z)e^{-\frac{r^2}{b(z)}}\right) = 0 \tag{9.53a}$$

$$A\frac{1}{r}\left(\frac{-4r}{b} + \frac{4r^3}{b^2}\right)e^{-\frac{r^2}{b}} - j2k\left(\frac{dA}{dz} + \frac{r^2}{b^2}A\frac{db}{dz}\right)e^{-\frac{r^2}{b}} = 0 \tag{9.53b}$$

$$A\frac{r^2}{b^2}\left(4 - j2k\frac{db}{dz}\right) = 2\left(\frac{2A}{b} + jk\frac{dA}{dz}\right) \tag{9.53c}$$

Note that the left side is proportional to r^2, while the right side has no r-dependence at all. For this to be satisfied for all r, we must have both sides equal to zero,

$$\frac{db}{dz} = \frac{2}{jk} \tag{9.54a}$$

$$\frac{2A}{b} = -jk\frac{dA}{dz} \tag{9.54b}$$

We may note at this point that while $A(z)$ and $b(z)$ were prescribed to be real at the waist, $z = 0$, the paraxial Helmholtz equation has dictated that they cannot remain so at all values of z. We shall see that the imaginary parts of these terms contribute to phase rotation beyond that which would be experienced by a simple plane wave. Solving (9.54a) for $b(z)$,

$$b(z) = b(0) + \frac{2}{jk}z = w_0^2 + \frac{2}{jk}z \tag{9.55}$$

Let us now substitute (9.54a) in (9.54b),

$$\frac{2A}{b} = -jk\frac{dA}{dz} = -2\frac{dA}{db} \tag{9.56a}$$

$$\frac{dA}{db} = -\frac{A}{b} \tag{9.56b}$$

$$\therefore A = \frac{C_0}{b} = \frac{C_0}{w_0^2 + \frac{2}{jk}z} \tag{9.56c}$$

Clearly, $1/b(z)$ is an important quantity in defining both the amplitude and phase variations along the beam axis. Let us rewrite this in a more useful form,

$$\frac{1}{b(z)} = \frac{1}{w_0^2 + \frac{2}{jk}z} = \frac{w_0^{-2}}{1 - j\frac{2}{kw_0^2}z} = \frac{w_0^{-2}}{1 - j\frac{z}{z_R}} \tag{9.57a}$$

$$= \frac{w_0^{-2}\left(1 + j\frac{z}{z_R}\right)}{1 + \left(\frac{z}{z_R}\right)^2} \tag{9.57b}$$

where

$$z_R = \frac{kw_0^2}{2} = \frac{\pi w_0^2}{\lambda} \tag{9.58}$$

is known as the *Rayleigh distance* [6]. Some other useful definitions are

$$w(z) = w_0\sqrt{1 + \left(\frac{z}{z_R}\right)^2} \tag{9.59a}$$

$$R(z) = z\left(1 + \left(\frac{z_R}{z}\right)^2\right) \tag{9.59b}$$

$$\psi(z) = \tan^{-1}\left(\frac{z}{z_R}\right) \tag{9.59c}$$

We thus have

$$\frac{1}{b(z)} = \frac{1}{w^2(z)} + j\frac{k}{2R(z)} \tag{9.60}$$

and

$$\frac{1}{b(z)} = \frac{1}{w_0 w(z)} e^{j\psi(z)} \tag{9.61}$$

Finally, we can put these expressions back into (9.51) to find the envelope of the propagating wave,

$$u(r, z) = E_0 \frac{w_0}{w(z)} e^{-\frac{r^2}{w^2(z)}} e^{-j\left(\frac{kr^2}{2R(z)} - \psi(z)\right)} \tag{9.62}$$

where we have let $C_0 = E_0 w_0^2$. Adding the normal phase term in (9.40), we find that the total electric-field phasor for an axially symmetric Gaussian beam may be written as

$$E(r, z) = E_0 \frac{w_0}{w(z)} e^{-\frac{r^2}{w^2(z)}} e^{-j\left(kz + \frac{kr^2}{2R(z)} - \psi(z)\right)} \tag{9.63}$$

Recall that we have assumed $\mathbf{E}$ is transverse and linearly polarized. For simplicity, let us write $\mathbf{E} = E(r, z)\mathbf{x}$. We may then calculate the magnetic field by application of Faraday's law,

$$\mathbf{H} = \frac{j}{\omega\mu} \nabla \times \mathbf{E} = \frac{j}{\omega\mu}\left(\frac{\partial E}{\partial z}\mathbf{y} - \frac{\partial E}{\partial y}\mathbf{z}\right) \tag{9.64}$$

Under the paraxial approximation, the longitudinal variation due to phase rotation is much faster than the radial amplitude variation,[4] allowing us to neglect the second term above,

$$\therefore \mathbf{H} = H(r, z)\mathbf{y} \approx \frac{j}{\omega\mu}\frac{\partial E}{\partial z}\mathbf{y} = \frac{k}{\omega\mu} E(r, z)\mathbf{y} = \frac{1}{\eta} E(r, z)\mathbf{y} \tag{9.65}$$

A plot of this general field solution is shown in Figure 9.5(a). The solution is valid in both the positive and negative z directions from the waist. We may thus

4 The opposite was true of the envelope, u. While the longitudinal variation in u is slow compared to its radial variation, the longitudinal variation in E is quite rapid, due to the propagating phase term, e^{-jkz}.

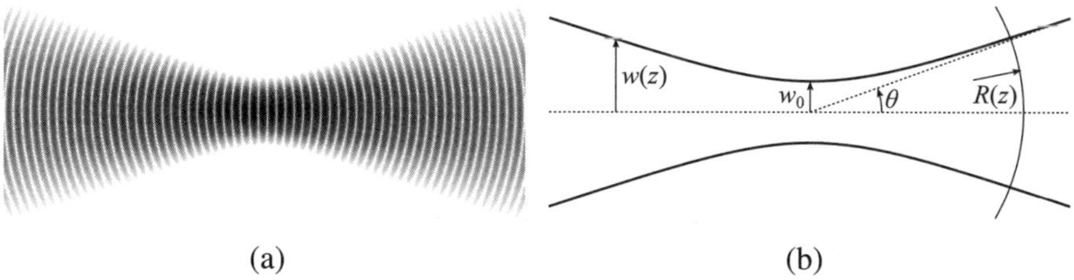

Figure 9.5 (a) Intensity map of a Gaussian beam at one instant in time. The brightness scale is logarithmic with full black representing peak intensity, while full white is 1,000 times weaker. (b) Diagram of the Gaussian beam annotated with key beam parameters.

envision a series of spherical wavefronts approaching from the left and collapsing toward a focal point at $z = 0$, then passing through the beam waist before expanding again to the right of the figure. The finite diameter of the waist represents the minimum spot size onto which a wave having this divergence angle and intensity profile can be focused. Figure 9.5(b) is a diagram of the beam envelope highlighting some of the key geometrical beam parameters which we have identified.

The power in a Gaussian beam is found as usual by integrating the Poynting vector over the entire cross-section (which in this case has infinite extent),

$$P_0 = \iint (\mathbf{E} \times \mathbf{H}^*) \cdot d\mathbf{A} = \int_0^{2\pi} \int_0^{\infty} (\mathbf{E} \times \mathbf{H}^*) \cdot \mathbf{z} r dr d\theta \tag{9.66a}$$

$$= 2\pi \frac{|E_0|^2}{\eta} \frac{w_0^2}{w^2(z)} \int_0^{\infty} r e^{-\frac{2r^2}{w^2(z)}} dr = \frac{\pi}{2\eta} |E_0|^2 w_0^2 = \frac{\pi}{k\eta} |E_0|^2 z_R \tag{9.66b}$$

where E_0 is assumed to be the rms amplitude.

9.3.3 Terms of the Gaussian Beam Equation

Equation (9.63) (and by extension (9.65)) can be understood as a combination of factors and terms that modify the basic plane-wave solution. Let us first look at the middle exponential having a real-valued argument, $-r^2/w^2(z)$. This term truncates the amplitude in the radial direction so that it falls off by a factor $1/e$ at the *beam width radius*, $w(z)$, given by (9.59a). The width radius reaches a minimum at the beam waist, where it has radius w_0. At the Rayleigh distance, z_R, the beam width

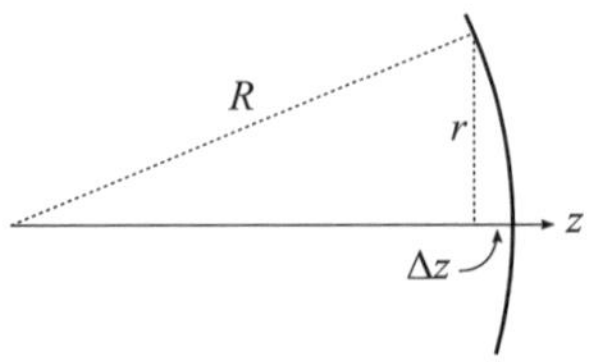

Figure 9.6 Geometric explanations of the radius of curvature, $R(z)$, based on the excess phase imparted at a given distance from the z axis for a spherical wavefront.

has increased (relative to the waist) by a factor of $\sqrt{2}$. The distance between $\pm z_R$ on the z axis is referred to as the *depth of focus* or *confocal parameter* of the beam.

To the left of that exponential in (9.63), we have a scalar ratio, $w_0/w(z)$. This attenuates the field amplitude far from the waist along the longitudinal axis consistent with the wavefront spreading out over a larger area. The angle of divergence may be found by considering the asymptotic behavior of (9.59a) as z becomes very large,

$$\theta = \lim_{z\to\infty} \tan^{-1}\left(\frac{w(z)}{z}\right) \approx \tan^{-1}\left(\frac{w_0}{z_R}\right) \tag{9.67}$$

The last exponential in (9.63) has a purely imaginary argument, thus affecting the phase of the wave only. It has three terms. The first, kz, is the usual plane-wave propagation term where k is the wavenumber.

The second term increases the phase along the radial axis and may be understood geometrically via the diagram in Figure 9.6. The curvature of the wavefront means that at a radius r away from the z axis, the wave has propagated a slightly greater distance than its z-coordinate alone would imply. That extra distance is given by

$$\Delta z = R - \sqrt{R^2 - r^2} = R\left(1 - \sqrt{1 - \frac{r^2}{R^2}}\right) \tag{9.68a}$$

$$\approx R\left(1 - 1 + \frac{r^2}{2R^2}\right) = \frac{r^2}{2R} \tag{9.68b}$$

where the approximation holds due to the paraxial assumption ($r \ll R$). This result is exactly the expression multiplying k in the middle term of the phase exponential of (9.63). It is therefore seen that the term in the denominator, $R(z)$,

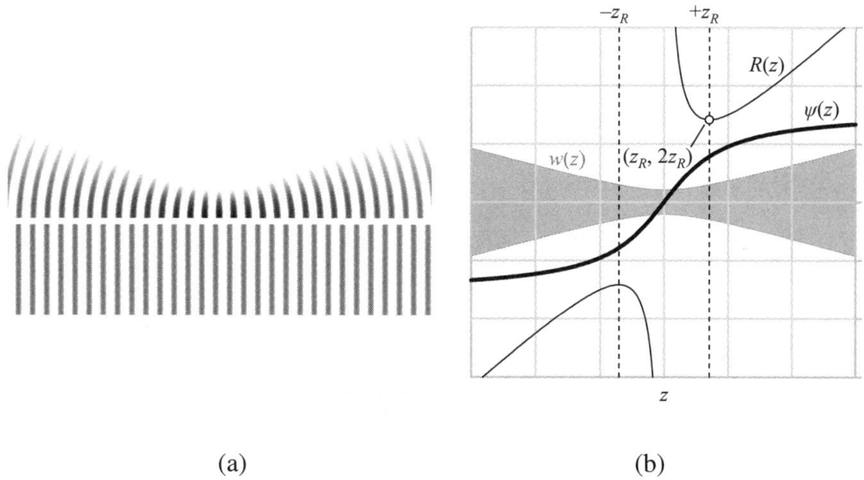

Figure 9.7 (a) Illustration of the Gouy phase shift by plotting a Gaussian beam (top) against a simple plane wave (bottom) having the same wavelength. Only half of the intensity peaks (corresponding to positive E field) are shown so that the 180° phase shift is more evident. (b) Plot of some key parameters of the Gaussian beam along the axis of propagation.

is the radius of curvature of the wavefront, given by (9.59b). Note that near the waist, the wavefronts are nearly planar, having a nearly infinite radius of curvature, but they become spherical as one moves away from the waist, diverging from it as though from a focal point. The point of greatest curvature (where $R(z)$ is minimum) occurs at the Rayleigh distance, z_R, where it has value $R(z_R) = 2z_R$.

The last phase term, $\psi(z)$, is called the *Gouy phase*, and is given by (9.59c). It imparts a total phase shift of π radians smoothly from one side of the beam waist to the other. This is illustrated in Figure 9.7(a), where the wavefronts of a Gaussian beam have been plotted along with those of a plane wave having the same free-space wavelength. In contrast with the intensity plot of Figure 9.5(a), only every other intensity peak is shown (e.g., those corresponding to the positive field amplitude), making it easier to compare the phase of the two waveforms. Note that both the Gaussian beam and the plane wave are in phase with one another at the left side of the plot, that is, the wavefronts align with one another vertically at the z axis, but nearer to the waist the wavefronts of the Gaussian beam slide further ahead of those belonging to the plane wave. At the far right of the plot, the Gaussian beam emerges from its beam waist 180° out-of-phase with the plane wave. This is entirely due to the Gouy phase term, which is plotted along with some other parameters in Figure 9.7(b). Note that this implies that wavefronts in the vicinity of the beam waist have longer effective wavelength and move with phase velocity faster than c, the plane-wave speed of light, just as the wavefronts do in the normal modes of a smooth-walled waveguide.

Finally, we may calculate the beam's numerical aperture as

$$\text{NA} = n \sin\theta \approx \frac{w_0}{z_R}\sqrt{\varepsilon_r} \tag{9.69}$$

According to Section 8.8.1, if a Gaussian beam is incident upon the end of an optical fiber, and the NA of the beam is less than that of the fiber, then the bulk of the energy in the Gaussian beam could be captured by a guided mode in the optical fiber (neglecting reflections from the surface).

It is worth noting that any one of the beam waist radius, the divergence angle, the Rayleigh distance, or the numerical aperture is sufficient to describe all the properties of a Gaussian beam for a given wavelength, since (9.63) can be derived in its entirety from knowledge of just one of these parameters. A summary of the equations describing the relationships between these parameters and the Gaussian beam field solution is given in Table 9.2.

Although we have focused in this section on the lowest-order axially symmetric solution of the paraxial Helmholtz equation, one should be aware that higher-order solutions and those not having axially symmetric profiles are also possible. In rectangular coordinates, the higher-order modes are known as *Hermite-Gaussian modes*, while in cylindrical coordinates they are *Laguerre-Gaussian modes*. In axially nonsymmetric cases, the beam waist width in one dimension differs from that in the other dimension. Notably, the dimension having the larger waist diameter has a narrower divergence angle, and vice versa; thus, a converging beam that is wide, horizontally, away from the waist, becomes narrow and tall at the waist, and then back again. Further, beams for which the waists in the two dimensions do not coincide longitudinally are called *astigmatic*.

9.3.4 Complex Beam Parameter

Later analyses can be simplified if we define a quantity known as the *complex beam parameter* as follows,

$$q(z) = z + jz_R \tag{9.70}$$

Although deceptively simple, a great deal of information is encoded in this quantity. Recall that the Rayleigh distance, the imaginary part, completely defines the global

Table 9.2
Equations for Axially Symmetric Gaussian Beams

Parameter	Symbol	Expression
E phasor amplitude	$E(r,z)$	$E_0 \frac{w_0}{w(z)} e^{-\frac{r^2}{w^2(z)}} e^{-j\left(kz+\frac{kr^2}{2R(z)}-\psi(z)\right)}$
H phasor amplitude	$H(r,z)$	$\frac{E_0}{\eta} \frac{w_0}{w(z)} e^{-\frac{r^2}{w^2(z)}} e^{-j\left(kz+\frac{kr^2}{2R(z)}-\psi(z)\right)}$
Wave impedance	$\frac{E(r,z)}{H(r,z)}$	$\eta=\sqrt{\frac{\mu}{\varepsilon}}$
Beam waist radius	w_0	$\frac{\lambda}{\pi \tan\theta}$
Rayleigh distance	z_R	$\frac{kw_0^2}{2}=\frac{\pi w_0^2}{\lambda}$
Beam width radius	$w(z)$	$w_0\sqrt{1+\left(\frac{z}{z_R}\right)^2}$
Wavefront radius of curvature	$R(z)$	$z\left(1+\left(\frac{z_R}{z}\right)^2\right)$
Gouy phase	$\psi(z)$	$\tan^{-1}\left(\frac{z}{z_R}\right)$
Beam divergence (half-angle)	θ	$\tan^{-1}\left(\frac{w_0}{z_R}\right)\approx\frac{\lambda_0}{\pi w_0\sqrt{\varepsilon_r}}$
Numerical aperture	NA	$\frac{w_0}{z_R}\sqrt{\varepsilon_r}$
Transmitted power	P_0	$\frac{\pi}{2\eta}\left\|E_0\right\|^2 w_0^2=\frac{\pi}{k\eta}\left\|E_0\right\|^2 z_R$

parameters of the Gaussian beam. The z coordinate, in turn, allows that information to be converted into a local field solution. Consider the inverse of this parameter,

$$\frac{1}{q(z)}=\frac{1}{z+jz_R}=\frac{z}{z^2+z_R^2}-\frac{jz_R}{z^2+z_R^2} \tag{9.71a}$$

$$=\frac{1}{z\left(1+\left(\frac{z_R}{z}\right)^2\right)}-\frac{j}{z_R\left(1+\left(\frac{z}{z_R}\right)^2\right)}=\frac{1}{R(z)}-\frac{jw_0^2}{z_R w^2(z)} \tag{9.71b}$$

$$=\frac{1}{R(z)}-\frac{j2}{kw^2(z)} \tag{9.71c}$$

and

$$\frac{1}{q(z)} = \frac{1}{z + jz_R} = \frac{-j}{z_R\left(1 - j\frac{z}{z_R}\right)} \tag{9.72a}$$

$$= \frac{-j}{z_R\sqrt{1 + \left(\frac{z}{z_R}\right)^2}} e^{-j\tan^{-1}\left(-\frac{z}{z_R}\right)} = \frac{-jw_0}{z_R w(z)} e^{j\psi(z)} \tag{9.72b}$$

This gives us a very simple expression for the envelope of the Gaussian beam in terms of $q(z)$,

$$u(r,z) = \frac{1}{q(z)} e^{\frac{-jkr^2}{2q(z)}} \tag{9.73a}$$

$$= \frac{-jw_0}{z_R w(z)} e^{j\psi(z)} e^{\frac{-jkr^2}{2R(z)} - \frac{r^2}{w^2(z)}} \tag{9.73b}$$

$$= \frac{-jw_0}{z_R w(z)} e^{-\frac{r^2}{w^2(z)}} e^{-j\left(\frac{kr^2}{2R(z)} - \psi(z)\right)} \tag{9.73c}$$

which is the same as (9.62) if $E_0 = -j/z_R$. As written, then, the transmitted power in the beam would be

$$P_0 = \frac{\pi}{k\eta} |E_0|^2 z_R = \frac{\pi}{k\eta z_R} \tag{9.74}$$

We can normalize this so that the power is independent of $q(z)$ if we multiply by $\sqrt{z_R}$, or

$$u(r,z) = \frac{\sqrt{\mathrm{Im}\{q(z)\}}}{q(z)} e^{\frac{-jkr^2}{2q(z)}} \tag{9.75}$$

9.4 RAY TRANSFER MATRICES

The power of the complex beam parameter is that it allows us to track the evolution of a Gaussian beam not only as it propagates through space, but as it interacts with quasioptical elements, simply by noting changes in $q(z)$. In a sense, $q(z)$ serves for quasioptics what the terminal parameters — voltage, current, or wave amplitudes — did for lumped-element, transmission-line, or waveguide networks. The role of the network parameters for those constructions will be replaced in this instance by what are known as *ray transfer matrices*.

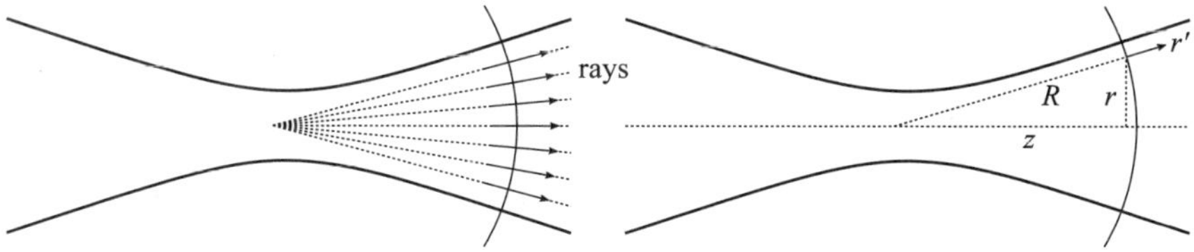

Figure 9.8 (a) Illustration of rays in a Gaussian beam and (b) their relationship to the wavefront radius of curvature.

9.4.1 Rays in Gaussian Beams

A *ray* is a line of propagation. It has a position and a direction. In the paraxial approximation for axially symmetric systems, the radius, r, of a point (its distance from the axis) is sufficient to define its position, and the radial slope, r', determines its direction. For propagating waves, that direction may be considered perpendicular to the wavefronts, as illustrated in Figure 9.8. Thus, under the paraxial approximation, the radius of curvature may be related to the ray parameters, r and r', as follows,

$$r \approx Rr' \tag{9.76a}$$

$$\therefore R \approx \frac{r}{r'} \tag{9.76b}$$

In the optical limit, characterized by high frequency and vanishingly small wavelength, the beam waist and Rayleigh distance shrink to zero, so that the radius of curvature becomes $R(z) = z$. Thus, all rays seem to pass through and/or emanate from a fixed focal point on the axis (the vanishing point where the infinitesimal beam waist was located), and are infinitely straight. In a true Gaussian beam, $R(z)$ is slightly larger than z, and thus the rays' point of origin moves, from a point on the opposite side of the beam waist to $z = 0$ as the wavefront propagates into the far field. Nonetheless, the straight-line approximation for rays may be considered accurate locally, and much of the optical formalism known as *ray tracing* can be applied successfully to Gaussian beams.

9.4.2 Matrices for Common Optical Elements

Ray transfer matrices relate the ray parameters on one side of an optical system (e.g., lenses, mirrors, and their combinations) to the ray parameters on the other

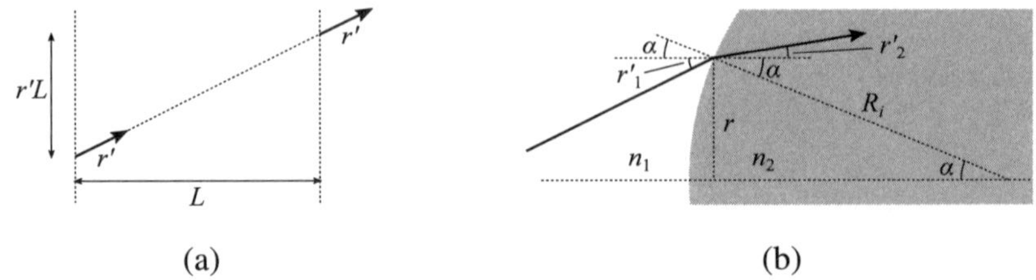

Figure 9.9 Ray transformation due to (a) empty space with length L, and (b) a dielectric interface at a curved boundary with radius R_i.

side [7]. Thus,

$$\begin{pmatrix} r_2 \\ r_2' \end{pmatrix} = \begin{pmatrix} A & B \\ C & D \end{pmatrix} \begin{pmatrix} r_1 \\ r_1' \end{pmatrix} \tag{9.77}$$

or

$$\mathbf{r}_2 = \mathbf{M}\mathbf{r}_1 \tag{9.78}$$

Although sometimes referred to as $ABCD$-parameters, one should be careful not to confuse them with the network parameters of the same name that relate the terminal voltages and currents of a two-port lumped-element and/or transmission-line network. Those $ABCD$-parameters and these do have one convenient feature in common, however: cascades of multiple elements (optical elements, in this case) are characterized by the product of the $ABCD$-matrices of the individual parts.

Let us therefore articulate the ray transfer matrices for some common cases, to use as building blocks for the matrices of more complex systems. The simplest optical element of all is empty space, over which a ray is free to propagate along its given slope. If the length of that space is L, then we may deduce that its position r has shifted by $r'L$, while its slope is unchanged,as illustrated in Figure 9.9. Thus,

$$r_2 = r_1 + r_1'L \tag{9.79a}$$

$$r_2' = r_1' \tag{9.79b}$$

Putting this in matrix form, we have

$$\begin{pmatrix} r_2 \\ r_2' \end{pmatrix} = \begin{pmatrix} 1 & L \\ 0 & 1 \end{pmatrix} \begin{pmatrix} r_1 \\ r_1' \end{pmatrix} \tag{9.80a}$$

$$\therefore \mathbf{M} = \begin{pmatrix} 1 & L \\ 0 & 1 \end{pmatrix} \tag{9.80b}$$

Note that the determinant of this matrix is $\det(\mathbf{M}) = 1$. In fact, it can be shown more generally that the determinant of any ray transfer matrix is given by $\det(\mathbf{M}) = n_1/n_2$, where n_1 and n_2 are the indices of refraction of the input and output media [7].

This fact is best illustrated by the following example, shown in Figure 9.9(b), where we have an interface between dielectric regions with a curved boundary having radius R_i. In the paraxial approximation, we may consider the incident ray slope, r_1', to be the same as its angle relative to the propagating axis. The angle of incidence upon the dielectric boundary is then given by $r_1' + \alpha$, where α is the angle of the dielectric interface itself (independent of the ray slope) relative to the axis. We may calculate that angle as

$$\alpha = \sin^{-1}\left(\frac{r}{R_i}\right) \approx \frac{r}{R_i} \tag{9.81}$$

Snell's law then gives us

$$n_1 \sin\left(r_1' + \alpha\right) = n_2 \sin\left(r_2' + \alpha\right) \tag{9.82a}$$

$$n_1\left(r_1' + \alpha\right) \approx n_2\left(r_2' + \alpha\right) \tag{9.82b}$$

$$n_1 r_1' + n_1 \frac{r}{R_i} = n_2 r_2' + n_2 \frac{r}{R_i} \tag{9.82c}$$

$$\therefore r_2' = \frac{n_1 - n_2}{n_2 R_i} r + \frac{n_1}{n_2} r_1' \tag{9.82d}$$

The radial position, r, after the interface is unchanged. Thus, the ray transfer matrix for this case is

$$\mathbf{M} = \begin{pmatrix} 1 & 0 \\ \frac{n_1 - n_2}{n_2 R_i} & \frac{n_1}{n_2} \end{pmatrix} \tag{9.83}$$

where $R_i > 0$ if the interface is convex toward the incident wave (as shown), and $R_i < 0$ if it is convex away from it. (This convention is not universally adopted, however [8].)

Note that we have neglected any reflection from the dielectric interface. If needed, an *antireflection coating*, essentially a thin impedance-matching layer analogous to a quarter-wave transformer in the transmission-line model, can be applied to the surfaces of high-quality, refractive, optical components to better realize this approximation. In the ray-matrix formalism, such a coating could be modeled as a pair of closely spaced dielectric interfaces having concentric curvature. Proof that this results in asymptotically zero change to the original matrix in the high frequency approximation will be left as a problem for the reader.

We may use the cascade-multiplication property of the ray transfer matrix to calculate the response of a thin lens. Note that, unlike the $ABCD$-matrices of lumped elements and transmission lines, the subsequent elements in this case *premultiply* the compound matrix formed from the preceding elements. Thus, the matrix for a thin lens is found by multiplying the matrix for the output interface times the matrix for the input interface,

$$\mathbf{M} = \begin{pmatrix} 1 & 0 \\ \frac{n_2-n_1}{n_1 R_2} & \frac{n_2}{n_1} \end{pmatrix} \begin{pmatrix} 1 & 0 \\ \frac{n_1-n_2}{n_2 R_1} & \frac{n_1}{n_2} \end{pmatrix} = \begin{pmatrix} 1 & 0 \\ \frac{n_2-n_1}{n_1}\left(\frac{1}{R_2} - \frac{1}{R_1}\right) & 1 \end{pmatrix} \tag{9.84}$$

It is customary to write the above expression as

$$\mathbf{M} = \begin{pmatrix} 1 & 0 \\ -\frac{1}{f} & 1 \end{pmatrix} \tag{9.85}$$

where f is the *focal length* of the lens, and

$$\frac{1}{f} = \frac{n_2 - n_1}{n_1}\left(\frac{1}{R_1} - \frac{1}{R_2}\right) \tag{9.86}$$

For a lens with thickness t, we must insert a space extension between the two interfaces,

$$\mathbf{M} = \begin{pmatrix} 1 & 0 \\ \frac{n_2-n_1}{n_1 R_2} & \frac{n_2}{n_1} \end{pmatrix} \begin{pmatrix} 1 & t \\ 0 & 1 \end{pmatrix} \begin{pmatrix} 1 & 0 \\ \frac{n_1-n_2}{n_2 R_1} & \frac{n_1}{n_2} \end{pmatrix} \tag{9.87a}$$

$$= \begin{pmatrix} 1 + \frac{t}{R_1}\left(\frac{n_1-n_2}{n_2}\right) & t\frac{n_1}{n_2} \\ \frac{n_2-n_1}{n_1}\left(\frac{1}{R_2} - \frac{1}{R_1}\right) - \frac{(n_2-n_1)^2 t}{n_1 n_2 R_1 R_2} & 1 + \frac{t}{R_2}\left(\frac{n_2-n_1}{n_2}\right) \end{pmatrix} \tag{9.87b}$$

For a thick lens in air ($n_1 = 1$), and letting $n = n_2$, we may write

$$\frac{1}{f} = (n-1)\left(\frac{1}{R_1} - \frac{1}{R_2} + \frac{(n-1)t}{nR_1R_2}\right) \tag{9.88}$$

This is known as the *Lensmaker's equation.*

Curved, reflective surfaces, or mirrors, may be treated as well, as indicated in Figure 9.10. For the case in which the mirror is convex toward the incident ray

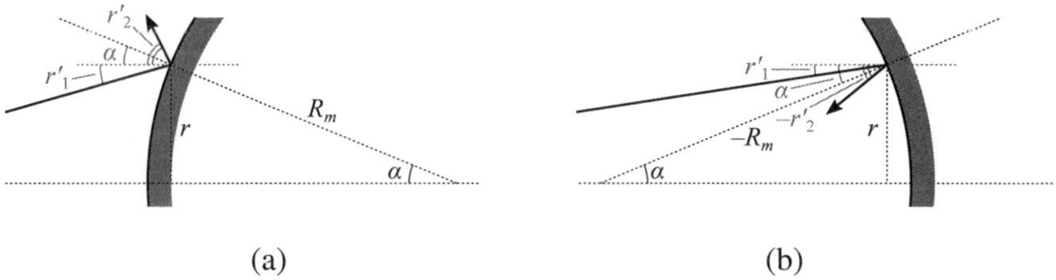

Figure 9.10 Ray transformation due to (a) a convex reflective surface and (b) a concave reflective surface.

($R_m > 0$), we have

$$r_2' - \alpha = r_1' + \alpha \tag{9.89a}$$

$$\therefore r_2' = r_1' + 2\alpha = r_1' + 2\frac{r}{R_m} \tag{9.89b}$$

If, instead, the mirror is concave ($R_m < 0$), we have

$$-r_2' - \alpha = \alpha - r_1' \tag{9.90a}$$

$$\therefore r_2' = r_1' - 2\alpha = r_1' + 2\frac{r}{R_m} \tag{9.90b}$$

In both cases, we have for the ray transfer matrix,

$$\mathbf{M} = \begin{pmatrix} 1 & 0 \\ \frac{2}{R_m} & 1 \end{pmatrix} \tag{9.91}$$

A summary of the ray transfer matrices for these common cases is given Table 9.3. The flat dielectric interface and slab are found from the formulae already derived for curved interfaces and lenses by letting $R \to \infty$ for the appropriate surfaces.

9.4.3 Application to Gaussian Beams

The classical straight-ray approximation for the propagation of light is only valid in the high-frequency limit, where the wavelength, beam waist, and Rayleigh distance all shrink to zero. Nevertheless, the ray transfer matrices just derived can be used, without modification, to track the evolution of Gaussian beams through a

Table 9.3
Common Ray-Transfer Matrices

Optical Element	Ray Transfer Matrix
Empty space	$\begin{pmatrix} 1 & L \\ 0 & 1 \end{pmatrix}$
Curved dielectric interface	$\begin{pmatrix} 1 & 0 \\ \frac{n_1-n_2}{n_2 R_i} & \frac{n_1}{n_2} \end{pmatrix}$
Flat dielectric interface	$\begin{pmatrix} 1 & 0 \\ 0 & \frac{n_1}{n_2} \end{pmatrix}$
Thin lens	$\begin{pmatrix} 1 & 0 \\ -\frac{1}{f} & 1 \end{pmatrix}$ where $\frac{1}{f} = (n-1)\left(\frac{1}{R_1} - \frac{1}{R_2}\right)$
Thick lens	$\begin{pmatrix} 1 + \frac{t}{R_1}\left(\frac{1-n}{n}\right) & \frac{t}{n} \\ -\frac{1}{f} & 1 + \frac{t}{R_2}\left(\frac{n-1}{n}\right) \end{pmatrix}$ where $\frac{1}{f} = (n-1)\left(\frac{1}{R_1} - \frac{1}{R_2} + \frac{(n-1)t}{nR_1R_2}\right)$
Flat dielectric slab	$\begin{pmatrix} 1 & \frac{t}{n} \\ 0 & 1 \end{pmatrix}$
Mirror	$\begin{pmatrix} 1 & 0 \\ \frac{2}{R_m} & 1 \end{pmatrix}$

In all cases, $R > 0$ if the surface is convex toward the incident ray.
n is the refractive index of the lens relative to the surrounding media.

quasioptical system as well [5, 9, 10], using the *Kogelnik ABCD law*,

$$q_2 = \frac{Aq_1 + B}{Cq_1 + D} \tag{9.92}$$

For example, consider a Gaussian beam with complex beam parameter $q_1 = z_1 + jz_R$ at position $z = z_1$ propagating through a distance L. Applying (9.92) to

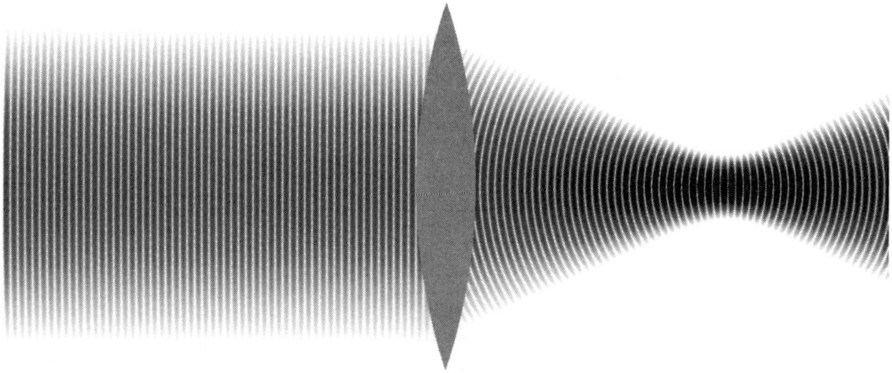

Figure 9.11 A nearly collimated Gaussian beam incident upon a biconvex lens. The waist of the outgoing beam is located near the focal point of the lens.

the empty space matrix in Table 9.3 gives for the outgoing beam parameter,

$$q_2 = \frac{Aq_1 + B}{Cq_1 + D} = \frac{1 \cdot q_1 + L}{0 \cdot q_1 + 1} = q_1 + L = (z_1 + L) + jz_R \tag{9.93}$$

As expected, the real part of the complex beam parameter has advanced by an amount L along the z axis, and the beam width, amplitude, and phase will evolve according to (9.71c) and (9.72b).

If, instead, the beam is incident upon a thin lens, we have

$$q_2 = \frac{Aq_1 + B}{Cq_1 + D} = \frac{q_1 + 0}{-\frac{1}{f}q_1 + 1} = \frac{q_1}{1 - q_1/f} \tag{9.94a}$$

$$\therefore \frac{1}{q_2} = \frac{1}{q_1} - \frac{1}{f} \tag{9.94b}$$

Since the focal length, f, is real-valued, only the real part of $1/q_1$ is modified. Thus, according to (9.71c), the radius of curvature is altered by the lens, while the beam width in the immediate vicinity of the lens is unchanged. This is illustrated in Figure 9.11. Here we show a beam with a large waist and very small divergence angle, almost a plane wave, incident upon a biconvex lens from the left. Refraction due to the lens material causes the wavefronts to curve inward, without altering the beam width itself. The outgoing beam, then, converges upon its waist at a distance from the lens approximately equal to its focal length.

More sophisticated optical systems may be constructed and system-level ray transfer matrices computed in the usual way by multiplication. The input-output behavior of a Gaussian beam propagating through such a system may then be determined by applying the Kogelnik $ABCD$ law with this compound matrix.

9.5 FIELD EQUATIONS FOR ANTENNAS

Let us now turn our attention away from the modes of propagating beams, and talk instead about the structures that excite them, namely *antennas*. Antennas may be thought of as a kind of transition, between the confined fields of guided waves, and the radiated far fields of free-space (which may or may not satisfy paraxial approximations). Prior to that discussion, it is useful to return once more to Maxwell's equations, and derive a new formulation expressed solely in terms of field potentials, as was done previously for electrostatics and magnetostatics, only this time with the effects of propagation delay taken into account.

9.5.1 Potential Formulation of Maxwell's Equations

The definition of the magnetic field as the curl of a vector potential, $\mathbf{A}$, had as its basis the nonexistence of magnetic charge, leading to the conclusion that the magnetic field should have no divergence. That has not changed, so the original definition remains intact,

$$\mathbf{B} = \nabla \times \mathbf{A} \tag{9.95}$$

The definition of the electric field as the gradient of a scalar potential was based on its lack of a curl in the static case. Once time derivatives are allowed, Faraday's law (1.19b) tells us that the electric field does indeed have a curl component, given by the rate of change of the magnetic flux density. A new field-potential definition is needed, preferably one that reverts back to the static form if time derivatives are once again reduced to zero. We thus propose the following,

$$\mathbf{E} = -\nabla\varphi - \frac{\partial \mathbf{A}}{\partial t} \tag{9.96}$$

where φ is the scalar potential. Note that (9.96) reverts to (1.32) when the time derivative equals zero, and (9.95) is the same as the previous (1.42b).

We must now verify (or derive additional conditions to ensure) that the remainder of Maxwell's equations are still satisfied in their entirety with these definitions. The nonexistence of magnetic charge is satisfied automatically, since

$$\nabla \cdot \mathbf{B} = \nabla \cdot (\nabla \times \mathbf{A}) = 0 \tag{9.97}$$

Faraday's law gives

$$\nabla \times \mathbf{E} = -\nabla \times \nabla\varphi - \frac{\partial}{\partial t}(\nabla \times \mathbf{A}) = -\frac{\partial}{\partial t}(\nabla \times \mathbf{A}) = -\frac{\partial \mathbf{B}}{\partial t} \tag{9.98}$$

and is also satisfied automatically. Ampère's law gives

$$\nabla \times \mathbf{H} = \frac{1}{\mu}\nabla \times \mathbf{B} = \mathbf{J} + \frac{\partial \mathbf{D}}{\partial t} \tag{9.99a}$$

$$\therefore \nabla \times \nabla \times \mathbf{A} = \mu\mathbf{J} + \mu\varepsilon\frac{\partial \mathbf{E}}{\partial t} \tag{9.99b}$$

$$\nabla(\nabla \cdot \mathbf{A}) - \nabla^2\mathbf{A} = \mu\mathbf{J} - \frac{1}{c^2}\frac{\partial}{\partial t}\left(\nabla\varphi + \frac{\partial \mathbf{A}}{\partial t}\right) \tag{9.99c}$$

$$\left(\nabla^2\mathbf{A} - \frac{1}{c^2}\frac{\partial^2 \mathbf{A}}{\partial t^2}\right) - \nabla\left(\nabla \cdot \mathbf{A} + \frac{1}{c^2}\frac{\partial \varphi}{\partial t}\right) = -\mu\mathbf{J} \tag{9.99d}$$

where c is the speed of light in the medium. Finally, Gauss's law gives

$$\nabla \cdot \mathbf{D} = \varepsilon\nabla \cdot \mathbf{E} = \varepsilon\left(-\nabla^2\varphi - \frac{\partial}{\partial t}(\nabla \cdot \mathbf{A})\right) = \rho \tag{9.100a}$$

$$\therefore \nabla^2\varphi + \frac{\partial}{\partial t}(\nabla \cdot \mathbf{A}) = -\frac{\rho}{\varepsilon} \tag{9.100b}$$

Equations (9.99d) and (9.100b) are the potential formulation of Maxwell's equations [11]. Although more complex in form, they have reduced the total number of vector equations from four to two, and the number of unknowns from six (the three components of both $\mathbf{E}$ and $\mathbf{H}$) to four (three components of $\mathbf{A}$ and one of φ). The general solution is still underdetermined, however; there is more than one set of field potentials, $\mathbf{A}$ and φ, which solve both of these equations, and thus indirectly determines $\mathbf{E}$ and $\mathbf{H}$, for a given set of boundary conditions.

This is an example of *gauge freedom*. We are free to set some additional constraints that simplifies our solutions. In Section 1.2.2, we chose (1.43), the

Coulomb gauge, which, if substituted above, yields

$$\nabla^2 \mathbf{A} - \frac{1}{c^2}\frac{\partial^2 \mathbf{A}}{\partial t^2} = -\mu \mathbf{J} + \frac{1}{c^2}\nabla\frac{\partial \varphi}{\partial t} \tag{9.101a}$$

$$\nabla^2 \varphi = -\frac{\rho}{\varepsilon} \tag{9.101b}$$

For antenna problems, it is more convenient to use the *Lorenz gauge*,[5]

$$\nabla \cdot \mathbf{A} = -\frac{1}{c^2}\frac{\partial \varphi}{\partial t} \tag{9.102}$$

which yields very similar expressions for the scalar and vector potentials,

$$\nabla^2 \mathbf{A} - \frac{1}{c^2}\frac{\partial^2 \mathbf{A}}{\partial t^2} = -\mu \mathbf{J} \tag{9.103a}$$

$$\nabla^2 \varphi - \frac{1}{c^2}\frac{\partial^2 \varphi}{\partial t^2} = -\frac{\rho}{\varepsilon} \tag{9.103b}$$

In fact, it is customary to write both expressions in terms of the *d'Alembertian operator*, which is defined as follows,[6]

$$\Box^2 = \nabla^2 - \frac{1}{c^2}\frac{\partial^2}{\partial t^2} \tag{9.104}$$

or, in the time-harmonic case,

$$\Box^2 = \nabla^2 + \frac{\omega^2}{c^2} = \nabla^2 + k^2 \tag{9.105}$$

Thus, we have for the potential formulation of Maxwell's equations under the Lorenz gauge condition,

$$\Box^2 \mathbf{A} = -\mu \mathbf{J} \tag{9.106a}$$

$$\Box^2 \varphi = -\frac{\rho}{\varepsilon} \tag{9.106b}$$

A summary of Maxwell's equations in the potential formulation is given in Table 9.4.

---

5 On an academic note, the Lorenz gauge is properly spelled here without a "t", since it honors the Danish physicist, Ludvig Valentin Lorenz [12], not the Dutch physicist, Hendrik Antoon Lorentz (with a "t"), for whom the Lorentz force law and Lorentz reciprocity theorem are named. The fact that the Lorenz gauge condition is itself a Lorentz invariant further adds to the confusion.

6 Some texts use the box symbol alone ($\Box$) as the d'Alembertian, instead of the box-squared ($\Box^2$).

Table 9.4
Potential Formulation of Maxwell's Equations

Gauge Condition	Equations
Gauge-free	$\left(\nabla^2\mathbf{A} - \frac{1}{c^2}\frac{\partial^2\mathbf{A}}{\partial t^2}\right) - \nabla\left(\nabla\cdot\mathbf{A} + \frac{1}{c^2}\frac{\partial\varphi}{\partial t}\right) = -\mu\mathbf{J}$ $\nabla^2\varphi + \frac{\partial}{\partial t}\left(\nabla\cdot\mathbf{A}\right) = -\frac{\rho}{\varepsilon}$
Coulomb gauge: $\nabla\cdot\mathbf{A} = 0$	$\nabla^2\mathbf{A} - \frac{1}{c^2}\frac{\partial^2\mathbf{A}}{\partial t^2} = -\mu\mathbf{J} + \frac{1}{c^2}\nabla\frac{\partial\varphi}{\partial t}$ $\nabla^2\varphi = -\frac{\rho}{\varepsilon}$
Lorenz gauge: $\nabla\cdot\mathbf{A} = -\frac{1}{c^2}\frac{\partial\varphi}{\partial t}$	$\Box^2\mathbf{A} = -\mu\mathbf{J}$ $\Box^2\varphi = -\frac{\rho}{\varepsilon}$

9.5.2 Retarded Potentials

Note that for the time-harmonic case in a source-free region, both field potentials satisfy the Helmholtz equation,

$$\Box^2\mathbf{A} = \left(\nabla^2 + k^2\right)\mathbf{A} = 0 \tag{9.107a}$$

$$\Box^2\varphi = \left(\nabla^2 + k^2\right)\varphi = 0 \tag{9.107b}$$

For electrostatics, we initially solved for the potential of a localized, infinitesimal source, a point charge, then wrote the field solution for more complex charge distributions as a superposition integral. We may do the same here by supposing there is an oscillating point charge having phasor amplitude q located at the origin.

First, let us expand the Helmholtz equation for φ in spherical coordinates. Since our excitation is a scalar point source, we may assume that the solution depends only on r. Thus,

$$\left(\nabla^2 + k^2\right)\varphi = \frac{1}{r}\frac{d}{dr}\left(r\frac{d\varphi}{dr}\right) + k^2\varphi = \frac{d^2\varphi}{dr^2} + \frac{1}{r}\frac{d\varphi}{dr} + k^2\varphi = 0 \tag{9.108}$$

Note that this is valid everywhere except at the origin. Solutions to this differential equation are given by

$$\varphi(r) = \frac{\varphi_1}{r}e^{-jkr} + \frac{\varphi_2}{r}e^{+jkr} \tag{9.109}$$

Like our previous solutions of the Helmholtz equation, these two terms take the form of traveling waves, only instead of traveling forward and backward, these travel inward and outward, respectively, with an amplitude that scales as $1/r$. We may neglect the inward-traveling wave in this situation, since it would require a source at infinity. Thus, we may set $\varphi_2 = 0$. To solve for the remaining constant, we must integrate (9.106b) throughout a spherical volume centered at the origin,

$$\iiint \Box^2 \varphi dV = -\iiint \frac{\rho}{\varepsilon} dV \tag{9.110a}$$

$$\iiint \left(\nabla^2 \varphi + k^2 \varphi\right) dV = -\frac{q}{\varepsilon} \tag{9.110b}$$

$$\iiint \nabla \cdot \nabla \varphi dV + k^2 \iiint \varphi dV = -\frac{q}{\varepsilon} \tag{9.110c}$$

$$\oint\!\!\oint \nabla \varphi \cdot d\mathbf{S} + k^2 \iiint \varphi dV = -\frac{q}{\varepsilon} \tag{9.110d}$$

where the last step follows from the divergence theorem. This can be simplified if we allow the region of integration to become vanishingly small ($r \to 0$). Since φ is proportional to $1/r$, and dV is proportional to $r^2 dr$, the second term on the left side vanishes entirely. The remaining closed surface integral may be evaluated using our general, outward-traveling wave solution,

$$\oint\!\!\oint \nabla \varphi \cdot d\mathbf{S} = \oint\!\!\oint \nabla \left(\frac{\varphi_1}{r}e^{-jkr}\right) \cdot d\mathbf{S} \approx \varphi_1 \oint\!\!\oint \nabla \left(\frac{1}{r}\right) \cdot d\mathbf{S} \tag{9.111a}$$

$$= -\varphi_1 \oint\!\!\oint \frac{1}{r^2} dS = -4\pi\varphi_1 = -\frac{q}{\varepsilon} \tag{9.111b}$$

$$\therefore \varphi_1 = \frac{q}{4\pi\varepsilon} \tag{9.111c}$$

and our final field potential due to the point source is

$$\varphi(r) = \frac{q}{4\pi\varepsilon r}e^{-jkr} \tag{9.112}$$

For antenna work, instead of a point charge, it is somewhat more convenient to use an infinitesimal bit of oscillating current as the source element, a purely mathematical construct known as a *Hertzian dipole* or *elemental doublet* [13]. We assume it is oriented in the z direction. Thus,

$$\Box^2 \mathbf{A} = -\mu \mathbf{J} \tag{9.113a}$$

$$\left(\nabla^2 + k^2\right)\left(A_x \mathbf{x} + A_y \mathbf{y} + A_z \mathbf{z}\right) = -\mu J_z \mathbf{z} \tag{9.113b}$$

$$\therefore \Box^2 A_z = -\mu J_z \tag{9.113c}$$

Note that we have temporarily used Cartesian coordinates in order to show that $\mathbf{A}$ has only a z component, or more generally, that the vector potential is everywhere directed parallel to the dipole orientation. Having reduced (9.113c) now to a scalar equation, we may return to spherical coordinates and write down its solution most simply by analogy to (9.112) (substituting $\varphi \to \mathbf{A}$, $\rho \to J_z$, $q \to idz$, and $\varepsilon \to 1/\mu$),

$$\mathbf{A}(r) = \frac{\mu i dz}{4\pi r} e^{-jkr} \mathbf{z} \tag{9.114}$$

Expressions for both potentials resulting from more complex source distributions may then be written as an integral of superposition over these elemental results,

$$\varphi(\mathbf{r}) = \frac{1}{4\pi\varepsilon} \iiint \frac{\rho\left(\mathbf{r}'\right)}{\left|\mathbf{r} - \mathbf{r}'\right|} e^{-jk\left|\mathbf{r}-\mathbf{r}'\right|} dV' \tag{9.115a}$$

$$\mathbf{A}(\mathbf{r}) = \frac{\mu}{4\pi} \iiint \frac{\mathbf{J}\left(\mathbf{r}'\right)}{\left|\mathbf{r} - \mathbf{r}'\right|} e^{-jk\left|\mathbf{r}-\mathbf{r}'\right|} dV' \tag{9.115b}$$

where the integrals are taken over the source distributions in primed coordinates. These are referred to as the *retarded field potentials* [14] since the exponential term indicates that the response due to a dynamic source distribution is delayed by an amount commensurate with the plane-wave speed of light.[7]

---

7 As an aside, if the Coulomb gauge was used here instead of the Lorenz gauge, we would have found that $\mathbf{A}$ is delayed, while φ responds instantly throughout space. This in itself is in no way a violation of Maxwell's equations, nor of causality under special relativity, since the only directly measurable quantities, such as $\mathbf{E}$ and $\mathbf{H}$, depend on $\mathbf{A}$ also, and are still delayed. Nevertheless, the Lorenz gauge is preferred in these situations for simplicity and because it is somehow more satisfying that both field potentials should experience an identical amount of delay.

9.6 WIRE ANTENNAS

In order to illustrate how radiation patterns for antennas are calculated, it is instructive to begin with a relatively simple example: the Hertzian dipole. This may then be used as an approximate model for more-realistic but electrically short dipoles.

9.6.1 Short Dipoles

The vector potential for a Hertzian dipole was already given in (9.114). We may calculate its magnetic field using (9.95),

$$\mathbf{B} = \nabla \times \mathbf{A} = \nabla \times \left(\frac{\mu i dz}{4\pi r} e^{-jkr} \mathbf{z} \right) \tag{9.116}$$

It is useful here to convert the unit vector into spherical coordinates, $\mathbf{z} = \cos\theta \mathbf{r} - \sin\theta \boldsymbol{\theta}$,

$$\mathbf{B} = \frac{\mu i dz}{4\pi} \nabla \times \left(\frac{e^{-jkr}}{r} \cos\theta \mathbf{r} - \frac{e^{\ jkr}}{r} \sin\theta \boldsymbol{\theta} \right) \tag{9.117a}$$

$$= \frac{\mu i dz}{4\pi r} \left(-\frac{\partial}{\partial\theta} \frac{e^{-jkr}}{r} \cos\theta - \frac{\partial}{\partial r} e^{-jkr} \sin\theta \right) \boldsymbol{\phi} \tag{9.117b}$$

$$= \frac{\mu i dz}{4\pi} e^{-jkr} \sin\theta \left(\frac{1}{r^2} + \frac{jk}{r} \right) \boldsymbol{\phi} \tag{9.117c}$$

Therefore,

$$\mathbf{H} = \frac{1}{\mu} \mathbf{B} = \frac{i dz}{4\pi} e^{-jkr} \sin\theta \left(\frac{jk}{r} + \frac{1}{r^2} \right) \boldsymbol{\phi} \tag{9.118}$$

The electric field is then given by Ampère's law,

$$\mathbf{E} = \frac{\nabla \times \mathbf{H}}{j\omega\varepsilon} = \frac{1}{j\omega\varepsilon} \left(\frac{1}{r\sin\theta} \frac{\partial}{\partial\theta} \left(H_\phi \sin\theta \right) \mathbf{r} - \frac{1}{r} \frac{\partial}{\partial r} \left(r H_\phi \right) \boldsymbol{\theta} \right) \tag{9.119a}$$

$$\therefore E_r = \frac{i dz}{j4\pi\omega\varepsilon} \frac{e^{-jkr}}{r\sin\theta} \left(\frac{1}{r^2} + \frac{jk}{r} \right) \frac{\partial}{\partial\theta} \left(\sin^2\theta \right) \tag{9.119b}$$

$$= \frac{i dz \eta}{2\pi} \left(\frac{1}{r^2} - \frac{j}{kr^3} \right) e^{-jkr} \cos\theta \tag{9.119c}$$

$$E_\theta = -\frac{i dz}{j4\pi\omega\varepsilon} \frac{\sin\theta}{r} \frac{\partial}{\partial r} \left(r e^{-jkr} \left(\frac{1}{r^2} + \frac{jk}{r} \right) \right) \tag{9.119d}$$

$$-j\frac{idz\eta}{4\pi}\left(\frac{k}{r}-\frac{j}{r^2}-\frac{1}{kr^3}\right)e^{-jkr}\sin\theta \tag{9.119e}$$

Note that $\mathbf{E}$ has both $\mathbf{r}$ and $\boldsymbol{\theta}$ components, while $\mathbf{H}$ has only a $\boldsymbol{\phi}$ component. Also, these fields have terms which decay as r^{-1}, r^{-2}, and r^{-3}. The latter two are near-field components that rapidly become insignificant in the far field. They contribute primarily to the driving-point impedance of the antenna and mutual coupling to adjacent antennas, but do not contribute to the total radiated power. It is therefore useful to isolate only the far-field terms,

$$\mathbf{H} = j\frac{idz}{4\pi}\frac{k}{r}e^{-jkr}\sin\theta\boldsymbol{\phi} \tag{9.120a}$$

$$\mathbf{E} = j\frac{idz\eta}{4\pi}\frac{k}{r}e^{-jkr}\sin\theta\boldsymbol{\theta} \tag{9.120b}$$

where it is seen that, in the far field, $E_\theta = \eta H_\phi$. The far-field radiated flux (power per unit area) is then given by the Poynting vector, S,

$$S = E_\theta H_\phi^* = \left(\frac{|i|dz}{4\pi}\frac{k}{r}\sin\theta\right)^2\eta = \frac{\eta}{4}|i|^2\left(\frac{dz}{\lambda}\right)^2\frac{1}{r^2}\sin^2\theta \tag{9.121}$$

where i is assumed to be an rms current amplitude. Note that, due to the dz/λ term, the radiated flux is vanishingly small. Hertzian dipoles are fictitious elements, meant only as infinitesimals to calculate the far-field patterns of more complex wire antennas by integration. Nonetheless, they can be used as an approximate solution for electrically short dipoles, provided that we use half the driving-point current, $i = i'/2$, as the effective average current over its length (since the current profile in any real dipole must fall to zero at the endpoints). Thus, for a short dipole with length $l \ll \lambda$,

$$S = \frac{\eta}{16}|i'|^2\left(\frac{l}{\lambda}\right)^2\frac{1}{r^2}\sin^2\theta \tag{9.122}$$

Several plots of the far-field pattern is shown in Figure 9.12. Since the radiation pattern in this case is azimuthally omnidirectional (uniform in ϕ), we can show all pertinent far-field information in a two-dimensional polar plot with θ as the independent variable and the radius corresponding to relative power, as in Figure 9.12(a). More simply, it is customary in some cases to simply show the relative power in decibels versus θ, as in Figure 9.12(b). If there were some dependence on ϕ, we could add a second curve to either of these plots, but one would have to specify from which cut angle is each curve taken. For more complex

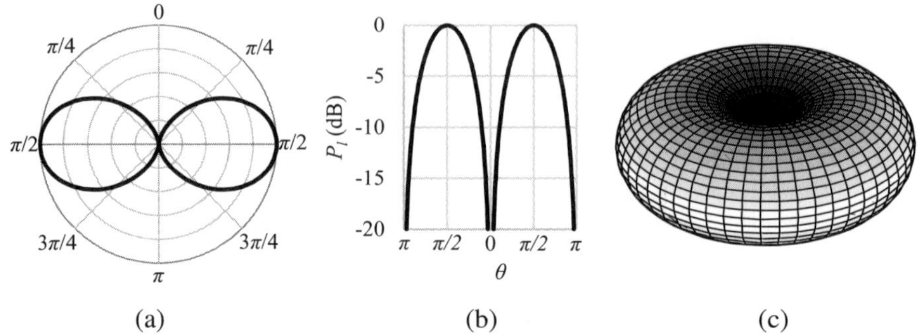

Figure 9.12 Far-field radiation pattern for a short dipole plotted (a) versus θ on a polar coordinate frame, (b) in a decibel scale versus θ on Cartesian axes, and (c) as a three-dimensional surface.

radiation patterns, visualization is greatly enhanced by a three-dimensional surface plot, such as that in Figure 9.12(c), where again the radius indicates relative flux in a given direction.

9.6.2 Half-Wave Dipoles

For larger dipoles, such as that shown in Figure 9.13(a), we must take into account more carefully the geometry and electrical extent of the arms. First, we assume that the dipole is fed from the center, that the arms are oriented in the z direction, and that the ends are open-circuited. Therefore, an excitation at the terminals sets up a current standing wave that must be zero at both end points. Although we have not derived the exact profile of the current distribution, it is reasonable to assume that it is nearly sinusoidal since the arms look very much like open-ended transmission-line stubs (but not exactly, since no clear return path, or ground, has been identified). Thus, we may write the rms current as

$$i(z) = \begin{cases} i \cos\left(\frac{\pi z}{L}\right) & |z| \leq \frac{L}{2} \\ 0 & \text{otherwise} \end{cases} \tag{9.123}$$

where i is the rms phasor amplitude and L is the total length of the dipole from end to end. Note that the propagation distance from any point z on the axis is

$$|\mathbf{r} - \mathbf{r}'| = \sqrt{r^2 + z^2 - 2rz\cos\theta} \tag{9.124}$$

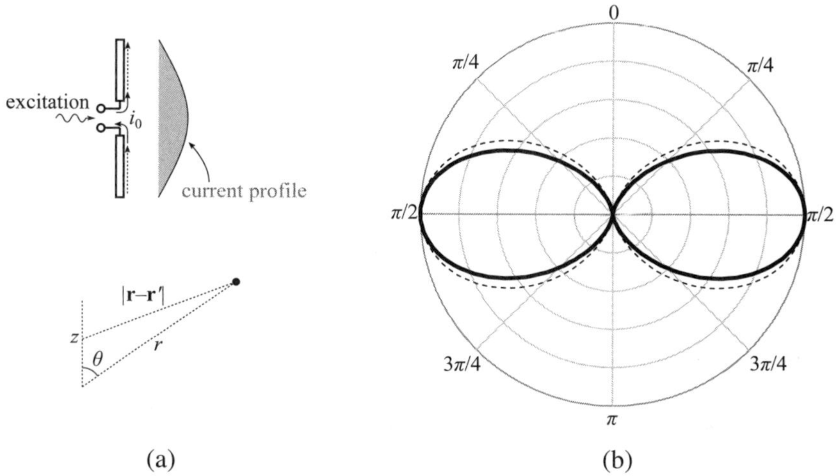

Figure 9.13 (a) Geometry of an electrically large dipole and (b) radiation pattern of a half-wave dipole (solid line) compared to that of Hertzian dipole (dashed line).

according to the law of cosines (see Section A.1). In the far field, we may assume that $z \ll r$; therefore,

$$|\mathbf{r} - \mathbf{r}'| = \sqrt{r^2 + z^2 - 2rz\cos\theta} = r\sqrt{1 + \left(\tfrac{z}{r}\right)^2 - 2\tfrac{z}{r}\cos\theta} \tag{9.125a}$$

$$\approx r\sqrt{1 - 2\tfrac{z}{r}\cos\theta} \approx r\left(1 - \tfrac{z}{r}\cos\theta\right) = r - z\cos\theta \tag{9.125b}$$

Therefore, if we apply (9.115b) in spherical coordinates to the source distribution in (9.123), we have for the vector potential in the far field,

$$\mathbf{A} = \frac{\mu}{4\pi}\iiint \frac{\mathbf{J}(\mathbf{r}')}{|\mathbf{r} - \mathbf{r}'|} e^{-jk|\mathbf{r}-\mathbf{r}'|} dV' \tag{9.126a}$$

$$= \frac{\mu i}{4\pi} e^{-jkr}\mathbf{z} \int_{-L/2}^{L/2} \frac{e^{+jkz\cos\theta}}{r - z\cos\theta} \cos\left(\frac{\pi z}{L}\right) dz \tag{9.126b}$$

$$\approx \frac{\mu i}{4\pi r} e^{-jkr} \cos\left(\tfrac{kL}{2}\cos\theta\right) \frac{2\pi/L}{\left(\frac{\pi}{L}\right)^2 - k^2\cos^2\theta}\mathbf{z} \tag{9.126c}$$

A useful special case occurs when $L = \lambda/2$. Known as a *half-wave dipole*, the vector potential simplifies somewhat,

$$\mathbf{A} = \frac{\mu i}{2\pi kr} e^{-jkr} \frac{\cos\left(\frac{\pi}{2}\cos\theta\right)}{\sin^2\theta}\mathbf{z} \tag{9.127}$$

It remains to calculate the electric and magnetic fields from this potential,

$$\mathbf{H} = \frac{1}{\mu}\mathbf{B} = \frac{1}{\mu}\nabla\times\mathbf{A} \tag{9.128a}$$

$$= \frac{i}{2\pi k}\nabla\times\left(\frac{e^{-jkr}}{r}\frac{\cos\left(\frac{\pi}{2}\cos\theta\right)}{\sin^2\theta}(\cos\theta\mathbf{r} - \sin\theta\boldsymbol{\theta})\right) \tag{9.128b}$$

$$= -\frac{i}{2\pi kr}\left(\frac{e^{-jkr}}{r}\frac{\partial}{\partial\theta}\frac{\cos\left(\frac{\pi}{2}\cos\theta\right)}{\sin\theta\tan\theta} + \frac{\partial}{\partial r}e^{-jkr}\frac{\cos\left(\frac{\pi}{2}\cos\theta\right)}{\sin\theta}\right)\boldsymbol{\phi} \tag{9.128c}$$

The first term in the parentheses above will clearly include an r^{-2} term, which can be neglected in the far field. Therefore,

$$\mathbf{H} \approx j\frac{i}{2\pi r}e^{-jkr}\frac{\cos\left(\frac{\pi}{2}\cos\theta\right)}{\sin\theta}\boldsymbol{\phi} \tag{9.129}$$

The electric field should then be given by

$$\mathbf{E} = \eta\mathbf{H}\times\mathbf{r} = j\frac{i\eta}{2\pi r}e^{-jkr}\frac{\cos\left(\frac{\pi}{2}\cos\theta\right)}{\sin\theta}\boldsymbol{\theta} \tag{9.130}$$

Finally, the radiated flux is

$$S_{\lambda/2} = E_\theta H_\phi^* = \frac{\eta}{(2\pi r)^2}|i|^2\frac{\cos^2\left(\frac{\pi}{2}\cos\theta\right)}{\sin^2\theta} \tag{9.131}$$

This radiation pattern is shown with a solid line in Figure 9.13(b). It is very similar to that of the Hertzian dipole, shown also in the figure with a dashed line, only with slightly narrower beam width in the azimuthal plane.

9.6.3 Gain and Directivity

As the above calculated far-field radiation patterns make obvious, a given antenna is typically more effective at radiating power in certain directions than others, and therefore, according to the principles of reciprocity which were discussed much earlier in Section 1.7, is more effective at receiving power from those directions also. Let us now begin to quantify this directionality of antenna transmission and reception using metrics which can be applied universally to all types of antennas.

First, let us define the total radiated power as the integral of the radiated flux over a sphere in the far field,

$$P_{tot} = \oiint S(r,\theta,\phi)dA = \int_0^{2\pi}\int_0^{\pi} S(r,\theta,\phi)r^2 \sin\theta d\theta d\phi \tag{9.132}$$

Conservation of energy ensures that the result of this calculation should not depend on r, so long as r is large enough to be in the far field. We may define the *radiant intensity*, U, as the power per unit solid angle, given simply by

$$U(\theta,\phi) = r^2 S(r,\theta,\phi) \tag{9.133}$$

For electrically short dipoles, we have from (9.122),

$$U(\theta,\phi) = \frac{\eta}{16}|i|^2\left(\frac{l}{\lambda}\right)^2 \sin^2\theta \tag{9.134a}$$

$$\therefore P_{tot} = \int_0^{2\pi}\int_0^{\pi} U(\theta,\phi)\sin\theta d\theta d\phi = \frac{\eta|i|^2}{16}\left(\frac{l}{\lambda}\right)^2 \int_0^{2\pi}\int_0^{\pi} \sin^3\theta d\theta d\phi \tag{9.134b}$$

$$= \frac{\eta|i|^2\pi}{8}\left(\frac{l}{\lambda}\right)^2 \int_0^{\pi} \sin^3\theta d\theta = \frac{\eta|i|^2\pi}{6}\left(\frac{l}{\lambda}\right)^2 \tag{9.134c}$$

The average radiated power per unit solid angle is then found by dividing this total by 4π, the number of *steradians* (sr) in a complete sphere,

$$P_{av} = \frac{P_{tot}}{4\pi} = \frac{\eta|i|^2}{24}\left(\frac{l}{\lambda}\right)^2 \tag{9.135}$$

This would be the constant radiant intensity of a theoretical isotropic, or omnidirectional, antenna radiating the same total power. The dipole's directive gain is the ratio of its radiant intensity to the average radiated power,

$$D(\theta, \phi) = \frac{U(\theta, \phi)}{P_{av}} = \tfrac{3}{2}\sin^2\theta \tag{9.136}$$

and its directivity, D, is the peak value of the directive gain. In this case, the directivity is $3/2$, or, expressed in decibels, about 1.76 dBi (where dBi stands for *decibels isotropic*). The gain of an antenna is closely related to the directive gain, but includes an extra factor known as the *antenna efficiency*, η_{ant}, associated with ohmic losses in the antenna structure,

$$G = \eta_{ant} D \tag{9.137}$$

Finally, the beam area or beam solid angle, Ω_A, is the total integrated power divided by the peak radiated flux [1],

$$\Omega_A = \frac{P_{tot}}{U_{max}} = \frac{P_{tot}}{D \cdot P_{av}} = \frac{4\pi}{D} \tag{9.138}$$

which, for the short dipole is $8\pi/3 \approx 8.13$ sr.

The radiated flux of the half-wave dipole, (9.131), is not as easy to integrate in closed form, but we can estimate its directivity by associating the beam solid angle with the narrowing of the far-field radiation pattern illustrated in Figure 9.13. Visually, the half-power beamwidth in declination, or θ_{HP}, appears to be about 10% smaller for the half-wave dipole than for the short dipole. Thus, we might expect its beam area to be smaller by a similar amount, and its directivity to be correspondingly larger at $3/2 \cdot 1.1 \approx 1.65$. In fact, careful numerical analysis shows that the directivity of a half-wave dipole is about 1.64, or 2.15 dBi.

9.6.4 Radiation Resistance

The total radiated power from an antenna represents a form of loss from the local circuit attached to its terminals. Since the terminals of a dipole are electrically small and close together, this loss can be accounted for in a lumped-element equivalent circuit as a load resistor. The value of that resistor is known as the *radiation resistance*, and is found simply by applying (2.36) to the total radiated power from the antenna with i given by its rms driving current. For the electrically short dipole, we have

$$R_{rad} = \frac{P_{tot}}{|i|^2} = \frac{\pi}{6}\eta\left(\frac{l}{\lambda}\right)^2 \tag{9.139}$$

The equivalent load impedance of an antenna will have a reactive component as well, but this is not so easily derived. Considering that electrically short dipoles closely resemble lumped-element open-circuits, or open-ended transmission-line stubs, it should not be surprising that the reactive impedance of such elements is capacitive. The following expression is given in [15],

$$X = \frac{-120}{\pi}\left(\frac{\lambda}{l}\right)\left(\ln\left(\frac{l}{d}\right) - 1\right) \tag{9.140}$$

where d is the diameter of the wire. We therefore see that since the radiation resistance diminishes at a rate proportional to l^2 and is combined in series with a capacitance approaching an open circuit, an electrically short dipole will be very difficult to impedance match, and can be done so only for very narrow bandwidths.

We might expect the half-wave dipole, in contrast, to be resonant at its center frequency, thus exhibiting only a real part and no reactance. This is approximately true, but not exact, due in part to the effects of finite wire diameter. Typically, a dipole that is truly a half-wavelength long will have a radiation resistance of 73Ω along with an inductive reactance of about $+j43\Omega$. The reactive portion of its feed point impedance can be nullified by shortening the dipole on average about 5%, again depending slightly on the wire diameter [13].

9.6.5 Other Wire Antennas

Dipoles are important as a representative baseline element for many kinds of antennas and antenna arrays. Their properties are therefore summarized in Table 9.5. Some of the numerous variations on, and derivatives from, the basic dipole design are shown in Figure 9.14. The folded dipole, Figure 9.14(b), has the two ends of a half-wave dipole connected end-to-end by another half-wavelength wire. The currents on both sides of the fold are in-phase, producing an antenna pattern very much the same as the original dipole, but with broader bandwidth and four times the driving-point impedance, matching it well to some common 300Ω balanced transmission lines. The halo antenna, Figure 9.14(c), has the arms of the dipole bent into a horizontal circle, producing a laterally omnidirectional radiation pattern with greater attenuation at zenith.

Dipole bandwidth can also be broadened by flaring the arms out into flat triangles, as in the bowtie antenna of Figure 9.14(d), or into full cones like the

Table 9.5
Summary of Dipole Antennas

Parameter	Symbol	Short Dipole ($l \ll \lambda$)	Half-Wave Dipole ($l = \lambda/2$)
Electric far field	E_θ	$j\frac{i\eta}{4r}\left(\frac{l}{\lambda}\right)e^{-jkr}\sin\theta$	$j\frac{i\eta}{2\pi r}e^{-jkr}\frac{\cos\left(\frac{\pi}{2}\cos\theta\right)}{\sin\theta}$
Magnetic far field	H_ϕ	$\frac{E_\theta}{\eta}$	$\frac{E_\theta}{\eta}$
Radiated flux	S	$\frac{\eta}{16}\lvert i\rvert^2\left(\frac{l}{\lambda}\right)^2\frac{1}{r^2}\sin^2\theta$	$\frac{\eta}{(2\pi r)^2}\lvert i\rvert^2\frac{\cos^2\left(\frac{\pi}{2}\cos\theta\right)}{\sin^2\theta}$
Directivity	D	1.5 (1.76 dBi)	1.64 (2.15 dBi)*
Beam area	Ω_A	$\frac{8\pi}{3} \approx 8.13$ sr	7.66 sr*
Radiation resistance	R_{rad}	$\frac{\pi}{6}\eta\left(\frac{l}{\lambda}\right)^2$	$73 + j43\Omega$*

*Derived from numerical approximation.

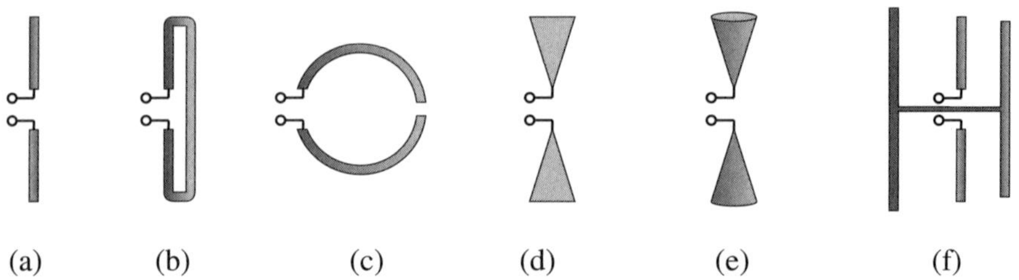

Figure 9.14 Various wire and wire-based antennas. (a) Basic dipole. (b) Folded dipole. (c) Halo. (d) Bowtie. (e) Biconic. (d) Yagi-Uda.

biconic antenna in Figure 9.14(e). Substantially improved directivity can also be obtained by adding parasitic elements to the main driven dipole, such as the three-element Yagi-Uda antenna shown in Figure 9.14(f), which has one reflector element behind the driven dipole and one director element in front. Few of these modifications lend themselves easily to closed-form analysis. The reader is referred to the references (especially [1, 16]) for an extensive review of the many types of antennas that are available.

One simple case that does warrant a brief mention, however, is the monopole antenna shown in Figure 9.15(a). Extending vertically from a flat, conductive

Figure 9.15 (a) Quarter-wavelength monopole antenna and (b) its model as a half-wave dipole where one arm is formed by the image in the ground plane.

ground plane, it is usually fed from the base with one terminal on the bottom end of the vertical shaft and the other connected to ground. Most often a quarter-wavelength tall, it can be modelled as a half-wave dipole (as in Figure 9.15(b)) where one arm of the dipole is formed as a virtual element by the image in the ground metal. The far-field radiation pattern is the same as that shown for the half-wave dipole in Figure 9.13(b) in the upper hemisphere ($\theta < \pi/2$). Thus, the total radiated power is half while the drive current is the same, from which we can conclude that the radiation resistance (and any reactive component) is half what it would be for the complete half-wave dipole.

Quarter-wave monopoles are common as radio broadcast towers or masts throughout much of the world. They are sometimes referred to as *Marconi antennas* in honor of the early pioneer of radio technology Guglielmo Marconi, who invented this type of antenna in 1895.

9.7 COMPLEMENTARY ANTENNAS

Just as lumped-element and transmission-line networks have duals, antenna structures have complementary forms as well, having their roots in both the duality of Maxwell's equations as described in 1.1.2, as well as a fundamental truth of optics known as *Babinet's principle*.

9.7.1 Half-Wave Slot

Let us start with an antenna that we already know well, the basic dipole, only this time, instead of a cylindrical wire, let us flatten the arms out into thin strips, as shown in Figure 9.16(a). If the width of the strips is still very small compared

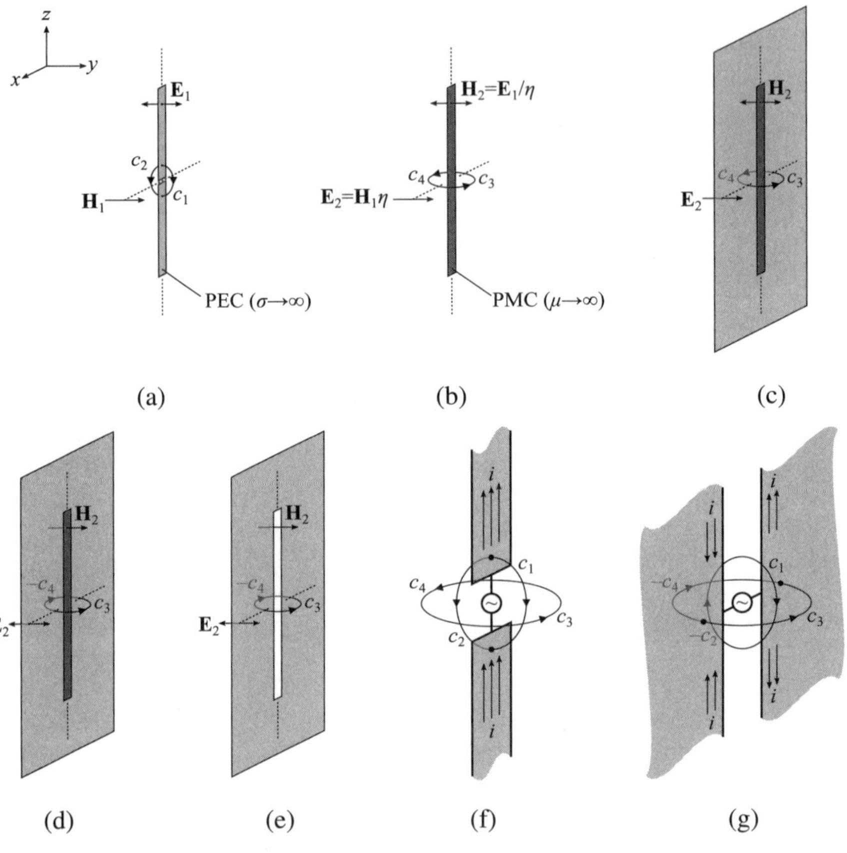

Figure 9.16 (a) Dipole of flat strips oriented along the z axis. (b) Dual structure comprising a strip of perfect magnetic conductor. (c) Addition of a flat electrically conducting plane surrounding the magnetically conducting strip. (d) Reversal of the magnetic excitation and fields on one side of the plane. (e) Removal of magnetically conducting strip, revealing the slot antenna. (f) Detail of generator and integration paths used for calculation of dipole impedance. (g) Detail of generator and integration paths for calculation of slot impedance.

to a wavelength, this minor detail should have negligible effect on the antenna's radiation pattern or driving-point impedance. Define two short paths of integration, c_1 and c_2, passing from the end of one strip to the other on each side of the xz plane. Although a single generator within the gap is sufficient to excite the structure, it will be useful for now to imagine two signal generators in parallel, cooriented with each of these two integration paths, c_1 and c_2. Note that the electric field, $\mathbf{E}_1$, is perpendicular to the surface of the metal strips and points in opposite directions on both sides. Further, the magnetic field, $\mathbf{H}_1$, anywhere in the xz plane (but not on the strips) is perpendicular to that plane and is continuous through it.

Now, imagine an alternate scenario in which we exchange all electric and magnetic quantities. For the electric and magnetic fields in particular, we have

$$\mathbf{E}_2 = A_0\mathbf{H}_1 \tag{9.141a}$$

$$\mathbf{H}_2 = A_1\mathbf{E}_1 \tag{9.141b}$$

The scaling factors, A_0 and A_1, are constrained by Maxwell's equations, specifically so that the new fields, $\mathbf{E_2}$ and $\mathbf{H}_2$, have the proper ratio given by the characteristic impedance of free space. That is, at least in the far field,

$$\frac{E_2}{H_2} = \frac{A_0}{A_1}\frac{H_1}{E_1} = \frac{A_0}{A_1}\frac{1}{\eta} = \eta \tag{9.142a}$$

$$\therefore A_0 = \eta^2 A_1 \tag{9.142b}$$

This still leaves one degree of freedom, which amounts to an arbitrary amplitude scaling factor. For simplicity, we will assume that $A_0A_1 = 1$, and consequently $A_0 = A_1^{-1} = \eta$.

The new configuration is shown in Figure 9.16(b). It is not sufficient just to exchange the electric and magnetic fields, we must swap electric and magnetic boundary conditions and stimuli as well. The metal strips of the dipole have therefore been replaced with a *perfect magnetic conductor* (PMC), which can be modeled as a material having infinite permeability.[8] The electric and magnetic fields have swapped places, and our excitation now comprises a loop in the xy plane surrounding the magnetic strip, where two signal sources may be connected antiparallel (the circulating electric current of these two signal sources acting together induces a magnetic current in the strip). Let us call that portion of the excitation loop in front of the strip c_3 and that portion on the back of the strip c_4.

Since the electric field is now continuous through and perpendicular to the xz plane, we may fill that plane with a thin conducting sheet that fully surrounds the

8 Recall that electrical conductivity appears in the time-harmonic form of Maxwell's equations identically to the imaginary part of permittivity. Thus, a perfect electric conductor can be thought of as having either infinite conductivity or infinite permittivity. Likewise, a perfect magnetic conductor can be thought of as having infinite permeability.

magnetically conducting strip and extends infinitely in all directions, as shown in Figure 9.16(c). The addition of this conducting plane has no effect on either field, but, along with the magnetically conducting strip, it effectively isolates the two sides from one another. We may therefore reverse the backside excitation path, c_4, along with the fields associated with it, as shown in Figure 9.16(d). The fields on the back side of the xz plane are thus inverted, but those on the front side remain unchanged.

Note that due to the symmetry of the resulting structure and the configuration of its excitation, the magnetic field is now continuous through the magnetically conducting strip, it has the same magnitude and direction on both sides. The strip may therefore be removed without altering the field, leaving a slot in the conducting plane, with two in-phase signal sources connected across the midpoint through paths c_3 and $-c_4$.

This exercise has shown that a radiating slot in an infinite conducting plane has the exact same radiation pattern as the complementary dipole of conducting strips, only with the electric and magnetic fields exchanged, and their orientation reversed on one side of the conducting plane versus the other [1, 17].

Further, the driving-point impedance of the slot antenna can be derived from what we know of the basic dipole and the relationships between the two antennas' fields. In the above derivation, we imagined parallel signal sources, one on each side of the xz plane, so that their orientations could be reversed at intermediate steps. However, for the both the original dipole and the final slot, the two parallel sources were oriented in the same direction. Thus, the structures may be excited by single generators, in the z-direction across the gap for the dipole as shown in Figure 9.16(f), and in the x-direction across the width of the slot for the slot antenna in Figure 9.16(g)

Still, the paths c_1, c_2, c_3, and c_4 are useful as paths of integration for determining voltage and current. The terminal voltage and current of the basic dipole, for example, are given by

$$v_1 = \int_{c_1} \mathbf{E}_1 \cdot d\mathrm{l} \tag{9.143a}$$

$$i_1 = \int_{c_3+c_4} \mathbf{H}_1 \cdot d\mathrm{l} = 2 \int_{c_3} \mathbf{H}_1 \cdot d\mathrm{l} \tag{9.143b}$$

and those of the complementary slot antenna are given by

$$v_2 = \int_{c_3} \mathbf{E_2} \cdot d\mathrm{l} \tag{9.144a}$$

$$i_2 = \int_{c_1 - c_2} \mathbf{H}_2 \cdot d\mathbf{l} = 2 \int_{c_1} \mathbf{H}_2 \cdot d\mathbf{l} \tag{9.144b}$$

We know from our derivation, however, that $\mathbf{E}_1 = \eta \mathbf{H}_2$ and $\mathbf{E}_2 = \eta \mathbf{H}_1$ on the front side of the xz plane — that is, for paths c_1 and c_3 (on c_2 and c_4 they have been inverted). Therefore,

$$v_2 = \int_{c_3} \mathbf{E_2} \cdot d\mathbf{l} = \eta \int_{c_3} \mathbf{H_1} \cdot d\mathbf{l} = \frac{\eta}{2} i_1 \tag{9.145a}$$

$$i_2 = 2 \int_{c_1} \mathbf{H}_2 \cdot d\mathbf{l} = \frac{2}{\eta} \int_{c_1} \mathbf{E}_1 \cdot d\mathbf{l} = \frac{2}{\eta} v_1 \tag{9.145b}$$

and the driving-point impedance of the slot antenna is

$$z_2 = \frac{v_2}{i_2} = \frac{\eta^2 i_1}{4 v_1} = \frac{(\eta/2)^2}{z_1} \tag{9.146}$$

Thus, we see that the impedance of the complementary slot antenna is the inverse, or dual, of the impedance of the original wire antenna with a normalizing impedance of $\eta/2$. Since a half-wave dipole has an impedance of about 73Ω (if shortened slightly to nullify the reactive part), a half-wave slot driven from the mid-point would have an impedance of

$$z_{slot} = \frac{(377\Omega/2)^2}{73\Omega} \approx 484\Omega \tag{9.147}$$

This is a rather high impedance and is often not compatible with the transmission-line impedances we might wish to match it to. One way to alleviate this issue is to shift the driving point off-center [1], as shown in Figure 9.17(a). It is also common to combine the antenna with a microstrip-to-slotline transition as the launcher, such as that shown in Figure 9.17(b) (compare this with the general transition shown earlier in Figure 8.5).

9.8 PLANAR ANTENNAS

The enormous growth in printed-circuit technologies has naturally driven along with it interest in antenna structures that can be printed on planar substrates along with microstrip transmission lines and integrated, active, electronic devices. Although

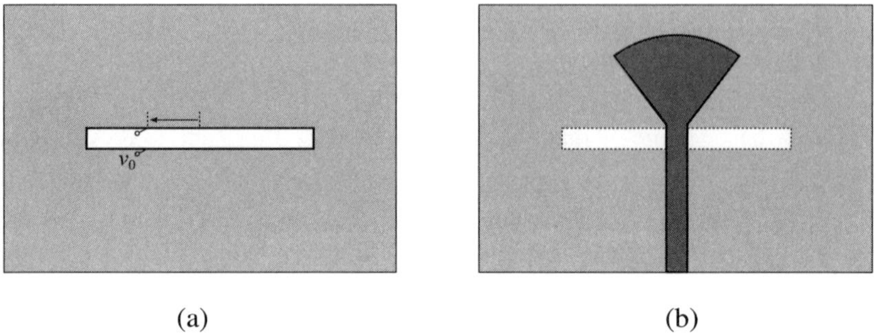

Figure 9.17 Excitation of a slot antenna by (a) offset terminals and (b) a microstrip-to-slotline transition.

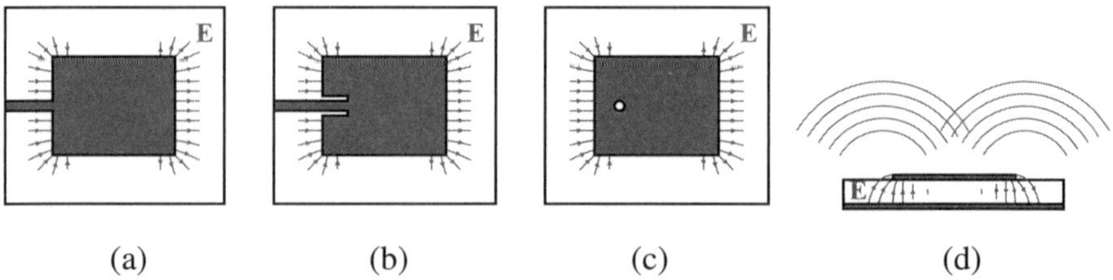

Figure 9.18 Microstrip patch antenna fed from (a) an edge-connected microstrip line, (b) a recessed microstrip line, and (c) a via hole. (d) Side view showing the resonant field pattern beneath the substrate and radiation from the fringing fields on the forward and backward edges of the patch.

accurate, closed-form, theoretical solutions will be rare for antennas of this type, modern computer-aided simulation techniques have made it possible to design practical antennas that are very effective in their intended applications. A brief summary of such antennas will be included here.

9.8.1 Microstrip Patch Antenna

One of the first practical, planar antennas to be developed is the *patch antenna* [18]. It comprises a large (approximately half-wavelength) square or rectangular patch printed on a dielectric substrate over a ground plane. The simplest style for feeding a patch antenna is from a microstrip line attached to the center of one edge, as shown in Figure 9.18(a). Alternatively, one may recess the microstrip line deeper into the patch, as shown in Figure 9.18(b), in order to achieve a more convenient antenna impedance. It is also common to feed patch antennas through a via hole from the bottom side of the substrate, as shown in Figure 9.18. The location of the via will effect the impedance match, bandwidth, and in some cases polarization properties of the radiated fields.

In all these cases, the patch operates much like a half-wave transmission-line resonator, with a low loaded Q due to the radiation losses. The open-circuit conditions create large fringing fields at both the forward and backward edges of the patch as indicated in the figure, and it is from these fields that most of the radiation takes place, as indicated in Figure 9.18(d). These fringing fields make the patch seem larger, electrically, than its physical dimensions, so that a properly designed patch is actually a bit less than half-wavelength in size (similar to the step and open-circuit discontinuity corrections for microstrip that we discussed in Section 3.5.2).

Note that the lateral component of the electric fields at the front and back edges are in phase (whereas the vertical components in the substrate are out-of-phase). Thus, the independent radiation patterns of the two edges add constructively toward zenith, straight above the substrate, while they interfere destructively at angles away from zenith. There will be declination angles at which the two radiation patterns exactly cancel, leading to a null in the field pattern on either side of the main lobe.

Unlike most printed circuits, patch antennas will typically be printed on substrates that are electrically thick and have low dielectric constant in order to improve radiation efficiency. If the patch is to be integrated on the substrate with an integrated circuit that would be negatively impacted by unintended leakage, some compromises shall have to be made.

9.8.2 Planar Inverted-F Antenna (PIFA)

Another, very popular form of planar antenna is a combination of the patch antenna and another type known as the *inverted-F antenna*. The inverted-F antenna itself may be understood as a modification of the simple quarter-wave monopole above a ground plane, as illustrated in Figure 9.19. The bent monopole in Figure 9.19, also known as an *inverted-L*, is more compact than the standard quarter-wave monopole, but has very low driving-point impedance. That can be improved by shifting the feed point further down the length of the wire, leaving the base end shorted to the ground plane as a matching stub, as in Figure 9.19(c). This completes the inverted-F antenna [19].

In the planar version of the inverted-F antenna, the wire of the monopole is widened dramatically into a flat patch, and the offset ground connection is made by

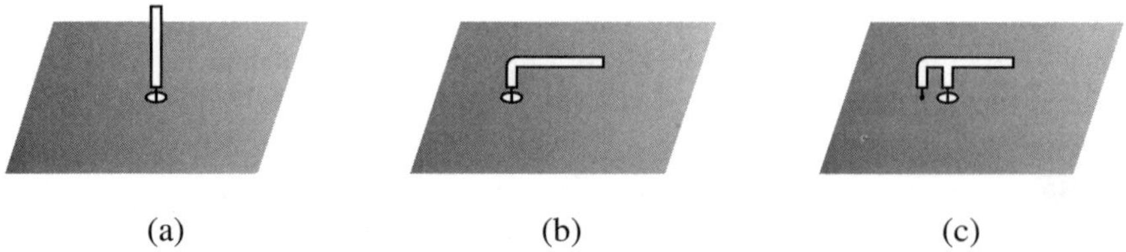

Figure 9.19 Evolution of the inverted-F antenna. (a) A quarter-wave monopole above a ground plane. (b) The monopole is bent over so that it is parallel to the ground plane. This makes the antenna shorter and also directs the energy more vertically where the monopole had a null. (c) The feed point is moved from the base to a point further along the length of the antenna. The base of the antenna is then shorted to the ground plane, acting as an inductive stub and helping with the impedance match.

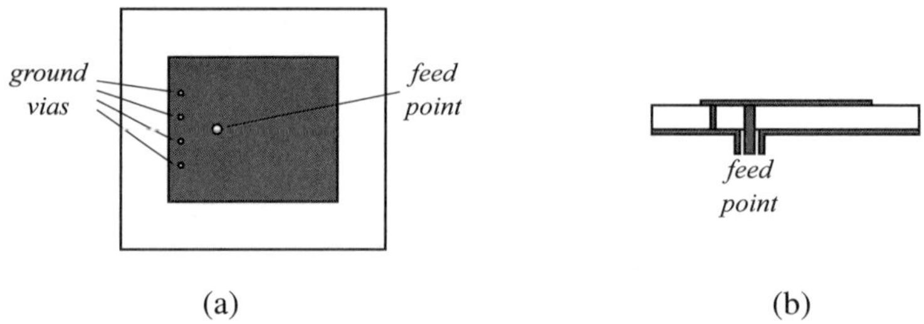

Figure 9.20 Planar inverted-F antenna (PIFA). (a) Top view. (b) Side view.

a via fence at one edge, as illustrated in Figure 9.20(a). Also called a *shorted patch*, this antenna may seem to bear little resemblance to the wire form of inverted-F antenna when viewed form the top, but its common features become more evident when viewed from the side, as shown in Figure 9.20(b). Planar inverted-F antennas (PIFAs) are very common in handheld wireless devices due to their compact size and broad radiation pattern.

9.9 HORN ANTENNAS

While it is best for broadcast antennas to be as omnidirectional as possible (at least in the horizontal plane), directed communication or reception over longer distances benefit from the higher gain and spatial selectivity of a narrow beam. The general principles of Fourier optics discussed in Section 9.2 tell us that to achieve a narrow beam we should have a wide (large compared to a wavelength) aperture. This

(a) (b)

Figure 9.21 Two common horn antennas. (a) Pyramidal horn fed from a rectangular waveguide. (b) Conical horn fed by a circular waveguide.

is further supported by our analysis of Gaussian beams for which a larger waist diameter is associated with smaller divergence angle.

The best way to illuminate an electrically large aperture is with a *horn antenna*, essentially an impedance-matching taper between a single-moded (or dual-polarization) waveguide and one that is much larger and open-ended. Two fairly simple examples are shown in Figure 9.21.

9.9.1 Pyramidal Horn

The first type, in Figure 9.21(a), is known as a *pyramidal horn*. It is generally fed by a rectangular waveguide in the dominant mode. A fully rigorous solution would require accounting for mode conversion in the taper, phase curvature at the resulting wavefront, and edge effects associated with currents that wrap around the lip and flow back down the outside of the horn (which further requires detailed knowledge of the exterior dimensions and surface quality), but for a sufficiently gradual taper and large aperture, these effects can be neglected. We thus assume the aperture fields are simply those of the incident mode in the rectangular waveguide, enlarged to the dimensions of the aperture. We may then take the two-dimensional inverse Fourier transform of that field distribution and apply (9.37) to recover an estimate of the far-field radiation pattern.

Let us assume the inner width and height of the horn aperture are A and B, respectively, and that the electric field in the aperture is given by the TE_{10} rectangular waveguide mode,

$$E(x, y) = E_0 \cos\left(\frac{\pi x}{A}\right) \tag{9.148}$$

where $-A/2 \leq x \leq A/2$ and $-B/2 \leq y \leq B/2$. (Note that we have shifted the origin of the coordinate system from the corner to the center of the aperture to simplify the math that follows.) The plane-wave decomposition of this aperture field distribution is given by the inverse Fourier transform,

$$\mathcal{E}\left(k_x, k_y\right) = \int_{-A/2}^{A/2} \int_{-B/2}^{B/2} E_0 \cos\left(\tfrac{\pi x}{A}\right) e^{j(k_x x + k_y y)} dy dx \tag{9.149a}$$

$$= 4E_0 \int_0^{A/2} \int_0^{B/2} \cos\left(\tfrac{\pi x}{A}\right) \cos\left(k_x x\right) \cos\left(k_y y\right) dy dx \tag{9.149b}$$

$$= \frac{4E_0}{k_y} \sin\left(\tfrac{k_y B}{2}\right) \int_0^{A/2} \cos\left(\tfrac{\pi x}{A}\right) \cos\left(k_x x\right) dx \tag{9.149c}$$

$$= 2\pi A B E_0 \operatorname{sinc}\left(\tfrac{k_y B}{2}\right) \frac{\cos\left(\frac{k_x A}{2}\right)}{\pi^2 - \left(k_x A\right)^2} \tag{9.149d}$$

The far-field radiation pattern, at least for small declination angle θ, is then given by (9.37),

$$E(r, \theta, \phi) = j2\pi(k \cos\theta) \frac{e^{-jkr}}{r} \mathcal{E}\left(k_x, k_y\right) \tag{9.150a}$$

$$= j(2\pi)^2 A B E_0 (k \cos\theta) \frac{e^{-jkr}}{r} \operatorname{sinc}\left(\tfrac{k_y B}{2}\right) \frac{\cos\left(\frac{k_x A}{2}\right)}{\pi^2 - \left(k_x A\right)^2} \tag{9.150b}$$

or, in terms of power,

$$P = |E_0|^2 \frac{(2\pi)^4 A^2 B^2 k^2}{r^2 \eta} \cos^2\theta \operatorname{sinc}^2\left(\tfrac{k_y B}{2}\right) \frac{\cos^2\left(\frac{k_x A}{2}\right)}{\left[\pi^2 - \left(k_x A\right)^2\right]^2} \tag{9.151}$$

where the wavenumber components may be expressed in spherical coordinates,

$$k_x = k \sin\theta \cos\phi \tag{9.152a}$$

$$k_y = k \sin\theta \sin\phi \tag{9.152b}$$

It is customary to take cuts through the radiation patterns at $\phi = 0$ (the H-plane) and at $\phi = \pi/2$ (the E-plane). This simplifies the expressions somewhat,

$$P_H = P_0 \pi^4 \frac{\cos^2\theta \cos^2\left(\frac{A}{2} k \sin\theta\right)}{\left[\pi^2 - \left(A k \sin\theta\right)^2\right]^2} \tag{9.153a}$$

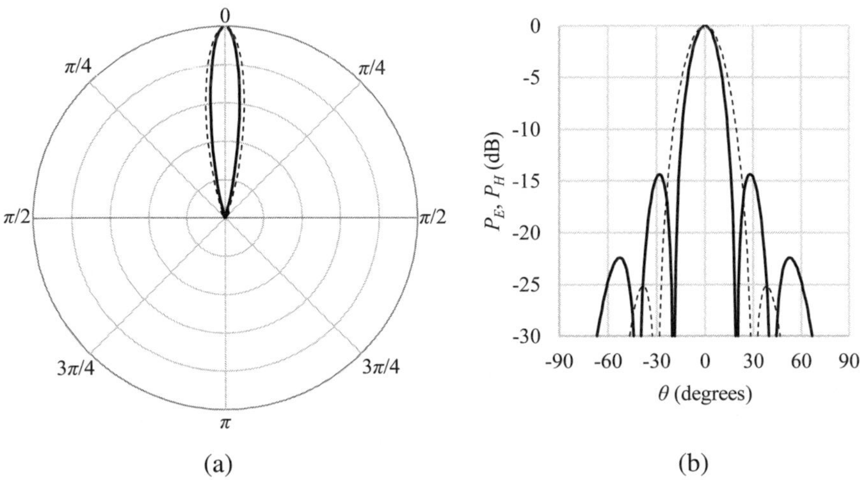

Figure 9.22 The E-plane (solid thick line) and H-plane (dashed thin line) far-field radiation patterns for a pyramidal horn with $A = B = 3\lambda$. (a) Polar plot. (b) Logarithmic plot.

$$P_E = P_0 \cos^2\theta \,\text{sinc}^2\left(\tfrac{B}{2} k \sin\theta\right) \tag{9.153b}$$

where the prefactor is

$$P_0 = |E_0|^2 \frac{16A^2B^2k^2}{r^2\eta} \tag{9.154}$$

A plot of these cut-plane curves is given in Figure 9.22. There is a narrow, main lobe indicating much greater directivity than the dipole-style antennas studied previously. Note that despite the aperture having equal dimensions in both planes, the H-plane beam pattern is slightly broader due to the sinusoidal taper of its field in the aperture. However, the E-plane pattern, while narrower, has relatively high sidelobes peaking at about -15 dB, due to the abrupt truncation of its otherwise uniform aperture field. These sidelobes are difficult to see in the polar plot of Figure 9.22(a), but nonetheless can be significant in high-performance applications.

9.9.2 Conical Horn

The conic horn in Figure 9.21(b) has similar directivity properties as the pyramidal horn, but with a slight twist. Assuming the aperture fields are those of the dominant, TE_{11} mode in circular waveguide, shown in Figure 9.23(a), then the aperture fields and the associated radiation pattern are no longer polarized along a single linear

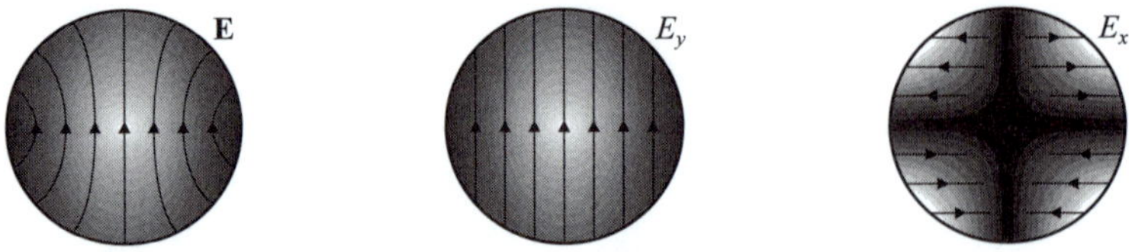

(a) (b) (c)

Figure 9.23 Decomposition of (a) TE_{11} aperture field distribution in a conical horn into (b) copolar and (c) cross-polar components.

dimension. The aperture fields may be broken down into vertical and horizontal components, shown in Figures 9.23(b, c), and each contributing independently to a *copolar* and *cross-polar* far-field beam patterns, respectively. The copolar, far-field pattern will be qualitatively similar to that from the pyramidal horn, with a single main lobe and smaller sidelobes. The cross-polar pattern will have a null on boresight, where the copolar main lobe peaks, but four equal lobes in the upper-left, upper-right, lower-left, and lower-right quadrants. The *cross-polar leakage* is therefore zero at the peak of the main beam, but has peaks of its own off-center.

9.9.3 Potter Horn

Despite having a rotationally symmetric aperture geometry, the aperture field pattern of a smooth-wall conical horn is still not circularly symmetric, even if cross-polar components are excluded, resulting in a far-field beam that is not the same width in the E and H-planes. A dual-polarization antenna system (one in which both orthogonal forms of the TE_{11} mode are excited) would therefore illuminate noncongruent areas of the far field in the two polarizations. Further, the cross-polar leakage described above would contaminate polarization-sensitive measurements.

A novel solution to these issues is known as a *Picket-Potter* or *dual-mode horn* [20], shown in Figure 9.24(a). Whereas the pyramidal and conical horn utilized gradual tapers to minimize conversion from the dominant modes into higher-order modes as the cross-section grew toward the aperture, the Potter horn has an abrupt step near the throat that is designed specifically to convert some energy from the dominant TE_{11} mode of circular waveguide into the TM_{11} mode. The step is followed by a short uniform section designed to bring these two modes into the proper phase (compensating in part for modal dispersion which follows in the long taper to the aperture). If properly balanced, the TE_{11} and TM_{11} modes combine to

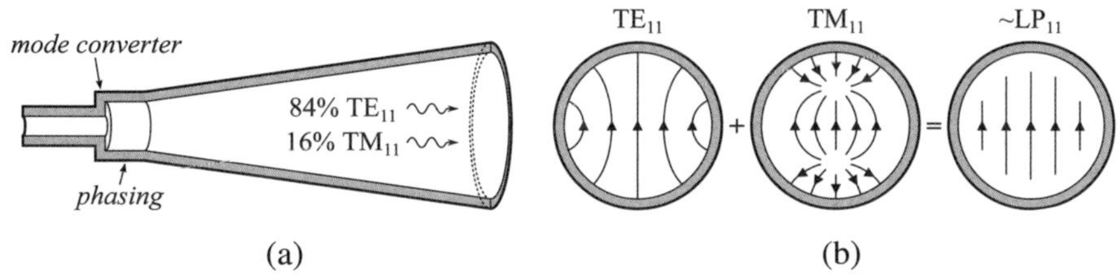

Figure 9.24 Dual-mode Potter horn (a) profile, and (b) combination of the TE_{11} and TM_{11} modes to synthesize an approximate linearly polarized (LP_{11}) mode.

form a good approximation of a linearly polarized mode with circularly symmetric intensity. This yields a far-field radiation pattern that is rotationally symmetric, relatively free of sidelobes and has low cross-polarization leakage. A comparison of the beam patterns (computed numerically by computer) for a smooth conical horn and a Potter horn having the same overall length and aperture diameter is shown in Figure 9.25. The E and H-planes are shown, along with a cross-polarization cut taken along a diagonal between these two planes (where the cross-polarization patterns peak). Note that the Potter horn's pattern is equal width in both E and H-planes (and is nearly circular), while the sidelobes are much lower and cross-polar rejection much deeper than the smooth conical horn.

9.9.4 Conical Corrugated Feedhorn

Unfortunately, while the Potter horn provides almost ideal beam patterns, the beneficial effect is narrowband, due to the frequency-sensitivity of the mode-conversion and subsequent dispersion in the critical TE/TM_{11} modes. To get around this, we must utilize a far more sophisticated kind of horn, known as a *conical corrugated feedhorn*, shown in Figure 9.26. Based on the corrugated circular waveguide which we studied in Section 7.7, the goal is to excite the HE_{11} mode in the balanced hybrid condition at the aperture (note that we have used a launcher like that shown in Figure 8.18(b)). This results in an aperture field pattern like that shown in Figure 7.27 where $\Lambda = 1$. It is linearly polarized with an approximately Gaussian intensity profile, the ideal conditions for launching a Gaussian beam.

The term *feedhorn* emphasizes a particular application for this kind of antenna, that of illuminating a large optical element such as a parabolic satellite dish. Since these structures are almost always circular, the circular beam makes the best use of the dish's collecting area, while minimizing the energy that misses the outer

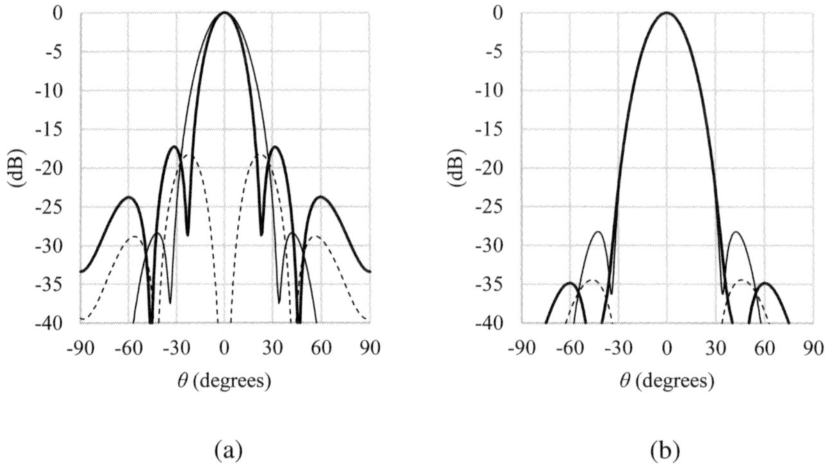

Figure 9.25 Comparison of the beam patterns for (a) a smooth-wall conical horn and (b) a Potter horn. In both cases, the aperture diameter $D = 3\lambda$, and the taper length $L = 30\lambda$. The E-plane pattern is shown with a thick solid line, the H-plane pattern with a thin solid line, and the cross polarization pattern in the diagonal plane with a dashed line.

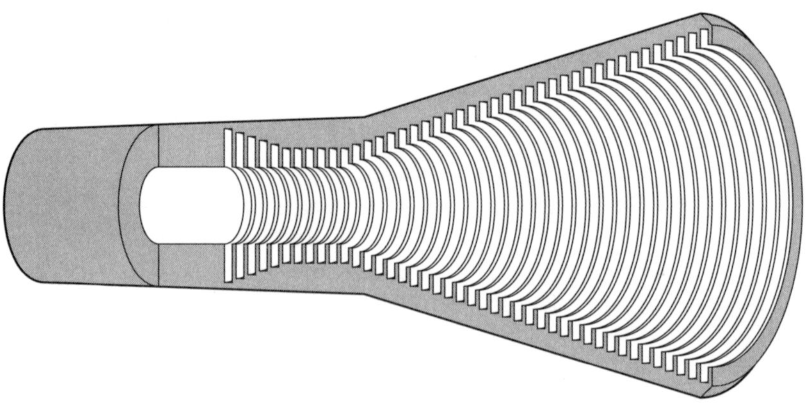

Figure 9.26 Cutaway illustration of a conical corrugated feedhorn.

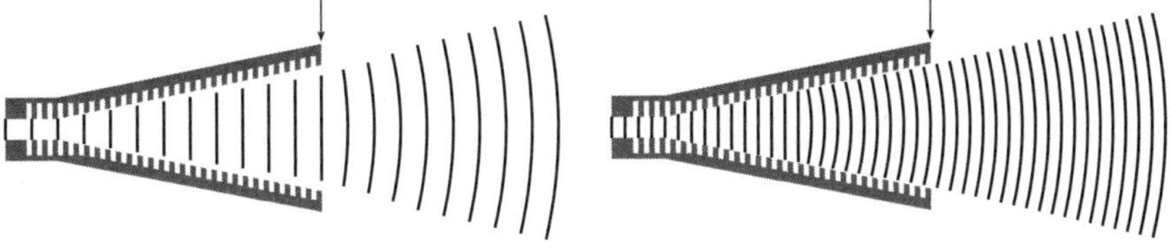

Figure 9.27 Wavefronts leaving a broadband corrugated feedhorn at (a) low and (b) high frequencies. The latter wavefronts have more curvature at the aperture, such that the divergence angle of the two beams is the same.

edge (known as *spillover*). To make the best use of the reflector over broad bandwidth, it is further desirable that the beamwidth is also constant with frequency. This can be achieved by matching the taper of the horn to the divergence angle of the (approximate) Gaussian beam launched from the aperture at the lowest frequency. Higher frequencies effectively launch from further down the throat of the horn, acquiring significant wavefront curvature by the time they reach the aperture, as illustrated in Figure 9.27. If this were not the case, then higher-frequency beams with a flat-phase at the aperture would launch as though a Gaussian beam with waste size equal to the aperture diameter, having much lower divergence angles than they do at lower frequencies.

9.10 METRICS FOR DIRECTIVE BEAMS

Having now seen a number of antennas with highly directive beam patterns, it is useful to define some performance metrics that are well suited to their characteristics. Like gain and directivity, defined previously, these metrics deal primarily with the ability of an antenna to direct most of its energy in a desired direction and in a single main beam without loss to wide-angle spillover and sidelobes.

9.10.1 Beam Area

Recall the beam area defined in (9.138). Although it is sufficiently general to apply to all antennas, it takes on a somewhat more intuitive meaning when applied to beams with high directivity. Specifically, it is the angular width of the cone

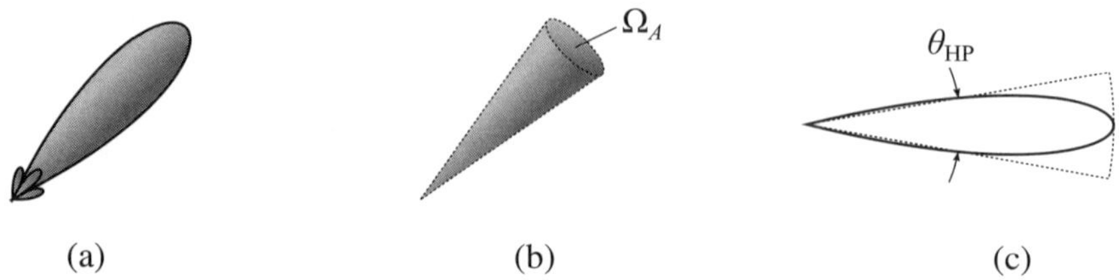

Figure 9.28 Comparison of (a) a realistic beam pattern and (b) the cone of emission representing its beam area. (c) Approximation of the beam area using the half-power beam width of the main lobe.

subtended by a fictitious far-field radiation pattern having uniform intensity equal to the peak of the true far-field radiation pattern and radiating the same total power. This is illustrated in Figures 9.28(a, b). The cone of emission associated with the beam area is a simple way to approximate a directed beam in rough calculations without having to account for the minute details of a more realistic radiation pattern. This utility is absent when applied to more omnidirectional patterns such as those from dipoles and slots, since the cone of emission bears such little resemblance to the actual patterns in those cases, and any preferred direction implied by drawing such a cone is misleading.

For directive beams, the beam area can be closely approximated as the product of the half-power beamwidths along any two orthogonal cuts of the beam, such as the E and H-planes,

$$\Omega_A \approx \theta_{HP}^{(E)} \theta_{HP}^{(H)} \tag{9.155}$$

as illustrated in Figure 9.28(c).

9.10.2 Beam Efficiency and Stray Factor

When a beam pattern consists of a main lobe and sidelobes, as it often does for horn antennas, the beam area may further be broken down into that portion which resides in the main lobe, Ω_M, and the remainder, which resides in the minor lobes, Ω_m,

$$\Omega_A = \Omega_M + \Omega_m \tag{9.156}$$

The *beam efficiency* is a measure of how much of the total radiated power is packed into the main lobe,

$$\eta_{beam} = \frac{\Omega_M}{\Omega_A} \tag{9.157}$$

and, conversely, the *stray factor* is a measure of how much is lost to the sidelobes,

$$\eta_{stray} = \frac{\Omega_m}{\Omega_A} = 1 - \eta_{beam} \tag{9.158}$$

When a beam pattern does not clearly exhibit nulls, the definition of what constitutes the main lobe versus the sidelobes or even a back lobe is somewhat subjective, and these quantities can only be approximately defined.

9.10.3 Effective Area

When thinking about a receiving antenna, we imagine that it is illuminated by a planar wavefront from a very distant source, such that the field intercepting the antenna is uniform with fixed power per unit area. Thus, from at least a standpoint of unit analysis, to recover the received power, we would like to simply multiply the power flux density of that wave by an area representing the flux which is intercepted by the antenna and transferred to its port or terminals. This will be called the *effective area* or *effective aperture*, and is denoted as A_e.

Logically, the larger the physical aperture, A_{phys}, of an antenna, such as a horn, the more power it can capture from an incident wave. However, since no antenna may couple perfectly to a uniform field distribution over its physical aperture, even with lossless, ideal materials, the effective aperture is less than the physical aperture area by a factor known as the *aperture efficiency*,

$$\eta_{ap} = \frac{A_e}{A_{phys}} \tag{9.159}$$

Horns, especially feedhorns paired with paraboloidal reflectors, typically have aperture efficiencies between about 0.5 and 0.8. Arrays of smaller antennas can be designed to have greater efficiencies, but usually at the cost of other aspects of beam quality, as the necessary uniform field distribution inevitably leads to unwanted sidelobes.

Conceptually, the effective area is a measure of how well an antenna receives power from a distant source in a particular direction, whereas directivity is a measure of how well an antenna delivers power to a distant point in that direction. Reciprocity tells us that these two quantities must be linked. The effective area of an antenna is thus directly related to its directivity, which, in turn, fixes the beam area according to (9.138). We may therefore derive a general relationship between the beam area and the antenna's effective area by considering the properties of the beam alone, without any knowledge of the structure of the antenna itself. It does not matter what beam we use, so long as it is physically accurate. We shall thus use a beam for which we already have a detailed analytical solution, namely, the Gaussian beam studied in Section 9.3, with the presumed aperture of the antenna at the beam's waist.

To calculate the beam area, we shall need the radiant intensity in the far field,

$$U(r) = \lim_{z\to\infty} z^2 S(r) = \lim_{z\to\infty} z^2 E(r) H^*(r) \tag{9.160a}$$

$$= \frac{|E_0|^2}{\eta} \lim_{z\to\infty} z^2 \frac{w_0^2}{w^2(z)} e^{-\frac{2r^2}{w^2(z)}} \tag{9.160b}$$

$$= \frac{|E_0|^2}{\eta} \lim_{z\to\infty} z^2 \frac{w_0^2}{w_0^2\left(1+\left(\frac{z}{z_R}\right)^2\right)} e^{-\frac{2r^2}{w_0^2\left(1+\left(\frac{z}{z_R}\right)^2\right)}} \tag{9.160c}$$

$$= \frac{|E_0|^2}{\eta} z_R^2 \lim_{z\to\infty} e^{-\frac{2r^2 z_R^2}{w_0^2 z^2}} = \frac{|E_0|^2}{\eta} z_R^2 e^{-2\frac{\tan^2\theta}{\tan^2\theta'}} \tag{9.160d}$$

where θ is the declination angle in the spherical coordinate system, and θ' is the divergence angle of the Gaussian beam. The beam area is then given by (9.138)

$$\Omega_A = \frac{P_{tot}}{U_{max}} = \frac{\frac{\pi}{\eta}|E_0|^2 w_0^2}{\frac{1}{\eta}|E_0|^2 z_R^2} = \pi \tan^2\theta' \tag{9.161}$$

Note that we have used $P_{tot} = 2P_0$ where P_0 was given in Table 9.2 for the power flow in a Gaussian beam. The factor of 2 is needed because we have assumed the excitation at an aperture, which, in the absence of any other antenna structure, would produce dual beams in both the $+z$ and $-z$ directions.[9]

The maximum effective aperture of the Gaussian beam (that is, the effective aperture of the presumed antenna, neglecting ohmic losses) may be calculated at its waist by dividing the total power by the peak flux at the origin (thus recovering the area that would be occupied by the waist if the fields were uniformly distributed),

$$A_{em} = \frac{P_{tot}}{S(r=z=0)} = \frac{\frac{\pi}{\eta}|E_0|^2 w_0^2}{\frac{1}{\eta}|E_0|^2} = \pi w_0^2 = \frac{\lambda^2}{\pi \tan^2\theta'} \tag{9.162}$$

9 A more realistic antenna, say, a horn, producing a near Gaussian beam in the forward direction would likewise have backward radiation as a consequence of currents flowing along the outside of the horn. The far-field radiation in the reverse direction would thus be more complex, but the overall energy balance would be preserved.

Note that the product of the beam area and the effective aperture is independent of the beam divergence,

$$A_{em}\Omega_A = \lambda^2 \tag{9.163}$$

This is a quite general result that applies to all beams and all antennas. One may even calculate the effective area of a theoretically lossless, isotropic antenna ($\Omega_A = 4\pi$) without having to worry about how to design one,

$$A_{iso} = \frac{\lambda^2}{4\pi} \tag{9.164}$$

This, in turn, gives us another general expression for the directivity,

$$D = \frac{A_{em}}{A_{iso}} = \frac{4\pi A_{em}}{\lambda^2} \tag{9.165}$$

One may even define the effective area of an antenna which has no physical area at all (or for which the physical area is negligibly small) such as a very thin, half-wave dipole. Recall that the directivity of such a dipole is about 1.64, making its effective area

$$A_{em}(\text{dipole}) = \frac{1.64}{4\pi}\lambda^2 = 0.1305\lambda^2 \tag{9.166}$$

When ohmic losses are taken into account, the actual effective area is reduced by a factor of η_{ant}, and the expression for directivity above becomes an expression for the actual gain,

$$G = \frac{4\pi A_e}{\lambda^2} \tag{9.167}$$

A summary of these results can be found in Table 9.6.

9.11 FRIIS TRANSMISSION EQUATION

Now that we know how to calculate the power received by an antenna as well as the power transmitted, we can derive an expression for the link gain between two antennas, so long as they are in each other's far fields and pointed such that each lies in the direction of maximum gain (and copolarized). Consider the power flux density generated by a transmitting antenna at a distance, d,

$$S_{max} = \frac{U_{max}}{d^2} = \frac{G' P_{av}}{d^2} = \frac{G' P'}{4\pi d^2} = \frac{A'_e P'}{\lambda^2 d^2} \tag{9.168}$$

Table 9.6

General Equations for Directive Antennas

Parameter	Symbol	Expressions
Directivity	D	$= \frac{4\pi}{\Omega_A} = \frac{4\pi A_{em}}{\lambda^2}$
Gain	G	$= \eta_{ant} D = \frac{4\pi A_e}{\lambda^2}$
Beam area (total)	Ω_A	$= \frac{4\pi}{D} = \frac{\lambda^2}{A_{em}} \approx \theta_{HP}^{(E)} \theta_{HP}^{(H)}$
Beam area (main lobe)	Ω_M	$= \eta_{beam} \Omega_A$
Beam area (minor lobes)	Ω_m	$= \eta_{stray} \Omega_A$
Antenna efficiency	η_{ant}	$= \frac{G}{D} = \frac{A_e}{A_{em}}$
Aperture efficiency	η_{ap}	$= \frac{A_e}{A_{phys}}$
Beam efficiency	η_{beam}	$= \frac{\Omega_M}{\Omega_A}$
Stray factor	η_{stray}	$= \frac{\Omega_m}{\Omega_A}$

where P' is the transmitted power and A'_e is the transmitting antenna's effective area (including ohmic losses). The received power at the second antenna is then given by $P'' = A''_e S_{max}$, from the definition of effective area. We may therefore write the net effective gain between the ports of the two antennas,

$$\frac{P''}{P'} = \frac{A'_e A''_e}{\lambda^2 d^2} \tag{9.169}$$

or, in terms of gain,

$$\frac{P''}{P'} = G'G'' \left(\frac{\lambda}{4\pi d} \right)^2 \tag{9.170}$$

These are both forms of the Friis transmission equation. Equation (9.169) is the original form, but (9.170) is more commonly used in modern times. It accurately predicts the free-path loss between two antennas — and more importantly, provides a context for the definition of antenna gain that does not require accurate knowledge of the full, spherical radiation pattern — but it fails to account for a number of factors that may come into play in real-world transmission scenarios. These include loss and scattering in the intervening medium, line-of-sight obstructions, reflections from the ground, and constructive/destructive interference due to multiple reflections off of nearby objects (known as *multipath)*.

Problems

9-1 What are the conductive and dielectric loss constants at 10 GHz of the TEM mode in parallel-plate waveguide formed by a 100-μm-thick GaAs substrate ($\varepsilon_r = 12.9$, $\tan\delta = 0.0065$) with gold metalization ($\sigma = 4.1 \times 10^7$ S/m) on both sides?

9-2 Do any of the modes in parallel-plate waveguide exhibit asymptotically zero conductive loss at high frequency in the same way that the TE_{0m} modes do in circular waveguide?

9-3 Write an expression for the ratio of the conductive loss tangents of the TE_m and TEM modes in parallel-plate waveguide. At what point do the loss tangents become equal?

9-4 A parallel-plate waveguide has a triangular excitation profile of width A, where the electric field amplitude is E_0 at $x = 0$, and tapers linearly to zero at $x = \pm A$. Calculate the far-field radiation pattern of this excitation.

9-5 How large would the beam waist of a Gaussian beam at 10 GHz have to be in order to have a divergence half-angle less than $1°$? Less than $0.1°$? Express the answer in both meters and wavelengths.

9-6 What diameter lens is required in order to intercept 95% of the power in a 10-GHz Gaussian beam at the waist with $10°$ divergence half-angle?

9-7 Calculate the ray transfer matrix of a quarter-wavelength antireflection coating having thickness t on the surface of a curved dielectric interface with radius R. Show that it reverts to the identity matrix in the high-frequency limit.

9-8 Imagine a lens made of glass with $n = 1.45$ which is convex on one side and concave on the other, where both sides have the same radius of curvature. How thick would it have to be in order for the focal length to be 20 times the radius of curvature?

9-9 What is the ray transfer matrix of a system comprising two thin lenses with focal lengths f_1 and f_2 spaced $d = f_1 + f_2$ apart?

9-10 Estimate the gain of a half-wave dipole at 100 MHz in which the arms have a resistance of 5 Ω/m. Assume the dipole is exactly a half-wavelength long, ignoring any reactive part of the terminal impedance.

9-11 What is the radiation resistance of 5-cm-long slot at 1 GHz? Qualitatively, what would its reactive part look like?

9-12 Assume that the aperture fields of a rectangular, pyramidal horn are those of the TE_{10} mode, as in the example given in Section 9.9.1. Recalculate the far-field pattern based on the magnetic field configuration instead of the electric field configuration. Are they different? Should they be? Discuss any discrepancies.

9-13 Write down an expression for the x and y-components of the electric field in the aperture of a circular Potter horn as a superposition of the TE_{11} and TM_{11} modes with arbitrary amplitude and phase relationship. Align the modes' polarizations so that the y-component can be maximized while the x-component may nearly cancel.

9-14 What is the maximum effective area of a short slot antenna with length $l \ll \lambda$?

9-15 What is the maximum directivity in decibels of a 12-m parabolic dish with an aperture efficiency of 80% at a wavelength of 1 mm?

References

[1] J. D. Kraus, *Antennas*, 2nd ed. New York: McGraw Hill, 1988.

[2] Wikipedia. (2018) Fourier optics. https://en.wikipedia.org/wiki/Fourier_optics.

[3] Wikipedia. (2018) Stationary phase approximation. https://en.wikipedia.org/wiki/Stationary_phase_approximation.

[4] W. Rotman, "Wide angle microwave lens for line source applications," *IEEE Transactions on Antennas and Propagation*, vol. 11, no. 6, pp. 623–632, 1963.

[5] P. F. Goldsmith, *Quasioptical Systems*. New York: IEEE Press, 1998.

[6] Wikipedia. (2018) Gaussian beam. https://en.wikipedia.org/wiki/Gaussian_beam.

[7] Wikipedia. (2018) Ray transfer matrix analysis. https://en.wikipedia.org/wiki/Ray_transfer_matrix_analysis.

[8] Wikipedia. (2018) Lens (optics). https://en.wikipedia.org/wiki/Lens_(optics).

[9] P. A. Bélanger, "Beam propagation and the *abcd* ray matrices," *Optics Letters*, vol. 16, no. 4, pp. 196–198, 1991.

[10] H. Kogelnik and T. Li, "Laser beams and resonators," *Proceedings of the IEEE*, vol. 54, no. 10, pp. 1312–1329, 1966.

[11] Wikipedia. (2018) Mathematical descriptions of the electromagnetic field. https://en.wikipedia.org/wiki/Mathematical_descriptions_of_the_electromagnetic_field.

[12] Wikipedia. (2018) Lorenz gauge condition. https://en.wikipedia.org/wiki/Lorenz_gauge_condition.

[13] Wikipedia. (2018) Dipole antenna. https://en.wikipedia.org/wiki/Dipole_antenna.

[14] Wikipedia. (2018) Retarded potential. https://en.wikipedia.org/wiki/Retarded_potential.

[15] P. J. Bevelacqua. (2011) The short dipole antenna. http://www.antenna-theory.com/antennas/shortdipole.php.

[16] C. A. Balanis, *Antenna Theory: Analysis and Design*, 3rd ed. New York: Wiley, 2005.

[17] H. G. Booker, "Slot aerials and their relation to complementary wire aerials (Babinet's principle)," *Journal of the Institute of Electrical Engineers – Part IIIA: Radiolocation*, vol. 93, no. 4, pp. 620–626, 1946.

[18] J. Q. Howell, "Microstrip antennas," *Antennas and Propagation Society International Symposium*, pp. 177–180, 1972.

[19] Wikipedia. (2018) Inverted-F antenna. https://en.wikipedia.org/wiki/Inverted-F_antenna.

[20] P. D. Potter, "A new horn antenna with suppressed sidelobes and equal beamwidths," *JPL Technical Report*, no. 32-354, February 1963.

Chapter 10

Flat-Frequency Components

Now armed with a wide variety of fundamental circuit elements suitable for applications at almost any wavelength, we are ready to begin discussing the next level of building blocks implemented using these technologies, which form the components of electronic systems. In this chapter we will focus on *flat-frequency components*, components for which the target operational characteristics are constant or near-constant with frequency. That is not to say that any particular implementation will not have limited operational frequency range, or that the performance does not change over that bandwidth; only that these variations are small, unintentional, and, generally, undesirable. Typically, the goal will be to design a component having the widest operational bandwidth with the most consistent performance characteristics possible.

10.1 TERMINATIONS

Perhaps the most trivial component we can define in this category is that of a simple *load* or *termination*, ideally, a single-port network that absorbs all incident power without reflection from the ports of an external network to which it is attached. For those networks defined by electrically small ports (usually built of lumped elements or transmission lines), the termination takes on the simple form of a resistor, matched to the characteristic impedance of the system. When discrete resistors are available and have the required frequency range of performance, then no further elaboration is needed.

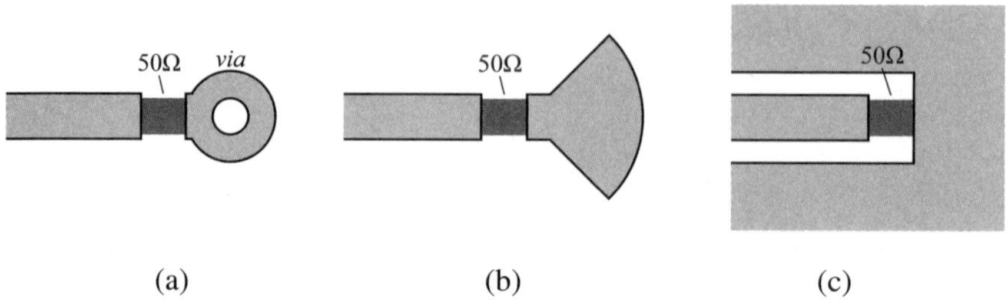

Figure 10.1 Example of printed-circuit terminations. (a) Microstrip grounded resistor. (b) Microstrip resistor with radial stub. (c) CPW grounded resistor.

10.1.1 Printed-Circuit Terminations

In the realm of planar, printed circuits, it is occasionally up to the system designer to fashion suitable terminations on their own using available materials and circuit patterns. Most printed-circuit processes will include the option of selectively depositing a thin film of material on the top surface, which has a known thickness and sheet resistance.[1] Several termination geometries which take advantage of this are shown in Figure 10.1. The first, Figure 10.1(a), is a simple resistor grounded with a via hole. This is generally a good termination down to DC, but is limited in its upper frequency range by the parasitic inductance of the via.

If the substrate is electrically thick, or the circuit is required to operate to very high frequencies, the simple via may be insufficient. One option is to replace the via with a radial stub, as shown in Figure 10.1(b). The stub provides a near short-circuit equivalent over a band-pass frequency range. This may be useful for high-frequency applications where the substrate is simply too thick for a via, but it does not work down to DC.

Another option is to use a resistor with coplanar waveguide, as shown in Figure 10.1(c). The easy, low-inductance access to ground on the top side of a substrate is one of the primary advantages of using CPW in integrated circuits.

Some processes and manufacturers give the user the option of selecting a sheet resistance to be used throughout an integrated circuit (controlled even with fixed materials by varying the thickness of the film). The resistors in the terminations may be implemented with any convenient sheet resistance, with some minor caveats. If

1 Recall that the concept of a surface, or sheet resistance, was first introduced in the context of skin effect losses in Section 1.6.4. The same mathematical framework is used here to describe materials which have intentionally dissipative characteristics.

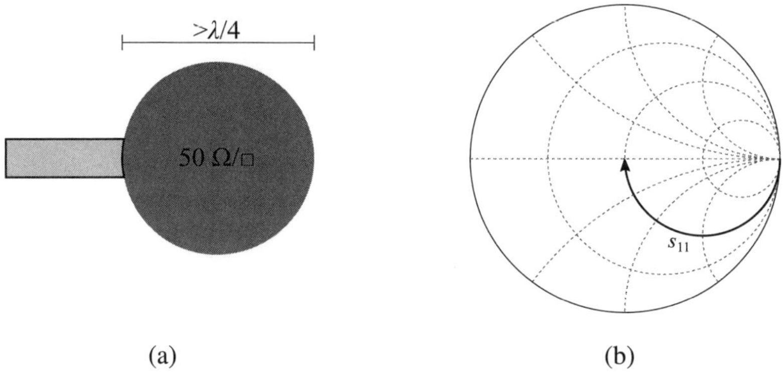

Figure 10.2 (a) Dot termination on a microstrip substrate. (b) Theoretical reflection coefficient on a Smith chart.

the sheet resistance is too low, then the resistor may have to be exceptionally long to achieve the desired 50Ω, but such a resistor itself would suffer from parasitic inductance, besides that of the via. However, if the sheet resistance is too high, then a 50Ω resistor would be unusually wide and, if wider than the microstrip trace leading up to it, would exhibit localized spreading resistance and inductance which will degrade the overall response of the termination. A sheet resistance of 50 Ω/□ is a common choice, giving the desired resistor in a 50Ω system a simple 1:1 aspect ratio.

A somewhat lesser-known solution, which is surprising, given its simplicity and practical advantages, is the *dot termination* [1] shown in Figure 10.2(a). It comprises a circular disc of resistive film at the end of a transmission line. No ground connection is needed. Unlike the previous terminations, the sheet resistance of the material is required to be 50 Ω/□, or, more accurately, one Z_0 per square, where Z_0 is the characteristic impedance of the system one wishes to terminate. The reflection response of this structure is shown on the Smith chart in Figure 10.2(b). Clearly, at low frequencies, it is an open circuit, but trends toward the center of the chart (a perfect match) as the frequency increases. Empirically, the return loss is better than 20 dB or so when the disc is greater than a quarter-wavelength in diameter, and remains very good so long as the microstrip feed line is not overmoded [2]. Although it does not work down to DC, it has a broader bandwidth response than the radial-stub terminated resistor in Figure 10.1(b), and will usually work to much higher frequencies than any via-grounded resistor could. It is especially worth remembering for those rare processes in which vias are not available. The primary disadvantages of this solution are that it is typically larger than most other solutions, and it lacks a two-terminal resistance the manufacturer can use to monitor and verify the sheet resistance of their film with a DC probe test (a point which could easily be addressed by including a dummy resistor elsewhere on the same substrate).

10.1.2 Waveguide Terminations

It is not so easy to simply attach a resistor to the end of a waveguide whose dominant mode is not even transverse-electric-and-magnetic (TEM). There are no obvious terminals, and if a resistor were bridged across the end of an open waveguide, say, from the center of one broad wall to the other, at least as much or more power would actually radiate out into space than couple into the resistor. In fact, the resistor is hardly needed; a simple open waveguide has return loss of approximately 10 dB, with most of the incident energy radiating away as though from an antenna. This is due to the large electrical size of the waveguide cross-section; in some ways, waveguides just below the cutoff of their first parasitic mode are as large as any structure can be while still supporting only a single mode of propagation. This feature is worthy of remembrance. While it is generally good practice to terminate all waveguides in any production equipment, in a quick laboratory experiment one can get away with simply leaving an unused waveguide port open, assuming that 10-dB return loss is sufficient for the test and there is no danger of leakage or cross-talk from the radiant signal contaminating the results.

More accurate waveguide terminations are most often made using magnetically loaded lossy material filling the waveguide (see, for example, [3]). Further, since the characteristic impedance of the material-loaded waveguide is highly reactive and very much different from that of the empty guide, these constructions almost invariably involve some kind of multisection-transformer or taper. Several common geometries are shown in Figure 10.3. The first, Figure 10.3(a), comprises a sharp, tapered cone of absorbing material protruding down the center of the waveguide toward the incident wave. Although it is capable of offering an excellent, broadband impedance match, the narrow tip is very delicate and difficult to machine or cast. There is also an issue in the case of a radiometric load where the physical temperature is significant; the tip has relatively poor connection to the rest of the waveguide housing, and may not stabilize at the same temperature as its surroundings, especially with external signal power incident upon it.

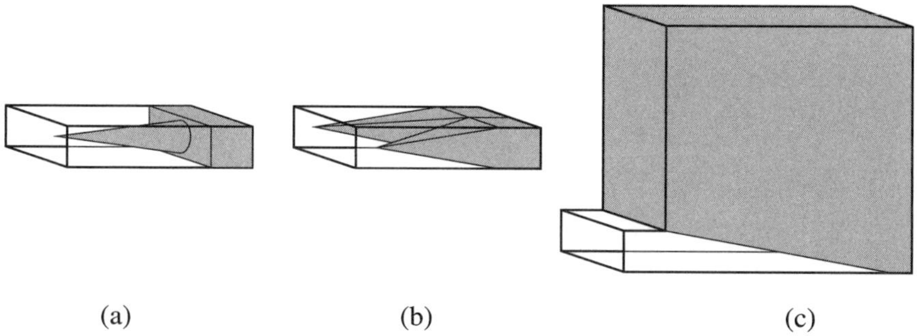

Figure 10.3 Waveguide termination geometries. (a) Central tapered cone. (b) Side-contacting tapered wedges. (c) Large cavity wedge.

A solution to the thermal problem may be to reshape the termination so that the taper makes contact with the waveguide walls (which are presumed to have very good thermal conductivity) all along its length, as in the pointed wedges of Figure 10.3(b). Note that such a load also splits easily into two identical pieces, each of which may be epoxied into one side of the housing in a split-block construction, as described earlier.

Typically, shaped absorbing elements such as these are machined from hard stock, or cast in the shapes desired. Either way, the very sharp features of either of the previous waveguide loads may be difficult to manufacture accurately when the frequency is high or too fragile for the delicate material to survive handling. In these cases, a large, cavity-filling wedge such as that shown in Figure 10.3(c) may be advantageous [4].

10.1.3 Absorber

In the quasioptical regime, terminations take the form of an absorbing boundary. Ideally, we would like to define a two-dimensional surface that is perfectly matched to the impedance of free space, but which absorbs 100% of the energy incident upon it, so that none passes through at any frequency. That is, it should have zero reflection but infinite attenuation. Unfortunately, Maxwell's equations do not permit such solution. To see why, consider the analogy of a termination at the end of a transmission line, as shown in Figure 10.4(a). As we know, a simple resistor is sufficient to realize a perfect termination. However, an implicit assumption of this model is that the transmission line, or the region supporting a propagating mode,

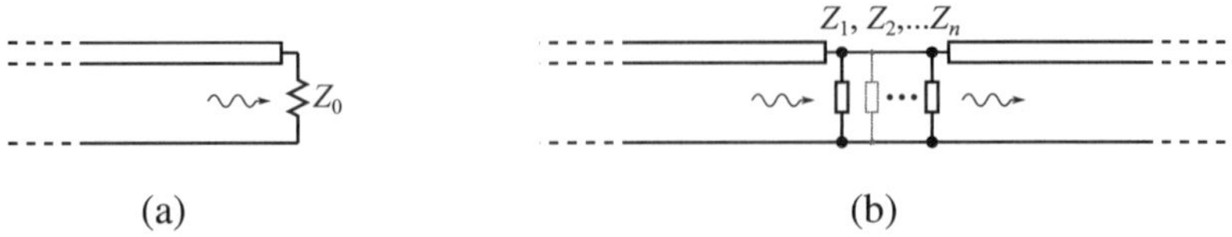

Figure 10.4 Transmission-line model of an absorbing boundary, for (a) truncated free space and (b) infinite free space.

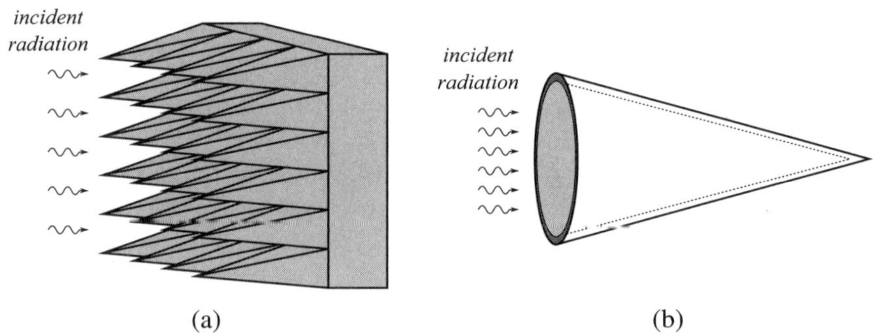

Figure 10.5 Two examples of quasioptical termination. (a) Pyramidal absorber. (b) Conical load.

ends at the terminals where the resistor is attached. In this case, the transmission line is a model for free space. Since space cannot end, per se, but can only be filled with some material, a more accurate model is one such as that in Figure 10.4(b), where one or more load impedances are connected in parallel with the transmission line (analogous to a multilayered material stackup in space). One may select impedances that provide a near-perfect impedance match over very broad frequencies, but some finite amount will always leak through, or else come to a point where the shunt impedance is zero (a short-circuit in the transmission-line case, or a PEC boundary in quasioptics), causing a reflection. As a matter of principle, we do not have such a thing as an absorbing surface, but rather absorbing regions, whose total absorption is dependent, in part, on their thickness (which is not to say that perfectly acceptable absorbers cannot be made physically quite thin).

Many practical absorbers use geometry in addition to material parameters to improve their characteristics. Two common forms are shown in Figure 10.5. The first, Figure 10.5(a), is known as pyramidal absorber, and comprises a two-dimensional array of tall pyramids cut from a lossy material. This is often the

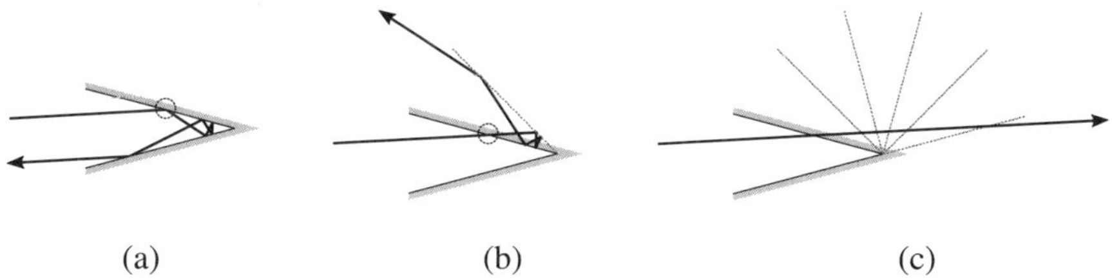

Figure 10.6 (a) A ray reflecting off the walls of a deep V-shaped corner, such as the inside of a cone or the sides of two pyramids in an array. The first reflection is circled with a dashed line. (b) New geometry after unfolding the first reflection. (c) Fully unfolded geometry revealing that the total number of reflections is increased with smaller opening angle.

geometry of choice for covering large surface areas, such as the inside of an *anechoic chamber*, an enclosed environment for testing antennas. In more compact setups, one may use the *conical load* shown in Figure 10.5(b), which has a layer of lossy material lining the inside of a deep, hollow cone.

The premise behind both of these geometries is essentially the same. It is difficult to manufacture a material that is simultaneously very lossy (absorbing) without also introducing impedance mismatch (reflections). We therefore adopt the approach of absorbing just a little energy with each reflection, and then create a geometry that maximizes the number of internal reflections before a wave can work its way back out to the source.[2] The tall pyramids and the deep cone ensure that waves incident near the primary axis impact the first surface at a glancing angle, and must then ricochet a large number of times to return back the way it came, as illustrated in Figure 10.6(a). In this example, a ray entering the V-shaped aperture of the cone (or the space between two pyramids) bounces off the walls a total of six times before reemerging. To better visualize this, we may unfold the geometry at each reflection by mirroring the ray's path as well as the other boundaries. The first step is shown in Figure 10.6(b). Note that the initial reflection of the ray has been rendered a straight line after mirroring. We follow the same procedure for each subsequent reflection, ultimately arriving at the diagram in Figure 10.6(c), where the ray is straight and passes through the image boundaries a total of six times. Thus, the total absorption coefficient and return loss of a correspondingly shaped absorber increases sixfold compared to a flat plane of the same material, and it is clear that smaller opening angles lead to a larger total number of reflections.

---

2 From another viewpoint, we maximize the lossy surface area with which an incoming wave must ultimately come in contact.

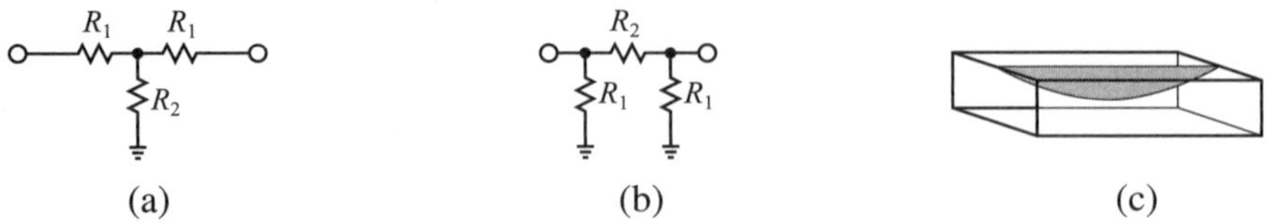

Figure 10.7 Attenuators. (a) Lumped-element tee-network. (b) Lumped-element pi-network. (c) Waveguide absorbing vane.

10.2 ATTENUATORS

Only slightly more sophisticated than a termination, *attenuators* are two-port devices which pass signals in both directions with some fixed amount of flat loss. These are most often placed in between other components in order to alleviate problems with impedance mismatch, since reflections are attenuated twice, once in the incident direction, and again upon the return. Thus, an attenuator having 6-dB loss, for example (sometimes called a 6-dB pad) will reduce the reflection coefficient of a device as seen from the other side of the attenuator by 12 dB. This usage makes it critical that the intrinsic return loss of the pad itself is very good.

For lumped and transmission-line circuits, attenuators usually take the form of tee- and pi-networks of resistors, as shown in Figures 10.7(a, b). In the first case, the resistor values are given by

$$R_1 = Z_0 \frac{1 - T}{1 + T} \tag{10.1a}$$

$$R_2 = Z_0 \frac{2T}{1 - T^2} \tag{10.1b}$$

where T is the transmission coefficient; in other words, the insertion loss, or attenuation, in decibels is $A = |20 \log T|$. The resistor values in the pi-network case may be derived from this using the delta-wye identity. More simply, we may recognize that the pi-network is the dual of the tee-network, so that each resistor is replaced by its normalized reciprocal,

$$R_1 = Z_0 \frac{1 + T}{1 - T} \tag{10.2a}$$

$$R_2 = Z_0 \frac{1 - T^2}{2T} \tag{10.2b}$$

The waveguide attenuator in Figure 10.7(c) uses an absorbing vane centered along the horizontal axis to provide the loss, while the ends are tapered to keep the reflection to a minimum. Specific design parameters may be optimized based on the available material thickness and resistivity.

10.3 SPLITTERS

A splitter or power divider is nothing more than a component that divides an incoming signal among two or more outputs (and, by reciprocity, it combines two or more inputs onto a common output path if operated in the reverse direction, in which case it is often referred to as a *power combiner*). Most often, the signal power is to be divided evenly among the outputs, and usually in-phase with one another, though there are exceptions. Thus, in the passive, lossless case, each path (whether forward or backward) experiences an attenuation in power of $1/N$, where N is the number of outputs, or a factor of $1/\sqrt{N}$ in wave amplitude. Ideally, also, the output ports should be isolated from one another; signals incident in reverse from one output port should couple back to the common input, but not to the other outputs.

Thus, the ideal scattering parameter matrix for a two-way splitter is

$$\mathbf{S} = \frac{e^{-j\phi}}{\sqrt{2}} \begin{pmatrix} 0 & 1 & 1 \\ 1 & 0 & 0 \\ 1 & 0 & 0 \end{pmatrix} \tag{10.3}$$

where ϕ is an arbitrary, common phase delay (which is typically considered unimportant in this device). Recall from Section 4.3.6 that a three-port network cannot be lossless, reciprocal, and matched at the same time. The matrix above is clearly reciprocal ($\mathbf{S}^T = \mathbf{S}$) and matched (the diagonal elements are zero), so it must not be lossless. This may be confirmed by multiplying $\mathbf{S}$ with its conjugate transpose,

$$\mathbf{S}^*\mathbf{S} = \frac{1}{2} \begin{pmatrix} 0 & 1 & 1 \\ 1 & 0 & 0 \\ 1 & 0 & 0 \end{pmatrix} \begin{pmatrix} 0 & 1 & 1 \\ 1 & 0 & 0 \\ 1 & 0 & 0 \end{pmatrix} = \frac{1}{2} \begin{pmatrix} 2 & 0 & 0 \\ 0 & 1 & 1 \\ 0 & 1 & 1 \end{pmatrix} \neq \mathbf{I} \tag{10.4}$$

At a minimum, this proves that any two-way splitter meeting the ideal conditions, even in principle at a single frequency point, must include lossy elements. This principle will be further demonstrated by exploring a number of different splitter topologies, some of which satisfy the ideal conditions, and others that do not (but are nonetheless useful).

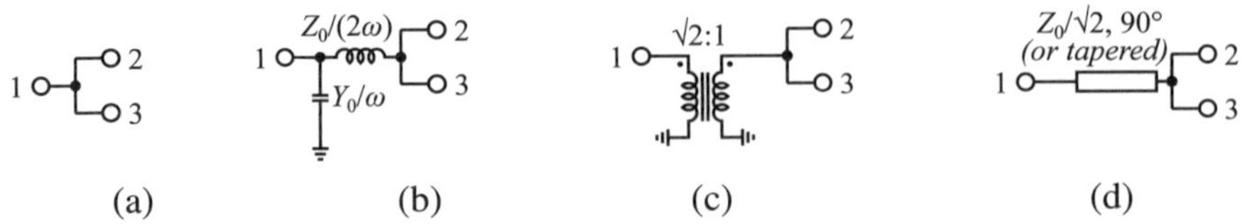

Figure 10.8 Tee-junction splitters, where port 1 is the common input port, and ports 2 and 3 are the outputs. (a) No matching network. (b) Lumped-element matching network. (c) Transformer matching network. (d) Transmission-line matching network.

10.3.1 Tee Junctions

A simple tee junction, as shown in Figure 10.8(a) may be considered a two-way splitter; however, it is not matched from any port. Assuming each port is terminated with the common characteristic impedance of the system, the common port sees an input admittance of $2Y_0$, and thus experiences a reflection coefficient of $\Gamma = -1/3$, or about 9.5 dB (independent of frequency). It also has poor isolation between the output channels, coupling just as well between outputs as it does from input to output. The impedance match, at least from the common port, may be improved with the addition of a matching network, as indicated in Figure 10.8(b). Here, we have taken the best, minimal lumped-element matching network indicated by Figure 6.15, with values given in Table 6.1. In this way, the input match at a single frequency point can be made perfect, but the bandwidth may be insufficient for some applications. That bandwidth may be improved by using a transformer with $\sqrt{2}$ turns ratio, as shown in Figure 10.8(c). The poor output match and lack of isolation remains.

Finally, we may match the tee junction with a quarter-wave transformer (or, even better, a tapered transformer), as shown in Figure 10.8(d). A plot of the scattering parameters for the lumped-element matched and quarter-wave transmission-line matched tee-junction splitters is shown in Figure 10.9. At the design frequency ($\omega = 1$), both achieve perfect match from the input port and an even 3-dB split to the output ports, but have only 6-dB isolation and 6-dB output return loss. This may be acceptable for some applications, but is far from the ideal laid out at the beginning of this section. Note also that the transmission line has rendered the response of the second junction periodic. For future commensurate-line networks, we will only plot the first octave, as this provides complete information about the network's behavior.

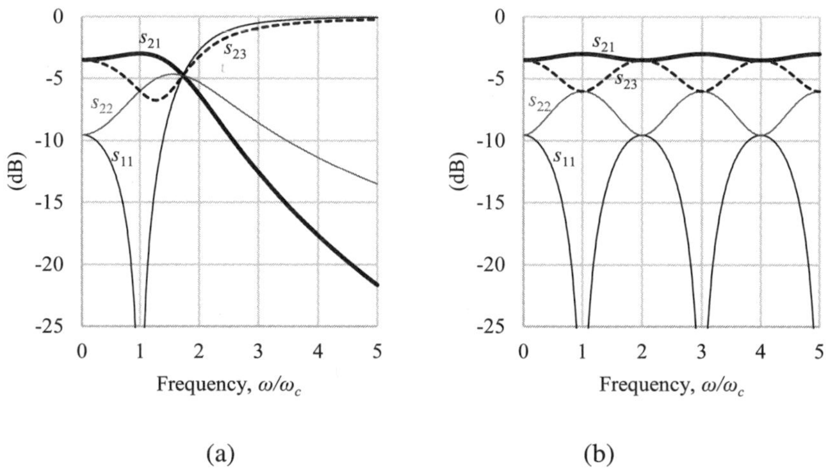

Figure 10.9 Scattering parameters for tee-junction splitters with (a) lumped-element matching and (b) quarter-wave transmission-line matching.

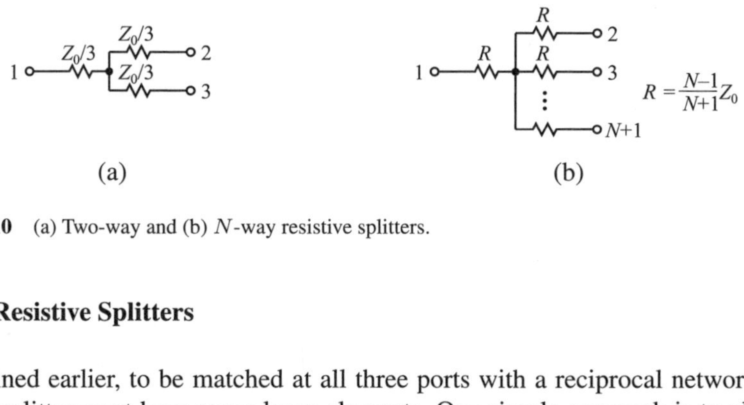

Figure 10.10 (a) Two-way and (b) N-way resistive splitters.

10.3.2 Resistive Splitters

As explained earlier, to be matched at all three ports with a reciprocal network, a two-way splitter must have some lossy elements. One simple approach is to place resistors in series with each port, as shown in Figure 10.10(a). The resistor value is chosen to provide a perfect impedance match to all three ports. A similar technique can be used for N-way splitters as well (having $N+1$ ports including the common port), where the resistor value is found as follows,

$$Z_{in} = R + \frac{(R + Z_0)}{N} = \frac{(1+N)R + Z_0}{N} \tag{10.5a}$$

$$\therefore \Gamma = \frac{Z_{in} - Z_0}{Z_{in} + Z_0} = \frac{(1+N)R + (1-N)Z_0}{(1+N)R + (1+N)Z_0} = 0 \tag{10.5b}$$

$$\therefore R = \frac{N-1}{N+1} Z_0 \tag{10.5c}$$

This solves the matching problem, but the output isolation is still poor, no better than the direct insertion loss, which is increased now as a consequence of the resistors (to 6 dB for a two-way splitter, for example). We will have to be a little more clever if we wish to realize a splitter closer to the ideal suggested in (10.3).

10.3.3 Wilkinson Power Dividers

One simple structure that actually does achieve ideal characteristics, at least in principle and at a single frequency point, is the *Wilkinson power divider* [5] shown in Figure 10.11(a). The resistor bridging between the two output ports is called an *isolation resistor*, for reasons that will soon become clear.

The easiest way to understand the Wilkinson power divider is through even and odd-mode analysis, first explained in Section 4.5. Drawing the symmetry line on the horizontal axis, we may identify the even and odd-mode equivalent circuits as shown in Figures 10.11(b, c). For simplicity, we will consider first only the circuit as seen from the output ports (2 and 3), and let port 1 be terminated. To facilitate division along the symmetry plane, we must first split the termination into parallel resistors, each having value $2z_0$. We thus have a resistor at the left end of the transmission line with value $2Z_0$ in the even-mode equivalent circuit. The isolation resistor is inactive in the even mode (since it would be open-circuited). Thus, the even-mode, port 2 impedance at the center frequency is

$$Z_{2,e} = \frac{\left(Z_0\sqrt{2}\right)^2}{2Z_0} = Z_0 \tag{10.6}$$

and thus the even-mode reflection coefficient is $s_{22,e} = 0$.

In the odd-mode circuit, the termination at what was port 1 is shorted out, leaving a short-circuited transmission-line stub in parallel with the half-circuit isolation resistor. The odd-mode equivalent impedance at center frequency is therefore also $Z_{2,o} = Z_0$ and the reflection coefficient is $s_{22,o} = 0$. Putting these together, we have

$$s_{22} = s_{33} = \tfrac{1}{2}\left(s_{22,e} + s_{22,o}\right) = 0 \tag{10.7a}$$

$$s_{23} = s_{32} = \tfrac{1}{2}\left(s_{22,e} - s_{22,o}\right) = 0 \tag{10.7b}$$

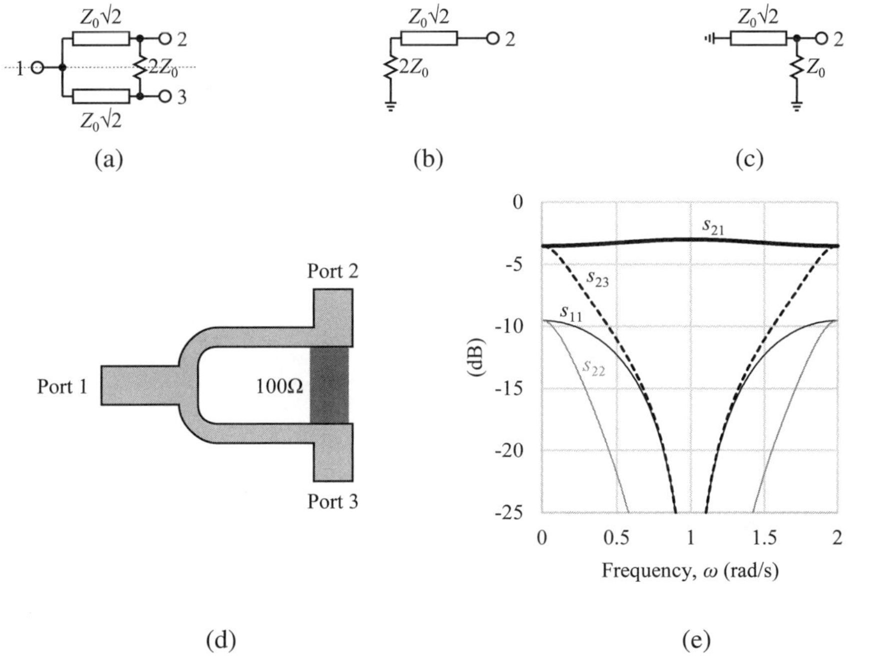

Figure 10.11 (a) A Wilkinson power divider with (b) even-mode and (c) odd-mode equivalent circuits as seen from the output ports. (d) Microstrip layout. (e) Simulated performance. All transmission lines are a quarter-wavelength long at $\omega = 1$ rad/s.

When a signal is incident at port 1, the isolation resistor is inactive by symmetry, and the signal delivered to ports 2 and 3 must be equivalent. We may therefore write the input impedance at port 1 as

$$Z_{in,1} = \frac{1}{2}\frac{\left(Z_0\sqrt{2}\right)^2}{Z_0} = Z_0 \tag{10.8}$$

Therefore, $s_{11} = 0$, and,

$$s_{21} = s_{12} = s_{31} = s_{13} = \frac{-j}{\sqrt{2}} \tag{10.9}$$

where the $-j$ denotes a 90° phase delay associated with the quarter-wave transmission lines. Thus, all reflection terms and the undesirable output-port coupling

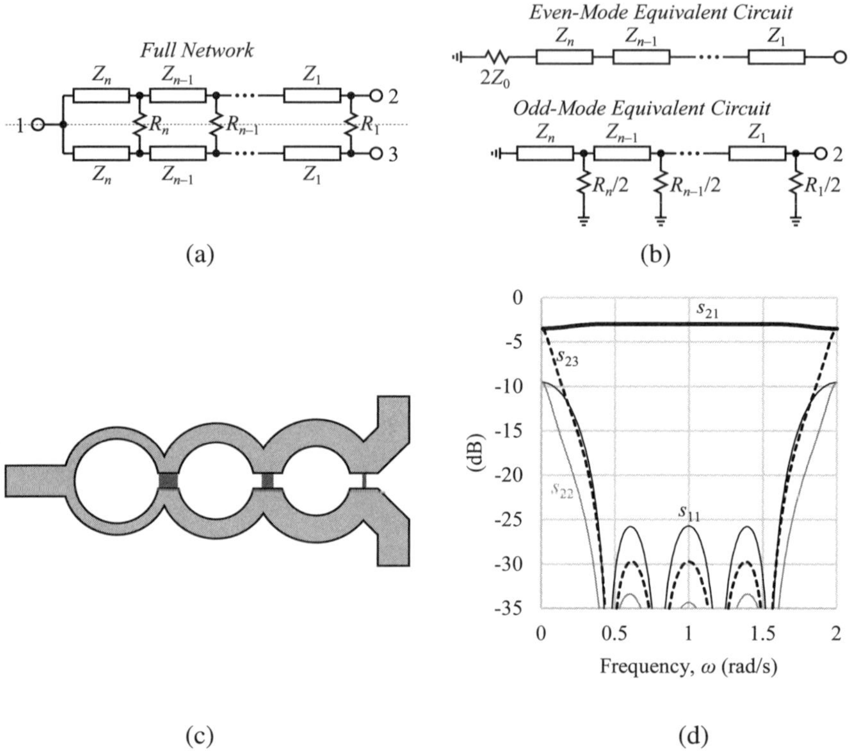

Figure 10.12 (a) A multisection Wilkinson power divider with (b) even and odd-mode equivalent circuits as seen from the output ports. (c) Microstrip layout. (d) Simulated performance.

reduce to zero at the center frequency, while the insertion loss achieves the theoretical minimum of 3 dB, as shown in Figure 10.11(e). The scattering parameters that we have derived at that frequency point exactly match the ideal matrix in (10.3) with $\phi = \pi/2$.

As Figure 10.11(e) shows, the bandwidth of this splitter is limited to about 35% (at roughly 20-dB impedance match and isolation). It may be broad-banded by introducing a multisection transformer into the even-mode equivalent circuit, as shown in Figure 10.12(a), with even and odd-mode equivalent circuits in Figure 10.12(b). The characteristic impedances for this splitter may be taken directly

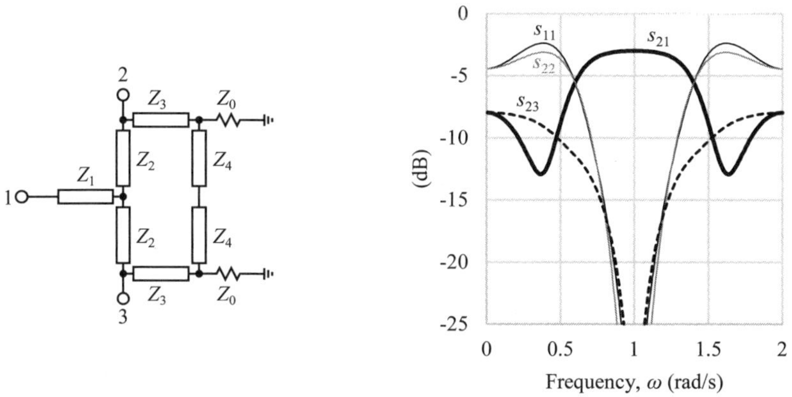

(a) (b)

Figure 10.13 (a) Gysel two-way power divider. (b) Simulated performance when $Z_1 = Z_0$ (omitted), $Z_2 = Z_0\sqrt{2}$, $Z_3 = Z_0$, and $Z_4 = Z_0/2$. All transmission lines are a quarter-wavelength long at $\omega = 1$ rad/s.

from matching transformer tables such as Tables 6.4 or 6.5, where the normalized load impedance is $z_L = 2$ (noting that the transformer sections are numbered from right to left). The values of the isolation resistors are not so readily determined closed form, although some guidance and tabulated values are given in [6]. Modern simulation software tools allow for easy numerical optimization to derive these resistor values. Figure 10.12(d) shows the simulated performance of a four-section Wilkinson power divider with impedances taken from Table 6.5 where $\Gamma_{min} = 0.05$.

10.3.4 Gysel Power Divider

Another form of transmission-line splitter is shown in Figure 10.13(a) and is called a *Gysel power divider* after its developer [7]. There are sufficient degrees of freedom in its design to optimize the performance in a number of different ways, that is, for bandwidth, flatness, isolation, or input/output return loss. No single set of standard characteristic impedances is given, but one reasonable selection is $Z_1 = Z_3 = Z_0$, $Z_2 = Z_0\sqrt{2}$, and $Z_4 = Z_0/2$, which guarantees the ideal performance (infinite isolation and no reflection from any port) at least at the center

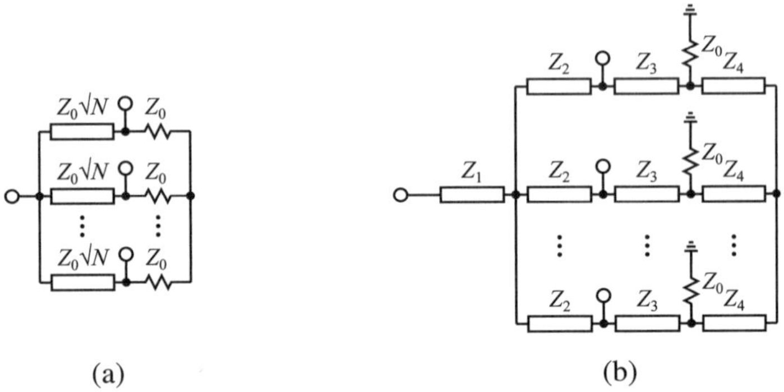

Figure 10.14 N-way (a) Wilkinson and (b) Gysel power dividers. All transmission lines are a quarter-wavelength long at the center frequency of operation.

frequency. Note that this selection allows the Z_1 transmission line to be omitted. Simulated characteristics using these values are given in Figure 10.13(b).

The advantage of the Gysel power divider over Wilkinson's power divider is that the isolation resistors are grounded. This is convenient in high-power applications since the resistors are better heat-sunk and are easier to implement in technologies where the Wilkinson's floating isolation resistors may be difficult to fabricate without having some parasitic electrical phase (it even suggests a possible waveguide implementation where the isolation resistors are replaced by one of the terminations discussed in Section 10.1.2).

Although two-way splitters are most common, both the Wilkinson and Gysel power dividers were originally conceived as arbitrary N-way splitters. Their higher port-number extensions are shown in Figure 10.14.

10.3.5 Rectangular Waveguide Combiners

Power dividers and combiners in rectangular waveguide typically take the form of tee-junctions. Without lossy elements, these necessarily have poor isolation between the divided output ports, but this is generally accepted in favor of simplicity. When isolation is required, one might try implementing a Gysel-type divider with terminations in place of the isolation resistors, or more simply to use a hybrid coupler, a type of four-port circuit, which will be discussed in Section 10.6.

The most prevalent form of rectangular waveguide combiner is shown in Figure 10.15(a). It is frequently seen in split-block constructions, where the pattern

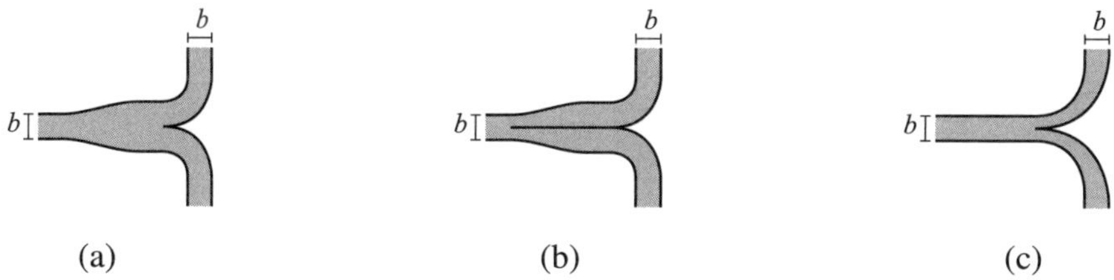

Figure 10.15 Two-way dividers/combiners in rectangular waveguide. (a) With a taper from full-height to double-height guide before division. (b) With tapers from half-height to full-height guide after division. (c) With tapers integrated into the bends after division.

shown is machined into each half of the housing. As such, the narrow wall having height b is apparent on the figure at each port, whereas the guide width, a, is perpendicular to the plane of the paper. Viewed as a divider, it begins with a stepped[3] or tapered transformer from regular full-height waveguide (where, typically, $b = a/2$) to double-height waveguide ($b' = a$) before splitting into two regular-height waveguides again.

There is a subtle issue with this geometry, however. It is obvious that by expanding the height of the waveguide before the split, the cutoff frequencies of higher-order modes are lowered, possibly into the band of operation. Some of these higher-order modes become cutoff again at in-band frequencies when the central wall is encountered that divides the now double-height waveguide into two. As described in Section 8.6.1, this is a perfect setup for a trapped mode resonance, and indeed combiners such as these are prone to them [8].

A better solution, shown in Figure 10.15, is to divide the waveguide first, creating two half-height waveguides, which can then be tapered (or stepped, if the designer insists) back to full height in the same space that was taken before. Proponents of the original geometry in Figure 10.15(a) like to point out that double-height waveguide is less lossy than the equivalent length of half-height waveguide in the second example. While true in principle, the difference is almost certainly negligible in all but the most extreme cases (extreme enough that most engineers will likely never see them in their entire careers).

Let us back this claim up with some numbers. Take a WR-10 waveguide having dimensions $a = 2b = 0.1$ in. Assume that a combiner is machined for

3 In my view, a stepped transformer is rather pointless here given the electrical length typically subtended by waveguide structures, and the inevitible rounding of corners, which results from the machining process. The latter requires extensive CAD modeling to be performed, which might just as well be expended on a more easily drawn and equally well-performing taper.

this size waveguide in a gold-plated split-block housing (thus, $\sigma = 4.1 \cdot 10^7$ S/m). From the expression given in Table 7.2, we may calculate the loss of a double-height waveguide in the dominant mode as

$$\alpha_c = \frac{a^3k^2 + 2b\pi^2}{a^3b\beta k\eta}\sqrt{\frac{\omega\mu'}{2\sigma}} = \frac{2a^2k^2 + 2\pi^2}{a^3\beta k\eta}\sqrt{\frac{\omega\mu'}{2\sigma}} \tag{10.10}$$

and the loss of a half-height waveguide as

$$\alpha_c' = \frac{a^3k^2 + 2b\pi^2}{a^3b\beta k\eta}\sqrt{\frac{\omega\mu'}{2\sigma}} = \frac{4a^2k^2 + 2\pi^2}{a^3\beta k\eta}\sqrt{\frac{\omega\mu'}{2\sigma}} \tag{10.11}$$

Selecting a frequency of 90 GHz, we find the difference is

$$\Delta\alpha_c = \frac{2k}{a\beta\eta}\sqrt{\frac{\omega\mu'}{2\sigma}} \tag{10.12a}$$

$$= \frac{2(47.9 \text{ rad/in})}{(0.1 \text{ in})(36.2 \text{ rad/in})(377\Omega)}\sqrt{\frac{(2\pi \cdot 90 \text{ GHz})(4\pi \cdot 10^{-7} \text{ H/m})}{2(4.1 \cdot 10^7 \text{ S/m})}} \tag{10.12b}$$

$$= 0.0065 \text{ Np/in} = 0.057 \text{ dB/in} \tag{10.12c}$$

If the length of the taper is $3a = 0.3$ in, we may calculate the excess loss (conservatively, since the taper will not be double-height over the entire length in the first case, nor half-height over the entire length in the latter case) as $0.3 \cdot 0.057 = 0.017$ dB, a truly minuscule difference.

Surface roughness may increase these numbers by a factor of a few, but this difference is still unlikely to be significant in almost any application. Better still, the overall length of the combiner (and with it, the loss) may be reduced by integrating the tapers into the subsequent bends, if present, as shown in Figure 10.15(c). This is perhaps the simplest version of all, where the bend in each leg of the divider comprises quarter turns with differing inner and outer radii of curvature designed to start the waveguide at half-height and end it at full-height.

10.4 PHASE SHIFTERS

A phase shifter is a device that introduces an offset in the transmitted phase of a signal, ideally without affecting its amplitude. Depending on the application,

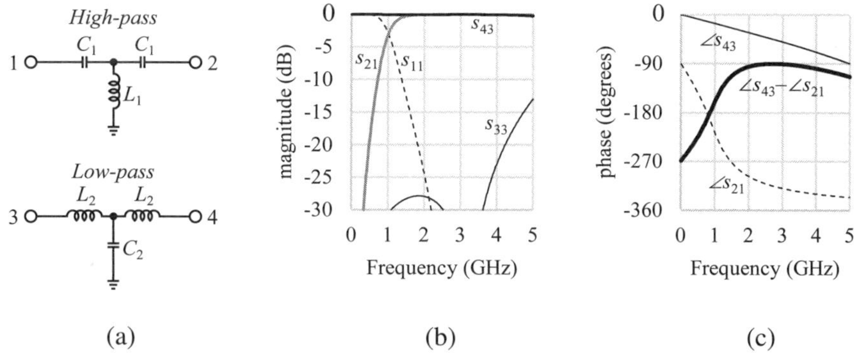

Figure 10.16 (a) High-pass/low-pass phase shifter networks. (b) Magnitude response. (c) Phase response.

one may desire to impart a constant phase change as a function of frequency, or a constant time delay (implying a frequency-increasing and linearly proportional phase offset). In the latter case, simple lengths of matched transmission line or waveguides operating in a quasi-TEM mode (to avoid dispersion) are usually the best passive options. At lower frequencies, where the needed propagation lengths are prohibitively large, active circuits that synthesize true time delay characteristics do exist, but are beyond the scope of this book. In keeping with the focus of this chapter on components having flat-frequency characteristics, we will devote our attention here to constant phase delay circuits.

10.4.1 Lumped-Element High-Pass and Low-Pass

Typically, it is not the absolute transmission phase of a two-port network that matters, but the difference in phase between two channels or between the network in one state versus another (i.e., by switching between alternate signal paths). One does not therefore need to synthesize a network where the phase delay is nearly constant, but rather two networks whose difference in phase is nearly constant. One such example is the lumped element network pair shown in Figure 10.16(a). Called a *high-pass/low-pass phase shifter* because the two networks take on the form of high-pass and low-pass filters, these can nevertheless be designed to have widely overlapping passbands, as indicated in Figure 10.16(b). The element values can be optimized for various criteria, in this case to achieve better than 20-dB return loss from 2–4 GHz, with a nearly constant phase difference of 90° over that range,

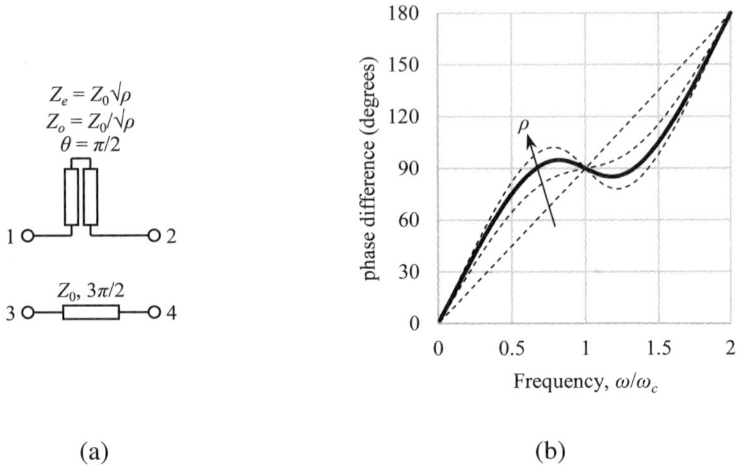

Figure 10.17 (a) Schiffman phase shifter. (b) Differential phase for coupling factors $\rho = 1, 2, 3$, and 4. The thick line corresponds to a coupling factor of $\rho = 3$.

shown in Figure 10.16(c). Alternatively, one may equate the $ABCD$-parameters of the lumped filter networks at a single frequency with those of an equivalent, matched transmission line having the desired phase delay, and thus solve for the element values.

Tee networks are shown in this example, but either tee or pi networks can be used in either position, or both, depending on topological desirability or the availability of components. Much larger lumped-element networks can also be used, to achieve bandwidths even up to several decades, if desired [9].

10.4.2 Schiffman Phase Shifters

Another very simple phase shifter is the coupled-line or Schiffman phase shifter shown in Figure 10.17(a). Described by Bernard Schiffman in 1958 [10], it comprises in one arm a single, quarter-wavelength, coupled transmission line with the ends linked, and in the other arm a straight, matched, three-quarter-wavelength transmission-line section. It is useful in networks like this having coupled lines to define a parameter, ρ, known as the *coupling factor*, which is the ratio of the even and odd-mode impedances ($\rho = Z_e/Z_o$). Note that if the lines in the upper arm are uncoupled ($\rho = 1$), Figure 10.17(b) indicates that the phase difference is a

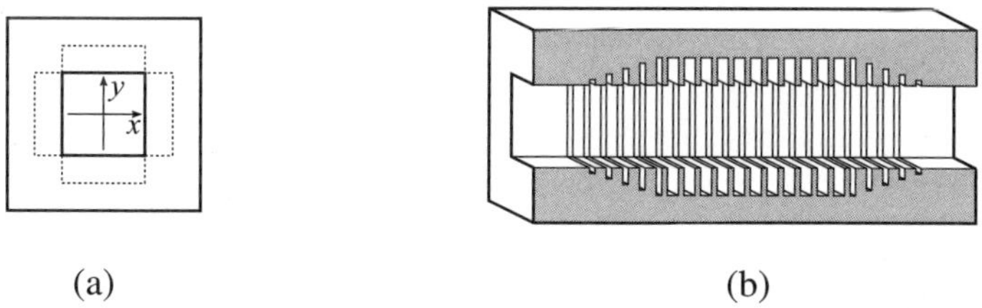

Figure 10.18 Square corrugated-waveguide phase shifter. (a) Front view showing x and y linear polarizations. (b) Cutaway isometric view showing interior corrugations.

straight, linear function of frequency, as one would expect for a simple time delay. As the arms become more tightly coupled, the phase difference takes on more of an S shape. When the coupling factor is $\rho = 3$, for example, the phase difference is within 5° of quadrature (that is, 90±5°) over more than an octave of bandwidth.

10.4.3 Corrugated Phase Shifters

An important special case for phase shifters is when the two channels correspond to two polarizations in an antenna transmit or receive system. The antenna structure or downstream hardware, for example, may intrinsically detect two linear polarizations, where one may wish to work with circular polarizations instead. A simple way to do this is to introduce a 90° differential phase shift in the two linear polarizations prior to detection. This is because the superposition of two linearly polarized waves in space that are 90° out of phase with one another exactly corresponds to left or right-circular polarized waves, depending on the sign of phase difference.

If the two polarizations feed into or from the antenna in a dual-mode waveguide (e.g., from a horn antenna into a circular or square waveguide), the differential phase shift can be accomplished in that waveguide without separating the two polarizations by introducing corrugations, as shown for example in Figure 10.18. Recall from Section 7.7 that the dispersion curves of the Floquet modes in corrugated waveguides depend on the depth and periodicity of the corrugations. One may take advantage of this to introduce slightly different corrugations in the vertical and horizontal directions, thereby engineering a differential phase between the two polarization modes that is constant over a very broad bandwidth [11]. The loss of these modes is also very low, making this structure especially important in applications where the reception sensitivity is crucial.

10.5 DIRECTIONAL COUPLERS

A directional coupler is a four-port device having two main ports that serve as the input and output of the primary signal path, while a portion of the forward-traveling wave is diverted, or coupled, to a third port and a portion of the backward-traveling wave is directed to the fourth port. The fraction of energy coupled to the third and fourth ports can take on any value, but is typically small, so that the forward and backward-traveling waves are only sampled, without incurring too much loss in the primary signal path. These are commonly used as a means of monitoring signals traveling forward and backward through a system without interrupting its operation.

10.5.1 Coupled-Line Coupler

A common and simple form of a directional coupler comprises a simple pair of coupled transmission lines, as shown in Figure 10.19(a). The general scattering parameters for such a structure were already given in (4.113) and (4.114). Let us assume for now that the propagation constants in the even and odd modes of the coupled lines are the same ($\gamma_e l = \gamma_o l = \gamma l = j\theta$) and that the normalized even and odd-mode impedances are given by

$$z_e = \sqrt{\rho} \tag{10.13a}$$

$$z_o = \frac{1}{\sqrt{\rho}} \tag{10.13b}$$

We thus have

$$\Delta_e = \Delta_o = 2\cos\theta + j\left(\sqrt{\rho} + \tfrac{1}{\sqrt{\rho}}\right)\sin\theta \tag{10.14a}$$

$$s_{11} = s_{22} = s_{33} = s_{44} = 0 \tag{10.14b}$$

$$s_{12} = s_{21} = s_{34} = s_{43} = \frac{2\sqrt{\rho}}{2\sqrt{\rho}\cos\theta + j(\rho+1)\sin\theta} \tag{10.14c}$$

$$s_{13} = s_{31} = s_{24} = s_{42} = \frac{j(\rho-1)\sin\theta}{2\sqrt{\rho}\cos\theta + j(\rho+1)\sin\theta} \tag{10.14d}$$

$$s_{14} = s_{41} = s_{23} = s_{32} = 0 \tag{10.14e}$$

Thus, at midband ($\theta = \pi/2$), we have

$$\mathbf{S} = \frac{1}{\rho+1}\begin{pmatrix} 0 & -j2\sqrt{\rho} & \rho-1 & 0 \\ -j2\sqrt{\rho} & 0 & 0 & \rho-1 \\ \rho-1 & 0 & 0 & -j2\sqrt{\rho} \\ 0 & \rho-1 & -j2\sqrt{\rho} & 0 \end{pmatrix} \tag{10.15}$$

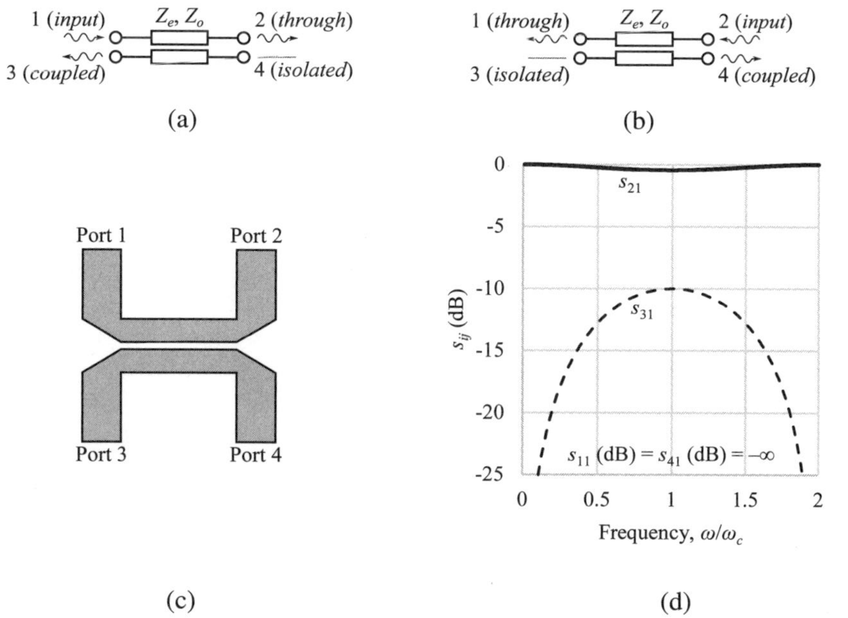

Figure 10.19 A coupled-line directional coupler labeled with (a) forward (rightward) traveling waves and (b) backward (leftward) traveling waves. (c) Microstrip layout. (d) Simulated performance for C = 10 dB.

For rightward traveling waves, we may define port 1 as the input, port 2 as the through or output port, port 3 as the coupled port, and port 4 as isolated. The peak coupling from port 1 to port 3 may be written as

$$C = \frac{\rho - 1}{\rho + 1} \tag{10.16}$$

It is customary to express this coupling as a positive number in decibels, or

$$C(\text{dB}) = 10 \log \left(\frac{P_{in}}{P_{cpl}} \right) = 20 \log \left(\frac{\rho + 1}{\rho - 1} \right) \tag{10.17}$$

and the insertion loss from port 1 to port 2 is given by

$$A = 10 \log \left(\frac{P_{in}}{P_{thru}} \right) = 20 \log \left(\frac{\rho + 1}{2\sqrt{\rho}} \right) \tag{10.18}$$

For leftward traveling waves, the roles of some of the ports are exchanged in accordance with the symmetry of the structure, as shown in Figure 10.19(b), and is why the coupler is called *directional.* A microstrip layout is shown in Figure 10.19(c), and simulated performance for a 10-dB coupler using ideal coupled lines is shown in Figure 10.19(b).

The above analysis assumes ideal coupled lines with identical propagation velocities in the even and odd modes. In practice, all coupled lines have some parasitics. In microstrip, especially, the field configuration of the even mode is more concentrated within the substrate than the odd mode, which has more field energy in the space above the substrate, such that the two modes have different effective dielectric constants and thus different phase velocities. The primary effect is to make the isolation terms in the above scattering parameter matrix not exactly zero. Isolation is thus measured in decibels as

$$Iso(\text{dB}) = 10\log\left(\frac{P_{in}}{P_{iso}}\right) = -20\log\left(|s_{41}|\right) \tag{10.19}$$

and the *directivity*, a critical performance metric for realistic couplers, is the difference between coupling and isolation,

$$D(\text{dB}) = Iso(\text{dB}) - C(\text{dB}) = 10\log\left(\frac{P_{cpl}}{P_{iso}}\right) = 20\log\left(\left|\frac{s_{31}}{s_{41}}\right|\right) \tag{10.20}$$

10.5.2 Multisection Couplers

Note that the input match and isolation of the (ideal) coupled-line coupler in Figure 10.19 are infinite at all frequencies while, even in the ideal case, the coupling is band-limited. The bandwidth can be improved by employing a multisection coupler, as shown in Figure 10.20(a). Let us solve (10.16) for the coupling factor, ρ_k, of each section in terms of its peak port-to-port coupling, C_k,

$$\rho_k = \frac{1+C_k}{1-C_k} \tag{10.21}$$

and then write the key frequency-dependent scattering parameters in terms of C_k using (10.14),

$$s_{21}^{(k)} = \frac{\sqrt{1-C_k^2}}{\sqrt{1-C_k^2}\cos\theta + j\sin\theta} \tag{10.22a}$$

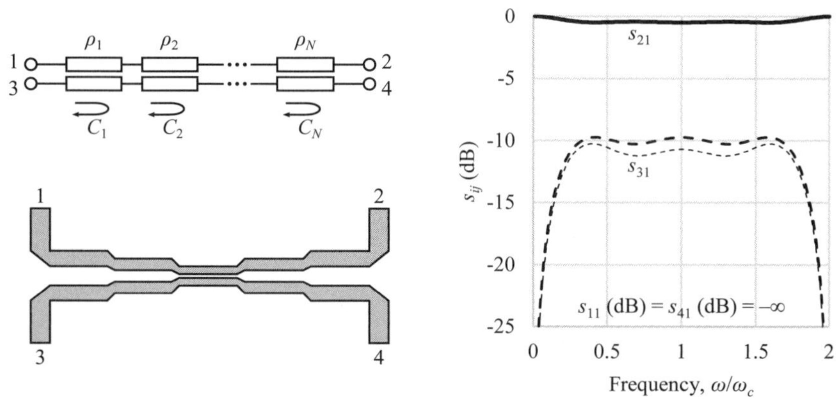

(a) (b)

Figure 10.20 (a) Multisection coupled-line coupler schematic and microstrip layout. (b) Simulated performance of a five-section 10-dB coupler. The thick dashed line shows the actual coupling from an ideal transmission-line model, while the thin dashed line shows the coupling derived from the small-coupling assumption in (10.24).

$$s_{31}^{(k)} = \frac{jC_k \sin\theta}{\sqrt{1 - C_k^2}\cos\theta + j\sin\theta} \tag{10.22b}$$

If we assume that the coupling of an individual section is small ($C_k \ll 1$), then we may simplify these expressions,

$$s_{21}^{(k)} \approx \frac{1}{\cos\theta + j\sin\theta} = e^{-j\theta} \tag{10.23a}$$

$$s_{31}^{(k)} \approx \frac{jC_k \sin\theta}{\cos\theta + j\sin\theta} = jC_k \sin\theta e^{-j\theta} \tag{10.23b}$$

The approximate expression for s_{21} above is simply a phase rotation, neglecting the small amount of signal which is lost to coupling. This is analogous to the same small violation of conservation of energy that occurred under the principle of small reflections when designing multisection impedance transformers in Section 6.3.1.

The total coupling from the input to the coupled port is therefore given by the sum of these individual couplings over all sections with the concomitant phase delay,

$$s_{31} = \sum_{k=1}^{N} \left(jC_k \sin\theta e^{-j\theta} \right) e^{-j2(k-1)\theta} = j\sin\theta e^{j\theta} \sum_{k=1}^{N} C_k e^{-j2k\theta} \tag{10.24}$$

where θ is the electrical length of a single section, all of which are assumed to be the same. As it was with multisection transformers, the summation in this expression has the form of a Fourier series, but it is combined in this case with an additional multiplicative shaping factor, $\sin\theta$. Nonetheless, it is possible to synthesize maximally flat and/or equiripple coupling functions with more than a decade of bandwidth, so long as the overall coupling is low [12]. An example is shown in Figure 10.20(b).

10.5.3 Codirectional Couplers

Note in Figure 10.19 that whichever direction the input signal comes from, the coupled port is always at the same end, propagating in the opposite direction as the input. For this reason, couplers of that type are often referred to as *backward-wave couplers*. This applies to the multisection coupler of Figure 10.20 as well, and the larger overall electrical length in that case makes it even more important to ensure that the even and odd-mode phase velocities are the same, to maintain the high directivity needed for a good directional coupler.

One may take a different approach, wherein the difference in phase velocities is maximized, such that significant coupling is achieved in the forward direction, that is, the adjacent port on the same end of the coupled lines as the through output port is taken as the coupled output, and that port on the same end as the input is taken to be isolated, as shown in Figures 10.21(a, b). This is known as a forward-wave or codirectional coupler [13, 14]. To maximize the directivity in a codirectional coupler, it is necessary to mitigate the coupling mechanism responsible for backward-wave coupling, namely, the even and odd-mode impedance mismatch.

In the layouts of Figures 10.19 and 10.20, the port transmission lines were turned away at right angles to the coupled lines themselves. Not simply a convenience, this was done in order to decouple the signal path from the coupled line section as abruptly as possible. In the codirectional coupler, to minimize the effect of impedance mismatch, we taper the port lines away gradually, as shown in Figure 10.21(c).

If the port tapers are sufficiently gradual, we can assume that both the even and odd modes of the coupled section are excited equally without reflection. Forward

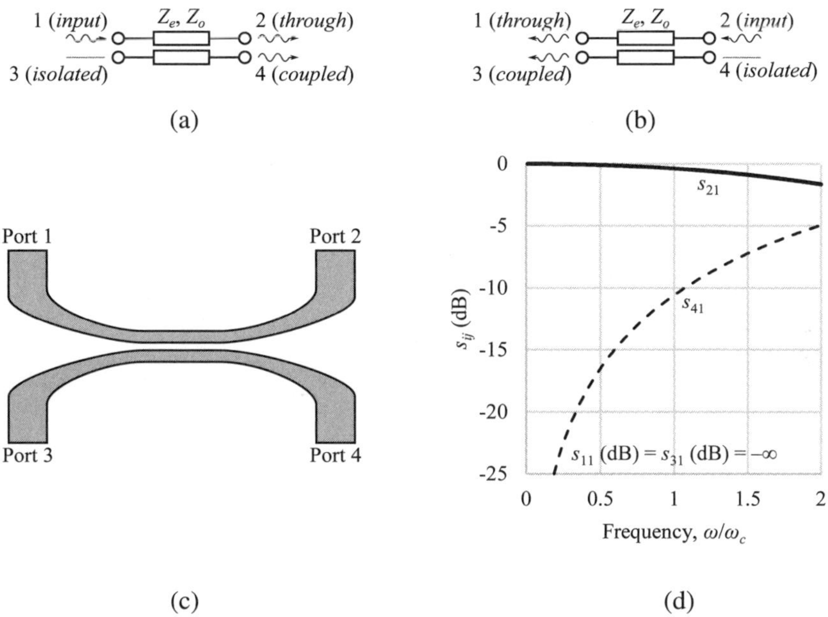

Figure 10.21 A codirectional coupler labeled with (a) forward (rightward) traveling waves and (b) backward (leftward) traveling waves. (c) Microstrip layout. (d) Simulated performance for C = 10 dB.

coupling is then given by (4.112d),

$$s_{41} = \tfrac{1}{2}\left(s_{21,e} - s_{21,o}\right) = \tfrac{1}{2}\left(e^{-j\theta_e} - e^{-j\theta_o}\right) \tag{10.25}$$

where θ_e and θ_o are the electrical lengths of the coupled-line section in the even and odd modes, respectively. Let us write each of these as a perturbation from a nominal electrical length, θ,

$$\theta_e = \theta + \tfrac{1}{2}\Delta\theta \tag{10.26a}$$

$$\theta_o = \theta - \tfrac{1}{2}\Delta\theta \tag{10.26b}$$

$$\Delta\theta = \frac{2\pi f L}{c}\Delta n_{eff} \tag{10.26c}$$

where L is the total length of the coupler and Δn_{eff} is the difference in effective index of refraction, given by

$$\Delta n_{eff} = \sqrt{\varepsilon_e^e} - \sqrt{\varepsilon_e^o} \tag{10.27}$$

Figure 10.22 Multi-aperture Beth hole coupler. (a) Bottom view showing two rows of circular coupling apertures. (b) Side view showing superposition of forward-wave and backward-wave coupling from subsequent apertures.

Therefore,

$$s_{41} = \tfrac{1}{2}e^{-j\theta}\left(e^{-j\frac{1}{2}\Delta\theta} - e^{j\frac{1}{2}\Delta\theta}\right) = -je^{-j\theta}\sin\left(\frac{\Delta\theta}{2}\right) \tag{10.28a}$$

$$= -je^{-j\theta}\sin\left(\frac{\pi f L}{c}\Delta n_{eff}\right) \tag{10.28b}$$

Unlike a backward-wave coupler, where each coupled section is a quarter-wavelength long at the center frequency and the peak magnitude of the coupling depends only on the ratio of even and odd-mode impedances, the magnitude of forward coupling increases with greater electrical length. In principle, if a forward-wave coupler is long enough, a point will be reached where the coupling is 100%, that is, all of the energy is transferred from the input line to the adjacent line. More commonly, weaker coupling is desired, but the dependence on electrical length means that the coupling will have some positive slope across a wide band (a slope that can be somewhat offset by increasing losses at higher frequencies), as shown in Figure 10.21(d).

In practice, the impedance-matching tapers at the ends of the coupler will not be perfect, and some backward-wave coupling will occur as a consequence of mode reflection, degrading the directivity. Nonetheless, codirectional couplers with high directivity can be realized over very wide bandwidths [2].

10.5.4 Multi-Aperture Waveguide Couplers

Directional couplers can be made in waveguide as well. Typically, this involves one or more coupling apertures in the wall between two adjacent waveguides, such as the circular holes shown in Figure 10.22. Called a *Bethe-hole coupler*, the circular tunnels in this case are clearly below cutoff, but each admits a small amount of energy to pass, via evanescent waves, from one waveguide to the other. As such,

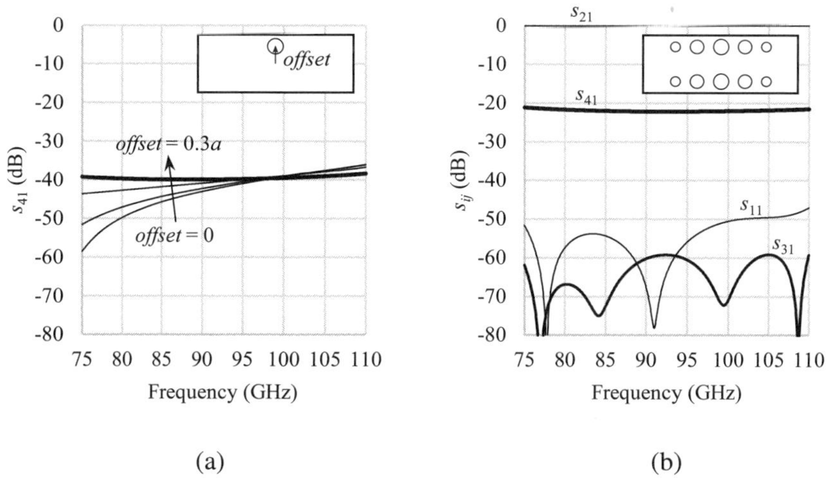

Figure 10.23 (a) Coupling response of a single circular aperture between two WR-10 waveguides with a diameter of 30 mils and a wall thickness of 5 mils. Coupling is shown for lateral offsets from center of 0, 10, 20, and 30 mils. (b) Multi-aperture coupler comprising 10 circular apertures. The geometry for each plot is shown in the inset.

the level of this coupling is strongly dependent on the thickness of the wall between the two waveguides. This feature is an advantage to manufacturing, as it allows the coupling apertures to be drilled into a machined housing from the outside [15, 16], leaving access holes in the much thicker exterior wall, which technically permit leakage to the outside of the module, but at levels that are many orders of magnitude weaker than the coupling itself and are negligible in almost any application.

Approximate and/or empirical theories for the magnitude and frequency dependence of coupling from various simple apertures do exist, but it is often much simpler with modern simulation software to simply optimize responses based on simple rules of thumb. For example, the coupling response of a single circular aperture through a 5-mil wall having 30-mil diameter in 50$\times$100-mil (WR-10) waveguide is shown in Figure 10.23(a). Curves are shown for several offset positions from center, revealing that the forward coupling slope is critically dependent on this parameter. In order to design a broadband coupler, we select an offset of $0.3a$, or 30 mils in this case, which minimizes this slope. There is backward-wave coupling also, but it is less by roughly an order of magnitude and will be further suppressed by the combined action of multiple coupling apertures in the final design.

Note that when we designed backward-wave couplers in Section 10.5.2, we focused our design effort on the realization of a desired coupling curve through constructive interference from multiple coupled sections. The isolation in the forward direction was essentially free, an intrinsic property of the coupled lines. The situation is reversed here, as the desired coupling is in the forward direction, and the combined response will be flat so long as the coupling from a single aperture is flat, since the contributions from each simply add coherently in this direction. However, the backward isolation must be realized by destructive interference, with the phasing between adjacent apertures designed to cancel one another out in the reverse direction.

To illustrate this point, let us write the intrinsic forward coupling of a single aperture as F, and the intrinsic backward coupling as B. We have selected an offset for the aperture such that the magnitude of F is nearly constant (in this case, $F \approx 0.01$). It turns out that its phase is nearly constant also (at about 90°), but this is not important. If two apertures are located a quarter-wavelength apart longitudinally, the combined forward coupling may be written

$$F' \approx -jF - jF = -j2F \tag{10.29}$$

since the phase delay between coupling apertures is the same in both channels (primary and coupled). For backward coupling, the coupled signal from the second aperture has undergone the quadrature phase delay twice, while coupling from the first aperture has none. Therefore,

$$B' \approx B - B = 0 \tag{10.30}$$

The precision of this cancellation deviates as you move away from the center frequency since the spacing is no longer strictly a quarter-wavelength, but it does not depend on the phase of the intrinsic coupling, B, itself. Nevertheless, the combined partial cancellation of the backward-coupling from a larger number of apertures at near quarter-wavelength can be quite significant, leading to couplers with exceptional directivity.

The response of an optimized 10-hole WR-10 coupler is thus shown in Figure 10.23(b). The 30-mil offset giving the flattest coupling response in Figure 10.23(a) has been used for all 10 apertures, but with diameters that vary from 25 mils at the ends of the coupler to 32 mils in the center. Each aperture was paired with a symmetric counterpart and spaced 45 mils longitudinally. Because the coupled waves from all 10 apertures add in voltage, the total coupled power (S_{41}) has increased from that of the single aperture by about $10^2 = 100$, or 20 dB (slightly

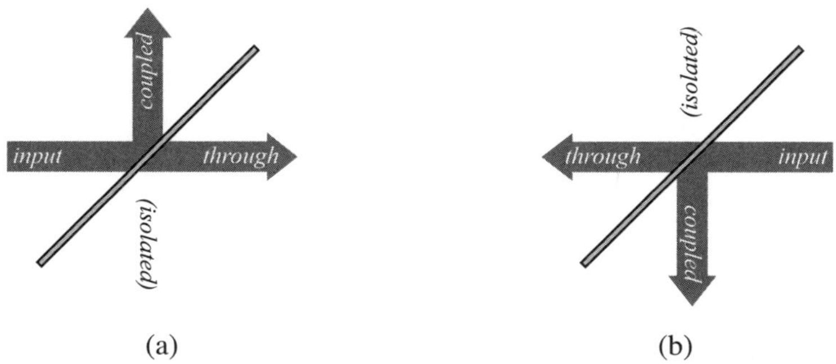

Figure 10.24 Beam splitter with (a) incident wave from the left and (b) incident wave from the right.

less, since the outermost apertures were smaller). Most significantly, however, the isolation term (s_{31}) is better than 60 dB across the entire waveguide bandwidth, for directivity of about 40 dB.

10.5.5 Beam Splitters

In the quasioptical realm, a directional coupler is most simply realized by a thin slab of dielectric material oriented at a 45° angle to the incident beam, as shown in Figure 10.24(a). A portion of the wave is reflected (coupled) toward the top of the page, while the remainder is passed through to the right. There is no mechanism for diverting energy downward in the figure, so this position may be considered isolated. However, with an incident wave from the right, as shown in Figure 10.24(b), the coupled energy is directed downward, while the upward position is isolated. We may thus consider this a directional coupler, just like the transmission-line and waveguide directional couplers previously described, where each of the four cardinal directions may be considered a port.

In order to calculate the coupling for a given slab of material, we must consider the interaction of multiple reflections between the two interfaces, as shown in Figure 10.25(a). The principal reflection from the first interface is $r_1 = r$ where r is simply the Fresnel reflection coefficient given in Table 1.3 (which obviously will depend on the polarization). To that we must add the secondary reflection from the opposite interface, which has magnitude $|r_2| = |tr't'|$, where r and t represent the Fresnel coefficients, and the primed coefficients refer to incidence from inside the material. The form of the Fresnel coefficients guarantees that $r' = -r$ and $tt' = 1 - r^2$; therefore, $tr't' = -r\left(1 - r^2\right)$, irregardless of polarization.

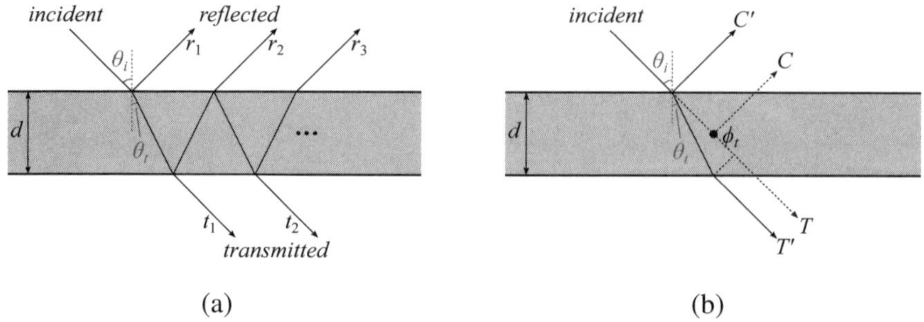

Figure 10.25 (a) Interaction of multiple reflected waves in a beam splitter. (b) Reference plane corrections.

The secondary reflection additionally has a phase term associated with the propagation delay through the slab. It is useful to break this down into two parts. The first part is the phase through the dielectric from one interface to the other and back, at angle θ_t,

$$2\phi_d = \frac{2d}{\cos\theta_t}\beta n = 2\beta d \frac{n^2}{\sqrt{n^2 - \frac{1}{2}}} \tag{10.31}$$

where d is the thickness of the slab, λ is the wavelength in the surrounding medium (usually air), n is the refractive index of the slab material relative to that medium, and we have assumed that the incident angle is $\theta_i = \pi/4$. The second part corresponds to a reference-plane correction, projecting the exit point of the ray back to the corresponding wavefront of the principal reflection,

$$\phi_a = (2d\tan\theta_t)\,\beta\sin\theta_i = 2\phi_d \frac{\sin\theta_i \sin\theta_t}{n} = \frac{\phi_d}{n^2} = \frac{\beta d}{\sqrt{n^2 - \frac{1}{2}}} \tag{10.32}$$

The total additional phase per reflection is then given by

$$\Delta\phi = 2\phi_d - \phi_a = 2\phi_d\left(1 - \frac{1}{2n^2}\right) = 2\beta d\sqrt{n^2 - \tfrac{1}{2}} \tag{10.33}$$

and the net secondary reflection is

$$r_2 = tr't'e^{-j\Delta\phi} = -r\left(1 - r^2\right)e^{-j\Delta\phi} \tag{10.34}$$

while the tertiary reflection is

$$r_3 = tr'r'r't'e^{-j2\Delta\phi} = (-r)^3(1-r)^2e^{-j2\Delta\phi} \tag{10.35}$$

and so on. We thus find ourselves with a geometric series for the total coupling, which has the solution,

$$C' = r - r\left(1-r^2\right)e^{-j\Delta\phi}\sum_{k=0}^{\infty}\left(r^2e^{-j\Delta\phi}\right)^k = \frac{r\left(1-e^{-j\Delta\phi}\right)}{1-r^2e^{-j\Delta\phi}} \tag{10.36}$$

Transmission through the slab derives from a similar geometric series,

$$T' = \left(1-r^2\right)e^{-j\phi_d}\sum_{k=0}^{\infty}\left(r^2e^{-j\Delta\phi}\right)^k = \frac{\left(1-r^2\right)e^{-j\phi_d}}{1-r^2e^{-j\Delta\phi}} \tag{10.37}$$

The above transmission and reflection coefficients assume the outer surfaces of the slab as their reference planes. For beam splitters, it is customary to shift the reference planes instead to a point in the center of the slab, as indicated in Figure 10.25(b). The reflection coefficient needs no adjustment, since rays C and C' have the same phase. However, the transmission coefficient must be advanced by

$$\phi_t = \frac{\beta d}{\cos\theta_t}\cos\left(\theta_i-\theta_t\right) - \frac{\beta d}{\sqrt{2}} = \frac{\beta d}{\sqrt{4n^2-2}} = \tfrac{1}{2}\phi_a \tag{10.38}$$

Therefore,

$$T = T'e^{j\phi_a/2} = \frac{1-r^2}{1-r^2e^{-j\Delta\phi}}e^{-j(\phi_d-\phi_a/2)} = \frac{1-r^2}{1-r^2e^{-j\Delta\phi}}e^{-j\Delta\phi/2} \tag{10.39}$$

and

$$C = C' = \frac{r\left(1-e^{-j\Delta\phi}\right)}{1-r^2e^{-j\Delta\phi}} = \frac{j2r\sin\left(\frac{1}{2}\Delta\phi\right)}{1-r^2e^{-j\Delta\phi}}e^{-j\Delta\phi/2} \tag{10.40}$$

We thus find that with the chosen reference plane position, the transmission and reflection coefficients are in quadrature (there is a 90° phase differential) independent of the frequency, dielectric material, or its thickness [17].

The coupling and insertion loss in decibels are then easily calculated,

$$C(\text{dB}) = 10\log\left(\frac{4r^2\sin^2\left(\frac{1}{2}\Delta\phi\right)}{\left(1-r^2\right)^2+4r^2\sin^2\left(\frac{1}{2}\Delta\phi\right)}\right) \tag{10.41a}$$

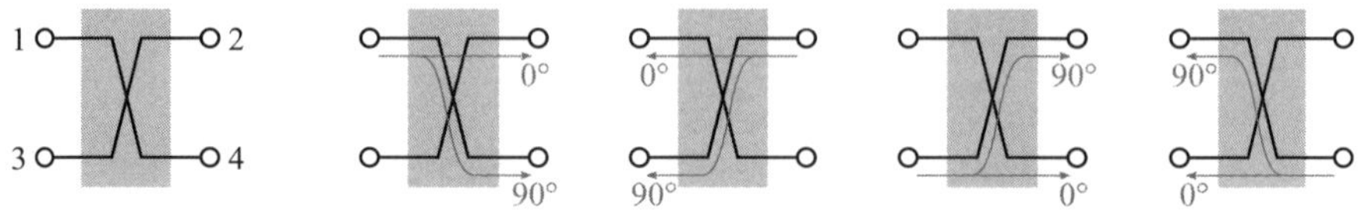

Figure 10.26 (a) Circuit symbol for a quadrature hybrid. Signal flow path for incidence (b) from port 1, (c) from port 2, (d) from port 3, and (e) from port 4.

$$A(\text{dB}) = 10\log\left(\frac{\left(1-r^2\right)^2}{\left(1-r^2\right)^2 + 4r^2\sin^2\left(\frac{1}{2}\Delta\phi\right)}\right) \tag{10.41b}$$

These are both periodic in frequency, with the peak coupling value determined solely by the material dielectric constant, and the periodicity determined by is thickness. Mylar is a popular material for beam splitters at millimeter-wave frequencies due to its mechanical strength, its low-loss properties, and its availability in thin sheets that can be stretched as a tight membrane over a mechanical frame [18]. At even higher, optical frequencies, the coupling dielectric may comprise a resin film or adhesive layer between two triangular, glass prisms assembled into a cube [19]. In that instance, the surfaces of the cube may be treated with antireflection coating, as was described for refractive lenses in Section 9.4.2.

10.6 QUADRATURE HYBRIDS

A special case of directional couplers is when the coupling is strong enough (3 dB, specifically) that equal amounts of energy flow out of both the coupled port and the through port. This is called a *hybrid coupler*. Hybrids fall into two general categories: *quadrature hybrids*, where the through and coupled outputs are 90° out-of-phase, and *180° hybrids*, which (not surprisingly) have 180° phase differential. Quadrature hybrids will be discussed in this section, while 180° hybrids will be discussed later in Section 10.7.

Quadrature hybrids are useful building blocks in a number of high-level devices such as balanced amplifiers (which will be discussed in Section 12.1.7) and sideband-separating mixers (which will be discussed in Section 13.3.5). A common circuit symbol for this component is shown in Figure 10.26, which highlights an important property of many implementations, namely, that it is symmetric about two planes, so that any of the four ports can be excited with identical results. It is

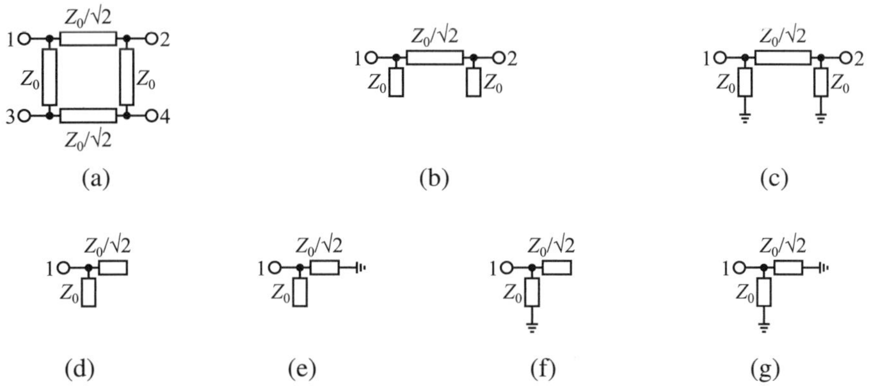

Figure 10.27 (a) Schematic of a branchline coupler. All lines are a quarter-wavelength long. (b) Even-mode equivalent circuit. (c) Odd-mode equivalent circuit. (d) Even-even-mode equivalent circuit. (e) Even-odd-mode equivalent circuit. (f) Odd-even-mode equivalent circuit. (g) Odd-odd-mode equivalent circuit.

typically drawn as shown here with a forward-coupling port orientation, although this is not strictly a requirement. Finally, the phase shifts indicated on the figure are not meant to suggest the absolute phase of each scattering parameter, only that the difference between the through and coupled outputs is $\pm 90°$.

As a point of interest, every one of the general directional couplers described in Section 10.5, whether backward-coupling or forward-coupling, provides a 90° phase differential, and could therefore be used as a quadrature hybrid if 3-dB coupling were selected. However, there are certain structures that achieve specifically this level of coupling by design, and will be discussed in more detail here.

10.6.1 Branchline Hybrids

One of the simplest and most popular quadrature hybrids is known as a *branchline coupler*, shown in Figure 10.27(a). A type of *ring circuit* in which the transmission lines are arranged in a loop, it exhibits the aforementioned two-plane symmetry and thus lends itself well to even and odd-mode analysis. We divide first along the horizontal symmetry line, arriving at the even and odd-mode equivalent circuits in Figures 10.27(b, c), respectively. Each of these may then be divided again into even and odd-mode equivalents as shown in Figures 10.27(d–g). The stubs in these circuits may then be considered to have electrical lengths of $\theta/2$, where θ was the electrical length of the lines in the original hybrid. At midband ($\theta = \pi/2$), the input admittances of these circuits are

$$Y_{ee} = jY_0 \left(1 + \sqrt{2}\right) \tag{10.42a}$$

$$Y_{eo} = jY_0 \left(1 - \sqrt{2}\right) \tag{10.42b}$$

$$Y_{oe} = -jY_0 \left(1 - \sqrt{2}\right) \tag{10.42c}$$

$$Y_{oo} = -jY_0 \left(1 + \sqrt{2}\right) \tag{10.42d}$$

Reconstruction of the original hybrid's admittance parameters is then a two-step process,

$$Y_{11} = \tfrac{1}{2}\left(Y_{11,e} + Y_{11,o}\right) = \tfrac{1}{4}\left(Y_{ee} + Y_{eo} + Y_{oe} + Y_{oo}\right) = 0 \tag{10.43a}$$

$$Y_{21} = \tfrac{1}{2}\left(Y_{21,e} + Y_{21,o}\right) = \tfrac{1}{4}\left(Y_{ee} - Y_{eo} + Y_{oe} - Y_{oo}\right) = jY_0\sqrt{2} \tag{10.43b}$$

$$Y_{31} = \tfrac{1}{2}\left(Y_{11,e} - Y_{11,o}\right) = \tfrac{1}{4}\left(Y_{ee} + Y_{eo} - Y_{oe} - Y_{oo}\right) = jY_0 \tag{10.43c}$$

$$Y_{41} = \tfrac{1}{2}\left(Y_{21,e} - Y_{21,o}\right) = \tfrac{1}{4}\left(Y_{ee} - Y_{eo} - Y_{oe} + Y_{oo}\right) = 0 \tag{10.43d}$$

Therefore, by symmetry, we have

$$\mathbf{Y} = jY_0 \begin{pmatrix} 0 & \sqrt{2} & 1 & 0 \\ \sqrt{2} & 0 & 0 & 1 \\ 1 & 0 & 0 & \sqrt{2} \\ 0 & 1 & \sqrt{2} & 0 \end{pmatrix} \tag{10.44}$$

and the scattering parameters are

$$\mathbf{S} = \left(Y_0\mathbf{I} - \mathbf{Y}\right)\left(Y_0\mathbf{I} + \mathbf{Y}\right)^{-1} = -\frac{1}{\sqrt{2}} \begin{pmatrix} 0 & j & 0 & 1 \\ j & 0 & 1 & 0 \\ 0 & 1 & 0 & j \\ 1 & 0 & j & 0 \end{pmatrix} \tag{10.45}$$

Clearly, all ports are matched, and all input signals are divided equally into the through and forward-coupled outputs, with quadrature phase differential, while the backward-coupled output is isolated.

A typical microstrip layout for a branchline coupler and its frequency response based on an ideal transmission-line model are given in Figure 10.28. The

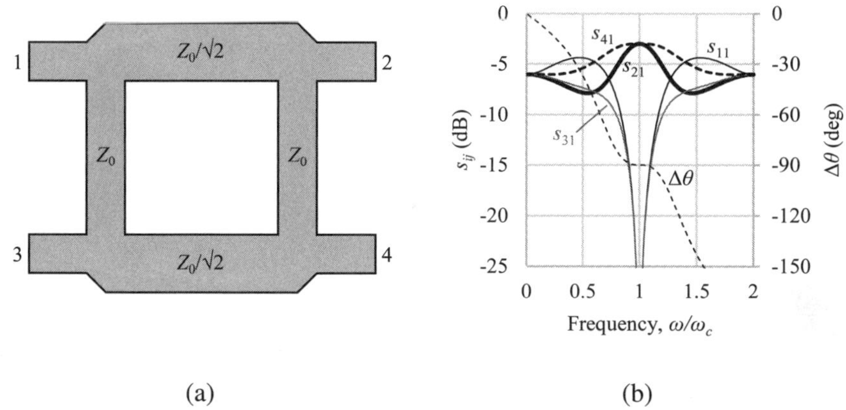

Figure 10.28 (a) Microstrip layout and (b) simulated performance of a branchline quadrature coupler.

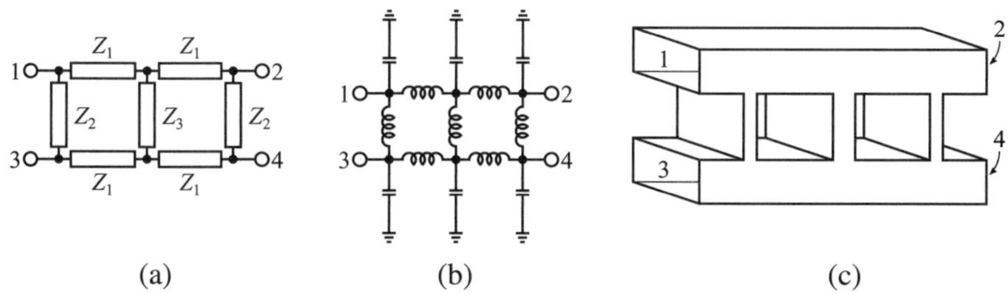

Figure 10.29 Variations on the branchline coupler. (a) Double-box transmission-line. (b) Lumped-element equivalent. (c) Rectangular waveguide.

bandwidth can be improved with additional branches, such as the double-box design [20] shown in Figure 10.29(a). In fact, the basic branchline concept can be extended beyond the transmission-line domain by substituting lumped-element equivalents for the branches (typically, pi-networks comprising series inductors and shunt capacitors), as indicated in Figure 10.29(b), or to rectangular waveguide as shown in Figure 10.29(c). There is no single, accepted way to tune these structures, but each can be optimized for various aspects of performance (impedance match, amplitude balance, and phase differential over frequency) in its particular application.

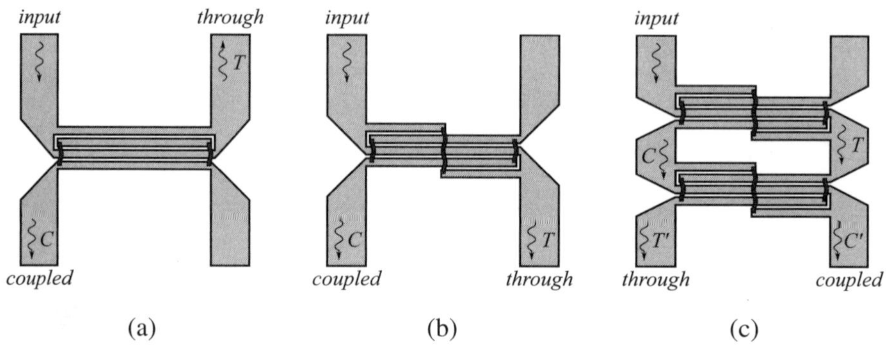

Figure 10.30 (a) Interdigitated coupler. (b) Lange coupler (note the bond wires or bridges connecting nonadjacent strips at the middle and end points). (c) Tandem Lange coupler. The coupled sections are a quarter-wavelength long in all cases.

10.6.2 Lange Couplers and Tandem Couplers

As was stated at the beginning of Section 10.6, any directional coupler that produces a 90° phase-differential (as most do) can be used to make a quadrature hybrid if the coupling can be made strong enough. In the case of single and multisection coupled lines, such as those shown in Figures 10.19 and 10.20, the required level of coupling can be a challenge. Broadside coupling is usually stronger than edge-coupling, but this is rarely an option in monoplanar technologies such as microstrip.

In these situations, other methods must be explored to increase the coupling. One popular method is to increase the number of edges that participate in the coupling by dividing each coupled-line section into parallel strips and then interleaving them, as shown in Figure 10.30(a). Generally known as an *interdigitated coupler*, it has the disadvantage of requiring bond wires or air bridges to connect the alternating strips electrically.

Recall that the coupled line section is a backward-wave coupler and as such, if used as a hybrid, has its two outputs diagonally opposed. Circuit layouts are typically simplified if the two outputs can be made adjacent to one another, so the interdigitated coupler is most often folded about the middle as shown in Figure 10.30(b). This is commonly referred to as a *Lange coupler* [21]. In fact, it works much better than its unfolded counterpart in Figure 10.30(a) due to the improved symmetry. Lange couplers are capable of fairly wide-bandwidth, low-loss performance and are a workhorse in microwave circuit designs. Some variants have even more fingers to further improve coupling, but the four-finger variety shown here is most common.

If the appropriate level of coupling is still lacking, the alignment of the output ports facilitates yet another technique for increasing it, known as the *tandem coupler* shown in Figure 10.30(c). By combining two couplers in cascade, the degree of coupling required of each stage is substantially relaxed (rarely do the individual couplers need to be interdigitated; however, it is helpful if the outputs are aligned as shown).

The coupling required of each stage in a tandem coupler to achieve a net 3-dB split is not immediately obvious — it is not 6 dB, nor $20\log(2\sqrt{2})$, as is often assumed. Let T be the transmission coefficient from the input to the through port of an individual coupler, and for simplicity let us assume that T is real (i.e., the phase delay is zero). The coupling coefficient is then $C = \pm j\sqrt{1-T^2}$. The net transmission coefficient for the tandem coupler is then given by

$$T' = T^2 + C^2 = T^2 - \left(1 - T^2\right) = 2T^2 - 1 \tag{10.46}$$

and must be equal to $1/\sqrt{2}$; therefore,

$$T = \sqrt{\frac{1+\sqrt{2}}{2\sqrt{2}}} \tag{10.47}$$

and

$$C(\text{dB}) = -10\log\left(1 - T^2\right) = -10\log\left(\frac{\sqrt{2}-1}{2\sqrt{2}}\right) = 8.34\text{ dB} \tag{10.48}$$

10.7 180° HYBRIDS

Another common type of hybrid is the 180° hybrid, where the input is divided into two equal parts with 180° phase shift, and again a fourth port is isolated. In contrast with the quadrature hybrid, which typically behaves in an identical (but mirrored) fashion when the isolated port is stimulated, the behavior of a 180° hybrid changes when driven from the isolated port. Specifically, the incident signal is split equally between the same two outputs as before, but in phase with one another. Thus, when used in reverse as a combiner, the originally isolated port takes the sum of the two

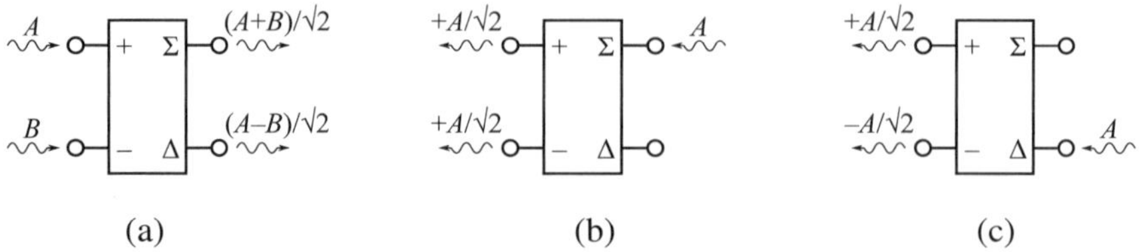

Figure 10.31 Port behavior of 180° hybrid (a) with simultaneous inputs on the plus and minus ports, (b) with a lone input on the sum port, and (c) with a lone input on the difference port.

incident signals, while the original input port takes the difference. These two ports are therefore conventionally labeled the sum (Σ) and difference (Δ) ports, while the remaining two are labeled plus ($+$) and minus ($-$). This nomenclature and some common input-output scenarios are illustrated in Figure 10.31.

Many practical implementations of 180° hybrids exhibit symmetry such that one may exchange the Σ port with the $+$ port and the Δ port with the $-$ port and achieve the same results, but this need not always be the case. Real implementations further may have an additional delay term that is common to all scattering parameters, but this amounts to nothing more than a reference phase and does not alter the overall behavior.

10.7.1 Rat-Race Hybrids

A very simple and common form of 180° hybrid is the *rat-race* or *hybrid ring* coupler, shown in Figure 10.32(a). All transmission lines have the same characteristic impedance, $Z_0\sqrt{2}$, and each line segment as drawn is a quarter-wavelength long at the center frequency, so the bottom portion of the loop between ports 1 and 4 is three-quarter-wavelengths long.

An ingenious method of broadbanding the response of the hybrid ring coupler replaces this three-quarter-wavelength section with a pair of short-circuited coupled lines only one-quarter-wavelength long [22], as shown in Figure 10.32(b). These coupled lines have the same phase between opposite ports as a three-quarter-wavelength section at the center frequency, but with less slope. This is sometimes called a *reverse-phase hybrid ring coupler*. The responses of both the original rat-race and reverse-phase couplers are shown in Figures 10.32(c, d). The bandwidth of the reverse-phase coupler depends on the coupling factor of the coupled lines, $\rho = Z_e/Z_o$, whose even and odd-mode characteristic impedances for perfect match,

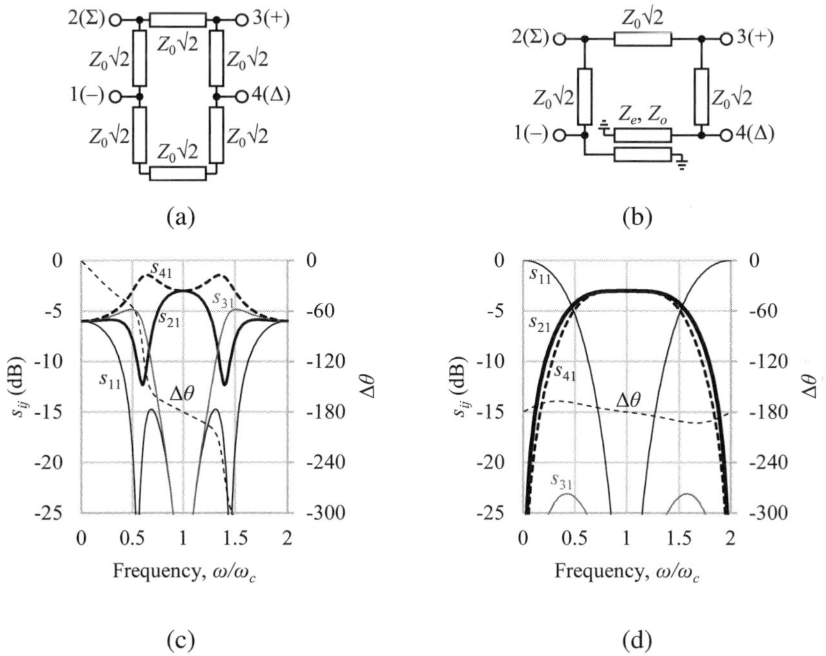

Figure 10.32 (a) Conventional rat-race hybrid coupler. (b) Reverse-phase hybrid ring coupler. All line segments are a quarter-wavelength long at the center frequency. Equations for the even and odd-mode characteristic impedances of the coupled lines are given in the text. (c) Simulated performance of conventional rat-race hybrid coupler. (d) Simulated performance of reverse-phase hybrid ring coupler with $\rho = Z_e/Z_o = 5$.

isolation, and amplitude balance at center frequency are then given by

$$Z_e = Z_0 \frac{\rho - 1}{\sqrt{2}} \tag{10.49a}$$

$$Z_o = \frac{Z_e}{\rho} = Z_0 \frac{\rho - 1}{\rho\sqrt{2}} \tag{10.49b}$$

The hybrid ring couplers are good examples of circuits that can be profitably realized using mixed transmission-line technologies. A variety of possible layouts using combinations of microstrip, coplanar waveguide (CPW), and slotline are shown in Figure 10.33.

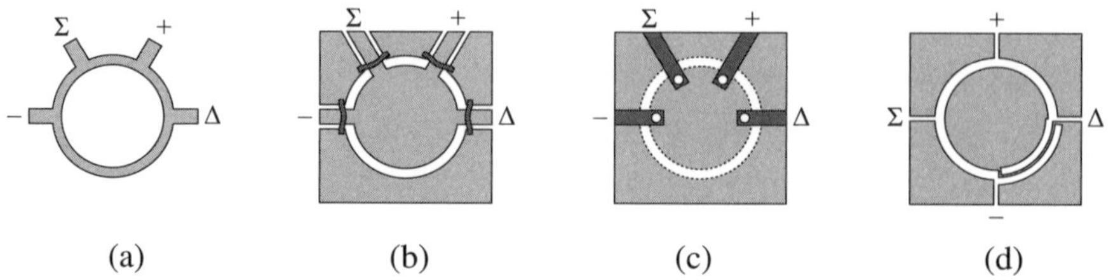

Figure 10.33 (a) Microstrip rat-race hybrid. (b) CPW-slotline rat-race hybrid. (c) Double-sided microstrip-CPW rat-race hybrid. (d) Slotline reverse-phase hybrid.

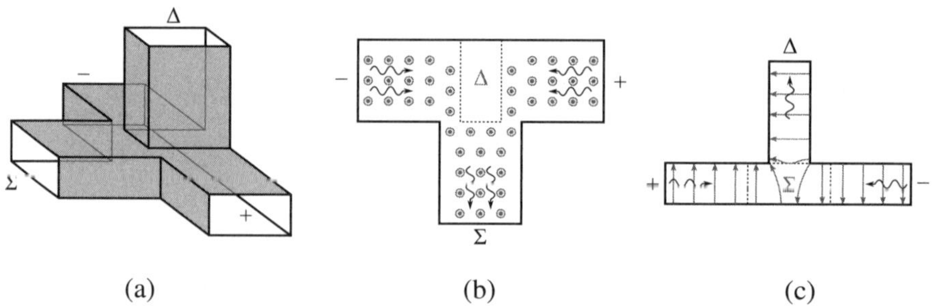

Figure 10.34 (a) Rectangular waveguide magic tee. (b) Top view showing the summation of waves from the two side ports. (c) Back view showing the subtraction of waves from the two side ports.

10.7.2 Waveguide Magic Tee

The hybrid ring structures of Section 10.7.1 do admit solutions in waveguide if the resulting tee junctions are appropriately compensated; however, there is another 180° hybrid construct that is specific to rectangular waveguide and warrants special mention. This structure is known as a *magic tee*, shown in Figure 10.34. Its operation can be understood in terms of symmetry, and the summation and subtraction, or in-phase and out-of-phase combination, of the input waves from the two side ports. In the first case, Figure 10.34(b), the two balanced input waves cancel one another on the Δ port, while adding constructively at the Σ port. In the latter case, Figure 10.34(c), the two incident waves are out-of-phase and thus cancel at the Σ port while combining at the Δ port.

In practice, impedance matching structures, such as tuning posts and/or irises, will be required to ensure good return loss, but so long as these symmetries are preserved, the isolation terms at least will have a very broadband response.

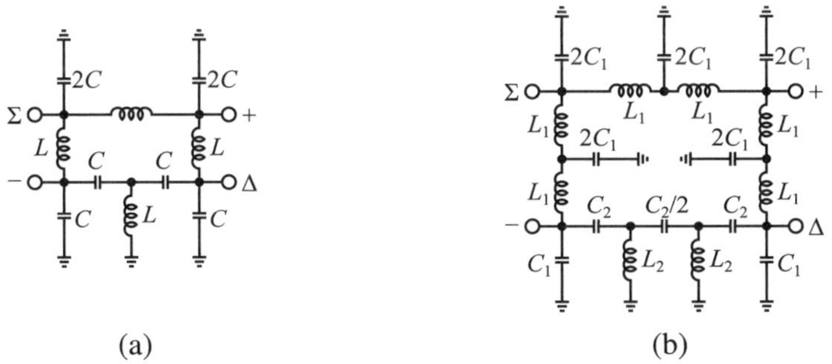

Figure 10.35 (a) Lumped-element 180° hybrid. (b) More sophisticated hybrid with increased bandwidth.

10.7.3 Lumped-Element 180° Hybrids

Like the quadrature coupler in Figure 10.29(b), lumped-element 180° hybrids are typically based on transmission-line prototypes (e.g., the branchline or hybrid ring), but with pi and tee-network approximations of the transmission-line branches. High-pass and low-pass sections can be used to realize the different phase shifts in each branch, as indicated in Figure 10.35, and typically the more elements used, the better the approximation will be [23].

10.8 NONRECIPROCAL COMPONENTS

Every circuit or component that we have discussed so far in this book has obeyed the reciprocity theorem first introduced in Section 1.7, with the exception of the gyrator described in Section 4.7.1, but that was only a mathematical construct. Nothing was said at the time how such an element might be realized.

Recall that the derivation of the Lorentz reciprocity theorem hinged on the assumption that the material constitutive parameters, ε and μ, were simple scalars, such that their multiplication with the electromagnetic field vectors was commutative. In anisotropic materials, these quantities are *tensors* (for present purposes, a matrix in three dimensions). For most materials, these matrices are diagonal, so the commutative property holds and Lorentz reciprocity is preserved, despite the anisotropy. In certain exotic materials, notably ferrites, the μ tensor may be rendered nondiagonal by an externally applied magnetic field, typically to the point of magnetic saturation, either from a permanent magnet or electromagnet,

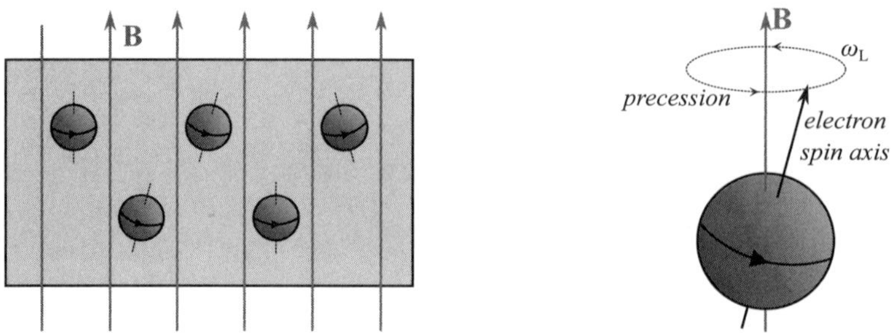

(a) (b)

Figure 10.36 Illustration of the source of nonreciprocity in ferrites. (a) Model of a ferrite material containing multiple magnetic dipoles exposed to an applied magnetic field. (b) Precession of the spin axis.

thus providing a mechanism for breaking reciprocity [12, 24]. Importantly, despite these exotic magnetic properties which can be exploited, ferrites are generally also nonconductive so that waves can propagate through them with relatively low loss.

10.8.1 Physical Mechanism of Nonreciprocity

The underlying physical mechanism of this departure from reciprocity is the interaction of the applied magnetic bias with the magnetic dipole moments within the ferrite, such as the spin axes of the electrons, as shown in Figure 10.36(a). The external magnetic field exerts a gyroscopic force on these particles, causing their spin axes to precess as shown in Figure 10.36(b). Called *Larmor precession*, it has a direction determined by the orientation of the DC magnetic field, and a rotation speed determined by the field's strength as well as the material properties, but, importantly, it does not depend on the declination angle of the spin axis from the applied magnetic field vector [25]. The angular frequency at which the spin axis precesses is called the *Larmor frequency* and is given by

$$\omega_L = \gamma B_0 = \gamma \mu_0 H_0 \tag{10.50}$$

where γ is the *gyromagnetic ratio* [26]. The gyromagnetic ratio for electrons, and thus for most ferrites, is $\gamma_e = 1.760859644 \times 10^{11}$ rad/(s·T).

The easiest way to see how this generates nonreciprocal behavior is to imagine, now, a circularly polarized wave traveling through such magnetically biased material. The circular polarization can either be left-handed or right-handed (i.e., clockwise or counterclockwise). Larmor precession creates a preferential direction for the torque applied to the internal magnetic dipole moments, such that the passing wave may either work with or against the ambient Larmor precession. Thus, circularly polarized waves of different handedness in a magnetically saturated material will have different effective permeabilities, and thus different phase velocities.

More rigorously, it can be shown [12, 24] that the permeability tensor in Cartesian coordinates at radial frequency ω for a z-directed magnetic bias field, $\mathbf{H} = H_0\mathbf{z}$, is

$$\boldsymbol{\mu} = \begin{pmatrix} \mu & j\kappa & 0 \\ -j\kappa & \mu & 0 \\ 0 & 0 & \mu_0 \end{pmatrix} \tag{10.51}$$

where

$$\mu = \mu_0 \left(1 + \frac{\omega_L \omega_m}{\omega_L^2 - \omega^2}\right) \tag{10.52a}$$

$$\kappa = \mu_0 \frac{\omega_L \omega_m}{\omega_L^2 - \omega^2} \tag{10.52b}$$

In these equations, ω_L is the now familiar Larmor frequency, and ω_m is similarly defined as $\omega_m = \gamma\mu_0 M_0$, where $M_0 = \chi_m H_0$ is the saturated bias magnetization (see Section 1.6.2).

10.8.2 Faraday Rotation Devices

One of the chief consequences of this is the *Faraday effect* or *Faraday rotation* [27]. Consider a linearly polarized wave passing through a biased ferrite medium. Such a wave may be written as the superposition of two circularly polarized waves, one left-handed and one right-handed, which as we have said above will have different phase velocities. The two circularly polarized waves will therefore emerge with a slight differential delay between them, which when they are recombined amount to a rotation of the original linear polarization. The angle of rotation, β (not to be confused with the phase constant), may be written in a fairly simple form,

$$\beta = \mathcal{V} B d \tag{10.53}$$

Here, B is the magnetic flux density applied, d is path length over which the passing wave and magnetic field interact, and the empirical scaling factor, $\mathcal{V}$, is known as the *Verdet constant* for the magnetic material.

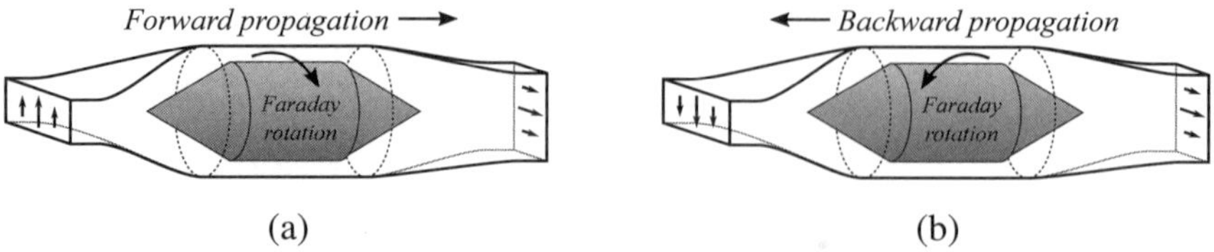

Figure 10.37 Waveguide gyrator based on Faraday rotation. (a) Forward propagation. (b) Backward propagation.

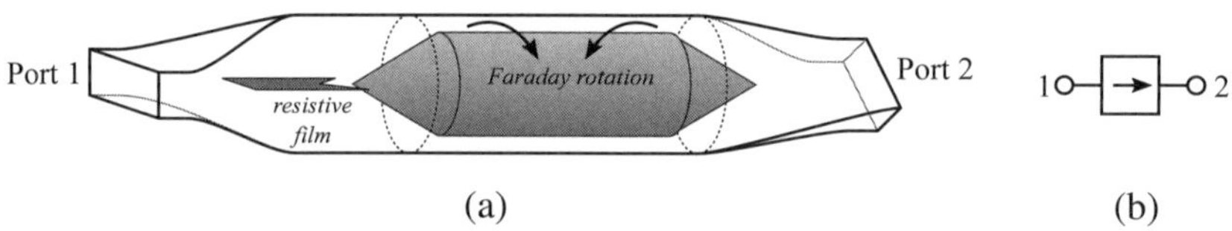

Figure 10.38 Faraday rotation isolator. (a) Structure. (b) Schematic symbol.

Faraday rotation is the basis of some useful nonreciprocal devices such as the gyrator shown in Figure 10.37. It comprises rectangular to circular waveguide tapers at each end with the broad walls in orthogonal planes, and a ferrite rod in the center. The rod may include matching tapers of its own, and longitudinal magnetic bias will typically be provided by a permanent magnet outside and surrounding the circular the waveguide (not shown). The ferrite rod would be designed to provide a precisely 90° rotation so that signals incident from the left waveguide emerge from the right waveguide as shown. In the reverse direction, signals incident with the exact same polarization undergo Faraday rotation in the opposite direction (as seen looking along the direction of propagation), resulting in an output that is flipped 180° from the original. We may thus infer a scattering parameter matrix identical to that of the original gyrator in (4.129).

A similar but somewhat conventionally more useful device is shown in Figure 10.38. Called an isolator, its internal structure is essentially the same, except that the angle of Faraday rotation in each direction is designed in this case to be 45 degrees, and a resistive film is included to absorb reflections coming from the right-side port. The film is oriented in such a way that it has no effect on input signals from the left side. The isolator therefore propagates signals only in one direction, but is well matched from both sides.

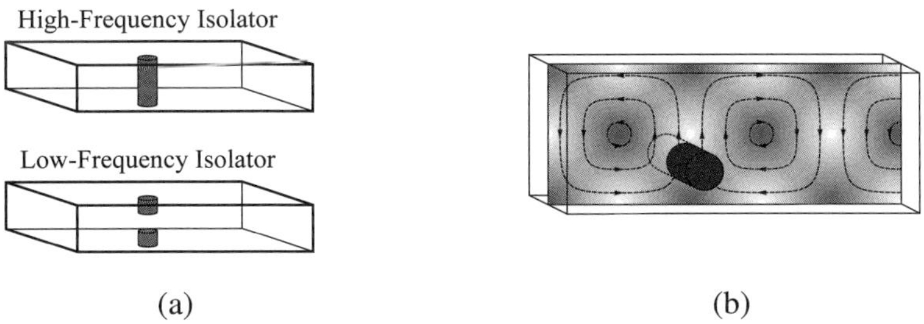

Figure 10.39 Resonant absorption isolator. (a) High-frequency and low-frequency variants. (b) High-frequency variant showing lateral position within the H-field of a passing TE_{10} wave.

Real-world devices may additionally include a 45° waveguide twist at one of the ports to bring the two waveguides back into alignment, but that is incidental to its operation. The schematic symbol for an isolator is shown in Figure 10.38(b).

10.8.3 Resonant Absorption Isolators

Note from (10.52) that the permeability passes through a singularity around the Larmor frequency, ω_L, indicative of resonant behavior. In reality, loss mechanisms which were not included in our simple model will dampen this gyromagnetic resonance, absorbing significant amounts of energy at or near the resonant frequency. This is the basis of another type of waveguide isolator, called the *resonant absorption isolator* [28], shown in Figure 10.39(a). A ferrite post or slab is located off-center in a rectangular waveguide, and magnetic bias is applied vertically, usually from a permanent magnet outside the waveguide (not shown). The location of the post is optimized so that a passing wave in the dominant mode exhibits a rotating magnetic field. If the direction of this rotation is aligned with the Larmor precession and close to that frequency, it will couple strongly to the gyromagnetic resonance and dissipate energy. This rotational alignment occurs only for waves passing in one direction. For waves traveling in the reverse direction, the magnetic field of the dominant mode rotates opposite to the Larmor precession. In this case, little coupling to the gyromagnetic resonance occurs, and the loss is very low.

Resonant absorption isolators are typically used for higher powers than the Faraday rotation isolators discussed previously, with power capacity limited by the temperature rise due to dissipation. One challenge associated with their design is that the resonant frequency is controlled, in part, by the field intensity, via (10.50), but for low frequencies that field might not be strong enough to saturate the ferrite.

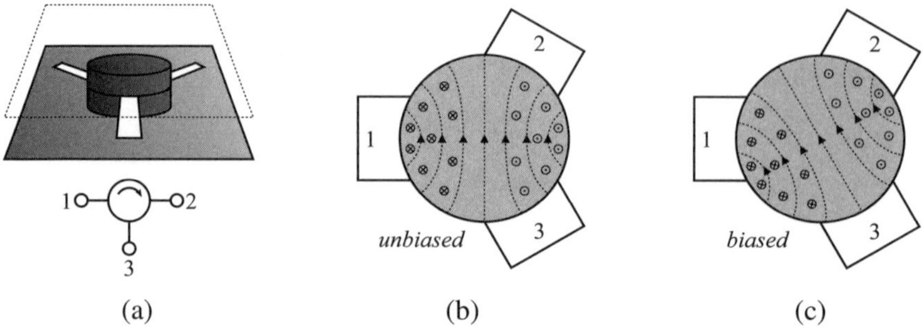

Figure 10.40 (a) Structure (above) and schematic symbol (below) for a stripline junction circulator. (b) Dielectric resonator mode without vertical magnetic bias. (c) Dielectric resonator mode with vertical magnetic bias. The dashed lines in the plane of the paper correspond to the magnetic field, while the vectors noted perpendicular to the paper correspond to the electric field.

Geometry may help to increase the ambient field required to achieve resonance, with one such solution shown at the bottom of Figure 10.39(a). It is also possible to stretch the ferrite out into a longitudinal slab to lengthen the interaction distance and increase absorption.

10.8.4 Stripline Circulators and Isolators

Ferrites can be used in planar, nonreciprocal circuits as well. Figure 10.40(a) shows a stripline junction circulator comprising a junction of three microstrip lines at 120° angles with ferrite pucks above and below. The pucks act as a dielectric resonator which couples to the field excitation from one of the waveguide. A diagram of that dielectric resonator mode is shown in Figure 10.40(b), excited in this case from port 1, and without an external magnetic bias applied. This mode may be considered a standing wave formed by the superposition of two counterrotating modes (like circularly polarized waves). When a vertical magnetic bias is applied, the permeability for one mode changes with respect to the other, differentiating their phase velocities such that the standing wave pattern rotates, as shown in Figure 10.40(c). If tuned properly, the field pattern rotation is 30°, aligning a null in the direction of port 3, while coupling port 1 to port 2, as shown. Symmetry shows that excitation from port 2 would likewise couple to port 3 while isolating from port 1, and so on.

We thus see that in a circulator, coupling is always from one port to the neighboring port, in this case in a clockwise direction, but not in reverse. Circulators are commonly used in transmit/receive duplexers, where a transmitting power amplifier on port 1 is coupled to an antenna on port 2, and that antenna is coupled back to a receiving low-noise amplifier on port 3. In this way, the sensitive low-noise amplifier is protected from damage and interference from the much stronger power amplifier.

A circulator can be made into an isolator simply by terminating one port, say port 3. Signals may then flow from port 1 to port 2, but signals reflecting back from port 2 are terminated in port 3 without coupling back to port 1. This is a common way to protect power amplifiers from damage due to back-reflections from poorly matched loads.

Problems

10-1 Use the ray-folding analogy illustrated in Figure 10.6 to estimate the absorption of a conical absorber with 30° opening angle as a function of incidence angle, assuming that the absorption per bounce is 5 dB.

10-2 What are the resistor values needed for a tee attenuator having 20-dB attenuation in a 50Ω system? For a pi attenuator?

10-3 What is the insertion loss penalty for an N-way splitter (beyond the factor of $1/N$ for power division) that is incurred by a resistive splitter designed for perfect impedance matching?

10-4 Derive frequency-dependent expressions for the complete list of s-parameters of the basic Wilkinson power divider in Figure 10.11.

10-5 List the characteristic impedances of the transmission-line sections in a five-section, maximally flat Wilkinson power divider matched to 50Ω.

10-6 Write an expression for the frequency-dependent differential phase shift of the Schiffman phase-shifter shown in Figure 10.17.

10-7 What are the even and odd-mode characteristic impedances required of a single-section coupled-line coupler to have 10-dB coupling in a system where $Z_0 = 75\Omega$? What about 3-dB coupling?

10-8 A single-section, 10-dB backward-wave coupler on microstrip has 30-dB directivity at mid-band. Estimate the difference in the effective index of refraction of the even and odd-modes.

10-9 How many coupling apertures of the type indicated in Figure 10.23 with $0.3a$ offset are needed to synthesize a coupler with 14-dB coupling?

10-10 What thickness of Mylar ($\varepsilon = 3.1$) is needed to make a 10-dB beam splitter at 100 GHz in the s polarization? What coupling will this same sheet of Mylar have in the p polarization?

10-11 Prove that a tandem coupler comprising two matched quadrature couplers is still a quadrature coupler, regardless of its overall coupling factor.

10-12 Illustrate that a quadrature hybrid cascaded with a differential, 90° phase shifter at its through and coupled ports behaves as a 180° hybrid.

10-13 Derive the s-parameters of the rate-race hybrid coupler in Figure 10.32(a) at mid-band using even and odd-mode analysis.

10-14 Compare the scattering parameters at mid-band of the three-quarter-wave-length line between ports 1 and 4 in Figure 10.32(a) and the quarter-wave coupled-line section between ports 1 and 4 in Figure 10.32(b).

10-15 What is the Larmor frequency for a typical ferrite biased with a magnetic field of 3×10^5 A/m?

10-16 Approximately how long would the ferrite rod within a Faraday rotation isolator have to be if it had a Verdet constant of $\mathcal{V} = 100$ rad/(T·m) in a 1T field?

References

[1] B. Oldfield, "Connector and termination construction above 50 GHz," *Applied Microwave & Wireless*, pp. 56–66, 2001.

[2] M. A. Morgan and S. Weinreb, "Octave-bandwidth high-directivity microstrip codirectional couplers," *IEEE MTT-S International Microwave Symposium Digest*, vol. 2, June 2003, pp. 1227–1230.

[3] Laird. (2018) ECCOSORB®MF – Emerson & Cuming Microwave Products. http://www.eccosorb.com/products-eccosorb-mf.htm.

[4] F. P. Mena and A. M. Baryshev, "Design and simulation of a waveguide load for ALMA-band 9," *ALMA Memo Series*, no. 513, 2005.

[5] E. J. Wilkinson, "An N-way hybrid power divider," *IRE Transactions on Microwave Theory and Techniques*, vol. 8, no. 1, pp. 116–118, January 1960.

[6] S. B. Cohn, "A class of broadband three-port TEM-mode hybrids," *IEEE Transactions on Microwave Theory and Techniques*, vol. 19, no. 2, pp. 110–116, February 1968.

[7] U. H. Gysel, "A new N-way power divider/combiner suitable for high-power applications," *IEEE MTT-S International Microwave Symposium*, May 1975.

[8] M. A. Morgan and S.-K. Pan, "Graphical prediction of trapped mode resonances in sub-mm and THz waveguide networks," *IEEE Transactions on Terahertz Science and Technology*, vol. 3, no. 1, pp. 72–80, January 2013.

[9] S. D. Bedrosian, "Normalized design of 90° phase-difference networks," *IRE Transactions on Circuit Theory*, vol. 7, no. 2, pp. 128–136, June 1960.

[10] B. M. Schiffman, "A new class of broadband microwave 90-degree phase shifters," *IRE Transactions on Microwave Theory and Techniques*, vol. 6, no. 2, pp. 232–237, May 1958.

[11] S. Srikanth, "A wide-band corrugated rectangular waveguide phase shifter for cryogenically cooled receivers," *IEEE Microwave and Guided Wave Letters*, vol. 7, no. 6, pp. 150–152, June 1997.

[12] D. M. Pozar, *Microwave Engineering*, 4th ed. New York: Wiley, 2011.

[13] P. K. Ikäläinen and G. L. Matthaei, "Wide-band, forward-coupling microstrip hybrids with high directivity," *IEEE Transactions on Microwave Theory and Techniques*, vol. 35, no. 8, pp. 719–725, August 1987.

[14] S. Uysal, *Nonuniform Line Microstrip Directional Couplers and Filters*. Norwood, MA: Artech House, 1993.

[15] E. W. Bryerton, "A cryogenic integrated noise calibration and coupler module using a MMIC LNA," *IEEE Transactions on Microwave Theory and Techniques*, vol. 59, no. 8, pp. 2117–2122, August 2011.

[16] N. R. Erickson, "High performance dual directional couplers for near-mm wavelengths," *IEEE Microwave and Wireless Components Letters*, vol. 11, no. 5, pp. 205–207, May 2001.

[17] P. F. Goldsmith, *Quasioptical Systems*. New York: IEEE Press, 1998.

[18] E. W. Weisstein, "Millimeter/submillimeter fourier transform spectroscopy of jovian planet atmospheres." Ph.D. dissertation, California Institute of Technology, Pasadena, CA, 1996.

[19] Wikipedia. (2018) Beam splitter. https://en.wikipedia.org/wiki/Beam_splitter.

[20] Unknown Editor. (2018) Double-box branchlines. https://www.microwaves101.com/encyclopedias/double-box-branchlines.

[21] J. Lange, "Interdigitated strip-line quadrature hybrid," *IEEE MTT-S International Microwave Symposium*, pp. 10–13, May 1969.

[22] K. Chang, *Microwave Ring Circuits and Antennas*. New York: Wiley, 1996.

[23] S. J. Parisi, "180 degrees lumped element hybrid," *IEEE MTT-S International Microwave Symposium*, pp. 1243–1246, June 1989.

[24] S. Ramo, J. Whinnery, and T. Van Duzer, *Fields and Waves in Communication Electronics*. New York: Wiley, 1984.

[25] Wikipedia. (2019, April) Larmor precession. https://en.wikipedia.org/wiki/Larmor_precession.

[26] Wikipedia. (2019, May) Gyromagnetic ratio. https://en.wikipedia.org/wiki/Gyromagnetic_ratio.

[27] Wikipedia. (2019, April) Faraday effect. https://en.wikipedia.org/wiki/Faraday_effect.

[28] Skyworks. (2007, March) Use of ferrimagnetic material in isolators. http://www.skyworksinc.com/uploads/documents/No6510.pdf.

Chapter 11

Frequency-Selective Components

In the previous chapter, we often sought ways to avoid or mitigate the frequency-dependent factors, to give the broadest possible response in frequency with minimal variations. In this chapter, we turn that aspiration around, focusing our efforts instead on circuits whose frequency dependence is a defining characteristic. The goal will typically be to control the variation across frequency, either to maximize selectivity or precisely compensate the amplitude and phase variations in other broadband components.

11.1 EQUALIZERS

Among the most prevalent imperfections in any electronic circuit is ohmic loss. Furthermore, although the frequency dependence of loss can take on almost any mathematical form, by far the most common is that loss increases monotonically with frequency, due in large part to the natural form of the skin effect resistance, (1.118), which applies broadly to virtually all good conductors. Thus, lumped elements and transmission lines (e.g., coaxial cables and microstrip traces) acquire loss slopes that increase with frequency.[1] A complex electronic system in which multiple components and interconnects are cascaded may therefore accumulate significant, negative slopes.

This variation in amplitude over frequency can be just as harmful, if not more, to the system performance as is the loss itself. To mitigate this problem, we may insert into the signal path a compensatory device known as an *equalizer*, a passive

1 A notable exception to this trend is hollow metallic waveguides, which often exhibit negatively sloping loss characteristics over their single-mode operating range, as indicated in Figure 7.9.

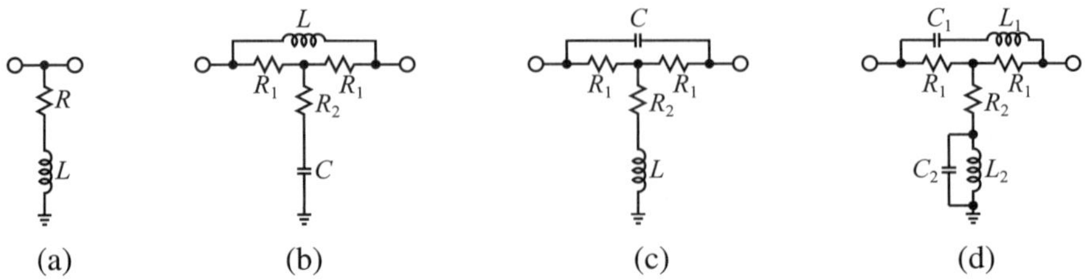

Figure 11.1 Lumped-element gain slope equalizers. (a) Simple, but not perfectly matched. (b) Matched and negatively sloping. (c) Matched and positively sloping. (d) Matched with a finite 0-dB loss frequency.

network having the opposite loss slope to the rest of the system. Thus, the combined response of the equalizer in cascade with the remaining components is rendered flat.

11.1.1 Lumped-Element Equalizers

In their simplest form, slope equalizers may comprise as little as two elements, such as the resistor and inductor combination shown in Figure 11.1(a). This may suffice for the correction of small gain slopes, but equalizers with larger slopes will typically have unacceptable return loss. Ideally, an equalizer should be matched at both ports, so that dissipative losses alone create the slope [1]. Attenuation due to reflections will create standing waves that render the circuit more sensitive to component variations, interconnect lengths, temperature effects, and nonlinearities.

A slightly more sophisticated topology shown in Figure 11.1(b) is capable of providing a controlled loss slope along with theoretically perfect impedance match at all points [2]. It may be understood by considering the limiting behavior at extreme frequencies. At DC, for example, the inductor shorts the two ports together while the capacitor leaves the resistors open-circuited. Thus, the network passes signals unattenuated at the lowest frequencies. At very high frequencies, the inductor becomes open while the capacitor grounds the resistor tee, resembling a simple attenuator. In fact, we may apply (10.1) for the resistor values, where A is the attenuation in decibels associated with the total loss slope from DC to infinity.

It remains to determine the reactance elements, L and C, so that the reflection coefficient is zero at intermediate frequencies as well as at the end points. Let us apply even and odd-mode analysis to this structure. In the even-mode equivalent circuit, looking at Figure 11.1(b), the inductor is left open while the central branch with the series capacitor splits in two,

$$Z_{even} = R_1 + 2R_2 + 2(j\omega C)^{-1} \tag{11.1}$$

In the odd-mode equivalent circuit, the inductor is split in two and then grounded, while R_2 and C are shorted out,

$$Y_{odd} = R_1^{-1} + 2(j\omega L)^{-1} \tag{11.2}$$

The conditions for perfect impedance match are then found by setting (4.100a) equal to zero,

$$s_{11} = 0 \tag{11.3a}$$

$$\therefore \Gamma_{even} = -\Gamma_{odd} \tag{11.3b}$$

$$\frac{Z_{even} - Z_0}{Z_{even} + Z_0} = -\frac{Y_0 - Y_{odd}}{Y_0 + Y_{odd}} \tag{11.3c}$$

$$\frac{Z_{even}}{Z_0} = \frac{Y_{odd}}{Y_0} \tag{11.3d}$$

$$\frac{R_1 + 2R_2}{Z_0} + \frac{2}{j\omega C Z_0} = \frac{1}{R_1 Y_0} + \frac{2}{j\omega L Y_0} \tag{11.3e}$$

The equivalence of the real parts above is confirmed by the resistor values found from the attenuator equations,

$$\frac{R_1 + 2R_2}{Z_0} = \frac{1-T}{1+T} + \frac{4T}{1-T^2} = \frac{1+T}{1-T} = \frac{1}{R_1 Y_0} \tag{11.4}$$

where $T < 1$ is the desired transmission coefficient at peak attenuation (that is, $A = 20 \log T$). To satisfy the imaginary part of (11.3e), we must have

$$\frac{L}{C} = Z_0^2 \tag{11.5}$$

The initial slope, or rather the range of frequencies over which most of the slope occurs, may be shifted by scaling the LC product. A plot of this equalizer's response when designed for slopes ranging from 1 to 10 dB is given in Figure 11.2(a). Note that since A (the attenuation in decibels) determines total slope from DC to infinity, it must be made somewhat larger than the total slope desired over a finite range of frequencies. Generally speaking, the most linear behavior at low frequency is

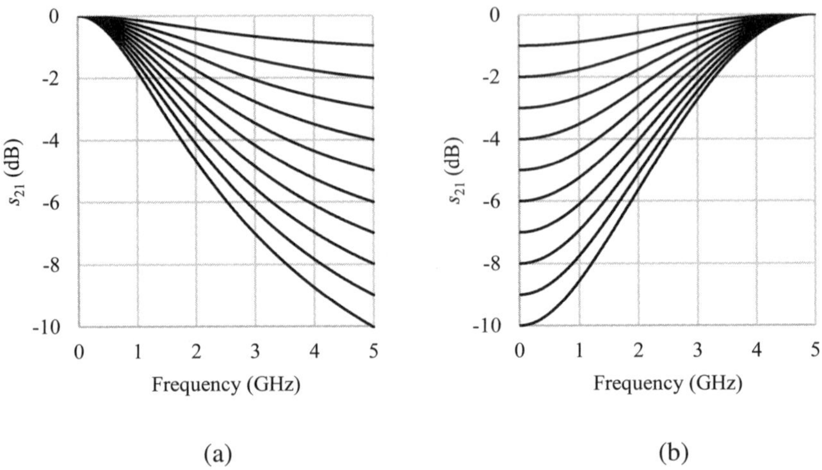

Figure 11.2 Simulated response of the lumped-element equalizers in (a) Figure 11.1(b), and (b) Figure 11.1(d) with slopes ranging from 1 to 10 dB.

achieved for very small LC product and very large A; however, some compromise must be made for parasitics.

For positively sloping equalizers, we could simply apply a high-pass transformation to this circuit, resulting in the equalizer from Figure 11.1(c). Unfortunately, this only reaches zero loss asymptotically as frequency approaches infinity, such that any equalizer using this topology must have some excess flat loss over a finite frequency range. It is better, therefore, to apply a band-pass transformation, resulting in the circuit from Figure 11.1(d). The series and parallel resonators should then be tuned so that they are resonant at the upper end of the desired frequency range. Responses for this circuit are shown in Figure 11.2(b).

Some empirical fitting of design parameters is generally required with these circuits to most closely approximate a linear slope from DC to ω. For Figure 11.1(a), we have let $L = Z_0 A/(5\omega)$ and then adjusted A until the intercept at ω (=5 GHz) was close to the desired total slope value. For Figure 11.1(b), we have applied the bandpass transformation given in Table 5.1 with bandwidth parameter $\Delta = 30/A$ to a low-pass prototype having $L = 2Z_0/\omega$.

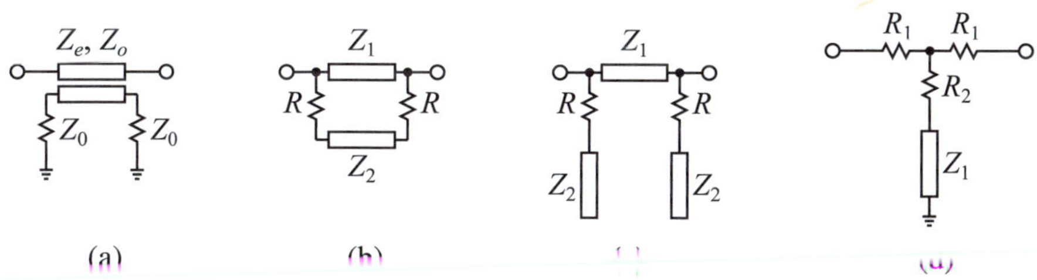

Figure 11.3 Transmission-line equalizers using (a) coupled lines, (b) two parallel transmission lines in a ring, (c) a quasi-pi network, and (d) a quasi-tee network [3].

11.1.2 Transmission-Line Equalizers

One of the simplest ways to make an equalizer with transmission lines is to design a matched coupler and terminate the coupled energy in either direction with resistors, as shown in Figure 11.3(a). The slope is determined by the coupling factor, $\rho = Z_e/Z_o$, where

$$Z_e = Z_0\sqrt{\rho} \tag{11.6a}$$

$$Z_o = \frac{Z_0}{\sqrt{\rho}} \tag{11.6b}$$

$$\rho = \frac{\left(2 - T^2\right) + 2\sqrt{1 - T^2}}{T^2} \tag{11.6c}$$

and T is the desired transmission coefficient at peak attenuation (the derivation will be left as an exercise for the reader). Note that because commensurate-line networks such as this have a periodic frequency response, equalizers built from them can be used for either positive or negative slopes, depending on the frequency range. The response for this network over one full period is shown in Figure 11.4(a). It produces a negative gain slope over the first half of the period, and a positive slope over the octave of the second half. The equations above ensure that the reflection coefficient is zero everywhere.

It is difficult, however, to generate large slopes with simple coupled lines as shown. The range plotted assumes $\rho < 3$, producing a meager 1.25-dB slope. An interdigitated (Lange) coupling section or tandem coupler could be used to improve this somewhat, but a slope of more than a few decibels would be difficult. A more practical solution is the circuit in Figure 11.3(b), having element values given by

$$Z_1 = \frac{Z_0}{T} \tag{11.7a}$$

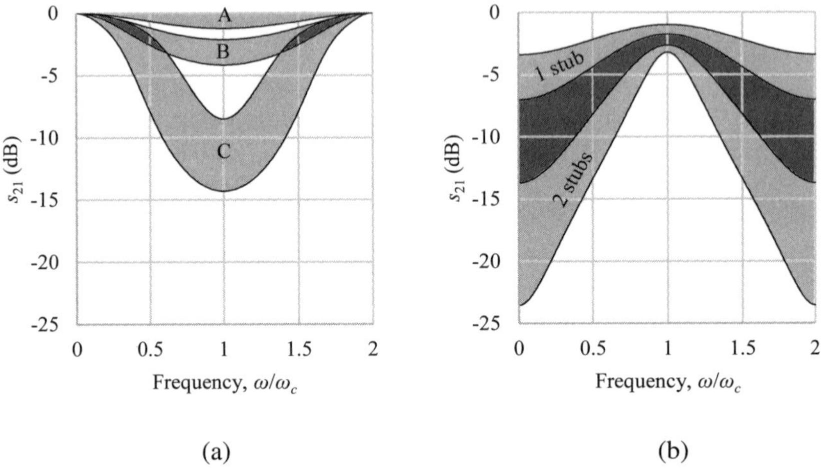

Figure 11.4 Performance of the transmission-line equalizers in (a) Figure 11.3(a) (A), Figure 11.3(b) (B), Figure 11.3(c) (C), and (b) Figure 11.3(d), with 1 and 2 stubs as indicated.

$$Z_2 = \frac{Z_0 T}{1 - T^2} \tag{11.7b}$$

$$R = \frac{Z_0}{1 - T^2} \tag{11.7c}$$

It, too, has perfect impedance match at all frequencies, and with line impedances ranging from $Z_0/2$ to $2Z_0$, it is capable of achieving slopes between 2.15 dB and 4.18 dB, with Z_2 taking on the most extreme values in both cases.

Larger slopes may be achieved with the quasi-pi network shown in Figure 11.3(c), where

$$Z_2 = \frac{Z_1}{(Z_1^2/Z_0^2) - 1} \tag{11.8a}$$

$$R = \sqrt{\tfrac{1}{2} Z_1 Z_2} \tag{11.8b}$$

$$T = \frac{2R\sqrt{2(2R^2 - Z_0^2)}}{(2R + Z_0)^2} \tag{11.8c}$$

and for the same range of line impedances reaches slopes of 8.5 dB to 14.3 dB. The disadvantage of this network is that it is not perfectly matched, but it nevertheless achieves return loss better than about 16.8 dB for slopes within that range.

All of the above transmission-line equalizers provide negative slope over the first half of the periodic response, and positive slope over the second half. The opposite characteristic may be achieved with the quasi-tee network in Figure 11.3(d), where

$$R_1 = \tfrac{1}{2} Z_0 (1 - T) \tag{11.9a}$$

$$R_2 = \tfrac{1}{\text{[illegible]}} Z_0 T \frac{3 - T}{1 - T} \tag{11.9b}$$

$$Z_1 = \tfrac{1}{4} Z_0 \sqrt{T} \frac{3 - T}{1 - T} \tag{11.9c}$$

With the standard 2:1 ratio of immittance (relative to the system characteristic impedance), this network as shown can achieve slopes between about 2.4 dB and 11.1 dB. If the resistor and stub are divided into two parallel branches, this can be extended down to 20.3 dB. Unfortunately, this network is neither perfectly matched nor zero loss at the center of the frequency response, as shown in Figure 11.4(b).

More complicated frequency variations may be compensated for by more sophisticated equalizers, but the returns are diminishing and equalizers having other than positive or negative slopes, or perhaps gentle concave-up or concave-down responses, are rare. More common are continuously adjustable or switched banks of equalizers [3], but even these are almost exclusively based on the fixed versions presented here, with integrated switches or active elements that simply provide control over the element values.

11.2 FOUNDATIONS OF ELECTRONIC FILTERS

The archetype of frequency selective components is the electronic filter, a circuit designed to pass a given range of frequencies (the *passband*) with minimum loss while blocking another range of frequencies (the *stopband*) with maximum attenuation. Filters are among the most well studied of passive electronic components, and there is a rich mathematical theory that has been developed to describe them.

11.2.1 Periodic Networks

Among the first types of filters to be developed were periodic structures. The mathematical basis of their design, known as the *image-parameter method*, has been largely supplanted by network synthesis, and will not be covered here in detail; however, it is useful to briefly discuss the properties of strictly periodic (infinitely long) structures before finite-length filters can be developed.

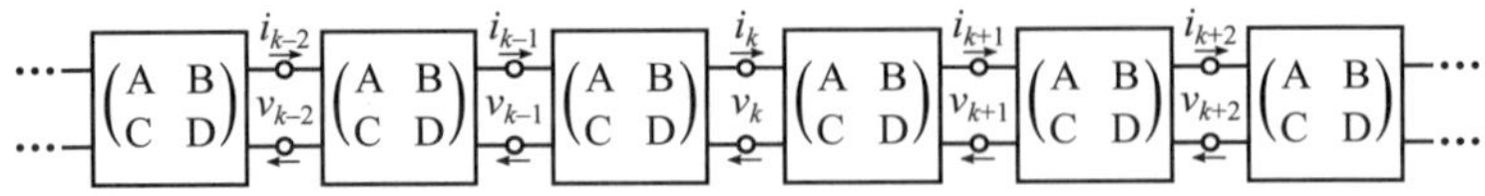

Figure 11.5 An infinite cascade of two-port network cells described by their $ABCD$-parameters.

Consider an infinite cascaded chain of two-port network cells, as shown in Figure 11.5. Each cell may be described by a common $ABCD$-matrix. The terminal voltage and current at any node between cells may therefore be written

$$\begin{pmatrix} v_k \\ i_k \end{pmatrix} = \begin{pmatrix} A & B \\ C & D \end{pmatrix} \begin{pmatrix} v_{k+1} \\ i_{k+1} \end{pmatrix} \tag{11.10}$$

(Recall that the definition of the $ABCD$-parameters has a negative sign for the current on one side, but this was based on the convention that currents were always directed inward at a network port. In this case, since each node is both the input of one cell and the output of another, we have for convenience defined the currents as all running in the same direction.)

We may also model this chain of cells as a kind of synthetic transmission line supporting a Floquet mode, where the voltage and current configuration repeats at every node between cells with a complex propagation delay,

$$\begin{pmatrix} v_{k+1} \\ i_{k+1} \end{pmatrix} = e^{-\gamma} \begin{pmatrix} v_k \\ i_k \end{pmatrix} \tag{11.11}$$

Putting these together, we find

$$\begin{pmatrix} v_k \\ i_k \end{pmatrix} = \begin{pmatrix} A & B \\ C & D \end{pmatrix} \begin{pmatrix} v_{k+1} \\ i_{k+1} \end{pmatrix} = e^{\gamma} \begin{pmatrix} v_{k+1} \\ i_{k+1} \end{pmatrix} \tag{11.12a}$$

$$\therefore \begin{pmatrix} A - e^{\gamma} & B \\ C & D - e^{\gamma} \end{pmatrix} \begin{pmatrix} v_{k+1} \\ i_{k+1} \end{pmatrix} = 0 \tag{11.12b}$$

For nontrivial solutions, then, the determinant of the 2×2 matrix in (11.12b) must be zero,

$$\det \begin{pmatrix} A - e^{\gamma} & B \\ C & D - e^{\gamma} \end{pmatrix} = 0 \tag{11.13a}$$

$$(A - e^{\gamma})(D - e^{\gamma}) - BC = 0 \tag{11.13b}$$

$$e^{2\gamma} - (A + D)e^{\gamma} + AD - BC = 0 \tag{11.13c}$$

If we assume the cells are reciprocal, then $AD - BC = 1$ according to Table 4.4, simplifying the expression,

$$e^{2\gamma} - (A + D)e^{\gamma} + 1 = 0 \tag{11.14a}$$

$$e^{\gamma} + e^{-\gamma} = A + D \tag{11.14b}$$

$$\cosh \gamma = \tfrac{1}{2}(A + D) \tag{11.14c}$$

or, in another form,

$$e^{\gamma} = \tfrac{1}{2}(A + D) \pm \sqrt{\tfrac{1}{4}(A + D)^2 - 1} \tag{11.15}$$

We may also determine the effective characteristic impedance of the chain, called the *Bloch impedance*, by solving (11.12b) for the ratio of voltage and current,

$$Z_B = \frac{v_{k+1}}{i_{k+1}} = \frac{B}{e^{\gamma} - A} = \frac{e^{\gamma} - D}{C} \tag{11.16a}$$

$$\therefore Z_B = \frac{-2B}{A - D \mp \sqrt{(A + D)^2 - 4}} \tag{11.16b}$$

Note that there are two solutions corresponding to forward- and backward-traveling waves. If the network cell is symmetric, then (referring once again to Table 4.4) we have $A = D$, and

$$Z_B = \pm \frac{B}{\sqrt{A^2 - 1}} \tag{11.17}$$

so the forward and backward-traveling waves' impedances differ only by sign, due to the reversal of the current reference direction relative to the direction of propagation.

As an example, consider the capacitively loaded transmission line in Figure 11.6(a), with an asymmetric unit cell given in Figure 11.6(b). (A symmetric unit cell could just as well have been chosen, and would have affected the apparent characteristic impedance as measured from different points within the periodic structure, but would have no other impact on the line's propagation characteristics.) The $ABCD$-parameters of the unit cell are

$$\begin{pmatrix} A & B \\ C & D \end{pmatrix} = \begin{pmatrix} 1 & 0 \\ j\omega C & 1 \end{pmatrix} \begin{pmatrix} \cos\theta & jZ_1 \sin\theta \\ jY_1 \sin\theta & \cos\theta \end{pmatrix} \tag{11.18a}$$

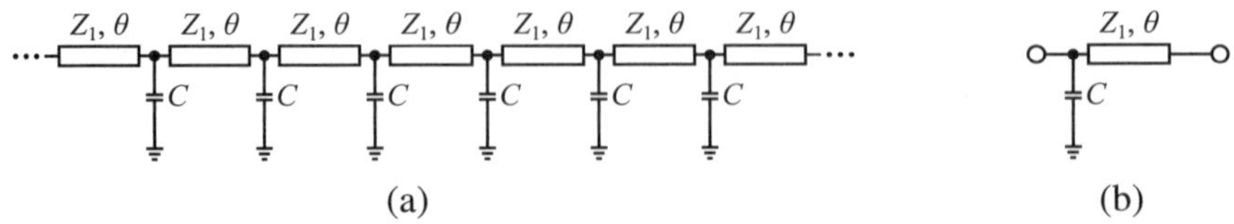

Figure 11.6 (a) Transmission line periodically loaded with capacitors. (b) Asymmetric unit cell.

$$= \begin{pmatrix} \cos\theta & jZ_1 \sin\theta \\ j\omega C \cos\theta + jY_1 \sin\theta & -\omega C Z_1 \sin\theta + \cos\theta \end{pmatrix} \tag{11.18b}$$

From (11.14c), we may write

$$\cosh\gamma = \cosh(\alpha + j\beta) = \cosh\alpha\cos\beta + j\sinh\alpha\sin\beta \tag{11.19a}$$

$$= \cos\theta - \tfrac{1}{2}\omega C Z_1 \sin\theta \tag{11.19b}$$

where we have applied some of the hyperbolic function identities in Section A.2. Note that since the final expression is real-valued, we must have either $\alpha = 0$ or $\beta = n\pi$, and

$$\cosh\alpha\cos\beta = \cos\theta - \tfrac{1}{2}\omega C Z_1 \sin\theta \tag{11.20}$$

Note also that $\cosh\alpha$ is always greater than or equal to 1, and the magnitude of $\cos\beta$ is always less than or equal to 1. Thus, $\alpha = 0$ when the magnitude of the right side above is less than 1, and $\beta = n\pi$ when the magnitude is greater than one (and $n = 0$ or 1 depending on its sign).

A dispersion diagram for this structure is given in Figure 11.7(a). For this example, we have loaded a 30Ω transmission line with 10-pF capacitors a quarter wavelength apart at 1 GHz. The structure has alternating passbands and stopbands, where each successive passband is narrower, and each stopband deeper, than the one before it (due in part to the capacitors asymptotically approaching a short-circuit).

The actual transmission and reflection coefficients of a finite length of this periodic structure will depend on the number of sections and will be affected almost as much by the impedance mismatch as by the propagation constant. To emphasize this point, the response for a length of four sections is given in Figure 11.7(b). Although the passbands and stopbands can be roughly discerned, the response is hardly optimal for a filter. Impedance-matching sections can be added to the input and output, but this would be wasteful of components. Better to incorporate the matching into the structure itself.

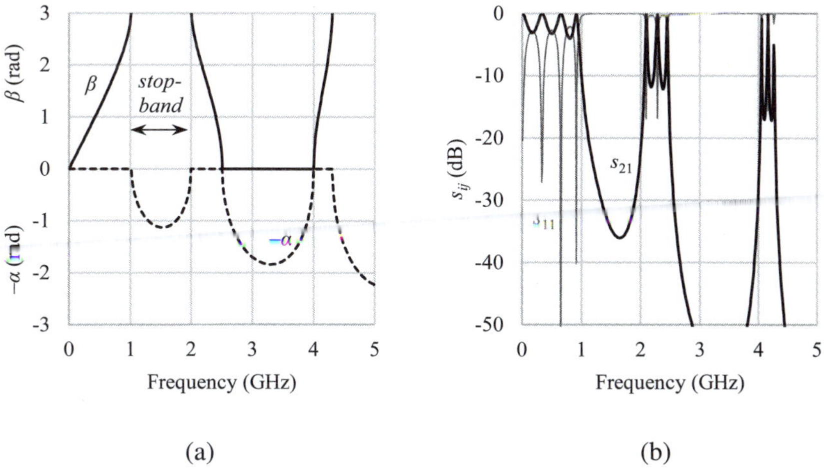

Figure 11.7 (a) Dispersion diagram showing attenuation constant, α, and phase constant, β, for a capacitively loaded transmission line with $Z_1 = 30\Omega$ and $C = 10$ pF, and where the spacing between capacitors is a quarter wavelength at 1 GHz. The first stopband is highlighted. (b) Transmission and reflection coefficient for four full sections.

11.2.2 Canonical Filter Responses

A great deal of mathematical effort has been expended over the years to determine exactly what is the optimal frequency-selecting response which can be achieved in principle using a finite number of elements. "Optimal" depends somewhat on which features are considered most important for a given application (e.g., loss slope, monotonicity, selectivity, group delay) but the boundaries of what is possible are now fairly well understood.

The principal drivers of realizability for a filter's transmission response, or *transfer function*, are causality and stability. Causality is the condition that a network's response to an input cannot precede that input in time. Stability is the condition that the output remains bounded for bounded inputs.

A reasonable baseline for what would constitute an ideal filter, absent the above constraints, is one that has zero loss in the passband, infinite attenuation in the stopband, zero (or at least linear) phase delay, and perfect impedance match everywhere. That is,

$$s_{11} = 0 \tag{11.21a}$$

$$s_{21} = \begin{cases} 1 & f < f_c \\ 0 & f \geq f_c \end{cases} \tag{11.21b}$$

where for convenience we have assumed a low-pass response (Section 5.1 taught us how to transform an existing low-pass network into high-pass, band-pass, and band-stop configurations) and f_c is the cutoff, or corner frequency. The transmission response of (11.21b), sometimes called a *brick wall filter*, is not realizable, however, for essentially the same reason that perfect impedance matching was ruled out in Section 6.2.4; the time-domain representation of such a response, a sync function having nonzero value over infinite lengths of time, is noncausal.

Most authors would cite difficulty with (11.21a) as well, as it has been implicitly assumed throughout most of the historical development of filters that the filter elements will be essentially lossless, with few exceptions, ensuring that $s_{11} \approx 1$ in the stopband due to conservation of energy. As we will see in Section 11.3, this need not be a limitation at all, as the same, optimal responses that can be realized using conventional techniques may be realized in a reflectionless form as well, at the cost of some additional elements [2].

For finite, lumped-element networks, the transfer function may be written as the ratio of two complex polynomials [2, 4, 5],

$$H(\omega) = a_0 \frac{\prod\limits_k (j\omega - z_k)}{\prod\limits_l (j\omega - p_l)} \tag{11.22}$$

where z_k are the *zeros* of the transfer function (where the transmission coefficient becomes identically zero), and p_l are the *poles* (where the transmission coefficient becomes infinite). The poles and zeros are complex-valued, and stability requires that all the poles must lie in the left half of the complex plane. The number of poles determines the order of the filter, which shall be designated N, and the number of elements required to realize it. The prefactor a_0 ensures that $|H(\omega)| \leq 1$ for all ω.

The consequence of this formulation is that realizable transfer functions will, in general, reach zero or one only at discrete frequency points, with ripples in between, and have finite slopes in the transition between passband and stopband. It has been proven mathematically that the steepest transition slopes for a given amplitude of ripples are achieved when the ripples have equal amplitude in either the passband or stopband or both. For some applications, however, ripples of any amplitude are undesirable, in which cases responses having a monotonic characteristic in the passband or stopband or both are preferred.

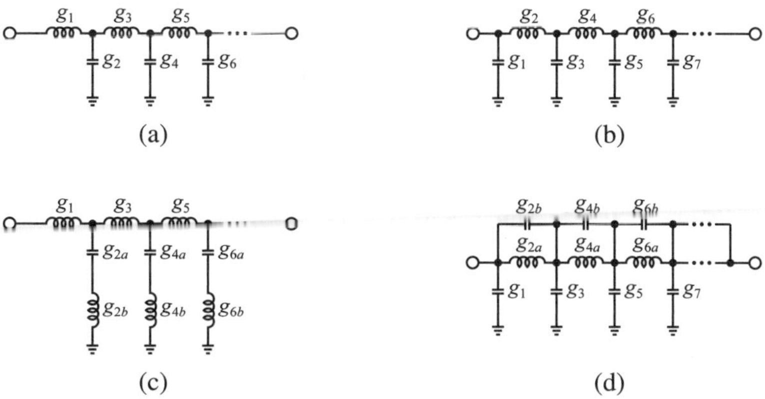

Figure 11.8 Low-pass ladder-topology prototypes, labeled with normalized element values. (a) All-pole ladder. (b) Dual all-pole ladder. (c) Ladder supporting finite transmission zeros. (d) Dual ladder supporting finite transmission zeros.

We thus may start with four canonical responses — known as *Butterworth* (also called *binomial* or *maximally flat*), *Chebyshev type I*, *Chebyshev type II* (also called *inverse Chebyshev*), and *elliptic* (also called *Cauer*) — corresponding to the four possible combinations of equiripple versus monotonic in the passband and stopband. For those that allow ripples, the amplitude of those ripples is a free parameter. These canonical responses are summarized in Table 11.1. There are a wide variety of other responses that have been identified by researchers that are optimal in other ways — *Legendre-Papoulis*, *Achieser-Zolotarev*, *Gaussian*, *Bessel-Thomson*, etc. — but these four are the most common (with the possible exception of Chebyshev type II). For a more thorough treatment of these and other responses as well as the mathematical bases for them, one is referred to the available literature [2, 4, 5].

11.2.3 Lumped-Element Ladders

Although there is usually more than one topology that can realize a given transfer function, the *ladder* topologies shown in Figure 11.8 are the de facto standard and have been proven to be minimal in the sense that no other topology can realize the same responses with fewer elements. Low-pass versions of these topologies are shown, but high-pass, band-pass, and band-stop versions may also be generated using the transformations found in Table 5.1.

Table 11.1

Canonical Filter Responses

Response	$\lvert H(\omega)\rvert$	Passband Ripple	Stopband Ripple	Low-Pass Plot
Butterworth	$\frac{1}{\sqrt{1+\omega^{2N}}}$	monotonic	monotonic	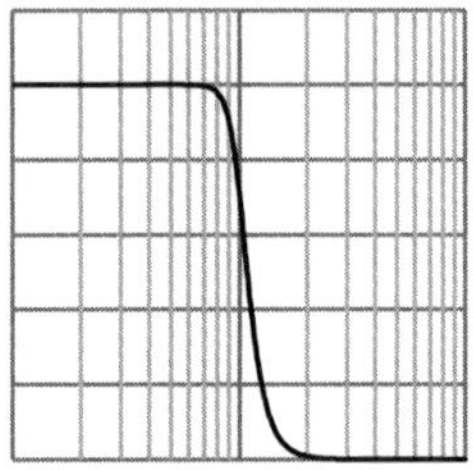
Chebyshev type I	$\frac{1}{\sqrt{1+\varepsilon^2 T_N^2(\omega)}}$	$\frac{1}{\sqrt{1+\varepsilon^2}}$	monotonic	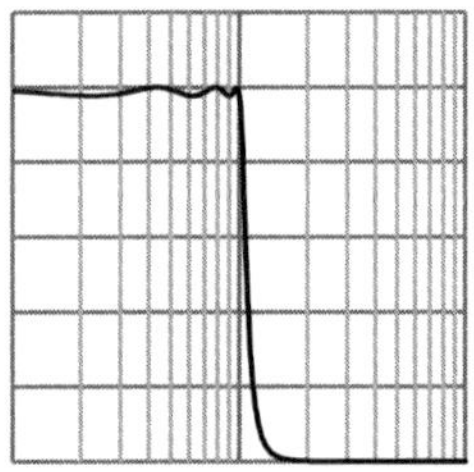
Chebyshev type II	$\frac{\varepsilon T_N(\omega^{-1})}{\sqrt{1+\varepsilon^2 T_N^2(\omega^{-1})}}$	monotonic	$\frac{\varepsilon}{\sqrt{1+\varepsilon^2}}$	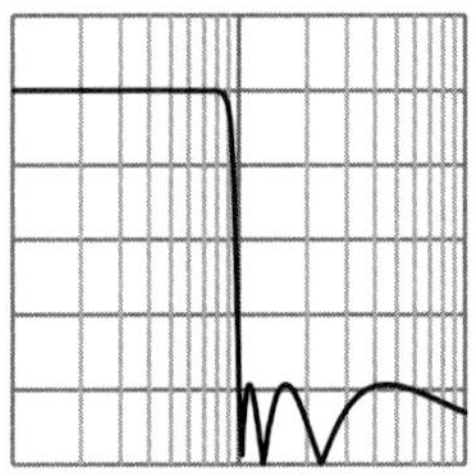
Elliptic	$\frac{1}{\sqrt{1+\varepsilon^2 R_N^2(\xi,\omega)}}$	$\frac{1}{\sqrt{1+\varepsilon^2}}$	$\frac{1}{\sqrt{1+\varepsilon^2 R_N^2(\xi,\xi)}}$	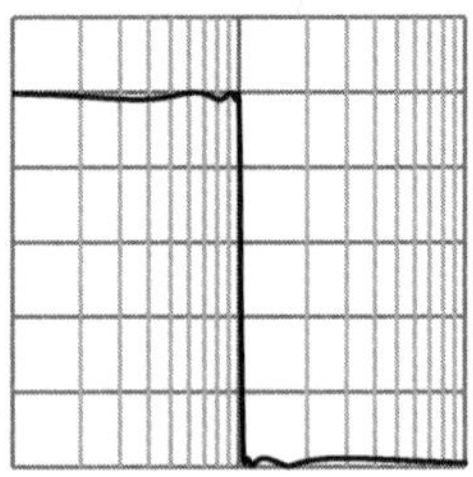

N = filter order, ε = ripple factor, ξ = selectivity factor, T_N = Nth-order Chebyshev polynomial, and R_N = Nth-order elliptic rational function.

Figures 11.8(a, b) are duals of one another. In accordance with the conclusions of Section 5.1.7, they have the same transmission response (in both magnitude and phase), while their reflection coefficients are additive inverses. The element values are conventionally labeled with quantities that are normalized in both frequency and impedance. That is, if the inductors are assumed to be labeled in henries, and the capacitors in farads, then a typical set of prototype parameters, g_k, will yield a filter with cutoff frequency at $\omega = 1$ rad/s and is impedance-matched (in the passband) to 1Ω. The impedance and frequency may then be scaled to any values desired using the transformations in Table 5.1. Thus, for a filter having cutoff frequency ω_c and matched to a system characteristic impedance of Z_0, the inductors and capacitors should be given the following values,

$$L_k = g_k \frac{Z_0}{\omega_c} \tag{11.23a}$$

$$C_k = g_k \frac{Y_0}{\omega_c} \tag{11.23b}$$

The filters in Figures 11.8(a, b) are also what are known as *all-pole filters*, which is to say that the numerator in (11.22) is merely a constant, there are no finite transmission zeros (or, said differently, all N transmission zeros lie at infinite frequency). Butterworth and Chebyshev type-I filters are examples of all-pole filters. Formulas for the prototype parameters yielding filters having both of these responses can be determined mathematically[2] [5]. For Butterworth filters having a 3-dB cutoff corner at $\omega_c = 1$ rad/s, we have

$$g_k = 2 \sin\left(\frac{2k-1}{2N}\pi\right) \text{ for } k = 1 \ldots N \tag{11.24}$$

where, again, N is the order of the filter, and also the number of reactive elements.

The formulas for the prototype parameters of a Chebyshev type-I response are slightly more complicated,

$$\beta = \ln\left(\coth\left(\tfrac{1}{4}\ln\left(1+\varepsilon^2\right)\right)\right) \tag{11.25a}$$

2 Historically, books would often include tables of prototype parameter values for filters of this type having a variety of orders and selected ripple factors. In my view, however, these tables were primarily a holdover from the days before desktop computers were readily available, which could evaluate the necessary trigonometric functions with ease. As we shall see in Section 11.3, there are some recently discovered nonladder topologies that benefit from specific ripple factors not often tabulated in other texts, so it is considered worthwhile for the next generation of engineers to begin developing the habit of using the formulas here within their simulation models, granting them total flexibility and control of the filter response.

$$\gamma = \sinh\left(\tfrac{\beta}{2N}\right) \tag{11.25b}$$

$$a_k = \sin\left(\tfrac{(2k-1)\pi}{2N}\right) \text{ for } k = 1 \ldots N \tag{11.25c}$$

$$b_k = \gamma^2 + \sin^2\left(\tfrac{k\pi}{N}\right) \text{ for } k = 1 \ldots (N-1) \tag{11.25d}$$

$$g_1 = \frac{2a_1}{\gamma} \tag{11.25e}$$

$$g_k = \frac{4a_{k-1}a_k}{b_{k-1}g_{k-1}} \text{ for } k = 1 \ldots N \tag{11.25f}$$

$$g_{N+1} = \begin{cases} 1 & \text{for } N \text{ odd} \\ \coth^2\left(\tfrac{\beta}{4}\right) & \text{for } N \text{ even} \end{cases} \tag{11.25g}$$

where the ripple factor, ε, is a free parameter previously described. (Take care that ε in this context and the intermediate parameters β and γ are not confused with permittivity or the phase and propagation constants using the same Greek letters in other contexts.)

For the Butterworth filter, we assumed that both ports were terminated with the system characteristic impedance, Z_0, regardless of order. For the Chebyshev type-I filter, this is only possible (at least, using the ladder topology) when N is odd. When N is even, the filter is necessarily impedance-transforming, and the second port must be terminated with a port impedance having nonunity normalized value. That normalized value is given by g_{N+1} in (11.25g).

Plots of the filter responses using the ladder topologies of Figures 11.8(a, b) with these element values are shown in Figure 11.9. Note that the passband for the Butterworth filter is smooth, while the Chebyshev filter has ripples in its passband with a steeper transition.

Let us work out the parameter values for the Chebyshev type-I filter as an example. First, we must calculate the prototype parameter values, using (11.25) with the order $N = 7$ and ripple factor $\varepsilon = 0.2187$,

$$\beta = \ln\left(\coth\left(\tfrac{1}{4}\ln\left(1 + 0.2187^2\right)\right)\right) \approx 4.45 \tag{11.26a}$$

$$\gamma = \sinh\left(\frac{\beta}{2N}\right) = \sinh\left(\frac{4.45}{2 \times 7}\right) \approx 0.3232 \tag{11.26b}$$

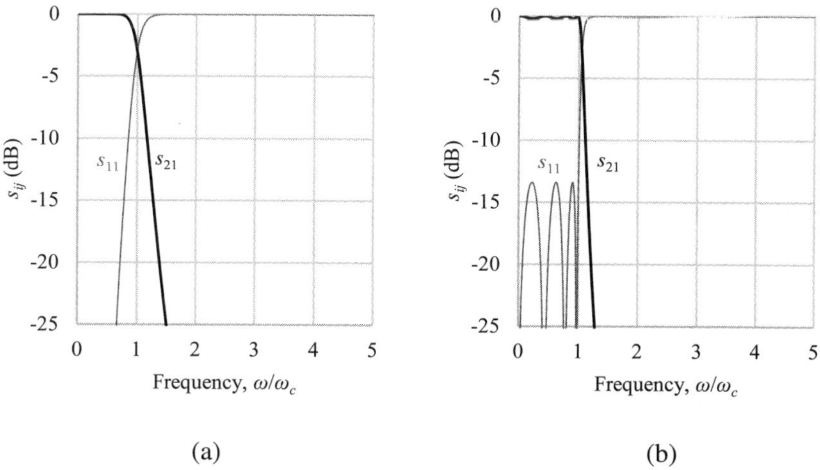

Figure 11.9 Seventh-order all-pole filters implemented using the ladder topologies. (a) Butterworth response. (b) Chebyshev type-I response with $\varepsilon = 0.2187$.

$$a_k = \sin\left(\frac{(2k-1)\pi}{2N}\right) \approx \begin{cases} 0.2225 & k=1 \\ 0.6235 & k=2 \\ 0.9010 & k=3 \\ 1.0000 & k=4 \\ 0.9010 & k=5 \\ 0.6235 & k=6 \\ 0.2225 & k=7 \end{cases} \tag{11.26c}$$

$$b_k = \gamma^2 + \sin^2\left(\frac{k\pi}{N}\right) \approx \begin{cases} 0.2927 & k=1 \\ 0.7157 & k=2 \\ 1.0549 & k=3 \\ 1.0549 & k=4 \\ 0.7157 & k=5 \\ 0.2927 & k=6 \end{cases} \tag{11.26d}$$

$$g_1 = \frac{2a_1}{\gamma} = \frac{2 \times 0.2225}{0.3232} \approx 1.377 \tag{11.26e}$$

$$g_k = \frac{4a_{k-1}a_k}{b_{k-1}g_{k-1}} \approx \begin{cases} 1.377 & k=2 \\ 2.280 & k=3 \\ 1.498 & k=4 \\ 2.280 & k=5 \\ 1.377 & k=6 \\ 1.377 & k=7 \end{cases} \tag{11.26f}$$

and finally, since N is odd, $g_8 = 1$. (These values are repeated again later in Table 11.2.)

Although plotted in Figure 11.9(b) over a normalized frequency scale, let us use for this example a cutoff corner of $f_c = 1$ GHz, and normalize it for a characteristic impedance of $Z_0 = 50\Omega$. We will select the topology in Figure 11.8(a), which starts with a series inductor (which we will call L_1, numbering all elements from left to right). The scaled element values are given by the formulas in Table 5.1,

$$L_1 = \frac{g_1 Z_0}{\omega_c} = \frac{(1.377)(50\Omega)}{2\pi(1\text{ GHz})} \approx 10.96\text{ nH} \tag{11.27a}$$

$$C_2 = \frac{g_2 Y_0}{\omega_c} = \frac{(1.377)(50\Omega)^{-1}}{2\pi(1\text{ GHz})} \approx 4.383\text{ pF} \tag{11.27b}$$

$$L_3 = \frac{g_3 Z_0}{\omega_c} = \frac{(2.280)(50\Omega)}{2\pi(1\text{ GHz})} \approx 18.14\text{ nH} \tag{11.27c}$$

$$C_4 = \frac{g_4 Y_0}{\omega_c} = \frac{(1.498)(50\Omega)^{-1}}{2\pi(1\text{ GHz})} \approx 4.768\text{ pF} \tag{11.27d}$$

$$L_5 = \frac{g_5 Z_0}{\omega_c} = \frac{(2.280)(50\Omega)}{2\pi(1\text{ GHz})} \approx 18.14\text{ nH} \tag{11.27e}$$

$$C_6 = \frac{g_6 Y_0}{\omega_c} = \frac{(1.377)(50\Omega)^{-1}}{2\pi(1\text{ GHz})} \approx 4.383\text{ pF} \tag{11.27f}$$

$$L_7 = \frac{g_7 Z_0}{\omega_c} = \frac{(1.377)(50\Omega)}{2\pi(1\text{ GHz})} \approx 10.96\text{ nH} \tag{11.27g}$$

and of course the output termination in this case is $R = g_8 Z_0 = 1 \times 50\Omega = 50\Omega$.

Chebyshev type-II and elliptic filters, in contrast, are *not* all-pole filters, and therefore cannot be implemented using the topologies in Figures 11.8(a, b). In order to generate transmission zeros, we must add elements such that the series immittances in the ladder are parallel resonators, or the shunt immittances are series resonators, as shown in Figures 11.8(c, d). The element values for filters having these topologies are rarely included in books, but some tables do exist [6].

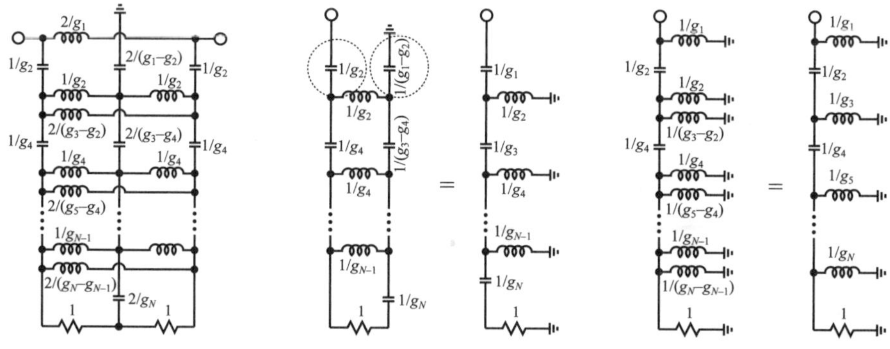

(a) (b) (c)

Figure 11.10 (a) Low-pass reflectionless filter topology. (b) Even-mode circuit equivalent to a terminated, high-pass ladder topology. The encircled capacitors are in series and may be combined. (c) Odd-mode circuit equivalent to the dual, terminated, high-pass ladder topology. Many of the shunt inductors are in parallel and may be combined.

11.3 REFLECTIONLESS FILTERS

Since the elements of a ladder network are all lossless, conservation of energy requires that $|s_{11}|^2 + |s_{21}|^2 = 1$, ensuring that the filter is almost totally reflective in the stopbands (see the thin gray lines that approach 0 dB at high frequency in Figure 11.9). Although commonly accepted as inevitable, experienced systems designers know that it is erroneous to dismiss these effects simply because they occur out-of-band; if out-of-band signals did not matter, why filter them at all? Reflections trap energy in standing waves that interact with broadband and/or nonlinear components in a wide variety of ways that degrade overall performance [7, 8].

11.3.1 Topological Basis of Reflectionless Filters

Fortunately, nonladder topologies do exist that implement the exact same optimal filter functions, but without reflection at any frequency. Consider the network shown in Figure 11.10(a). Like the ladder topologies in Figure 11.8, the elements are labeled with normalized prototype values. As a symmetric circuit, we may analyze its behavior using even and odd-mode analysis (some inductors and capacitors may need to be split into series or parallel pairs first). The even-mode equivalent circuit is shown in Figure 11.10(b). Note that the two capacitors at the top are effectively in series and may be combined. This, in turn, leaves the next two capacitors in series, and so on, until we arrive at a ladder topology, terminated with unit normalized resistance. In fact, this is the high-pass transformation of the ladder topology shown in Figure 11.8(a). The odd-mode circuit, Figure 11.10(c), also simplifies to a terminated ladder network, the high-pass transformation of Figure 11.8(b). The even and odd-mode equivalent circuits are therefore duals of each other, $\Gamma_{even} = -\Gamma_{odd}$, and thus, according to (4.100a), $s_{11} = 0$. The network is reflectionless.

What of its transmission response, however? This will depend on the prototype parameters (g_k) used, but according to (4.100b) and considering the duality condition, we know that $s_{21} = \Gamma_{even}$. Thus, if Butterworth or Chebyshev parameters are applied to this topology, then we may expect the transmission response to be a frequency-inverted form of the gray reflection curves shown in Figures 11.9(a) and (b), respectively. Referring to the magnitude response curves in Table 11.1, for the Butterworth parameters we have

$$|H'(\omega)| = \sqrt{1 - |H(\omega^{-1})|^2} = \sqrt{1 - \frac{1}{1+\omega^{-2N}}} = \frac{1}{\sqrt{1+\omega^{2N}}} \tag{11.28}$$

which is once again a Butterworth response. For the Chebyshev (type I) parameters, we have

$$|H'(\omega)| = \sqrt{1 - |H(\omega^{-1})|^2} = \sqrt{1 - \frac{1}{1+\varepsilon^2 T_N^2(\omega^{-1})}} \tag{11.29a}$$

$$= \frac{\varepsilon T_N(\omega^{-1})}{\sqrt{1+\varepsilon^2 T_N^2(\omega^{-1})}} \tag{11.29b}$$

which is a Chebyshev type-II response. We thus see that the Chebyshev type-I and type-II responses are in one sense inverses of one another, while the Butterworth is its own inverse.

As an example, let us once again revisit the seventh-order Chebyshev response with ripple factor $\varepsilon = 0.2187$. Derivation of the prototype parameters as previously worked out in (11.26) is unchanged. This time, we will apply those prototype parameters to the topology in Figure 11.10(a). Again, let us choose a corner frequency of 1 GHz, and 50Ω normalizing characteristic impedance. Numbering the inductors from top to bottom, we have,

$$L_1 = \frac{2}{g_1} \cdot \frac{Z_0}{\omega_c} = \frac{2}{1.377} \cdot \frac{50\Omega}{2\pi(1\text{ GHz})} \approx 11.56\text{ nH} \tag{11.30a}$$

$$L_2 = \frac{1}{g_2} \cdot \frac{Z_0}{\omega_c} = \frac{1}{1.377} \cdot \frac{50\Omega}{2\pi(1\text{ GHz})} \approx 5.779 \text{ nH} \tag{11.30b}$$

$$L_3 = \frac{2}{(g_3 - g_2)} \cdot \frac{Z_0}{\omega_c} = \frac{2}{(2.280 - 1.377)} \cdot \frac{50\Omega}{2\pi(1\text{ GHz})} \approx 17.63 \text{ nH} \tag{11.30c}$$

$$L_4 = \frac{1}{g_4} \cdot \frac{Z_0}{\omega_c} = \frac{1}{1.498} \cdot \frac{50\Omega}{2\pi(1\text{ GHz})} \approx 5.312 \text{ nH} \tag{11.30d}$$

$$L_5 = \frac{2}{(g_5 - g_4)} \cdot \frac{Z_0}{\omega_c} = \frac{2}{(2.280 - 1.498)} \cdot \frac{50\Omega}{2\pi(1\text{ GHz})} \approx 20.35 \text{ nH} \tag{11.30e}$$

$$L_6 = \frac{1}{g_6} \cdot \frac{Z_0}{\omega_c} = \frac{1}{1.377} \cdot \frac{50\Omega}{2\pi(1\text{ GHz})} \approx 5.779 \text{ nH} \tag{11.30f}$$

$$L_7 = \frac{2}{(g_7 - g_6)} \cdot \frac{Z_0}{\omega_c} \to \infty \tag{11.30g}$$

Note that the last inductor diverges to infinity on account of the fact that $g_7 = g_6$ (the ripple factor was selected expressly for this purpose). In practice, we may therefore leave this inductor as open circuit. The capacitors, numbered from the top middle to bottom left, may be found as follows,

$$C_1 = \frac{2}{(g_1 - g_2)} \cdot \frac{Y_0}{\omega_c} \to \infty \tag{11.31a}$$

$$C_2 = \frac{1}{g_2} \cdot \frac{Y_0}{\omega_c} = \frac{1}{1.377} \cdot \frac{(50\Omega)^{-1}}{2\pi(1\text{ GHz})} \approx 2.312 \text{ pF} \tag{11.31b}$$

$$C_3 = \frac{2}{(g_3 - g_4)} \cdot \frac{Y_0}{\omega_c} = \frac{2}{(2.280 - 1.498)} \cdot \frac{(50\Omega)^{-1}}{2\pi(1\text{ GHz})} \approx 8.141 \text{ pF} \tag{11.31c}$$

$$C_4 = \frac{1}{g_4} \cdot \frac{Y_0}{\omega_c} = \frac{1}{1.498} \cdot \frac{(50\Omega)^{-1}}{2\pi(1\text{ GHz})} \approx 2.125 \text{ pF} \tag{11.31d}$$

$$C_5 = \frac{2}{(g_5 - g_6)} \cdot \frac{Y_0}{\omega_c} = \frac{2}{(2.280 - 1.377)} \cdot \frac{(50\Omega)^{-1}}{2\pi(1\text{ GHz})} \approx 7.050 \text{ pF} \tag{11.31e}$$

$$C_6 = \frac{1}{g_6} \cdot \frac{Y_0}{\omega_c} = \frac{1}{1.377} \cdot \frac{(50\Omega)^{-1}}{2\pi(1\text{ GHz})} \approx 2.312 \text{ pF} \tag{11.31f}$$

$$C_7 = \frac{2}{g_7} \cdot \frac{Y_0}{\omega_c} = \frac{2}{1.377} \cdot \frac{(50\Omega)^{-1}}{2\pi(1\text{ GHz})} \approx 4.623 \text{ pF} \tag{11.31g}$$

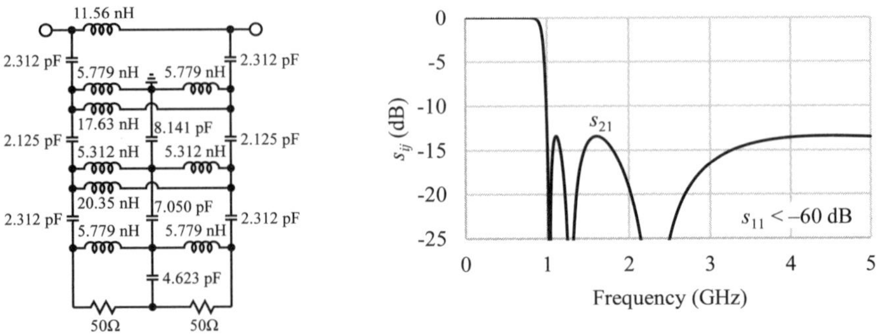

(a) (b)

Figure 11.11 Seventh-order, reflectionless, low-pass filter with values calculated in the text. (a) Circuit schematic. (b) Frequency response.

and in this case, the infinite capacitor may be considered a short circuit. The final circuit and its frequency response are thus shown in Figures 11.11(a, b). Note that the reflection coefficient in decibels is very large (it would be infinite if not for roundoff errors). On a Smith chart, which is not shown, the immittance locus would be little more than a dot in the very center. This, as described, is a Chebyshev type-II response and may be considered a frequency-inverted replica of the reflection coefficient for the ladder prototype shown earlier in Figure 11.9(b).

11.3.2 Mitigation of Negative Elements

There is an issue with the foregoing reflectionless topology, namely, that some of the elements have values based on the difference between two prototype parameters (specifically, an odd-numbered parameter minus an even-numbered parameter), and there is no guarantee that this difference will be positive. In fact, the ripple factor that we chose in our worked example saw a couple of the filter elements become infinite (and were thus omitted) as a consequence of perfect cancellation of neighboring prototype parameters. If we examine the list of parameters for Chebyshev filters over a range of ripple factors, as in Table 11.2, we see that beyond this point ($\varepsilon < 0.2187$ when $N = 7$) the subtraction would have a negative result. At ripple factors lower than this, then, a negative inductor and/or capacitor would

Table 11.2

Prototype Parameters for Seventh-Order Chebyshev Filters

ε	g_1	g_2	g_3	g_4	g_5	g_6	g_7
0.25	1.465	1.350	2.366	1.461	2.366	1.350	1.465
0.2187	**1.377**	**1.377**	2.280	1.498	2.280	**1.377**	**1.377**
0.2	1.323	1.392	2.229	1.520	2.229	1.392	1.323
0.15	1.173	1.424	2.089	1.576	2.089	1.424	1.173
0.1	1.008	1.437	1.940	1.622	1.940	1.437	1.008
0.05	0.807	1.396	1.757	1.634	1.757	1.396	0.807
0.01891	0.621	1.274	**1.569**	**1.569**	**1.569**	1.274	0.621
0.01	0.535	1.179	1.464	1.500	1.464	1.179	0.535

Boldface indicates ripple factors for which adjacent prototype parameters become equal.

be required. More generally [2], the threshold occurs for filters of any order, N, at

$$\varepsilon_{min} = \sqrt{e^{4\tanh^{-1}(e^{-\beta})} - 1} \tag{11.32}$$

where

$$\beta = 2N\sinh^{-1}\left(\sqrt{\frac{1}{2}\sin\left(\frac{\pi}{N}\right)\tan\left(\frac{\pi}{N}\right)}\right) \tag{11.33}$$

For Chebyshev type-II filters, this limits the stopband rejection per stage to approximately 13.4 dB. More seriously, it precludes Butterworth filters entirely, since a Butterworth response is essentially a limiting case of a Chebyshev response wherein the ripple factor is reduced to zero.

There is a way around this problem, however. Recall the current-reversing transformer identities we identified in Figure 5.15. These resulted in equivalent pi- and tee-networks having negative central elements, and the negative elements present within our topology either already have this form or can be manipulated until they do.

Consider the repeating rings of inductors in the low-pass topology, such as the bottom group indicated in Figure 11.12(a). Each of these groupings can be drawn as a pi-network, Figure 11.12(b), where the central element is negative when $g_k < g_{k-1}$. If the two shunt inductors in that group are then split into parallel pairs, as shown in Figure 11.12(c), we have a form to which we can apply the identity in Figure 5.15(a), resulting in the equivalent subcircuit shown in Figure 11.12(d). An

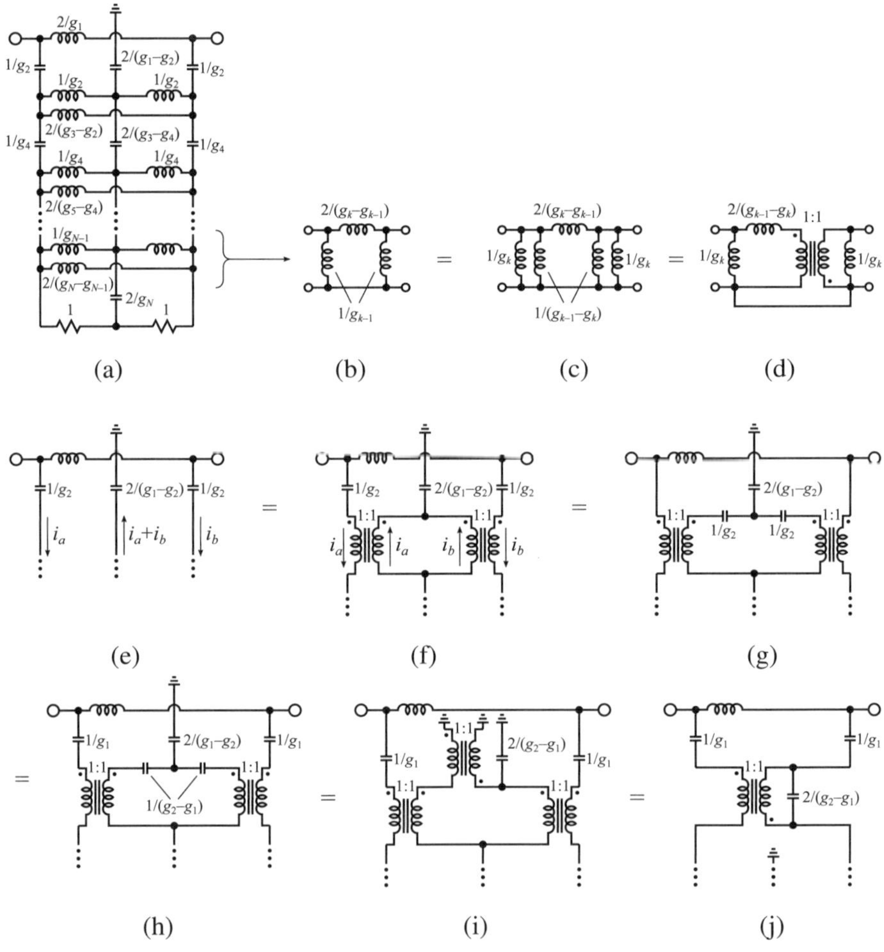

Figure 11.12 Elimination of negative elements in reflectionless filters. (a) Original topology. (b) Pi-network of inductors where the central element is assumed to be negative ($g_k < g_{k-1}$. (c) Division of outer inductors into parallel pairs. (d) Application of transformer identity from Figure 5.15(a). (e) Trio of series capacitors where the central element is assumed to be negative. (f) Temporary incorporation of transformers to hold current sum condition. (g) Relocation of outer inductors to central path. (h) Division of outer capacitors into series pairs, and relocation of residual capacitors back to outer branches. (i) Application of transformer identity from Figure 5.15(b). (j) Elimination of redundant transformers.

element with a parameter subtraction remains, but the order of that subtraction has been reversed, now $2/\left(g_{k-1}-g_k\right)$.

The negative capacitors in this topology take a bit more effort to eliminate. Take the set of three capacitors at the top, as indicated in Figure 11.12(e), and assume that $g_1 < g_2$ so the central capacitor is negative. The outer capacitors are effectively in series with it, but this is difficult to see since they are not directly connected. To aid in that visualization, we note that the entirety of the circuit below the capacitors has no ground sink, so any currents flowing down in the outer branches, i_a and i_b, must be balanced by a sum current, $i_a + i_b$, flowing up in the central branch. We may therefore add a pair of 1:1 transformers to enforce this condition while we make changes, as shown in Figure 11.12(f). The transformers alone do not alter the response, but they allow us to move the outer-branch capacitors into the center, as shown in Figure 11.12(g). Now we have a tee-network with a negative central element. This time we split the positive capacitors into series pairs, and relocate the residual capacitors back to the outer branches, as shown in Figure 11.12(h), leaving two capacitors of value $1/\left(g_2-g_1\right)$ in the middle. This is the form we need to apply the identity from Figure 5.15(b), giving us the result in Figure 11.12(i) [9]. At this time, it is possible to remove the outer two transformers since their function of duplicating currents in various branches is redundant, leaving us the single-transformer equivalent in Figure 11.12(j).

The final topology for a Chebyshev type-II filter having ripple factor below the threshold given in (11.32), after replacing two negative elements with transformer equivalents, is shown in Figure 11.13(a). As Table 11.2 shows, if the ripple factor is made too small ($\varepsilon < 0.01891$ for $N = 7$) another critical point is eventually reached where the next pair of elements shall become negative. These negative elements may be eliminated in the very same way. It is actually convenient, when designing equiripple filters, to select a ripple factor that exactly matches one of these critical points since the divergence of the associated element values to infinity allows them to be eliminated from the circuit entirely, slightly simplifying the overall topology.

11.3.3 All-Pole Reflectionless Filters

The reader should be reminded once again that the topologies in Figure 11.10(a) and Figure 11.13(a) achieve their canonical transmission responses (Butterworth, Chebyshev type II) without any reflection whatsoever — $s_{11} = 0$ at all frequencies. It turns out that all of the energy which is incident at one port in the stopband is terminated in the unit-value resistor on the same side as that port. We may thus

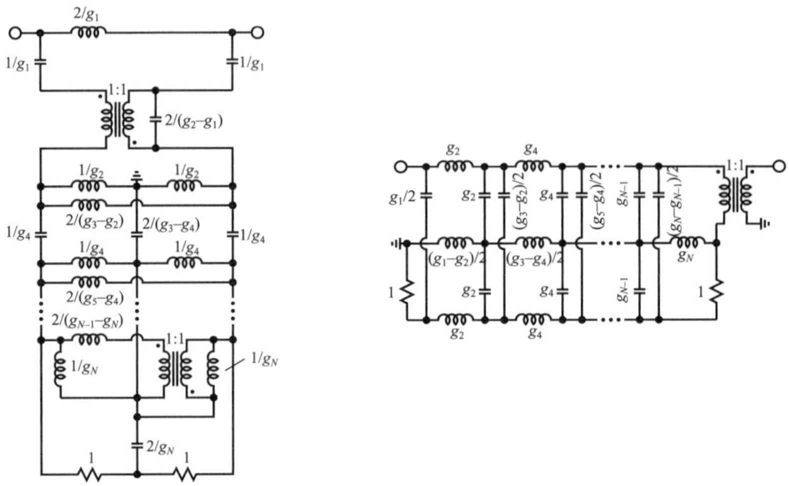

(a) (b)

Figure 11.13 (a) Reflectionless filter topology with two substitutions for negative elements, suitable for a Chebyshev type-II response having ripple factor smaller than threshold given in (11.32). (b) Reflectionless filter topology suitable for an all-pole frequency response such as Chebyshev type I.

pick up the inverse response (e.g., Chebyshev type I) by coupling out this energy through a balun or transformer, as indicated in Figure 11.13(b). Inasmuch as the balun transformer is ideal, and the former output is appropriately terminated, this topology is likewise reflectionless from both remaining ports. For this example, we have applied a frequency-inverting (high-pass) transformation to the circuit first so that the inverse response is still low-pass. No negative-element substitutions were made in this instance, but could be if they were needed.

11.4 DISTRIBUTED FILTERS

Many common forms of transmission-line filters are based on lumped-element equivalents to which Richard's transformation and various transmission-line identities have been applied (as was illustrated initially in Figures 5.18 and 5.19). As this procedure generally results in commensurate-line networks with periodic response, those deriving from a low-pass lumped-element prototype effectively yield a band-stop response, while those deriving from a high-pass prototype yield a band-pass response, with the lines being a quarter-wavelength in the center of the stopband or passband, respectively. However, there are other methods of generating filters at high frequency, which will be discussed in this section.

11.4.1 Stepped-Impedance Filters

While Richard's transformation yields stub-only topologies (which can then be separated by cascade lines with a port extension and the help of Kuroda's identity), an approximate low-frequency equivalence between simple lumped elements and cascade transmission lines can also be established. Consider the $ABCD$-parameters of a lossless transmission line as given in Table 4.1,

$$ABCD = \begin{pmatrix} \cos(\beta l) & jZ_k \sin(\beta l) \\ jY_k \sin(\beta l) & \cos(\beta l) \end{pmatrix} \tag{11.34}$$

where $Z_k = 1/Y_k$ is the characteristic impedance of the transmission line. Let us now take the limit when the frequency is low ($\beta \to 0$) and when Z_k is either very large or very small (compared to Z_0),

$$\lim_{\substack{Z_k \to \infty \\ \beta l \to 0}} ABCD = \begin{pmatrix} 1 & jZ_k \beta l \\ 0 & 1 \end{pmatrix} \tag{11.35a}$$

$$\lim_{\substack{Z_k \to 0 \\ \beta l \to 0}} ABCD = \begin{pmatrix} 1 & 0 \\ jY_k \beta l & 1 \end{pmatrix} \tag{11.35b}$$

The first case is the same as a series inductor having value $L_k = Z_k \beta l/\omega$ and the second is equivalent to a shunt capacitor having value $C_k = Y_k \beta l/\omega$. To illustrate, the equivalent series and shunt impedance of both kinds of networks is compared versus frequency in Figure 11.14.

To make a low-pass filter, we may take a lumped-element ladder prototype, as in Figure 11.8(a), and replace each element with its approximately equivalent cascade transmission line. The resulting structure has alternating high and low-impedance cascade transmission lines and is known as a *stepped-impedance filter*. It is generally best to maximize the ratio between high and low impedances, such that one usually takes the largest and smallest characteristic impedance that can practically be fabricated (and avoids overmoding), then evaluating the line length, βl, to match the desired inductance or capacitance,

$$\theta_k = \beta l_k = g_k \begin{cases} Z_0/Z_{hi} & \text{series inductor} \\ Z_{lo}/Z_0 & \text{shunt capacitor} \end{cases} \tag{11.36}$$

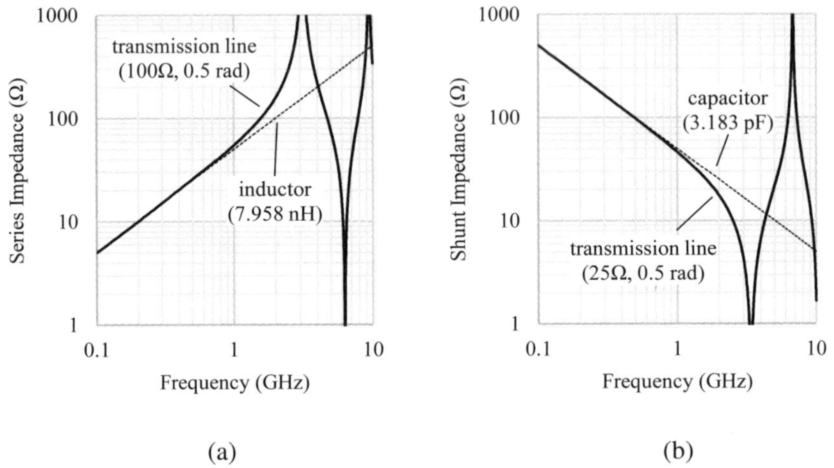

Figure 11.14 Comparison of (a) a high-impedance transmission line against a series inductor, and (b) a low-impedance transmission line against a shunt capacitor. The line electrical lengths (0.5 rad) are evaluated at 1 GHz.

Note that we have implicitly evaluated each line length at $\omega = \omega_c$, the corner frequency, for best results under the above approximations. An example of a fifth-order filter using this technique is shown in Figure 11.15. Note that since the line lengths are not commensurate in this case, the resulting low-pass filter does not have a periodic response (nonetheless, it will still have parasitic passbands at sufficiently high frequency where the small-length approximation fails).

11.4.2 Coupled Resonators

In modern times, engineers have begun to take a somewhat different view of the way filters are constructed, not as a network of fundamental elements like inductors, capacitors, or transmission lines, but as an ensemble of generic resonators which are coupled to one another by circuits approximating ideal impedance inverters. Since resonators can be constructed using almost any kind of element (lumped, transmission lines, waveguides, general cavities, dielectric bricks); this facilitates the translation of a generic filter prototype into an implementation using almost any technology that is desired.

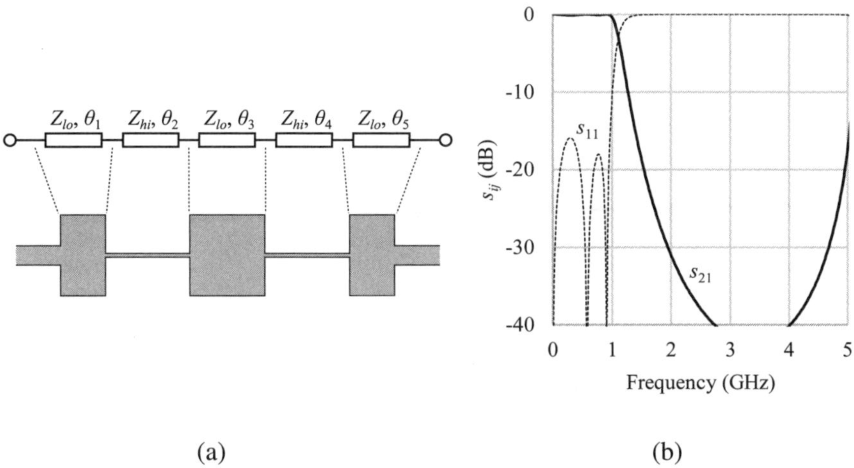

Figure 11.15 (a) Schematic and layout of a stepped-impedance low-pass filter. (b) Simulated response when line lengths are based on a Chebyshev prototype.

To illustrate, let us start from the simple ladder topology in Figure 11.8(a) having only three elements ($N = 3$). To this, we add a chain of admittance inverters as shown in Figure 11.16(a). Provided that the normalizing admittance of these inverters is equal to the system characteristic admittance ($J = Y_0$, or unit-value in this example), these have no affect on the amplitude response. The overall phase may change depending on how many we add, but that is considered unimportant in this context. Repeated application of the inverter-dual identity from Figure 5.11 yields a circuit containing only inverters and capacitors, as shown in Figure 11.16(b). Note that the capacitors in this case have normalized values equal to the prototype parameters from the original ladder. Alternatively, we could have let the capacitors all have the same normalized value, say, unity, and modified the inverter parameters to preserve the response, specifically by dividing each by a factor equal to the square root of the product of the adjacent prototype parameter values, as indicated in Figure 11.16(c).

Conceptually, each of the shunt capacitors may be considered a kind of resonator with a resonant frequency of zero[3] [10]. We may therefore generalize

3 Alternatively, one could apply a band-pass transformation first to the lumped-element ladder in Figure 11.16(a), in which case the capacitors in Figures 11.16(b, c) would become parallel LC resonators in accordance with Table 5.1. As shown here, the circuit represents a low-pass prototype.

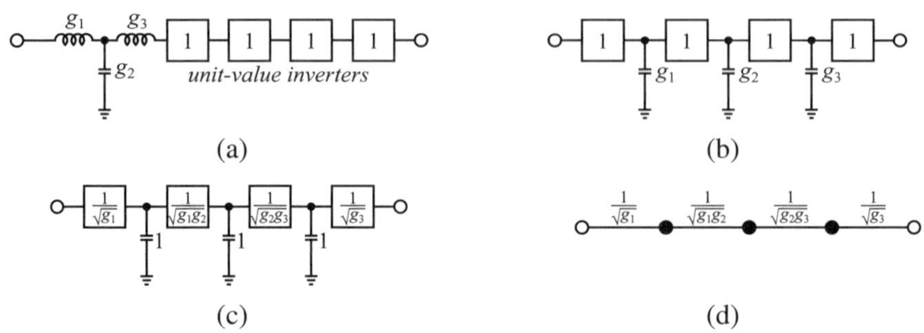

Figure 11.16 Evolution from a lumped-element ladder topology to a coupled-resonator topology. (a) Lumped-element ladder with $N + 1$ admittance inverters of unit normalized value. (b) Repeated application of the inverter-dual identity. (c) Renormalization of resonators to unit value. (d) Final coupling-routing diagram.

this schematic across a wide variety of technologies by redrawing the capacitors as *resonant nodes* using a simple filled circle as shown in Figure 11.16(d). The ports, in contrast, become *nonresonant nodes*, and the nodes are connected by lines representing the admittance inverters, labeled with their normalized admittance parameters. The normalized inverter parameters are referred to as the *couplings* between the resonators, and this graphical representation is called a *coupling-routing diagram*.

Recall that the inverter-dual identity applies for inverters having positive and negative phase. Thus, any number of couplings in this topology, in any combination, may be negated and the amplitude response of the filter will remain the same. However, the phase response will be different, so the same general statement will not be true of more complex topologies in which the couplings do not form a simple linear array.

Although we already know how this circuit behaves from its origins as a lumped-element ladder, it is useful to work through the analysis of its response, as we will soon generalize this to more complex coupling-routing diagrams which do not derive from previously studied lumped-element topologies. Consider the *nodal admittance matrix* for this structure, essentially the admittance matrix where every node is considered a potential port. By extension of the admittance parameters for a two-port pi-network shown in Table 4.1, the diagonal elements of the nodal admittance matrix are simply the sums of the admittances connected to each node, while the off-diagonal elements are the negatives of the admittances connected between nodes.

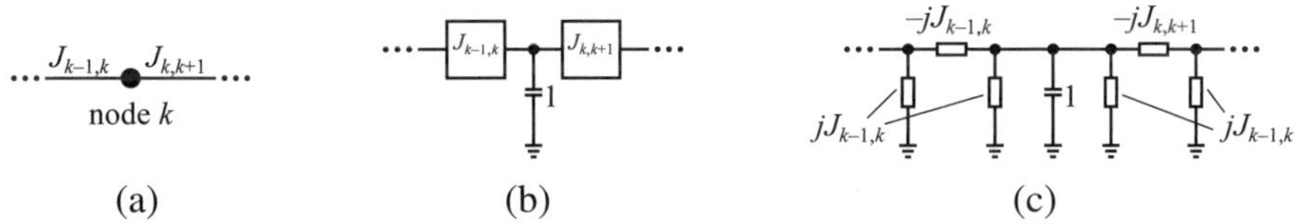

Figure 11.17 (a) Detail of node k in a sequential coupling-routing diagram. (b) Low-pass prototype representation using a unit-valued capacitor as the resonator and ideal admittance inverters with values $J_{k-1,k}$ and $J_{k,k+1}$. (c) Expansion of each admittance inverter into a pi-network of constant admittances.

As shown in Figure 11.17, the vicinity around a particular node, k, may be drawn as a unit-valued capacitor for the resonator along with the adjoining ideal admittance inverters. Those admittance inverters have been expanded here using pi-networks of constant admittance elements, as first suggested in Figure 4.27(b). Note that since each inverter comprises series and shunt elements with the same value and opposite sign, they add nothing to the diagonal elements of the nodal admittance matrix, while the couplings themselves form the off-diagonal elements. For the entire third-order filter, then,

$$\mathbf{Y} = \begin{pmatrix} 1 & jJ_{0,1} & 0 & 0 & 0 \\ jJ_{0,1} & j\omega & jJ_{1,2} & 0 & 0 \\ 0 & jJ_{1,2} & j\omega & jJ_{2,3} & 0 \\ 0 & 0 & jJ_{2,3} & j\omega & jJ_{3,4} \\ 0 & 0 & 0 & jJ_{3,4} & 1 \end{pmatrix} = \mathbf{G} + \mathbf{U} + j\mathbf{M} \tag{11.37}$$

where

$$\mathbf{G} = \begin{pmatrix} 1 & 0 & 0 & 0 & 0 \\ 0 & 0 & 0 & 0 & 0 \\ 0 & 0 & 0 & 0 & 0 \\ 0 & 0 & 0 & 0 & 0 \\ 0 & 0 & 0 & 0 & 1 \end{pmatrix} \tag{11.38a}$$

$$\mathbf{U} = \begin{pmatrix} 0 & 0 & 0 & 0 & 0 \\ 0 & j\omega & 0 & 0 & 0 \\ 0 & 0 & j\omega & 0 & 0 \\ 0 & 0 & 0 & j\omega & 0 \\ 0 & 0 & 0 & 0 & 0 \end{pmatrix} \tag{11.38b}$$

$$\mathbf{M} = \begin{pmatrix} 0 & J_{0,1} & 0 & 0 & 0 \\ J_{0,1} & 0 & J_{1,2} & 0 & 0 \\ 0 & J_{1,2} & 0 & J_{2,3} & 0 \\ 0 & 0 & J_{2,3} & 0 & J_{3,4} \\ 0 & 0 & 0 & J_{3,4} & 0 \end{pmatrix} \tag{11.38c}$$

representing the matrix of termination conductances (**G**), resonant-node reactances (**U**), and the *coupling matrix* (**M**), respectively. By convention, the input port is node number 0, while the output port is node $N + 1$ (in this case, node 4). The couplings to these nodes are known as *external couplings*.[4] Since **G** and **U** are known a priori, the coupling matrix, **M**, alone is sufficient to completely describe the filter response.

Let us generalize now for an arbitrary filter of order N, and inject a test current of 2A into the network at the input node 0 (and no currents anywhere else). By the definition of the admittance matrix, we have

$$\mathbf{v} = \begin{pmatrix} v_0 \\ v_1 \\ \vdots \\ v_{N+1} \end{pmatrix} = \mathbf{Y}^{-1}\mathbf{i} = \mathbf{Y}^{-1} \begin{pmatrix} 2 \\ 0 \\ \vdots \\ 0 \end{pmatrix} \tag{11.39}$$

Note that, in this scenario, the total admittance from node 0 to ground is given by $2/v_0$. This includes both the input admittance of the filter, Y_{in}, as well as the input termination resistor (which has unit value). Subtracting the latter, we have

$$Y_{in} = \frac{2}{v_0} - 1 \tag{11.40}$$

The input reflection coefficient at port 1 (node 0) is therefore given by

$$s_{11} = \frac{1 - Y_{in}}{1 + Y_{in}} = \frac{2 - \frac{2}{v_0}}{\frac{2}{v_0}} = v_0 - 1 \tag{11.41}$$

Since v_0 embodies the combined voltage of both the incident and reflected waves, we further have

$$v_0 = v_{in}\left(1 + s_{11}\right) \tag{11.42a}$$

4 In some texts, the external couplings are not included in the main coupling matrix, which then has only $N \times N$ elements. In those contexts, the larger matrix shown here would more specifically be called the $N + 2$ *coupling matrix*, for clarity.

$$\therefore v_{in} = \frac{v_0}{1 + s_{11}} = 1 \tag{11.42b}$$

Thus, the transmission coefficient from port 1 to port 2 (node 0 to the node $N + 1$) is given by

$$s_{21} = \frac{v_{N+1}}{v_{in}} = v_{N+1} \tag{11.43}$$

More directly, in terms of the matrix $\mathbf{Y}$, we may simply write

$$s_{11} = 2\frac{C_{00}}{\det \mathbf{Y}} - 1 \tag{11.44a}$$

$$s_{21} = 2\frac{C_{N+1,0}}{\det \mathbf{Y}} \tag{11.44b}$$

where C_{kl} is the *cofactor* or *signed minor* of matrix $\mathbf{Y}$.

11.4.3 Susceptance Slope and Band-Pass Scaling

This derivation of coupled-resonator technology has made extensive use of normalized quantities. If the resonators comprise simple unit-value (1 Farad) capacitors as shown, and the inter-resonator couplings are based on the prototype parameters (g_k) taken from canonical filter tables, then the circuits shown, given 1Ω terminations, will yield a low-pass filter response with a cutoff frequency of 1 rad/s. To implement a band-pass filter, we must replace the simple capacitors with actual resonators which are resonant at the desired frequency. We must also scale the resonator couplings to account for the desired fractional bandwidth, and the rate at which the immittance of that particular resonator changes around its resonant frequency.

Consider three common types of resonators: a parallel LC circuit, a shorted transmission line with electrical length θ, and an open transmission line with electrical length θ', all resonant at the same frequency. The susceptance of each of these three resonators types is listed here:

$$B(\omega) = \begin{cases} \omega C - \frac{1}{\omega L} & \text{lumped } LC \\ -Y_0 \cot\theta & \text{shorted quarter-wave line} \\ Y_0 \tan\theta' & \text{open half-wave line} \end{cases} \tag{11.45}$$

and is plotted in Figure 11.18(a). In this case, we have assumed that $\sqrt{L/C} = Z_0 = 50\Omega$, and that both transmission-line resonators have characteristic impedance Z_0. Although all three have the same resonant frequency of 1 GHz, the width of the

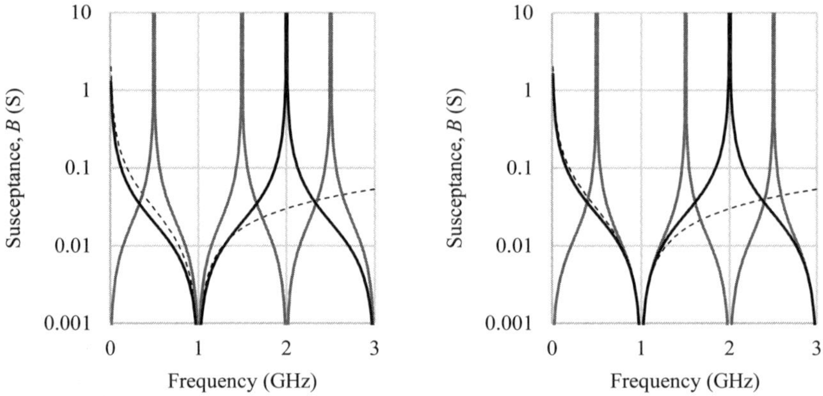

(a) (b)

Figure 11.18 Susceptance of lumped-element LC (dashed line), quarter-wave transmission-line (solid black), and half-wave transmission-line (solid gray) resonators, with resonant frequency at 1 GHz. (a) $\sqrt{L/C} = Z_0 = 50\Omega$, and both transmission-line resonators have characteristic impedance Z_0. (b) $\sqrt{L/C} = Z_0 = 50\Omega$, the quarter-wave resonator has characteristic impedance $Z_0\pi/4$, and the half-wave resonator has characteristic impedance $Z_0\pi/2$.

susceptance curve around that resonance is different in all three cases. This would directly affect the achieved bandwidth of a filter made with these resonators, and must be accounted for. We may calculate the rate of change of the susceptance curve around the resonant frequency, ω_0, for the lumped-element resonator

$$\left.\frac{dB}{d\omega}\right|_{\omega=\omega_0} = C + \frac{1}{\omega_0^2 L} = \frac{2}{\omega_0^2 L} = 2\frac{Y_0}{\omega_0} \tag{11.46}$$

and for the quarter-wave transmission line,

$$\left.\frac{dB}{d\omega}\right|_{\omega=\omega_0} = Y_0 \csc^2(\pi/2) \left.\frac{d\theta}{d\omega}\right|_{\omega=\omega_0} = Y_0\frac{\pi}{2\omega_0} \tag{11.47}$$

Note that the susceptance slopes of the two resonators differ by a factor of $\pi/4$. To normalize out this condition, we define a resonator coefficient known as the *susceptance slope parameter* [11], labeled b, where

$$b = \frac{\omega}{2}\left.\frac{dB}{d\omega}\right|_{\omega=\omega_0} \tag{11.48}$$

The susceptance slope parameter has units of siemens, and as shown above is equal to Y_0 for the lumped-element resonator with $\sqrt{L/C} = Z_0$. For the quarter-wave transmission line, it is $b = Y_0\pi/4$, and a similar analysis for the half-wave resonator would show in that case that $b = Y_0\pi/2$.

One could account for this variation by scaling the transmission-line resonator impedances, as shown in Figure 11.18(b). Thus, all three types of resonators would have the same slope and would produce filters with the same relative bandwidth. The impedances of some types of resonators, however — cavity or dielectric slab, for example — are not so easily modified if the basic geometry is fixed (it might, for example, require fine control of the slab's dielectric constant). It is therefore conventional to absorb the susceptance slope parameter into the coupling coefficients themselves. For a linear resonator array based on a list of prototype parameters, g_k, we have

$$J_{k,k+1} = \begin{cases} \sqrt{\dfrac{Y_0 b\Delta}{g_0 g_1}} & k = 0 \\[2ex] \dfrac{b\Delta}{\sqrt{g_k g_{k+1}}} & 1 \leq k < N \\[2ex] \sqrt{\dfrac{Y_0 b\Delta}{g_N g_{N+1}}} & k = N \end{cases} \tag{11.49}$$

where Δ is the desired fractional bandwidth of the filter. Note that g_0 and g_{N+1} represent the normalized termination immittances and may implicitly be equal to 1 in most cases, especially when N is odd.

11.4.4 Edge-Coupled Band-Pass Filters

A broad class of microwave band-pass filter topologies utilizes a series of sequentially coupled half-wave resonators, as indicated schematically in Figure 11.19(a). Unique implementations of this basic concept are distinguished from one another primarily by the resonator technology used (transmission-line or waveguide) and by their method of coupling. Several common forms include capacitively end-coupled transmission-line resonators (Figure 11.19(b)), edge-coupled transmission-line resonators (Figure 11.19(c)), and inductive-iris coupled waveguide cavities (Figure 11.19(d)).

Let us take the edge-coupled band-pass filter in Figure 11.19(c) as an example. In order to relate it to the conceptual prototype in Figure 11.19(a), we must

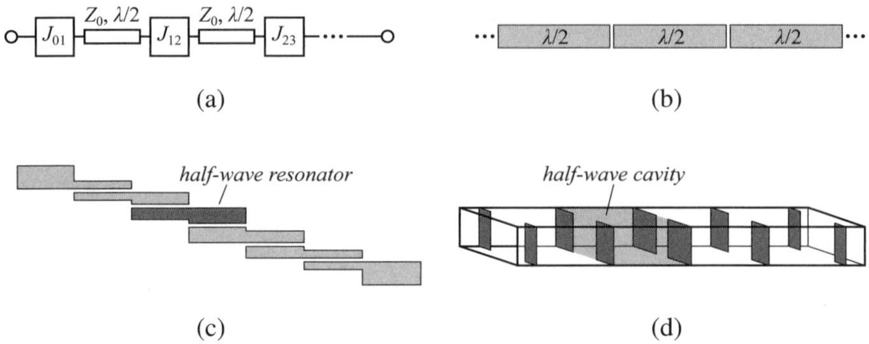

Figure 11.19 Several common forms of coupled half-wave resonator filters. (a) General schematic diagram. (b) Capacitively end-coupled transmission lines. (c) Edge-coupled transmission lines. (d) Waveguide cavities coupled by inductive irises.

Figure 11.20 A coupled-line section and its approximate equivalent circuit at mid-band.

first derive an approximate equivalent circuit for a single coupled-line section. The equivalent circuit that we wish to use is shown in Figure 11.20, and is valid near the center frequency of the filter when the coupled lines are a quarter wavelength long ($\theta = \pi/2$). To determine the even and odd-mode characteristic impedances of the coupled lines, we must compare the network parameters of both circuits. The impedance parameters of the coupled lines were given in (4.115) (terms associated with the unused ports may simply be omitted since the impedance parameters treat these as open-circuit terminations anyway),

$$z_{11} = z_{22} = -\tfrac{j}{2}\left(Z_e + Z_o\right)\cot\theta \tag{11.50a}$$

$$z_{12} = z_{21} = -\tfrac{j}{2}\left(Z_e - Z_o\right)\csc\theta \tag{11.50b}$$

$$\therefore \mathbf{Z} = \frac{1}{j2\sin\theta}\begin{pmatrix}(Z_e + Z_o)\cos\theta & Z_e - Z_o \\ Z_e - Z_o & (Z_e + Z_o)\cos\theta\end{pmatrix} \tag{11.50c}$$

Near mid-band, where $\theta \approx \pi/2$, we have $\cos\theta \approx 0$ and

$$\det\mathbf{Z} \approx \frac{\left(Z_e - Z_o\right)^2}{4\sin^2\theta} = -z_{21}^2 \tag{11.51}$$

The $ABCD$-parameters of the coupled lines are then given by the conversion formula,

$$\begin{pmatrix} A & B \\ C & D \end{pmatrix} = \frac{1}{z_{21}} \begin{pmatrix} z_{11} & \det \mathbf{Z} \\ 1 & z_{22} \end{pmatrix} \approx \begin{pmatrix} \frac{Z_e+Z_o}{Z_e-Z_o} \cos\theta & j\frac{Z_e-Z_o}{2} \\ j\frac{2}{Z_e-Z_o} & \frac{Z_e+Z_o}{Z_e-Z_o} \cos\theta \end{pmatrix} \tag{11.52}$$

The $ABCD$-parameters of the approximate equivalent circuit are given by cascade multiplication,

$$\begin{pmatrix} A & B \\ C & D \end{pmatrix} = \begin{pmatrix} \cos\theta & jZ_0 \sin\theta \\ jY_0 \sin\theta & \cos\theta \end{pmatrix} \cdot \begin{pmatrix} 0 & -jJ^{-1} \\ -jJ & 0 \end{pmatrix} \begin{pmatrix} \cos\theta & jZ_0 \sin\theta \\ jY_0 \sin\theta & \cos\theta \end{pmatrix} \tag{11.53a}$$

$$= \begin{pmatrix} \left(J^{-1}Y_0 + JZ_0\right) \sin\theta\cos\theta & jJZ_0^2 \sin^2\theta - jJ^{-1}\cos^2\theta \\ jJ^{-1}Y_0^2 \sin^2\theta - jJ\cos^2\theta & \left(J^{-1}Y_0 + JZ_0\right) \sin\theta\cos\theta \end{pmatrix} \tag{11.53b}$$

$$\approx \begin{pmatrix} \left(J^{-1}Y_0 + JZ_0\right) \cos\theta & jJZ_0^2 \\ jJ^{-1}Y_0^2 & \left(J^{-1}Y_0 + JZ_0\right) \cos\theta \end{pmatrix} \tag{11.53c}$$

Comparing (11.52) and (11.53c), we have

$$\frac{Z_e + Z_o}{Z_e - Z_o} = J^{-1}Y_0 + JZ_0 \tag{11.54a}$$

$$\frac{Z_e - Z_o}{2} = JZ_0^2 \tag{11.54b}$$

which can then be solved for Z_e and Z_o,

$$Z_e = Z_0 \left(1 + JZ_0 + J^2 Z_0^2\right) \tag{11.55a}$$

$$Z_o = Z_0 \left(1 - JZ_0 + J^2 Z_0^2\right) \tag{11.55b}$$

Thus, to design an edge-coupled band-pass filter, we may first select an order and filter type (e.g., Butterworth or Chebyshev), calculate the prototype parameter values using formulas such as (11.24) or (11.25), determine the resonator couplings from (11.49), and finally the coupled line impedances using (11.55).

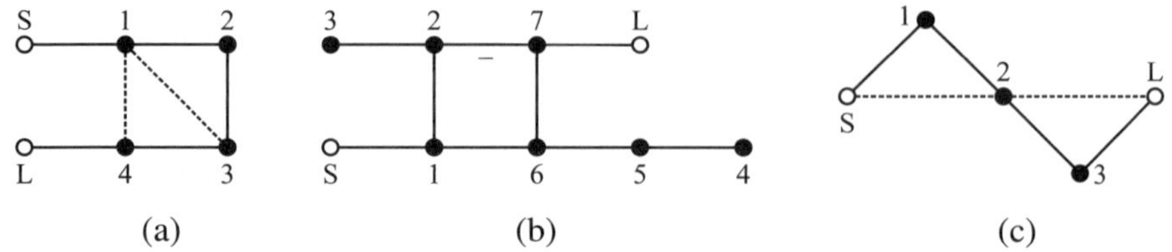

Figure 11.21 A variety of coupling-routing diagrams besides the linear array. (a) Canonical folded topology. (b) Cul-de-sac configuration. (c) Conjoined trisections. Resonant nodes are labeled with a number, while S and L denote the source and load, respectively.

11.4.5 General Coupling-Routing Topologies

There are many more possible ways to arrange and couple resonators than a simple linear array. To give an idea of just how general this theory can be, coupling-routing diagrams for several possibilities are shown in Figure 11.21. One very popular choice is to fold the linear array in half, as shown in Figure 11.21(a), then include additional couplings between nonadjacent nodes, such as between nodes 1–3 and 1–4. By convention, couplings between sequential nodes are called *mainline couplings* and are drawn in the coupling-routing diagram with a solid line, while those between nonadjacent nodes are called *cross couplings* and are drawn with a dashed line. The mainline couplings would appear one position away from the main diagonal of the coupling matrix, whereas the cross couplings are represented by nonzero terms further off the main diagonal. This distinction is merely notional, as a mainline coupling can be turned into a cross coupling, and vice-versa, simply by renumbering the nodes. Whereas the sequential array typically results in all-pole filter responses, the addition of nonadjacent couplings tends to introduce transmission zeros, allowing Chebyshev type-II and elliptic or quasi-elliptic filters to be realized.

In general, couplings can be either positive or negative, typically associated with magnetic coupling and electric coupling (or inductive coupling and capacitive coupling), but reversing this convention yields the same results. When the two are mixed, as in the cul-de-sac configuration of Figure 11.21(b), one can identify which couplings have the opposite-of-normal coupling sign by labeling the associated edge with a minus sign, as shown. There is also no intrinsic requirement that the source and load be coupled to the filter by a single coupling, either, as suggested by the cascaded trisection filter in Figure 11.21(c).

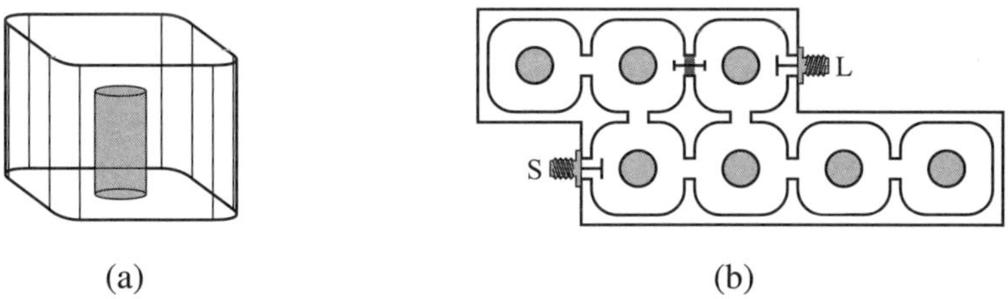

Figure 11.22 (a) Coaxial resonator geometry. (b) Example implementation of cul-de-sac topology from Figure 11.21(b), using apertures in the sidewalls for positive couplings, and an electric field probe for the negative coupling.

The range of topologies and desirable responses is virtually endless, and as such, optimal solutions for filters of this type do not lend themselves well to tabulation, as did the linear array. A comprehensive review is beyond the scope of this book; however, numerous methods have been developed to synthesize or numerically optimize the coupling matrices that implement filters subject to all kinds of specifications [4]. Once synthesized, the matrix may be modified using *similarity transforms* (also known as *rotations*) to nullify certain cross-coupling terms, yielding a topology which is more convenient to implement [12]. Mathematically, this works because the solutions in (11.44) depend only on the determinant of the $N \times N$ matrix of couplings (excluding the rows and columns associated with the terminations). Thus, any multiplication by a transformation matrix that has a determinant of unity has no effect on the reflection or transmission response of the filter.

11.4.6 Empirical Extraction of Couplings

As we have seen, resonators can take on many forms, not all of which lend themselves ideally to theoretical analysis, despite their very practical advantages. One particularly popular choice is the coaxial resonator shown in Figure 11.22(a), where a large cavity, which needs not necessarily be circular, is loaded with a conducting post, like the center conductor of a coaxial line. The post often does not extend the full height of the resonator, forming a capacitive gap at the end. Even without an equivalent circuit, it is not difficult to convince ourselves that such a structure should have resonant frequencies, and that coupling between two such resonators could take the form of apertures in the sidewall, or electrical field probes extending into the cavity. An illustration of what a filter might look like using such

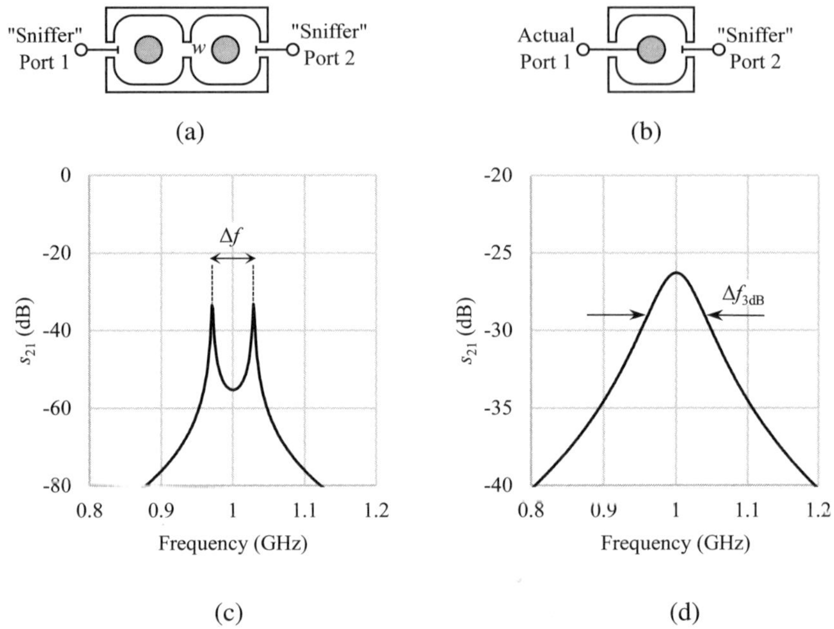

Figure 11.23 Extraction of normalized couplings. (a) Test circuit for inter-resonator couplings. (b) Test circuit for source/load couplings. (c) Measurement of inter-resonator coupling. (d) Measurement of source/load coupling.

resonators is shown in Figure 11.22(b), in this case utilizing the cul-de-sac coupling-routing diagram from Figure 11.21(b). All the resonators are identical, only the coupling features vary, in this case, inductive irises in the sidewalls for positive couplings, and a capacitive field probe for the negative coupling.

How do we determine the coupling coefficient between two resonators whose equivalent circuit is unknown, and for a coupling mechanism that cannot easily be associated with terminal parameters upon which to define a normalizing admittance, J? With modern CAD tools, one may answer that question by simulating standard test circuits and observing their behavior for a range of geometric variations.

Consider the pair of resonators coupled through an iris in the sidewall as shown in Figure 11.23(a). We wish to extract the coupling coefficient as a function of the width, w, of the sidewall aperture. To do so, we add a pair of sniffer ports which are only very loosely coupled to the structure, not enough to affect it, but enough that we can observe the resonance frequencies. The result is shown in Figure 11.23(c). Note that although both resonators were tuned to resonate at 1 GHz, their interaction has caused the resonant peak to split into two. It can be shown that the separation between these two resonances is proportional to the normalized coupling coefficient [13]. That is,

$$m_{12} = \frac{J_{12}}{b\Delta} = \frac{\Delta f}{BW} \tag{11.56}$$

where BW is the *ripple bandwidth* of the filter, or the width of the passband in the desired frequency response (whereas Δ is the fractional ripple bandwidth). As a rule of thumb, to ensure that the sniffer ports do not contaminate the results, the peaks of the two resonances in this test should be kept below -30 dB.

To extract the source and load coupling, we use a single resonator with the proposed port coupling mechanism on one side and again a loosely coupled sniffer port on the other side as shown in Figure 11.23(b). In this case, the response should look something like that shown in Figure 11.23(d). This time, we measure the 3-dB bandwidth around the peak. This can be associated with the external quality factor, Q_e, of the resonator. In most cases, the unloaded Q_U will be much larger than the loaded Q_L, so that we may assume $Q_e = Q_L$, and

$$Q_e = \frac{f_0}{\Delta f_{\text{3dB}}} = \frac{f_0}{BW \cdot m_{01}^2} \tag{11.57}$$

where m_{01} is again the normalized coupling coefficient, given by

$$m_{01} = \frac{J_{01}}{\sqrt{b\Delta}} = \sqrt{\frac{\Delta f_{\text{3dB}}}{BW}} \tag{11.58}$$

Once again, to avoid altering the result, the sniffer port should be sufficiently loosely coupled to keep the peak below about $25 - 30$ dB [13].

11.5 WAVEGUIDE FILTERS

Filters in waveguide are typically based either upon coupled-resonator theory or upon approximate transmission-line models. In the first case, the half-wave resonator topology is a particularly common choice, with many varieties having similar equivalent circuits as their basis and differing primarily by their method of construction.

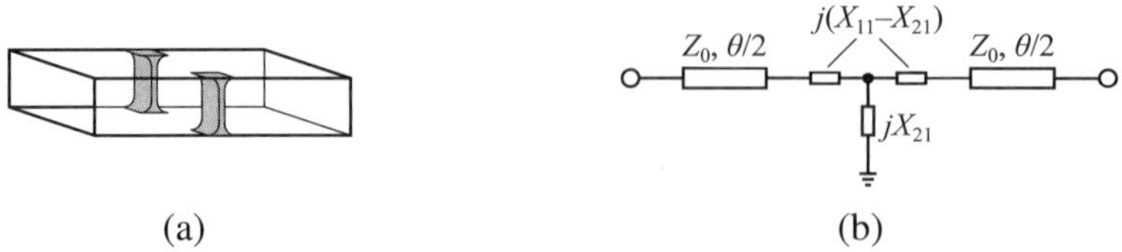

Figure 11.24 (a) Thick iris with radiused corners such as one might get in a machined split block. (b) Circuit model with port extensions needed to create an impedance inverter.

11.5.1 Half-Wave Resonator Waveguide Filters

Recall the iris-coupled half-wave resonator filter first shown in Figure 11.19(d). Table 7.3 showed that such an iris has a shunt inductor as its equivalent circuit. This may then be converted into an impedance inverter with the proper port extensions as shown in Figure 4.27(q). For a positive inductance, the port extensions will typically be negative, and are easily subtracted from the half-wavelength resonators on either side of the iris.

Although electromagnetic theory-based circuit models exist for irises that are very thin [14], thicker irises such as one might get in a machined split block, shown with radiused corners in Figure 11.24(a), are perhaps easier dealt with by computer-aided simulation [15]. In this case, a somewhat more sophisticated inverter model is needed, shown in Figure 11.24(b). The tee-network in the middle of this circuit represents the equivalent z-parameters of the discontinuity as determined by the simulation. To make it an impedance inverter, we must add port extensions [5] given by

$$\theta = -\tan^{-1}\left(\frac{X_{11}+X_{21}}{Z_0}\right) - \tan^{-1}\left(\frac{X_{11}-X_{21}}{Z_0}\right) \tag{11.59}$$

The normalizing impedance of the inverter is then

$$K = Z_0\left|\tan\left(\frac{\theta}{2} + \tan^{-1}\left(\frac{X_{11}-X_{21}}{Z_0}\right)\right)\right| \tag{11.60}$$

In most cases, the simulator will natively produce scattering parameters as the primary result, which can then be converted to effective impedance parameters, jX_{11} and jX_{21}, given a choice of normalizing impedance, Z_0. The dominant mode wave impedance, Z_{TE}, is a natural and common choice, although in reality this selection is arbitrary (note that (11.59) and (11.60) depend only on the normalized values anyway).

(a) (b)

Figure 11.25 Variants of the cascaded half-wave resonator waveguide filter. (a) With thin conducting vanes. (b) With posts.

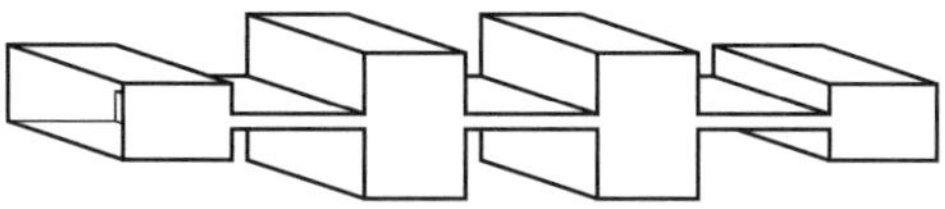

Figure 11.26 Stepped-height waveguide low-pass filter.

Other geometries having a similar basis in the linear half-wave resonator array are shown in Figures 11.25(a, b), where the inverters are realized using thin vanes or a fence row of posts as the coupling apertures, respectively. The inverter model in Figure 11.24(b) and its associated equations in (11.59) and (11.60) are quite general, and may be applied to these structures as well. The former geometry is especially handy in the microwave lab, because a crude, adjustable filter can be quickly assembled from readily available materials simply by pinching thin strips of metal between the two halves of an existing split block. Two strips are sufficient for a low-order filter, and rough tuning can be accomplished by adjusting the spacing between them.

11.5.2 Stepped-Height Waveguide Filter

Although no rectangular waveguide filter can truly be considered low-pass, since the cutoff of the dominant mode will always impart a finite lower limit to the passband, the stepped-impedance filter in Figure 11.15, which had a low-pass characteristic in transmission lines, can be emulated in rectangular waveguide to create an upper frequency limit as well. If the width of the waveguide sections are kept constant, then the cutoff frequency of the dominant mode (and hence the lower passband edge) will be uniform throughout the filter. Instead we vary the relative height of the waveguide, as shown in Figure 11.26.

Strictly speaking, the wave impedance, Z_{TE}, of the TE_{10} mode is the same in each section regardless of this height, and it should not therefore be considered a

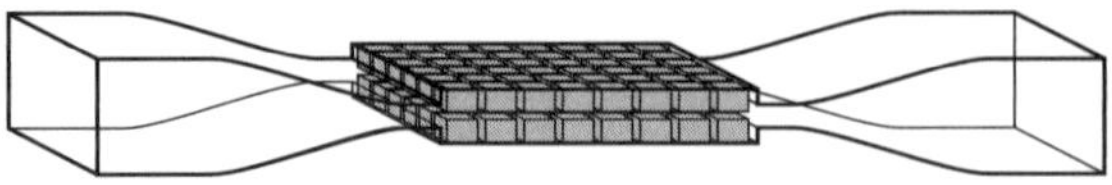

Figure 11.27 Conceptual drawing of a waffle-iron, low-pass waveguide filter.

substitute for the characteristic impedance of the transmission-line sections, but the step changes in height have a similar effect in terms of the magnitude and phase of the reflections at each discontinuity as the changes in transmission-line impedance do. Typically, one would select a minimum height based on losses and/or fabrication limitations, while the maximum would be determined by overmoding. Further, a large height step can impart significant phase to a passing waveform, which must be accounted for in the length of the adjoining sections.

11.5.3 Waffle-Iron Low-Pass Filter

The stopband of a stepped-height waveguide filter will generally only extend as high in frequency as the cutoff of the first higher-order mode. Mode conversion at the discontinuities invalidates the analogy to a simple transmission-line model. In order to construct a low-pass waveguide filter with a stopband that extends beyond its single-mode frequency range, we must consider and control the complex modal properties of the waveguide at these higher frequencies.

To that end, we first narrow the height of the waveguide at both ports using a stepped transformer or taper, as shown in Figure 11.27. The dominant mode is thus impedance-matched into the reduced height middle section, while the higher-order modes capable of propagating over a very wide range of frequencies is limited to the TE_{m0} modes, modes that can be decomposed into a superposition of finitely many TEM waves traveling in different directions. We next add a two-dimensional array of "bosses" in the top and bottom of the waveguide throughout the middle section, a bandgap-producing, periodic structure known as a *waffle-iron* due to its similarity in appearance to the kitchen appliance. So long as the periodicity of the bosses is shorter than half a free-space wavelength, the propagation of the elemental TEM waves that make up all TE_{m0} modes is isotropic, and thus independent of mode number [5]. This allows the waffle-iron filter to exhibit a stopband well beyond the normal operating range of the waveguide.

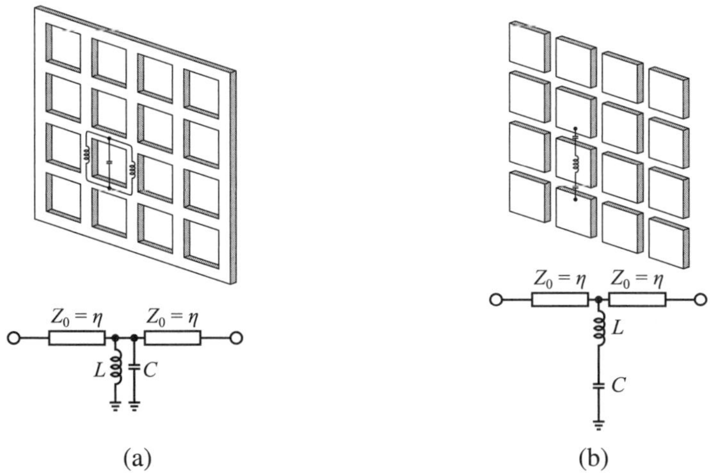

Figure 11.28 Frequency-selective surfaces. (a) Band-pass mesh. (b) Band-stop grid.

11.6 FREQUENCY-SELECTIVE SURFACES

In the quasioptical domain, filters take on the form of *frequency-selective surfaces*, two-dimensional structures perpendicular to the direction of propagation of a plane wave or Gaussian beam, and whose interactions with those wavefronts can be modelled as an element in an electromagnetic circuit. Two basic forms of frequency-selective surfaces are shown in Figure 11.28. In both cases, the space ahead and behind the surface along the direction of propagation may be modeled as a transmission line with characteristic impedance given by the free-space wave impedance, η. We assume for simplicity that the structures are electrically thin, so that their effect on the passing wavefront may be modeled as a shunt element across the transmission line [16].

In Figure 11.28(a), the surface is a conducting mesh of closed loops which may be approximately modeled over a finite range of frequencies as a parallel LC resonator. It therefore has a first-order band-pass response. The inductance of that resonator may be associated with the vertical current path around each mesh, as indicated in the figure, while the capacitance may be associated with the electrical coupling between the horizontal members.

Note that we have implicitly assumed in this drawing that the incident polarization is vertical. Because the structure is identical under a 90° rotation, the same equivalent circuit applies to horizontally polarized waves. Further, since any polarization may be considered a superposition of horizontal and vertical polarization components, the electrical behavior of this surface is fully independent of polarization.[5]

In Figure 11.28(b), the surface comprises an array of disconnected, metallic islands. In practice, a substrate or membrane will be needed for mechanical support of such a structure, but we assume here that any such supports are thin enough to be electrically neglected. In this case, the circuit model comprises a series LC resonator in shunt across the effective transmission line of free space. It therefore exhibits a first-order band-stop response.

More sophisticated structures may be fashioned to control the overall response, but the basic principles of frequency selective surfaces can be understood from these two examples. The specific values of the equivalent circuit elements in simple cases may be derived from a closed-form analysis of each unit cell using periodic boundary conditions. For more sophisticated forms, equivalent circuit models are now more readily derived using computer-aided numerical simulation.

Higher-order responses may be achieved by cascading several sheets like these with short gaps between them. Say, for example, we wish to create a low-pass filter modelled after the lumped-element ladder in Figure 11.8(b). We note that at low frequencies, significantly below resonance, the shunt admittance of the circuit model for the mesh surface in Figure 11.28(a) is dominated by the equivalent inductor, while the circuit model in Figure 11.28(b) is most accurately represented by a shunt capacitor. We may therefore select the band-stop grid in Figure 11.28(b) to represent the shunt capacitors in our desired low-pass ladder. The series inductances may then be approximated by the short length of free-space between sheets. Recall that an approximate equivalence between a series inductor and short length of transmission line was already derived earlier in the context of stepped-impedance low-pass filters in Section 11.4.1.

11.7 TIME-DOMAIN OR PULSE-SHAPING FILTERS

Although frequency selection is the defining characteristic of an electronic filter, not all filters are designed to maximize selectivity, at least not directly. Instead, the goal in *time-domain* or *pulse-shaping* filters is to control or limit *edge rates* (slopes),

5 However, projection effects will render the behavior of the surface dependent upon the angle of incidence.

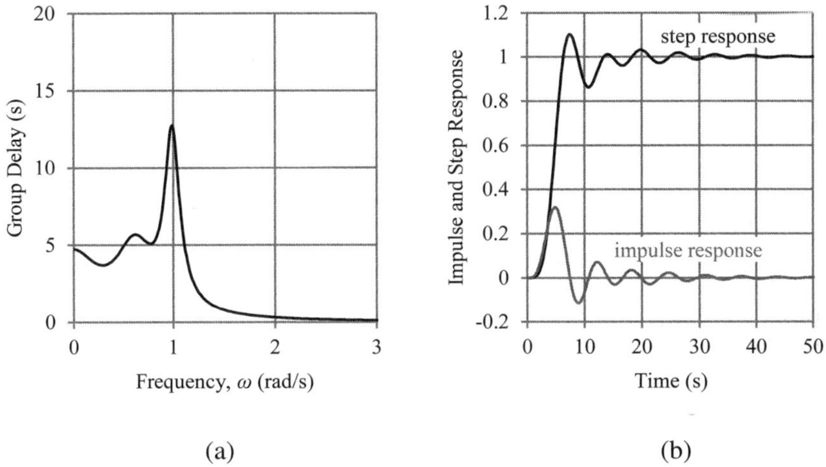

Figure 11.29 (a) Group delay and (b) impulse/step responses of a fifth-order, low-pass Chebyshev type-I filter having normalized cutoff, $\omega_c = 1$ rad/s.

ringing, overshoot, and other waveform characteristics that are sensitive to the phase as well as amplitude in the frequency domain.

The frequency-domain characteristic which is most telling in this sense is the *group delay*, related to the group velocity which was described in Section 7.2.6. Whereas the *phase delay* describes the amount by which a purely sinusoidal signal having frequency ω is delayed from the input to the output of a filter, and is given by

$$\tau_\theta = -\frac{\phi}{\omega} \tag{11.61}$$

the group delay describes the effective time interval for which a wave packet or enveloped is delayed, and is given by

$$\tau_g = -\frac{d\phi}{d\omega} \tag{11.62}$$

In these equations, ϕ is the phase angle of the transmission coefficient at frequency ω, or $\phi = \angle s_{21}$. The group delay of a fifth-order Chebyshev type-I filter is shown in Figure 11.29(a) for illustration. To minimize distortion of a modulated waveform, the group delay should be as flat as possible, so that waveform packets of all shapes and duration are delayed equally. Clearly, the Chebyshev type-I filter falls short in this regard, having not only a substantial slope across the passband but noticeable oscillations as well.

The impact of this on time-domain signals is even more clearly illustrated by the filter's *impulse response* and *step response*, which are shown in Figure 11.29(b). The impulse response, as the name implies, is the output of the filter given a very short impulse, or Dirac delta function, as its input. The impulse response may be calculated as the *inverse Laplace transform* [17–19] of the filter's transfer function when expressed in terms of complex frequency, $s = \sigma + j\omega$. Similarly, the step response characterizes the filter's output when driven by a unit step, or *Heaviside step function*, at its input, and may be calculated as the time integral of the impulse response (or, equivalently, as the inverse Laplace transform of the transfer function multiplied by s^{-1}).

The impulse response of the Chebyshev type-I filter exhibits noticeable ringing that continues long after the input waveform has passed. In modulated waveforms, this lasting output from a short-term input would lead to harmful *intersymbol interference* [20]. The step response further reveals an overshoot of the step height that could artificially reduce the system's dynamic range.

11.7.1 Gaussian Filters

Shifting our focus to these measures of distortion in the time domain, we may articulate an entirely different ideal filter response in contrast with the brick wall response of (11.21b). Known as a *Gaussian filter*, having its roll-off in amplitude given by a Gaussian or normal distribution curve, it has the property of exhibiting no overshoot in the presence of step functions while minimizing the rise and fall times, a feature that is closely related to its having the minimum group delay for all possible filters of comparable selectivity [21].

Unfortunately, like the brick-wall filter before it, a true Gaussian filter is noncausal (and thus unrealizable), having very small but nonzero temporal response at all times from minus infinity to plus infinity. Nevertheless, causal finite-order approximations do exist that closely replicate the Gaussian filter's key characteristics. Once such approximation is known as a Bessel-Thomson filter.

11.7.2 Bessel-Thomson Filters

Based on Bessel functions, the amplitude response of a Bessel-Thomson filter approaches that of a Gaussian filter as the order approaches infinity, but even fairly low-order filters are a close match to the Gaussian ideal, and while the overshoot is not technically zero, it can be quite negligible in almost any realistic application [22]. Additionally, Bessel-Thomson filters are distinguished by having maximally flat group delay among all filters of equivalent order, thus preserving the shape of transient waveforms, which are entirely contained within the filter's passband. These advantages may outweigh the filter's very poor selectivity in distortion-sensitive applications.

One very simple and convenient implementation of a Bessel-Thomson filter is shown in Figure 11.30(a). It comprises a resistively loaded ladder network of lumped elements that mimics a lossy transmission line. As a result, all of the interior elements are identical, like a chain of cascaded unit cells, with only minor tweaks at the end-points for impedance-matching into and out of the structure. The lossy elements render the filter fairly well matched (though not totally reflectionless) even into the stopband. More than a convenience, the absorptive properties of this filter are especially critical to the applications in which a Gaussian or Bessel-Thomson filter would typically be used, as signal reflections can be just as damaging to waveform integrity as direct-transmission ringing or overshoot.

The frequency responses of this filter network for several orders are plotted in Figure 11.30(b) on a linear (not decibel) scale. The nearly Gaussian profile is evident, and almost independent of order, while the reflection coefficient is below 0.2 (or about 15-dB return loss).

The group delay is then given in Figure 11.30(c) and the impulse and step responses in Figure 11.30(d). Clearly superior to the Chebyshev type-I filter, the group delay is smooth, the impulse response is nearly free of ringing, and the step response exhibits almost no overshoot. The overall delay, or *latency*, increases almost proportionally to order, but otherwise the characteristics are all about the same.

11.8 MULTIPLEXERS

The last frequency-selective device we will discuss is known as a *multiplexer*, in many ways, the marriage of a filter and a splitter. Multiplexers take an input signal and distribute it to two or more outputs according to their frequency content, or, operated in reverse, combine multiple inputs at different frequencies into a spectral composite at a single output. Most multiplexers simply take on the form of two or more standard filters — whether implemented with lumped elements, transmission lines, or waveguide — combined in parallel at a common junction, as shown in Figure 11.31. The number of outputs, or frequency channels, is arbitrary.

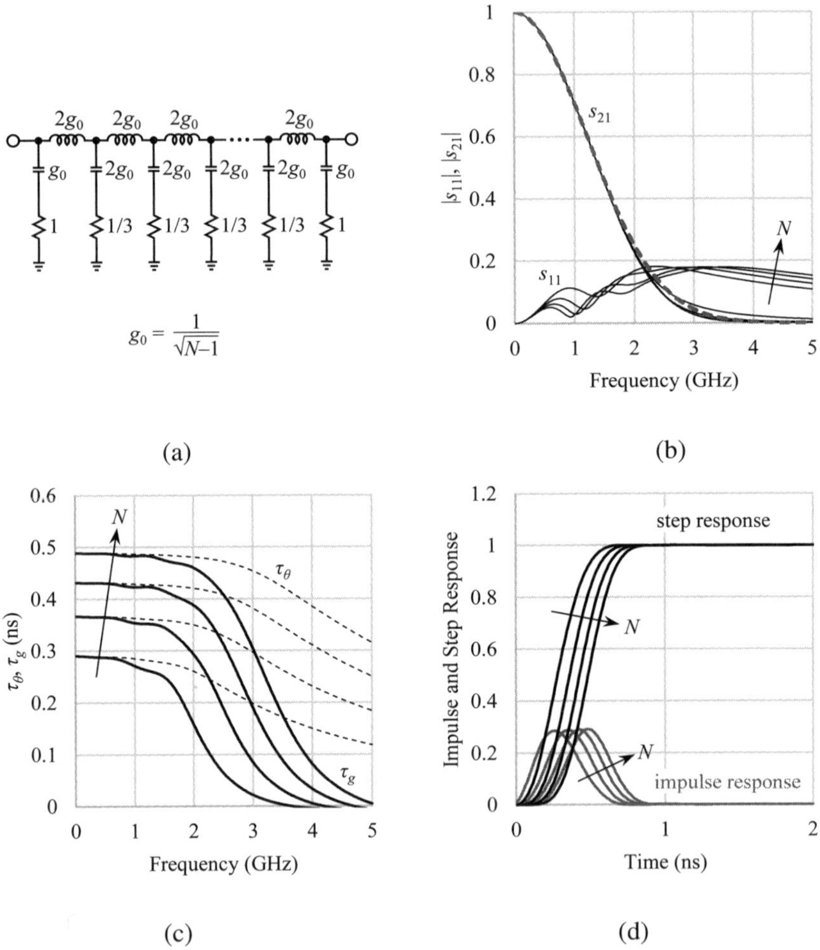

Figure 11.30 (a) Schematic of an absorptive, low-pass Bessel-Thomson filter. (b) Transmission and reflection coefficients for orders $N = 5, 7, 9$, and 11, all normalized such the the 3-dB bandwidth is 1 GHz. The dashed gray line is for an ideal Gaussian frequency response. (c) Phase and group delay. (d) Step responses.

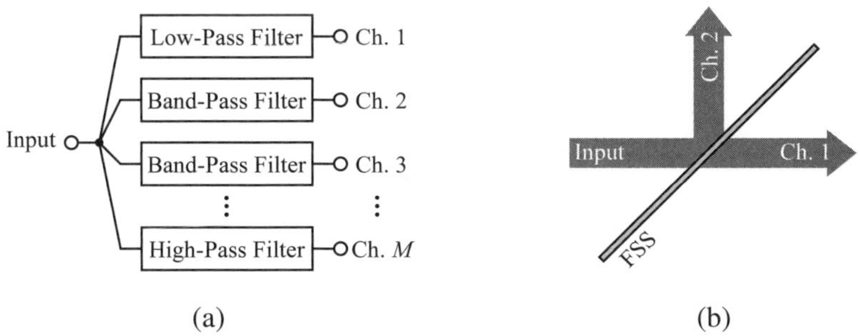

Figure 11.31 (a) General M-port multiplexer. (b) Quasioptical diplexer based on a frequency-selective surface (FSS).

A multiplexer with only two outputs is commonly referred to as a *diplexer*, while those with three outputs are called *triplexers*, and so on. It is generally assumed that the passbands of the individual filters are nonoverlapping.

In the quasioptical realm, a diplexer may be formed from a frequency-selective surface oriented at a 45° angle relative to the incoming wave, as shown in Figure 11.31(b). Higher channel counts may be constructed by cascading several such quasioptical diplexers, one after the other.

While it is not strictly necessary that the collective passbands of all output channels be contiguous, this is a common goal. The insertion loss of adjacent channels, then, intersect at 3 dB. As a general rule, otherwise well-designed filters that are indiscriminately combined in parallel interact poorly with one another at these band-edge frequencies, often resulting in return-loss peaks and distortion of the passband shape [2]. Classically, the only filter response that combined seamlessly with itself in such a manner was the Butterworth response; that is, appropriately designed[6] low-pass and high-pass filters of the same order could be combined in parallel if designed to have identical 3-dB corner frequencies and would exhibit no transition-edge reflections or passband distortions as a consequence of the combination. The Butterworth response is, in that sense, self complementary. It turns out that a Chebyshev type-II filter, say, high-pass, is in the same way complementary to a low-pass Chebyshev type-I filter. A rigorous derivation for designing such complementary, or *susceptance annulling*, filter pairs is given in [23, 24]. (In these

6 In this context, an "appropriate design" would consist of a pair of low-pass and high-pass *singly terminated* Butterworth filter ladder prototypes, which are distinct from the doubly terminated ladder prototypes described in Section 11.2.3.

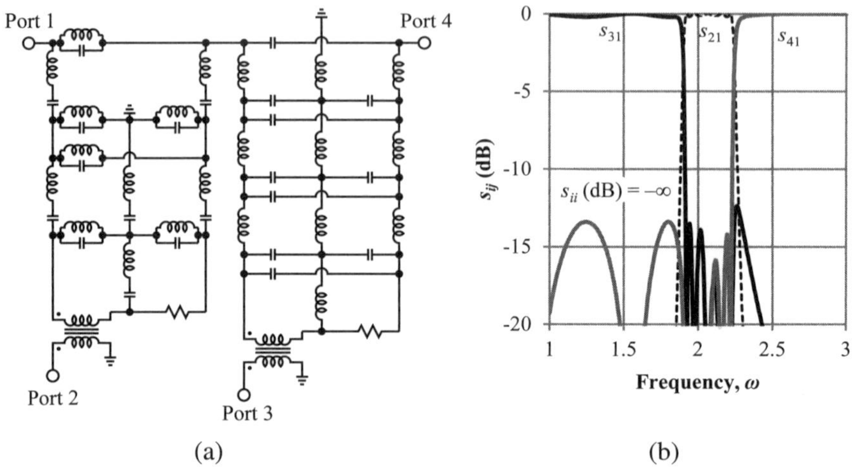

Figure 11.32 Reflectionless triplexer made from (a) one band-stop filter and one high-pass filter, exhibiting (b) low-pass, band-pass, and high-pass output channels.

references, the complementary network is not always presented as a separate, terminated filter in its own right, but a classical theorem by Sidney Darlington guarantees that any such network can be transformed into one.)

The reflectionless filters from Section 11.3 offer a unique and convenient approach to building contiguous-channel multiplexers, since the internal terminations can be replaced with external ports, thus transforming the filter into (potentially) a dual-directional diplexer without any further modification necessary. These can then be cascaded to create larger multiplexers having an arbitrary number of channels [2], all while maintaining the reflectionless property from all of its ports, input and output. An example of a multiplexer designed in this way, having three contiguous channels (low-pass, band-pass, and high-pass) is shown in Figure 11.32.

Problems

11-1 Calculate the element values for the equalizer in Figure 11.1(d) so that it has positive 6 dB of slope from DC to 5 GHz in a 50Ω system. (For the low-pass prototype, let $L = 2Z_0/\omega_0$ and $C = 2Y_0/\omega_0$, and then perform a band-pass transformation with $\Delta = 30/A$.)

11-2 Derive (11.6c) for the coupled-line equalizer, where T is the transmission coefficient at peak attenuation.

11-3 Derive the scattering parameters for the network in Figure 11.3(b) where the element values are given by (11.7).

11-4 Calculate the element values for the equalizer in Figure 11.3(c) to achieve a slope of 10 dB in a 50Ω network.

11-5 Calculate the mid-band insertion loss in decibels of the equalizer in Figure 11.3(d) in terms of the slope ratio T when the element values are given by (11.9).

11-6 Write an expression for the forward and backward-traveling wave Bloch impedances of the capacitively loaded transmission line in Figure 11.6(a), using the asymmetric unit cell in Figure 11.6(b). Rewrite for a symmetric unit cell with half the capacitance at each end. What can be observed about their relationship?

11-7 Prove that Chebyshev type-I and type-II responses of the same order have the same selectivity for a given passband insertion loss and stopband rejection level, where the selectivity is defined as the ratio of the passband and stopband corner frequencies at those levels.

11-8 What is the ripple factor for a Chebyshev type-II filter having a minimum stopband attenuation of 20 dB?

11-9 Calculate the ladder prototype parameters for a ninth-order Butterworth filter. What are actual element values for a high-pass implementation normalized to 50Ω and having a cutoff corner at 1 GHz?

11-10 Draw a schematic for a lumped-element reflectionless filter having a fifth-order Butterworth response with a cutoff frequency of 1 GHz and 50Ω characteristic impedance.

11-11 Derive the critical ripple factor given in (11.32) and (11.33) for a reflectionless filter for which $g_1 = g_2$. What is the limit as $N \to \infty$?

11-12 Calculate the electrical line lengths, θ_k, at the corner frequency, ω_c, for a seventh-order, stepped-impedance Butterworth filter in two cases: one based on the ladder prototype in Figure 11.8(a), and another based on Figure 11.8(b). Assume that the highest-impedance transmission line available is $3Z_0$, and the lowest is $Z_0/2$.

11-13 Write down the normalized transmission response of a two-resonator filter whose coupling coefficients are $J_{02} = 1$ and $J_{12} = J_{13} = -1$. Assume that $\mathbf{G}$ and $\mathbf{U}$ have the form given in (11.38).

11-14 Calculate the susceptance slope parameter for a resonator formed by the parallel combination of two equal-length transmission-line stubs, one short-circuited and one open-circuited, both having characteristic impedance Z_0.

11-15 Calculate the coupling coefficients, $J_{k,k+1}$, for a seventh-order Chebyshev filter with ripple factor $\varepsilon = 0.1$ and fractional bandwidth $\Delta = 0.1$, for system characteristic impedance $Z_0 = 50\Omega$, using a linear array of half-wave resonators.

11-16 Calculate the even and odd-mode impedances of a fifth-order edge-coupled bandpass filter having a Chebyshev type-I response with ripple factor $\varepsilon = 0.1$. Assume that the system characteristic impedance is 50Ω, and design for a fractional bandwidth of 10%.

11-17 If one were to attempt an empirical extraction of the couplings identified in the edge-coupled filter above with a center frequency of 10 GHz, what would be the desired resonant peak spacings for the interior couplings, and the 3-dB bandwidths of the source and load couplings?

References

[1] D. J. Mellor, "On the design of matched equalizers of prescribed gain versus frequency profiles," *IEEE International Microwave Symposium Digest*, pp. 308–311, June 1977.

[2] M. A. Morgan, *Reflectionless Filters*. Norwood, MA: Artech House, 2017.

[3] M. A. Morgan, T. Newton, R. Hayward, and T. A. Boyd, "Non-reflective transmission-line filters for gain slope equalization," *IEEE/MTT-S International Microwave Symposium*, June 2007, pp. 545–548.

[4] R. J. Cameron, C. M. Kudsia, and R. R. Mansour, *Microwave Filters for Communication Systems*. New York: Wiley, 2007.

[5] G. L. Matthaei, L. Young, and E. M. T. Jones, *Microwave Filters, Impedance-Matching Networks, and Coupling Structures*. Dedham, MA: Artech House, 1980.

[6] M. E. Van Valkenburg and W. M. Middleton, Eds., *Reference Data for Engineers: Radio, Electronics, Computer, and Communications*. Boston: Newnes, 2002.

[7] Mini-Circuits Inc. (2015) Pairing mixers with reflectionless filters to improve system performance. https://www.minicircuits.com/app/AN75-007.pdf.

[8] M. A. Morgan, "Think outside the band: Design and miniaturization of absorptive filters," *IEEE Microwave Magazine*, vol. 19, pp. 54–62, November 2018.

[9] M. A. Morgan, W. M. Groves, and T. A. Boyd, "Reflectionless filter topologies supporting arbitrary low-pass ladder prototypes," *IEEE Transactions on Circuits and Systems I*, vol. 66, no. 2, pp. 594–604, February 2019.

[10] H. C. Bell, "The coupling matrix in low-pass prototype filters," *IEEE Microwave Magazine*, vol. 17, no. 2, pp. 70–76, April 2007.

[11] R. Levy, R. V. Snyder, and G. L. Matthaei, "Design of microwave filters," *IEEE Transactions on Microwave Theory and Techniques*, vol. 50, no. 3, pp. 783–792, March 2002.

[12] R. J. Cameron, "Advanced filter synthesis," *IEEE Microwave Magazine*, vol. 12, no. 6, pp. 42–61, October 2011.

[13] M. Hagensen, "Narrowband microwave bandpass filter design by coupling matrix synthesis," *Microwave Journal*, vol. 53, no. 4, pp. 218–226, April 2010.

[14] N. Marcuvitz, *Waveguide Handbook*. Stevenage, UK: Peter Peregrinus Ltd., 1986.

[15] M. A. Morgan and S. Weinreb, "Techniques for the integration of high-Q millimeter-wave filters in multi-function mmic modules," *Microwave Journal*, vol. 48, no. 5, pp. 182–196, May 2005.

[16] Wikipedia. (2018, November) Frequency selective surface. https://en.wikipedia.org/wiki/Frequency_selective_surface.

[17] F. F. Kuo, *Network Analysis and Synthesis*, 2nd ed. New York: Wiley, 1962.

[18] G. Arfken and H. Weber, *Mathematical Methods for Physicists*, 4th ed. San Diego, CA: Academic Press, 1995.

[19] B. Lathi, *Linear Systems and Signals*. Carmichael, CA: Berkeley-Cambridge Press, 1992.

[20] T. Wong and T. Lok, "Intersymbol interference and equalization," *EEL6535 course notes*, pp. 4.1–4.24. http://wireless.ece.ufl.edu/twong/Notes/Comm/ch4.pdf.

[21] Wikipedia. (2018, October) Gaussian filter. https://en.wikipedia.org/wiki/Gaussian_filter.

[22] Wikipedia. (2018, March) Bessel filter. https://en.wikipedia.org/wiki/Bessel_filter.

[23] M. Khalaj-Amirhosseini and A. Khalaj-Amirhosseini, "Reflectionless filters with arbitrary transfer functions," *Journal of Telecommunications*, vol. 34, no. 2, October 2016.

[24] M. Khalaj-Amirhosseini and M. M. Taskhiri, "Twofold reflectionless filters of inverse-Chebyshev response with arbitrary attenuation," *IEEE Transactions on Microwave Theory and Techniques*, vol. 65, no. 11, pp. 4616–4620, 2017.

Chapter 12

Amplifiers

Thus far, we have entirely focused our analysis on passive, or dissipative, components, for which the total outgoing signal power is strictly less than or equal to the input signal power. At some point in almost any real electronic system, it is necessary to increase a signal's amplitude. This is the work of amplifiers.

It is beyond the scope of this book to elucidate the underlying physics that allow a device (usually a transistor) to amplify. Instead we investigate the external behavior of those devices in the context of a larger electromagnetic system, with their amplifying characteristics represented by effective network parameters or an equivalent circuit.

The first part of this chapter deals with the fundamental gain, stability, and noise analyses that apply to virtually any amplifier. Implicitly assumed is that the signals under consideration obey a *small-signal approximation*, comprising only slight perturbations from an operating DC, or *quiescent*, bias point. In Section 12.3, we will talk about *power amplifiers*, which violate the small-signal approximation, and discuss some of the special considerations which to apply to large-signal operation instead.

12.1 GAIN AND STABILITY

In principle, a device can amplify in reflection or transmission, in one or both directions, to one or any number of ports. Most often, amplifiers are two-port devices which exhibit gain in one preferred direction from input to output. The common circuit symbol for an amplifier, which emphasizes this preferred direction, is shown in Figure 12.1(a). By convention, the input is port 1 and the output is

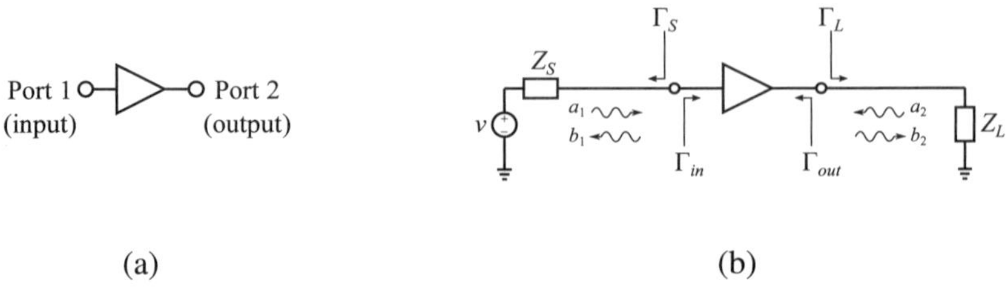

Figure 12.1 (a) Schematic symbol for a two-port amplifier. (b) Amplifier with embedding impedances.

port 2. We shall use this symbol equally to represent single active devices, such as transistors, and complete amplifiers.

12.1.1 Power Gain Factors

Ideally, a device would exhibit useful gain characteristics with both ports terminated in the system characteristic impedance, Z_0, but this is rarely the case without the aid of a matching network. We therefore show the amplifier in Figure 12.1(b) as generally having mismatched terminations at its ports, source and load, denoted as Z_S and Z_L, respectively. Although there may be zero electrical length between the amplifier ports and their terminations, we may still articulate effective forward- and backward-traveling waves at each port which are related by the amplifier's scattering parameters,

$$\begin{pmatrix} b_1 \\ b_2 \end{pmatrix} = \begin{pmatrix} s_{11} & s_{12} \\ s_{21} & s_{22} \end{pmatrix} \begin{pmatrix} a_1 \\ a_2 \end{pmatrix} \tag{12.1}$$

Further, the outgoing and returning waves at port 2 are related by the load's reflection coefficient,

$$\frac{a_2}{b_2} = \Gamma_L = \frac{Z_L - Z_0}{Z_L + Z_0} \tag{12.2}$$

Thus, the effective input reflection coefficient is given by Mason's rule,

$$\Gamma_{in} = \frac{b_1}{a_1} = s_{11} + \frac{s_{21}s_{12}\Gamma_L}{1 - s_{22}\Gamma_L} \tag{12.3}$$

It is tempting to also write the ratio a_1/b_1 as being equal to Γ_S; however, this is not the case when the source is active (at the frequency under consideration). The voltage generator, v, imposes a different overall boundary condition upon the combined wave amplitudes at port 1 than does Z_S by itself. Nevertheless, we may write Γ_S as the equivalent reflection coefficient which would occur if $v = 0$,

$$\Gamma_S = \frac{Z_S - Z_0}{Z_S + Z_0} \tag{12.4}$$

Similarly, the reflection coefficient seen looking into the output port of the amplifier, Γ_{out}, does not give the ratio b_2/a_2 unless $v = 0$, in which case

$$\Gamma_{out} = s_{22} + \frac{s_{21}s_{12}\Gamma_S}{1 - s_{11}\Gamma_S} \tag{12.5}$$

We learned from Chapter 6 that the power delivered from a source to a load is maximized when their immittances are conjugately matched. The effectiveness of an amplifier as a gain block may thus be articulated in various ways in terms of the power flows going into and out of its ports relative to that which would flow under the conjugate matching conditions.

The actual power accepted by the amplifier at its input, P_{in}, is given by the incident power minus the reflected power, or, in terms of wave amplitudes,

$$P_{in} = |a_1|^2 - |b_1|^2 = |a_1|^2 \left(1 - |\Gamma_{in}|^2\right) \tag{12.6}$$

where rms amplitudes have been assumed. It is useful to express the incident wave amplitude, a_1, in terms of the voltage source, v. To do so, we consider the effective voltage divider seen by the voltage source,

$$v\frac{Z_{in}}{Z_S + Z_{in}} = (a_1 + b_1)\sqrt{Z_0} \tag{12.7a}$$

$$v\frac{\frac{1+\Gamma_{in}}{1-\Gamma_{in}}}{\frac{1+\Gamma_S}{1-\Gamma_S} + \frac{1+\Gamma_{in}}{1-\Gamma_{in}}} = a_1 (1 + \Gamma_{in})\sqrt{Z_0} \tag{12.7b}$$

$$\therefore a_1 = \frac{v}{2\sqrt{Z_0}} \cdot \frac{1 - \Gamma_S}{1 - \Gamma_S\Gamma_{in}} \tag{12.7c}$$

Thus, the accepted input power in terms of v is

$$P_{in} = \frac{|v|^2}{4Z_0} \cdot \frac{|1 - \Gamma_S|^2}{|1 - \Gamma_S\Gamma_{in}|^2} \left(1 - |\Gamma_{in}|^2\right) \tag{12.8}$$

The available power at the input, in contrast, is that which would be accepted by the amplifier under the conjugate matching condition, that is, if $Z_{in} = Z_S^*$, or equivalently $\Gamma_{in} = \Gamma_S^*$,

$$P_{av,in} = P_{in}|_{\Gamma_{in}=\Gamma_S^*} = \frac{|v|^2}{4Z_0} \cdot \frac{|1-\Gamma_S|^2}{1-|\Gamma_S|^2} \tag{12.9}$$

Similarly, the actual output power delivered to the load is

$$P_{out} = |b_2|^2 - |a_2|^2 = |b_2|^2 \left(1 - |\Gamma_L|^2\right) \tag{12.10}$$

The wave amplitude b_2 may be expressed once again in terms of the source voltage by noting that

$$b_2 = a_1 s_{21} + a_2 s_{22} = a_1 s_{21} + b_2 \Gamma_L s_{22} \tag{12.11a}$$

$$\therefore \frac{b_2}{a_1} = \frac{s_{21}}{1 - s_{22}\Gamma_L} \tag{12.11b}$$

Therefore,

$$P_{out} = \frac{|v|^2}{4Z_0} \cdot \frac{|s_{21}|^2 |1-\Gamma_S|^2 \left(1-|\Gamma_L|^2\right)}{|1-\Gamma_S\Gamma_{in}|^2 |1-s_{22}\Gamma_L|^2} \tag{12.12}$$

However, the available power at the output of the amplifier is that which would be delivered to a load when $Z_L = Z_{out}^*$ ($\Gamma_L = \Gamma_{out}^*$),

$$P_{av,out} = \frac{|v|^2}{4Z_0} \cdot \frac{|s_{21}|^2 |1-\Gamma_S|^2 \left(1-|\Gamma_{out}|^2\right)}{|1-\Gamma_S\Gamma_{in}|^2 |1-s_{22}\Gamma_{out}^*|^2} \tag{12.13}$$

Importantly, Γ_{in} in (12.13) is the reflection coefficient seen at the input of the amplifier when $\Gamma_L = \Gamma_{out}^*$. Applying this to (12.3), we have

$$|1-\Gamma_S\Gamma_{in}|^2\Big|_{\Gamma_L=\Gamma_{out}^*} = \frac{|1-s_{11}\Gamma_S|^2 \left(1-|\Gamma_{out}|^2\right)^2}{|1-s_{22}\Gamma_{out}^*|^2} \tag{12.14}$$

The derivation of this result will be left as an exercise for the reader. Substituting this into (12.13), we have

$$P_{av,out} = \frac{|v|^2}{4Z_0} \cdot \frac{|s_{21}|^2 |1-\Gamma_S|^2}{|1-s_{11}\Gamma_S|^2 \left(1-|\Gamma_{out}|^2\right)} \tag{12.15}$$

Now, having calculated the actual and potential power flows at each port, we can describe the performance of a general amplifier in terms of its power gains [1, 2]. First, the realized, operating power gain is simply the ratio of actual power delivered to the load relative to the power accepted by the amplifier

$$G_P = \frac{P_{out}}{P_{in}} = \frac{|s_{21}|^2 \left(1 - |\Gamma_L|^2\right)}{\left(1 - |\Gamma_{in}|^2\right) |1 - s_{22}\Gamma_L|^2} \tag{12.16}$$

The available power gain is given by the ratio of the available output power to the available input power,

$$G_A = \frac{P_{av,out}}{P_{av,in}} = \frac{|s_{21}|^2 \left(1 - |\Gamma_S|^2\right)}{|1 - s_{11}\Gamma_S|^2 \left(1 - |\Gamma_{out}|^2\right)} \tag{12.17}$$

Finally, the *transducer power gain* is defined as the ratio of actual power delivered to the load relative to the power available from the source,

$$G_T = \frac{P_{out}}{P_{av,in}} = \frac{|s_{21}|^2 \left(1 - |\Gamma_S|^2\right) \left(1 - |\Gamma_L|^2\right)}{|1 - \Gamma_S\Gamma_{in}|^2 |1 - s_{22}\Gamma_L|^2} \tag{12.18}$$

Note that when both ports are terminated in the system characteristic impedance ($\Gamma_S = \Gamma_L = 0$), the transducer power gain is $G_T = |s_{21}|^2$.

The transducer power gain is perhaps the most accurate description of an amplifier's effectiveness at increasing power in a system, since it directly compares what level of power transfer was achieved versus what was available, and accounts for both port terminations (no other expression above depends on both Γ_S and Γ_L, even implicitly). It is customary in some texts to break it down into three components,

$$G_T = G_S G_0 G_L \tag{12.19}$$

where

$$G_S = \frac{1 - |\Gamma_S|^2}{|1 - \Gamma_S\Gamma_{in}|^2} \tag{12.20a}$$

$$G_0 = |s_{21}|^2 \tag{12.20b}$$

$$G_L = \frac{1 - |\Gamma_L|^2}{|1 - s_{22}\Gamma_L|^2} \tag{12.20c}$$

and where the significance is that G_S is a factor embodying the input mismatch, G_L describes the output mismatch, and G_0 is an intrinsic gain-factor of the amplifier or device itself, neglecting mismatch effects.

It should be clear from this derivation that neither port can be conjugately matched in isolation; the impedance presented at one port affects the impedance realized at the other, as a consequence of the feedback and feedforward terms of the scattering-parameter matrix (s_{12} and s_{21}, respectively). This creates a kind of two-dimensional matching problem which complicates amplifier design. Fortunately, most common amplifying devices, such as transistors, are strongly nonreciprocal, having large reverse isolation (which is to say that s_{12} is very small while s_{21} is large). The *unilateral transducer power gain*, G_{TU}, is defined under the assumption that s_{12} is negligible (and consequently $\Gamma_{in} = s_{11}$),

$$G_{TU} = \frac{1 - |\Gamma_S|^2}{|1 - s_{11}\Gamma_S|^2} |s_{21}|^2 \frac{1 - |\Gamma_L|^2}{|1 - s_{22}\Gamma_L|^2} \tag{12.21}$$

A summary of the reflection, power, and gain terms for a general, linear, amplifying device is given in Table 12.1.

12.1.2 Oscillations

Gain is the provenance of all oscillations; that is, beneath every oscillation is a device having gain, and in every gain-bearing device there is the potential to oscillate. At its core, gain is the introduction of new signal power into a circuit. When that new power is self-generated, without any coherent relationship to signals that are supplied to the device, we call it an oscillation.

To put it another way, oscillations occur as a result of regenerative amplification. Small fluctuations — present as noise that is due, at least, to thermal excitation of charge carriers within a circuit, even when no external signal is applied — are amplified and reamplified until either the fluctuation self-limits due to nonlinear effects, or the active device destroys itself.

Any gain-bearing device can be made to oscillate with positive feedback, from output back to input, having the right phase. Such feedback is usually unintentional and should be carefully avoided whenever one is working with a high-gain device. Even without direct feedback, the reverse leakage (s_{12}) of a practical device such as a transistor can sustain an oscillation if the port terminations are right. A device or circuit which cannot oscillate for any arbitrarily chosen pair of source and load

Table 12.1
Linear Amplifier Power and Gain Factors

Factor	Symbol	Expression	Termination Dependencies
Input reflection coefficient	Γ_{in}	$s_{11} + \frac{s_{12}s_{21}\Gamma_L}{1-s_{22}\Gamma_L}$	Γ_L
Output reflection coefficient	Γ_{out}	$s_{22} + \frac{s_{12}s_{21}\Gamma_S}{1-s_{11}\Gamma_S}$	Γ_S
Operating input power	P_{in}	$\frac{\lvert v\rvert^2}{4Z_0} \cdot \frac{\lvert 1-\Gamma_S\rvert^2(1-\lvert\Gamma_{in}\rvert^2)}{\lvert 1-\Gamma_S\Gamma_{in}\rvert^2}$	Γ_S, Γ_L
Operating output power	P_{out}	$\frac{\lvert v\rvert^2}{4Z_0} \cdot \frac{\lvert 1-\Gamma_S\rvert^2}{\lvert 1-\Gamma_S\Gamma_{in}\rvert^2} \lvert s_{21}\rvert^2 \frac{1-\lvert\Gamma_L\rvert^2}{\lvert 1-s_{22}\Gamma_L\rvert^2}$	Γ_S, Γ_L
Available input power	$P_{av,in}$	$\frac{\lvert v\rvert^2}{4Z_0} \cdot \frac{\lvert 1-\Gamma_S\rvert^2}{1-\lvert\Gamma_S\rvert^2}$	Γ_S
Available output power	$P_{av,out}$	$\frac{\lvert v\rvert^2}{4Z_0} \cdot \frac{\lvert 1-\Gamma_S\rvert^2}{\lvert 1-s_{11}\Gamma_S\rvert^2} \lvert s_{21}\rvert^2 \frac{1}{1-\lvert\Gamma_{out}\rvert^2}$	Γ_S
Operating power gain	G_P	$\frac{P_{out}}{P_{in}} = \frac{1}{1-\lvert\Gamma_{in}\rvert^2} \lvert s_{21}\rvert^2 \frac{1-\lvert\Gamma_L\rvert^2}{\lvert 1-s_{22}\Gamma_L\rvert^2}$	Γ_L
Available power gain	G_A	$\frac{P_{av,out}}{P_{av,in}} = \frac{1-\lvert\Gamma_S\rvert^2}{\lvert 1-s_{11}\Gamma_S\rvert^2} \lvert s_{21}\rvert^2 \frac{1}{1-\lvert\Gamma_{out}\rvert^2}$	Γ_S
Transducer power gain	G_T	$\frac{P_{out}}{P_{av,in}} = \frac{1-\lvert\Gamma_S\rvert^2}{\lvert 1-\Gamma_S\Gamma_{in}\rvert^2} \lvert s_{21}\rvert^2 \frac{1-\lvert\Gamma_L\rvert^2}{\lvert 1-s_{22}\Gamma_L\rvert^2}$	Γ_S, Γ_L
Unilateral transducer power gain	G_{TU}	$\frac{P_{out}}{P_{av,in}} = \frac{1-\lvert\Gamma_S\rvert^2}{\lvert 1-s_{11}\Gamma_S\rvert^2} \lvert s_{21}\rvert^2 \frac{1-\lvert\Gamma_L\rvert^2}{\lvert 1-s_{22}\Gamma_L\rvert^2}$	Γ_S, Γ_L

v is the source rms voltage.

terminations, so long as they are passive, is called *unconditionally stable.*[1] If a device is not unconditionally stable, then it is *conditionally stable* or, equivalently, *potentially unstable.*

There is a rich body of literature regarding methods of analyzing or testing a circuit model to see if it might oscillate. In the strictest mathematical terms, a circuit may oscillate if it has a transfer function with a pole in the right half of the complex frequency plane [3]. One especially versatile method for testing if such a pole exists

1 This is a slight misnomer since, as we have just said, any gain-bearing device can be made to oscillate given appropriate feedback. "Unconditional," in this context, refers simply to the unlimited selection of source and load impedances to which it can be subjected without fear of oscillations.

is known as the *normalized determinant function* [4–6]. These methods may require more sophisticated models or measurement data than are readily available, and are beyond the scope of this book. Instead, we will focus on relatively straightforward tests on a circuit's scattering parameters. These are often the first line of defense in any new amplifier design.

12.1.3 Stability Circles

The basis of s-parameter stability factors is the understanding that a stable device will exhibit passive input and output reflection coefficients with passive source and load impedances. That is,

$$|\Gamma_{in}| = \left| s_{11} + \frac{s_{12}s_{21}\Gamma_L}{1 - s_{22}\Gamma_L} \right| < 1 \qquad (12.22a)$$

$$|\Gamma_{out}| = \left| s_{22} + \frac{s_{12}s_{21}\Gamma_S}{1 - s_{11}\Gamma_S} \right| < 1 \qquad (12.22b)$$

Note that if the device is unilateral ($s_{12} = 0$), these expressions simply imply that $|s_{11}| < 1$ and $|s_{22}| < 1$, respectively, independent of the source and load. If the device is not unilateral, these inequalities define a complex range of source and load reflection coefficients, Γ_S and Γ_L, for which the device is stable. Instability could occur, for example, when the product $s_{22}\Gamma_L$ is close to 1; then the denominator (12.22a) would be close to zero, making Γ_{in} very large and violating the inequality. Unconditional stability would simply require that the region of stability for each termination covers the entire Smith chart.

The boundary of the stable region for the output termination may may be found by allowing the input reflection coefficient to have exactly unit magnitude [1],

$$|\Gamma_{in}| = \left| s_{11} + \frac{s_{12}s_{21}\Gamma_L}{1 - s_{22}\Gamma_L} \right| = 1 \qquad (12.23a)$$

$$\therefore |s_{11} - \Gamma_L\Delta| = |1 - s_{22}\Gamma_L| \qquad (12.23b)$$

where

$$\Delta = \det \mathbf{S} = s_{11}s_{22} - s_{21}s_{12} \qquad (12.24)$$

Squaring both sides of (12.23b), we have

$$|s_{11} - \Gamma_L\Delta|^2 = |1 - s_{22}\Gamma_L|^2 \qquad (12.25a)$$

$$|s_{11}|^2 - s_{11}\Gamma_L^*\Delta^* - s_{11}^*\Gamma_L\Delta + |\Gamma_L\Delta|^2 = 1 - s_{22}\Gamma_L - s_{22}^*\Gamma_L^* + |s_{22}\Gamma_L|^2 \quad (12.25b)$$

$$\left(|s_{22}|^2 - |\Delta|^2\right)|\Gamma_L|^2 - 2\text{Re}\left\{(s_{22} - s_{11}^*\Delta)\Gamma_L\right\} = |s_{11}|^2 - 1 \quad (12.25c)$$

$$|\Gamma_L|^2 - 2\text{Re}\left\{\frac{(s_{22} - s_{11}^*\Delta)\Gamma_L}{|s_{22}|^2 - |\Delta|^2}\right\} = \frac{|s_{11}|^2 - 1}{|s_{22}|^2 - |\Delta|^2} \quad (12.25d)$$

Completing the square on the left side, we have

$$\left|\Gamma_L - \frac{(s_{22} - s_{11}^*\Delta)^*}{|s_{22}|^2 - |\Delta|^2}\right|^2 = \frac{|s_{11}|^2 - 1}{|s_{22}|^2 - |\Delta|^2} + \frac{|s_{22} - s_{11}^*\Delta|^2}{\left(|s_{22}|^2 - |\Delta|^2\right)^2} \quad (12.26a)$$

$$= \frac{|s_{11}|^2|s_{22}|^2 + |\Delta|^2 - 2\text{Re}\left\{s_{11}s_{22}\Delta^*\right\}}{\left(|s_{22}|^2 - |\Delta|^2\right)^2} = \left|\frac{s_{12}s_{21}}{|s_{22}|^2 - |\Delta|^2}\right|^2 \quad (12.26b)$$

which can be written as

$$|\Gamma_L - C_L| = r_L \quad (12.27)$$

where

$$C_L = \frac{(s_{22} - s_{11}^*\Delta)^*}{|s_{22}|^2 - |\Delta|^2} \quad (12.28a)$$

$$r_L = \left|\frac{s_{12}s_{21}}{|s_{22}|^2 - |\Delta|^2}\right| \quad (12.28b)$$

These equations describe the boundary of a circle in the complex reflection coefficient plane, having center C_L and radius r_L, and is known as *output stability circle* since it describes the set of reflection coefficients, Γ_L, for which the circuit remains stable.

A similar derivation yields the *input stability circle*, having center and radius given by

$$C_S = \frac{(s_{11} - s_{22}^*\Delta)^*}{|s_{11}|^2 - |\Delta|^2} \quad (12.29a)$$

$$r_S = \left|\frac{s_{12}s_{21}}{|s_{11}|^2 - |\Delta|^2}\right| \quad (12.29b)$$

Note in both cases that some or all of the stability circle may extend beyond the passive boundaries of the Smith chart. In addition, the circle only defines the

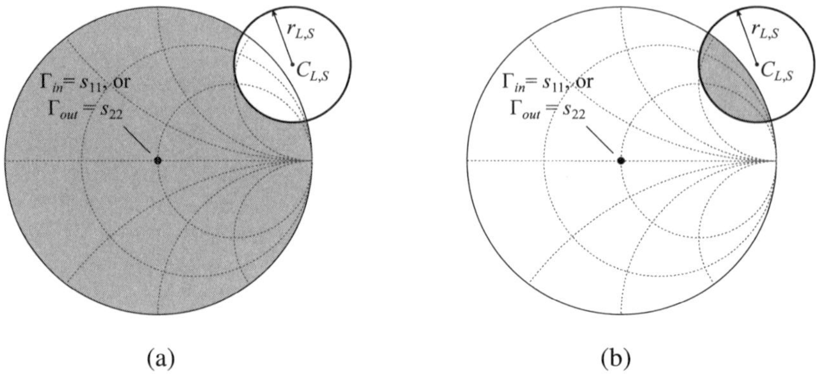

Figure 12.2 Input (or output) stability circles where (a) $|s_{22}| < 1$ (or $|s_{11}| < 1$) and (b) $|s_{22}| > 1$ (or $|s_{11}| > 1$). The shaded region represents stable source (load) impedances.

boundary of the stable region, with stable points lying either outside or inside the circle, as indicated in Figures 12.2(a, b). The simplest way to test which side of the boundary circle is stable is to consider the center of the Smith chart, for which $\Gamma_{in} = s_{11}$ and $\Gamma_{out} = s_{22}$. If $s_{11} < 1$, then the center of the Smith chart lies within the output stability region, otherwise it lies outside the stability region. Similarly, if $s_{22} < 1$, then the center of the chart lies within the input stability region; otherwise, it is excluded.

Thus, unconditional stability is simply the condition that the stable regions bounded (either inside or outside) by the stability circles encompass the entire Smith chart. Fortunately, it can be shown that if the region bounded by one stability circle includes the entire Smith chart, then the region bounded by the other will as well [7]. Therefore, only one stability circle needs to be checked for unconditional stability.

12.1.4 Rollet Stability Factor

The test for unconditional stability may be simplified even further by reducing the equations to two simple expressions. The first, known as the *Rollet stability factor* [7–10] is given below

$$K = \frac{1 - |s_{11}|^2 - |s_{22}|^2 + |\Delta|^2}{2\,|s_{21}s_{12}|} > 1 \tag{12.30}$$

Necessary and sufficient conditions for unconditional stability are that $K > 1$ and any one of the following auxiliary conditions are met,

$$B_1 = 1 + |s_{11}|^2 - |s_{22}|^2 - |\Delta|^2 > 0 \quad (12.31a)$$

$$B_2 = 1 - |s_{11}|^2 + |s_{22}|^2 - |\Delta|^2 > 0 \quad (12.31b)$$

$$|\Delta| < 1 \quad (12.31c)$$

$$|s_{12}s_{21}| < 1 - |s_{11}|^2 \quad (12.31d)$$

$$|s_{12}s_{21}| < 1 - |s_{22}|^2 \quad (12.31e)$$

It is important to emphasize that $K > 1$ by itself is not strictly sufficient to prove unconditional stability, although it receives the most attention since the auxiliary conditions above are typically met by most practical active devices anyway.

12.1.5 Geometric Stability Factors

A further drawback of having to combine two inequalities to form a stability criterion is that the relative magnitude of either one does not, strictly speaking, indicate a higher margin of stability. That is, we cannot rigorously say that an amplifier with $K = 1.00001$ is any more or less stable than an amplifier with $K = 10^6$.

To resolve this issue, researchers have identified a number of single-parameter stability conditions that are equivalent to Rollet's, the most famous of which is called μ. It comes in two forms, but either one alone is both necessary and sufficient to prove unconditional stability,

$$\mu_1 = \frac{1 - |s_{11}|^2}{|s_{22} - s_{11}^*\Delta| + |s_{12}s_{21}|} > 1 \quad (12.32a)$$

$$\mu_2 = \frac{1 - |s_{22}|^2}{|s_{11} - s_{22}^*\Delta| + |s_{12}s_{21}|} > 1 \quad (12.32b)$$

These are sometimes called *geometric stability factors* because they effectively measure the distance from the center of the Smith chart to the nearest point on the corresponding stability circle in the complex plane.

12.1.6 Maximum Gains

Another question that bears asking is what is the maximum gain that can be derived from an active device given complete freedom to choose its termination impedances (that is, with appropriate matching networks). Stability plays a role in answering this question, since a device which is unstable has, at least theoretically, infinite available gain. This can easily be verified by considering (12.20a), the source gain factor of the transducer power gain. The fact that a network is unstable implies that there is a passive load termination, $|\Gamma_L| \leq 1$, for which the input reflection coefficient has magnitude larger than unity, or $|\Gamma_{in}| > 1$. It is therefore possible to select a passive source reflection coefficient, $|\Gamma_S| < 1$, such that the denominator, $|1 - \Gamma_S\Gamma_{in}|$, is identically zero. Thus, the transducer power gain with the selected (passive) source and load terminations is infinite.

When the device is unconditionally stable, the maximum available gain is finite and is given by

$$G_{max} = \left(K - \sqrt{K^2 - 1}\right)\left|\frac{s_{21}}{s_{12}}\right| \quad \text{for } K > 1 \tag{12.33a}$$

$$= \frac{1}{K + \sqrt{K^2 - 1}}\left|\frac{s_{21}}{s_{12}}\right| \quad \text{for } K > 1 \tag{12.33b}$$

where K is the Rollett stability factor. Although the expression in (12.33a) is more commonly cited, computational precision errors may lead to incorrect results when K is very large (necessitating the subtraction of two very large numbers to obtain a small result). The latter expression (12.33b) is strictly equivalent yet immune to this problem [9].

Note that when $K < 1$, the calculation for G_{max} is imaginary, but as K approaches 1 from above, the maximum available gain converges on a finite value, $|s_{21}/s_{12}|$. It can be shown that this limit is the *maximum stable gain (MSG)* of a potentially unstable device. Putting both regions together, we have for the maximum stable gain

$$MSG = \begin{cases} \frac{1}{K+\sqrt{K^2-1}}\left|\frac{s_{21}}{s_{12}}\right| & K > 1 \\ \\ \left|\frac{s_{21}}{s_{12}}\right| & K \leq 1 \end{cases} \tag{12.34}$$

Any attempt to extract more gain from a device than this by adjusting the port impedances will result in instability.

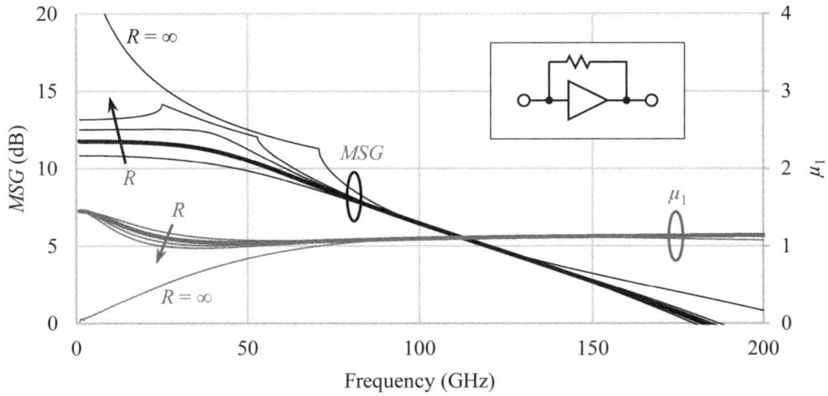

Figure 12.3 Maximum stable gain and geometric stability factor for a millimeter-wave transistor with resistive feedback (see inset). The feedback resistance was varied from $R = \infty$ to $R = 400\Omega$, 350Ω, 300Ω, and 250Ω. The thick solid lines are for $R = 300\Omega$.

Note that all stability criterion discussed in this section are derived from the scattering parameters, which are themselves functions of frequency. As an example, consider the plot in Figure 12.3, which shows the maximum stable gain and geometric stability factor, μ_1, for a typical millimeter-wave transistor. The intrinsic device (for which $R = \infty$ in this plot) will typically increase in available gain at low frequency and will become potentially unstable below a certain threshold, in this case, about 70 GHz. The stability coefficient, μ_1, drops below unity at this frequency, and the maximum stable gain goes through a slope discontinuity as a consequence of the piecewise nature of (12.34). To stabilize the device, we have added a feedback resistor with progressively smaller values (some devices may require a DC block in the feedback path to allow biasing). The low-frequency end will stabilize first, leaving an unstable band at mid-frequencies, until the feedback is strong enough to stabilize the entire curve.

Since a device is only stable if it is stable at all frequencies, DC to ∞, it is reasonable to ask over how wide a frequency range must the stability factor be considered? Clearly, low frequencies are important, but more complex amplifier circuits than this lone transistor do not always exhibit such obvious monotonicity. Ultimately, a device's internal parasitics will limit its gain at higher frequencies to the point that it can no longer be made to oscillate. The frequency at which this occurs, the *maximum frequency of oscillation*, is called f_{max}, and is an important figure of merit for high-frequency transistors.

The basis of f_{max} is a measure known as the *maximum unilateral gain*, also called *Mason's invariant* [11], given by

$$U = \frac{|s_{12} - s_{21}|^2}{\det(\mathbf{I} - \mathbf{SS}^*)} \tag{12.35}$$

It is important to make a distinction between the maximum unilateral gain, U, the maximum available gain, G_{max}, and the unilateral transducer gain, G_{TU}. G_{max} is the maximum gain available from a (generally nonunilateral) device given total freedom of termination impedance selection, and could be infinite if the device is unstable. G_{TU} is the achieved gain of a unilateral device given specific terminations. U is the gain that an otherwise nonunilateral device can achieve given arbitrary terminations after it has been rendered unilateral by a carefully crafted feedback network (think of the feedback network as cancelling out the intrinsic reverse leakage of the device).

The frequency at which U becomes unity is f_{max}. It has also been shown that G_{max} and MSG become unity at the same point [11]. Beyond this frequency, there is no gain available — via any passive embedding network — and hence no ability to oscillate. This provides an upper bound to the frequency range over which a device or circuit (or their models) must be tested to ensure stability. It is usually not possible (or is it at least impractical) to measure U accurately all the way out f_{max}, but for most active devices, it eventually settles into a monotonic decline that can be extrapolated (or, if it is easier, G_{max} can be used for extrapolation as well, since it reaches unity at the same point).

12.1.7 Balanced Amplifiers

By definition, an amplifier conjugately matched to the system characteristic impedance through lossless, reciprocal impedance-matching networks will have zero reflection at the external ports of those matching networks. In practice, this can be difficult to ensure over a wide frequency range, and the peak level of the return loss may be compromised in favor of bandwidth. Furthermore, as we will see in the following Sections 12.2 and 12.3, there are reasons to provide a termination impedance to the amplifier which is something other than the conjugate. In these cases, the return loss of the amplifier may suffer at the expense of other metrics.

A fairly clever way of solving these issues, at least over moderate bandwidths, is a configuration known as a *balanced amplifier*, shown in Figure 12.4. Two

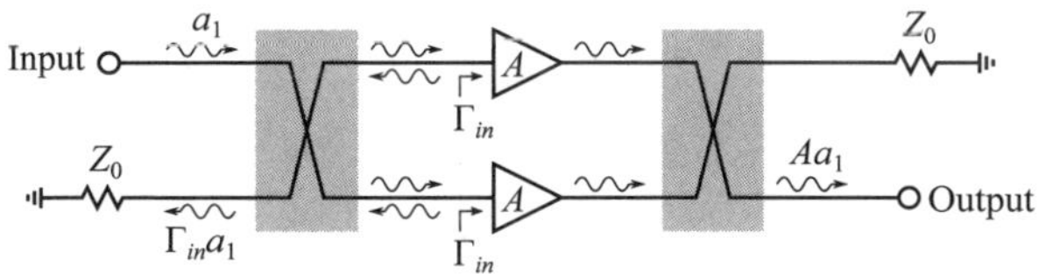

Figure 12.4 Balanced amplifier configuration showing an input wave (having phasor amplitude a_1), an amplified output wave (Aa_1), and a reflected wave ($\Gamma_{in}a_1$) which is redirected to the input-side termination.

identical amplifiers having voltage gain A and input reflection coefficient Γ_{in} are combined with quadrature hybrids at both input and output. In the hybrids' operating range, waves traversing from one side to the other experience an excess phase of 90° when crossing diagonally compared to when they cross directly. Thus the direct reflection at the input port comprises the superposition of two reflected waves, one of which has experienced a total phase shift (diagonally and back) 180° greater than the other (directly across and back). Thus, these two reflected components cancel.

In contrast, the two reflected waves add constructively at the adjacent termination (after both reflected waves make one direct crossing and one diagonal). The amplified output waves add constructively at an output port diagonally opposed to the input port, as shown.

Backward-traveling waves incident at the output port (not shown) are similarly redirected to a termination on the output side. In principle, this would allow the balanced amplifier to have ideal return loss at both ports. In practice, this depends on the amplitude and phase balance of the hybrids and on the homogeneity of characteristics between the two amplifiers.

12.2 LOW-NOISE AMPLIFIERS

The ultimate sensitivity of an electronic system to weak inputs, say, the wireless transmission from a very distant transmitter, is not actually a function of gain, but of noise. Recall from Section 1.6.3 that the random, thermally driven motions of charge carriers in any conductor creates small, broad-spectrum fluctuations in voltage and current which appears across the conductor's terminals as noise. Active devices have additional sources of noise, not all of which are thermal in origin. A received signal, no matter how weak, can always be amplified, but the background

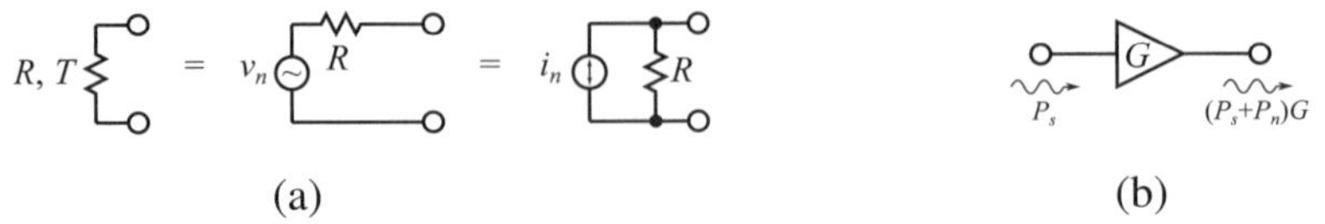

Figure 12.5 (a) Resistor at a physical temperature, T, and its associated Thévenin and Norton equivalent noise sources. (b) An amplifier with power gain, G, which adds input-referred noise power, P_n, to the signal power, P_s, prior to amplification.

noise from the external environment and from the internal electronics will be amplified along with it. If this noise is too strong, it will obscure the input signal, and no amount of electronic gain will recover it.

This makes it critically important for some amplifiers that the noise power they add to the signals passing through them is as small as possible. This, it turns out, requires an entirely different impedance be presented to an amplifier's input than the conjugate match required for optimum gain. To understand why, we must first look at how electronic noise is generated and propagates through a linear network.

12.2.1 Noise Temperature and Power

Let us consider a single resistor at a finite physical temperature, T (expressed in absolute temperature units of kelvin), as indicated in Figure 12.5(a). Although passive, we may consider this resistor a stochastic signal generator with Thévenin and Norton equivalent sources. The available power spectral density from this generator is given by *Planck's blackbody radiation law*,

$$S_n = \frac{P_n}{B} = \frac{hf}{e^{hf/kT} - 1} \tag{12.36}$$

where B is the bandwidth under consideration, P_n is the total noise power available in that bandwidth, h is *Planck's constant* ($= 6.62607015 \times 10^{-34}$ J·s), and k is *Boltzmann's constant* ($= 1.380649 \times 10^{-23}$ J/K)[2].

2 When the writing of this book began, the values of both Planck's and Boltzmann's constants were approximate quantities determined by physical measurements. However, the 26th General Conference on Weights and Measures held in November 2018 approved a redefinition of certain SI units, such as the kelvin and the kilogram, in terms of fundamental physical quantities. Thus, as these new definitions came into effect on May 20, 2019, the above values for both Planck's constant and Boltzmann's constant are now exact [12].

For most practical frequencies and temperatures, we have $hf \ll kT$; thus, the above may be simplified,

$$P_n \approx \frac{hfB}{1 + hf/kT - 1} = kTB \tag{12.37}$$

This, known as the *Rayleigh-Jeans approximation*, expresses the available noise power per unit bandwidth as being simply proportional to temperature. For a simple resistor (or termination) at the standard temperature, $T_0 = 290\text{K}$, the Rayleigh-Jeans approximation gives a noise spectral power density of about -84 dBm/GHz. We may furthermore write the noisy equivalent voltage or equivalent current noise sources attached to our resistor as follows,

$$\overline{v_n^2} = 4kTBR \tag{12.38a}$$

$$\overline{i_n^2} = \frac{4kTB}{R} \tag{12.38b}$$

where the overline represents the long-term mean, or expected value, and the factors of four are needed since P_n was written as the available power (e.g., the rms voltage or current is divided equally between the source and load resistor).

For devices other than simple resistors, it is often convenient to express the noise power that they deliver in terms of their *equivalent noise temperature*, T_e, in reference to (12.37), which for active devices may not be the same as their physical temperature. The ratio of the noise generated by such a source to the standard-temperature resistor above is termed the *excess noise ratio* (ENR) and is often given in decibels.

An important performance metric for amplifiers (or of any networks for that matter) is the amount of noise they add to the signals passing through them, as illustrated in Figure 12.5(b). In the case of amplifiers, it is customary to quantify that noise power as though it were added as the signal enters the input, before being amplified. Thus, the signal power and the noise power undergo the same gain factor, G, after passing through the device. This is termed the *input-referred noise power*, or, if expressed as a temperature via the Rayleigh-Jeans approximation, the *input-referred equivalent noise temperature*.

The degradation in signal-to-noise ratio for signals passing through an amplifier, when the input is assumed to be at the standard temperature, T_0, is the amplifier's *noise factor*, F (commonly called *noise figure* when expressed in decibels). The noise factor is therefore directly related to the equivalent noise temperature,

$$F = \frac{SNR_{in}}{SNR_{out}} = \frac{P_s/P_0}{P_s/\left(P_0 + P_n\right)} = 1 + \frac{P_n}{P_0} = 1 + \frac{T_e}{T_0} \tag{12.39}$$

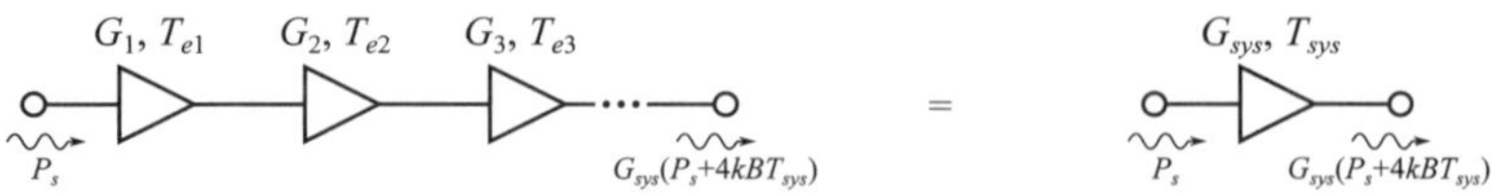

Figure 12.6 A cascade of low noise amplifiers each having different power gains and equivalent noise temperatures, and its single-stage equivalent.

where P_s is the input signal power, P_n is the noise power added by the network, and P_0 is the noise power associated with a source termination at temperature T_0. SNR stands for signal-to-noise ratio.

It is interesting to consider the equivalent noise temperature or noise factor of a cascade of amplifiers, each having potentially different characteristics, as shown in Figure 12.6. First, note that the total system power gain is simply the product of all the single stage power gains,

$$G_{sys} = \prod_{i=1}^{N} G_i \tag{12.40}$$

The total signal and noise power at the output of the system is then

$$P_{sys} = P_s G_{sys} + 4kB \left(T_{e1} \prod_{i=1}^{N} G_i + T_{e2} \prod_{i=2}^{N} G_i + T_{e3} \prod_{i=3}^{N} G_i + \cdots \right) \tag{12.41a}$$

$$= G_{sys} \left(P_s + 4kBT_{sys} \right) \tag{12.41b}$$

where

$$T_{sys} = T_{e1} + \frac{T_{e2}}{G_1} + \frac{T_{e3}}{G_1 G_2} + \frac{T_{e4}}{G_1 G_2 G_3} + \cdots \tag{12.42}$$

Applying (12.39), we may also write

$$F_{sys} = 1 + \frac{T_{sys}}{T_0} = F_1 + \frac{F_2 - 1}{G_1} + \frac{F_3 - 1}{G_1 G_2} + \frac{F_4 - 1}{G_1 G_2 G_3} + \cdots \tag{12.43}$$

Note that the first amplifier in the cascade makes a dominant contribution to the overall system noise performance.

Still, if the gain of the input stage is not sufficiently large, the second, and possibly later stages will contribute more to the system equivalent noise temperature than one may consider acceptable. It is convenient to define a secondary metric of performance that takes into account the amplifier's gain by expressing the input-referred equivalent noise temperature of an infinitely long cascade of identical devices,

$$T_{cas} = T_e \left(1 + \frac{1}{G} + \frac{1}{G^2} + \frac{1}{G^3} + \cdots \right) = \frac{T_e}{1 - 1/G} \tag{12.44}$$

and the closely related *noise measure*, N_m, which is simply the noise factor of an infinitely long cascade, or $N_m = 1 + T_{cas}/T_0$.

A real system has devices other than amplifiers. The cascade rules in (12.42) and (12.43) still apply, where the power gains for lossy, passive devices will be less than one; however, we must derive an expression for the noise factor, or equivalent noise temperature, of a passive two-port device. The simplest way to do that is to recognize that a matched, passive device with loss, connected to a matched input termination, say, a lossy transmission line with characteristic impedance Z_0, emits the same noise at room temperature as a simple resistor. In other words, while the line may attenuate some of the noise delivered by its input termination, it replaces that noise with an equal amount from its own, internal lossy elements. Driven, non-noise, input signals are simply attenuated (and not replaced). Therefore, the signal-to-noise ratio is degraded by the loss factor, $G = 1/L \leq 1$. By the definition of noise factor and equivalent noise temperature,

$$F = L = \frac{1}{G} \tag{12.45a}$$

$$T_e = T_0(F - 1) = T_0 \left(\frac{1}{G} - 1\right) \tag{12.45b}$$

In fact, the equivalent noise temperature of a passive component at any ambient temperature is $T_e = T_{amb}(1/G - 1)$. However, the noise factor $F = 1/G$ is only correct at the standard room temperature, $T_0 = 290\text{K}$.

12.2.2 Noise Models of Two-Port Networks

Fundamentally, an N-port device will generate noise at each of its ports as though an otherwise noiseless network had generators attached, as shown for the two-port networks in Figure 12.7. These generators may take the form of Thévenin equivalent voltage sources, Norton equivalent current courses, or even wave sources, as shown in Figures 12.7(a–c), where the source impedance is determined by the network. In general, the stochastic signals generated by these sources will be partially correlated

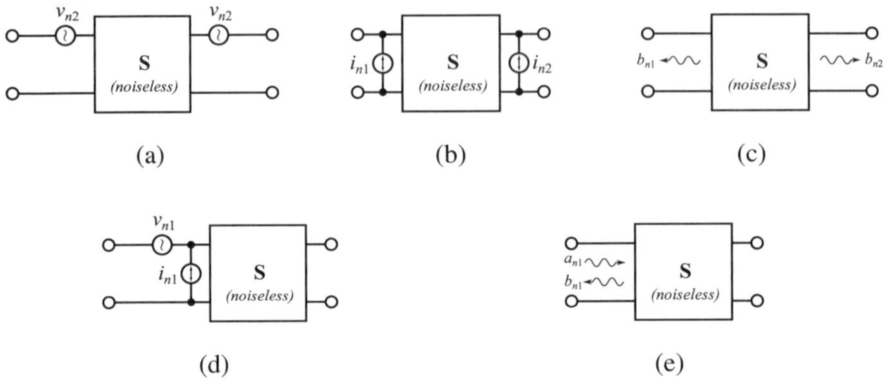

Figure 12.7 Several representations of a noisy two-port network, comprising an equivalent noiseless network with scattering matrix **S** plus (a) a voltage noise source at each port, (b) a current noise source at each port, (c) a wave noise source at each port, (d) a voltage and a current noise source at one port, and (e) a forward-traveling and backward-traveling wave noise source at one port. The noise sources will, in general, be partially correlated.

with one another. It therefore takes four independent parameters at any given frequency to describe the noise properties of a two-port network (two real-valued source amplitudes and a complex correlation coefficient) aside from the network immittance or scattering parameters themselves.

It is also possible to move the equivalent noise sources from one port to another, through an appropriate transformation given by the network parameters. Thus, we may gather both noise-sources for a two-port network at the input as shown for the voltage and current sources in Figure 12.7(d) and for the forward-traveling and backward-traveling wave sources in Figure 12.7(e) (keeping in mind that these wave amplitudes are not related by the network's reflection coefficient, since they are generated independently by internal sources). Each model is a complete and equivalent representation of a network and the noise that it generates.

12.2.3 Noise Parameters and Matching

It was stated in Section 12.2.2 that four noise parameters are needed to describe the noise-emission properties of a two-port network, aside from the network port parameters themselves. This set may take on many forms, but an especially convenient form expresses the minimum equivalent noise temperature (or noise figure) which a network can be made to have, the complex-valued source impedance or reflection

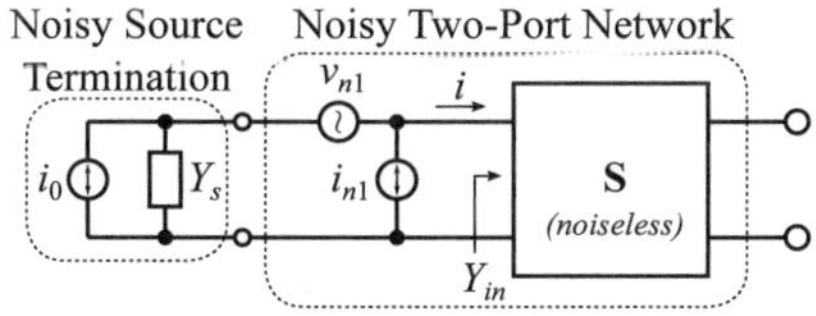

Figure 12.8 Model of a noisy two-port network with noisy source termination, used for calculation of the noise parameters.

coefficient that achieves it, and an additional parameter describing how rapidly the achieved noise temperature deviates from that minimum with nonoptimal source impedance.

Consider the noise model in Figure 12.7(d) having two noise generators at the input, one voltage source and one current source. Say attached to the input is an arbitrary source admittance, $Y_s = 1/Z_s$, which, at the standard temperature, T_0, delivers an additional noise current, i_0, into the network, as shown in Figure 12.8. The total current entering the device may then be written,

$$i = \frac{Y_{in}\left(i_0 + i_{n1}\right)}{Y_{in} + Y_s} + \frac{v_{n1}}{Z_{in} + Z_s} = \frac{Y_{in}}{Y_{in} + Y_s}\left(i_0 + i_{n1} + Y_s v_{n1}\right) \tag{12.46}$$

where $Y_{in} = 1/Z_{in}$ is the input admittance of the two-port network. The noise factor is given by (12.39) in terms of the noise powers,

$$F = 1 + \frac{P_n}{P_0} = 1 + \frac{\overline{\left(i_{n1} + Y_s v_{n1}\right)^2}}{\overline{i_0^2}} \tag{12.47}$$

Note that since these are stochastic signals, we have used the rms values of the currents to calculate the noise powers. The numerator can be simplified by noting that the network noise generators, v_{n1} and i_{n1} are partially correlated. Let us write the current noise source as having an uncorrelated part, i_u, and a part that is correlated to v_{n1},

$$i_{n1} = i_u + Y_c v_{n1} \tag{12.48}$$

where Y_c is called the *correlation admittance* [13]. Thus,

$$F = 1 + \frac{\overline{\left(i_u + \left(Y_c + Y_s\right) v_{n1}\right)^2}}{\overline{i_0^2}} = 1 + \frac{\overline{i_u^2} + \left|Y_c + Y_s\right|^2 \overline{v_{n1}^2}}{\overline{i_0^2}} \tag{12.49}$$

where the last step follows because the remaining terms in the numerator are, by definition, uncorrelated. The Rayleigh-Jeans approximation gives $\overline{i_0^2} = 4kT_0BG_s$ (where $G_s = \text{Re}\{Y_s\}$). Let us express i_u and v_{n1} in similar forms,

$$\overline{i_u^2} = 4kT_0BG_u \tag{12.50a}$$

$$\overline{v_{n1}^2} = 4kT_0BR_n \tag{12.50b}$$

Therefore, the noise factor may be written

$$F = 1 + \frac{G_u}{G_s} + \frac{R_n}{G_s}|Y_c + Y_s|^2 \tag{12.51a}$$

$$= 1 + \frac{G_u}{G_s} + \frac{R_n}{G_s}\left((G_c + G_s)^2 + (B_c + B_s)^2\right) \tag{12.51b}$$

Let us now select a source admittance, $Y_s = G_s + B_s$, that minimizes this expression. The imaginary part is obviously given by $B_s = -B_c$. For the real part, we set the partial derivative equal to zero,

$$\frac{\partial F}{\partial G_s} = -\frac{G_u}{G_s^2} + R_n\frac{2G_s(G_c + G_s) - (G_c + G_s)^2}{G_s^2} = 0 \tag{12.52a}$$

$$\therefore R_n\left(G_s^2 - G_c^2\right) = G_u \tag{12.52b}$$

$$G_s = \pm\sqrt{G_c^2 + \frac{G_u}{R_n}} \tag{12.52c}$$

Taking the positive root, the optimum passive source admittance is

$$Y_{opt} = G_{opt} + jB_{opt} = \sqrt{G_c^2 + \frac{G_u}{R_n}} - jB_c \tag{12.53}$$

and the minimum noise factor given this source termination is

$$F_{min} = 1 + \frac{G_u + R_n(G_c + G_s)^2}{G_s} \tag{12.54a}$$

$$= 1 + \frac{G_u + R_n\left(G_c + \sqrt{G_c^2 + \frac{G_u}{R_n}}\right)^2}{\sqrt{G_c^2 + \frac{G_u}{R_n}}} \tag{12.54b}$$

$$= 1 + 2R_nG_c + 2\sqrt{(R_nG_c)^2 + R_nG_u} \tag{12.54c}$$

More importantly, the achieved noise factor for any source admittance is

$$F = 1 + \frac{G_u}{G_s} + \frac{R_n}{G_s}\left((G_c + G_s)^2 + (B_c + B_s)^2\right) \tag{12.55a}$$

$$= 1 + 2R_nG_c + 2R_nG_{opt} + \frac{R_n}{G_s}\left((G_s - G_{opt})^2 + (B_s - B_{opt})^2\right) \tag{12.55b}$$

$$= F_{min} + \frac{R_n}{G_s}\left|Y_s - Y_{opt}\right|^2 \tag{12.55c}$$

We therefore see once again that the noise properties of a linear, two-port network may be completely described by four noise parameters: F_{min}, R_n, and the real and imaginary parts of Y_{opt}.

Alternatively, since $T_e = T_0(F - 1)$, we may write

$$T_e = T_{min} + T_0\frac{R_n}{G_s}\left|Y_s - Y_{opt}\right|^2 \tag{12.56}$$

R_n is called the *noise resistance*, and describes how rapidly the noise factor and noise temperature degrade as the source admittance, Y_s, deviates from the optimum. A superior formulation uses an alternative to this parameter, $N = R_nG_{opt}$, where

$$T_e = T_{min} + NT_0\frac{\left|Y_s - Y_{opt}\right|^2}{G_sG_{opt}} \tag{12.57a}$$

$$= T_{min} + NT_0\frac{\left|Z_s - Z_{opt}\right|^2}{R_sR_{opt}} \tag{12.57b}$$

Known as the *Lange noise parameter*, N serves the same function as the noise resistance, but is furthermore invariant under lossless, reciprocal embedding [14]. In other words, when a matching network is added to a device for which T_{min}, N, and the optimum source immittance are known, the optimum source immittance changes, but N stays the same (as does T_{min}).

It is useful also to write the equivalent noise temperature in terms of the reflection coefficients,

$$T_e = T_{min} + NT_0\frac{\left|Z_s - Z_{opt}\right|^2}{\text{Re}\left\{Z_s\right\}\text{Re}\left\{Z_{opt}\right\}} \tag{12.58a}$$

$$= T_{min} + 4NT_0 \frac{\left| \frac{1+\Gamma_s}{1-\Gamma_s} - \frac{1+\Gamma_{opt}}{1-\Gamma_{opt}} \right|^2}{\left(\frac{1+\Gamma_s}{1-\Gamma_s} + \frac{1+\Gamma_s^*}{1-\Gamma_s^*} \right) \left(\frac{1+\Gamma_{opt}}{1-\Gamma_{opt}} + \frac{1+\Gamma_{opt}^*}{1-\Gamma_{opt}^*} \right)} \quad (12.58b)$$

$$= T_{min} \left(1 + M \frac{\left| \Gamma_s - \Gamma_{opt} \right|^2}{\left(1 - |\Gamma_s|^2 \right) \left(1 - |\Gamma_{opt}|^2 \right)} \right) \quad (12.58c)$$

where

$$M = \frac{4NT_0}{T_{min}} \quad (12.59)$$

Although the alternative noise parameter M is not commonly articulated in textbooks, it is in some ways more fundamental than either R_n or N. Like both N and T_{min}, it is invariant under lossless, reciprocal embedding, and is as well unchanged by the parallel connection of identical devices (which is effectively the case when one scales a transistor's base or gate width). Unlike N, the parameter M eliminates the meaningless scaling factor of T_0, an arbitrary reference temperature which may have nothing at all to do with the current application (say, that of a cryogenically cooled front-end).

Since M (like R_n and N) describes the sensitivity of an amplifier's noise performance to suboptimal impedance-matching, designers are strongly motivated to identify active devices for which M is as small as possible, thus easing the design of broadband systems. Unfortunately, there is a fundamental physical limit on how small M may be. Consider how the value of M depends on the correlation of the network noise sources, v_{n1} and i_{n1},

$$M = \frac{4NT_0}{T_{min}} = \frac{4R_n G_{opt}}{F_{min} - 1} = \frac{2G_{opt}}{G_c + G_{opt}} = 2\left(1 + \frac{G_c}{G_{opt}}\right)^{-1} \quad (12.60a)$$

$$= 2\left(1 + \frac{G_c}{\sqrt{G_c^2 + \frac{G_u}{R_n}}}\right)^{-1} = 2\left(1 + \frac{G_c}{\sqrt{G_c^2 + \frac{\overline{i_u^2}}{\overline{v_{n1}^2}}}}\right)^{-1} \quad (12.60b)$$

Note that the second term in parentheses must always have magnitude less than 1. The minimum value of M occurs when the two noise sources are 100% correlated (making $i_u = 0$). Since no two sources can be more than 100% correlated, we must have $M \geq 1$. This is a fundamental property of any linear, two-port network, irregardless of the technology.

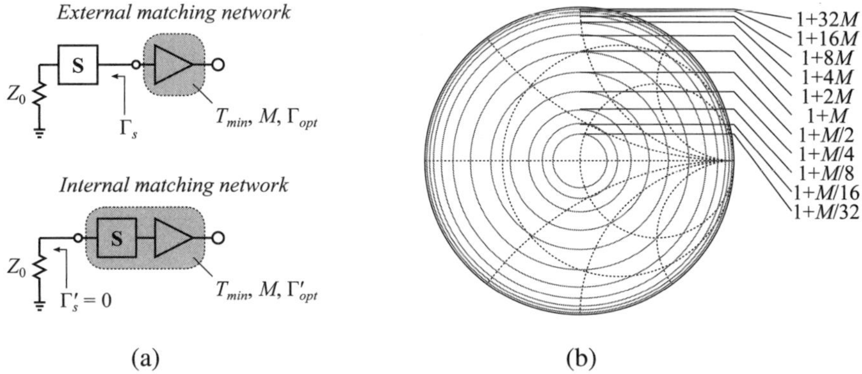

Figure 12.9 (a) Two viewpoints for designing a low-noise amplifier: as a device with intrinsic noise parameters to be matched by an external matching network, and as an embedded system with internal matching designed to yield $\Gamma'_{opt} \approx 0$. Note that T_{min} and M are the same in both scenarios, only Γ'_{opt} is modified by the matching network. (b) Contours for Γ'_{opt} on a Smith chart yielding noise temperature ratios $T_e/T_{min} = 1 + M/32$ to $1 + 32M$.

Incidentally, when the two sources are completely uncorrelated, we have $G_c = 0$, and thus $M = 2$. Strictly speaking, it is possible for M to be larger than 2 for partially correlated sources if $G_c < 0$, but this has been shown not to be the case for most common forms of devices like transistors [15]. We therefore find that, in most practical cases, M is constrained such that

$$1 \le M \le 2 \tag{12.61}$$

The typical perspective one takes with an expression like (12.58c) is that shown at the top in Figure 12.9(a), where an external matching network transforms the system characteristic impedance into one having reflection coefficient as close as possible to Γ_{opt}. An alternative point of view, shown at the bottom in Figure 12.9(a) considers the matching network as integral to the device, so that the optimum reflection coefficient of the combined network, Γ'_{opt}, is as close as possible to 0. Since the external terminations in this view are always Z_0, we have $\Gamma'_s = 0$ and (12.58c) may be simplified,

$$T_e = T_{min}\left(1 + M\frac{\left|\Gamma'_{opt}\right|^2}{1 - \left|\Gamma'_{opt}\right|^2}\right) \tag{12.62}$$

Table 12.2
Reflection Coefficient Contours for Constant Gain, Constant Noise, and Stability

Metric	Reflection Plane	Contour Parameter	Center	Radius
Unilateral gain	Γ_S	$g_s = \frac{G_S}{G_{S,max}}$	$\frac{g_s s_{11}^*}{1-(1-g_s)\lvert s_{11}\rvert^2}$	$\frac{\sqrt{1-g_s}(1-\lvert s_{11}\rvert^2)}{1-(1-g_s)\lvert s_{11}\rvert^2}$
Unilateral gain	Γ_L	$g_l = \frac{G_L}{G_{L,max}}$	$\frac{g_l s_{22}^*}{1-(1-g_l)\lvert s_{22}\rvert^2}$	$\frac{\sqrt{1-g_l}(1-\lvert s_{22}\rvert^2)}{1-(1-g_l)\lvert s_{22}\rvert^2}$
Available gain	Γ_S	$g_a = \frac{G_A}{\lvert s_{21}\rvert^2}$	$\frac{g_a(s_{11}^*-\lvert\Delta\rvert^2 s_{22})}{1+g_a(\lvert s_{11}\rvert^2-\lvert\Delta\rvert^2)}$	$\frac{\sqrt{1-2K\lvert s_{12}s_{21}\rvert g_a+\lvert s_{12}s_{21}\rvert^2 g_a^2}}{\lvert 1+g_a(\lvert s_{11}\rvert^2-\lvert\Delta\rvert^2)\rvert}$
Operating gain	Γ_L	$g_p = \frac{G_P}{\lvert s_{21}\rvert^2}$	$\frac{g_p(s_{22}^*-\lvert\Delta\rvert^2 s_{11})}{1+g_p(\lvert s_{22}\rvert^2-\lvert\Delta\rvert^2)}$	$\frac{\sqrt{1-2K\lvert s_{12}s_{21}\rvert g_p+\lvert s_{12}s_{21}\rvert^2 g_p^2}}{\lvert 1+g_p(\lvert s_{22}\rvert^2-\lvert\Delta\rvert^2)\rvert}$
Noise*	Γ_S	X	$\frac{\Gamma_{opt}}{X+1}$	$\frac{\sqrt{X(X+1-\lvert\Gamma_{opt}\rvert^2)}}{X+1}$
Stability	Γ_S	$K=1$	$\frac{(s_{11}-s_{22}^*\Delta)^*}{\lvert s_{11}\rvert^2-\lvert\Delta\rvert^2}$	$\left\lvert\frac{s_{12}s_{21}}{\lvert s_{11}\rvert^2-\lvert\Delta\rvert^2}\right\rvert$
Stability	Γ_L	$K=1$	$\frac{(s_{22}-s_{11}^*\Delta)^*}{\lvert s_{22}\rvert^2-\lvert\Delta\rvert^2}$	$\left\lvert\frac{s_{12}s_{21}}{\lvert s_{22}\rvert^2-\lvert\Delta\rvert^2}\right\rvert$

$^*X = \left(\frac{T_e}{T_{min}} - 1\right)\frac{1-\lvert\Gamma_{opt}\rvert^2}{M}$

This allows us to show contours on a Smith chart in Figure 12.9(b) corresponding to the excess noise temperature ratio, T_e/T_{min}, which applies universally to all linear, two-port networks. The insight provided by this representation is simply that the noise performance of an actual amplifier, relative to the best that it could theoretically achieve, is simply a function of the distance Γ_{opt} lies form the origin, scaled by M. Whether Γ_{opt} is inductive or capacitive, for example, makes no difference.

For the more general case when $\Gamma_s \neq 0$, constant noise contours are still circles, but their centers and radii depend on the device parameters. Formulas for these circles [1] as well as constant gain circles and stability circles (derived in Section 12.1.3) are shown in Table 12.2.

)Stability, gain, and noise circles can now be easily calculated and plotted across frequency using commonly available computer-aided design (CAD programs, but it is helpful for the designer to under their origin and limitations. To illustrate their use, let us consider the design of low-noise amplifier as an example,

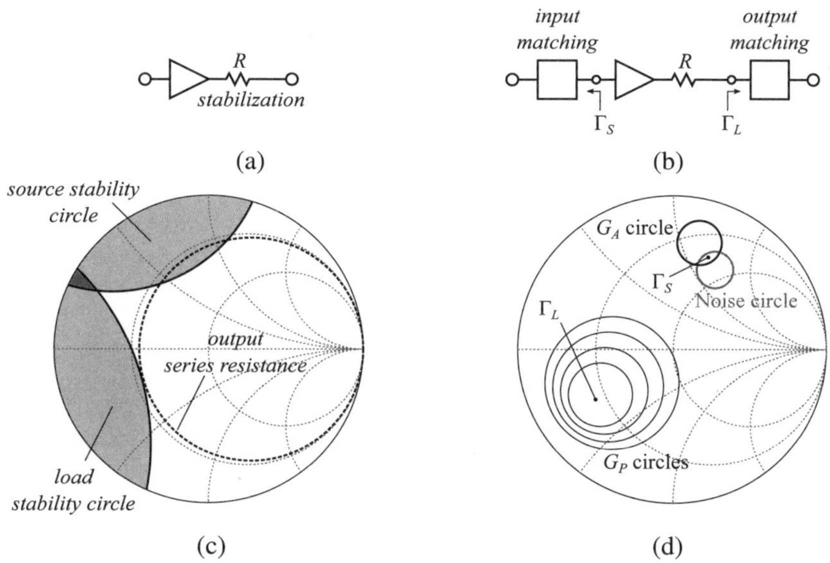

Figure 12.10 (a) Amplifying device with output stabilization resistor. (b) Full amplifier with input and output matching networks. (c) Source and load stability circles for the amplifying element. The shaded regions denote potentially unstable source and load impedances. (d) Available gain (1-dB contour), operating gain (0.2-dB contours), and noise figure (0.1-dB contour) circles used for designing the matching networks.

shown in Figure 12.10(a). The scattering parameters for a high-frequency transistor were simulated at 20 GHz. It has a maximum stable gain at that frequency of about 13 dB, and minimum noise figure of about 1.1 dB, but it is not unconditionally stable. The source and load stability circles are plotted on the Smith chart in Figure 12.10(c).

The location of the stability circles towards the left edge of the chart suggest that a series resistance could help to stabilize it. A circle of constant resistance that is tangent to the load stability circle is shown with a dashed line. If a resistor at least as large as that represented by this contour is added in series at the output, all other passive loads will present an impedance to the amplifier that lies within this circle, within the stable region (we typically select a slightly larger resistor to provide a safety margin). Note that the intersection with the source stability circle is unimportant, since we are considering output resistances here. If the stabilization resistor was added to the input instead of the output, then a different (even larger) resistance value would be needed. We will choose to stabilize at the output, since an input resistor would have greater impact on the achievable noise figure. Whichever stabilization resistor is used, ensuring a stable termination at one port will simultaneously move the stability circle for the other port beyond the passive boundary of the Smith chart.[3]

With stability now assured, we must now design the input and output matching networks as shown in Figure 12.10(b). To that end, we calculate and plot the gain and noise circles for the transistor-resistor combination in Figure 12.10(d). For the output match, we use the operating gain circles, since these depend only on the load impedance. We have plotted contours representing a 0.2-dB penalty each. Since there is no other constraint on the load impedance, we select a value in the center of these contours.

The input match is a little more difficult. We must compromise between gain and noise matching. For the gain, we plot a 1-dB available gain contour (since available gain depends only on the source impedance). For noise we plot a contour corresponding to a 0.1-dB noise figure penalty. We select a source impedance that lies in the overlap region of these two circles, as shown. Ultimately, we achieve a gain of 11 dB at the selected frequency with an associated noise figure of 1.5 dB (the degradation from optimum in both cases is due to the stabilization resistor which introduces some loss). This moderate level of gain may not be sufficient in some applications, as later, lesser-performing components may still make a significant contribution to the overall system noise. If more gain is needed, additional gain stage may be required. While there is no universal recipe for optimizing multistage amplifiers, a brief discussion of useful strategies is given in Section 12.4.

12.3 POWER AMPLIFIERS

When amplifiers are used not to receive a weak signal, but to transmit, or to generate a powerful coherent signal for energetic applications, it is necessary to ensure that the amplifier delivers as much output power as possible. It may further be important that the amplifier remain linear in its behavior, and/or to do so most efficiently (that is, maximize the signal output while minimizing the DC bias power consumed). Amplifiers for which these metrics are most important are called *power amplifiers*.

---

3 We are only considering a single frequency for this example, but in the real world it is necessary to ensure stability at all frequencies from DC to f_{max}. In practice, stabilization resistors such as this stabilize best near the operating band and above, while much lower frequencies must be stabilized using lossy elements in the bias tees or power supply, where they do not affect in-band performance.

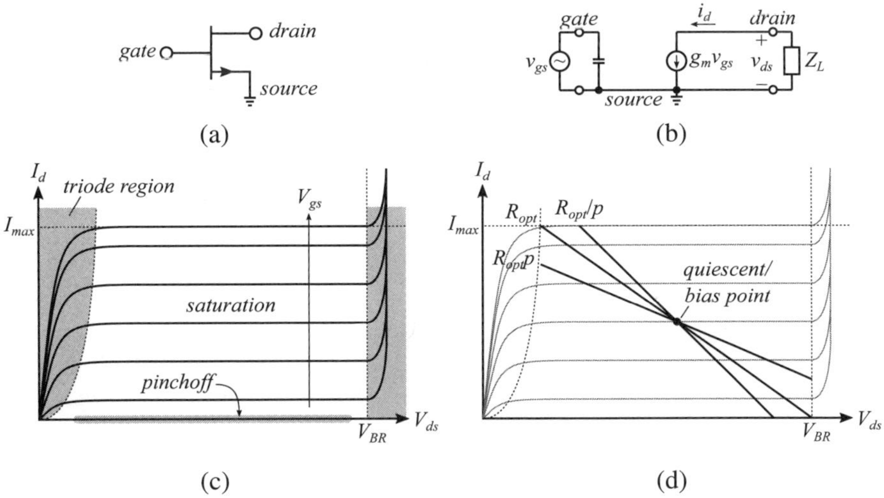

Figure 12.11 (a) Schematic symbol for a field-effect transistor (common-source configuration). (b) Simplified small-signal model with input voltage generator and output load impedance. (c) Intrinsic current-voltage (IV) curves. Each curve represents a different gate-source voltage, and V_{BR} denotes the drain breakdown voltage. (d) Resistive loadlines for maximum power (R_{opt}), suboptimal voltage swing (R_{opt}/p) and suboptimal current swing ($R_{opt}p$).

Small-signal characteristics such as noise figure are rarely a concern for power amplifiers, but the stability criterion discussed previously remain important.

12.3.1 Saturated Power

We have not yet had much to say about nonlinearity or nonlinear devices in this book. The physical mechanisms that govern nonlinear behavior are generally device-specific and beyond the current scope; however, it should not be too hard to convince oneself that no real-world device, no matter what its construction, can maintain linear (and nondestructive) behavior over all amplitude scales. Dynamic range is limited; at some point, all devices cease to be linear.

For power amplifiers, the aspects of design that one must consider when attempting to maximize output power are best illustrated by an example. Figure 12.11(a) shows the schematic symbol for a field-effect transistor (FET), and a (vastly simplified) circuit model is given in Figure 12.11(b), where we have attached a voltage signal generator at the gate as the input and an arbitrary load impedance,

Z_L at the drain as the output. FETs are generally transconducting devices, wherein the voltage applied between the gate and source terminals (v_{gs}) is sensed and a proportional amount of current is drawn into the drain terminal (i_d). The proportionality factor, g_m, is the transconductance, and has units of siemens, or inverse-ohms. It must be emphasized that this circuit represents the small-signal behavior only; in reality, the transistor will need a DC bias at each port to center it within its linear operating range. The DC bias sources and associated bias tees (comprising chokes and DC blocks) are not shown.

The full current-voltage (IV) relationship for a typical FET — including both the bias and small-signal components — is shown in Figure 12.11(c). The horizontal and vertical axes represent the drain-source voltage and the drain current, respectively, while each curve is for a different gate-source voltage. The limits over which the device operates as a linear transconductor are drawn with dotted lines. That linear operating region is somewhat misleadingly called the *saturation region*, not to indicate that the power is saturated, but that the drain current has reached a point where it is no longer dependent on drain-source voltage (note the lines are horizontal in this region).

At the left edge of the plot, a curved line marks the boundary between this region and the *triode region*, where the drain-source channel is not fully turned on. Also, as the gate voltage is lowered, a point is reached where the drain current drops to zero and the device is turned off. This is called *pinchoff*.

Both the current and the voltage have maximum values as well. At the top edge of the plot, a limit is reached beyond which further increases in V_{gs} can no longer add to the current. We will call this limit I_{max}. Finally, at the right edge of the plot, an ultimate drain-source voltage (V_{BR}) is reached wherein the gate can no longer exercise complete control over the channel, and the drain current rises sharply. This is called *breakdown* and is usually associated with damage to the device.

We thus find that output voltage and current associated with the drain terminal have a limited range over which they can swing. If the signal perturbations away from the quiescent point are small, say, less than one-tenth the bias level, then small-signal operation can be assumed and the results of the previous section apply. If, however, the signal swing carries the instantaneous operating point of the device near any of these limits, we must consider the large-signal behavior. A current-voltage trajectory which maximizes the swing in both dimensions is shown in Figure 12.11(d), labeled R_{opt}. R_{opt} represents the real-valued impedance of the load, Z_L, for which this trajectory is traced out given adequate drive power at the gate. The negative slope is due to the reference direction of the current in this plot, i_d, which is into the drain terminal and out of the (passive) load. The dot in the middle represents the DC bias, or quiescent point. Ideally, to maximize power delivered, the quiescent point should be located halfway between the minimum and maximum voltage and current limits. Trajectories such as these are called *loadlines*.

Two other loadlines are shown on the plot, one for which the voltage swing is suppressed by a factor p, and the other for which the current swing is suppressed by a factor p. These are labeled R_{opt}/p and $R_{opt}p$, respectively, as these are the resistive load impedances which achieve them. Note that in both cases, the power delivered to the load is suppressed by exactly the same factor,

$$P_{opt} = \frac{1}{2} v_{sw} i_{sw} \tag{12.63a}$$

$$P_{sub-V} = \frac{1}{2} \left(\frac{v_{sw}}{p} \right) i_{sw} = \frac{P_{opt}}{p} \tag{12.63b}$$

$$P_{sub-I} = \frac{1}{2} v_{sw} \left(\frac{i_{sw}}{p} \right) = \frac{P_{opt}}{p} \tag{12.63c}$$

where $\pm v_{sw}$ and $\pm i_{sw}$ are the maximum voltage and current swings available, respectively, from the quiescent point to the appropriate limits on both axes.

For complex loads, having both real and imaginary parts, we must be a little more careful. Note that if we start from R_{opt}/p and add some series reactance, the current through the resistive part is unchanged, but an additional reactive component is added to the total voltage trajectory seen by the device, increasing the voltage swing without delivering any extra power. However, if we start from $R_{opt}p$ and add a susceptance in parallel, the voltage across the resistive part will stay fixed, while the current swing at the device increases, and again the power is unchanged. The loadline in these two cases appears elliptical, as shown in Figure 12.12(a), and is traced either clockwise or counterclockwise depending on the sign of the reactance added. Clearly a point will be reached in either case where the voltage and/or current swings once again reach the practical limits of the device.

We thus obtain a closed boundary on the Smith chart, delineated by constant-resistance contours on the left and constant-conductance contours on the right, which corresponds to output power delivery of P_{opt}/p as shown in Figure 12.12(b). One may follow the same procedure for any number of power reduction factors, p, thus creating a contour diagram on the Smith chart which shows the expected maximum output power deliverable from an amplifier as a function of the output load impedance. Unlike the gain, noise, and stability contours of the previous

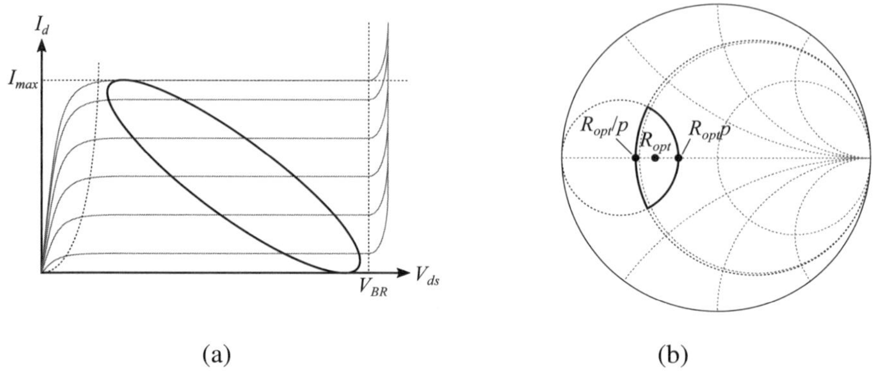

Figure 12.12 (a) A reactive loadline trajectory where some series reactance has been added to a suboptimal resistance, R_{opt}/p. The current swing is still maximized, but the voltage excursion has not yet reached device limits. (b) Locus of impedances on a Smith chart corresponding to a reduced output power capacity of P_{opt}/p.

sections, these contours are not circular, but instead are almond-shaped, formed by the intersection of two incongruent circular discs.

For this analysis, we have chosen a very simple transistor model with an ideal current source as its output stage. A real transistor will have parasitic elements as well, such as a drain-source capacitance, and a lead terminal inductance, as shown in Figure 12.13(a). These may be subsumed into the load termination, having the effect of shifting the contours around the Smith chart in opposition to that added by the parasitics themselves (in other words, since a parasitic series inductor adds reactance to the load impedance, that load must now have an equivalent negative reactance added to it in order to cancel out the parasitic's effect). While the original contours were frequency-independent, the shifted contours will generally drift across the Smith chart as a function of frequency [16], as shown in Figure 12.13(b).

12.3.2 Linearity

In the preceding analysis, it was generally assumed that the active device would behave linearly right up to the point where the voltage and current swings of the signal carried it past certain hard limits, causing the waveform to immediately *clip*, or else damage the device. In practice, the onset of nonlinearity is more gradual. When the energetic delivery of power is the application's only concern, these subtle

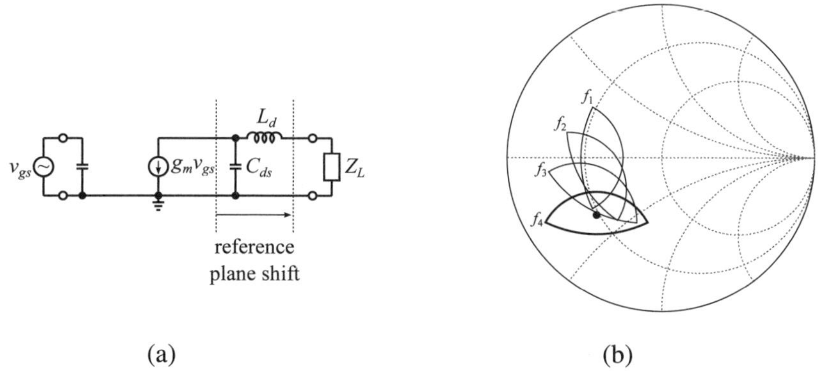

Figure 12.13 (a) Shift in reference plane from the transistor's internal current source to the drain terminal after parasitics. (b) Resulting, frequency-dependent load impedance contour.

effects can be largely ignored. In more sensitive applications, such as wireless data transmission, the distortion and generation of extraneous spectral components that tend to accompany these nonlinearities can be quite harmful.

A fairly typical plot of output power (P_{out}) versus input power (P_{in}) for an amplifier is shown in Figure 12.14(a). For sufficiently small signals, at the lower left of the plot where the device is linear, this curve is a straight line. The slope when plotted on decibel scales is 1:1, and its offset from the main diagonal (where $P_{out} = P_{in}$) is the operating power gain. As the power is increased beyond a certain limit, that offset will decrease. This is known as *gain compression*. The output power at which the gain has compressed by 1 dB is known as the *1-dB compression point*, denoted P_{1dB}. Additional increases in power will eventually drive the amplifier to saturation, P_{sat}.

While P_{1dB} emphasizes the loss of gain, other metrics directly relate to the growth of noninput spectral tones. Variously called *harmonics*, *spurs*, *mixing products*, or *intermodulation products* depending on the context, those which are further described as *third-order* exhibit 3:1 power growth slopes when the signal is sufficiently weak (in contrast with the directly amplified input, which has a 1:1 slope).[4] It is typical for the actual measured powers of these products (and of the input tone) to deviate from the simple linear slopes at high powers, but

4 The mathematical basis for the appearance and growth slopes of noninput spectral tones will be described in more detail in Section 13.1.

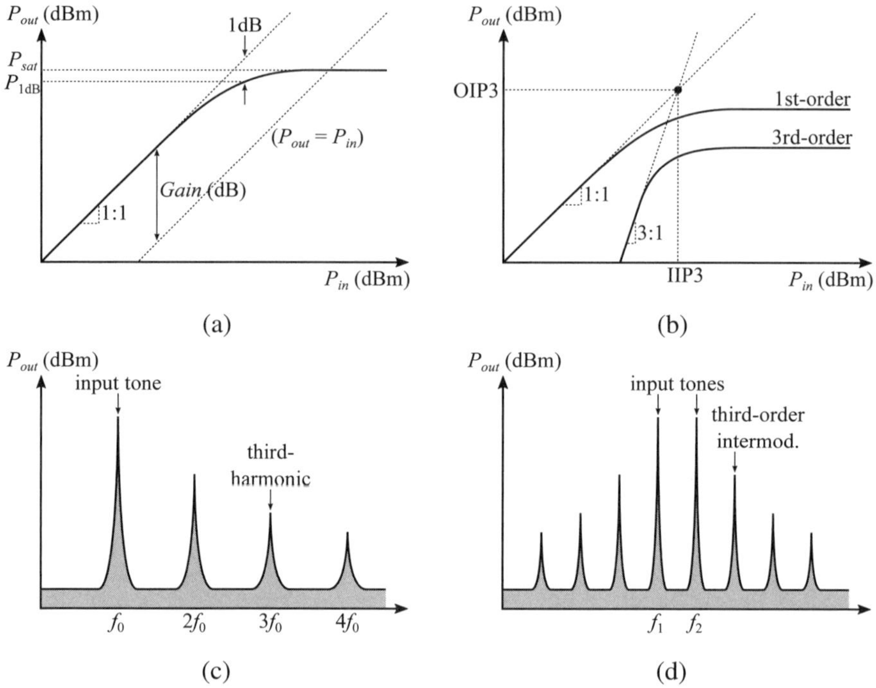

Figure 12.14 Definition of linearity performance metrics. (a) Output-referred 1-dB gain compression ($P_{1\text{dB}}$). (b) Third-order intercept (IIP3 or OIP3 for input- and output-referred, respectively). (c) Key spectral tones for third-order intercept measurement based on harmonics. (d) Key spectral tones for third-order intercept measurement based on intermodulation products.

the initial slopes can be extrapolated to where they intersect, known as the *third-order intercept point* (IP3). The input and output powers at which this intercept occurs are labeled IIP3 and OIP3, respectively. A general rule of thumb is that 1-dB compression occurs about 10 dB below the third-order intercept.

One should be careful that there are two different kinds of third-order products in common use with these definitions, but which do not give exactly the same results. The first is based on harmonics and is tested simply by driving a single input tone through the device. The output powers at that same frequency and at three times the input frequency are measured and their intercept point extrapolated. The issue with this measurement is that it typically requires that both the amplifier under test and the measurement system have sufficiently broad bandwidth to capture

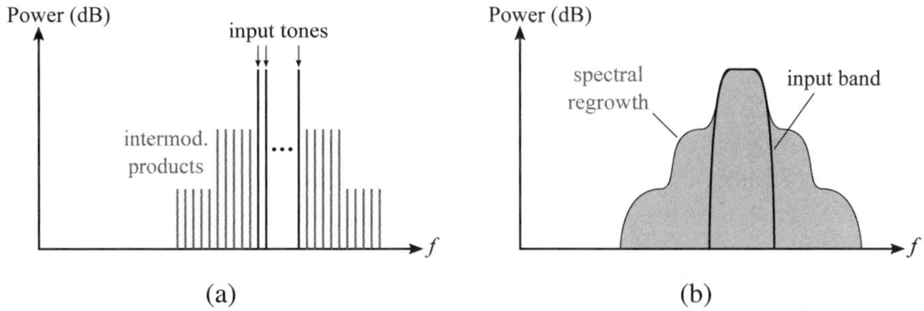

Figure 12.15 (a) Two-tone intermodulation distortion giving rise undesired, in-band or near-band spurious products. (b) Extension to bandlimited signals comprising (conceptually) an infinite number of tones.

both the fundamental tone and the third harmonic. An alternative test is based on intermodulation products. In this case, two closely spaced input tones, f_1 and f_2, are driven at equal powers into the amplifier. Nonlinear mixing in the amplifier generates third-order spectral tones at $2f_2 - f_1$ and $2f_1 - f_2$. The output powers at these frequencies are compared to those at the input frequencies, and again their intercept point is extrapolated [17].

One might wonder why the third-order products are used instead of second-order. Some standards for second-order intercepts do exist, but are less common. One possible reason is that intentional symmetries in certain nonlinear devices inherently suppress second-order products (some examples will be given in Chapter 13). More importantly, there are no second-order intermodulation products that are close in frequency to the input tone(s) — even for the two-tone test described above, the second-order intermodulation products occur at $2f_1$, $2f_2$, and $|f_1 \pm f_2|$, all of which are near to the second harmonic or DC, requiring the device and the measurement system to exhibit broadband behavior.

The third-order intermodulation products also play an important role in spectral regrowth, which is illustrated in Figure 12.15. Signals that comprise not just one or two tones but an entire band of spectral energy (Figure 12.15(b)) are still subject to intermodulation distortion, and as described above, it is the third-order (and higher) nonlinearities that lead to mixing products near to the band of interest.

12.3.3 Efficiency

Whenever working with devices that handle large amounts of power, efficiency becomes an important factor, in this case, the efficiency with which DC bias power is converted into signal energy at the desired output. For field-effect transistors, since power is almost exclusively consumed at the drain terminal (more accurately, between the drain and source terminals), we define the *drain efficiency* as the ratio of output power (at the fundamental harmonic), P_{out}, to the DC power,

$$\eta_d = \frac{P_{out}}{P_{DC}} \tag{12.64}$$

For an amplifier biased as in Section 12.3.1, such that the voltage and current swings, $\pm v_{sw}$ and $\pm i_{sw}$, just reach zero at the sinusoidal minimum, that is, the quiescent voltage and current are $v_q = v_{sw}$ and $i_q = i_{sw}$, respectively, the maximum theoretical drain efficiency is 50%,

$$\eta_d = \frac{P_{out}}{P_{DC}} \leq \frac{v_{sw} i_{sw}/2}{v_q i_q} = \frac{v_{sw} i_{sw}/2}{v_{sw} i_{sw}} = \frac{1}{2} \tag{12.65}$$

In practice, internal losses in the device as well as additional bias that must be supplied to stay out of the triode region will degrade this efficiency further. It should also be mentioned that modulated communication signals have much more complex behaviors than the simple sinusoids used here; an important metric for such a signal is the *peak-to-average-power ratio* (PAPR), which characterizes how much stronger the largest amplitude swings are compared to those which are typical. This has a direct impact on efficiency, since the bias point must be far enough from the limits to support the larger swings while, most of the time, the current-voltage trajectory is nearer to the center.

In the above scenario, the device is turned on, or conducting, throughout the entire sinusoidal cycle. We thus say that it has a *conduction angle* of 2π. This is known as a *Class A* amplifier. A *Class B* amplifier, in contrast, is biased to have a conduction angle of only π, or half the sine-wave, as shown in Figure 12.16(a). Clearly, a clipped waveform such as this will comprise significant harmonic content as well as the fundamental. The load will typically comprise a resonant circuit that isolates the fundamental tone while presenting a near short-circuit to the higher harmonics. This allows the clipped current waveform, including harmonics, to flow freely from the active device's terminals, while the voltage harmonic terms will be reduced to zero, leaving behind a perfect sinusoidal voltage with amplitude $\pm v_{sw}$.

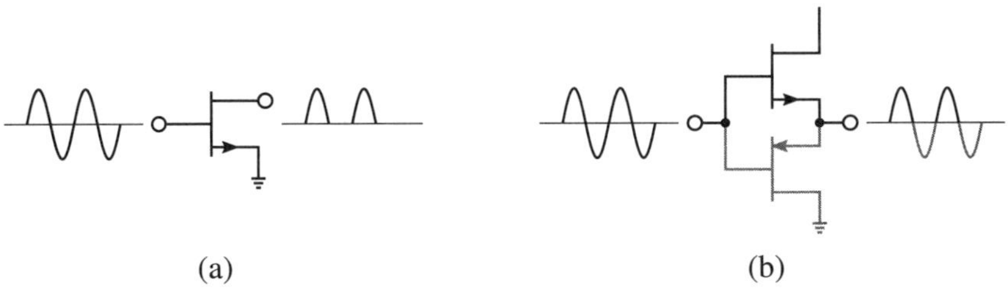

Figure 12.16 Class B amplifier conceptual waveforms. (a) Single-ended configuration. (b) Push-pull configuration. Note that devices having complementary polarity (e.g., nMOS and pMOS) are used.

Fourier sine series analysis of the current waveform reveals that its amplitude at the fundamental frequency is reduced by half,

$$i_1 = \frac{1}{\pi}\int_0^{2\pi} i(\theta)\sin\theta d\theta = \frac{i_{sw}}{\pi}\int_0^{\pi}\sin^2\theta d\theta = \frac{i_{sw}}{2} \tag{12.66}$$

However, the DC current draw from the supply is found by taking the time average of the current waveform,

$$i_{DC} = \frac{i_{sw}}{2\pi}\int_0^{\pi}\sin\theta d\theta = \frac{i_{sw}}{\pi} \tag{12.67}$$

We therefore have for the drain efficiency,

$$\eta_d = \frac{P_{out}}{P_{DC}} = \frac{v_{sw}i_1/2}{v_{sw}i_{DC}} = \frac{v_{sw}i_{sw}/4}{v_{sw}i_{sw}/\pi} = \frac{\pi}{4} \approx 78.5\% \tag{12.68}$$

A common variation for Class B amplifiers uses two active devices combined in parallel, each conducting over opposite halves of the cycle, as shown in Figure 12.16(b). Distortion due to harmonics is reduced, while the overall efficiency remains the same, since each device consumes and contributes an equal amount of power. This is known as the push-pull configuration [16].

Of course it is possible to bias an amplifier with conduction angles between that of the Class A and Class B, a compromise which is naturally referred to as *Class AB*, and having linearity and efficiencies which are partway between the

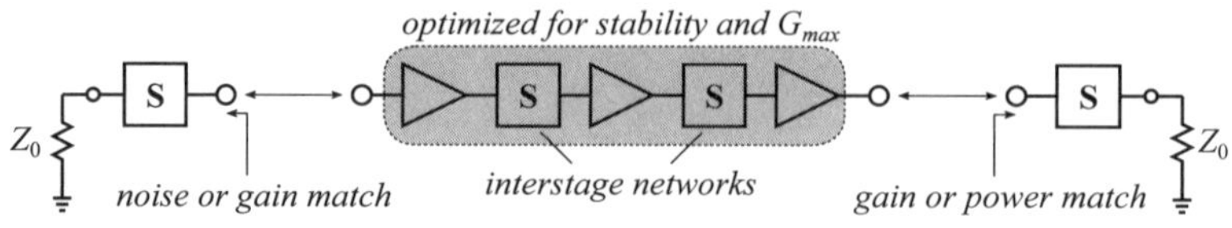

Figure 12.17 General multistage amplifier.

two. *Class C* amplifiers operate with even further reduced conduction angle, and other classes exist also which specify harmonic loading conditions and other aspects to improve device performance. Obviously, operating an amplifier in these modes results in significant harmonic generation and distortion. Some of these effects can be mitigated with appropriate output filtering, while others cannot. For applications which demand good linearity, it is typically best to stick with Class A operation.

Finally, it is worth noting that a high-power device with low gain requires a significant amount of signal drive power at its input. Strictly speaking, therefore, the drain efficiency as it has been defined need not be constrained to less than unity; a simple pass-through network that bypasses the transistor entirely could have infinite efficiency, while the gain is less than 1. In the same way that the noise measure and T_{cas} defined in (12.44) account for an amplifier's gain in qualifying its noise performance, a new metric has been defined for amplifier efficiency which counts only the increase in power from input to output. Known as the *power-added efficiency*, it is given by

$$PAE = \frac{P_{out} - P_{in}}{P_{DC}} \tag{12.69}$$

12.4 MULTISTAGE AMPLIFIERS

A single active device, say, a transistor, will typically have sufficient gain by itself for an acceptable amplifier only at the lowest frequencies. At higher frequencies, multiple devices must be cascaded, as shown in Figure 12.17. In fact, in the millimeter-wave range, it is not unusual for amplifier stages to be so gain-starved that the appropriate cascade performance measures (T_{cas}, noise measure, and PAE) emphasize conjugate/gain-matching even more heavily than noise and/or power matching.

It is not generally optimal to match each individual stage to 50Ω, as though independent amplifiers, and then cascade them. Better results, and especially broader bandwidth, can be achieved by optimizing the interstage networks to provide the appropriate conjugate, noise optimal, or power-maximizing impedance match to both sides directly. As there are many compromises to be made, this leads to a multidimensional optimization problem for which any number of design strategies can be devised. One technique preferred by the author proceeds in the following approximate sequence,

- First, the active device is selected and scaled to suit the application. In addition to key aspects of performance, such as T_{min} and power capacity, it is advantageous at this point to select a device whose intrinsic impedance (or perhaps Z_{opt}) is as close as possible to Z_0 near the target operating frequencies, for this will ease the development of matching networks that provide good performance over the widest possible bandwidths. Note the devices of different sizes may be selected for the input, intermediate, and output stages.
- The individual active devices are next stabilized (unconditionally) with appropriate lossy feedback or series/shunt elements.
- All stabilized active device cells are then combined with interstage networks which are optimized for maximum available gain, G_{max}. We are not yet ready to consider input and output-matching or the realized transducer gain.
- Finally, input and output matching networks are added. For a low-noise, small-signal amplifier, a noise match is provided at the input while a conjugate gain-match is provided at the output, while for a power amplifier the input is gain-matched and the output is power-matched. This is also the point at which input and output reflection coefficients are considered.
- Iterate over the previous steps as necessary until acceptable overall performance is achieved.

Problems

12-1 For a given set of two-port s-parameters, derive the load termination, Γ_L, which causes the input to be perfectly matched. Under what conditions is it realizable with a passive load?

12-2 Consider a transistor for which $s_{11} = -j0.8$, $s_{12} = 0.16+j0.08$, $s_{21} = j2.2$, and $s_{22} = 0.3 - j0.6$ at a particular frequency in a 50Ω environment. What is its input reflection coefficient given a load of $\Gamma_L = j0.5$?

12-3 Prove the relation given in (12.5).

12-4 For the transistor in Problem 12-2, what is the operating power gain, the available power gain, and the transducer power gain, assuming 50Ω terminations?

12-5 Which is lowest for a given amplifier and port terminations, the operating power gain, the available power gain, or the transducer power gain?

12-6 For the transistor in Problem 12-2, what is the center and radius of the input stability circle? The output stability circle?

12-7 Is it possible to define source and load terminations for an unconditionally stable linear amplifier such that the operating power gain is larger than the available power gain? Why or why not?

12-8 For the transistor in Problem 12-2, what is the Rollet stability factor? What are the input and output geometric stability factors?

12-9 Prove the equivalence of (12.33a) and (12.33b).

12-10 What is the maximum available gain of the transistor in Problem 12-2? The maximum stable gain?

12-11 Derive an expression for the return loss in decibels of a balanced amplifier where the reflection coefficients of the component amplifiers are slightly mismatched, with voltage amplitude ratio X and phase difference ϕ. Assume quadrature hybrid is perfect, and the nominal (average) return loss of the amplifiers is RL_0.

12-12 What is Mason's invariant for the transistor in Problem 12-2?

12-13 At what frequency is the Rayleigh-Jeans approximation in error by 1% at room temperature ($T_0 = 290$K)? At what temperature is it in error by the same amount at 100 GHz?

12-14 An amplifier has a noise figure of 3 dB. How much does it degrade the signal-to-noise ratio of a matched source at an ambient temperature of 40 °C?

12-15 A given transistor has a minimum equivalent noise temperature of 50K and a maximum associated gain of 10 dB. What is the best noise temperature that a three-stage amplifier made using these devices could have? How does this compare to the best noise-measure that an amplifier of any number of stages could have?

12-16 What is the equivalent noise temperature as a function of frequency of a reflectionless Butterworth filter at a physical temperature, T_{phys}?

12-17 A transistor has a minimum equivalent noise temperature of 100K. What is the minimum value of its associated Lange noise parameter, N?

12-18 Write expressions for the current-voltage trajectory at the terminals of a power transistor driving complex load immittances of $Z_1 = R_{opt}/p + jX$ and $Y_2 = G_{opt}/p + jB$. In both cases, solve for the imaginary parts, X and B, such that the voltage swing is constrained to $v_q \pm v_{sw}$ and the current is constrained to $i_q \pm i_{sw}$, where $R_{opt} = v_{sw}/i_{sw}$.

12-19 What is the maximum input power that an amplifier having 30 dB of gain and $P_{1dB} = 20$ dBm can accept at its input before compressing by 1 dB?

12-20 Write an expression for the maximum theoretical drain efficiency of a transistor-like device as a function of conduction angle. Assume that the voltage range of the device is 0 to $2v_{sw}$ and the current range is 0 to $2i_{sw}$.

References

[1] D. M. Pozar, *Microwave Engineering*, 4th ed. New York: Wiley, 2011.

[2] G. Gonzalez, *Microwave Transistor Amplifiers: Analysis and Design*, 2nd ed. Upper Saddle River, NJ: Prentice Hall, 1996.

[3] H. Nyquist, "Regeneration theory," *Bell System Technical Journal*, vol. 11, pp. 126–147, January 1932.

[4] A. Platzker, W. Struble, and K. Hetzler, "Instabilities diagnosis and the role of K in microwave circuits," *IEEE MTT-S Digest*, vol. 3, pp. 1185–1188, June 1993.

[5] W. Struble and A. Platzker, "A rigorous yet simple method for determining stability of linear N-port networks," *15th Annual GaAs IC Symposium Digest*, pp. 251–254, October 1993.

[6] A. Platzker and W. Struble, "Rigorous determination of the stability of linear N-node circuits from network determinants and the appropriate role of the stability factor K of their reduced two-ports," *3rd International Workshop on Integrated Nonlinear Microwave and Millimeterwave Circuits*, pp. 93–107, October 1994.

[7] G. Lombardi and B. Neri, "Criteria for the evaluation of unconditional stability of microwave linear two-ports: a critical review and new proof," *IEEE Transactions on Microwave Theory and Techniques*, vol. 47, no. 6, pp. 746–751, June 1999.

[8] J. M. Rollett, "Stability and power-gain invariants of linear twoports," *IRE Transactions on Circuit Theory*, vol. 9, pp. 29–32, March 1962.

[9] Unknown Editor. (2019) Stability factor. https://www.microwaves101.com/encyclopedias/stability-factor.

[10] R. P. Meys, "Review and discussion of stability criteria for linear 2-ports," *IEEE Transactions on Circuits and Systems*, vol. 37, no. 11, pp. 1450–1452, November 1990.

[11] M. S. Gupta, "Power gain in feedback amplifiers, a classic revisited," *IEEE Transactions on Microwave Theory and Techniques*, vol. 40, pp. 864–879, May 1992.

[12] Wikipedia. (2019, March) 2019 redefinition of si basae units. https://en.wikipedia.org/wiki/2019_redefinition_of_SI_base_units.

[13] A. M. Niknejad. (2019) Two-port noise. http://rfic.eecs.berkeley.edu/~niknejad/ee242/pdf/eecs242_lect8_twoportnoise.pdf.

[14] J. Lange, "Noise characterization of linear twoports in terms of invariant parameters," *IEEE Journal of Solide-State Circuits*, vol. 2, no. 2, pp. 37–40, June 1967.

[15] M. Pospieszalski, "Interpreting transistor noise," *IEEE Microwave Magazine*, vol. 11, no. 6, pp. 61–69, November 2010.

[16] S. C. Cripps, *RF Power Amplifiers for Wireless Communications*. Norwood, MA: Artech House, 1999.

[17] Wikipedia. (2019, January) Third-order intercept point. https://en.wikipedia.org/wiki/Third-order_intercept_point.

Chapter 13

Frequency Conversion

So far in this book we have covered the fundamental elements of electromagnetic circuits — whether lumped, distributed, or quasi-optical — and then used them to develop system building blocks that are frequency-independent or frequency-selective, reciprocal or nonreciprocal, and passive or power-adding. In this final chapter we discuss circuit building blocks which convert signal power from one frequency to another. This, necessarily, involves circuit elements that are nonlinear. As with gain-bearing elements, the physical processes that lead to nonlinear behavior are specific to the actual devices, and are generally beyond the intended scope of this text. We will therefore treat the nonlinear elements in most cases that follow as black boxes, focusing instead on their external characteristics and how to leverage them in the larger circuits of which they are a part to achieve a particular goal. For convenience, we will often show diodes as the nonlinear elements in circuit diagrams, but this is for illustration only, as the concepts presented apply more generally.

13.1 GENERAL NONLINEARITY

Although nonlinear devices can have any number of terminals, most of the important features for the discussion that follows can be captured in a simple two-terminal element. The most recognizable example is a diode, whose schematic symbol and typical current-voltage characteristics are shown in Figure 13.1. At its simplest, the diode may be considered a nonlinear resistor. Under most operating conditions, it conducts only in the forward direction (from the anode to the cathode when $v > 0$)

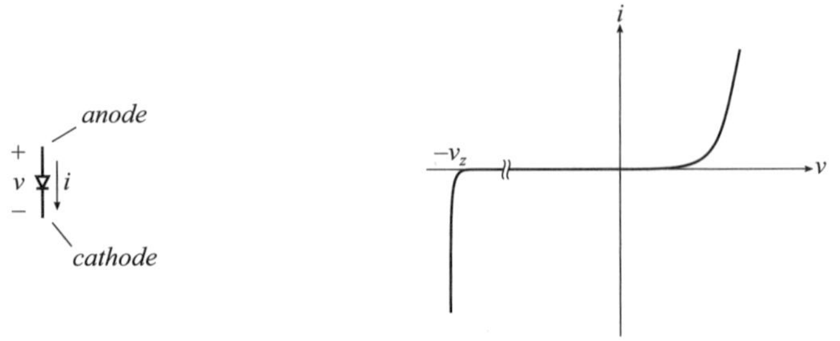

(a) (b)

Figure 13.1 (a) Schematic symbol and (b) typical current-voltage characteristic of a diode.

as shown by the upturn of the curve, or *knee*, in the upper right quadrant of Figure 13.1(b). Only very little leakage current flows in the reverse direction, until a reverse breakdown voltage is reached, $v = -v_z$, typically much further away from the origin than the knee at positive voltage. For many diodes, this reverse breakdown is unintentional, and can be damaging. For certain diodes, called *Zener diodes*, the reverse breakdown is carefully engineered and exploited, as it is usually a much sharper transition into conduction than the positive one. For these diodes, v_z is called the *Zener voltage*.

Nearer to the origin, the diode conduction may be represented by an exponential curve called the *Shockley ideal diode equation*,

$$i = I_s \left(e^{\frac{v}{nV_T}} - 1 \right) \tag{13.1}$$

where I_s is the reverse bias saturation current, or *scale current*, n is a fitting parameter called the *ideality factor* (typically between 1 and 2), and V_T is the *thermal voltage*, given by

$$V_T = \frac{kT}{q_e} \approx 25.85 \text{ mV at } T = 300\text{K} \tag{13.2}$$

In this equation, k is Boltzmann's constant, and $q_e = 1.602176634 \times 10^{-19}$ C (exactly) is the *elementary charge* — the charge of a single proton (or the equivalent magnitude of charge of a single electron).

What matters at this stage is simply that the current may be expanded as a Taylor series about the origin,

$$i = a_0 + a_1 v + a_2 v^2 + a_3 v^3 + \cdots = \sum_{m=0}^{\infty} a_m v^m \tag{13.3}$$

(Strictly speaking, a diode cannot have finite current flowing with zero voltage applied, but the constant term, a_0 is carried forward nonetheless to account for possible DC bias on top of the signal voltage, v.) If the voltage applied is sinusoidal with time, $v(t) = V\cos(\omega t)$, the current may then be written

$$i(t) = \sum_{m=0}^{\infty} a_m V^m \cos^m(\omega t) \tag{13.4}$$

The trigonometric power reduction identities in Section A.1.5 show that this expression will have pure sinusoidal components at frequencies $m\omega$ for all integers $m \geq 0$. These are called *harmonics* of the fundamental driven frequency, ω. Specifically, 2ω is the second harmonic, 3ω is the third harmonic, and so on.

Furthermore, if the applied voltage contains two sinusoidal components,

$$v(t) = V_1 \cos(\omega_1 t) + V_2 \cos(\omega_2 t) \tag{13.5}$$

then the current may be written as follows

$$i(t) = \sum_{m=0}^{\infty} a_m \left(V_1 \cos(\omega_1 t) + V_2 \cos(\omega_2 t)\right)^m \tag{13.6a}$$

$$= \sum_{m=0}^{\infty} \sum_{r=0}^{m} a_m \binom{m}{r} V_1^r V_2^{m-r} \cos^r(\omega_1 t) \cos^{m-r}(\omega_2 t) \tag{13.6b}$$

where the binomial theorem has been used. In this case, power reduction identities along with the prosthaphaeresis formulas (see Section A.1.6) show that the current will contain sinusoidal components having frequency $m\omega_1 \pm n\omega_2$ for all integers m and n. These are called *mixing products*.

For small V (or V_1 and V_2), only the first few terms in the Taylor series will be significant, because higher powers of V will be vanishingly small. Coupled with the fact that the mth term in the Taylor series is the lowest-order term to generate an mth harmonic (as born out by the power reduction identities), we know that, for sufficiently small V, the amplitude of the mth harmonic is proportional to V^m. The power of the mth harmonic in decibels will then be proportional to m times the power in the fundamental frequency (albeit with a potentially very small proportionality constant). Similarly, kth-order mixing products of small signals (where $k = |m| + |n|$) will have logarithmic power proportional to m times the power of the first-order inputs. This is why the slopes of the fundamental and third-order harmonics/intermodulation products for weak signals in Section 12.3.2 were known to be 1:1 and 3:1, respectively, which is the basis of the third-order intercept extrapolation.

Nonlinearities need not be resistive. Most diodes will also exhibit some form of nonlinear capacitance as well, for example. More sophisticated expressions of nonlinearity and the associated computational algorithms exist to more generally describe nonlinear behavior (e.g., *Volterra series* and *harmonic balance* [1]) and computer codes to aid in using them are now readily available with many CAD programs. It is not our intention to provide a thorough exposition of these formulations here. Instead, we focus on some basic key principles that are apparent from the relatively simple examples above,

- For inputs which comprise only sinusoidal tones (ω_1, ω_2, and so forth), nonlinear elements generate output frequencies that are sums and differences of integer multiples of the input frequencies ($m\omega_1 \pm n\omega_2$), but no others.

- Even-order nonlinearities generate only even harmonics ($m\omega_1$ where m is even) or even-order mixing products ($m\omega_1 \pm n\omega_2$ where $k = |m|+|n|$ is even) of the input frequencies (plus DC), while odd-order nonlinearities generate only odd-order harmonics ($m\omega_1$ where m is odd) and mixing products ($m\omega_1 \pm n\omega_2$ where $k = |m| + |n|$ is odd) plus the fundamental tones.

- When all input signals are relatively weak and equal in strength, the relative power of each harmonic or mixing product is proportional to its order times the power of the inputs or $P\left(m\omega_1 + n\omega_2\right) \propto \left(|m| + |n|\right) P\left(\omega_1\right)$.

- If one input tone (ω_1) is much stronger than another (ω_2), the relative power of any mixing product involving the weaker tone is proportional to the coefficient (i.e., harmonic number) of that tone in the mixing product times its input power or $P\left(m\omega_1 + n\omega_2\right) \propto |n| P\left(\omega_2\right)$.

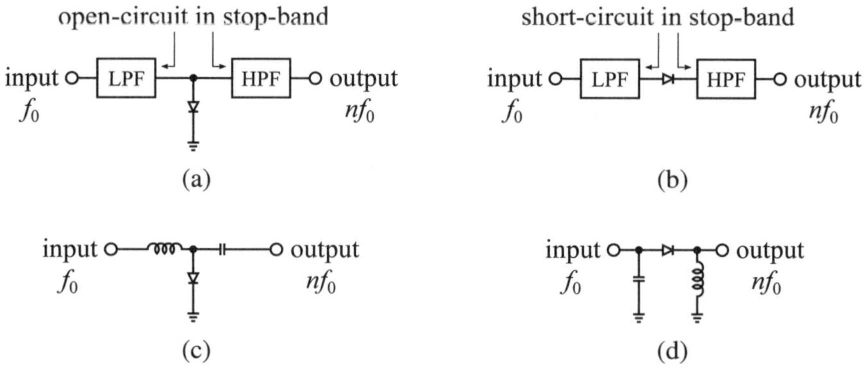

Figure 13.2 Harmonic multipliers using single diodes. (a) Shunt configuration. (b) Series configuration. (c) With single, series lumped-element filters. (d) With single, shunt lumped-element filters.

13.2 MULTIPLIERS

Perhaps the simplest component, at least in principle, to exploit these properties of nonlinear elements is the harmonic generator, or *multiplier*. Typically, these are used to generate sinusoidal tones at frequencies that are too high to generate directly with the desired amplitude or the desired coherence to a lower frequency reference.

13.2.1 Single-Element Multipliers

Typically, one will want to drive, or *pump*, a nonlinear element, such as a diode, with a strong signal at the fundamental frequency, f_0, and extract the desired harmonic, nf_0, from the output. The ratio of output power at the harmonic frequency to the input at the fundamental frequency is called the *conversion efficiency*, or, alternatively, the *conversion gain* or *conversion loss*, depending on the sign convention that is to be implied. For the greatest efficiency, one should use at least a high-pass (or band-pass) filter at the output port to ensure that all of the input power is used to pump the diode, as shown for a shunt diode in Figure 13.2(a) and for a series diode in Figure 13.2(b). A low-pass filter would also be used on the input side to ensure that all the power from the desired harmonic couples to the output and does not leak back through the input.

One should be mindful of the impedance presented by each of these filters in their stopbands. It is not sufficient to simply block the undesired signals. When a shunt diode is used, one must be careful that the filters in either case do not short out the diode in the stopband, preventing either the input or output signals from coupling to it. Ideally, an open-circuit would be provided by both filters in their respective stopbands. Similarly when a series diode is used, the far end of the diode must be grounded so that it can act as a proper load for the input signal and a proper source for the output. Thus, the filter stopbands should approximate a short-circuit in that case. The filters in one case will be duals of the filters in the other.

The filters can be virtually any type that we discussed in Chapter 11, whether lumped-element, distributed, or quasi-optical. They could be very sophisticated, or almost trivially simple, as shown with the single lumped-elements in Figures 13.2(c, d). These filters may also incorporate impedance-matching networks, to match the diode or other nonlinear element to the input port at the fundamental frequency, and to the output port at the desired harmonic.

It is also common to incorporate bias tees so that DC power can be applied to center the quiescent point of the nonlinear element where its nonlinearity is strongest (i.e., at the diode knee) to improve its conversion efficiency at low input powers. This may further be used to control what kind of nonlinearity is exploited: a diode under reverse bias usually exhibits a strong nonlinear capacitance (in which case it is called a *varactor*, an abbreviation for *variable reactor*), whereas one under forward bias exhibits a strong nonlinear resistance (a *varistor*) as previously shown. *Varactor* multipliers are typically more efficient, while *varistor* multipliers are easier to impedance-match over broad bandwidths.

13.2.2 Selective Harmonic Suppression

Basic nonlinear elements will typically generate lower-order harmonics more strongly than higher-order ones, but still the presence of the undesired harmonics can unacceptably contaminate the applications in which they are used. Good filtering is one option for suppressing these harmonics, but this may not be possible if the desired tuning range of a multiplier is sufficiently wide — the harmonics will simply be too close together and even overlap the desired output frequency range.

Take, for example, a tripler (a multiplier designed to generate the third harmonic) which is required to generate outputs between 12 and 18 GHz. Its input frequency range would then have to be 4–6 GHz. One could design a very sharp filter at the output that passes 12–18 GHz exclusively, but when the filter is driven at the low end of its input range, 4 GHz, it will generate not only the third harmonic, 12 GHz, but a fourth harmonic at 16 GHz as well. As 16 GHz is within the output passband, not only will the filter allow it through, but one can also expect the impedance to be well matched at that frequency. Similarly, when the tripler is driven

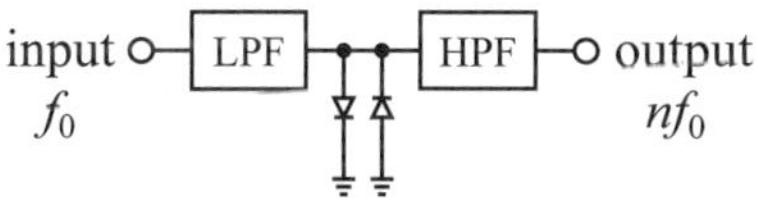

(a)

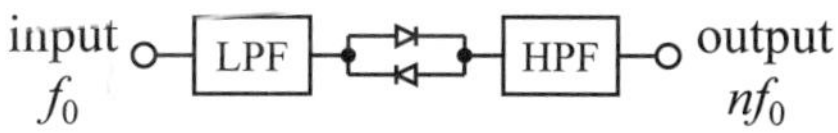

(b)

Figure 13.3 Odd-harmonic multipliers utilizing antiparallel diodes (a) in shunt configuration, and (b) in series configuration.

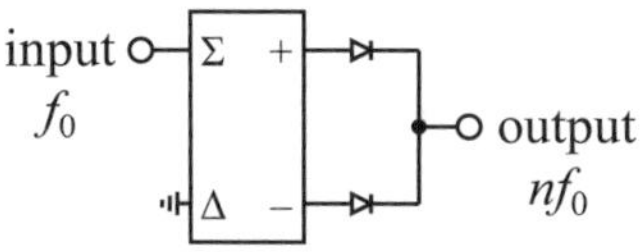

(a)

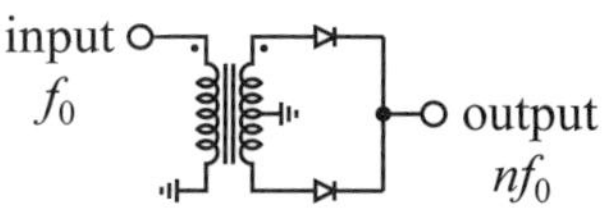

(b)

Figure 13.4 Even-harmonic multipliers. (a) Conceptual diagram with a 180° hybrid. (b) Implemented with a transformer having a center tap.

at the upper end of its range, 6 GHz, a second harmonic will be present at 12 GHz, right at the lower edge of the output passband.

In this case, it would be advantageous if the second and fourth harmonics were not present at all, intrinsically, before filtering. The analysis of Section 13.1 suggests that what we need is a device that exhibits odd-order nonlinearity, but not even-order nonlinearity, which shall be the case only if the nonlinearity is antisymmetric on either side of the quiescent point. If this is not already a feature of the nonlinearity element selected, one can be engineered simply by putting two identical elements in an antiparallel configuration [2], as shown in Figure 13.3. If the two diodes in these examples have the same characteristics, then the overall conduction curve of the pair will be antisymmetric about the origin (that is, the same current will flow in one direction as it does in the other when equivalent voltages are applied in opposite polarity). This is sufficient to ensure that the Taylor series expansion has no even-order terms.

What if we wish to make an even-order multiplier, and suppress the odd harmonics instead? This can be accomplished by driving two diodes out-of-phase and then extracting their outputs in-phase, as shown conceptually in Figure 13.4(a). Almost any kind of 180° splitter can be used: a rat-race hybrid, a magic tee, or a simple transformer (or one might just use a balanced/differential transmission line on the input, while the output is single-ended). In any case, the same considerations as before apply to the impedance presented to the diodes at the input and output frequencies. In Figure 13.4(b), a transformer is used with a center tap that grounds the diodes for coupling to the output.

13.3 MIXERS

Recall, in Section 13.1, we said that multiple tones incident upon a nonlinearity lead to numerous mixing products comprising the sums of integer multiples of those tones, and that further if one tone is much stronger than another, the relative power of a particular mixing product is proportional to the input power times the coefficient of that tone within the mixing product. Let us consider a two-tone excitation where one tone is very strong and the other is very weak. Let us call the strong tone the *local oscillator*, or LO, and denote its frequency ω_{LO}, and the weak tone the *radio frequency*, RF, denoted ω_{RF}. Let us further consider only the lowest-order mixing products involving both tones, $|\omega_{LO} \pm \omega_{RF}|$. Note that, mathematically, a simple product of the two input tones (which is inevitable in any nonlinearity having a Taylor series expansion extending to the second degree or higher) is sufficient to explain the appearance of these mixing products,

$$\cos(\omega_{LO}t)\cos(\omega_{RF}t) = \tfrac{1}{2}\cos((\omega_{LO}+\omega_{RF})t) + \tfrac{1}{2}\cos((\omega_{LO}-\omega_{RF})t) \quad (13.7)$$

According to the previous analysis, since the coefficient of ω_{RF} in the frequencies of these mixing products is ± 1, the power of these output tones will be directly proportional to the input RF power. Thus, despite passing through a strong nonlinearity, the translation of spectral frequencies from one place to another, a process generally described as *mixing*, can nonetheless be considered linear for small signals.

In some cases, the higher of these two outputs will be excluded by filtering. The resultant frequency spectrum associated with these input and output tones, then, is illustrated in Figure 13.5(a). The weaker tone is shifted from a relatively high frequency (the RF) to a relatively lower frequency called the *intermediate frequency*, or IF. This, a special case of mixing, is called *downconversion*. Note that if the weaker signal is not only a tone, but a band-pass spectrum as shown in Figure 13.5(b), then the entire spectrum is downconverted intact.

There is really no requirement that the RF band of frequencies lie entirely above the LO in the frequency spectrum. If there is significant signal energy below the LO tone, but at the same distance from it as the original RF band, then that band

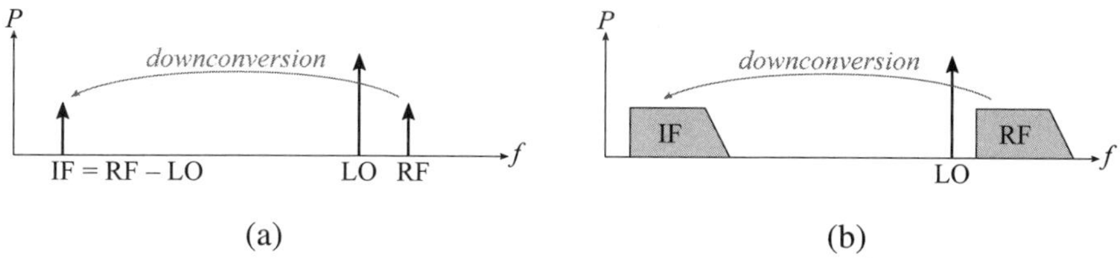

Figure 13.5 Downconversion mixing (a) from an RF tone to an IF tone, and (b) from an RF band to an IF band.

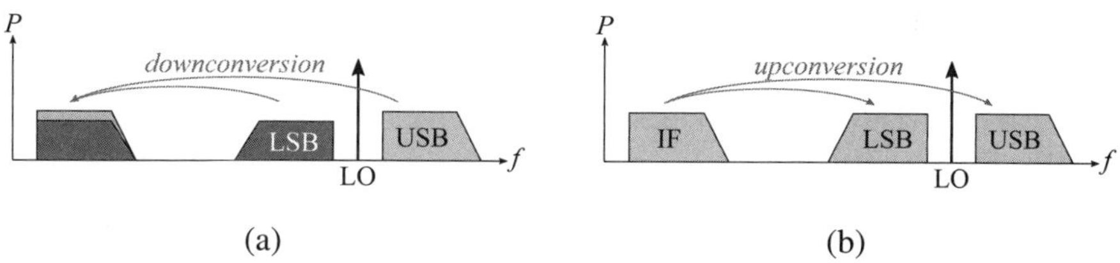

Figure 13.6 (a) Double-sideband downconversion. (b) Upconversion.

is translated into the exact same output frequency range, only reversed and with any residual phase negated, since $\cos(-\omega_{IF}t + \phi) = \cos(\omega_{IF}t - \phi)$, as shown in Figure 13.6(a). The RF spectrum above the LO frequency is called the *upper sideband*, or USB, while the RF spectrum below it is called the *lower sideband*, or LSB. Often, the RF spectrum will be filtered prior to mixing to select one or the other sideband so that one does not contaminate the other after downconversion. When both sidebands are desired, there are special techniques for recovering both independently, which are described later in Section 13.3.5.

Alternatively, one may inject the IF as an input at a relatively low frequency, as shown in Figure 13.6(b) to achieve an upward shift to higher frequencies, called *upconversion*. In this case, the two RF sidebands appear as separate outputs, one of which may be selected by filtering.

13.3.1 Single-Diode Mixers

The component that implements this upconversion or downconversion is called a *mixer*, and is customarily indicated by the symbol in Figure 13.7(a), which emphasizes its function as a multiplier of input signals. Specific implementations are widely varying, but in the case of a single diode, as shown for a shunt configuration

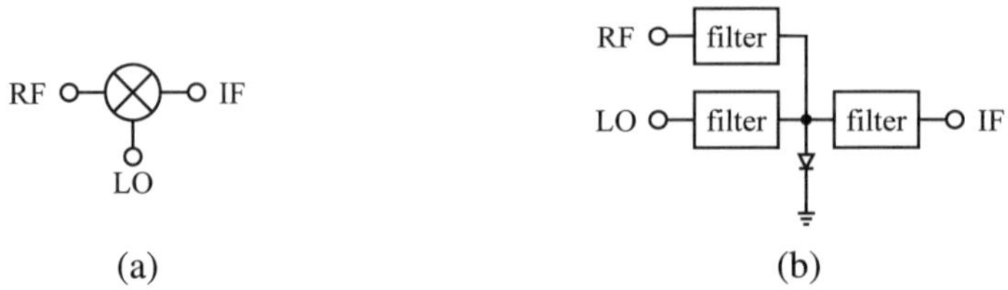

Figure 13.7 (a) Mixer symbol. (b) Generic single-diode mixer with independent RF, LO, and IF filters/matching networks.

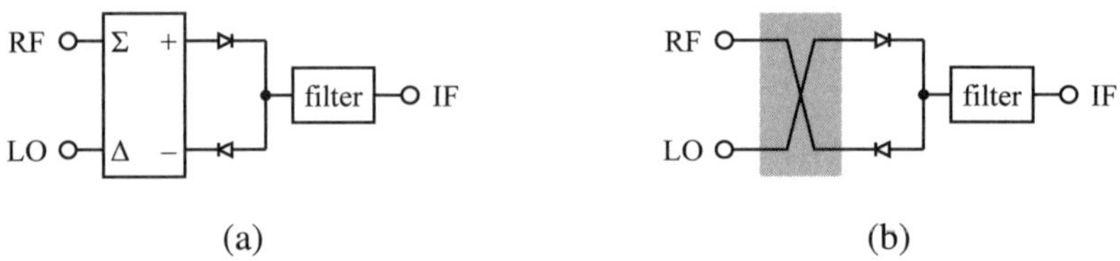

Figure 13.8 Single-balanced mixers using (a) a 180° hybrid, which can be either upconverting or downconverting, and (b) a quadrature hybrid, which can only be downconverting as shown (to make it an upconverter, one of the diodes must be reversed).

in Figure 13.7(b), it will typically require individual filters and matching networks for the RF, LO, and IF ports. As the RF and LO are often close to one another in frequency, their filters may be implemented in a unified way as a diplexer. In any case, the same considerations of grounding and coupling that accompanied the frequency-dependent matching networks for multipliers will apply here.

13.3.2 Single-Balanced

More commonly, mixers exhibit *balanced* configurations, where the inputs and outputs are applied to or extracted from multiple nonlinear elements through hybrids or transformers with phase relationships that allow for the cancellation of undesired mixing products and LO noise contributions. This may also help to eliminate the need for a difficult RF/LO diplexer, and further allow the RF/LO ranges to overlap, since these ports are topologically isolated via the associated hybrid.

Single-balanced mixers use a single hybrid, which can either be 180° or quadrature, as shown in Figure 13.8. Most often, the diodes (or other nonlinear elements) are oriented in opposing directions, as shown. It is possible to orient the

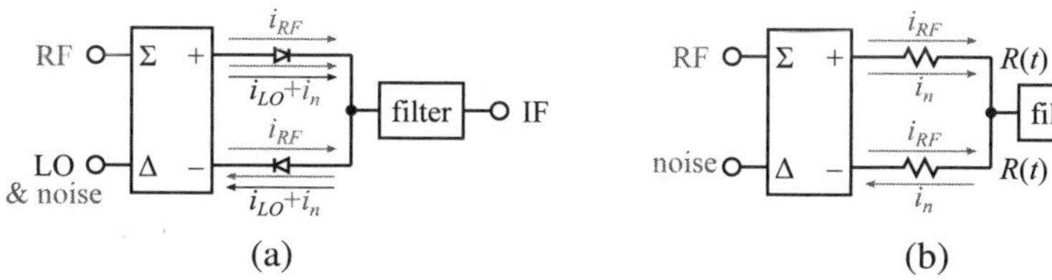

Figure 13.9 LO noise rejection in a single-balanced mixer. (a) Directed currents comprising RF, LO, and noise components. (b) Weak signals applied to equivalent time-varying resistors.

diodes in the same direction, but then an additional 180° hybrid will be needed at the IF to subtract the resulting currents.

To visualize how the balanced mixer works, it is useful to separate, conceptually, the functions of the LO and RF signals. The LO is presumed to have a large enough amplitude to move the operating point along the IV curve of a nonlinear element, such as the diode's IV curve in Figure 13.1, substantially changing its conductance from one moment to the next. The RF, in contrast, is a small-amplitude perturbation on top of that trend, one that only exercises the conduction curve locally around the momentary operating point determined by the LO. For a given instant in time, we may perceptualize the diodes as variable resistors that are controlled by the LO, and then modulated by the RF.

See Figure 13.9(a) for an illustration of a downconverting, single-balanced mixer with 180° hybrid. We consider three distinct inputs: the RF signal, the LO signal, and any amplitude-modulated or additive noise present on the LO port. Note that the reference direction of the currents is consistent with the hybrid operation: in-phase for the RF applied to the Σ port, and out-of-phase for the LO plus noise applied to the Δ port. The LO is considered a strong enough signal to activate the diodes' nonlinearity, and is shown in black, whereas the RF and noise signals are weak, shown in gray. Thus, we may redraw the circuit as shown in Figure 13.9(b), where the diodes have been replaced with time-variable resistors, $R(t)$, implicitly controlled by the now-absent LO signal. Since the LO reference direction is forward-biased for both diodes, the equivalent resistances are the same at all times, t. The weak signals can now be applied across those resistors. Note from the figure that the RF signals (downconverted by the variable resistance) combine in-phase at the IF port, while the LO noise components cancel.

To be a bit more rigorous, let us write the phases of the RF, LO, and noise signals at the "+" and "–" ports of the hybrid,

$$\phi_{RF} = 0 \text{ (at both ports)} \tag{13.8a}$$

$$\phi_{LO} = \begin{cases} 0 & \text{at the "+" port} \\ \pi & \text{at the "–" port} \end{cases} \tag{13.8b}$$

$$\phi_n = \begin{cases} 0 & \text{at the "+" port} \\ \pi & \text{at the "–" port} \end{cases} \tag{13.8c}$$

Since the diodes have opposite orientation, the signals impressed upon the diode at the "–" port with respect to its forward reference direction are reversed.

$$\phi'_{RF} = \begin{cases} 0 & \text{at the "+" port} \\ \pi & \text{at the "–" port} \end{cases} \tag{13.9a}$$

$$\phi'_{LO} = 0 \text{ (at both ports)} \tag{13.9b}$$

$$\phi'_n - 0 \text{ (at both ports)} \tag{13.9c}$$

Since the LO and noise signals have the same phase in both diodes, the downconverted noise currents at frequency $\omega_n - \omega_{LO}$ in both diodes will also have the phase, and therefore cancel when the final IF is extracted from the difference of these two currents. This proves that downconverted noise is suppressed by the single-balanced mixer.

Similarly, any specific mixing product $\omega_{IF} = m\omega_{RF} + n\omega_{LO}$ will arise from diode currents having phase $\phi_{IF} = m\phi'_{RF} + n\phi'_{LO}$, or

$$\phi_{IF} = \begin{cases} 0 & \text{at the "+" port} \\ m\pi & \text{at the "–" port} \end{cases} \tag{13.10}$$

If m is even, the IF currents in the diodes are in-phase, and cancel when the currents from the diodes are subtracted, otherwise, they combine. We thus see that the single-balanced mixer in Figure 13.9 rejects all mixing products for which m is even. Similar analysis shows that if the RF and LO ports are swapped — so that the LO now enters the Σ port of the hybrid, while the RF enters the Δ port — the same LO noise rejection is achieved, but the spurious mixing products are rejected this time if n is even, not m.

One can also orient both diodes in the same direction (rather than opposed, as shown here) but then a 180° output IF hybrid is needed. In that case, the spurious response rejection is the opposite of that described in the previous paragraph. Orienting the nonlinear elements in the same direction may ease the introduction of DC bias and could be useful for certain nonlinear elements that are not so easy to reverse, such as transistors [3].

Analysis of the single-balanced mixer using a quadrature hybrid, shown in Figure 13.8(b), shows that it too rejects downconverted LO noise, but only rejects spurious mixing products for which the difference between m and n is a multiple of four. It also suffers from relatively poor RF-to-LO isolation, being simply equivalent to the individual diode's return loss.

Single-balanced mixers using quadrature hybrids do have one somewhat surprising property. All previously described mixers work in either direction, as both upconverters and downconverters, and reject both upconverted and downconverted LO noise, but the single-balanced mixer with a quadrature hybrid shown in Figure 13.8(b) only works as a downconverter, not an upconverter. If one of the diodes is reversed (and no IF hybrid is added, as is usually done with other types of single-balanced mixers), then it will only work as an upconverter, not a downconverter. Further, the upconverting mixer rejects upconverted LO noise, while the downconverting mixer rejects downconverted LO noise.[1] Proof of this behavior is left as an exercise for the reader.

13.3.3 Double-Balanced

Even more spurious mixing products can be rejected if a mixer is *double-balanced*, employing two separate 180° hybrids on the RF and LO ports, as shown in Figure 13.10. Also called a *ring mixer*, this configuration uses a four-diode ring, or *quad*, which is pumped by the LO at diagonally opposed corners 180° out-of-phase, and modulated at the remaining two corners 180° out-of-phase by the RF. Typically drawn with center-tapped transformers, as shown in Figure 13.10(a), any generic 180° hybrid will suffice, where the IF signal and ground are taken from the common-mode hybrid ports, indicated in Figure 13.10(b). Thus, double-balanced ring mixers may be implemented effectively even in the millimeter-wave range where transformers are not practical [4].

Double-balanced mixers have the same LO noise rejection properties as single-balanced mixers, while also rejecting spurious mixing products $m\omega_{RF} + n\omega_{LO}$ if either m or n is even, not just one or the other, as was the case for single-balanced mixers. It should be noted that this rejection is not dependent on the hybrid

1 It is worth noting that most authors define balanced mixers as those which reject downconverted LO noise, making the upconverter-type described here not technically a balanced mixer. In my view, what matters is the noise that is rejected at the same or similar output frequencies as the RF signal, making the more common definition erroneous.

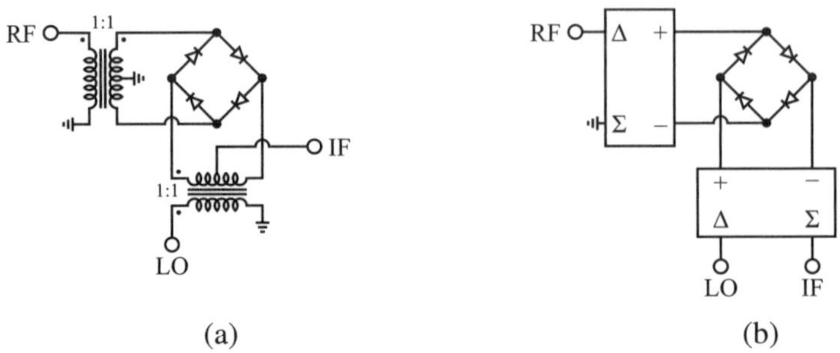

Figure 13.10 Double-balanced ring mixers with (a) center-trapped transformers and (b) generic 180° hybrids.

having operable bandwidth at the second or third harmonic of the LO or RF; these harmonics are generated by the nonlinear elements. Thus, the hybrids need only have good performance in the fundamental LO and RF bands to realize the spur-rejection benefits. (The IF will typically be much too low in frequency to be in the valid operating range of the hybrid, making Figure 13.10(b) more notional than rigorous, and one must take care that the IF is properly grounded and coupled to the output through the hybrids at low frequency.) The ring mixer is an especially popular configuration, and diode quads are readily available from manufacturers in a single package, often fabricated monolithically on the same bit of semiconductor to ensure near-identical characteristics.

An alternative configuration for the double-balanced mixer is shown in Figure 13.11(a), known as a *star mixer*. Appearing almost like a dual of the ring mixer, the star mixer employs a diode quad, this time emanating from a common node, and being driven differentially from an orthogonal pair of Marchand baluns. The IF is extracted from the central common node of the quad.

13.3.4 Triple-Balanced

In addition to the rejection of LO amplitude noise and of selected spurious responses, balanced mixers have the added benefit of providing inherent topological isolation between the RF and LO ports without the need of filters. This makes it easier, for example, for the RF and LO frequency bands to overlap. In many practical cases, however, the IF will still have to be significantly lower in frequency in order to provide adequate grounding and isolation. One solution is to employ a *triple-balanced configuration*, also called a *doubly double-balanced mixer*, wherein

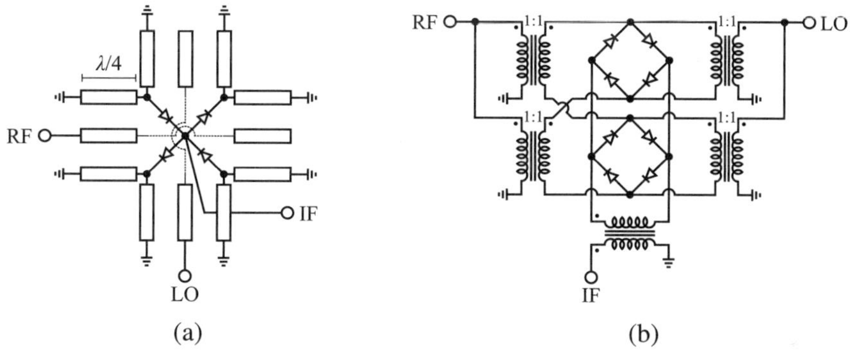

Figure 13.11 (a) Double-balanced star mixer. (b) Triple-balanced mixer.

each of the three mixer ports are topologically isolated using differential hybrids or baluns, as shown in Figure 13.11(b). Having the same LO noise and spur rejection properties as a double-balanced mixer, the RF, LO, and IF ranges of the triple-balanced mixer may be mutually overlapping.

In progressing from single-diode mixers, to singe-balanced mixers, double-balanced mixers, and finally triple-balanced mixers, we have also gone from using a single diode or nonlinear element, to two, four, and finally eight diodes or nonlinear elements, along with the necessary supporting circuitry. It should not be surprising that each progressively more complex mixer also requires more LO power to drive those nonlinear elements to an appropriate level of nonlinearity. However, we have also assumed that the RF must be significantly weak compared to the LO for the RF-to-IF signal path (or its reverse) to behave linearly; by increasing the LO pump power, and spreading it out over a larger number of nonlinear elements, we have also increased the RF/IF dynamic range over which the mixer may be treated as linear in the primary signal path.

It should also be noted here that the spurious product and noise suppression characteristics that we have described are based on the assumption that all nonlinear elements have identical characteristics, and that the associated hybrids and transformers are well balanced. In practice, small amplitude and phase imbalances can degrade the isolation of these terms considerably.

A summary of the critical properties of single-diode (or unbalanced) and balanced mixers is given in Table 13.1.

Table 13.1

Properties of Unbalanced and Balanced Mixers

Balance Type[1]	Ports[2] RF	LO	IF	Element Orient.[3]	Up/Down-Converts?	Rel. LO Power[4]	Rejects LO Noise?	Rejects Spur $m\omega_{RF} + n\omega_{LO}$?
Unbal.				either	0 dB	no	no	
Single	Σ	Δ	com.	opp.	either	+3 dB	yes	if m is even
Single	Δ	Σ	com.	opp.	either	+3 dB	yes	if n is even
Single	Σ	Δ	Δ	same	either	+3 dB	yes	if n is even
Single	Δ	Σ	Δ	same	either	+3 dB	yes	if m is even
Single	90°	90°	com.	opp.	down	+3 dB	down	if $(m-n) \bmod 4 = 0$
Single	90°	90°	Δ	same	down	+3 dB	down	if $(m-n) \bmod 4 = 0$
Single	90°	90°	com.	same	up	+3 dB	up	if $(m-n) \bmod 4 = 2$
Single	90°	90°	Δ	opp.	up	+3 dB	up	if $(m-n) \bmod 4 = 2$
Double	Δ	Δ	Σ	ring	either	+6 dB	yes	if either is even
Double	Δ	Δ	Σ	star	either	+6 dB	yes	if either is even
Triple	Δ	Δ	Δ	2-quad	either	+9 dB	yes	if either is even

[1]Mixers may be unbalanced (unbal.), single-balanced, double-balanced, or triple-balanced.
[2]Signals are marked Σ or Δ if connected to the ports of a 180° hybrid, or 90° if connected to a quadrature hybrid. IFs marked com. are extracted from a common node between the diodes.
[3]Diode or element orientation may be opposite (opp.) or the same for single-balanced mixers, ring or star for double-balanced mixers, or two quads (2-quad) for triple-balanced mixers.
[4]LO power relative to an unbalanced, single-diode mixer.

13.3.5 Sideband-Separating

Recall from Figure 13.6 that, during downconversion, spectra on both sides of the local oscillator known as the upper and lower sidebands are downconverted to the same IF. Usually, one sideband or the other will be considered the desired signal, while the other is considered interference, or else both sidebands contain useful information but that information is devalued if the two are combined. In the former case, the desired sideband can be isolated by prefiltering at the RF, provided the separation between the two is wide enough. However, when the IF is to be downconverted near to baseband or when both sidebands are desired as separate outputs, we must rely on a special configuration known as a *sideband-separating mixer*, shown in Figure 13.12. It relies on two identical submixers (which can be unbalanced or single-, double-, or triple-balanced, as desired) driven in-phase by one signal, the RF or the LO, and in quadrature by the other. After downconversion, the upper and lower sidebands appear combined at both submixers' outputs, but

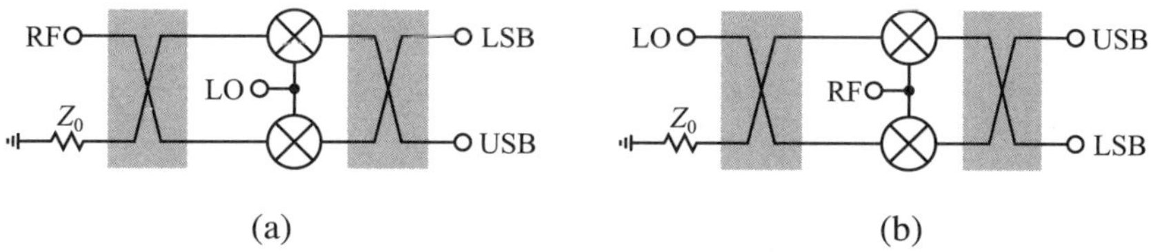

Figure 13.12 Sideband-separating mixers with (a) RF divided in quadrature and LO divided in-phase, or (b) LO divided in quadrature and RF divided in-phase.

with quadrature phase relationships of opposite polarity, such that an additional quadrature hybrid at the IF isolates them onto separate outputs.

To see how this works, let us write the upper and lower sidebands as sinusoidal signals at frequencies $\omega_{LO} + \omega_{IF}$ and $\omega_{LO} - \omega_{IF}$, respectively,

$$v_{USB}(t) = U\cos(\omega_{LO}t + \omega_{IF}t) \tag{13.11a}$$

$$v_{LSB}(t) = L\cos(\omega_{LO}t - \omega_{IF}t) \tag{13.11b}$$

U is the amplitude of the upper-sideband signal and L is the amplitude of the lower-sideband signal. Using the diagram in Figure 13.12(a), we may write the total signal at the RF inputs to both submixers (after the first quadrature hybrid) as

$$RF_1 = \frac{U}{\sqrt{2}}\cos(\omega_{LO}t + \omega_{IF}t) + \frac{L}{\sqrt{2}}\cos(\omega_{LO}t - \omega_{IF}t) \tag{13.12a}$$

$$RF_2 = \frac{U}{\sqrt{2}}\cos\left(\omega_{LO}t + \omega_{IF}t - \frac{\pi}{2}\right) + \frac{L}{\sqrt{2}}\cos\left(\omega_{LO}t - \omega_{IF}t - \frac{\pi}{2}\right) \tag{13.12b}$$

where RF_1 is the signal input to the top submixer in the figure, and RF_2 is the input to the bottom submixer. First-order mixing of these inputs with the LO generates sum and difference frequencies. We will assume that the submixers include a low-pass filter, so that only the difference frequencies survive at their IF outputs, or

$$IF_1 = \frac{L_cU}{\sqrt{2}}\cos(\omega_{IF}t) + \frac{L_cL}{\sqrt{2}}\cos(-\omega_{IF}t) \tag{13.13a}$$

$$= \frac{L_cU}{\sqrt{2}}\cos(\omega_{IF}t) + \frac{L_cL}{\sqrt{2}}\cos(\omega_{IF}t) \tag{13.13b}$$

for the top submixer, and

$$IF_2 = \frac{L_c U}{\sqrt{2}} \cos\left(\omega_{IF} t - \frac{\pi}{2}\right) + \frac{L_c L}{\sqrt{2}} \cos\left(-\omega_{IF} t - \frac{\pi}{2}\right) \tag{13.14a}$$

$$= \frac{L_c U}{\sqrt{2}} \cos\left(\omega_{IF} t - \frac{\pi}{2}\right) + \frac{L_c L}{\sqrt{2}} \cos\left(\omega_{IF} t + \frac{\pi}{2}\right) \tag{13.14b}$$

for the bottom submixer, where the conversion loss of both submixers is L_c. What is most important to take away from this is that the phase differential in the lower sideband changes sign after downconversion. Thus, the two signal paths carry a phase differential of $\pi/2$ for spectra downconverted from the upper sideband, and a differential of $-\pi/2$ for spectra downconverted from the lower sideband, a difference of π. A single sideband may thus be isolated by combining the two channels with a $\pi/2$ phase shift. Which sideband is recovered and which is canceled out are determined by which signal path receives the extra phase shift. One final quadrature hybrid at the IF accomplishes both simultaneously, thus giving the final outputs,

$$OUT_1 = \frac{1}{\sqrt{2}} IF_1 \angle 0 + \frac{1}{\sqrt{2}} IF_2 \angle(-\pi/2) \tag{13.15a}$$

$$= \frac{1}{\sqrt{2}} \left(\frac{L_c U}{\sqrt{2}} \cos(\omega_{IF} t) + \frac{L_c L}{\sqrt{2}} \cos(\omega_{IF} t) \right) + \frac{1}{\sqrt{2}} \left(\frac{L_c U}{\sqrt{2}} \cos(\omega_{IF} t - \pi) + \frac{L_c L}{\sqrt{2}} \cos(\omega_{IF} t) \right) \tag{13.15b}$$

$$= L_c L \cos(\omega_{IF} t) \tag{13.15c}$$

$$OUT_2 = \frac{1}{\sqrt{2}} IF_1 \angle(-\pi/2) + \frac{1}{\sqrt{2}} IF_2 \angle 0 \tag{13.15d}$$

$$= \frac{1}{\sqrt{2}} \left(\frac{L_c U}{\sqrt{2}} \cos\left(\omega_{IF} t - \frac{\pi}{2}\right) + \frac{L_c L}{\sqrt{2}} \cos\left(\omega_{IF} t - \frac{\pi}{2}\right) \right) + \frac{1}{\sqrt{2}} \left(\frac{L_c U}{\sqrt{2}} \cos\left(\omega_{IF} t - \frac{\pi}{2}\right) + \frac{L_c L}{\sqrt{2}} \cos\left(\omega_{IF} t + \frac{\pi}{2}\right) \right) \tag{13.15e}$$

$$= L_c U \cos\left(\omega_{IF} t - \frac{\pi}{2}\right) \tag{13.15f}$$

Therefore, we find that the top output delivers the lower sideband downconverter with a conversion loss of L_c, while the bottom output delivers the upper sideband with same loss (and an inconsequential residual phase shift of $\pi/2$). In practice, the mixers and various interconnects will introduce additional amplitude loss and phase shifts of their own, but these do not affect the general conclusion so long as they are the same in both channels.

Although most authors show the quadrature hybrid on the RF port and the in-phase splitter on the LO port, as in Figure 13.12(a), the sideband separation is just as effective with the roles reversed, as shown in Figure 13.12(b). Proponents of the initial configuration argue that the RF return loss is better when viewed through the quadrature hybrid (since reflected waves are rerouted to the resistive termination), and the LO return loss is usually less critical. However, the isolation between the two downconverted sidebands, called *image rejection*, depends on the balance of the two signal paths, that is, on how well the hybrids, mixers, and interconnects combine to produce the ideal $\pi/2$ phase shift and matched amplitudes. Quadrature hybrids inevitably deviate from these ideals away from the center frequency. In-phase splitters, in contrast, are usually physically symmetric, such that the amplitude and phase-imbalance are independent of frequency (although impedance matching and isolation may suffer away from the design frequency). Since the RF port in a sideband-separating downconverter must always sweep out a wider frequency range than the LO, this gives the configuration in Figure 13.12(b) a significant bandwidth advantage.

To quantify the image rejection, IR, we may consider the additive construction and subtractive cancellation of sinusoids given a total amplitude error, A, and phase error, ϕ,

$$\sin\theta \pm A\sin(\theta+\phi) = \sqrt{1+A^2 \pm 2A\cos\phi}\,\sin(\theta+\cdots) \tag{13.16}$$

where we have used the outphase summation rule from Section A.1.9. The image rejection is simply the ratio of the above sinusoids' amplitudes,

$$IR = 10\log\left(\frac{1+A^2+2A\cos\phi}{1+A^2-2A\cos\phi}\right) \tag{13.17}$$

Contours of image rejection as a function of amplitude imbalance on the horizontal axis and phase imbalance on the vertical axis are shown in Figure 13.13(a).

Included in the amplitude and phase balance calculations is the IF hybrid, which may be most difficult of all considering that the IF will often extend over a wider fractional bandwidth, approaching infinity if a baseband IF is desired. It will

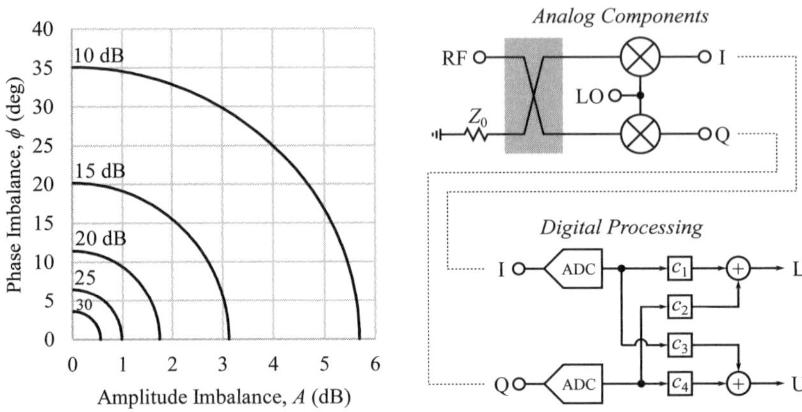

(a) (b)

Figure 13.13 (a) Contours of constant image rejection for amplitude imbalance, A, and phase imbalance, ϕ. (b) Numerical implementation of IF hybrid for precision sideband separation and bandwidth.

almost certainly be the largest hybrid in the system and may be totally impractical if the specifications are tight. A fairly elegant solution in applications where the IF signal is destined to be digitized (as is most often the case these days) is to move the IF hybrid into the digital domain, where it can be implemented numerically with almost unlimited precision[2] [5]. In fact, the numerical coefficients of the hybrid, labeled c_1–c_4 in Figure 13.13(b), can be tuned not only to mimic a perfect hybrid, but one that directly compensates the amplitude and phase errors of the preceding analog components, allowing one to achieve extraordinarily high image-rejection ratios [6]. The intermediate analog signals after downconversion but prior to digitization are conventionally called I and Q for *in-phase* and *quadrature*, respectively, and the mixer which generates them, no longer fully sideband-separating on its own, is called an *I/Q mixer*.

2 On a historical note, prior to the advent of crystal filters for sideband selection, essentially this same structure was used to achieve *single-sideband suppressed carrier* (SSBSC) modulation, where a baseband signal is upconverted into a single RF sideband with both the image band and the LO tone substantially absent from the final output.

Figure 13.14 Subharmonic mixers. (a) Shunt configuration. (b) Series configuration.

If the RF is low enough, the analog mixer can be eliminated in favor of a direct digital downconverter — a sampler configuration in which the individual analog-to-digital converters (ADCs) act very much like mixers downconverting their inputs via aliasing. These are not practical yet at millimeter-wave frequencies, however.

13.3.6 Subharmonic

In some instances, a local oscillator may not be readily available at a frequency sufficiently close to the RF band for direction downconversion. In these instances, it may be advantageous to enhance higher-order mixing products based on harmonics of the LO, and to suppress mixing with the LO fundamental frequency which would, under most circumstances, be stronger. This would be called a *subharmonic mixer*. There are many ways to accomplish this, but perhaps the simplest and most widely adopted are those shown in Figure 13.14(a), where a single diode has been replaced with an antiparallel diode pair. Looking quite similar to the odd-harmonic multipliers in Figure 13.3, which suppressed even-order harmonics, in this case mixing with the even harmonics of the LO is enhanced. These seemingly contradictory statements are actually both true for the same reasons. The antiparallel diode pair, assuming identical diodes, appears from the outside to have no second-order nonlinearity; thus, second harmonics of the inputs are suppressed at the output (in both cases). However, second-order currents are still present; they simply circulate between the two diodes without coupling to the outside. That allows internal mixing with the second-order LO currents to occur, and the phase is such that mixing products from both diodes with the RF fundamental (that is, $2\omega_{LO} \pm \omega_{RF}$) combine in-phase at the IF output, while those with LO fundamental ($\omega_{LO} \pm \omega_{RF}$) cancel.

The concepts of harmonic mixing, balanced mixing, and sideband separation are not mutually exclusive. The nonlinear elements in a single-balanced mixer, for example, could be replaced with antiparallel diode pairs in order to enhance mixing with the LO second-harmonic while suppressing mixing products with the

LO fundamental; one simply must be careful that the chosen configuration allows the mixing product that one is targeting. In other words, a single-balanced mixer that suppresses products of the form $m\omega_{RF} + n\omega_{LO}$ when n is even will not suddenly start generating those mixing products simply because an antiparallel diode pair is used. In that case, all mixing products would be suppressed, and the circuit would not be very useful. One should also carefully analyze the behavior of these compound constructions using the techniques illustrated in this chapter to determine whether, for example, the appropriate subharmonically upconverted or downconverted LO noise is suppressed in the manner that one is expecting.

13.4 FUNDAMENTAL LIMITS UPON CONVERSION EFFICIENCY

One might ask if there is any fundamental limit to the efficiency that can be achieved by frequency-converting circuits. In fact, several limits have been derived based on the assumptions of different kinds of nonlinear elements.

Perhaps the most famous are the *Manley-Rowe relations* [7, 8], which apply to arbitrary nonlinear reactors (varactors),

$$\sum_{m=0}^{\infty}\sum_{n=-\infty}^{\infty}\frac{mP_{mn}}{m\omega_1 + n\omega_2} = \frac{H}{\omega_1} \tag{13.18a}$$

$$\sum_{m=-\infty}^{\infty}\sum_{n=0}^{\infty}\frac{nP_{mn}}{m\omega_1 + n\omega_2} = 0 \tag{13.18b}$$

where P_{mn} is the time-average power flowing into the nonlinear reactor at frequency $m\omega_1 + n\omega_2$, and H is the average power dissipated by hysteresis in the varactor, if any. Extraordinarily, these expressions hold so long as the nonlinearity is reactive, no assumption is made about its specific form (e.g., its Taylor series).

Ideal reactors (without hysteresis) are energy-storage devices only — a linear reactor (be it an inductor or a capacitor) stores energy for a time and then releases it back to the circuit at the same frequency. A nonlinear reactor, in contrast, may store energy at one frequency and then release it at another, but no reactor without hysteresis, not even a nonlinear one, may dissipate that energy internally. The Manley-Rowe relations represent a fundamental conservation principle that has found widespread use in many applications involving energy conservation in nonlinear systems, beyond just those of electronic circuits.

To see how the Manley-Rowe relations can be applied, consider the case of an upper-sideband upconversion or downconversion where $\omega_{LO} = \omega_1$, $\omega_{IF} = \omega_2$, and $\omega_{RF} = \omega_{LO} + \omega_{IF}$ (that is $m = n = 1$). To maximize efficiency, we allow only the desired inputs and outputs to couple in or out of the device (other output terms would only reduce the portion that remains for the desired output in the above summation). Therefore, (13.18b) becomes

$$\frac{P_{11}}{\omega_1 + \omega_2} + \frac{P_{01}}{\omega_2} = \frac{P_{RF}}{\omega_{RF}} + \frac{P_{IF}}{\omega_{IF}} = 0 \tag{13.19a}$$

$$\therefore -\frac{P_{RF}}{P_{IF}} = \frac{\omega_{RF}}{\omega_{IF}} \tag{13.19b}$$

The minus sign simply indicates that one power term is negative while the other is positive (one is an input and the other an output). When upconverting, an optimally designed varactor mixer could have conversion gain equal to the ratio of the RF and IF. Effectively, the weak IF input modulates the much stronger LO in such a way that substantial energy from the LO pump is converted into RF energy. When operating as a downconverter, the LO takes energy away from the RF input, transferring only a small portion to the IF, with the loss factor given by the same ratio.

The situation for lower-sideband upconversion or downconversion ($\omega_{RF} = \omega_{LO} - \omega_{IF}$) is rather more interesting,

$$\frac{-P_{1,-1}}{\omega_1 - \omega_2} + \frac{P_{01}}{\omega_2} = -\frac{P_{RF}}{\omega_{RF}} + \frac{P_{IF}}{\omega_{IF}} = 0 \tag{13.20a}$$

$$\therefore \frac{P_{RF}}{P_{IF}} = \frac{\omega_{RF}}{\omega_{IF}} \tag{13.20b}$$

In this case, P_{RF} and P_{IF} have the same sign. Since power cannot be extracted from passive terminations, it must be that LO power delivered to the varactor emerges at both the RF and IF frequencies simultaneously (with proportionately more coming out at the higher frequency). The real part of the immittance presented to the RF and IF circuits is negative; therefore, the device is potentially unstable, and depending on the source and load impedances, conversion gain anywhere from zero to infinity may be achieved. This is the basis of a device called a *parametric amplifier*, wherein a pumped varactor generates a negative impedance that can be exploited for reflective gain at a frequency lower than the pump.

Harmonic generation with varactors can be quite efficient,

$$\frac{P_{10}}{\omega_1} + \frac{mP_{m0}}{m\omega_1} = \frac{H}{\omega_1} \tag{13.21a}$$

$$\therefore -\frac{P_{m0}}{P_{10}} = 1 - \frac{H}{P_{10}} \tag{13.21b}$$

which approaches 100% when hysteresis is absent ($H \to 0$).

Pantell [9] extended Manley and Rowe's analysis to nonlinear resistors (varistors) and found that positive varistors (for which $\partial i/\partial v > 0$) obey the following inequalities,

$$\sum_{m=0}^{\infty} \sum_{n=-\infty}^{\infty} m^2 P_{mn} \geq 0 \tag{13.22a}$$

$$\sum_{m=-\infty}^{\infty} \sum_{n=0}^{\infty} n^2 P_{mn} \geq 0 \tag{13.22b}$$

These equations show that upconversion or downconversion with varistors may approach 100% conversion efficiency. For example, in the case of an upper-sideband upconverter,

$$P_{01} + P_{11} = P_{IF} + P_{RF} \geq 0 \tag{13.23a}$$

$$\therefore -\frac{P_{RF}}{P_{IF}} \geq 1 \tag{13.23b}$$

Harmonic generators (multipliers) using varistors are found by Pantell's equations to have efficiency limited by the harmonic number,

$$-\frac{P_{m0}}{P_{10}} \leq \frac{1}{m^2} \tag{13.24}$$

confirming a similar result by Page [10], who analyzed the special case of an *ideal rectifier*, an extreme form of nonlinear resistance that transitions instantly from an open circuit at negative voltages to a short-circuit under positive bias. We may thus conclude that varactors are potentially more efficient for multiplication and upconversion (though potentially unstable), while varistors may be more efficient for downconversion.

Other limits may apply to more specific cases. For example, Kelly [11] has shown that varistor-based mixers supporting operation in either sideband have a theoretical minimum conversion loss (up or down) of 3 dB. In the case where matched loads are presented to all idler frequencies (other mixing products), the theoretical minimum loss is $20\log(\pi/2) = 3.92$ dB.

Problems

13-1 What is the small-signal resistance of a diode obeying the Shockley equation as a function of the DC bias point?

13-2 What is the temperature coefficient of the current in a diode obeying the Shockley equation assuming fixed voltage bias?

13-3 A device which generates a DC output voltage or current proportional to an AC input power, which may be considered a special case of frequency conversion, from a finite frequency to zero frequency, is called a *detector*. Calculate the *responsivity* of a Shockley diode detector, defined for the purposes of this problem as the ratio of detected DC current to accepted input power. Hint: the integrals in Section A.4.2 might be useful.

13-4 What is the maximum fractional bandwidth that a single-element, Nth-harmonic multiplier can have such that its output can be filtered without any undesired harmonic overlapping in the passband? What if antiparallel elements are used, rejecting even harmonics? What if a 180° hybrid is used, rejecting odd harmonics?

13-5 Suppose we wish to design a mixer that does not translate an input RF band, but simply reverses it in frequency space. That is, signal components at the lower band edge, ω_1, are shifted to the upper band edge, ω_2, and vice versa. What LO frequency is needed?

13-6 Analyze the single-balanced mixer with quadrature hybrid in Figure 13.8(b) to show that it has the LO noise and mixing product rejection properties listed in Table 13.1.

13-7 Explain why the single-balanced mixer with quadrature hybrid and opposite diode orientation, as shown in Figure 13.8(b), works only as an downconverter.

13-8 What is the image rejection achieved by a sideband-separating mixer that has a 1.2-dB amplitude imbalance and 15° phase imbalance?

13-9 Show that the mixer in Figure 13.12(b) is also sideband-separating, despite the exchange of RF and LO ports from the example given in the text.

13-10 A sideband-separating mixer is made with two imperfect quadrature hybrids. Assume that the RF hybrid has amplitude and phase imbalance such that it alone would limit the image rejection to 20 dB, even if the IF hybrid was perfect, and that the IF hybrid has imbalances that would limit image rejection to 15 dB, even if the RF hybrid was perfect. Give approximate upper and lower bounds for the image rejection of the combination.

13-11 Say we wish to design an upconverting, second-harmonic, single-balanced mixer by replacing the single diodes in a fundamental single-balanced mixer with antiparallel diode pairs. Select a configuration from Table 13.1 that is appropriate to the task, draw the circuit, and then confirm that LO noise undergoing the same second-harmonic upconversion is suppressed (if it is not, select another configuration).

References

[1] S. A. Maas, *Nonlinear Microwave Circuits*. New York: IEEE Press, 1997.

[2] M. A. Morgan and S. Weinreb, "A full waveguide band MMIC tripler for 75-110 GHz," *IEEE MTT-S International Microwave Sympsoium Digest*, vol. 1, May 2001, pp. 103–106.

[3] S. A. Maas, *Microwave Mixers*, 2nd ed. Norwood: Artech House, 1993.

[4] M. A. Morgan, E. W. Bryerton, P. Cesarano, T. A. Boyd, D. Thacker, K. Saini, and S. Weinreb, "A millimeter-wave diode-MMIC chipset for local oscillator generation in the ALMA telescope," *IEEE MTT-S International Microwave Symposium Digest*, June 2005.

[5] M. A. Morgan and J. R. Fisher, "Experiments with digital sideband-separating downconversion," *Publications of the Astronomical Society of the Pacific*, vol. 122, no. 889, pp. 326–335, March 2010.

[6] R. Finger, P. Mena, N. Reyes, R. Rodriguez, and L. Bronfman, "A calibrated digital sideband separating spectrometer for radio astronomy applications," *Publications of the Astronomical Society of the Pacific*, vol. 125, pp. 263–269, March 2013.

[7] J. M. Manley and H. E. Rowe, "Some general properties of nonlinear elements — part I: general energy relations," *Proceedings of the IRE*, vol. 44, no. 7, pp. 904–913, July 1956.

[8] H. E. Rowe, "Some general properties of nonlinear elements — part I: general energy relations," *Proceedings of the IRE*, vol. 46, no. 5, pp. 850–860, May 1958.

[9] R. H. Pantell, "General power relationships for positive and negative nonlinear resistive elements," *Proceedings of the IRE*, vol. 46, no. 12, pp. 1910–1913, December 1958.

[10] C. Page, "Harmonic generation with ideal rectifiers," *Proceedings of the IRE*, vol. 46, no. 10, pp. 1738–1740, October 1958.

[11] A. J. Kelly, "Fundamental limits on conversion loss of double sideband resistive mixers," *IEEE Transactions on Microwave Theory and Techniques*, vol. 25, no. 11, pp. 867–869, November 1977.

Appendix A

Mathematical Identities

The following sections list a variety mathematical results that are useful in solving electromagnetic circuit problems.

A.1 TRIGONOMETRIC IDENTITIES

A.1.1 Pythagorean Identities

$$\sin^2\theta + \cos^2\theta = 1 \qquad \tan^2\theta + 1 = \sec^2\theta$$

$$1 + \cot^2\theta = \csc^2\theta \qquad \sec^2\theta + \csc^2\theta = \sec^2\theta\csc^2\theta$$

A.1.2 Sum-Angle Formulas

$$\sin(\alpha+\beta) = \sin\alpha\cos\beta + \sin\beta\cos\alpha \qquad \tan(\alpha+\beta) = \tfrac{\tan\alpha+\tan\beta}{1-\tan\alpha\tan\beta}$$

$$\cos(\alpha+\beta) = \cos\alpha\cos\beta - \sin\alpha\sin\beta \qquad \cot(\alpha+\beta) = \tfrac{\cot\alpha\cot\beta-1}{\cot\alpha+\cot\beta}$$

A.1.3 Double-Angle and Triple-Angle Formulas

$$\sin(2\theta) = 2\sin\theta\cos\theta \qquad \sin(3\theta) = 3\sin\theta - 4\sin^3\theta$$

$$\cos(2\theta) = \cos^2\theta - \sin^2\theta \qquad \cos(3\theta) = 4\cos^3\theta - 3\cos\theta$$

$$\tan(2\theta) = \frac{2\tan\theta}{1-\tan^2\theta} \qquad \tan(3\theta) = \frac{3\tan\theta-\tan^3\theta}{1-3\tan^2\theta}$$

$$\cot(2\theta) = \frac{\cot^2\theta-1}{2\cot\theta} \qquad \cot(3\theta) = \frac{3\cot\theta-\cot^3\theta}{1-3\cot^2\theta}$$

$$\sec(2\theta) = \frac{\sec^2\theta}{2-\sec^2\theta} \qquad \sec(3\theta) = \frac{\sec^3\theta}{4-3\sec^2\theta}$$

$$\csc(2\theta) = \tfrac{1}{2}\sec\theta\csc\theta \qquad \csc(3\theta) = \frac{\csc^3\theta}{3\csc^2\theta-4}$$

A.1.4 Half-Angle Formulas

$$\tan\left(\tfrac{1}{2}\theta\right) = \csc\theta - \cot\theta = \frac{\sin\theta}{1+\cos\theta} = \frac{1-\cos\theta}{\sin\theta} = \frac{\tan\theta}{1+\sec\theta}$$

$$\cot\left(\tfrac{1}{2}\theta\right) = \csc\theta + \cot\theta = \frac{\sin\theta}{1-\cos\theta} = \frac{1+\cos\theta}{\sin\theta}$$

$$\tan\left(\frac{\alpha+\beta}{2}\right) = \frac{\sin\alpha+\sin\beta}{\cos\alpha+\cos\beta} = -\frac{\cos\alpha-\cos\beta}{\sin\alpha-\sin\beta}$$

$$\tan\left(\tfrac{1}{2}\theta + \tfrac{\pi}{4}\right) = \sec\theta + \tan\theta$$

A.1.5 Power Reduction

$$\sin^2\theta = \tfrac{1}{2}\left(1-\cos(2\theta)\right) \qquad \sin^3\theta = \tfrac{1}{4}\left(3\sin\theta - \sin(3\theta)\right)$$

$$\cos^2\theta = \tfrac{1}{2}\left(1+\cos(2\theta)\right) \qquad \cos^3\theta = \tfrac{1}{4}\left(3\cos\theta + \cos(3\theta)\right)$$

$$\cos^n\theta = \begin{cases} \left(\tfrac{1}{2}\right)^{n-1}\sum\binom{n}{(n-k)/2}\cos(k\theta) + \tfrac{1}{2^n}\binom{n}{n/2} & n \text{ is even} \\ \left(\tfrac{1}{2}\right)^{n-1}\sum\binom{n}{(n-k)/2}\cos(k\theta) & n \text{ is odd} \end{cases}$$

$$\sin^n\theta = \begin{cases} \left(\tfrac{1}{2}\right)^{n-1}\sum(-1)^{k/2}\binom{n}{(n-k)/2}\cos(k\theta) + \tfrac{1}{2^n}\binom{n}{n/2} & n \text{ is even} \\ \left(\tfrac{1}{2}\right)^{n-1}\sum(-1)^{(k-1)/2}\binom{n}{(n-k)/2}\sin(k\theta) & n \text{ is odd} \end{cases}$$

In the above formulae, the summations are taken over all $0 < k \leq n$, where all k are even if n is even, and all k are odd if n is odd.

A.1.6 Product-to-Sum Formulas (Prosthaphaeresis)

$$2\sin\alpha\sin\beta = \cos(\alpha-\beta) - \cos(\alpha+\beta)$$
$$2\cos\alpha\cos\beta = \cos(\alpha-\beta) + \cos(\alpha+\beta)$$
$$2\sin\alpha\cos\beta = \sin(\alpha+\beta) + \sin(\alpha-\beta)$$
$$\tan\alpha\tan\beta = \frac{\cos(\alpha-\beta) - \cos(\alpha+\beta)}{\cos(\alpha-\beta) + \cos(\alpha+\beta)}$$

A.1.7 Sum-to-Product Formulas

$$\sin\alpha + \sin\beta = 2\sin\left(\tfrac{\alpha+\beta}{2}\right)\cos\left(\tfrac{\alpha-\beta}{2}\right)$$
$$\cos\alpha + \cos\beta = 2\cos\left(\tfrac{\alpha+\beta}{2}\right)\cos\left(\tfrac{\alpha-\beta}{2}\right)$$
$$\cos\alpha - \cos\beta = -2\sin\left(\tfrac{\alpha+\beta}{2}\right)\sin\left(\tfrac{\alpha-\beta}{2}\right)$$

A.1.8 Composite-Angle Formulas

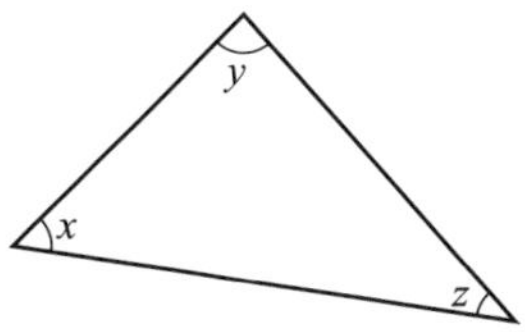

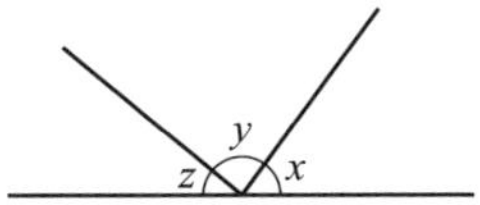

If $x+y+z=\pi$, then,

$$\sin(2x) + \sin(2y) + \sin(2z) = 4\sin x\sin y\sin z$$

$$\tan x + \tan y + \tan z = \tan x\tan y\tan z$$
$$\cot x\cot y + \cot y\cot z + \cot z\cot x = 1$$

If $x+y+z=\frac{\pi}{2}$, then,

$$\cot x + \cot y + \cot z = \cot x\cot y\cot z$$

A.1.9 Outphase Summation

$$a\sin\theta + b\cos\theta = \sqrt{a^2+b^2}\cdot\sin\left(\theta + \tan^{-1}\left(\tfrac{b}{a}\right)\right)$$

$$\begin{aligned} &a\sin\theta + b\sin(\theta+\phi) \\ &\qquad = \sqrt{a^2+b^2+2ab\cos\phi}\cdot\sin\left(\theta+\tan^{-1}\left(\tfrac{b\sin\phi}{a+b\cos\phi}\right)\right) \end{aligned} \tag{A.1}$$

Note, it may be necessary to use a four-quadrant implementation of the arctangent in the above formulas to get a result with the correct sign.

A.1.10 Trigonometric Products

$$\sin(nx) = 2^{n-1}\prod_{k=0}^{n-1}\sin\left(x+\frac{k\pi}{n}\right)$$

$$\prod_{k=1}^{n-1}\sin\left(\frac{k\pi}{n}\right) = \frac{n}{2^{n-1}}$$

$$\prod_{k=1}^{n-1}\cos\left(\frac{k\pi}{n}\right) = \frac{\sin\left(\frac{n\pi}{2}\right)}{2^{n-1}}$$

$$\prod_{k=1}^{n-1}\tan\left(\frac{k\pi}{n}\right) = \frac{n}{\sin\left(\frac{n\pi}{2}\right)}$$

$$\prod_{k=1}^{n}\sin\left(\frac{(2k-1)\pi}{4n}\right) = \prod_{k=1}^{n}\cos\left(\frac{(2k-1)\pi}{4n}\right) = \frac{\sqrt{2}}{2^n}$$

$$\prod_{k=1}^{n}\sin\left(\frac{k\pi}{2n+1}\right) = \frac{\sqrt{2n+1}}{2^n}$$

$$\prod_{k=1}^{n}\cos\left(\frac{k\pi}{2n+1}\right) = \frac{1}{2^n}$$

A.1.11 Exponential Forms (Euler's Formula)

$$e^{j\theta} = \cos\theta + j\sin\theta$$

or more generally

$$e^{\alpha+j\beta} = e^{\alpha}e^{j\beta} = e^{\alpha}\cos\beta + je^{\alpha}\sin\beta$$

and

$$\cos\theta = \tfrac{e^{j\theta}+e^{-j\theta}}{2} \qquad \sin\theta = \tfrac{e^{j\theta}-e^{-j\theta}}{j2}$$

A.2 HYPERBOLIC FUNCTIONS AND IDENTITIES

A.2.1 Relationships to Standard Trigonometric Functions

$$j\sinh x = \sin(jx) \qquad j\sin x = \sinh(jx)$$

$$\cosh x = \cos(jx) \qquad \cos x = \cosh(jx)$$

$$j\tanh x = \tan(jx) \qquad j\tan x = \tanh(jx)$$

$$\coth x = j\cot(jx) \qquad \cot x = j\coth(jx)$$

$$\operatorname{sech} x = \sec(jx) \qquad \sec x = \operatorname{sech}(jx)$$

$$\operatorname{csch} x = j\csc(jx) \qquad \csc x = j\operatorname{csch}(jx)$$

A.2.2 Pythagorean Identities

$$\cosh^2 x - \sinh^2 x = 1$$

$$1 - \tanh^2 x = \operatorname{sech}^2 x$$

$$\coth^2 x - 1 = \operatorname{csch}^2 x$$

A.2.3 Sum-Argument Formulas

$$\sinh(x+y) = \sinh x\cosh y + \sinh y\cosh x$$

$$\cosh(x+y) = \cosh x\cosh y + \sinh x\sinh y$$

$$\tanh(x+y) = \frac{\tanh x + \tanh y}{1 + \tanh x\tanh y}$$

A.2.4 Double-Argument Formulas

$$\sinh(2x) = 2\sinh x \cosh x$$
$$\cosh(2x) = \sinh^2 x + \cosh^2 x$$
$$\tanh(2x) = \frac{2\tanh x}{1+\tanh^2 x}$$

A.2.5 Half-Argument Formulas

$$\tanh\left(\tfrac{1}{2}x\right) = \coth x - \operatorname{csch} x = \frac{\sinh x}{1+\cosh x} = \frac{\cosh x - 1}{\sinh x}$$
$$\coth\left(\tfrac{1}{2}x\right) = \coth x + \operatorname{csch} x$$

A.2.6 Sum-to-Product Formulas

$$\sinh x - \sinh y = 2\cosh\left(\tfrac{x+y}{2}\right)\sinh\left(\tfrac{x-y}{2}\right)$$
$$\cosh x - \cosh y = 2\sinh\left(\tfrac{x+y}{2}\right)\sinh\left(\tfrac{x-y}{2}\right)$$

A.2.7 Exponential Forms

$$e^x = \cosh x + \sinh x$$
$$e^{-x} = \cosh x - \sinh x$$
$$e^x = \sqrt{\frac{1+\tanh x}{1-\tanh x}} = \frac{1+\tanh\left(\frac{x}{2}\right)}{1-\tanh\left(\frac{x}{2}\right)}$$
$$\cosh x = \frac{e^x + e^{-x}}{2}$$
$$\sinh x = \frac{e^x - e^{-x}}{2}$$

A.3 OTHER THEOREMS

A.3.1 Binomial Theorem

$$(x+y)^n = \sum_{k=0}^{n} \binom{n}{k} x^{n-k} y^k = \sum_{k=0}^{n} \binom{n}{k} x^k y^{n-k}$$

A.3.2 Law of Sines and Cosines

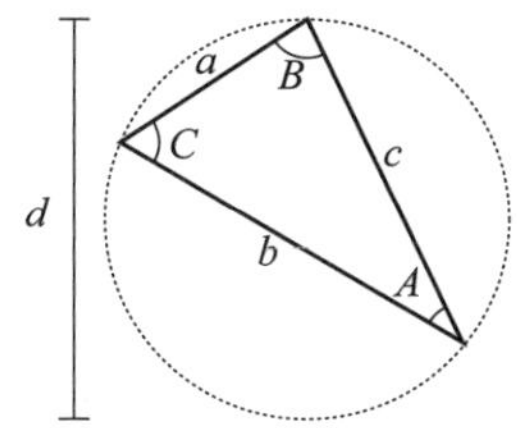

The law of sines:

$$\frac{a}{\sin A} = \frac{b}{\sin B} = \frac{c}{\sin C} = d$$

The law of cosines:

$$c^2 = a^2 + b^2 - 2ab\cos C$$

A.4 INTEGRALS

A.4.1 Trigonometric Integrals

$$\int_0^a \cos\left(\frac{m\pi x}{a}\right)\cos\left(\frac{n\pi x}{a}\right)dx = \frac{a}{2}\begin{cases} 2 & m = n = 0 \\ 1 & m = n \neq 0 \\ 0 & m \neq n \end{cases}$$

$$\int_0^a \sin\left(\frac{m\pi x}{a}\right)\sin\left(\frac{n\pi x}{a}\right)dx = \frac{a}{2}\begin{cases} 0 & m = n = 0 \\ 1 & m = n \neq 0 \\ 0 & m \neq n \end{cases}$$

Or, more succinctly,

$$\int_0^a \cos\left(\frac{m\pi x}{a}\right)\cos\left(\frac{n\pi x}{a}\right)dx = \frac{a}{2}\left(1 + \delta_m\right)\delta_{mn}$$

$$\int_0^a \sin\left(\frac{m\pi x}{a}\right)\sin\left(\frac{n\pi x}{a}\right)dx = \frac{a}{2}\left(1 - \delta_m\right)\delta_{mn}$$

and always

$$\int_0^a \cos\left(\frac{m\pi x}{a}\right)\sin\left(\frac{n\pi x}{a}\right)dx = 0$$

$$\int_0^a \cos^2\left(\frac{m\pi x}{a}\right)dx = \frac{a}{2}(1+\delta_m)$$

$$\int_0^a \sin^2\left(\frac{m\pi x}{a}\right)dx = \frac{a}{2}(1-\delta_m)$$

where δ_{mn} or just δ_m is the Kronecker delta,

$$\delta_{mn} = \begin{cases} 1 & m = n \\ 0 & m \neq n \end{cases}$$

$$\delta_m = \delta_{m0} = \begin{cases} 1 & m = 0 \\ 0 & m \neq 0 \end{cases}$$

A.4.2 Other Useful Integrals

$$\int_0^{2\pi} e^{x\cos\theta}d\theta = \int_0^{2\pi} e^{x\sin\theta}d\theta = 2\pi I_0(x)$$

$$\int_0^{2\pi} e^{x\cos\theta}\cos\theta d\theta = \int_0^{2\pi} e^{x\sin\theta}\sin\theta d\theta = 2\pi I_1(x)$$

$$\int_0^{2\pi} e^{x\cos\theta+y\sin\theta}d\theta = 2\pi I_0\left(\sqrt{x^2+y^2}\right)$$

where $I_n(x)$ is the nth-order modified Bessel function of the first kind (see Appendix G).

Appendix B

Integral Theorems

Here we review two very important integral theorems in multivariable calculus known as the *divergence theorem* and *Stokes' theorem*. These apply to any continuously differentiable vector field, **F**, such as the electric and magnetic fields, the vector magnetic potential, and the volume current density.

B.1 THE DIVERGENCE THEOREM

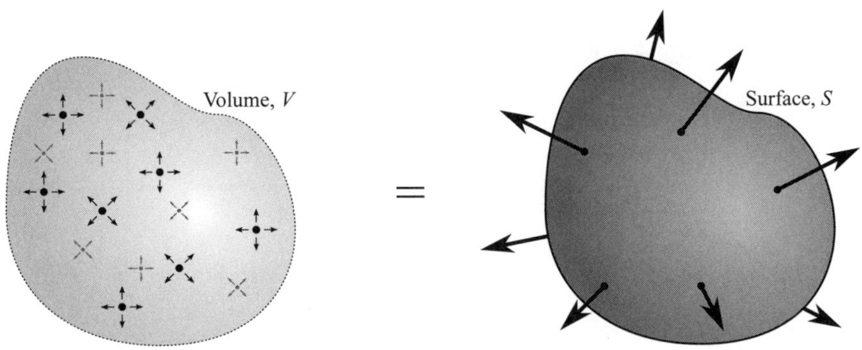

The divergence theorem (also called *Gauss's theorem* or *Ostrogradsky's theorem*) states that the total divergence in any arbitrary volume, V, is equal to the net outgoing flux through the surface, S, which bounds that volume,

$$\iiint_V (\nabla \cdot \mathbf{F}) dV = \oiint_S \mathbf{F} \cdot d\mathbf{S} \tag{B.1}$$

where $d\mathbf{S} = \mathbf{n} dS$ and $\mathbf{n}$ is the surface's unit normal vector.

B.2 STOKES' THEOREM

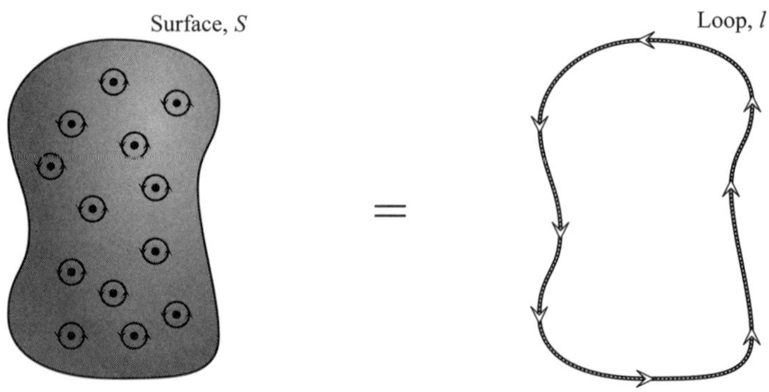

Stokes' theorem (also known as the *Kelvin-Stokes theorem*) states that the total curl on a surface, S (which need not be planar), is equal to the circulation around the perimeter, l, bounding that surface,

$$\iint_S (\nabla \times \mathbf{F}) \cdot d\mathbf{S} = \oint_l \mathbf{F} \cdot d\mathbf{l} \tag{B.2}$$

where the direction of integration around the perimeter, $d\mathbf{l}$, is related to the surface normal, $d\mathbf{S}$, by the right-hand rule.

Appendix C

Vector Identities

For any vectors $\mathbf{A}$, $\mathbf{B}$, $\mathbf{C}$, $\mathbf{D}$, and scalars f and g,

$$\mathbf{A} \cdot \mathbf{B} \times \mathbf{C} = \mathbf{B} \cdot \mathbf{C} \times \mathbf{A} = \mathbf{C} \cdot \mathbf{A} \times \mathbf{B} \tag{C.1}$$

$$\mathbf{A} \times (\mathbf{B} \times \mathbf{C}) = \mathbf{B}(\mathbf{A} \cdot \mathbf{C}) - \mathbf{C}(\mathbf{A} \cdot \mathbf{B}) \tag{C.2}$$

$$(\mathbf{A} \times \mathbf{B}) \cdot (\mathbf{C} \times \mathbf{D}) = \mathbf{A} \cdot \mathbf{B} \times (\mathbf{C} \times \mathbf{D}) \tag{C.3a}$$

$$= \mathbf{A} \cdot (\mathbf{B} \cdot \mathbf{D}\mathbf{C} - \mathbf{B} \cdot \mathbf{C}\mathbf{D}) \tag{C.3b}$$

$$= (\mathbf{A} \cdot \mathbf{C})(\mathbf{B} \cdot \mathbf{D}) - (\mathbf{A} \cdot \mathbf{D})(\mathbf{B} \cdot \mathbf{C}) \tag{C.3c}$$

$$(\mathbf{A} \times \mathbf{B}) \times (\mathbf{C} \times \mathbf{D}) = (\mathbf{A} \times \mathbf{B} \cdot \mathbf{D})\mathbf{C} - (\mathbf{A} \times \mathbf{B} \cdot \mathbf{C})\mathbf{D} \tag{C.4}$$

$$\nabla(fg) = f\nabla g + g\nabla f \tag{C.5}$$

$$\nabla \cdot (f\mathbf{A}) = f\nabla \cdot \mathbf{A} + \mathbf{A} \cdot \nabla f \tag{C.6}$$

$$\nabla \times (f\mathbf{A}) = f\nabla \times \mathbf{A} + \nabla f \times \mathbf{A} \tag{C.7}$$

$$\nabla(\mathbf{A} \cdot \mathbf{B}) = (\mathbf{A} \cdot \nabla)\mathbf{B} + (\mathbf{B} \cdot \nabla)\mathbf{A} + \mathbf{A} \times (\nabla \times \mathbf{B}) + \mathbf{B} \times (\nabla \times \mathbf{A}) \tag{C.8}$$

$$\nabla \cdot (\mathbf{A} \times \mathbf{B}) = \mathbf{B} \cdot (\nabla \times \mathbf{A}) - \mathbf{A} \cdot (\nabla \times \mathbf{B}) \tag{C.9}$$

$$\nabla \times (\mathbf{A} \times \mathbf{B}) = \mathbf{A}(\nabla \cdot \mathbf{B}) - \mathbf{B}(\nabla \cdot \mathbf{A}) + (\mathbf{B} \cdot \nabla)\mathbf{A} - (\mathbf{A} \cdot \nabla)\mathbf{B} \tag{C.10}$$

$$\nabla \cdot \nabla f = \nabla^2 f \tag{C.11}$$

$$\nabla \times \nabla \times \mathbf{A} = \nabla(\nabla \cdot \mathbf{A}) - \nabla^2 \mathbf{A} \tag{C.12}$$

$$\nabla \times \nabla f = 0 \tag{C.13}$$

$$\nabla \cdot (\nabla \times \mathbf{A}) = 0 \tag{C.14}$$

Note that in the above formulas, $\mathbf{A} \cdot \nabla \neq \nabla \cdot \mathbf{A}$. For example, in Cartesian coordinates,

$$(\mathbf{A} \cdot \nabla)\,\mathbf{B} = \left(A_x \frac{\partial}{\partial x} + A_y \frac{\partial}{\partial y} + A_z \frac{\partial}{\partial z}\right)(B_x\mathbf{x} + B_y\mathbf{y} + B_z\mathbf{z}) \tag{C.15a}$$

$$= \left(A_x \frac{\partial B_x}{\partial x} + A_y \frac{\partial B_x}{\partial y} + A_z \frac{\partial B_x}{\partial z}\right)\mathbf{x} \tag{C.15b}$$

$$+ \left(A_x \frac{\partial B_y}{\partial x} + A_y \frac{\partial B_y}{\partial y} + A_z \frac{\partial B_y}{\partial z}\right)\mathbf{y} \tag{C.15c}$$

$$+ \left(A_x \frac{\partial B_z}{\partial x} + A_y \frac{\partial B_z}{\partial y} + A_z \frac{\partial B_z}{\partial z}\right)\mathbf{z} \tag{C.15d}$$

whereas

$$(\nabla \cdot \mathbf{A})\,\mathbf{B} = \left(\frac{\partial A_x}{\partial x} + \frac{\partial A_y}{\partial y} + \frac{\partial A_z}{\partial z}\right)(B_x\mathbf{x} + B_y\mathbf{y} + B_z\mathbf{z}) \tag{C.16a}$$

$$= \left(\frac{\partial A_x}{\partial x} + \frac{\partial A_y}{\partial y} + \frac{\partial A_z}{\partial z}\right) B_x\mathbf{x} \tag{C.16b}$$

$$+ \left(\frac{\partial A_x}{\partial x} + \frac{\partial A_y}{\partial y} + \frac{\partial A_z}{\partial z}\right) B_y\mathbf{y} \tag{C.16c}$$

$$+ \left(\frac{\partial A_x}{\partial x} + \frac{\partial A_y}{\partial y} + \frac{\partial A_z}{\partial z}\right) B_z\mathbf{z} \tag{C.16d}$$

In the first case, it is the vector $\mathbf{B}$ that is differentiated, while in the latter case it is $\mathbf{A}$ that is differentiated.

Appendix D

Vector Operator Forms

D.1 CARTESIAN COORDINATES (x, y, z)

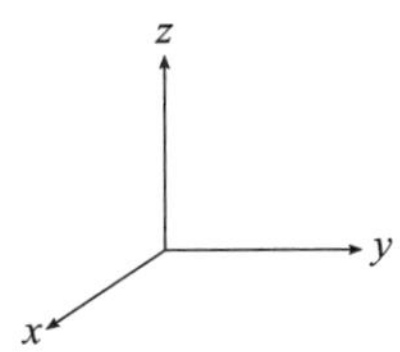

$$\nabla f = \frac{\partial f}{\partial x}\mathbf{x} + \frac{\partial f}{\partial y}\mathbf{y} + \frac{\partial f}{\partial z}\mathbf{z} \tag{D.1}$$

$$\nabla \cdot \mathbf{A} = \frac{\partial A_x}{\partial x} + \frac{\partial A_y}{\partial y} + \frac{\partial A_z}{\partial z} \tag{D.2}$$

$$\nabla \times \mathbf{A} = \begin{vmatrix} \mathbf{x} & \mathbf{y} & \mathbf{z} \\ \frac{\partial}{\partial x} & \frac{\partial}{\partial y} & \frac{\partial}{\partial z} \\ A_x & A_y & A_z \end{vmatrix} \tag{D.3a}$$

$$= \left(\frac{\partial A_z}{\partial y} - \frac{\partial A_y}{\partial z}\right)\mathbf{x} + \left(\frac{\partial A_x}{\partial z} - \frac{\partial A_z}{\partial x}\right)\mathbf{y} + \left(\frac{\partial A_y}{\partial x} - \frac{\partial A_x}{\partial y}\right)\mathbf{z} \tag{D.3b}$$

$$\nabla^2 f = \frac{\partial^2 f}{\partial x^2} + \frac{\partial^2 f}{\partial y^2} + \frac{\partial^2 f}{\partial z^2} \tag{D.4}$$

$$\nabla^2 \mathbf{A} = \left(\nabla^2 A_x\right)\mathbf{x} + \left(\nabla^2 A_y\right)\mathbf{y} + \left(\nabla^2 A_z\right)\mathbf{z} \tag{D.5}$$

D.2 CYLINDRICAL COORDINATES (r, θ, z)

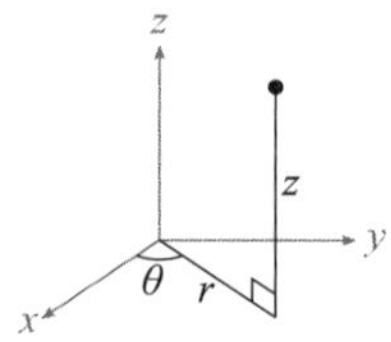

$$\nabla f = \frac{\partial f}{\partial r}\mathbf{r} + \frac{1}{r}\frac{\partial f}{\partial \theta}\boldsymbol{\theta} + \frac{\partial f}{\partial z}\mathbf{z} \tag{D.6}$$

$$\nabla \cdot \mathbf{A} = \frac{1}{r}\frac{\partial}{\partial r}(A_r) + \frac{1}{r}\frac{\partial A_\theta}{\partial \theta} + \frac{\partial A_z}{\partial z} \tag{D.7}$$

$$\nabla \times \mathbf{A} = \frac{1}{r}\begin{vmatrix} \mathbf{r} & r\boldsymbol{\theta} & \mathbf{z} \\ \frac{\partial}{\partial r} & \frac{\partial}{\partial \theta} & \frac{\partial}{\partial z} \\ A_r & rA_\theta & A_z \end{vmatrix} = \left(\frac{1}{r}\frac{\partial A_z}{\partial \theta} - \frac{\partial A_\theta}{\partial z}\right)\mathbf{r} \tag{D.8a}$$

$$+ \left(\frac{\partial A_r}{\partial z} - \frac{\partial A_z}{\partial r}\right)\boldsymbol{\theta} + \frac{1}{r}\left(\frac{\partial}{\partial r}(rA_\theta) - \frac{\partial A_r}{\partial \theta}\right)\mathbf{z} \tag{D.8b}$$

$$\nabla^2 f = \frac{1}{r}\frac{\partial}{\partial r}\left(r\frac{\partial f}{\partial r}\right) + \frac{1}{r^2}\frac{\partial^2 f}{\partial \theta^2} + \frac{\partial^2 f}{\partial z^2} \tag{D.9}$$

$$\nabla^2 \mathbf{A} = \nabla(\nabla \cdot \mathbf{A}) - \nabla \times \nabla \times \mathbf{A} \tag{D.10a}$$

$$= \left(\nabla^2 A_r - \frac{2}{r^2}\frac{\partial A_\theta}{\partial \theta} - \frac{A_r}{r^2}\right)\mathbf{r} \tag{D.10b}$$

$$+ \left(\nabla^2 A_\theta + \frac{2}{r^2}\frac{\partial A_r}{\partial \theta} - \frac{A_\theta}{r^2}\right)\boldsymbol{\theta} \tag{D.10c}$$

$$+ \left(\nabla^2 A_z\right)\mathbf{z} \tag{D.10d}$$

Note that, unlike in Cartesian coordinates, the Laplacian of a vector field in cylindrical coordinates is not the sum of the Laplacians of its individual components (i.e., $\nabla^2\mathbf{A} \neq \left(\nabla^2 A_r\right)\mathbf{r} + \left(\nabla^2 A_\theta\right)\boldsymbol{\theta} + \left(\nabla^2 A_z\right)\mathbf{z}$). This is because the direction of the unit vectors $\mathbf{r}$ and $\boldsymbol{\theta}$ are themselves dependent on the θ and r coordinates, respectively.

D.3 SPHERICAL COORDINATES (r, θ, ϕ)

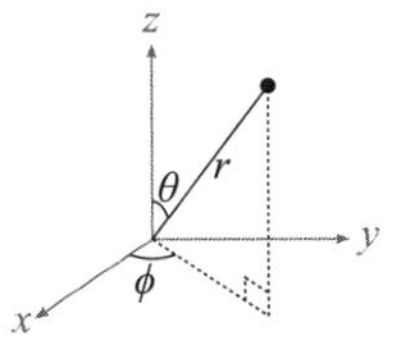

$$\nabla f = \frac{\partial f}{\partial r}\mathbf{r} + \frac{1}{r}\frac{\partial f}{\partial \theta}\boldsymbol{\theta} + \frac{1}{r\sin\theta}\frac{\partial f}{\partial \phi}\boldsymbol{\phi} \tag{D.11}$$

$$\nabla \cdot \mathbf{A} = \frac{1}{r^2}\frac{\partial}{\partial r}\left(r^2 A_r\right) + \frac{1}{r\sin\theta}\frac{\partial}{\partial \theta}\left(A_\theta \sin\theta\right) + \frac{1}{r\sin\theta}\frac{\partial A_\phi}{\partial \phi} \tag{D.12}$$

$$\nabla \times \mathbf{A} = \frac{1}{r^2\sin\theta}\begin{vmatrix} \mathbf{r} & r\boldsymbol{\theta} & (r\sin\theta)\,\boldsymbol{\phi} \\ \frac{\partial}{\partial r} & \frac{\partial}{\partial \theta} & \frac{\partial}{\partial \phi} \\ A_r & rA_\theta & (r\sin\theta)\,A_\phi \end{vmatrix} \tag{D.13a}$$

$$= \frac{1}{r\sin\theta}\left[\frac{\partial}{\partial \theta}\left(A_\phi \sin\theta\right) - \frac{\partial A_\theta}{\partial \phi}\right]\mathbf{r} \tag{D.13b}$$

$$+ \frac{1}{r}\left[\frac{1}{\sin\theta}\frac{\partial A_r}{\partial \phi} - \frac{\partial}{\partial r}\left(rA_\phi\right)\right]\boldsymbol{\theta} \tag{D.13c}$$

$$+ \frac{1}{r}\left[\frac{\partial}{\partial r}\left(rA_\theta\right) - \frac{\partial}{\partial \theta}A_r\right]\boldsymbol{\phi} \tag{D.13d}$$

$$\nabla^2 f = \frac{1}{r^2}\frac{\partial}{\partial r}\left(r^2\frac{\partial f}{\partial r}\right) + \frac{1}{r^2\sin\theta}\frac{\partial}{\partial \theta}\left(\sin\theta\frac{\partial f}{\partial \theta}\right) + \frac{1}{r^2\sin^2\theta}\frac{\partial^2 f}{\partial \phi^2} \tag{D.14}$$

$$\nabla^2 \mathbf{A} = \nabla\left(\nabla \cdot \mathbf{A}\right) - \nabla \times \nabla \times \mathbf{A} \tag{D.15a}$$

$$= \left[\nabla^2 A_r - \frac{2}{r^2}\left(A_r + \cot\theta A_\theta + \csc\theta\frac{\partial A_\phi}{\partial \phi} + \frac{\partial A_\theta}{\partial \theta}\right)\right]\mathbf{r} \tag{D.15b}$$

$$+ \left[\nabla^2 A_\theta - \frac{1}{r^2}\left(\csc^2\theta A_\theta - 2\frac{A_r}{\partial \theta} + 2\frac{\cot\theta}{\sin\theta}\frac{\partial A_\phi}{\partial \phi}\right)\right]\boldsymbol{\theta} \tag{D.15c}$$

$$+ \left[\nabla^2 A_\phi - \frac{1}{r^2}\left(\csc^2\theta A_\phi - 2\csc\theta\frac{\partial A_r}{\partial \phi} - 2\frac{\cot\theta}{\sin\theta}\frac{\partial A_\theta}{\partial \phi}\right)\right]\boldsymbol{\phi} \tag{D.15d}$$

Once again, the Laplacian of a vector field in spherical coordinates is not the sum of the Laplacian's of its individual components ($\nabla^2\mathbf{A} \neq \left(\nabla^2 A_r\right)\mathbf{r} + \left(\nabla^2 A_\theta\right)\boldsymbol{\theta} + \left(\nabla^2 A_\phi\right)\boldsymbol{\phi}$) because the direction of the unit vectors $\mathbf{r}$, $\boldsymbol{\theta}$, and $\boldsymbol{\phi}$ are themselves dependent on the r, θ, and ϕ coordinates.

Appendix E

Delta-Wye Identities

Although the delta-wye identity formulas given in Section 5.2.1 are mathematically general, only ten networks starting with three simple elements yield counterparts with no unphysical elements in any branch. These 10 are listed below.

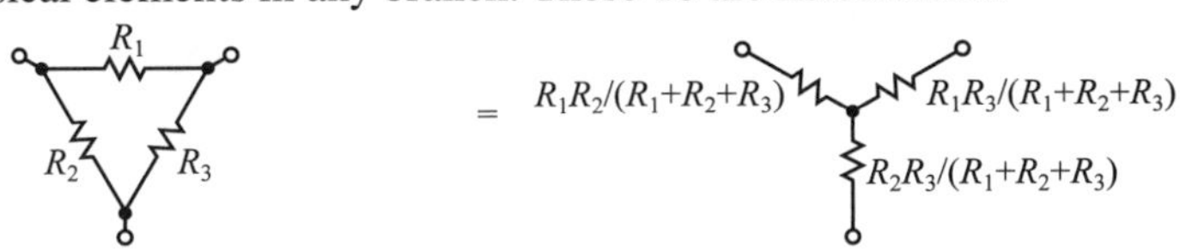

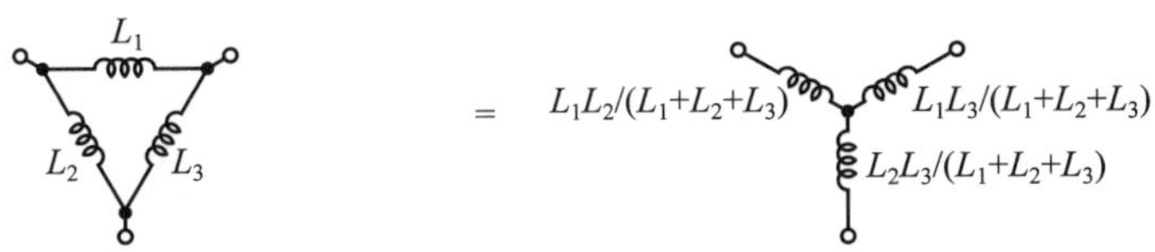

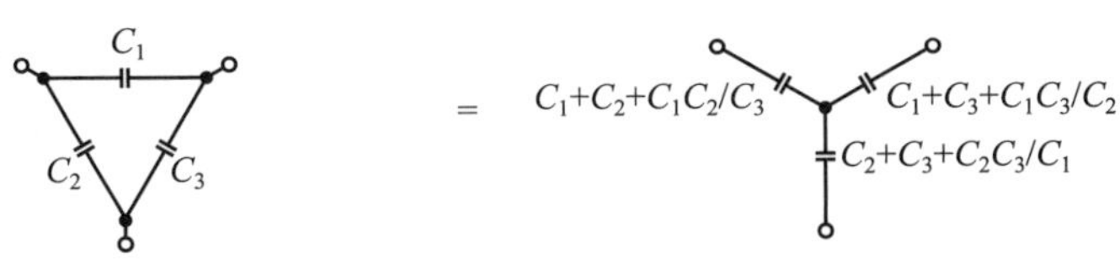

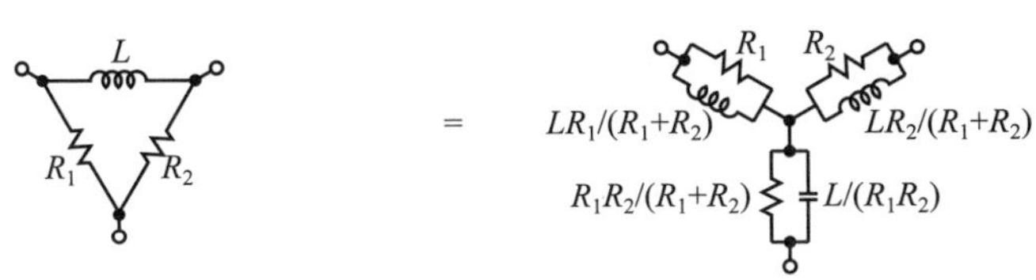

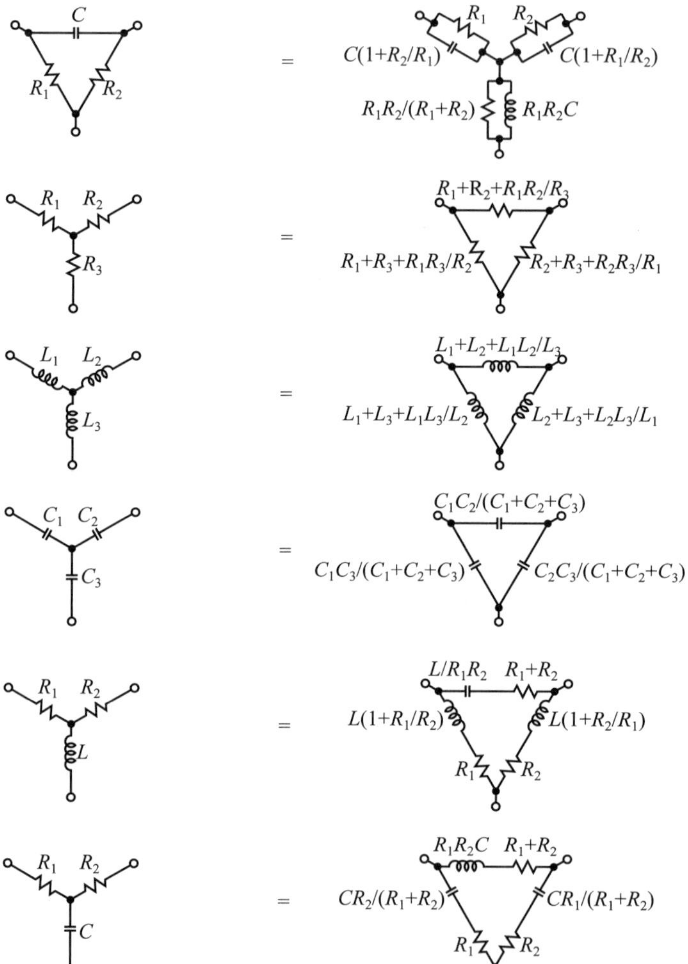
C
R1
R2
=
R1 R2
C(1+R2/R1)
C(1+R1/R2)
R1R2/(R1+R2)
R1R2C
R1 R2
R3
=
R1+R2+R1R2/R3
R1+R3+R1R3/R2
R2+R3+R2R3/R1
L1 L2
L3
=
L1+L2+L1L2/L3
L1+L3+L1L3/L2
L2+L3+L2L3/L1
C1 C2
C3
=
C1C2/(C1+C2+C3)
C1C3/(C1+C2+C3)
C2C3/(C1+C2+C3)
R1 R2
L
=
L/R1R2 R1+R2
L(1+R1/R2)
L(1+R2/R1)
R1 R2
R1 R2
C
=
R1R2C R1+R2
CR2/(R1+R2)
CR1/(R1+R2)
R1 R2

Appendix F

Transmission-Line Identities

The following transmission-line identities are useful in the construction of RF and microwave networks. Each table shows the schematics of several subcircuits which can be made equivalent across all frequencies, followed by the mathematical relationships between the element values under which said equivalence applies [1–5].

$$
\begin{aligned}
z_1 &= z_a\left(1+\frac{z_a}{z_b}\right) \\
&= \tfrac{1}{2}z_{e1}\left(1+\frac{z_{e1}}{z_{o1}}\right) \\
&= 2z_{e2}z_{o2}\frac{z_{e2}+z_{o2}}{\left(z_{e2}-z_{o2}\right)^2} \\
z_2 &= z_a+z_b \\
&= \tfrac{1}{2}\left(z_{e1}+z_{o1}\right) \\
&= \tfrac{1}{2}\left(z_{e2}+z_{o2}\right)
\end{aligned}
$$

$$
\begin{aligned}
z_a &= \frac{z_1z_2}{z_1+z_2} \\
&= \tfrac{1}{2}z_{e1} \\
&= \frac{2z_{e2}z_{o2}}{z_{e2}+z_{o2}} \\
z_b &= \frac{z_2^2}{z_1+z_2} \\
&= \tfrac{1}{2}z_{o1} \\
&= \frac{1}{2}\frac{\left(z_{e2}-z_{o2}\right)^2}{z_{e2}+z_{o2}}
\end{aligned}
$$

$$
\begin{aligned}
z_{e1} &= 2\frac{z_1z_2}{z_1+z_2} \\
&= 2z_a \\
&= \frac{4z_{e2}z_{o2}}{z_{e2}+z_{o2}} \\
z_{o1} &= 2\frac{z_2^2}{z_1+z_2} \\
&= 2z_b \\
&= \frac{\left(z_{e2}-z_{o2}\right)^2}{z_{e2}+z_{o2}}
\end{aligned}
$$

$$
\begin{aligned}
z_{e2} &= z_2\left(1+\sqrt{\tfrac{z_2}{z_1+z_2}}\right) \\
&= z_a+z_b+\sqrt{z_b\left(z_a+z_b\right)} \\
&= \tfrac{1}{2}\left(z_{e1}+z_{o1}\right)+\tfrac{1}{2}\sqrt{z_{o1}\left(z_{e1}+z_{o1}\right)}
\end{aligned}
$$

$$
\begin{aligned}
z_{o2} &= z_2\left(1-\sqrt{\tfrac{z_2}{z_1+z_2}}\right) \\
&= z_a+z_b-\sqrt{z_b\left(z_a+z_b\right)} \\
&= \tfrac{1}{2}\left(z_{e1}+z_{o1}\right)-\tfrac{1}{2}\sqrt{z_{o1}\left(z_{e1}+z_{o1}\right)}
\end{aligned}
$$

$$\begin{aligned} z_1 &= 2\frac{z_{e1}z_{o1}}{z_{e1}+z_{o1}} \\ &= \tfrac{1}{2}\left(z_{e2}+z_{o2}\right) \\ &= \tfrac{1}{2}z_{e3} \end{aligned}$$

$$\begin{aligned} z_{e3} &= 2z_1 \\ &= 4\frac{z_{e1}z_{o1}}{z_{e1}+z_{o1}} \\ &= z_{e2}+z_{o2} \end{aligned}$$

for any z_{o3}

$$\begin{aligned} z_{e1} &= \tfrac{1}{2}z_1\left(1+\rho_1\right) \\ &= \tfrac{1}{4}\left(z_{e2}+z_{o2}\right)\left(1+\rho_1\right) \\ &= \tfrac{1}{4}z_{e3}\left(1+\rho_1\right) \\ z_{o1} &= \tfrac{1}{2}z_1\left(1+\rho_1^{-1}\right) \\ &= \tfrac{1}{4}\left(z_{e2}+z_{o2}\right)\left(1+\rho_1^{-1}\right) \\ &= \tfrac{1}{4}z_{e3}\left(1+\rho_1^{-1}\right) \end{aligned}$$

for any ρ_1

$$\begin{aligned} z_{e2} &= 2z_1\frac{\rho_2}{1+\rho_2} \\ &= 4\frac{z_{e1}z_{o1}}{z_{e1}+z_{o1}}\frac{\rho_2}{1+\rho_2} \\ &= z_{e3}\frac{\rho_2}{1+\rho_2} \\ z_{o2} &= 2z_1\frac{1}{1+\rho_2} \\ &= 4\frac{z_{e1}z_{o1}}{z_{e1}+z_{o1}}\frac{1}{1+\rho_2} \\ &= z_{e3}\frac{1}{1+\rho_2} \end{aligned}$$

for any ρ_2

ρ is the ratio of even-mode to odd-mode impedance of the coupled lines, in some cases a free parameter.

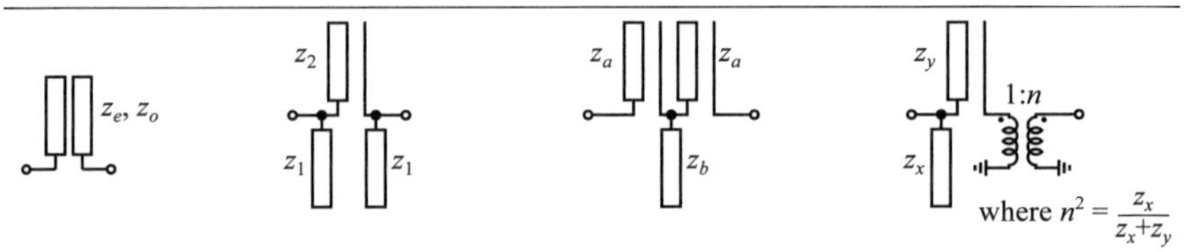

$$\begin{aligned} z_e &= z_1 \\ &= z_a+2z_b \\ &= z_x(n+1) \\ z_o &= \frac{z_1z_2}{2z_1+z_2} \\ &= z_a \\ &= z_y\frac{n^2}{n+1} \end{aligned}$$

$$\begin{aligned} z_1 &= z_e \\ &= z_a+2z_b \\ &= z_x(n+1) \\ z_2 &= \frac{2z_ez_o}{z_e-z_o} \\ &= \left(z_a+2z_b\right)\frac{z_a}{z_b} \\ &= z_yn \end{aligned}$$

$$\begin{aligned} z_a &= z_o \\ &= \frac{z_1z_2}{2z_1+z_2} \\ &= z_y\frac{n^2}{n+1} \\ z_b &= \tfrac{1}{2}\left(z_e-z_o\right) \\ &= \frac{z_1^2}{2z_1+z_2} \\ &= z_xn \end{aligned}$$

$$\begin{aligned} z_x &= \tfrac{1}{2}\left(z_e+z_o\right) \\ &= \frac{z_1}{n+1} \\ &= \frac{z_b}{n} \end{aligned}$$

$$\begin{aligned} z_y &= 2z_ez_o\frac{z_e+z_o}{\left(z_e-z_o\right)^2} \\ &= \frac{z_2}{n} \\ &= z_a\frac{n+1}{n^2} \end{aligned}$$

$$\begin{aligned} n &= \frac{z_e-z_o}{z_e+z_o} \\ &= \frac{z_1}{z_1+z_2} \\ &= \frac{z_b}{z_a+z_b} \end{aligned}$$

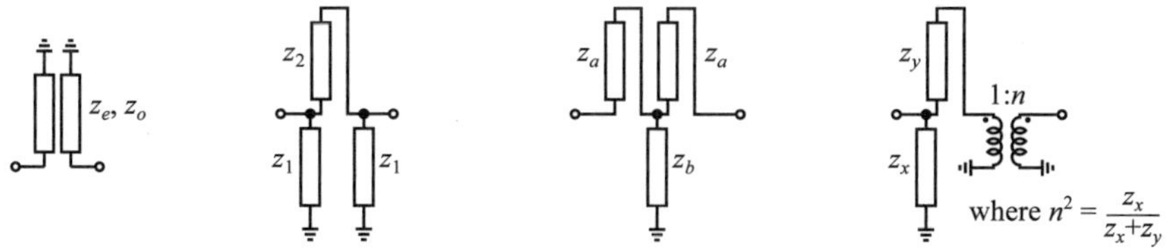

$$\begin{aligned} z_e &= z_1 \\ &= z_a + 2z_b \\ &= z_x(n+1) \\ z_o &= \frac{z_1 z_2}{2z_1 + z_2} \\ &= z_a \\ &= z_y \frac{n^2}{n+1} \end{aligned}$$

$$\begin{aligned} z_1 &= z_e \\ &= z_a + 2z_b \\ &= z_x(n+1) \\ z_2 &= \frac{2z_e z_o}{z_e - z_o} \\ &= (z_a + 2z_b)\frac{z_a}{z_b} \\ &= z_y n \end{aligned}$$

$$\begin{aligned} z_a &= z_o \\ &= \frac{z_1 z_2}{2z_1 + z_2} \\ &= z_y \frac{n^2}{n+1} \\ z_b &= \tfrac{1}{2}(z_e - z_o) \\ &= \frac{z_1^2}{2z_1 + z_2} \\ &= z_x n \end{aligned}$$

$$\begin{aligned} z_x &= \tfrac{1}{2}(z_e + z_o) \\ &= \frac{z_1}{n+1} \\ &= \frac{z_b}{n} \end{aligned}$$

$$\begin{aligned} z_y &= 2z_e z_o \frac{z_e + z_o}{(z_e - z_o)^2} \\ &= \frac{z_2}{n} \\ &= z_a \frac{n+1}{n^2} \end{aligned}$$

$$\begin{aligned} n &= \frac{z_e - z_o}{z_e + z_o} \\ &= \frac{z_1}{z_1 + z_2} \\ &= \frac{z_b}{z_a + z_b} \end{aligned}$$

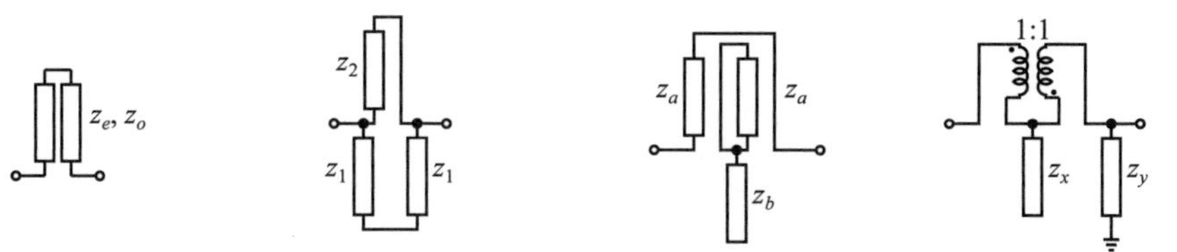

$$\begin{aligned} z_e &= z_1 \\ &= z_a + 2z_b \\ &= 2z_x \\ z_o &= \frac{z_1 z_2}{2z_1 + z_2} \\ &= z_a \\ &= 2z_y \end{aligned}$$

$$\begin{aligned} z_1 &= z_e \\ &= z_a + 2z_b \\ &= 2z_x \\ z_2 &= 2\frac{z_e z_o}{z_e - z_o} \\ &= \frac{z_a}{z_b}(z_a + 2z_b) \\ &= 4\frac{z_x z_y}{z_x - z_y} \end{aligned}$$

$$\begin{aligned} z_a &= z_o \\ &= \frac{z_1 z_2}{2z_1 + z_2} \\ &= 2z_x \\ z_b &= \tfrac{1}{2}(z_e - z_o) \\ &= \frac{z_1^2}{2z_1 + z_2} \\ &= z_x - z_y \end{aligned}$$

$$\begin{aligned} z_x &= \tfrac{1}{2} z_e \\ &= \tfrac{1}{2} z_1 \\ &= \tfrac{1}{2}(z_a + 2z_b) \\ z_y &= \tfrac{1}{2} z_o \\ &= \tfrac{1}{2}\frac{z_1 z_2}{2z_1 + z_2} \\ &= \tfrac{1}{2} z_a \end{aligned}$$

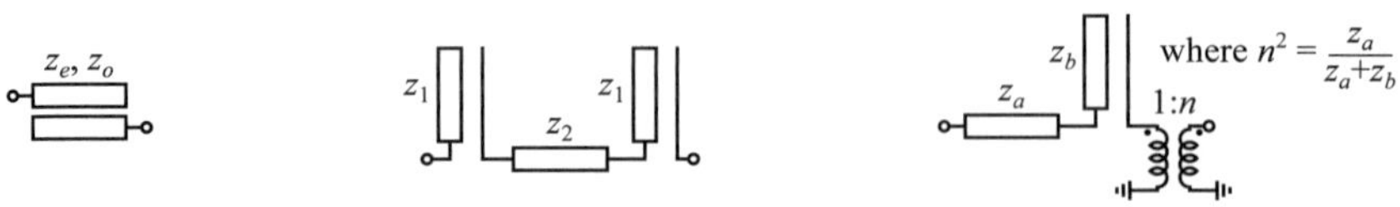

$$z_e = z_1 + 2z_2 = z_a(1+n)$$
$$z_o = z_1 = z_a(1-n)$$

$$z_1 = z_o = z_a(1-n)$$
$$z_2 = \tfrac{1}{2}(z_e - z_o) = z_a n$$

$$z_a = \tfrac{1}{2}(z_e + z_o) = \frac{z_1}{1-n}$$
$$z_b = 2z_e z_o \frac{z_e + z_o}{(z_e - z_o)^2} = z_1 \frac{1+n}{n^2}$$
$$n = \frac{z_e - z_o}{z_e + z_o} = \frac{z_2}{z_1 + z_2}$$

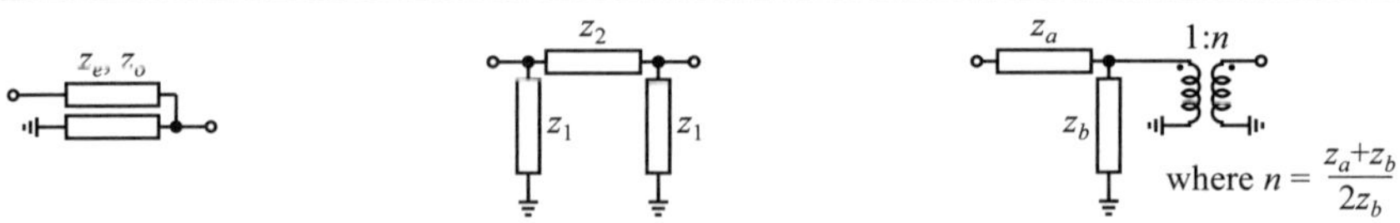

$$z_e = z_2 = z_a n$$
$$z_o = \tfrac{1}{2}\frac{z_1 z_2}{z_1 + 2z_2} = z_b n$$

$$z_1 = 2\frac{z_e z_o}{z_e - z_o} = 2n\frac{z_a z_b}{z_a - z_b}$$
$$z_2 = z_e = z_a n$$

$$z_a = \tfrac{1}{n} z_e = \tfrac{1}{n} z_2$$
$$z_b = \tfrac{1}{n} z_o = \tfrac{1}{n}\frac{z_1 z_2}{z_1 + 2z_2}$$
$$n = \tfrac{1}{2}\left(1 + \frac{z_e}{z_o}\right)$$

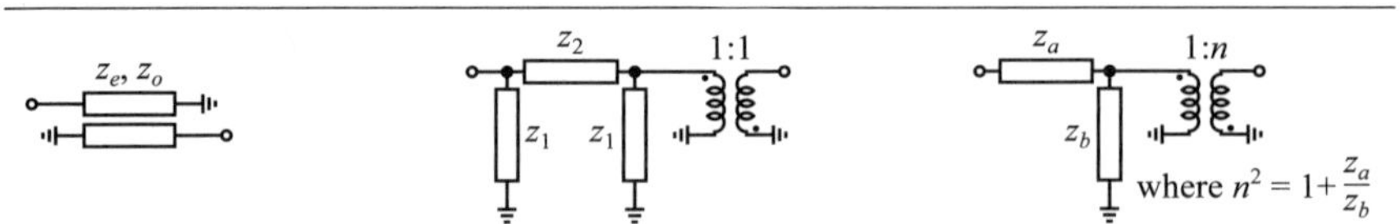

$$z_e = z_1 = z_a \frac{n}{n-1}$$
$$z_o = \frac{z_1 z_2}{2z_1 + z_2} = z_a \frac{n}{n+1}$$

$$z_1 = z_e = z_a \frac{n}{n-1}$$
$$z_2 = 2\frac{z_e z_o}{z_e - z_o} = z_a n$$

$$z_a = 2\frac{z_e z_o}{z_e + z_o} = \frac{z_1 z_2}{z_1 + z_2}$$
$$z_b = \tfrac{1}{2}\frac{(z_e - z_o)^2}{z_e + z_o} = z_1 \frac{1}{n(1+n)}$$
$$n = \frac{z_e + z_o}{z_e - z_o} = 1 + \frac{z_2}{z_1}$$

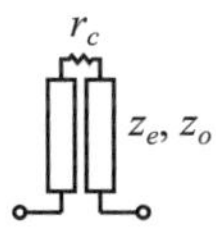

$$z_e = z_1 = z_a + 2z_b$$
$$z_o = \frac{z_1 z_2}{2z_1 + z_2} = z_a$$
$$r_c = \frac{r_1 r_2}{r_1 + r_2} = r_a$$

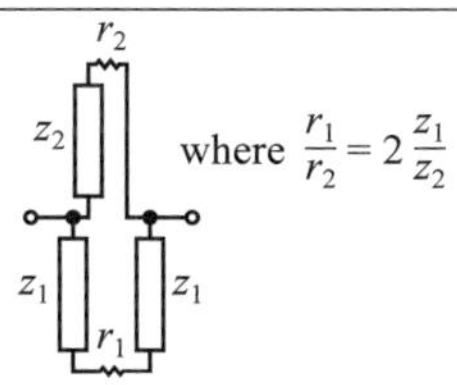

$$z_1 = z_e = z_a + 2z_b$$
$$z_2 = 2\frac{z_e z_o}{z_e - z_o} = \frac{z_a}{z_b}(z_a + 2z_b)$$
$$r_1 = \frac{z_e}{z_o} r_c = \left(1 + 2\frac{z_b}{z_a}\right) r_a$$
$$r_2 = \frac{z_e r_c}{z_e - z_o} = \left(1 + \frac{z_a}{2z_b}\right) r_a$$

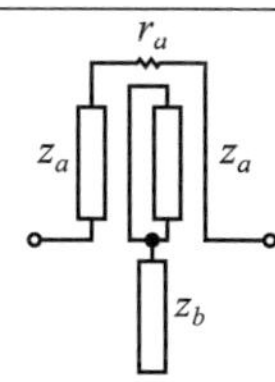

$$z_a = z_o = \frac{z_1 z_2}{2z_1 + z_2}$$
$$z_b = \tfrac{1}{2}(z_e - z_o) = \frac{z_1}{z_2}\frac{z_1 z_2}{2z_1 + z_2}$$
$$r_a = r_c = \frac{r_1 r_2}{r_1 + r_2}$$

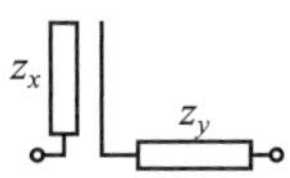

$$z_x = z_b n$$
$$z_y = z_a n$$

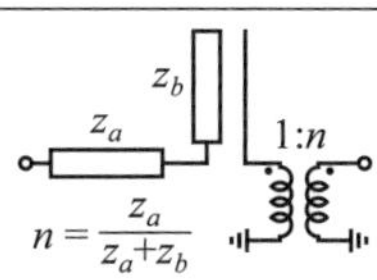

$$z_a = z_y \frac{1}{n}$$
$$z_b = z_x \frac{1}{n}$$
$$n = \frac{z_x}{z_x + z_y}$$

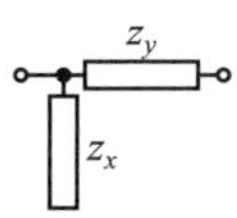

$$z_x = z_b n$$
$$z_y = z_a n$$

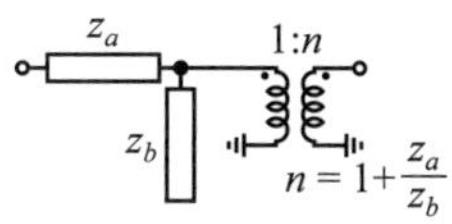

$$z_a = z_y \frac{1}{n}$$
$$z_b = z_x \frac{1}{n}$$
$$n = 1 + \frac{z_y}{z_x}$$

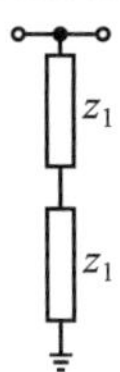

$$z_1 = \tfrac{1}{2} z_x$$

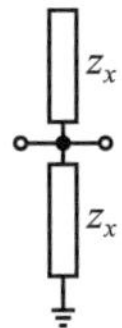

$$z_x = 2z_1$$

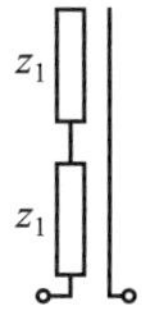

$$z_1 = 2z_x$$

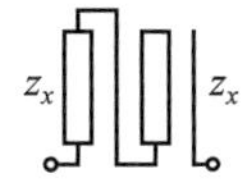

$$z_x = \tfrac{1}{2} z_1$$

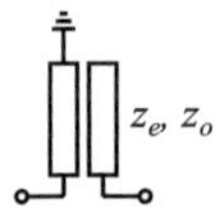

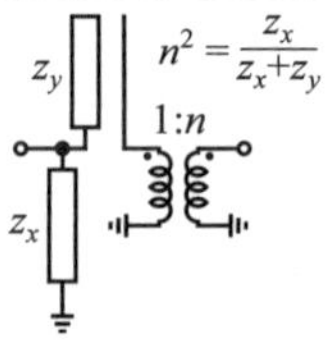

$$z_e = z_x(n+1) \qquad z_x = \tfrac{1}{2}(z_e + z_o)$$

$$z_o = z_y \frac{n^2}{n+1} \qquad z_y = \frac{2 z_e z_o (z_e + z_o)}{(z_e - z_o)^2}$$

$$n = \frac{z_e - z_o}{z_e + z_o}$$

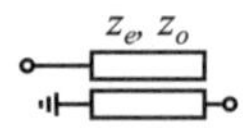

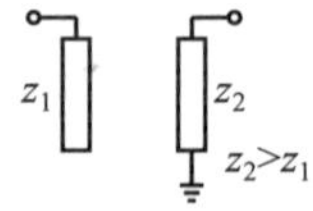

$$z_e = z_2 + \sqrt{z_2 (z_2 - z_1)}$$

$$z_o = z_2 - \sqrt{z_2 (z_2 - z_1)}$$

$$z_1 = 2 \frac{z_e z_o}{z_e - z_o}$$

$$z_2 = \tfrac{1}{2}(z_e + z_o)$$

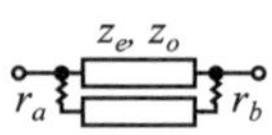

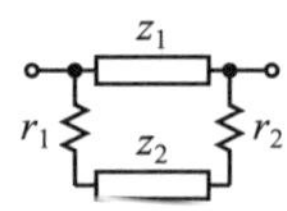

$$z_e = 2 \frac{z_1 z_2}{z_1 + z_2} \qquad z_1 = \tfrac{1}{2}(z_e + z_o)$$

$$z_o = 2 \frac{z_1^2}{z_1 + z_2} \qquad z_2 = \tfrac{1}{2} \frac{z_e}{z_o}(z_e + z_o)$$

$$r_a = 4 r_1 \left(\frac{z_1}{z_1 + z_2} \right)^2 \qquad r_1 = \tfrac{1}{4} r_a \left(\frac{z_e}{z_o} + 1 \right)^2$$

$$r_b = 4 r_2 \left(\frac{z_1}{z_1 + z_2} \right)^2 \qquad r_2 = \tfrac{1}{4} r_b \left(\frac{z_e}{z_o} + 1 \right)^2$$

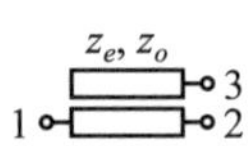

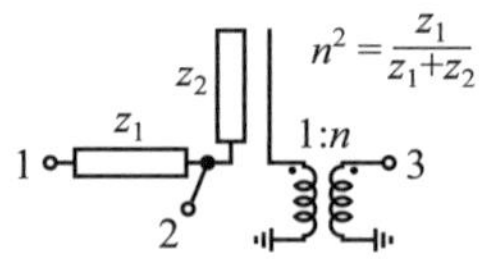

$$z_e = z_1(1+n)$$

$$z_o = z_1(1-n)$$

$$z_1 = \tfrac{1}{2}(z_e + z_o)$$

$$z_2 = 2 z_e z_o \frac{z_e + z_o}{(z_e - z_o)^2}$$

$$n = \frac{z_e - z_o}{z_e + z_o}$$

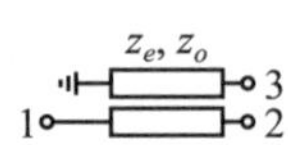

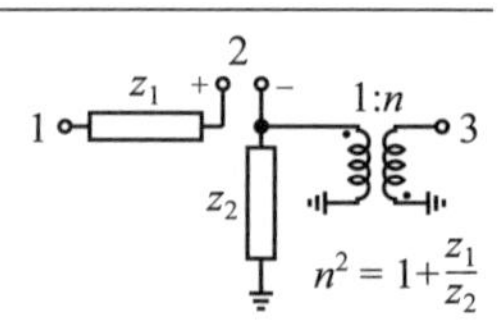

$$z_e = z_2 n(n+1)$$

$$z_o = z_2 n(n-1)$$

$$z_1 = 2 \frac{z_e z_o}{z_e + z_o}$$

$$z_2 = \tfrac{1}{2} \frac{(z_e - z_o)^2}{z_e + z_o}$$

$$n = \frac{z_e + z_o}{z_e - z_o}$$

	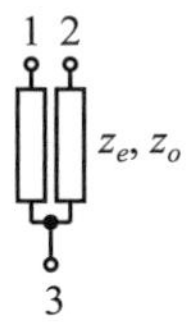	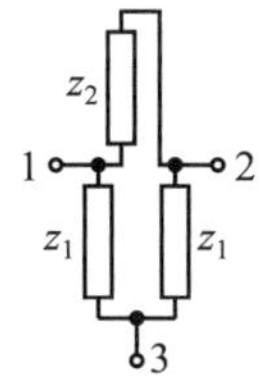	
$z_e = z_1$	$z_o = \dfrac{z_1 z_2}{2z_1 + z_2}$	$z_1 = z_e$	$z_2 = 2\dfrac{z_e z_o}{z_e - z_o}$
	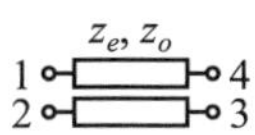	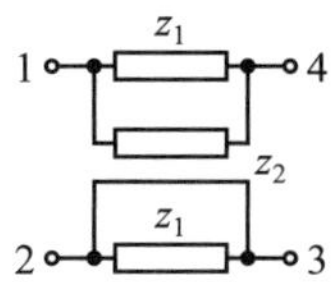	
$z_e = z_1$	$z_o = \dfrac{z_1 z_2}{2z_1 + z_2}$	$z_1 = z_e$	$z_2 = 2\dfrac{z_e z_o}{z_e - z_o}$

References

[1] G. L. Matthaei, L. Young, and E. M. T. Jones, *Microwave Filters, Impedance-Matching Networks, and Coupling Structures*. Dedham, MA: Artech House, 1980.

[2] H. Ozaki and J. Ishii, "Synthesis of a class of strip-line filters," *IEEE Transactions on Circuit Theory*, vol. 5, pp. 104–109, June 1958.

[3] A. Podcameni and L. F. M. Conrado, "A new transmission-line identity," *Microwave and Optical Technology Letters*, vol. 23, no. 1, pp. 62–63, October 1999.

[4] R. Wenzel, "Exact design of TEM microwave networks using quarter-wave lines," *IEEE Transactions on Microwave Theory and Techniques*, vol. 12, no. 1, pp. 94–111, January 1964.

[5] M. A. Morgan and T. A. Boyd, "Reflectionless filter structures," *IEEE Transactions on Microwave Theory and Techniques*, vol. 63, no. 4, pp. 1263–1271, April 2015.

Appendix G

Special Functions

G.1 CHEBYSHEV POLYNOMIALS

The Nth-order Chebyshev polynomial is given by

$$T_N(x) = \cos\left(N\cos^{-1}x\right) = \cosh\left(N\cosh^{-1}x\right) \tag{G.1}$$

Either definition is valid (and equivalent) over the full range of x, so long as one is prepared to utilize the full complex-valued definitions of $\cos(\cdot)$ and $\cosh(\cdot)$. If one wishes to work strictly in real numbers, than one should use

$$T_N(x) = \begin{cases} \cos\left(N\cos^{-1}x\right) & |x| \leq 1 \\ \cosh\left(N\cosh^{-1}x\right) & x > 1 \\ (-1)^N\cosh\left(N\cosh^{-1}(-x)\right) & x < 1 \end{cases} \tag{G.2}$$

To see that these do indeed describe a polynomial for integer N we observe the following,

$$T_0(x) = \cos\left(0\cdot\cos^{-1}x\right) = \cos(0) = 1 \tag{G.3a}$$

$$T_1(x) = \cos\left(1\cdot\cos^{-1}x\right) = \cos\left(\cos^{-1}x\right) = x \tag{G.3b}$$

and

$$T_{N\pm1}(x) = \cos\left(N\cos^{-1}x \pm 1\cdot\cos^{-1}x\right) \tag{G.4a}$$

$$= \cos\left(N\cos^{-1}x\right)\cos\left(\cos^{-1}x\right) \mp \sin\left(N\cos^{-1}x\right)\sin\left(\cos^{-1}x\right) \tag{G.4b}$$

$$= xT_N(x) \mp \sin\left(N\cos^{-1}x\right)\sin\left(\cos^{-1}x\right) \tag{G.4c}$$

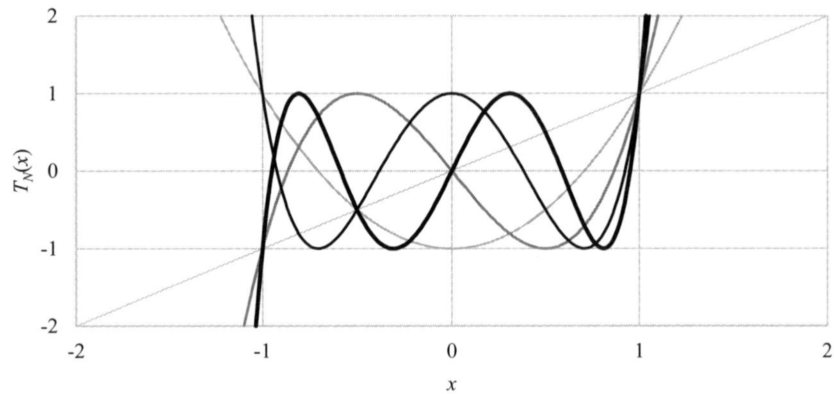

Figure G.1 Chebyshev polynomials for $N = 1 \ldots 5$.

$$\therefore T_{N+1}(x) + T_{N-1}(x) = 2xT_N(x) \tag{G.4d}$$

$$\therefore T_{N+1}(x) = 2xT_N(x) - T_{N-1}(x) \tag{G.4e}$$

Thus, the recurrence relation (G.4e) along with the initial values (G.3) allow us to define the Chebyshev polynomial recursively for any integer N,

$$T_0(x) = 1 \tag{G.5a}$$

$$T_1(x) = x \tag{G.5b}$$

$$T_2(x) = 2x^2 - 1 \tag{G.5c}$$

$$T_3(x) = 4x^3 - 3x \tag{G.5d}$$

$$T_4(x) = 8x^4 - 8x^2 + 1 \tag{G.5e}$$

$$T_5(x) = 16x^5 - 20x^3 + 5x \tag{G.5f}$$

The first five Chebyshev polynomials are plotted in Figure G.1. Note that all of them have less than or equal to unit magnitude for $|x| \leq 1$, and the higher the order, the more ripples there are in this range. Chebyshev polynomials are special for having the largest leading coefficient (and thus the steepest slope as x becomes large) for any polynomial of the same order that is bounded in this way. These properties lend themselves well to a number of circuit design problems where one wishes to maximize the slope out-of-band (such as in a filter or matching network) while keeping the ripples bounded in-band. Electromagnetic circuits based on Chebyshev polynomials will often be referred to as *equiripple* designs in reference to these properties.

The roots of the Chebyshev polynomials (which must all be inside the unit interval) are easily derived,

$$T_N(x_k) = \cos\left(N \cos^{-1} x\right) = 0 \tag{G.6a}$$

$$N \cos^{-1} x_k = k\pi - \tfrac{\pi}{2} \tag{G.6b}$$

$$x_k = \cos\left(\frac{2k-1}{N}\pi\right) \tag{G.6c}$$

for $k = 1 \ldots N$. Similarly, the extrema (which again lie inside the unit interval, and are always equal to ± 1), are derived,

$$\frac{d}{dx} T_N(x_m) = \frac{N}{\sqrt{1-x_m^2}} \sin\left(N \cos^{-1} x_m\right) = 0 \tag{G.7a}$$

$$N \cos^{-1} x_m = m\pi \tag{G.7b}$$

$$x_m = \cos\left(\frac{m\pi}{N}\right) \tag{G.7c}$$

for $m = 1 \ldots N-1$. (To include the endpoints on the interval $-1 \leq x \leq 1$ in this list of extrema, as some problems do, we may take $m = 0 \ldots N$.)

G.2 BESSEL FUNCTIONS

Bessel functions [1, 2] are defined as the solutions to Friedrich Bessel's differential equation,

$$x^2 \frac{d^2 y}{dx} + x \frac{dy}{dx} + \left(x^2 - n^2\right) y = 0 \tag{G.8}$$

where n is the order of the Bessel function. For the purposes of the most pertinent problems in electromagnetic circuits, x will be real-valued and positive, while n is a nonnegative integer. As a second-order equation, there must be two independent solutions, or kinds. The nth-order Bessel function of the first kind, denoted $J_n(x)$,

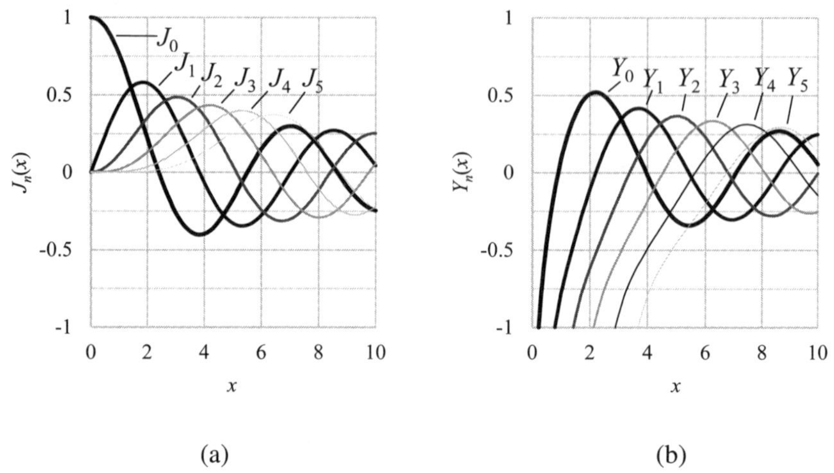

Figure G.2 Bessel functions of the (a) first and (b) second kinds, for orders $n = 1 \ldots 5$.

is given by the series

$$J_n(x) = \sum_{m=0}^{\infty} \frac{(-1)^m}{m!\Gamma(n+m+1)} \left(\frac{x}{2}\right)^{n+2m} \tag{G.9}$$

where Γ is the Gamma function, a generalized form of the factorial that takes complex arguments. It is defined by the integral

$$\Gamma(z) = \int_0^{\infty} x^{z-1} e^{-x} dx \tag{G.10}$$

If its argument is a positive integer, then $\Gamma(n) = (n-1)!$, so that the formula for the Bessel function of the first kind with integer order is

$$J_n(x) = \sum_{m=0}^{\infty} \frac{(-1)^m}{m!(n+m)!} \left(\frac{x}{2}\right)^{n+2m} \tag{G.11}$$

This is plotted for several orders in Figure G.2(a). It has the appearance of a decaying sinusoid, which is finite at the origin.

In contrast, the nth-order Bessel function of the second kind, denoted $Y_n(x)$, is divergent at the origin, as shown in Figure G.2(b). In its most general form,

$$Y_n(x) = \lim_{\alpha \to n} \frac{J_\alpha(x)\cos(\alpha\pi) - J_{-\alpha}(x)}{\sin(\alpha\pi)} \tag{G.12}$$

where the limit is required in the event that n is an integer (as is most often the case) due to the indeterminate form.

There is a rather complex series form that applies when n is a nonnegative integer,

$$\begin{aligned} Y_n(x) = \frac{2}{\pi}\left(\gamma + \ln\frac{x}{2}\right) J_n(x) - \frac{1}{\pi}\sum_{m=0}^{n-1} \frac{(n-m-1)!}{m!}\left(\frac{2}{x}\right)^{n-2m} \\ - \frac{1}{\pi}\sum_{m=0}^{\infty} \frac{(-1)^m (x/2)^{n+2m}}{m!(n+m)!}\left(\sum_{k=1}^{m}\frac{1}{k} + \sum_{k=1}^{n+m}\frac{1}{k}\right) \end{aligned} \tag{G.13a}$$

$$\gamma = \lim_{m\to\infty}\left(\sum_{k=1}^{m}\frac{1}{k} - \ln m\right) = 0.577215665\ldots \tag{G.13b}$$

The term γ is known as the *Euler-Mascheroni* constant.

G.2.1 Recurrence Formulas

If at least two orders of the Bessel functions have been calculated using the above methods, the rest may be obtained via the recurrence formula,

$$Z_{n+1}(x) = \frac{2n}{x} Z_n(x) - Z_{n-1}(x) \tag{G.14}$$

where $Z = J$ or Y. The derivatives are also conveniently given by recurrence formulas,

$$\frac{dZ_n}{dx} = Z_{n-1} - \frac{n}{x} Z_n \tag{G.15a}$$

$$= \frac{n}{x} Z_n - Z_{n+1} \tag{G.15b}$$

$$= \frac{1}{2}\left(Z_{n-1} - Z_{n+1}\right) \tag{G.15c}$$

Table G.1
Zeros of the Bessel Functions of the First Kind ($0 < x < 12$)

n	z_{n1}	z_{n2}	z_{n3}	z_{n4}
0	2.4048	5.5201	8.6537	11.7915
1	3.8317	7.0156	10.1735	
2	5.1356	8.4172	11.6198	
3	6.3802	9.7610		
4	7.5883	11.0647		
5	8.7715			
6	9.9361			
7	11.0864			

Higher-order derivatives may be found through the even more general expressions,

$$\left(\frac{1}{x}\frac{d}{dx}\right)^m x^n Z_n(x) = x^{n-m} Z_{n-m}(x) \tag{G.16a}$$

$$\left(\frac{1}{x}\frac{d}{dx}\right)^m x^{-n} Z_n(x) = \frac{(-1)^m}{x^{n+m}} Z_{n+m}(x) \tag{G.16b}$$

G.2.2 Zeros and Extrema

For many engineering problems, it is necessary to know the zeros and extrema of the Bessel functions of the first kind on the real axis. These are tabulated for convenience in Tables G.1 and G.2. Note that the first extremum of J_1 at $x = 0$ is omitted by convention.

Two integrals involving the zeros and extrema of the Bessel function of the first kind are also useful in some problems,

$$\int_0^{z_{nm}} \left(\frac{n^2}{x^2} J_n^2(x) + J_n'^2(x)\right) x dx = \frac{z_{nm}^2}{2} J_n'^2(z_{nm}) \tag{G.17a}$$

$$\int_0^{e_{nm}} \left(\frac{n^2}{x^2} J_n^2(x) + J_n'^2(x)\right) x dx = \frac{e_{nm}^2 - n^2}{2} J_n^2(e_{nm}) \tag{G.17b}$$

where z_{nm} is the mth root and e_{nm} the mth extremum of the nth-order Bessel function.

Table G.2
Extrema of the Bessel Functions of the First Kind ($0 < x < 12$)

n	e_{n1}	e_{n2}	e_{n3}	e_{n4}
0	3.8317	7.0156	10.1735	
1	1.8412	5.3314	8.5363	11.7060
2	3.0542	6.7061	9.9695	
3	4.2012	8.0512	11.3459	
4	5.3175	9.2824		
5	6.4156	10.5199		
6	7.5013	11.7349		
7	8.5778			
8	9.6474			
9	10.7144			
10	11.7709			

G.2.3 Asymptotic Approximations

It is convenient in some cases to have simpler formulas which approximate the Bessel functions (for real x and integer n) at small and large values of x. For $x \ll \sqrt{n+1}$,

$$J_n(x) \approx \frac{1}{n!}\left(\frac{x}{2}\right)^n \tag{G.18a}$$

$$Y_n(x) \approx \begin{cases} \frac{2}{\pi}\left(\ln\left(\frac{x}{2}\right)+\gamma\right) & n=0 \\ -\frac{1}{\pi}(n-1)!\left(\frac{2}{x}\right)^n & n>0 \end{cases} \tag{G.18b}$$

and for $x \gg n^2$,

$$J_n(x) \approx \sqrt{\frac{2}{\pi x}}\cos\left(x-\frac{\pi}{4}-\frac{n\pi}{2}\right) \tag{G.19a}$$

$$Y_n(x) \approx \sqrt{\frac{2}{\pi x}}\sin\left(x-\frac{\pi}{4}-\frac{n\pi}{2}\right) \tag{G.19b}$$

As an example, the second-order Bessel functions of the first and second kinds are shown in Figure G.3, along with these approximations in dashed lines. Note that although the Bessel functions have the appearance of damped sinusoids, the amplitude of neither kind decays to zero at infinity. This makes them unsuitable as solutions for open-boundary problems where a compact result is expected. For that, we need one of the modified Bessel functions of the following section.

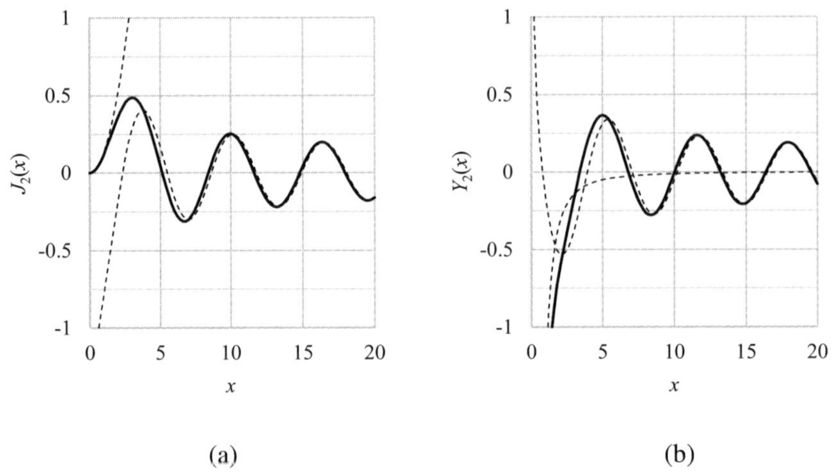

Figure G.3 Second-order Bessel functions of the (a) first and (b) second kind. Approximate forms are shown with dashed lines.

G.2.4 Modified Bessel Functions

Bessel functions have been generalized for complex arguments, x, and even for noninteger or complex orders, n. The forms in (G.9) and (G.12) are valid for all of these cases. A useful special case is when the argument is purely imaginary. These lead to the *modified Bessel functions of the first and second kind*, denoted $I_n(x)$ and $K_n(x)$, respectively, which are defined as

$$I_n(x) = j^{-n} J_n(jx) = \sum_{m=0}^{\infty} \frac{1}{m!\Gamma(n+m+1)} \left(\frac{x}{2}\right)^{n+2m} \tag{G.20a}$$

$$K_n(x) = \frac{\pi}{2} \left(j^{n+1} J_n(jx) - j^n Y_n(jx)\right) = \lim_{\alpha \to n} \frac{\pi}{2} \frac{I_{-\alpha}(x) - I_\alpha(x)}{\sin(\alpha\pi)} \tag{G.20b}$$

These are plotted for real x and integer n in Figure G.4. The derivatives in this case are given by recurrence formulas similar (but not all identical) to those for the original Bessel functions,

$$\frac{dI_n}{dx} = I_{n-1} - \frac{n}{x} I_n \tag{G.21a}$$

$$= \frac{n}{x} I_n + I_{n+1} \tag{G.21b}$$

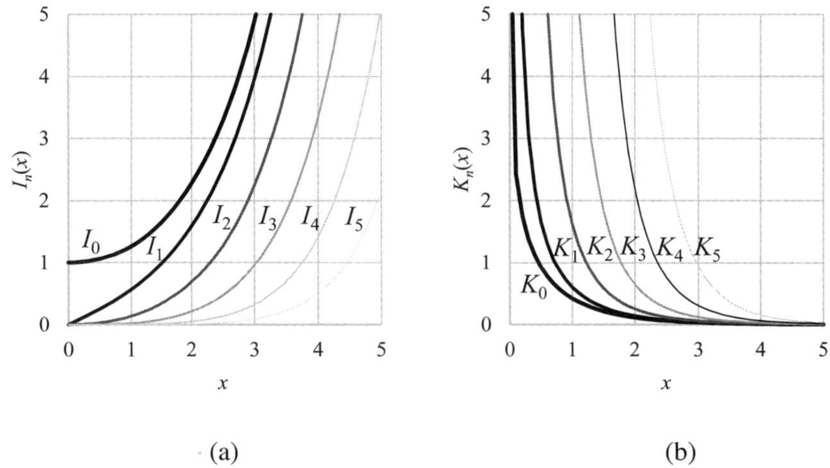

Figure G.4 Modified Bessel functions of the (a) first and (b) second kinds for real x and integer n.

$$= \frac{1}{2}\left(I_{n-1} + I_{n+1}\right) \tag{G.21c}$$

and

$$\frac{dK_n}{dx} = -K_{n-1} - \frac{n}{x}K_n \tag{G.22a}$$

$$= \frac{n}{x}K_n - K_{n+1} \tag{G.22b}$$

$$= -\frac{1}{2}\left(K_{n-1} + K_{n+1}\right) \tag{G.22c}$$

References

[1] Wikipedia. (2018) Bessel function. https://en.wikipedia.org/wiki/Bessel_function.

[2] G. Arfken and H. Weber, *Mathematical Methods for Physicists*, 4th ed. San Diego, CA: Academic Press, 1995.

About the Author

Matthew A. Morgan received his B.S. in electrical engineering from the University of Virginia in 1999, and his M.S. and Ph.D. in electrical engineering from the California Institute of Technology in 2001 and 2003, respectively. He has authored over 60 papers and holds 13 patents in the areas of microwave monolithic integrated circuit (MMIC) design, millimeter-wave system integration and packaging techniques, reflectionless filter development, high-speed serial communication, and ultrawideband millimeter-wave antennas.

During the summers of 1996 through 1998, he worked for Lockheed Martin Federal Systems in Manassas, Virginia, as an associate programmer, where he wrote code for acoustic signal processing, mathematical modeling, data simulation, and system performance monitoring. In 1999, he became an affiliate of NASA's Jet Propulsion Laboratory in Pasadena, California. There, he conducted research in the development of MMICs and MMIC-based receiver components for atmospheric radiometers, laboratory instrumentation, and the deep-space communication network.

In 2003, Dr. Morgan joined the Central Development Lab (CDL) of the National Radio Astronomy Observatory (NRAO) in Charlottesville, Virginia, where he now holds the position of scientist/research engineer. He was project engineer for the K-Band Focal Plane Array development on the Green Bank Telescope, and technical lead for Band 6 (211–275 GHz) cryogenic IF amplifier production, Band 6 receiver cartridge testing, and Band 3 (84–116 GHz) and Band 6 Orthomode Transducer (OMT) production for the Atacama Large Millimeter Array (ALMA). Dr. Morgan is currently the head of the CDL's Integrated Receiver Development program, and is involved in the design and development of low-noise receivers, components, and novel concepts for radio astronomy instrumentation in the centimeter-wave, millimeter-wave, and submillimeter-wave frequency ranges.

Index

D

E

F

G

H

I

J

K

L

M

N

S

T

U

V

W

Y

Z

Artech House Microwave Library

Behavioral Modeling and Linearization of RF Power Amplifiers, John Wood

Chipless RFID Reader Architecture, Nemai Chandra Karmakar, Prasanna Kalansuriya, Randika Koswatta, and Rubayet E-Azim

Control Components Using Si, GaAs, and GaN Technologies, Inder J. Bahl

Design of Linear RF Outphasing Power Amplifiers, Xuejun Zhang, Lawrence E. Larson, and Peter M. Asbeck

Design Methodology for RF CMOS Phase Locked Loops, Carlos Quemada, Guillermo Bistué, and Iñigo Adin

Design of CMOS Operational Amplifiers, Rasoul Dehghani

Design of RF and Microwave Amplifiers and Oscillators, Second Edition, Pieter L. D. Abrie

Digital Filter Design Solutions, Jolyon M. De Freitas

Discrete Oscillator Design Linear, Nonlinear, Transient, and Noise Domains, Randall W. Rhea

Distortion in RF Power Amplifiers, Joel Vuolevi and Timo Rahkonen

Distributed Power Amplifiers for RF and Microwave Communications, Narendra Kumar and Andrei Grebennikov

Electric Circuits: A Primer, J. C. Olivier

Electronics for Microwave Backhaul, Vittorio Camarchia, Roberto Quaglia, and Marco Pirola, editors

EMPLAN: Electromagnetic Analysis of Printed Structures in Planarly Layered Media, Software and User's Manual, Noyan Kinayman and M. I. Aksun

An Engineer's Guide to Automated Testing of High-Speed Interfaces, Second Edition, José Moreira and Hubert Werkmann

Envelope Tracking Power Amplifiers for Wireless Communications, Zhancang Wang

Essentials of RF and Microwave Grounding, Eric Holzman

Frequency Measurement Technology, Ignacio Llamas-Garro, Marcos Tavares de Melo, and Jung-Mu Kim

FAST: Fast Amplifier Synthesis Tool—Software and User's Guide, Dale D. Henkes

Feedforward Linear Power Amplifiers, Nick Pothecary

Filter Synthesis Using Genesys S/Filter, Randall W. Rhea

Foundations of Oscillator Circuit Design, Guillermo Gonzalez

Frequency Synthesizers: Concept to Product, Alexander Chenakin

Fundamentals of Nonlinear Behavioral Modeling for RF and Microwave Design, John Wood and David E. Root, editors

Generalized Filter Design by Computer Optimization, Djuradj Budimir

Handbook of Dielectric and Thermal Properties of Materials at Microwave Frequencies, Vyacheslav V. Komarov

Handbook of RF, Microwave, and Millimeter-Wave Components, Leonid A. Belov, Sergey M. Smolskiy, and Victor N. Kochemasov

High-Efficiency Load Modulation Power Amplifiers for Wireless Communications, Zhancang Wang

High-Linearity RF Amplifier Design, Peter B. Kenington

High-Speed Circuit Board Signal Integrity, Second Edition, Stephen C. Thierauf

Integrated Microwave Front-Ends with Avionics Applications, Leo G. Maloratsky

Intermodulation Distortion in Microwave and Wireless Circuits, José Carlos Pedro and Nuno Borges Carvalho

Introduction to Modeling HBTs, Matthias Rudolph

An Introduction to Packet Microwave Systems and Technologies, Paolo Volpato

Introduction to RF Design Using EM Simulators, Hiroaki Kogure, Yoshie Kogure, and James C. Rautio

Introduction to RF and Microwave Passive Components, Richard Wallace and Krister Andreasson

Klystrons, Traveling Wave Tubes, Magnetrons, Crossed-Field Amplifiers, and Gyrotrons, A. S. Gilmour, Jr.

Lumped Elements for RF and Microwave Circuits, Inder Bahl

Lumped Element Quadrature Hybrids, David Andrews

Microstrip Lines and Slotlines, Third Edition, Ramesh Garg, Inder Bahl, and Maurizio Bozzi

Microwave Circuit Modeling Using Electromagnetic Field Simulation, Daniel G. Swanson, Jr. and Wolfgang J. R. Hoefer

Microwave Component Mechanics, Harri Eskelinen and Pekka Eskelinen

Microwave Differential Circuit Design Using Mixed-Mode S-Parameters, William R. Eisenstadt, Robert Stengel, and Bruce M. Thompson

Microwave Engineers' Handbook, Two Volumes, Theodore Saad, editor

Microwave Filters, Impedance-Matching Networks, and Coupling Structures, George L. Matthaei, Leo Young, and E. M. T. Jones

Microwave Imaging Methods and Applications, Matteo Pastorino and Andrea Randazzo

Microwave Material Applications: Device Miniaturization and Integration, David B. Cruickshank

Microwave Materials and Fabrication Techniques, Second Edition, Thomas S. Laverghetta

Microwave Materials for Wireless Applications, David B. Cruickshank

Microwave Mixer Technology and Applications, Bert Henderson and Edmar Camargo

Microwave Mixers, Second Edition, Stephen A. Maas

Microwave Network Design Using the Scattering Matrix, Janusz A. Dobrowolski

Microwave Power Amplifier Design with MMIC Modules, Howard Hausman

Microwave Radio Transmission Design Guide, Second Edition, Trevor Manning

Microwave and RF Semiconductor Control Device Modeling, Robert H. Caverly

Microwave Transmission Line Circuits, William T. Joines, W. Devereux Palmer, and Jennifer T. Bernhard

Microwaves and Wireless Simplified, Third Edition, Thomas S. Laverghetta

Modern Microwave Circuits, Noyan Kinayman and M. I. Aksun

Modern Microwave Measurements and Techniques, Second Edition, Thomas S. Laverghetta

Modern RF and Microwave Filter Design, Protap Pramanick and Prakash Bhartia

Neural Networks for RF and Microwave Design, Q. J. Zhang and K. C. Gupta

Noise in Linear and Nonlinear Circuits, Stephen A. Maas

Nonlinear Microwave and RF Circuits, Second Edition, Stephen A. Maas

On-Wafer Microwave Measurements and De-Embedding, Errikos Lourandakis

Passive RF Component Technology: Materials, Techniques, and Applications, Guoan Wang and Bo Pan, editors

Practical Analog and Digital Filter Design, Les Thede

Practical Microstrip Design and Applications, Günter Kompa

Practical Microwave Circuits, Stephen Maas

Practical RF Circuit Design for Modern Wireless Systems, Volume I: Passive Circuits and Systems, Les Besser and Rowan Gilmore

Practical RF Circuit Design for Modern Wireless Systems, Volume II: Active Circuits and Systems, Rowan Gilmore and Les Besser

Principles of RF and Microwave Design, Matthew A. Morgan

Production Testing of RF and System-on-a-Chip Devices for Wireless Communications, Keith B. Schaub and Joe Kelly

Q Factor Measurements Using MATLAB®, Darko Kajfez

QMATCH: Lumped-Element Impedance Matching, Software and User's Guide, Pieter L. D. Abrie

Radio Frequency Integrated Circuit Design, Second Edition, John W. M. Rogers and Calvin Plett

Reflectionless Filters, Matthew A. Morgan

RF Bulk Acoustic Wave Filters for Communications, Ken-ya Hashimoto

RF Design Guide: Systems, Circuits, and Equations, Peter Vizmuller

RF Linear Accelerators for Medical and Industrial Applications, Samy Hanna

RF Measurements of Die and Packages, Scott A. Wartenberg

The RF and Microwave Circuit Design Handbook, Stephen A. Maas

RF and Microwave Coupled-Line Circuits, Rajesh Mongia, Inder Bahl, and Prakash Bhartia

RF and Microwave Oscillator Design, Michal Odyniec, editor

RF Power Amplifiers for Wireless Communications, Second Edition, Steve C. Cripps

RF Systems, Components, and Circuits Handbook, Ferril A. Losee

Scattering Parameters in RF and Microwave Circuit Analysis and Design, Janusz A. Dobrowolski

The Six-Port Technique with Microwave and Wireless Applications, Fadhel M. Ghannouchi and Abbas Mohammadi

Solid-State Microwave High-Power Amplifiers, Franco Sechi and Marina Bujatti

Spin Transfer Torque Based Devices, Circuits, and Memory, Brajesh Kumar Kaushik and Shivam Verma

Stability Analysis of Nonlinear Microwave Circuits, Almudena Suárez and Raymond Quéré

Substrate Noise Coupling in Analog/RF Circuits, Stephane Bronckers, Geert Van der Plas, Gerd Vandersteen, and Yves Rolain

System-in-Package RF Design and Applications, Michael P. Gaynor

Technologies for RF Systems, Terry Edwards

Terahertz Metrology, Mira Naftaly, editor

TRAVIS 2.0: Transmission Line Visualization Software and User's Guide, Version 2.0, Robert G. Kaires and Barton T. Hickman

Understanding Microwave Heating Cavities, Tse V. Chow Ting Chan and Howard C. Reader

Understanding Quartz Crystals and Oscillators, Ramón M. Cerda

Vertical GaN and SiC Power Devices, Kazuhiro Mochizuki

The VNA Applications Handbook, Gregory Bonaguide and Neil Jarvis

For further information on these and other Artech House titles, including previously considered out-of-print books now available through our In-Print-Forever® (IPF®) program, contact:

Artech House Publishers
685 Canton Street
Norwood, MA 02062
Phone: 781-769-9750
Fax: 781-769-6334
e-mail: artech@artechhouse.com

Artech House Books
16 Sussex Street
London SW1V 4RW UK
Phone: +44 (0)20 7596 8750
Fax: +44 (0)20 7630 0166
e-mail: artech-uk@artechhouse.com

Find us on the World Wide Web at: www.artechhouse.com